SOLDER JOINT RELIABILITY

SOLDER JOINT RELIABILITY

Theory and Applications

Edited by John H. Lau

VNR VAN NOSTRAND REINHOLD
New York

Library of Congress Catalog Card Number 90-12968
ISBN 0-442-00260-2

I(T)P Van Nostrand Reinhhold is a division of International Thomson Publishing. ITP logo is a trademark under license.

Printed in the United States of America

Van Nostrand Reinhold
115 Fifth Avenue
New York, NY 10003

International Thomson Publishing
Berkshire House
168-173 High Holborn
London WC1V 7AA, England

Thomas Nelson Australia
102 Dodds Street
South Melbourne 3205
Victoria, Australia

Nelson Canada
1120 Birchmount Road
Scarborough, Ontario
M1K 5G4, Canada

16 15 14 13 12 11 10 9 8 7 6 5 4 3 2

Library of Congress Cataloging-in-Publication Data

Solder joint reliability—theory and applications / [edited by] John H. Lau.
p. cm.
Includes index.
ISBN 0-442-00260-2
1. Welded joints. 2. Solder and soldering. I. Lau, John H.
TA492.W4S63 1991
671.5′2042—dc20 90-12968
CIP

Contents

Foreword

Solders have given the designer of modern consumer, commercial, and military electronic systems a remarkable flexibility to interconnect electronic components. The properties of solder have facilitated broad assembly choices that have fueled creative applications to advance technology. Solder is the electrical and mechanical "glue" of electronic assemblies.

This pervasive dependency on solder has stimulated new interest in applications as well as a more concerted effort to better understand materials properties. We need not look far to see solder being used to interconnect ever finer geometries. Assembly of micropassive discrete devices that are hardly visible to the unaided eye, of silicon chips directly to ceramic and plastic substrates, and of very fine peripheral leaded packages constitute a few of solder's uses.

There has been a marked increase in university research related to solder. New electronic packaging centers stimulate applications, and materials engineering and science departments have demonstrated a new vigor to improve both the materials and our understanding of them. Industrial research and development continues to stimulate new application, and refreshing new packaging ideas are emerging. New handbooks have been published to help both the neophyte and seasoned packaging engineer.

A critical element in the continued widespread use of solder is the reliability of the solder joint. Will solder provide the characteristics necessary to allow the world to depend on it in the future? This book, *Solder Joint Reliability—Theory and Applications,* edited by Dr. Lau and written by experts in the field, provides a focal point of current understanding. This will help all participants in the soldering world better plan for the future. It is an exciting time for solder, and reliability will be a key element in its development.

Donald W. Rice
Hewlett-Packard Company

Preface

Soldering has a history of several thousand years. It was considered an art until the electronic age, when it was recognized as a technology. Until very recently, however, research efforts into soldering technology have not been particularly aggressive because solder joints in plated-through hole technology do not cause serious reliability problems.

The 1980s have witnessed an explosive growth in the research efforts devoted to soldering science as a direct result of the rapid development of surface mount technology and the growing interest of miltichip module technology. (For a list of papers published after 1985 see references 1 through 202 in chapter 15; for a list of solder-joint papers published before 1985 see reference 185 in chapter 15.) Soldering is the joining method of choice for attaching component to printed circuit board or chip to substrate. In both cases the solder-joint functions simultaneously as the electrical and mechanical attachment medium. Thus, solder-joint reliability is one of the most critical issues in the development of these technologies.

We have now begun to obtain useful insights into the mechanical properties and microstructure of bulk solders and joints under fatigue, creep, and stress-relaxation conditions. Some effective methods of accelerated testing, failure analysis, and life prediction of solder joints have also been reported. We are also achieving a new understanding of how solder-joint reliability is influenced by flux reactions, solder paste, solderability, reflow methods, wave soldering, and cleaning. These results have been disclosed in diverse journals or more incidentally in the proceedings of many conferences, symposia, and workshops whose primary emphasis is material science or electronic packaging and interconnection. Consequently, there is no single source of information devoted to the state of the art of soldering science and technology. This book aims to remedy this deficiency and to present, in one volume, a timely summary of progress in all aspects of this fascinating field.

This book begins with the concerns of solder joint formation and ends with concerns over long term solder joint reliability. The book is divided into four basic parts. Chapters 1 and 2 describe the application of flux, solder paste, and solderability to solder joints. Chapters 3 through 5 review the various soldering and cleaning methods and their effects on solder joint reliability. Chapters 6 through 12 explain the failure mechanisms of bulk solders and joints under fatigue, creep, and stress-relaxation conditions. Chapters 13 through 18 provide the data and methods for the test, design, analysis, and life prediction of solder joints subjected to mechanicals thermal, shock, and vibration conditions.

Some duplication of material among chapters has been necessary if each chapter is to offer the reader all the information essential for understanding the subject matter. An attempt has been made to provide a degree of uniformity in perspectives, but diverse views on certain aspects of solder joint reliability are a reality. I hope that their inclusion here may be seen as an accurate reflection of the state of the art and a useful feature of the book.

For whom is this book intended? Undoubtedly, it will be of interest to three groups of specialists: (1) those who are active or intend to become active in research on soldering science and technology; (2) those who have encountered a practical soldering problem and wish to understand and learn more methods of solving such problems; and (3) students and professors at universities, in view of the fact that today's engineer receives on average less than one hour of instruction on soldering science during his or her university studies. I hope that this book will serve as a valuable reference to all those faced with the challenging problems created by the ever expanding use of solders in engineering practice, and that it will aid in stimulating further research on solder materials, testing and analytical methods, and the sounder use of solders. Reliability of the solder joint is limited only by the ingenuity and imagination of researchers, engineers, and management.

John H. Lau
Hewlett-Packard Company

Acknowledgments

Development and preparation of the manuscript was facilitated by the efforts of a number of dedicated people at Van Nostrand Reinhold. I would like to thank them all, with special mention to Stefania Taflinska for her coordination of the publication process and to Stephen Chapman for his unswerving support and advocacy. My special thanks to Marjorie Spencer who made my dream of this book come true by effectively sponsoring the project and solving many problems that arose during the book's preparation. It has been a great pleasure and fruitful experience to work with them.

Most of the materials in this book have been presented at various ASM/ASME/IEEE/ISHM conferences, symposia, and workshops in the past few years. I want to thank these respected societies for allowing me to organize technical sessions so I can have the privilege to invite the contributing authors to present their research results.

Each chapter of the book was reviewed by at least three individuals who are experts in soldering areas. According to their specialties, each individual reviewed at least three chapters of the book. These reviewers are: Dr. Birendra N. Agarwala of IBM, Professor Donald Barker of the University of Maryland, Dr. Kirk Bonner of Allied-Signal Inc., Professor Hans Conrad of North Carolina State University, Dr. T. Dixon Dudderar of AT&T Bell Laboratories, Dr. Darrel R. Frear of Sandia National Laboratories, Dr. Sung K. Kang of IBM, Dr. Joseph Kevra of Alpha Metals, Dr. Ken Kinsman of Intel Inc., Dr. Larry Moresco of Fujitsu Computer Packaging Technologies, Inc., Dr. Yi-Hsin Pao of Ford

Motor Company, Dr. Donald Rice of Hewlett-Packard Company Dr. Charles G. Schmidt of the Stanford Research Institute, Professor Karl K. Stevens of Florida Atlantic University, Dr. Boon Wong of Hughes Aircraft Company and Dr.Chee C.Wong of AT&T Bell Laboratories. I want to thank them for their many helpful comments and constructive suggestions that added significantly to this book.

I express my deep appreciation to all 30 contributing authors, experts in their respective fields, for their many helpful suggestions and cooperation in responding to requests for revisions. Their depth of knowledge, dedication, and patience have been demonstrated throughout the process of preparing this book.

I thank Dr. Donald W. Rice for bringing me to this wonderful world of electronic packaging and interconnection and for many fruitful discussions and much strong support in the past six years at Hewlett-Packard (first HP Laboratories and now HP Corporate Manufacturing). I also want to thank my daughter Judy and my wife Teresa for their consideration and patience by allowing me to work on this project on many weekends.

John H. Lau
Hewlett-Packard Company

SOLDER JOINT RELIABILITY

1

Flux Reactions and Solderability

Colin MacKay

This chapter attempts, on the basis of experimental results available, to explain the mechanism, for those fluxes that have been studied, by which they appear to work. This chapter is not a discussion of what fluxes are available for electronics soldering, of how they are classified, or how they are used or applied. It does not comment about the perceived relative merits of the various types available.

1.1 FLUX HISTORY

Originally the idea of a flux as a soldering aid developed from the metal coatings industry where the essential nature of having a clean surface before tin or zinc coating was well known. Whether iron or copper, sheet metal was immersed in an acid pickle solution prior to dipping in the molten metal bath. When soldering of two components was required, it was obvious that similar cleanliness requirements applied. Consequently it became the practice to use the same chemicals as for hot dipping, applied dropwise onto the joint area, to clean the joining components of oxide or other contaminating reaction product.

The other consideration when hot dipping was to protect the molten metal bath surface from oxidation. This was done with a "pot cover" of mixed fused salts that floated on the molten metal. The second property required of a solder flux is to provide a similar protective covering to the cleaned metal of the joint components between the time that the surfaces are cleaned and the molten solder wets the joint interface. This is exactly analogous to the pot cover operation and it is not surprising that a mixture of the materials used for the two operations developed for engineering soldering applications. The liquid acid component cleaned the joint surface, and as the temperature continued to rise and the liquid evaporated, the remnant salts fused in situ to provide the pot cover. Wetting of the joint metals by a solder with subsequent capillary penetration completed the soldering operation. Copious washing to remove the fused salt residues was, at that time, no disadvantage.

Under these conditions it was easy to show that as the acid concentration dropped, the time taken to clean the surface increased; consequently it took longer for the soldering operation, capillary rise, and penetration to initiate and complete. Soldering in the fledgling electrical industry used the same technology, but it soon became apparent that washing to remove the residues had to be more discrete, localized, and thorough, since water and salt residues represented a far more serious hazard in electrical circuitry than did corrosion in a heavy, over-engineered mechanical assembly. The natural precautions of only fluxing a small area around the joint and carefully washing the residues from that area carried over into the earliest electronics assemblies where heavy wiring harnesses of individual wires and large components were involved. With the advent of the printed circuit board and the associated wave soldering operation, residues were distributed over the whole surface of the electronics assembly and their presence became a serious consideration.

It was at this time that rosin-based fluxes became serious candidates for development and rapidly took over the market as electronics fluxes. The assumptions about how fluxes worked and what controlled their efficacy were simply carried over from the earlier ideas. As the complexity of the formulations of fluxes developed, so did their proprietary nature. Black magic pays when there aren't many witch doctors, so there was not a great effort to explain how these improved fluxes worked. Flux classification became a matter of considering the aftereffects of soldering; therefore tests and activity classifications became increasingly concerned with the activity of the residues and not with the effectiveness of a flux in promoting the soldering process. Little concern and less explanation was available about how these complex formulations worked or what controlled the magnitude of their effectiveness.

1.2 SOLDERABILITY TESTS

The most important requirement of any attempt to measure flux effectiveness is a quantitative solderability test that would allow discrimination between different formulations. Many tests have been reported in the literature, most of them based upon some aspect of the soldering process. These fall under the headings:

1. Visual assessment
2. Area of spread tests
3. Edge dip tests
4. Capillary tests
5. Globule tests
6. Rotary dip tests
7. Surface tension balance tests

1.2.1 Visual Assessment

Visual assessment relies upon inspector experience and memory. It offers no quantatitive evaluation but with experienced personnel offers a rapid comparison of a result with previously agreed visual standards.

1.2.2 Area of Spread Test

The area of spread test is a quantitative test that involves the spread, when melted, of a weighed cylindrical tablet of solder, fluxed and heated on a square coupon of substrate. The area is usually determined by planimetric methods. This test does not have a time element; therefore it cannot indicate anything about wetting rates. The test is allowed to continue until the spreading action has stopped. It indicates the combination of the maximum substrate-solder interaction with the minimum solder-air or solder-flux surface tension.

1.2.3 Edge Dip and Capillary Rise Tests

Both the edge dip and capillary rise tests involve the edgewise dipping of either a single flat piece of substrate material or a piece bent so that it creates a capillary gap. Assessment is usually qualitative, the edge dip test in particular being used to indicate solderability quality aspects (i.e., nonwetting, dewetting, pinholes, etc.) The capillary rise variant could be used quantitatively if the meniscus rise height were measured. These tests, like the previous one, measure an equilibrium situation, stressing to a different extent the same interfacial forces.

1.2.4 Globule Test

The globule test is a very geometry-specific test designed for solderability assessment of wires (preferably round wires). In this a weighed cylindrical tablet of solder is heated on an anvil. When the solder is molten, a wire is lowered into the solder bead. The time from initial contact until the bead coalesces around the wire is a measure of wetting time. This test also includes an element of applied pressure to the wetting interface resulting from the hydrostatic pressure of immersion and is slightly biaxial in that the bead may wet along the wire as well as around it. This is a rate measuring test.

1.2.5 Rotary Dip Test

Rotary dip testing is designed to measure the wetting time of flat sheet material in a system that emulates the wave soldering process. It copies the relative motion of solder and substrate by sweeping the test coupon tangentially across the surface of a solder bath (as opposed to the solder wave sweeping the circuit board surface). A contact timer indicates the wetting time for each test and a series of samples is required for a series of different rotation speeds of the sample arm. For a determination of optimum rotary dip solderability, the minimum solder contact time that gives complete coverage of the test coupon is determined. This test involves an element of squeegee action, a complex combination of applied pressure, and enhanced thermal contact behavior that aids and speeds the wetting process. Clearly this is a dynamic test.

1.2.6 Surface Tension Balance Test

The surface tension balance or wetting balance test measures the vertical force acting on a test coupon from the time wetting is initiated until it is completed or the force profile up until the time exceeds some preset maximum. Both dynamic and equilibrium parameters are measured by this test in that the time to reach any specified position in the wetting process can be determined. Also, provided the wetting interface reaches equilibrium within the duration of the test, the wetting force at the equilibrium wetting condition can likewise be determined. Commonly the endpoints chosen are the time for the wetting force to equal the upthrust force for a particular set of immersion depth and immersion rate conditions, the time to initiate wetting, and the time for wetting to be completed. Wetting rates at any position in the process can be calculated from the profile and the final wetting force is measured directly.

1.3 FLUX ACTION FROM SOLDERABILITY MEASUREMENTS

A quantitative test is required in order to make an effective measurement of flux action, and the type of test used determines which aspect of flux action is being quantified. The area of spread test determines factors independent of time that relate to the advance of the wetting interface across a surface. The surface tension balance test determines rate factors that relate to the speed of wetting. Both the rotary dip and the globule tests are more complex tests and are probably not good tests for determining flux mechanisms because the results tend to be test parameter dependent. The globule test is influenced by the size of the wire and the size of the solder pellet, and the rotary dip test is influenced by the rotation speed of the test arm.

1.4 FLUX TYPES

1.4.1 Mechanistic Studies for Inorganic Fluxes

On the basis of the historical development of fluxes, the purpose and action of the flux was to clean the substrate of all contaminating films and then to protect the clean surface from reoxidation until the coating metal or solder melts and, by reacting with it, wets the surface. According to this description of the process, any resulting response to a solderability test describes the effectiveness of the cleaning (and/or the protection). Implicit in this would be that once the surface was fully clean, the solderability would be optimum and could not be improved.

To test this, Turkodan and Zador[1] used a reducing atmosphere furnace to reduce the oxide film on a degreased and cleaned steel strip. After flooding the vessel with a protective, nonreducing atmosphere, a weighed quantity of tin was placed on the strip and the area of spread measured. Optimization of the reducing atmosphere and experiment parameters showed that the spread area reached a maximum and thereafter stayed at this value, which represented the ultimate spreading value possible with fully cleaned steel. When a weighed quantity of inorganic flux salts was applied with the tin or solder under the same optimum cleaning conditions, the spread area increased to being greater than the value that had been obtained for clean steel. Clearly flux in this form provided some additional aid to the wetting operation beyond what was possible from simply applying tin to a clean metal surface.

The explanation originally offered was on the basis of a classical triangle of forces equilibrium involving the three surface tension components, $\gamma_{H2/solid}$, $\gamma_{solid/liquid}$, and $\gamma_{H2/liquid}$ (Fig. 1-1):

$$\gamma_{(H2/solid)} = \gamma_{(solid/liquid)} + \gamma_{(H2/liquid)} \mathrm{Cos}\theta$$

where θ is the contact angle measured up from the solid interface to the surface of the liquid. The area of spread for the fluxed system was greater than for the hydrogen atmosphere system because $\gamma_{(flux/liquid)}$ was smaller than $\gamma_{H2/liquid)}$.

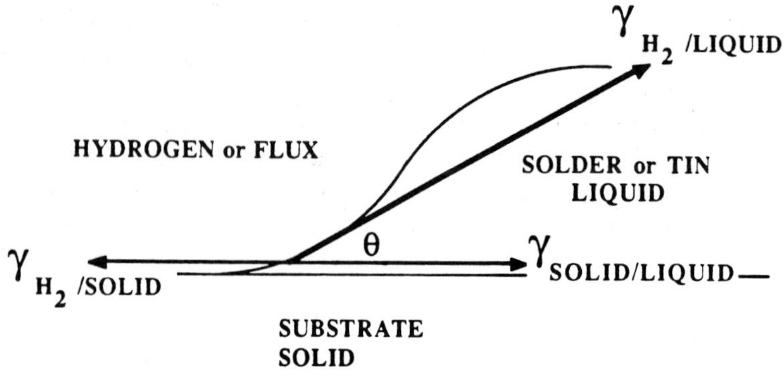

Figure 1-1 Wetting force equilibrium.

Effect of Impurities on Solderability

Later workers[2–9] who used the area of spread test attributed the performance improvement that they found to the reaction of the flux salts with the molten solder, creating an alloy of the metal of the flux salt with the tin or solder, and suggested that the alloy produced spread more readily than the tin or solder. This, however, ignored results of other workers[10] who had studied the effects

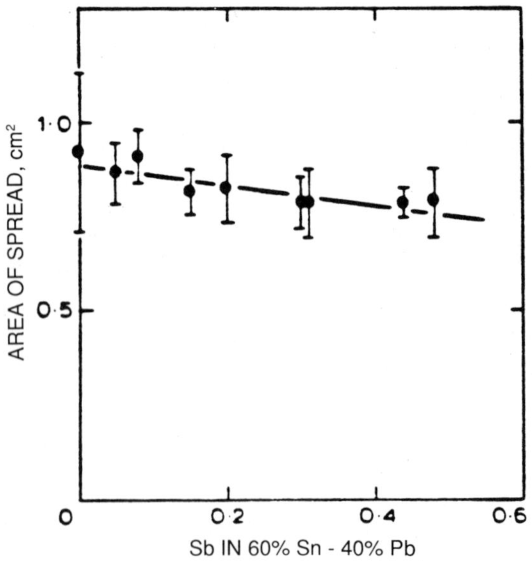

Figure 1-2 Effect of antimony on area of spread of 60% Sn-40% Pb solder on brass base metal (zinc ammonium chloride flux).

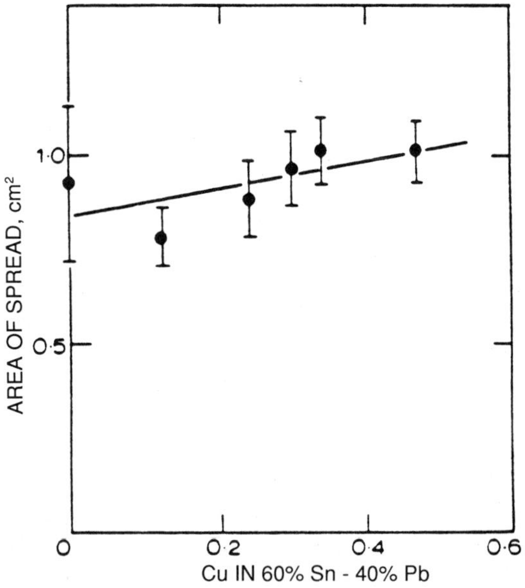

Figure 1-3 Effect of copper on area of spread of 60% Sn-40% Pb solder on brass base metal (zinc ammonium chloride flux).

of deliberate additions of a range of impurities to 63% Sn-37% Pb and 60% Sn-40% Pb alloys and had shown that in all cases this resulted in a decrease in solderability. Some effects of impurities on area of spread results are shown in Figures 1-2, 1-3, and 1-4 for the cases of low levels of antimony, copper, and zinc impurity on the solderability of brass with 60% Sn-40% Pb when a zinc ammonium chloride flux was used.

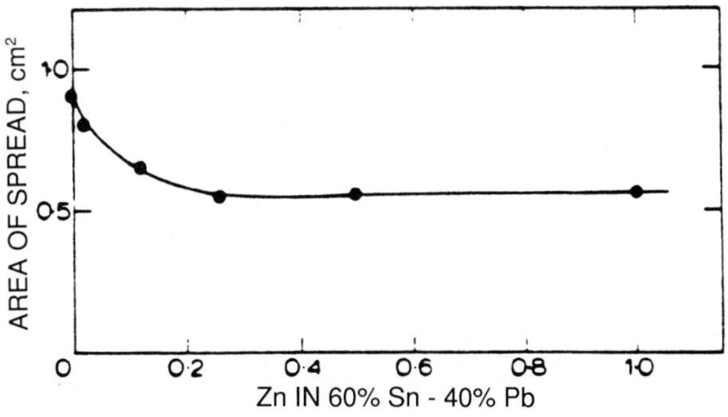

Figure 1-4 Effect of zinc on area of spread of 60% Sn-40% Pb solder on brass base metal (zinc ammonium chloride flux).

Only for the case of copper with brass as the substrate was any increase in spread area observed. This was attributed to reaction of the copper with the zinc dissolved from the brass surface to form small particles of copper-zinc compound, which was observed in microstructure of the joint, and not the formation of a modified solder alloy. The copper actually removed impurity arising from dissolution of the base material. In all other cases, for the brass, as shown, for copper, and for steel substrates the area of spread was decreased. Similar effects were also observed for wetting balance tests.

Table 1-1 shows a synopsis of these results for all impurities when fused salt fluxes were used. The percentage column gives the impurity level percentage that caused a 25 percent reduction in the area of spread value. Decreased areas were observed both in systems that reacted with the solder to produce intermetallic compounds with the impurity, as with copper, and in systems that dissolved the impurity to form a solution in the melt. Only 0.4 percent copper was required to initiate the formation of Cu_6Sn_5 intermetallic in 60% Sn-40% Pb alloy, in which it formed a fine suspension of needles of the compound. This in turn caused the viscosity of the melt to increase with the corresponding inability to flow and spread. Silver reacted in a similar manner, forming tiny platelets of Ag_3Sn. The effect of this compound was less marked than for copper. Antimony compound does not form until the composition reaches about 8 percent, so at the low levels used in these experiments it appeared to effect the spreading properties only as a fully dissolved element.

Table 1-1 Effect of Impurity on Area of Spread Values on Various Substrates Fluxed with Zinc Ammonium Chloride

Impurity	Test Piece	Percentage of Impurity to Produce 25% Reduction in Area of Spread
Antimony	Brass	0.9
	Steel	No effect
Arsenic	Brass	No effect
	Steel	0.2
Bismuth	Brass	5.1
	Steel	2.8
Cadmium	Steel	1.9
Copper	Steel	0.5
Zinc	Brass	0.04

An Electrochemical Mechanism

From the results mentioned earlier[2–9] it was clear that a large range of inorganic salts could be formulated into fused-salt-type fluxes and would demonstrate very significant improvements in solder flow characterisitics. The most probable mechanism for the fluxing effect with inorganic fluxes, as put forward by Latin[11] and supported by both Bailey and Watkins[12] and Turkodan and Zador,[1] was electrochemical. All of these studies detected an electrochemical potential between either tin or solder and the substrate material in a cell with the fused salt as electrolyte. Experiments[13] were made with a cell of the type shown in Figure 1-5, using fused salt mixes of ammonium chloride with stannous chloride and ferric chloride, with tin or solder as one electrode and iron or copper as the other. The gray film referred to by Latin[11] and Bailey[12] was analyzed by scanning electron microscopy and was found to be crystallites of Cu_6Sn_5 on the substrates, when either formulation was used at a temperature above the melting point of the solder or tin. When lower temperatures were used, a thin film of solid tin was observed. With steel as the substrate a thin film of $FeSn_2$ was found. This would indicate that the flux acts as an electrolyte of a fused salt electrochemical cell.

The positive effects of a tin coating layer on solderability are well known, so the deposition of tin had a clear beneficial effect since the substrate surface was replaced by one that was more solderable. The reason for other materials having beneficial action could be attributed to the extreme reactivity of freshly reacted materials. Nascent hydrogen, freshly evolved, is a well-known example of this phenomenon and freshly deposited metals also show this property. In the case of a soldering situation any freshly deposited metal would be laid down immediately before the advancing solder front at the wetting interface, ideally situated for establishment of a metal-metal interaction layer and essential for surface wetting.

When the work with a wide range of inorganic salts was examined, clearly some large effects were observed. Although the explanation given was faulty,

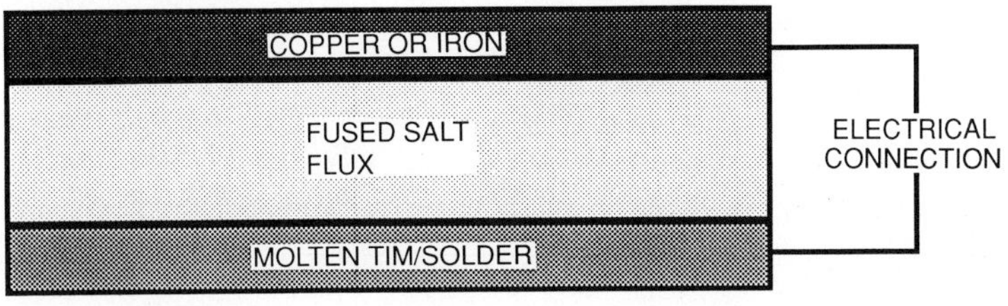

Figure 1-5 Test cell demonstrating electrochemical effect with fused salt fluxes.

the effects were real. Scrutiny of these results in the light of the material properties of the compounds used showed a clear correlation between the metal ion potentials and the free energies of the compounds, with the observed wetting responses. For all the systems tested (chloride, sulphate, and nitrate compounds), those with metal ion potentials below about −0.75 V had little, if any, effect on area of spread. Metal ion potentials above this showed a progressively larger effect (Fig. 1-6).

Similarly, compounds with high free energies were found to be less effective than those with low values, with the lower energies showing a progressively larger area of spread. The free energies of chlorides and nitrates, at which an effect began to be seen, were similar, at about 100 kcals/mol (Fig. 1-7). Sulphates in general had high free energy values and were effective at values below about 250 kcals/mol. Other properties included in the analysis were the valency of the metal, the compound melting point, and the atomic weight. All these showed no obvious correlation.

It seems, therefore, that for spreading capability of solder with inorganic salt fluxes, the prime controlling factors are the combination of the electropotential

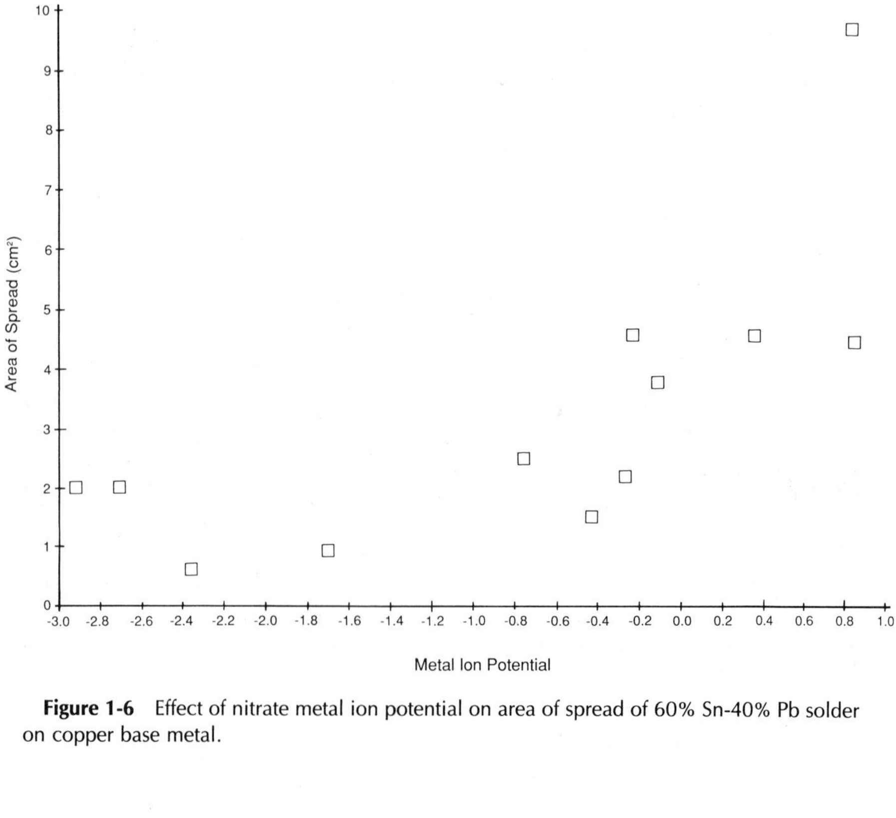

Figure 1-6 Effect of nitrate metal ion potential on area of spread of 60% Sn-40% Pb solder on copper base metal.

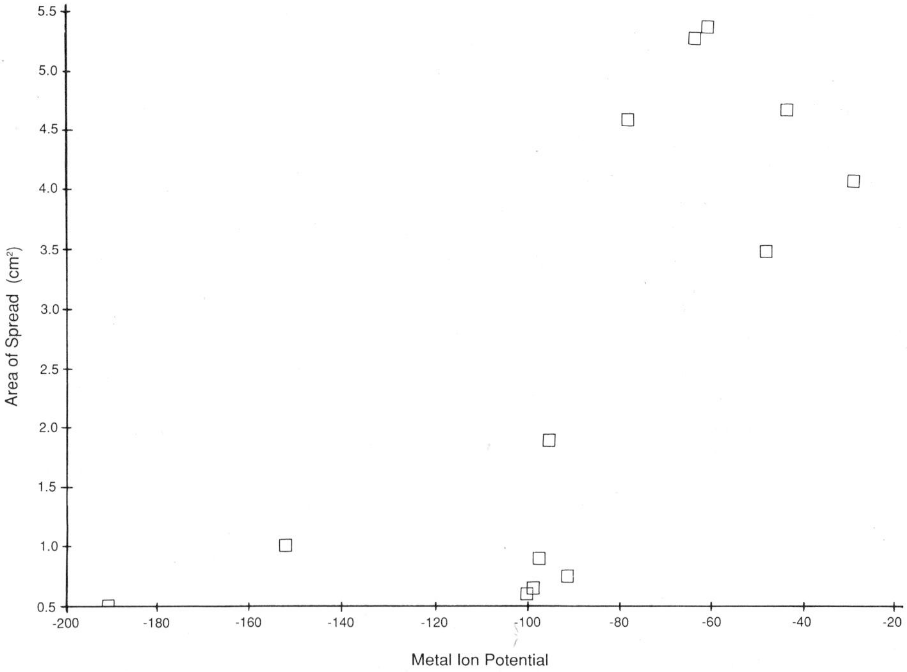

Figure 1-7 Effect of chloride metal ion potential on area of spread of 60% Sn-40% Pb solder on copper base metal.

of the metal ion and the ability of the compound to dissociate in a fused state. This agrees with an electrochemical mechanism being involved in the deposition of a metal ion onto the substrate in front of the advancing solder.

1.4.2 Mechanistic Studies for Rosin-based Fluxes

As mentioned previously, the reasons for using fluxes are well known. First, they clean the surface to be soldered by removing the oxidation and/or other chemical contamination. Second, they form a protective blanket over the clean surface to prevent reoxidation until the temperature becomes high enough for soldering. However, fluxes for electronics applications rarely consist of simple chemical solutions; usually they contain complex formulations designed to be both as wide ranging and process specific as possible. For example, a flux might be chosen so that it can cope with a particularly difficult surface or perform more quickly.

For electronics applications, fluxes are based on an organic rosin vehicle and contain the minimum possible activity so that any residues are noncorrosive.

Assessment of a flux's activity and assignment to a designated activity level among the classes acceptable for electronics applications depend upon a combination of criteria that often reduces soldering performance to a secondary consideration. Some classifications are by percentage of activator, others distinguish fluxes by the activator used, while still others combine water extract resistivity and flux corrosiveness on a thin copper film deposited onto glass as the definitive classifier.

Nevertheless, it is not realized that equivalent formulations conforming to all activation levels can be compounded from different amounts of activators that are not necessarily related. These fluxes could behave differently on a series of clean metals or on a range of increasingly contaminated or oxidized copper surfaces. Indeed, fluxes are called upon to operate on a wide range of oxidation levels, not simply on a set of carefully controlled surfaces. Both the speed and quality with which a flux renders such a range of surface solderable reflect the flux's usefulness.

Rosin

The principal constituent of an electronic flux is rosin. This is a natural product derived from the resinous excretion of trees. It is a glassy mix of abietic acid with some polymeric compounds of abietic acids and numerous hydrogenated modifications of that acid. The basic chemical structure is shown in Figure 1-8.

Since it is a natural product, the proportions of the various constituents will vary from genus to genus and also for different variations within a botanical group. In order to obtain as consistent a material as possible and thus produce a consistent, reliable flux product, rosin from a single species is used. For

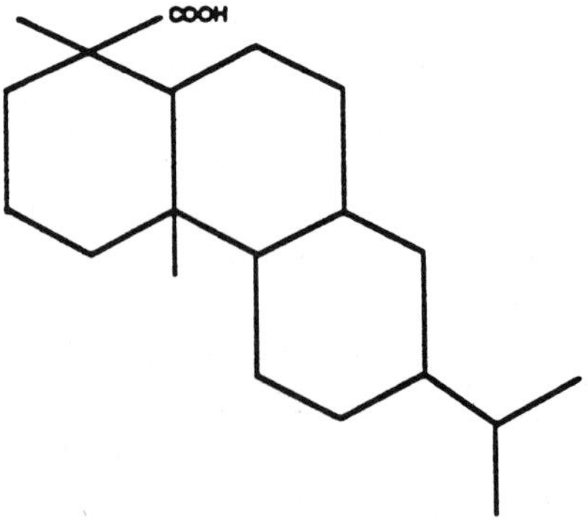

Figure 1-8 Basic structure of abietic acid.

example, in Europe, the material commonly used is derived from a specific type of Portuguese pine; in America, a North Carolina variant is specified. In addition to the gross differences in product constitution found between types of trees, trees of the same type grown in different locations, even material from the same woods in different years, can show significant variations. This obviously affects product uniformity and repeatability.

Because of its solid state and glassy structure the acid action of rosin is not apparent until it melts. Only then do the ions of the material have sufficient mobility to enter into reactions with substrate surfaces. The material has a pK rather than a pH number; that is, it exhibits ionic character only in the fused state.

Rosin Flux Formulations

Although commercial fluxes are formulated as solutions in alcohols so that the dispersed ions do have some mobility at room temperature, there is very little evidence of room temperature reaction on an oxidized copper surface. The main purpose of the solvent is to allow the application of a thin uniform film of the flux or, with further formulation, to allow a specific process application method (for example, foaming), as in the foam fluxing method. Preheating of the fluxed substrate, while serving to reduce thermal shock as the substrate contacts the molten solder surface, also serves to remove the solvent of the flux system and create a thin uniform film of rosin together with whatever activators have been added.

The rosin starts to soften at between 50° and 70°C and is fully fluid at about 120°C. In this condition it is weakly acidic and, without any augmentation, will clean lightly oxidized copper. However, printed circuit boards and components do not come in only one grade of oxidation. Storage conditions and controls produce a range of differently oxidized surfaces and the soldering assembly process is expected to easily cope with all of these and produce 99.99+ percent yields.

Another aspect of the soldering process is the production rate. Line speeds that allow economical production rates have to be achieved, so that although a plain rosin flux might be able to fully clean a surface eventually, the time required to do this could be unacceptably long. In order to improve both aspects of a plain rosin flux, ionic additives (usually chlorides) have been formulated on the basis of the same idea about fluxing mechanisms as applied for inorganic fluxes. Therefore the flux series designated rosin mildly activated (RMAs) contains less activator than does the rosin activated (RAs) series, and so on up to rosin super activated (RSA).

The problem with unbridled application of the most activated materials was that the residues remaining after soldering were corrosive to varying degrees. This led in turn to specification of materials according to the amount and activity of the residue and the inclusion of tests such as the copper mirror test of residue extracts within the activity specification for electronics fluxes.

Technological developments within the formulation industry also served to make any soldering performance-composition correlation more difficult. By preparing formulations with two-part activators that reacted to form the active constituent in situ, the temperature at which reaction with the substrate began was raised from being at room temperature to being above that of the reaction temperature of the two parts. Activators in which the active halogen is hydrogen bonded to a second molecule (usually amine) are also used. The temperature required to break the bond and release the halogen compound represented the minimum activity temperature. Yet higher activity temperatures could be realized with compounds that incorporated halogens covalently; the more stable bond required higher dissociation temperatures. Implied in these developments was the idea that because they required higher temperatures to initiate activity, the residues would somehow be less active after soldering. This completely ignored the fact that often the compound used as a carrier was volatile and was not available for recombination after soldering; therefore, one-half of the initial compound was no longer available for recombination after soldering. Only in the case of covalently bonded compounds was there evidence that residues were less active.

Wetting Balance Performance of Rosin Fluxes

So how well do rosin fluxes of different types perform? Wetting time tests using the standard surface tension balance procedures, for a range of different surface oxidation levels, achieved by immersion in an oxidizing solution, show the following results.[13]

The curves of Figure 1-9 essentially exhibited the same form, consisting of an initial sharp increase in wetting times followed by a plateau that remained nearly horizontal as oxidation levels, represented here by immersion times in an oxidizing solution, increased. Further increases produced an abrupt rise in wetting time immediately prior to the surfaces becoming nonwettable. In this region the results diverge considerably.

Certain features and the shape of the wetting time-immersion time curves become significant if it is assumed that all fluxes deposit an equal film over the test coupon so that the same volume of active constituent is involved in all cases. In this way a controlled quantity of rosin plus activator was available to react

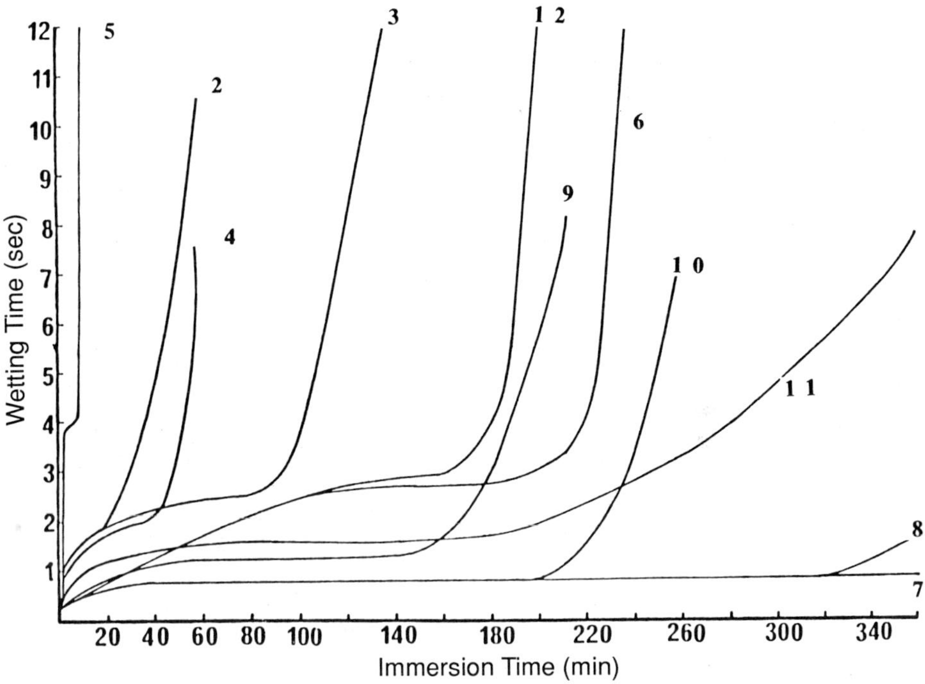

Figure 1-9 Wetting time curves on oxidized copper prepared by immersion in an oxidizing manganate solution, for the range of fluxes shown in Table 1-2.

with the oxide layer and made each sample representative of the bulk flux from which it was taken. Results then became comparable.

The fluxes tested were all of the rosin type with grades RMA and RA. They were also all from a single commercial supplier and the numbers quoted are for that generic product series. Details of the application and performance of these fluxes is given in Table 1-2. Plain rosin flux with no activator was also examined, but since the maximum immersion time treatment that this flux was able to solder was only 1 min, this would not fit on the scale of Figure 1-9.

All fluxes exhibited essentially the same features, namely an initial zone in which the wetting time increased sharply, followed by a level plateau, and then a second sharply increasing region that rose to an infinite wetting time (i.e., nonwetting). In some cases, the plateau was vestigial, as for RMA-1, while for others, such as RA-12, it was extremely extended. In general, the longest plateaus were for the most active (i.e., RA) fluxes. The plateau levels also tended to mirror the activity levels, occurring at the shortest wetting times for the RA

Table 1-2 Properties of Fluxes with Test Results in Figure 1-9

Flux Ref. No.	Classification	Properties
1	R	40% pure water white rosin in alcohol. Not activated, used for electronic soldering where residues are left on boards.
2	RMA	Rosin in alcohol, 37% solids activated with a nonionic agent. Used for electronic soldering by both foam dip and wave fluxing. Residue removal necessary only in most critical applications.
3	RMA	Most active rosin in alcohol flux in its class. 37% solids activated to higher level than (2) with different nonionic activating agent. For electronic soldering; also for dip and wave applications. Residue removal as above.
4	RMA	Rosin in chlorinated solvent. Medium activity flux as (2). Activated with low level of ionic activator (37% solids).
5	RMA	Special low solids content rosin in alcohol activated with nonionic agent (25% solids).
6	RA	Rosin in alcohol, organic acid activated flux. Completely halide free. Reduced residue corrosivity (15% solids).
7	RA	Rosin in alcohol, high solids content (57%) with an ionic activating agent. Does not degrade during extended soldering times.
8	RA	Rosin in alcohol, fully activated with an ionic activator. Residues generally must be removed (25% solids).
9	RA	Rosin in alcohol, low solids (13%) with an ionic activator. Easy cleaning flux.
10	RA	35% rosin in glycol ether solvent with an ionic activator. Low volatility, low maintenance flux.
11	RA	Rosin in alcohol (25% solids), moderate activity. Residues efficiently removed by chlorinated solvents.
12	RA	Rosin in alcohol spirit, special low wicking formulation with an ionic activating agent (25% solids).

fluxes and highest for the RMA and R grades. Some RA fluxes, however, exhibited the same plateau wetting times as the RMA materials; compare, for example, RA-12 and RA-6 with RMA-3.

An Active Constituent, Concentration Dependent Mechanism

From classical chemical studies it is known that the typical oxidation process follows an exponential growth curve and is particularly flat for copper (refer to, for example, *Metals Handbook,* vol. 6) over an extended thickness region. The immersion time scale on these graphs would therefore transpose to an essentially linear increase in oxide thickness. With permanganate treatment, it seemed that any initial nonlinear oxidation part was short, on the order of 1-min immersion time, as indicated by the similarity of the wetting time profile for plain rosin flux with those for the activated series. The treatment still gave evidence of increasing linear oxidation out to 6-h immersion time.

While these results were indicative and comparing them with normal soldering should be done with some caution, they do serve to clarify how the wetting rates are modified by the various compositions and they do contain solid evidence about how these fluxes work. Electron spectroscopy for chemical analysis (ESCA) clearly showed that the oxide film produced by permanganate solution differed from the natural oxide. In the test specimens, proportions of Cu_2O and CuO appeared to be nearly equal, whereas no CuO occurs in a natural film grown at ambient temperatures. The test films contained more CuO than would be expected from a film formed in air at high temperature. Precisely how this disparity affected fluxing action was difficult to assess.

Effect of Impurities in Solder with Rosin Flux

As with fused salt fluxes the general effect of impurities in solders was to decrease the area of spread results, as for example Figure 1-10. The effects on surface tension balance results were mixed and depended upon the phase diagram of the ternary systems that the impurity elements formed with the tin-lead alloy. For cadmium, whose results for both wetting rate and equilibrium wetting force are shown in Figure 1-11, wetting times increase and the equilibrium wetting force decreases. When antimony was the impurity (Fig. 1-12), there was no significant change in wetting time, while the equilibrium wetting force showed a significant decrease. In this situation the area of spread would have shown a decrease. By comparison, when bismuth was present (Fig. 1-13), wetting times actually decreased while the wetting force increased. This represented a smaller area of spread, which was reached more quickly.

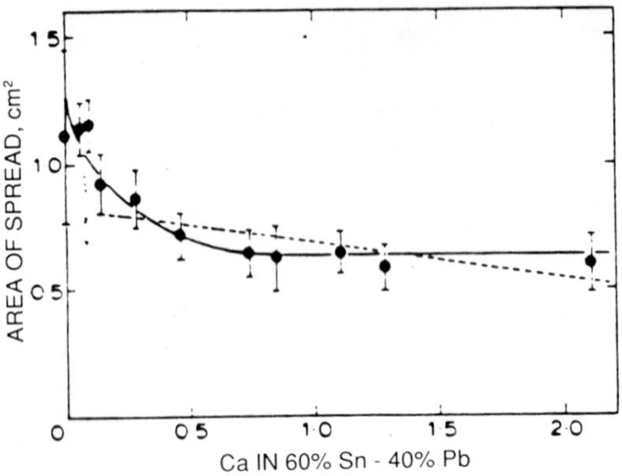

Figure 1-10 Effect of cadmium (area of spread of 60% Sn-40% I solder on copper base metal (nona tivated rosin flux).

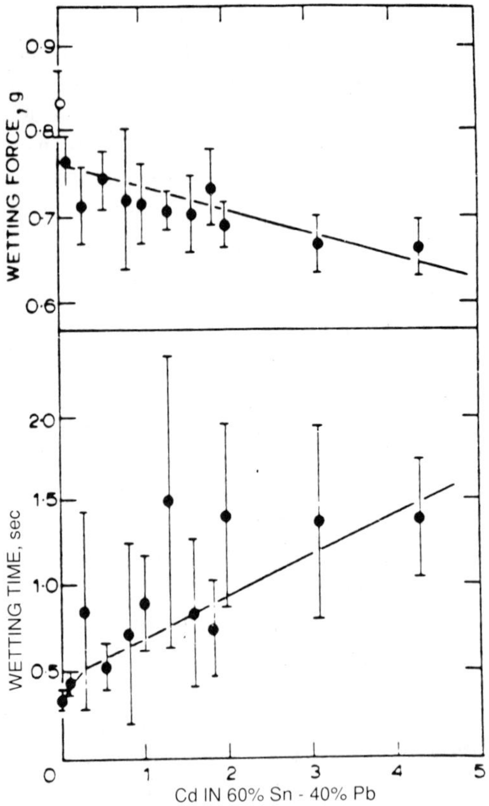

Figure 1-11 Effect of cadmium or wetting properties of 60% Sn-40% Pb solder on copper base meta (nonactivated rosin flux).

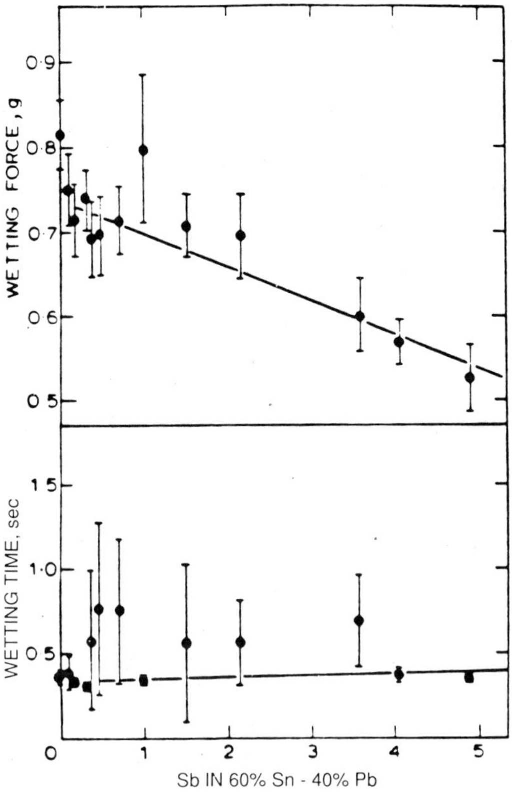

Figure 1-12 Effect of antimony on wetting properties of 60% Sn-40% Pb solder on copper base metal (nonactivated rosin flux).

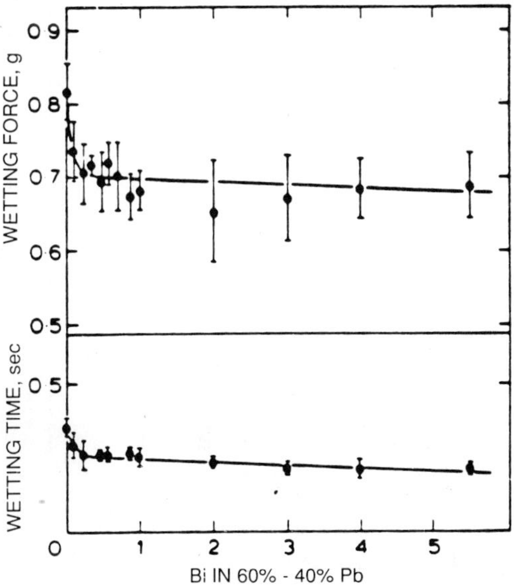

Figure 1-13 Effect of bismuth on wetting properties of 60% Sn-40% Pb solder on copper base metal (nonactivated rosin flux).

Thermodynamics of Flux Action

Thermodynamic studies of the processes taking place have been reported using differential scanning calorimetry (DSC),[14] thermogravimetric analysis (TGA),[14] and Fourier transform infrared analysis.[16] When the DSC and TGA results were examined in combination, so that a peak in the DSC scan, that occurred at the same temperature as an arrest in the DTA scan, could be assigned to a weight loss mechanism, and comparisons were made of results from a progressive series of mixes of components used in a typical flux formulation until a complete model formulation was assembled, certain features of the fluxing-soldering process were seen.

Starting Materials

Abietic Acid The basis material used for the flux vehicle was abietic acid, which was examined by both thermogravimetric analysis and differential scanning calorimetry with traces as shown in Figure 1-14.

The decomposition of this material began at about 150°C, as shown, with most of the reaction as shown by the sudden, sharp weight decrease, taking place

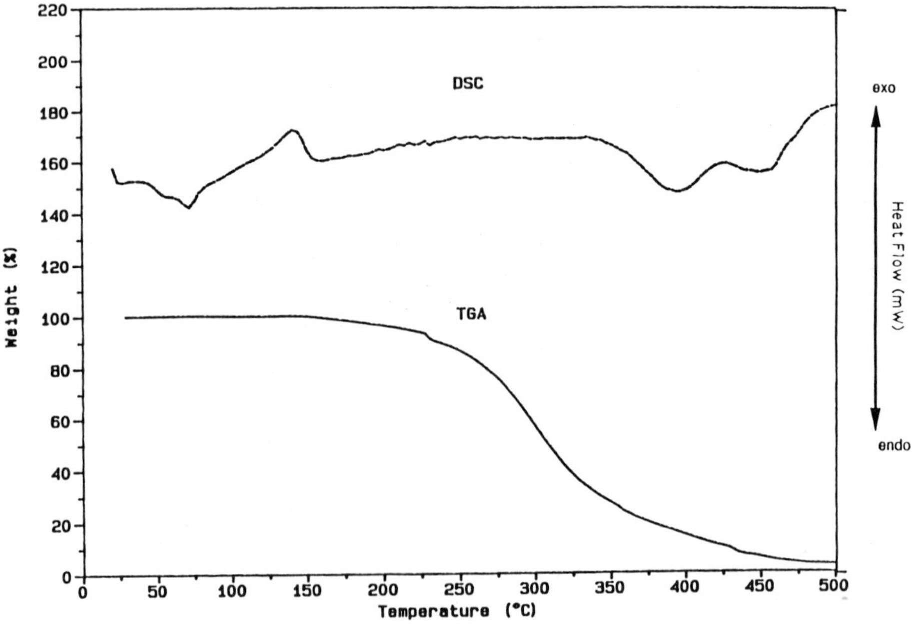

Figure 1-14 TGA and DSC scans of abietic acid.

at about 275°C. By about 475°C the material had fully decomposed, yielding a residue weighing only about 3 percent of the original sample weight.

DSC scans of samples sealed in aluminum pans (Fig. 1-14) showed a twin endothermic peak, the first beginning at 49°C, the second at 52°C, and both persisting to 148°C. These did not correspond to any weight loss temperatures indicated by the TGA scan; therefore they must have represented the complex melting behavior of abietic acid. The twin peaks at about 250°C corresponded to the position of a maximum rate of weight loss in the TGA and represented the decomposition of abietic acid into various molecular fragments. The DSC scan also showed two well-resolved endotherms at 380°C and at 425°C, indicating reaction between fragments of the molecule, possible evidence of polymerisation to dimeric or trimeric products.

When a sample with a vent hole in the lid was scanned in the DSC module (Fig. 1-15) the twin melting peaks were once again apparent. A similar rise was observed at 149°C in the scan for the sealed sample, but instead of decaying gradually, this peak persisted until suddenly decaying at 275°C. This sudden decay coincided exactly with the complete decomposition of the abietic acid in the TGA scans. This supports the contention that the presence of volatiles in the pan (sealed samples) produced further interaction peaks in the DSC scan rep-

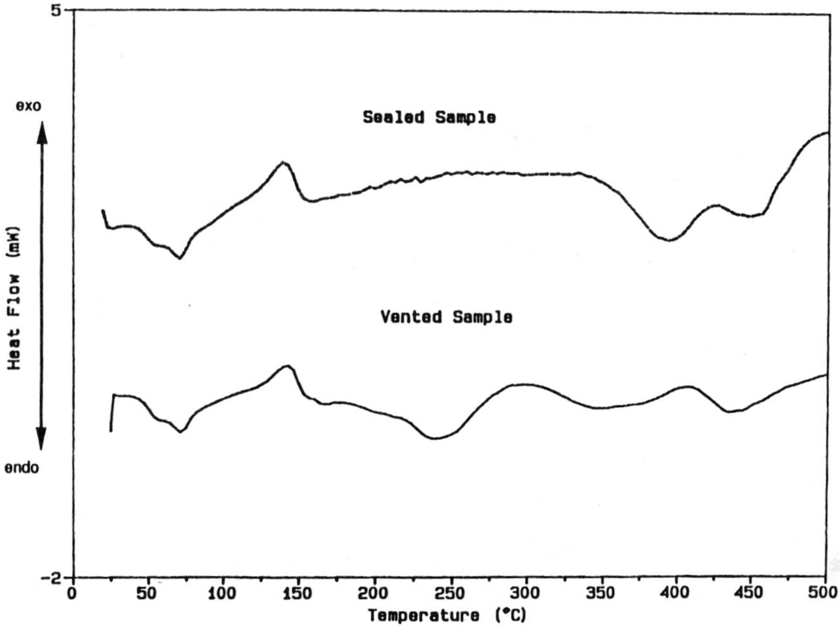

Figure 1-15 Comparison of DSC scans of abietic acid with sealed and vented sample pans.

resenting some other interactions occurring among the volatiles or with the resulting residues.

Isopropyl Alcohol (IPA) A TGA scan of IPA, using a 10°C/min heating rate, starting at room temperature, resulted in a smooth, continuous weight loss curve, ending with a 0 percent weight residue at 90°C. The vented sample in the DSC results showed a curve having one single endothermic peak beginning at room temperature and ending at about 80°C. The two curves were consistent with a simple evaporation process.

Triethanolamine-Hydrochloride (TEA-HCl) Using a 10°C/min heating rate on the TGA, the TEA-HCl weight loss scan of triethanolamine-hydròchloride showed a complex arrest starting at 51°C, with possible side reactions starting at 100°C and ending at 185°C. This arrest represented a weight loss of about 17.5 percent, which agreed relatively well with the theoretical percentage of hydrochloric acid (19 percent) in TEA-HCl. The slightly low value could be caused by adsorbed moisture in the deliquescent TEA-HCl, making the apparent molecular weight higher than 185.7. Indeed, the first, small arrest between 51 and 100°C could be the evolution of water vapor. The major arrest at 275°C

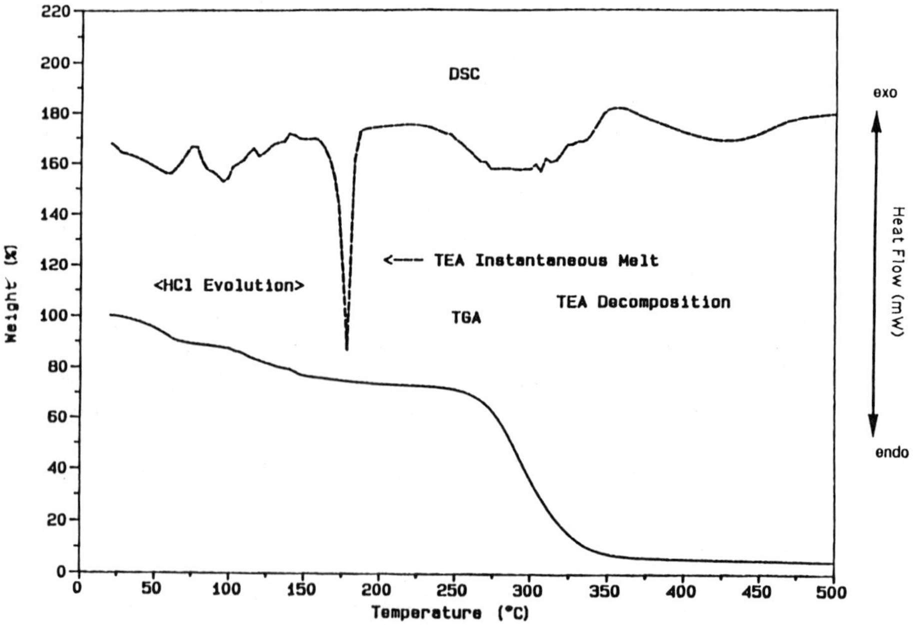

Figure 1-16 TGA and DSC scans of triethanolamine hydrochloride.

represented either the boiling or decomposition (or both) of the triethanolamine that had formed immediately following hydrochloric acid evolution.

The DSC scan for this material (Fig. 1-16) showed an endotherm starting at room temperature, probably associated with adsorbed water vapor, followed by a second, which occurred between 75°C and 90°C. This latter peak was probably caused by the evolution of any water of crystallization. The large peak at 177°C, while indicative of the evolution of hydrogen bonded HCl molecules, was caused by the spontaneous melting of triethanolamine, which was formed at above its melting point when the HCl molecule broke free. This mechanism must take place if the reported physical properties of triethanolamine and its HCl adduct were correct. The dissociation temperature of the adduct was 177°C, while the melting point of the parent compound was only 25°C. The boiling point of triethanolamine, 177°C, was shown in the endotherm starting at about 250°C. A superimposed fine structure on this peak indicated that this material also decomposed at these high temperatures.

Copper Oxide (CuO) The TGA scan for this material, over the entire temperature range, was a continuous horizontal line at 100 percent weight, showing no weight loss. The DSC scan was also a featureless horizontal line, indicating no reactions taking place.

Solder Powder The TGA scan for solder powder was essentially a continuous horizontal line, as for copper oxide. The DSC scan at a 10°C/min heating rate showed a single melting point endotherm at 179°C, which was in excellent agreement with the 179°C melting point of the 62% Sn-36% Pb-2% Ag alloy. As with all other scans, higher heating rates shifted the apparent melting point to higher temperatures so, using a 50°C/min heating rate, the melting point appeared to be 183°C.

Materials Combinations

Abietic Acid Plus IPA In combination, the abietic acid and isopropyl alcohol system represented the solvation of the basic rosin vehicle in a typical flux solvent. These components are considered inert and unreactive with one another. For plain rosin (R grade) flux, which is this system, the active constituent is the rosin, which when molten, behaves as an organic acid. The solution examined in this and subsequent experiments is a 45 percent solution of abietic acid in IPA.

The TGA scans were performed initially with open (no lids) pans. In all cases the scans showed an intermediate arrest, indicating the system was not simply one of classical evaporation. The presence of the arrest was evidence of some type of bonding between the abietic acid and the solvent, resulting in additional thermal energy being required to disrupt the association. This showed up in the

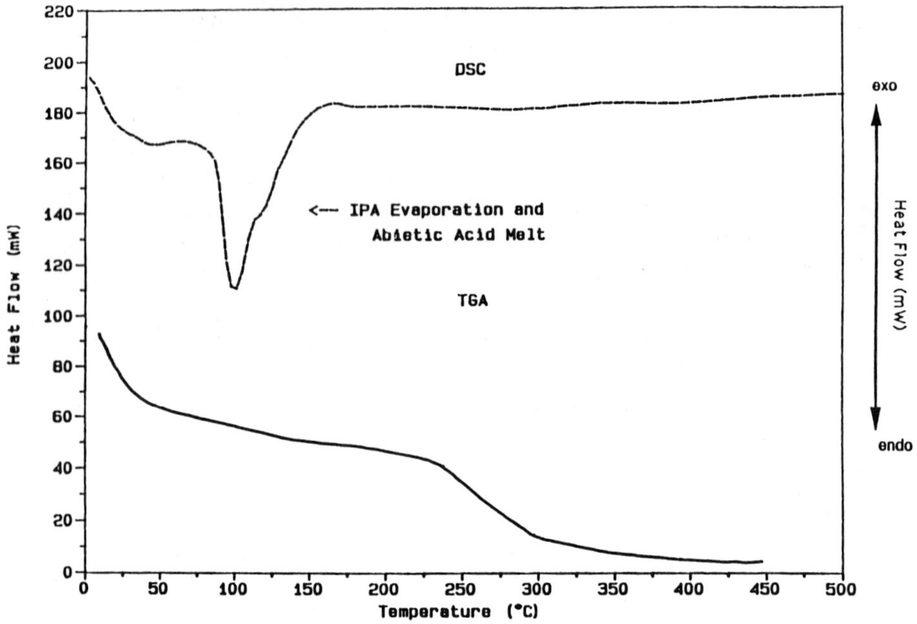

Figure 1-17 DSC and TGA scans of abietic acid plus isopropyl alcohol.

traces as an arrest. When the bonds were broken, the solvent was released and a sudden weight loss was registered. The precise temperature at which this occurred was 150°C. The final residue in all cases was only 3 percent to 5 percent of the original sample.

The associated DSC scans for this system, illustrated in Figure 1-17, indicated a large composite endothermic reaction peak consisting of at least two components. The first initiated at a temperature corresponding to the first weight loss arrest in the TGA scan and represented the evaporation of solvent. The second, which merged progressively with the first, occurred in the middle of the weight loss arrest and represented the melting curve of abietic acid. This was followed by an extended, more shallow peak corresponding with the decomposition of abietic acid. The melting peak also showed a small satellite at the two lowest heating rates, indicating reactions occurring at ~120°C.

Abietic Acid Plus IPA Plus Activator The next stage in the sequence involved adding a typical activator, in this case 0.3% TEA-HCl. The TGA scans were almost identical in comparison to the trace with no activator. The slight subsidiary arrest observed corresponded with the total evaporation of the IPA solvent. The only effect of the activator addition appeared to be that the decomposition temperature of the abietic acid was delayed by 5°C to 10°C. The only

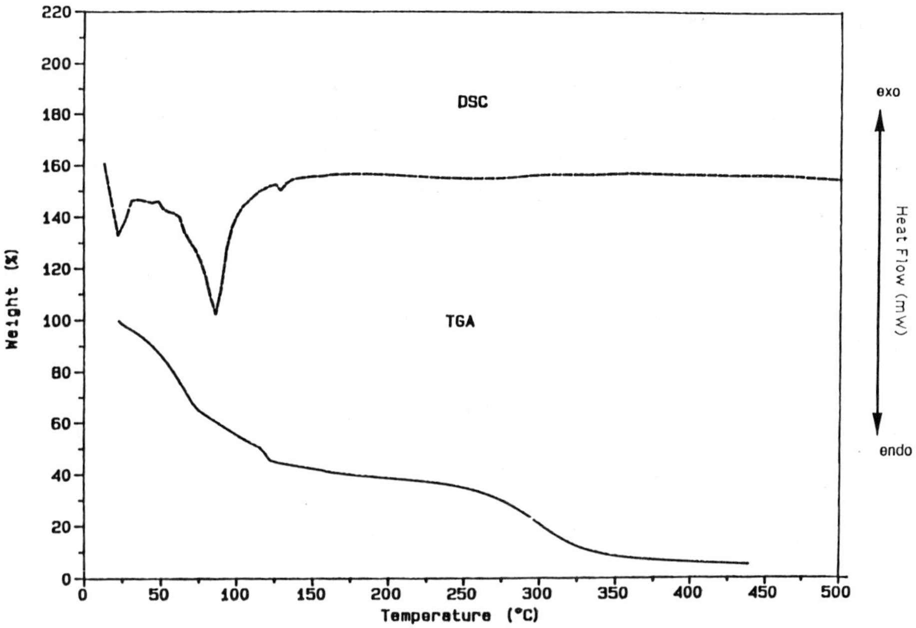

Figure 1-18 DSC and TGA scans of abietic acid plus isopropyl alcohol plus triethanolamine hydrochloride.

other evidence of the small addition of activator was seen on the 10°C/min heating rate curve where a noticeable change in the slope of the weight loss was indicated at ~150°C, which would correspond to the weight loss from the evolution of HCl, as shown in the TEA-HCl scan.

On the DSC scans, illustrated in Figure 1-18, the IPA evaporation peak merged into the abietic acid melt peak. The effect of the TEA-HCl addition could be seen only on the 10 and 50°C/min scans, where the TEA-HCl appeared as a satellite initiating at 130°C, in agreement with the temperature assigned to the evolution of HCl on the pure TEA-HCl curve and the associated large melting peak of tetraethanolamine, which formed spontaneously.

Abietic Acid Plus IPA Plus Activator Plus Copper Oxide Copper oxide was then added to the mixture as a representation of the oxide normally present on a copper surface of an assembly. It was added in excess in an effort to maximize any reaction occurring. When a major problem associated with splattering of the slurry that formed as the solvent evaporated had been overcome, continuous curves were obtained, typically showing three arrests. The first ended with a 68 percent weight loss, corresponding to complete evaporation of the IPA solvent. The second corresponded to the decomposition of the abietic acid and the third, occurring at ~420°C, was associated with the presence of copper oxide in the sample.

Calculations using values from the traces with and without CuO indicated that probably the CuO had not been reduced to copper metal during this experiment.

DSC scans of this formulation with vented samples (Fig. 1-19) showed the same trends as both abietic acid plus IPA and abietic acid plus IPA plus TEA-HCl traces. The twin peaks of the IPA evaporation and the abietic acid melting were the same as for the above two systems with the melting peak being larger than the evaporation. The abietic acid decomposition endotherms were also similar to those observed previously. One additional large feature was the exotherm at 400 to 450°C associated with the presence of copper oxide. There was also a small additional satellite peak at ~140°C.

Sealed DSC samples, while essentially exhibiting the same features as the vented, did reveal some differences. In these, the IPA evaporation endotherms were larger than the abietic acid melting endotherms. The exotherm associated with copper oxide was not so pronounced and occurred at slightly lower temperatures. These two series of DSC scans also indicated that there were differences in the reactions that took place when the volatile components of the reactions were allowed to escape and when they were contained.

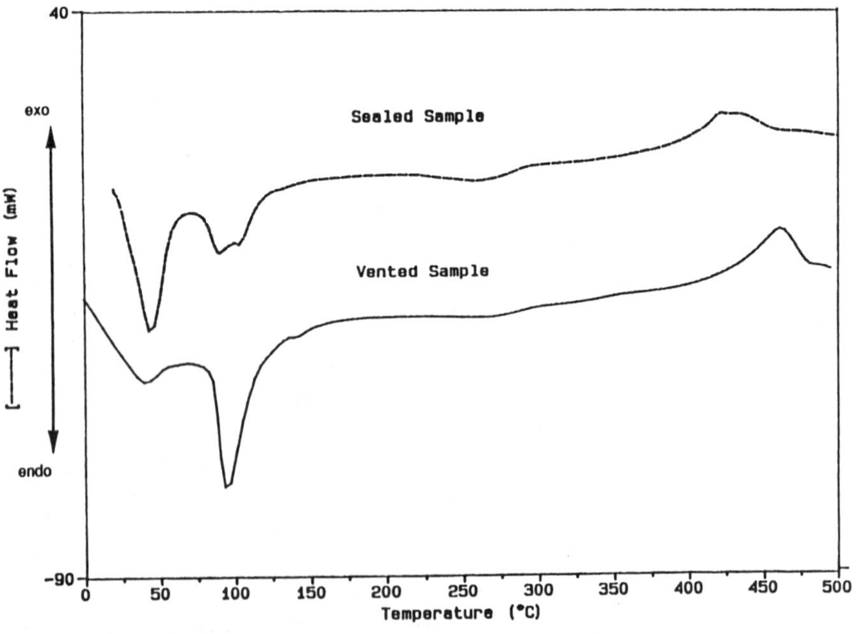

Figure 1-19 Comparison of DSC scans of abietic acid plus isopropyl alcohol plus triethanolamine hydrochloride plus cupric oxide in sealed and vented sample pans.

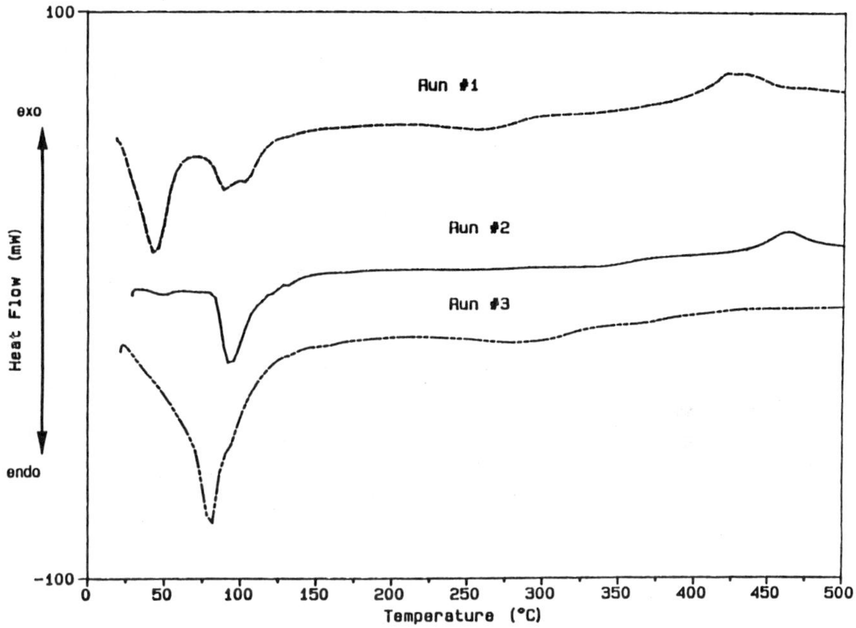

Figure 1-20 DSC scans indicating the progressive reduction of copper oxide with repeated application of flux.

The presence of the peaks associated specifically with the copper oxide, although absent from the raw copper oxide scans, indicate a reaction taking place. The exothermic nature of these peaks suggests a re-oxidation process, although the weight loss seen in the TGA scans seemed to show continued reduction. When the pans were opened after performing the scans, the black CuO had changed to a coppery red color. Microscopic examination of this showed gray-black films over parts of the residue, which might confirm the reoxidation theory.

In an additional set of experiments, the residues from a DSC scan were repeatedly refluxed and rerun. The successive scans (Fig. 1-20) show the gradual reduction of the peak at 400°C and its eventual disappearance. Visual examination of the residue after the final scan showed a sintered coppery mass indicative of complete reduction of the oxide. From these observations it appeared that the reaction progressed in two steps. Copper oxide was first reduced to cuprous oxide, which is reddish in color, and then was further reduced to copper metal.

Abietic Acid Plus IPA Plus Activator Plus Solder Powder As an intermediate for the full simulation of the complete soldering process, small additions of 62% Sn-36% Pb-2% Ag alloy solder powder (~325 mesh) with a melting point at 179°C were added to the flux formulation without any copper oxide being present. The TGA traces of this system showed a minor arrest occurring at a slightly higher weight percent than the weight loss corresponding to loss of all the solvent and a major arrest corresponding to the decomposition of the abietic acid when TEA was present. The weight loss percentage at this point corresponded exactly with that expected for decomposition of the abietic acid when allowance was made for the solder powder.

The main features of the DSC scans, both vented and sealed, were the abietic acid melting and decomposition peaks together with the solder melting peak. In the traces for the two lowest heating rates, vestigial peaks at ~130°C were observed, indicating the evolution of HCl from the small amount of activator present. The interaction of the flux with oxide on the solder powder surface did not show up as a thermodynamic peak in these traces.

Abietic Acid Plus IPA Plus Activator Plus Copper Oxide Plus Solder Powder When vented samples of these full soldering process simulation samples were scanned in the DSC, the major features were again the melting and decomposition of abietic acid and the melting peak of solder. A peak at ~400°C associated with CuO was also present. At the lowest heating rate there was a side peak at ~125°C, as seen in Figure 1-21.

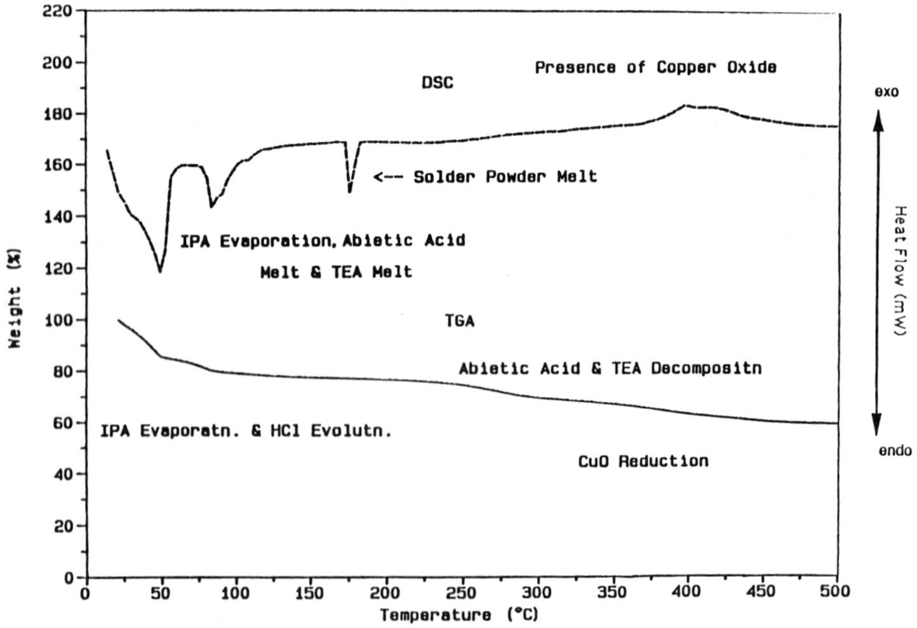

Figure 1-21 DSC and TGA scans of a simulated soldering process abietic acid plus isopropyl alcohol plus triethanolamine hydrochloride plus cupric oxide plus solder.

The scans from samples with sealed pans were more complex than those from vented pan samples. A complex of satellite peaks were observed between ~120 and 150°C; also the single exothermic peak from CuO was split into two, presumably due to an effect of reaction with the constrained vapors.

It was clear from comparison of the results of the vented and sealed DSC experiments that the exact sample conditions have an effect on the reactions taking place. In a real soldering situation obviously the assembly to be joined would not be contained, and volatiles would be able to escape. However, the volume of flux used would be considerably more than used in these experiments; therefore it would be entirely possible that with some of the flux acting as a blanket, conditions similar to those existing in our sealed tests could occur at the flux-metal interface.

The testing using CuO as the simulated surface oxide was clearly too severe for comparison with practical soldering reactions, but it showed some interesting features of the fluxing mechanism, in particular a stepwise process for reducing higher oxides. Exoelectron studies by Momose and Tami[15] had suggest that the lower oxide, Cu_2O, could have been reduced at ~180°C so that Cu_2O could have been a better choice of surface oxide simulation. These results indicated that CuO had a somewhat more complex behavior, possibly associated with reoxidation taking place at ~400°C.

Abietic Acid Plus IPA Plus Activator Plus Copper Powder Plus Solder Powder The results of experiments using copper powder are indicated in Figure 1-22. In these, the peak at ~400°C associated with the CuO was absent. Two shallow extended peaks corresponding to the formation of Cu_6Sn_5 plus Cu_3Sn formed at temperatures consistent with the formation temperatures of these compounds as shown in the phase diagram for this system.

The study was able to show many aspects of the sequence of events taking place during a complete soldering operation. However, it had not been able to show all the sequential details of the reactions involved. In part this was because of the magnitude of the thermodynamic energies involved. The large energies involved in change of state reactions such as melting and evaporation tended to mask the smaller energies associated with chemical reactions such as the formation of copper abiate. The glassy nature of the molten rosin that produced the melting peak over such a broad temperature range was obviously important since it tended to mask any sign of the other reactions taking place in that temperature range in the mixtures studied.

A formulation with only 0.3% activator addition, while it corresponded to the practical situation of an RMA grade flux, did not really allow observation of activator reactions in the flux systems. A higher level of activation, although not necessarily a practical composition, could have been more informative.

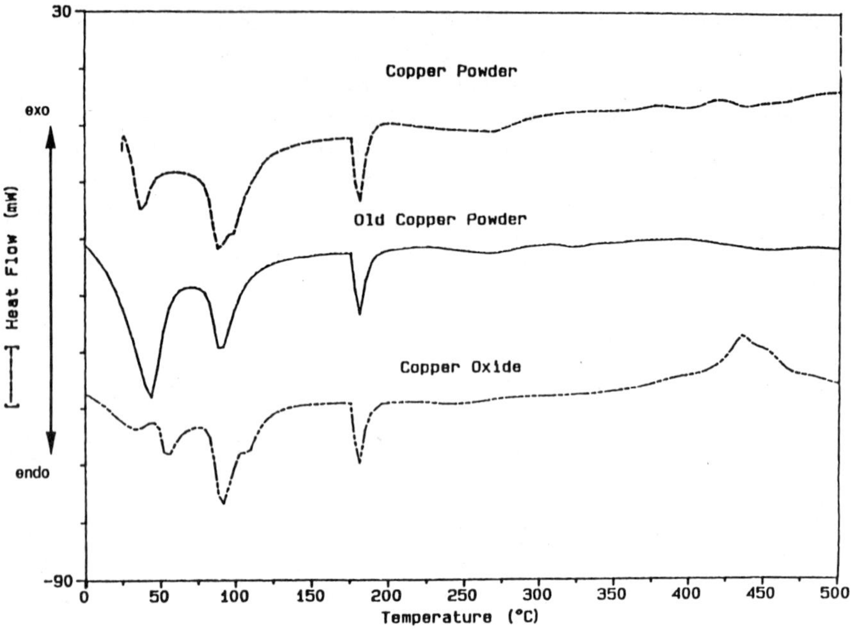

Figure 1-22 Comparison of DSC scans with fresh and aged copper powder with copper oxide.

Fourier Transform Infrared Analysis

Diffuse reflectance Fourier transform infrared analysis (DRIFT) has been used to study the surface reactions of rosin with a variety of metal oxides common to soldering systems.[16] By measuring the diffuse scattering from the surface of a sample and comparing it to the scattering from a reference surface (usually KBr), the absorbance characteristics of the sample can be measured. In the case of flux-oxide powder interactions the diffuse reflectance spectra showed changes arising from reactions within the flux film on the surface of the oxide powder. Copper, tin, and lead oxides were studied, together with salts of copper and lead. In the case of copper and stannous oxide (Sn_2O) the absorption band at 1600 cm^{-1}, arising from the acid group $[COO]^{-1}$, was monitored. From these spectra the presence of copper and tin carboxylate compounds were detected at the lowest test temperatures used, 260°C. With stannic oxide (SnO), no reaction was observed, even up to the highest temperatures used (350°C). This is not really surprising considering the stability of this last compound. The same generalized conclusions did not hold for the lead compounds, in that Pb_3O_4, a highly stable oxide, showed evidence of reaction at the lowest temperature (260°C) but with the absorption band shifted to 1520 cm^{-1}. Other lead compounds reacted differently, in that lead chloride did not react below 300°C, while lead sulphate was unreactive even at 350°C. Lead sulphate is highly stable while lead chloride has a free energy that is similar to stannous oxide.

It seemed from these studies that the reactivity with the flux and the solubility of the oxide or the reaction product with the flux were important factors and that these reactions in the case of lead compounds were complicated.

Rosin Flux Action Inferred from Solderability Measurements

The wetting time experiments mentioned earlier[13] allow some direct inferences as to the mechanism of flux action in a rosin-based system. If flux action were dependent upon reactant concentration, either rosin itself or the activator or both, as has been supposed, from classical considerations the reaction rate of the flux with an oxide film would be simply dependent upon the concentration of the active species in the solution. In physical chemical terms, for a system that depends only upon a single reactant, which is a first order reaction, this can be expressed as

$$\frac{dc}{dt} = kc$$

where c is the concentration of the reacting substance, which is a mathematical representation stating that at any instant the rate of disappearance of the reactant is proportional to the concentration at any instant.

If c_i is the starting concentration and this has reduced to c_t after time t, the remaining reactant concentration is $c_i - c_t$. In that case the above equation becomes

$$\frac{d(c_i - c_t)}{dt} = k(c_i - c_t)$$

or

$$d(c_t) = k(c_i - c_t)\, dt$$

The reaction rate constant shown here is called the specific reaction rate. From this it can be seen that the concentration of the reactant at any time is a function of the form

$$c_t = c_i\, e^{-kt}$$

which is a curve of the form shown in Figure 1-24.

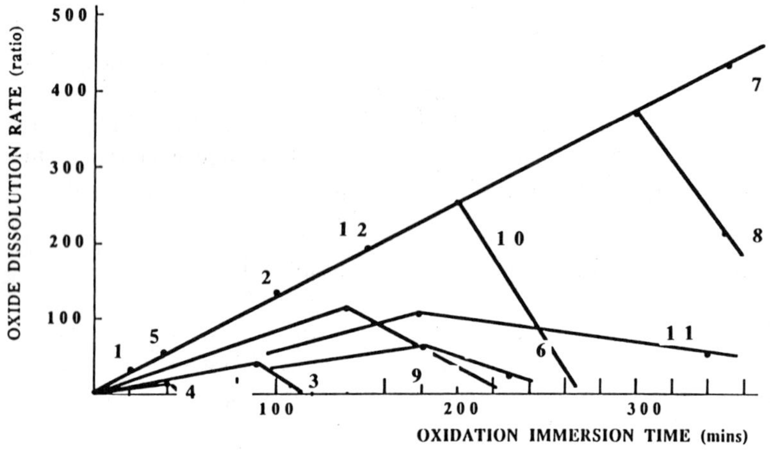

Figure 1-23 Oxide dissolution rates calculated from wetting time curves on oxidized copper prepared by immersion in an oxidizing manganate solution, for a range of fluxes.

For a reaction involving two components A and B the basic equation becomes

$$dx = k(A - x)(B - x)\, dt$$

where A = the concentration of A at $t = 0$
B = the concentration of B at $t = 0$
x = the reduction in concentration of A and B at $t = t$

In the case of a solid film dissolving into a solution being one of the constituents, the concentration of the dissolved solid remains constant until all of the solid has dissolved. Only then does the concentration of that component begin to decrease.

In this case the first part of the reaction behaves like a first-order reaction, while the last part behaves as a second-order reaction. In either case the general form of the relationship between concentration and time is a negative exponential, as before.

The effect of having only a restricted flux volume is to accelerate the rate of decrease of concentration so that the reaction rate drops more sharply. Higher order reactions would have the same general negative exponential form

No matter which precise order of reaction is occurring, the expected wetting time curve under this type of regime would be a continuously increasing wetting time as oxide thickness increased. But this is clearly not what is observed.

If we consider what is actually happening for the points across the wetting time versus oxidation curves (Fig. 1-9), the oxide is steadily being moved until a clean metal surface is exposed. When the metal is clean, the solder metal reacts

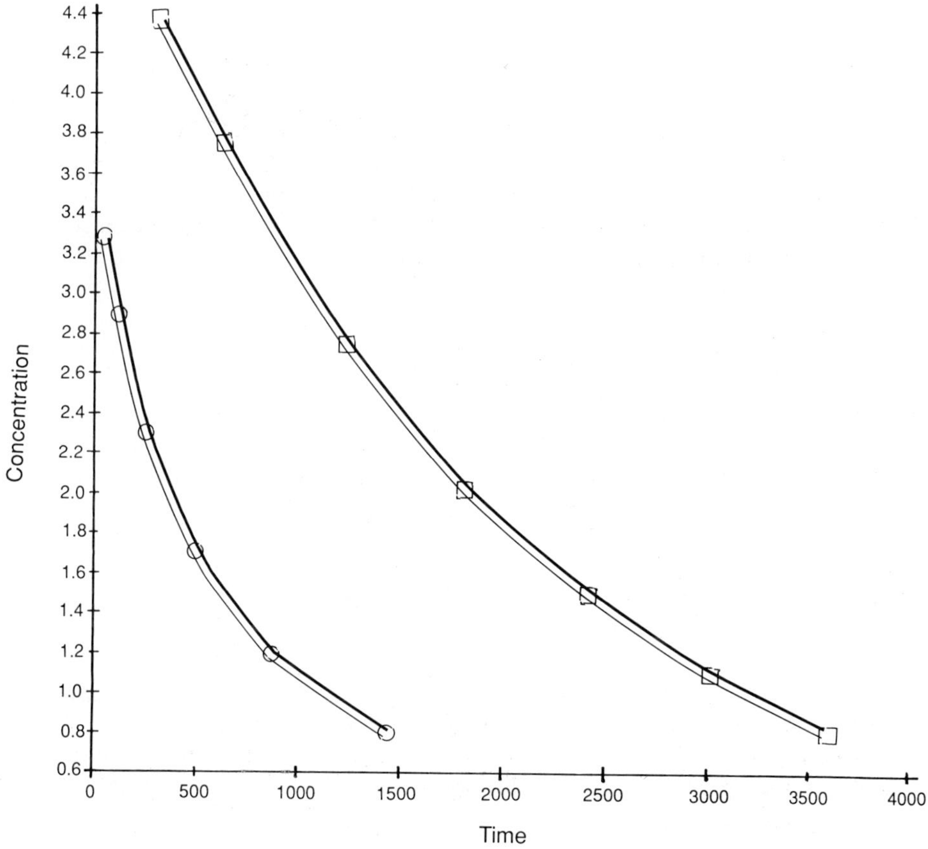

Figure 1-24 Concentration profiles for first and second order reaction mechanisms. ═□═ first-order reaction; ═○═ second-order reaction.

with it and wets the clean copper. The basic wetting time for clean copper is indicated by the intercept of the wetting curves with the wetting time axis when oxidation immersion time is zero. The graph shows this to be ~0.5 sec. However, other experiments indicated that the time shown here was limited by the rise time of the equipment used and that a time between 0.1 and 0.15 sec was probably the true value for clean copper with 60% tin-40% lead solder. The increase in wetting time above this minimum value would then be due to the time taken with the test flux, for the oxidation to be removed. The ratio of the wetting time to the oxidation time at any point on the curve would then be a measure of the reaction rate for dissolution of the oxide in the flux. When these ratios were measured for all points on the curves, it was found that they all fell on a single straight line as shown in Figure 1-23. As can be seen, this shows a positive slope with increasing reaction rate with increasing time, completely the opposite of what was indicated by the first- or second-order concentration driven mechanisms indicated by Figure 1-24.

Autocatalysis and Chain Reaction Equations

As a general rule the rate of a catalytic reaction is proportional to the concentration of the catalyst present. An example of this would be the catalysis of ether by iodine vapour where the equation for the rate of decrease in the amount of ether in the presence of iodine could be written

$$\frac{-d[\text{ether}]}{dt} \; {\scriptstyle [I_2]} \; k_1 \, [c_{I_2}] \, k_2 \, [c_{\text{ether}}]$$

With the large acceleration factors that catalysts impart, k_1 is very much bigger than k_2 and becomes the controlling factor in the reaction. When one of the products of a reaction is itself a catalyst for part of the reaction, the process is called autocatalysis. The effect of this process on the rate of a reaction, $A \Rightarrow P$ say, that obeys the law

$$\frac{d[P]}{dt} = k[c_A]\,[c_P]$$

would be that if the concentration of A at $t = t$ was $A_o - x$ and of P was $P_o - x$, where $x = 0$ at $t = 0$, then

$$\frac{dx}{dt} = k[A_o - x]\,[P_o - x]$$

and the fraction of P produced by time t would be

$$\frac{x}{P_o} = \frac{(e^{at} - 1)}{(1 + be^{at})}$$

where $a = (A_o + P_o) / k$ and $b = P_o / A_o$. This is a steadily increasing function with time and not one that decreases exponentially.

An alternative possibility would be a chain reaction in which an intermediate product of the reaction reacted with one of the constituents to create yet more of the intermediate. This could be represented as

$$A_2 + B_2 \Rightarrow 2X$$

which could proceed via the intermediate reactions

$$\begin{aligned} A_2 &\Rightarrow 2A \\ A + B_2 &\Rightarrow AB + B \\ B + A_2 &\Rightarrow AB + A \end{aligned}$$

with the reaction quenched by the process

$$B + AB \Rightarrow B_2 + A$$

and

$$2A \Rightarrow A_2$$

This gives eventually a relationship of the form

$$\frac{d[AB]}{dt} = \frac{K_a\,[c_{A,2}]\,[c_{B,2}]}{K_b + [c_{AB}]\,[c_{B,2}]}$$

This also represents a continuously increasing function.

On the basis of the form of the reaction rate curve that is observed in the surface tension balance experiments (Fig. 1-23), it seems clear that the mechanisms that would support a process as indicated here would be either autocatalysis or chain reactions. In the first case the reaction could be catalyzed by the same ions, either copper or hydrogen, that were being generated during oxide removal. A combination involving the halogen ions from the activator system would also be possible. Such driven catalytic reactions are not unknown.

A chain reaction system could be such that a secondary reaction involving copper ions, possibly even involving transitions from valence state Cu^{++} to Cu^{+}, assumes the rate determining function. Combinations of these could be possible together with the possibility that an exothermic reaction could further accelerate reaction rates.

The possibility that the increase in temperature arising from an exothermic reaction increased reaction rates to the extent indicated would tend to be discounted, since calorimetric results do not show large exothermic peaks.

Implications of Alternative Mechanisms

The implication of the differences in fluxing mechanism from a simple concentration dependence could be wide reaching. Perhaps injections of less corrosive compounds than currently used activators, but which contain the necessary ions, could replace the corrosive residue inorganic halides currently used. A source of copper ions, for example, could show a significant activity increase. In this way very highly activated systems that have no reactive residues could be formulated. In a similar vein questions involving the use of other ionic species, possibly other metal systems, certainly deserve consideration and study.

An example one can cite of where this type of study could lead, is the problem currently looming in the field of solder creams for surface mount applications where finer and finer detail requires finer solder powders for satisfactory screen printing performance. This in turn means higher oxide percentages in the powder requiring, in turn, more flux activity to allow solder reflow during the soldering operation. Finer printed circuit board detail is usually associated with smaller printed surface board-component joint gaps making it more and more difficult to clean residues from increasingly more activated solder creams. This seems to be a classical vicious circle that only a more enlightened approach to studies on how flux systems work can break.

REFERENCES

1. Turkdogan, E. T., and S. Zador, *J. Iron and Steel Inst.*, March, 1961, pp. 233–239.
2. Onishi, I., I. Okamoto, and A. Omori, *Trans. Japan Welding Inst.*, **1 (1),** 1972, pp. 23–27.
3. Ibid., **2 (1),** 1973, pp. 113–119.
4. Ibid., **2 (2),** 1973, pp. 97–102.
5. Ibid., **2 (2),** 1973, pp. 103–110.
6. Ibid., **3 (1),** 1974, pp. 99–103.
7. Ibid., **3 (1),** 1974, pp. 105–109.
8. Ibid., **4 (1),** 1975, pp. 79–84.
9. Ibid., **4 (1),** 1975, pp. 85–90.
10. Ackroyd, M. L., C. A. MacKay, and C. J. Thwaites, *Metals Technology,* February 1975, pp. 73–85.
11. Latin, A., *Trans. Faraday Soc.*, **34,** 1938, pp. 1384–1395.
12. Bailey, G. L. J., and H. C. Watkins, *J. Inst. Metals,* **19,** 1951, pp. 57–76.

13. MacKay, C. A., *Proc. ASM Conf.* "Microelectronics Packaging Technology," April 1989, Philadelphia.
14. Sherman, K., and C. A. MacKay, *Proc. ASM Conf.* "Microelectronics Packaging Technology," April 1989, Philadelphia.
15. Momose, Y., and Y. Tamia, *J.I.M.*, **93,** 1970.
16. Snyder, R. W., *Applied Spectroscopy,* **14(3),** 1987, pp. 460–463.

2

Solder Paste Technology and Applications

Jennie S. Hwang

The electronics industry, since the invention of the transistor in 1947, has continued to grow and change through evolutionary and revolutionary developments. Most recently, the strongest trend in the design and manufacture of electronics packages is the application of surface mount technology in lieu of conventional through-hole technology in populating printed circuit boards. As the name implies, surface mount technology involves placing components and devices on the surface of the circuit board, instead of through the board, by using engineering and scientific principles. This concept has been adopted in hybrid assemblies since the 1960s, yet its actual application in printed circuit assemblies was not fully developed until the early 1980s.

The aggregate merit of surface mount technology is superior performance-cost ratio for printed circuit board manufacture, in relation to conventional through-hole process. This is a result of increased circuit density, decreased component and board size, reduced weight, shorter interconnection and therefore improved electrical performance, and adaptability to automation.

For surface mounting, the most practical interconnecting material used today is solder paste because of its virtues of "deformable" viscoelastic form, readiness for automation, and tacky characteristic. The deformable form renders it applicable in any selected size and shape, and the tacky nature provides the capability of holding components and devices in position without the need of adhesive before the permanent metallurgical bonds are formed. The bonds are metallic in nature, providing electrical, thermal, and mechanical functions. Therefore, performance and quality of solder paste are critical to the integrity of solder joints, which, in turn, are critical to the overall reliability of the electronics assembly.

Solder paste, by one definition, is a homogeneous and kinetically stable mixture of solder powder, flux, and vehicle, which is capable of forming metallurgical bonds at a given set of soldering conditions and can be readily adapted to automated production in making reliable and consistent solder joints.

From the technological point of view, solder paste involves the interplay of several scientific disciplines. Sciences and technologies used include metallurgy and particle technology, chemistry and physics, rheology, and formulation technology as shown in Figure 2-1.[1] The three functional components of a solder paste are solder powder, flux system, and vehicle system. The solder powder is the only permanent component forming the metallurgical bond; the flux causes the solder powder and the substrate surfaces to be joined in a clean and metallic state so that good wetting and metallic continuity between solder and the surfaces can be formed; the vehicle is a carrier for solder powder and flux to provide a designated rheology. The final performance of a solder paste can be grouped as applicability, solderability, residue characteristics, and solder joint integrity. Figure 2-2 illustrates the correlation between the major components, the final performance, and the fundamental sciences and technologies. It is indicative that the solder paste is a genuine example of an interdisciplinary product and therefore demands an interdisciplinary approach.

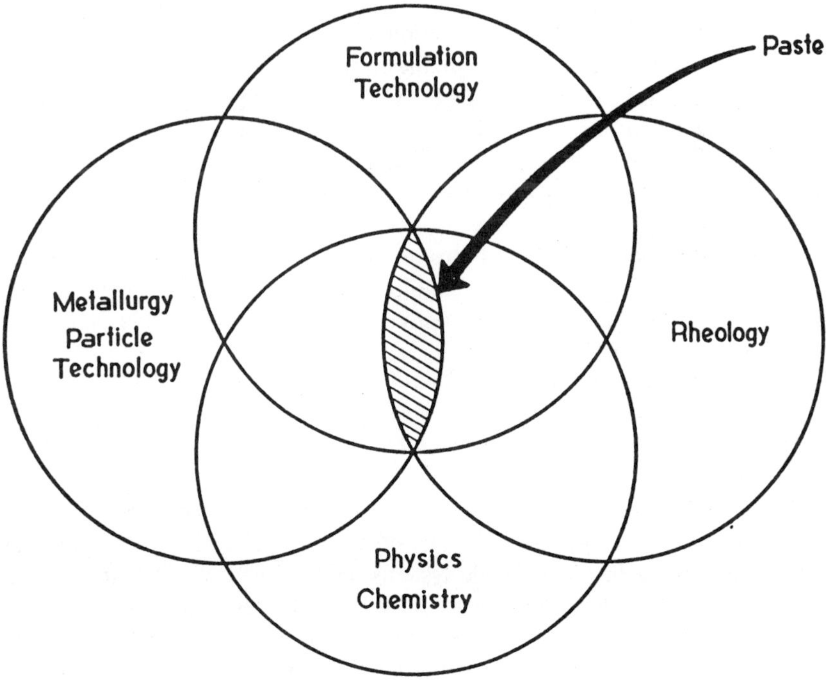

Figure 2-1 Interdisciplinary technology.

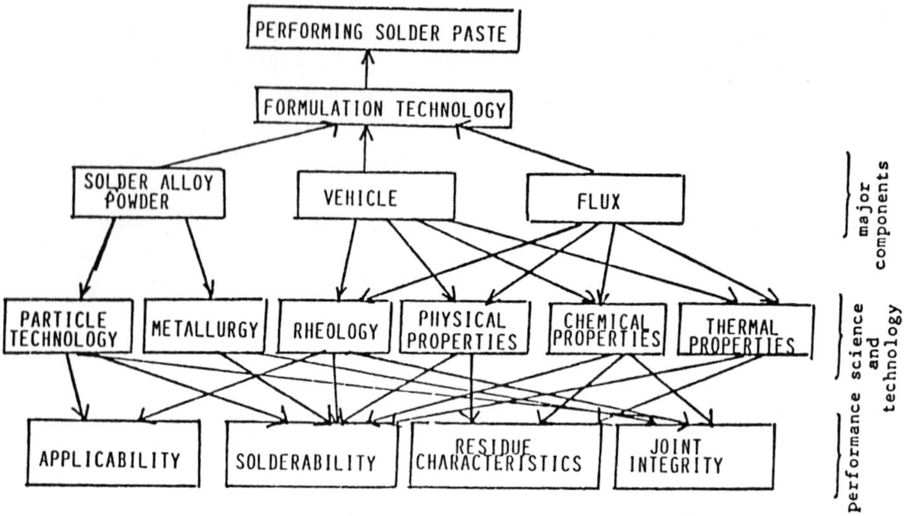

Figure 2-2 Solder paste technology.

Under the main thrust of surface mount technology, specific trends and driving forces prevalent in the electronics industry include increasingly complex and dense integrated circuitry, interconnection reliability, the chlorofluorocarbon-ozone issue, and cleanability and cleanliness measurement. These forces call for increased knowledge and continued new product development in all aspects of design, material, process, and equipment. For solder paste, finer and finer pitch paste, aqueous-cleaning paste, no-clean paste, and controlled atmosphere soldering are the new products and new processes to be developed and/or implemented.

With this demand in mind, this chapter is intended to provide an integrated knowledge of a full spectrum of technologies, techniques, and application of solder paste with emphasis on the new products, new processes, and new issues, as previously mentioned.

2.1 CHEMICAL AND PHYSICAL CHARACTERISTICS

In addition to the characteristics of solder powder, which are discussed later, the performance of a solder paste equally depends on the chemical and physical characteristics of the flux and vehicle systems. In this regard, the flux and vehicle have to be treated as one entity because in solder paste the functions of flux and vehicle are closely related and inseparable. A typical composition of the flux-vehicle system consists of a range of 5 to 20 ingredients. Each ingredient or group of ingredients provides a necessary function, such as binding, fluxing, controlling rheology, suspending, or other targeted functions. The specific characteristics to be considered are:

Boiling point of component
Boiling point of liquid phase
Melting point of component
Melting point of solid phase
Softening point of resin
Vapor pressure of component
Vapor pressure of system as a whole
Chemical functional groups and reactivities
Chemical reactivity versus temperature
Tackiness of resin versus surrounding medium
Environmental stability of component
Environmental stability of the system as a whole
Hygroscopicity of component
Hygroscopicity of system as a whole
Thermal stability of component
Thermal stability of system as a whole
Cohesive force of system
Viscosity of system
Metal load acceptability
Surface tension of liquid phase

Among the performance parameters of solder paste listed in Table 2-1, the relationship between the performance parameters and the above-listed characteristics is not always straightforward because of the potential interaction among ingredients and the interrelationships among the characteristics. For example, the boiling point of the component and the liquid phase, chemical functional groups and reactivities, tackiness of resin, and thermal stability of component and system all contribute to the residue properties in physical form (hard or soft), level of corrosivity, and cleanability. The tack and open times are highly dependent on the boiling point and vapor pressure of component and system, softening point and tackiness of resin, chemical functional groups and reactivities, environmental stability, and hygroscopicity of component and system. The rheology, wetting ability, and solder balling rely on the delicate balance of all the above characteristics. The cohesiveness of the flux-vehicle system is particularly important to the printing and dispensing efficiency. The metal load acceptability is most associated with paste appearance and the relationship between viscosity and printability-dispensability, as well as with the tack and open times.

Table 2-1 Solder Paste Performance Parameters

Stage	Parameter
Before soldering	Physical appearance
	Stability and shelf life
	Cold slump
	Dispensability through fine needle
	Screen and stencil printability
	Tack time
	Open time
	Adhesion
	Quality and consistency
During soldering	Compatibility with surfaces to be joined
	Flow property before becoming molten
	Flow property at and before molten
	Wettability
	Dewetting phenomenon
	Solder balling phenomenon
	Bridging phenomenon
	Wicking phenomenon
	Leaching phenomenon
Post soldering	Residue quantity
	Residue cleanability
	Residue corrosivity
	Electromigration
	Joint appearance
	Joint voids
	Joint strength
	Joint microstructure
	Joint integrity versus mechanical fatigue
	Joint integrity versus thermal fatigue
	Joint integrity versus thermal expansion coefficient differential
	Joint integrity versus intrinsic thermal expansion anisotropy
	Joint integrity versus creep
	Joint integrity versus corrosion-enhanced fatigue
	Joint integrity versus interfacial intermetallics
	Joint integrity versus bulk intermetallics

2.2 FLUXING AND FLUXES

The function of fluxes in solder paste is to chemically clean the substrate surfaces to be joined, to clean the surface of solder powder, and to maintain the cleanliness of both substrates and solder powder during soldering so that a metallic continuity at the interface and a complete coalescence of solder powder can be achieved.

Broadly, chemicals that contain functional groups that are active or become active at soldering temperature can be considered as fluxes. Organic chemicals containing active carboxylic groups ($-COOH$), amine groups ($-NH_2$, $-NHR$, $-NR_2$), and organic salts, such as hydrochloride or hydrobromide salts of amines, are common fluxing agents.

In addition to functional groups, the fluxing strength depends on the molecular structure, the surrounding medium, and the temperature in accordance with the organic chemistry principles.[2] For practical applications, this leads to the "flux activation temperature" and "flux activation time" for fluxes individually and fluxes in solder paste. It should be noted that activation temperature and time for the flux alone may be different when the same flux is incorporated in the solder paste composition. Some fluxes are active at ambient room temperature and others need elevated temperature to be activated. The normal activation temperature can range from 20 to 200°C, and the activation time can range from instantaneous to several minutes. Kinetically, flux activation temperature and activation time are always interrelated.

Rosin-based flux has been established as a reliable chemical and is widely used in the industry. The beauty of rosin is that it can provide multiple functions as a binder, a tackifier, a protecting layer, and a flux in the same system. The fact that its fluxing function is activated at an elevated temperature, but not at ambient room temperature, makes it a most desirable flux ingredient.

Rosin, a natural resin obtained from pine trees, is classified as gum, wood, and tall oil rosin, depending on its origin and processing. The unmodified rosin acids consist of abietic acid, isopimaric acid, neoabietic acid, pimaric acid, dihydroabietic acid, and dehydroabietic acid, with the abietic acid as a major component. The abietic acid, which is structurally characterized by one carboxylic group, a condensed three-ring structure, and a conjugated double bond, can be readily modified through different processes.[1]

With numerous grades of rosin available, rosin can be conveniently characterized by the softening point, acid number, saponification number, and color. Federal Specification LLL-R-626D calls for the following specifications for Class A Type I Grade WW or WG:

Characteristic	Specification
Insoluble matter in toluene (maximum percent by weight)	0.05
Softening point (0°C, minimum)	70
Acid number (minimum)	160
Saponification number (minimum)	166

For soldering, rosin is normally accompanied by an active chemical system, which is called activator, to enhance the flux strength. Customarily, several flux types have been used in the industry. They are Type R, Type RMA, Type RA, Type OA, and Type SA with the following definitions and properties:

Type R: Rosin without the presence of activator; mildest flux
Type RMA: Rosin mildly activated; containing both rosin and activator; no halides
Type RA: Rosin activated; higher flux strength than RMA; containing halides
Type OA: Organic acid; high fluxing strength and highly corrosive; aqueous-cleaning
Type SA: Synthetic activator; equivalent fluxing activity to organic acid; chlorofluorocarbon and alcohol cleaning

The flux activity and its potential corrosivity can be measured by water extract resistivity, copper mirror test, halide test as specified in Federal Specification QQ-S-571E, and surface insulation test as specified in Military Specification MIL-P-55110D and IPC specification IPC-TM-650 method 2.6.3.

2.3 SOLDER ALLOYS

An alloy is a material formed from a combination of two or more elements by means of dissolving and mixing them in the liquid state and then freezing them into the solid state through a proper solidification process. Solder generally refers

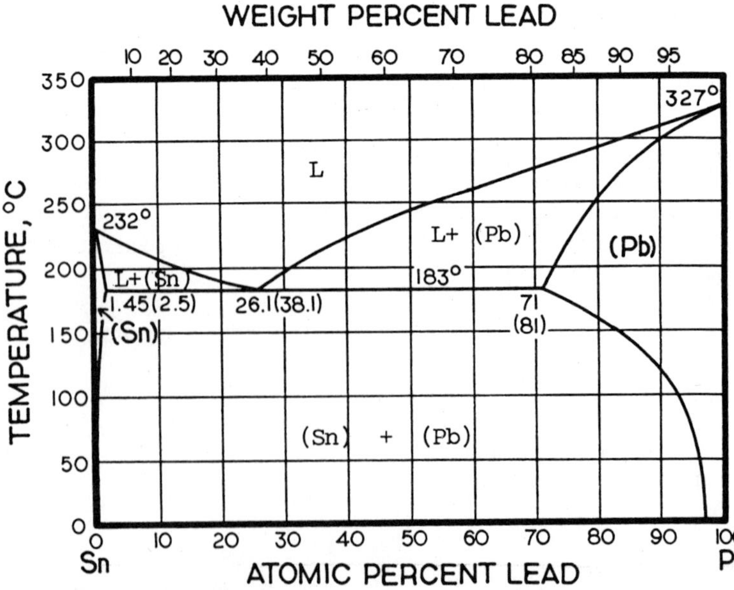

Figure 2-3 Phase diagram of lead-tin.

Table 2-2 Physical Properties of Solder Alloys

Alloy Composition	Liquid (°C)	Solidus (°C)	Ultimate Tensile Strength (ksi)	0.2% Yield Strength (ksi)	0.01% Yield Strength (ksi)	Uniform Elongation (%)
42% Sn-58% Bi	138	138	9.71	6.03	3.73	1.3
43% Sn-43% Pb-14% Bi	163	144	5.60	3.60	2.77	2.5
30% In-70% Sn	175	117	4.67	2.54	1.50	2.6
60% In-40% Sn	122	113	1.10	0.67	0.53	5.5
5% In-95% Pb	314	292	3.66	2.01	1.79	33.0
30% In-70% Pb	253	240	4.83	3.58	3.08	15.1
60% In-40% Pb	185	174	4.29	2.89	2.06	10.7
80% Sn-20% Pb	199	183	6.27	4.30	2.85	0.82
63% Sn-37% Pb	183	183	5.13	2.34	1.91	1.38
60% Sn-40% Pb	190	183	4.06	2.06	2.19	5.3
25% Sn-75% Pb	266	183	3.35	2.06	1.94	8.4
10% Sn-90% Pb	302	268	3.53	2.02	1.98	18.3
5% Sn-95% Pb	312	308	3.37	1.93	1.83	26.0
15% Sn-82.5% Pb-2.5% Ag	280	275	3.85	2.40	1.94	12.8
10% Sn-88% Pb-2% Ag	290	268	3.94	2.25	2.02	15.9
5% Sn-93.5% Pb-1.5% Ag	301	296	6.75	3.85	2.40	1.09
1% Sn-97.5% Pb-1.5 Ag	309	309	5.58	4.34	3.36	1.15
96.5% Sn-3.5% Ag	221	221	8.36	7.08	5.39	0.69
95% Sn-5% Ag	240	221	8.09	5.86	3.95	0.84
95% Sn-5% Sb	240	235	8.15	5.53	3.47	1.06
85% Sn-10% Pb-5% Sb	230	188	6.45	3.63	2.62	1.40
5% Sn-85% Pb-10% Sb	255	245	5.57	3.67	2.26	3.50
95% Pb 5% Sb	295	252	3.72	2.45	1.98	13.70

to fusible alloys with a liquidus temperature below 400°C (750°F). The elements constituting solder alloys include tin (Sn), lead (Pb), silver (Ag), bismuth (Bi), indium (In), antimony (Sb), and cadmium (Cd). The commonly used solder alloys are binary systems: tin-lead, tin-silver, tin-antimony, tin-indium, tin-bismuth, lead-indium; and ternary systems: tin-lead-silver, tin-lead-bismuth, tin-lead-indium, and tin-lead-antimony.

Regarding an alloy, a phase diagram, represented by composition versus temperature, is a useful tool for anticipating the metallurgical behavior of the alloy over a range of temperatures. Depending on the mutual solubility and miscibility of the elements involved in both the solid and liquid states, and on other metallurgical interactions, a phase diagram with different features will result. For example, Figure 2-3 shows the phase diagram of a tin-lead binary alloy showing the solid solution regions (α or β), the combined liquid solution and solid solution region ($L + \alpha$ or $L + \beta$), a solid solution mixture region ($\alpha + \beta$), and a liquid region (L). The presence of solid solution regions indicates that a tin-lead alloy will respond to designed heat treatment or to unpurposed thermal excursion.

Some mechanical properties of selected alloy composition are summarized in Table 2-2.

2.4 SOLDER POWDER

The technique of producing solder powders is largely based on an atomization process; that is, the disintegration of molten alloy into particulates by means of an energy-rich stream of inert gas or a centrifugal rotating disc. The process involves a number of operating parameters. To produce good quality powders with reasonable yield requires the understanding of each parameter in relation to the powder properties and also requires a balance among the operating parameters. The major properties of solder powder to be considered are:

Melting range	Particle size distribution
Flow rate	Oxide content
Apparent density	Dross content
Particle morphology	Elemental impurities
Particle size	Solderability

The melting, or mushy, range is the temperature difference between solidus and liquidus temperatures. For eutectic compositions where the solidus temperature coincides with liquidus temperature, there is no melting range. The melting range of tin-lead solder compositions can range from 0°C for eutectic composition 63%

Sn-37% Pb to 83°C for 25% Sn-75% Pb. The melting range can readily be determined by using a differential scanning calorimeter (DSC). However, considering that a phase diagram is derived at equilibrium conditions and that most processes and techniques used are under nonequilibrium conditions, the thermogram obtained should be referenced to the corresponding phase diagram for thermodynamic phase transitions and liquidus and solidus temperatures. The thermograms as shown in Figures 2-4 and 2-5 can serve as thermal fingerprints for a specific alloy.

The flow rate of solder powder is measured by a Hall flowmeter, in accordance with ASTM Method B-213. The flow rate can have a range of 8 to 14 sec/50 g, depending on the particle size and distribution, as well as on the surface condition of the powder. Generally, very fine powder size may tend to agglomerate and, therefore, hinder the flow efficiency. A "free flowing" condition is a minimum criteria for an acceptable powder. Apparent density is measured according to ASTM Method B-212; its value is significantly lower than that of bulk solder and falls in the range of 4.5 to 7.0 g/cc, varying with alloy composition.

The desirable particle morphology of solder powder includes a spherical shape and a smooth surface without satellites as shown in Figures 2-6 and 2-7. The particle size and distribution should be tailored to fit the needs of applications

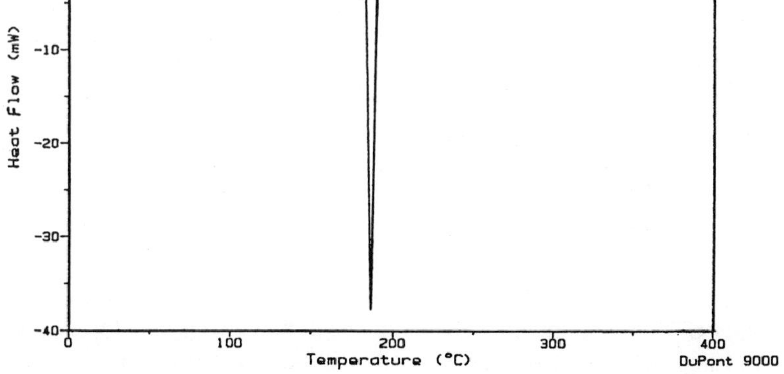

Figure 2-4 DSC thermogram of 63% Sn-37% Pb solder powder.

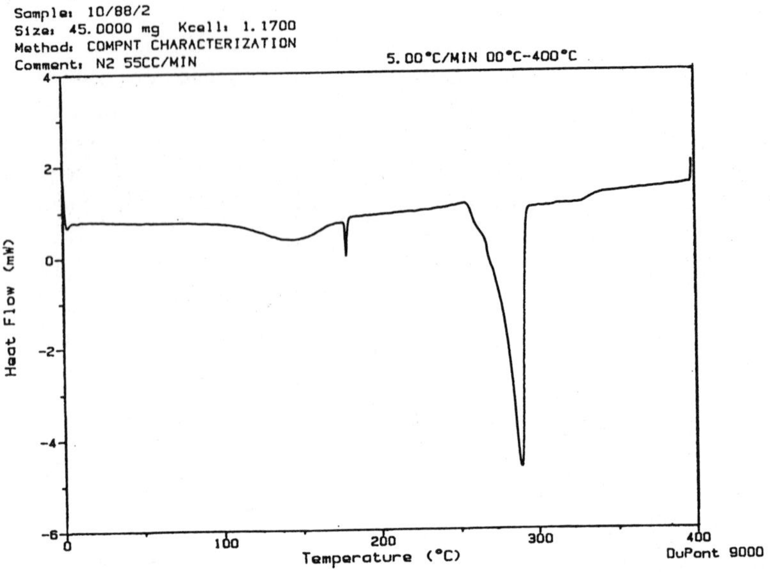

Figure 2-5 DSC thermogram of 105%-88% Pb-2% Ag solder powder.

with respect to rheology and solder paste disposition land pattern. The required particle size for different groups of applications is discussed later. For particles coarser than 44 μm (i.e., +325 mesh), the conventional sieve analysis according to ASTM Method B-214 is adequate for separation and specification. Below 325 mesh, a special technique is needed to separate different ranges of powder particle size.

The oxide content of solder powder can be determined by an oxygen detection apparatus. Truly oxide-free powder is hardly attainable. Although excessive oxide content is detrimental to solderability performance, with the state-of-the-art technology, powder containing oxygen in the range of 50 to 200 ppm is considered satisfactory. The dross content, a measurement of cleanliness of the powder, is another indicator of the quality of powder; a content below 0.5 wt %, as determined by wet chemistry technique, is a satisfactory level.

The limits of elemental impurities for tin (for nonantimony solder), bismuth, copper, iron, zinc, aluminum, arsenic, and cadmium are specified under Federal Specification QQ-S-571E as shown in Table 2-3. Their measurements can be carried out by means of wet chemistry techniques in conjunction with spectroscopy.

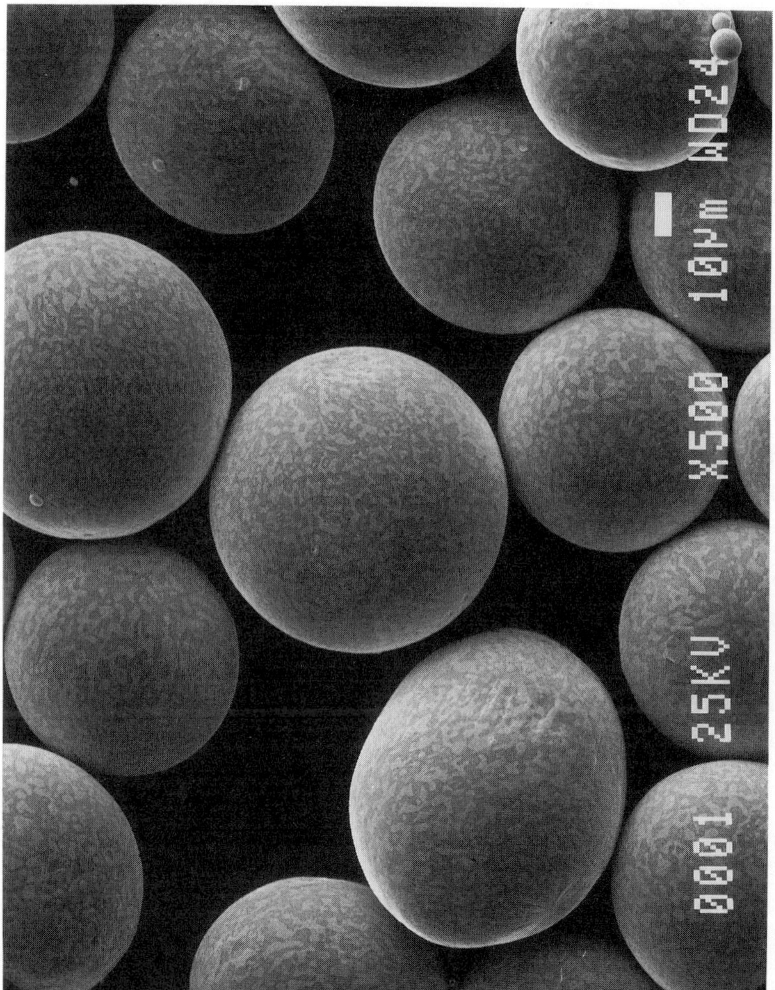

Figure 2-6 SEM micrograph of 63% Sn-37% Pb, −200/+325 mesh solder powder. (Courtesy of AMT, Inc.)

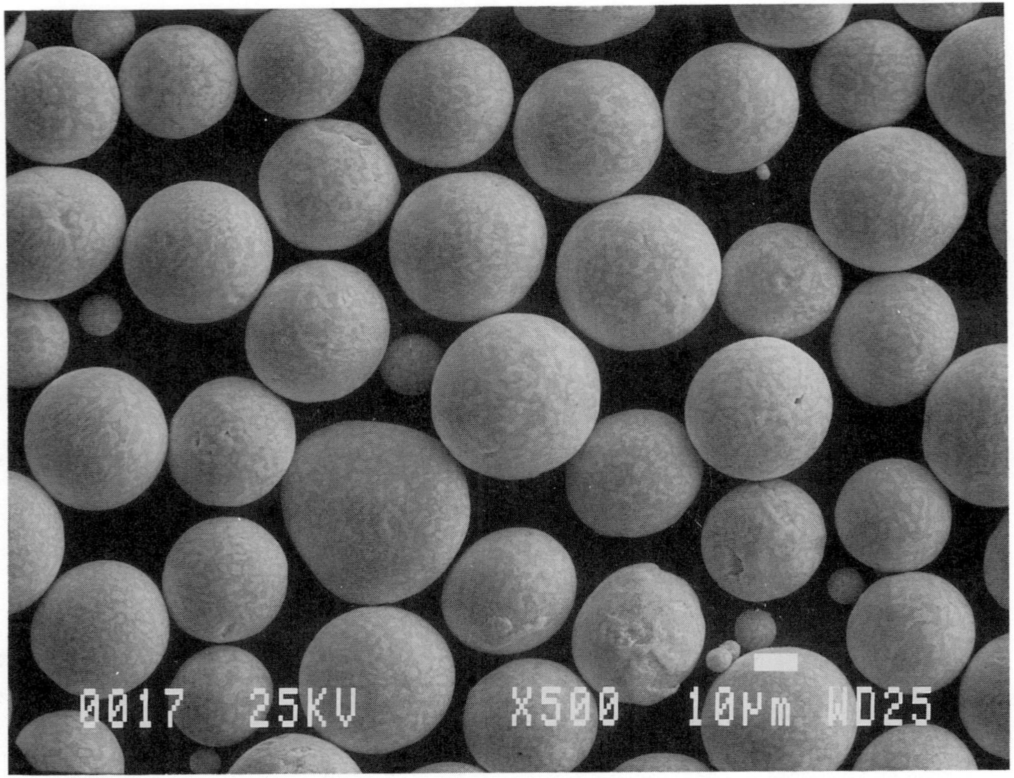

Figure 2-7 SEM micrograph of 63% Sn-37% Pb, −325/+500 mesh solder powder. (Courtesy of AMT, Inc.)

2.5 PASTE FORMULATION

Based on the flux activity and residue cleaning requirements, today's solder paste for electronics packaging may be classified as:

R-grade	Water saponifiable
RMA-grade	Water soluble
RA-grade	No-clean

The first three grades are rosin based and solvent cleaning and have been used in the industry for many years. The water-cleaning and no-clean pastes are discussed near the end of the chapter.

Solder paste is a homogeneous and kinetically stable mixture of solder alloy powder, flux, and vehicle that is capable of forming metallurgical bonds at a set of soldering conditions and can be readily adapted to automated production in making reliable and consistent solder joints. In terms of functionality, a solder paste can be considered as being composed of three major components. They are solder alloy powder, vehicle system, and flux system. The vehicle is primarily a carrier for alloy powder, as a compatible matrix for flux system, and as a basis for desirable rheology. The flux works to render the alloy powder and substrates to be joined in a clean and metallic state so that a metallic continuity and good wetting can be obtained. Both vehicle and flux are fugitive in nature at the completion of soldering; nevertheless, they are crucial to the formation of reliable, permanent bonds. On a permanent basis, the alloy powder part is the only functional component in forming a metallurgical bond.

As an example, a typical RMA solder paste formula may contain 12 ingredients, providing various designated functions:

Ingredient	Function
Rosin 1 Rosin 2 Rosin 3	Rosin system for designated softening point, acid number, thermal stability, fluxing, tackiness
Nonhalogen activator 4 Nonhalogen activator 5 Nonhalogen activator 6	Activator system for accomplishing fluxing action over a wide range of temperature, rheology, open time
Solvent 7 Solvent 8 Solvent 9	Solvent system to accommodate solubility, rheology, temperature compatibility, chemical compatibility, tack time
Binder 10	Providing compatible viscosity, rheology, tackiness
Fluxing modifier 11	Stabilizing and modifying flux activity and strength
Rheology modifier 12	Contributing to targeted rheology

Table 2-3 Solder Paste Trace Elements

Composition	Tin %	Lead %	Antimony %	Bismuth, max. %	Silver %	Copper, max. %	Iron, max. %	Zinc, max. %	Aluminum, max. %	Arsenic, max. %	Cadmium, max. %	Total of all Others, max. %
Sn96	Remainder	0.10, max.	—	—	3.6 to 4.4	0.20	—	0.005	—	0.05	0.005	—
Sn70	69.5 to 71.5	Remainder	0.50	0.25	0.1	0.08	0.02	0.005	0.005	0.03	—	0.08
Sn63	62.5 to 63.5	Remainder	0.50	0.25	0.1	0.08	0.02	0.005	0.005	0.03	—	0.08
Sn62	61.5 to 62.5	Remainder	0.50	0.25	1.75 to 2.25	0.08	0.02	0.005	0.005	0.03	—	0.08
Sn60	59.5 to 61.5	Remainder	0.50	0.25	0.1	0.08	0.02	0.005	0.005	0.03	—	0.08
Sn50	49.5 to 51.5	Remainder	0.50	0.25	0.1	0.08	0.02	0.005	0.005	0.025	—	0.08
Sn40	39.5 to 41.5	Remainder	0.50	0.25	0.1	0.08	0.02	0.005	0.005	0.02	—	0.08
Sn35	34.5 to 36.5	Remainder	1.6 to 2.0	0.25	0.1	0.08	0.02	0.005	0.005	0.02	—	0.08
Sn30	29.5 to 31.5	Remainder	1.4 to 1.8	0.25	0.1	0.08	0.02	0.005	0.005	0.02	—	0.08

Sn20	19.5 to 21.5	Remainder	0.80 to 1.2	0.25	0.1	0.08	0.02	0.005	0.005	0.02	—	0.08
Sn10	9.0 to 11.0	Remainder	0.20, max	0.03	0.17 to 2.4	0.08	—	0.005	0.005	0.02	—	0.10
Sn5	4.5 to 5.5	Remainder	0.50, max	0.25	0.1	0.08	0.02	0.005	0.005	0.02	—	0.08
Sb5	94.000 min.	0.20, max	4.0 to 6.0	—	0.1	0.08	0.08	0.03	0.03	0.05	0.03	0.03
Pb80	Remainder	78.5 to 80.5	0.20 to 0.50	0.25	0.1	0.08	0.02	0.005	0.005	0.02	—	0.08
Pb70	Remainder	68.5 to 70.5	0.20 to 0.50	0.25	0.1	0.08	0.02	0.005	0.005	0.02	—	0.08
Pb65	Remainder	63.5 to 65.5	0.20 to 0.50	0.25	0.1	0.08	0.02	0.005	0.005	0.02	—	0.08
Ag1.5	0.75 to 1.25	Remainder	0.40, max	0.25	1.3 to 1.7	0.30	0.02	0.005	0.005	0.02	—	0.08
Ag2.5	0.25, max	Remainder	0.40, max	0.25	2.3 to 2.7	0.30	0.02	0.005	0.005	0.02	—	0.03
Ag5.5	0.25, max	Remainder	0.40, max	0.25	5.0 to 6.0	0.30	0.02	0.005	0.005	0.02	—	0.03

The formula may appear to be somewhat straightforward. However, designing a viable product capable of delivering all desired performance characteristics as shown in Table 2-3 is complex and requires understanding of the technologies and skills in formulation. A product design can be accomplished by following design and manufacturing steps:

Define performance objective
Use fundamental technologies
Select raw materials
Understand and anticipate potential synergistic or antagonistic interactions
Balance performance parameters
Fine tune the formula to meet designated specification
Develop production process
Produce consistent product

Among the performance parameters, some are trade-offs. For example, high metal content is beneficial to a solder joint with respect to solder volume and void and to residue yet may make the paste more prone to crusting and drying and difficult to apply. High viscosity paste may improve flow control in resisting to rising temperature, yet the paste is difficult to apply. Using highly active fluxing chemicals may improve solderability in some cases, yet its use may leave more corrosive residue and render the paste less stable to shelf storage and to external forces. In such a case, improving the solderability by selecting proper ingredients without the use of highly active fluxing chemicals requires skill. It should be noted that increasing flux content does not always improve solderability in terms of wetting and/or solder balling elimination.

While the product is being designed, developing a practical and reproducible process for making the paste with consistent characteristics is equally important. It is not an exaggeration, but an indication of the importance of the role of process, to state that identical composition formulas can produce different products when the process is allowed to vary.

2.6 PASTE RHEOLOGY

Rheology of solder paste[1,3] is an important characteristic with respect to application, since the flow and deformation behavior of solder paste directly affect the quality of paste deposition on the intended solder pads. Therefore, the correlation of paste rheology with the application and its specific parameters is crucial to the control of rheological quality when manufacturing paste, as well as to the setting and control of application parameters when using paste.

In many rheological aspects, there are similarities between coating products where fine oxide particles are dispersed in an organic carrier phase and solder pastes. Yet, in solder pastes, different inherent factors are involved, making the solder paste rheology more of a challenge. Factors such as the following are influential:

Drastic difference in density between organic matrix and metallic particles
Suspended particles of a metallic nature
Relatively large particle size
Spherical particles
High load of particles

The drastic difference in density between the organic matrix (0.7 to 1.2 g/cm^3) and the metallic particles (7 to 12 g/cm^3) facilitates the settling of particles, and the large size (up to 75 μm) of particles reduces the interactions between the particle and matrix, which further facilitates settling. The metallic nature of particles renders the mixing or dispersion technique more delicate. A high load of particles, commonly more than 40 vol %, contributes to the complexity of the rheological behavior.

The primary driving forces underlying the rheology of solder paste include both kinetic and thermodynamic contribution. In principle, these cover thermal motion with magnitude determined by the mean free path of molecules or particles and by temperature, in conjunction with intermolecular or interparticle forces. The forces can be viewed as electrostatic repulsion and attraction between particles and the affinity between the chemical matrix and the particle surface. The particles will not aggregate when the affinity of particle surface for the chemical matrix exceeds the attractive forces between particles or when steric effects through absorption hinder the aggregation. Therefore, the rheology of solder paste may be affected by the following factors:

The composition, shape, and size of suspended particles
The chemical composition of suspending matrix
The relative concentration of effective ingredients in the matrix
The structure of ingredients in the matrix
The interactions between matrix and suspended particles, either of a physical or chemical nature, including wetting and solvation
The volume fraction occupied by the suspended particles; usually, the higher amount of particles, the more deviation from viscous flow
The internal structure and its response to external forces
The interactions among particles and resulting aglomeration and flocculants
Temperature

It is difficult to predict the rheology of such a system because of the lack of knowledge of the exact structure and the nature of forces exerted by molecules or particles. However, rheological behavior can be characterized, and the characterization can then be correlated to performance, as discussed in the next section.

The ideal rheology of solder paste for both dispensing and printing applications from an application point of view is expected to deliver the following properties:

Specified shelf stability
No separation or settling
No dripping
No stringiness but with enough tackiness to hold components
No slumping or sagging but with capability of leveling
Perfect transfer through printing pattern
Clean release from the screen
Clean break off through fine needles
Nearly instant recovery after transfer
Temperature insensitive

In practice, true instant recovery after paste transfer and temperature insensitivity is difficult to attain. It is thus obvious that trade-off and balance are essential to the design of pastes having a rheology able to accommodate the application processing parameters.

2.7 RHEOLOGY BEHAVIOR CHARACTERIZATION

Solder paste is neither a purely viscous nor a purely elastic material; it exhibits viscoelastic properties. Therefore, solder paste is characterized by its flow behavior as well as its elastic properties.

For viscous materials, Newton's law defines

$$\tau = \eta\dot{\gamma}$$

where $\dot{\gamma}$ is the shear stress, η is the shear rate, and τ is the apparent viscosity. Viscous flow described by this relationship is called newtonian flow, as shown in Figure 2-8.

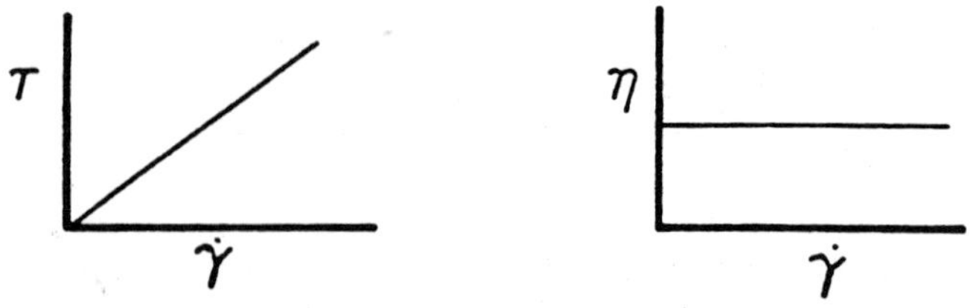

Figure 2-8 Newtonian flow.

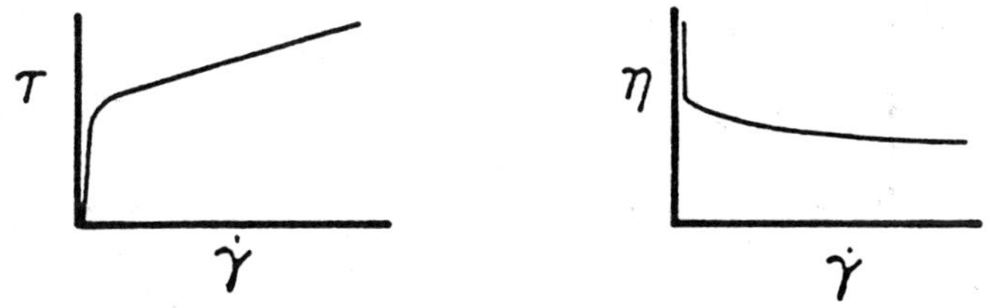

Figure 2-9 Plastic flow.

In addition to idealized newtonian flow, nonnewtonian types are exemplified by plastic, pseudoplastic, dilatant, thixotropic, and rheopectic flow. As illustrated in Figures 2-9 through 2-13, the viscosity of plastic or pseudoplastic material decreases with increasing shear rate, due primarily to the alignment of molecules or particles along the direction of shearing. The dilatant material displays increased viscosity as shear rate increases. It is believed that such flow resistance is attributed to the formation of a close packed structure. It often occurs in systems containing very small solid particles suspended in a highly defluocculated manner. The plasticity of materials is characterized by the presence of a yield value, which is the threshold shear stress required to overcome the forces of internal structure to start flow. In addition to shear thinning effect, thixotropic material exhibiting a time-dependent reestablishment in viscosity results in a hysteresis loop in the ramping up and down cycle. Rheopectic flow is not well understood, but it is postulated that a perfect structural orientation is associated with this phenomenon, assuming in-site chemical reaction and evaporation are absent.

Parallel to Newton's law for viscous fluids is Hooke's Law for solids. Hooke's law relates shear stress τ and shear strain γ in the following equation:

$$\tau = G\gamma$$

where G is the shear modulus and γ is the strain. Under stress, the deformation of most materials differs substantially from the idealized case described by Hook's law.

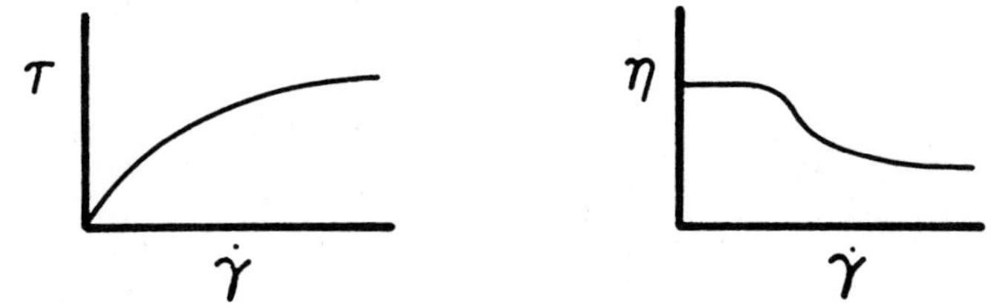

Figure 2-10 Pseudoplastic flow.

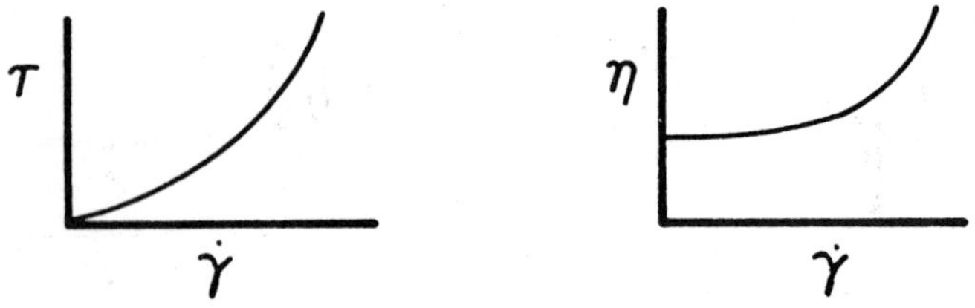

Figure 2-11 Dilatant flow.

Flow behavior can be characterized by several readily measured parameters: shear stress versus shear rate, yield point, viscosity versus shear rate, viscosity versus time, and shear stress versus time.

The relationship of shear stress to shear rate indicates the type of flow as shown in Figures 2.8 through 2-13. There is more than one way to determine the yield point. For convenience, the yield point may be estimated by the extrapolation of the flow curve to intercept the shear stress axis or may be taken as the shoulder point (inflection) of the curve, if present. These techniques are appropriate only for comparison purposes. When using these techniques to derive yield point, it is more reliable to measure flow at low shear rate conditions (<50 sec^{-1}). The relationship of viscosity to shear rate is equivalent to that of shear stress to shear rate. However, it provides direct readings on viscosity change in response to shear rate change.

Figure 2-14 shows the flow profiles of a dispensing solder paste, indicating very slight shear rate dependency. The shear rate ramping up and down curves indicate a very small amount of thixotropy. It also shows a very low yield point.

In comparison, Figure 2-15 is the flow profile of a printing paste, showing a relatively large yield point, moderate thixotropy, and a high shear rate depen-

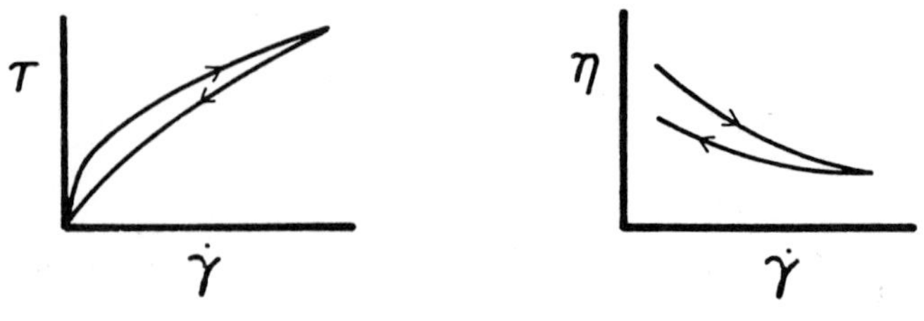

Figure 2-12 Thixotropic flow.

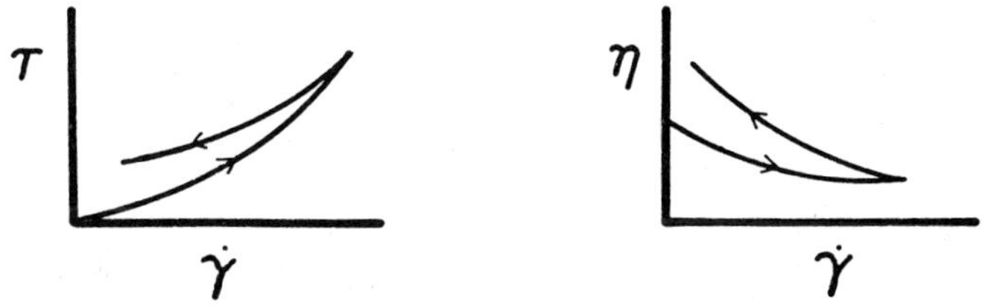

Figure 2-13 Rheopectic flow.

dency. The viscosity changes with time, after a constant shear rate is applied to a dispensing paste, as is shown in Figure 2-16. The stress relaxation curve of a printing paste also reflects the characteristics of a solder paste such as the magnitude of yield value, leveling viscosity, and the internal structure change of the paste.

At a constant temperature, the dynamic mechanical measurements that characterize the elastic properties include storage modulus, loss modulus, and complex viscosity in response to operating parameters of frequency, strain amplitude, and time.

Under dynamic measurements, the stress is applied sinusoidally at a given frequency, and the strain also alternates sinusoidally but is out of phase with the stress at a given phase angle. The theoretical and mathematical treatment is covered in the literature.[4]

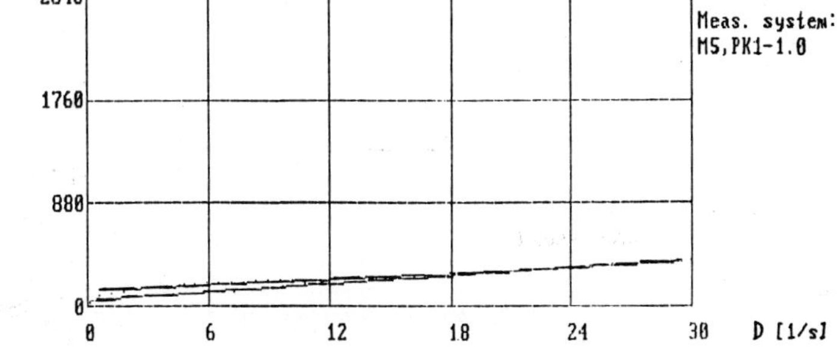

Figure 2-14 Flow curve of a dispensing solder paste.

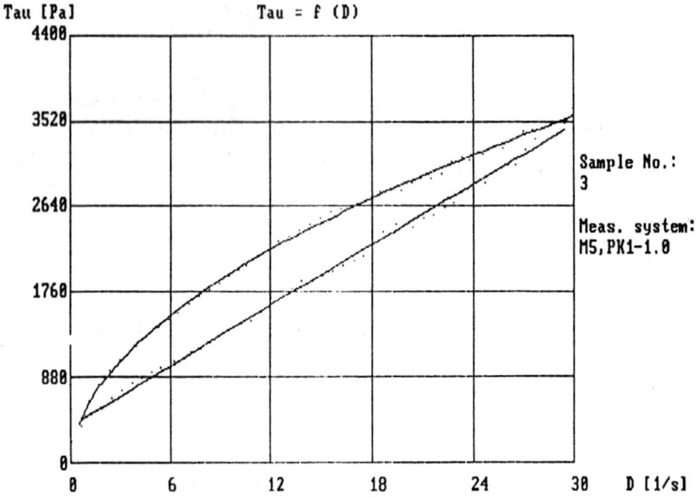

Figure 2-15 Flow curve of a printing solder paste.

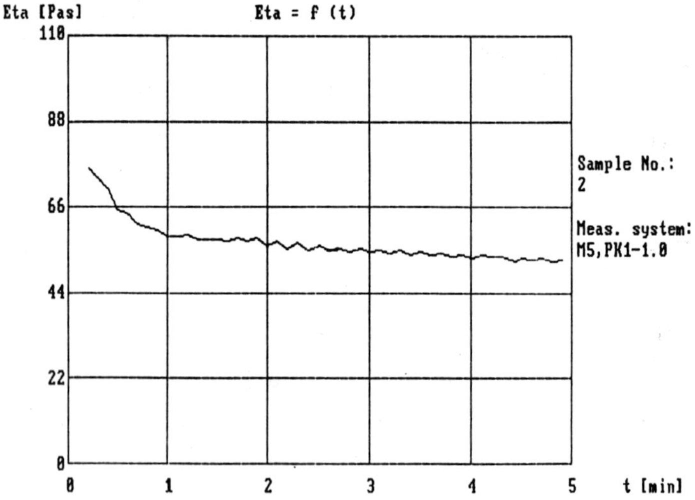

Figure 2-16 Viscosity versus time curve of a dispensing solder paste.

The elastic characteristics of paste may be defined by analyzing the following features:

Magnitude of the storage modulus, loss modulus, and complex viscosity
Relative magnitude of loss and storage modulus
Relative changes of loss and storage modulus in relation to frequency change, strain amplitude, or time
Relative rate of change in complex viscosity versus frequency change, strain amplitude, or time

The storage modulus (G'), loss modulus (G''), and complex viscosity (η^*) can be monitored in relation to frequency sweep under constant strain. These three properties can also be characterized in relation to strain sweep under constant frequency. Figure 2-17 is a measurement of storage modulus, loss modulus, and complex viscosity of a printing paste in relation to frequency under a constant strain of 5 percent. The relative weight of viscous and elastic portions in solder paste can be directly measured in loss tangent value, as shown in Figure 2-18.

Combining flow measurement and elastic measurement, a solder paste can be

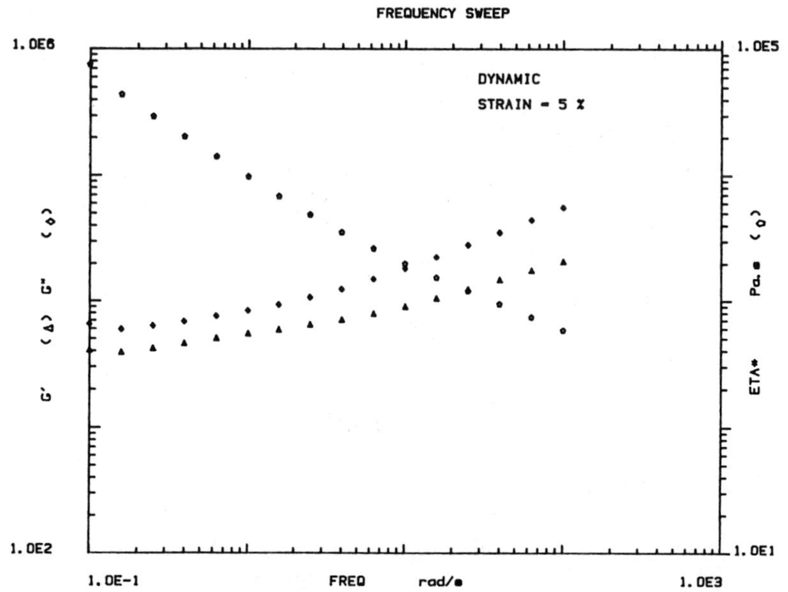

Figure 2-17 Measurement of storage modulus (G'), loss modulus (G''), and complex viscosity (η^*) in relation to frequency for a printing solder paste.

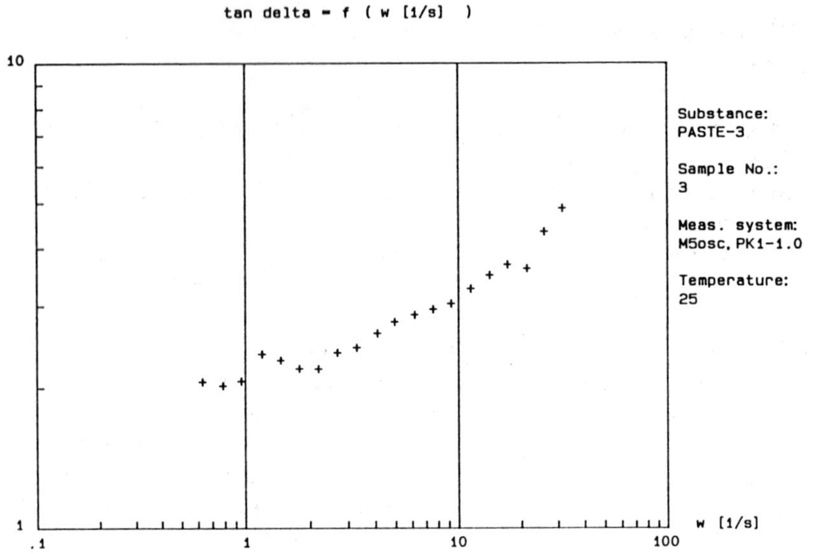

Figure 2-18 Loss tangent of a printing solder paste in relation to frequency.

adequately characterized. The characterization of the paste has to be correlated with the processing parameters as well as with the paste performance under a specific set of conditions.

Under two main paste application methods, fine dot dispensing and pattern printing, a paste possessing a low yield point and slight plastic behavior is found most suitable for dispensing applications. This is consistent with the formulation requirement that a dispensing paste normally has a lower upper limit in volume fraction of metal load as well as a lower viscosity range, in comparison with pastes designed for stencil or screen printing applications.

Low yield value physically signifies that paste will start to flow under low shear stress. Too high a yield point may result in poor leveling, which leaves an uneven surface, and too low a yield point may cause running and slumping.

Printing paste normally demands a higher viscosity range. A moderate yield point and moderate thixotropy are found to be suitable for printing mechanics. Too high a degree of thixotropy may cause slumping and sagging. However, shear thinning is desirable, especially at high shear rate, as encountered while a paste is being transferred through the opening during a printing process.

2.8 VISCOSITY AND MEASUREMENT

Table 2-4 lists the typical viscosity range and metal load for the dispensing and printing application techniques, although some exceptions may exist.

A single-point viscosity measurement as a criterion of the rheology of solder paste is prevalent in the industry. It is thus appropriate to discuss some practical aspects of single-point viscosity measurement. One of the common techniques adopted is to use the Brookfield viscometer with a helipath spindle TF at 5 rpm in accordance with Federal Specification QQ-S-571E. Measurement obtained by this technique is simple and convenient. However, single-point viscosity is valid only when the following conditions are met:

The rheology of the paste is known.
The paste handling technique immediately prior to measurement is standardized.
The measurement technique of the paste is standardized.
The paste container is specified in size and dimensions.

To illustrate the point, Table 2-5 is a summary of viscosity data measured using a Brookfield model RTV viscometer with helipath spindle TF at 5 rpm for two pastes, paste A and paste B. The variables in this table are paste handling prior to descending spindle, as expressed by the mixing time, and the measuring time (time to read) (i.e., the time interval from the start of the spindle descending to obtaining a dial reading). This time interval also indicates the spindle position in the paste container.

The data clearly indicate the effect of premixing. Longer mixing generates a lower viscosity number across the full range of reading time, from ½ to 3 min. Under the test conditions, after the paste is mixed thoroughly for 2 min, further mixing does not significantly affect viscosity. The data also indicate that the viscosity reading increases with increasing time to read, because of increased friction as the spindle is descending further down.

An equivalent test is carried out for paste A. The comparison between pastes A and B indicates that paste A has much less variation in response to mixing

Table 2-4 Viscosity Range of Dispensing and Printing Pastes

Solder Paste Type	Viscosity* (CPS)	Metal Load (wt %)
Fine dot dispensing	200,000–450,000	Up to 88
Screen printing	450,000–750,000	Up to 92
Stencil printing	650,000–1,400,000	Up to 94

*Brookfield Model RVT, helipath TF spindle, 5 rpm, 3 min mixing and 2 min reading.

Table 2-5 Single-Point Viscosity of Pastes A and B in Relation to Premixing Time and Time to Read[1]

	Viscosity (KCPS), time to read (sec)						
Mixing Time	30	45	60	90	120	100	180
Paste B:							
0	—	—	1440	1550	1700	—	—
1	—	—	840	950	1100	—	—
2	—	600	660	720	800	900	—
3		580	640	700	800	880	900
Paste A:							
0	200	220	220	280	400	430	—
1	190	210	220	240	270	290	—
2	190	200	210	220	230	250	—
3	170	180	200	200	220	230	—

time and time to read. Again, as the paste is mixed for 2 min, further mixing has no significant effect.

The comparative results demonstrate the relative sensitivity of pastes to handling (mixing) and measuring techniques. With these variations in mind, it is not surprising to see the inconsistency and lack of agreement from laboratory to laboratory and operator to operator in viscosity data. Nonetheless, with the proper technique and control, viscosity data can be consistently obtained. In practice, realizing such variations, as well as being able to control them, is important to the establishment of specifications and to communication between users and vendors, which in turn is important to overall quality and production yield.

Single-point viscosity can also be used to develop an indicator for shear thinning factor by measuring viscosities at two significantly different shear rates (revolutions per minutes, or rpm), such as viscosities at 1 and 20 rpm. The ratio of these two viscosities indicates the shear thinning factor. Again, the paste handling has to be kept consistent in order to obtain meaningful results.

2.9 PRINTING TECHNIQUE

Today, the printing technique is most widely used for transferring solder paste onto the intended solder pads on the surface mount production floor. Whether using mesh screen or metal mask stencil, the forces contributing to the paste transfer process are the relative adhesive between paste and screen interface and the paste and substrate interface, the push force by the squeegee along the direction parallel to screen, the pressure exerted by the squeegee along the direction perpendicular to the screen, the hydraulic pressure created in the paste when it is compressed between the squeegee and the screen, and the gravity force.

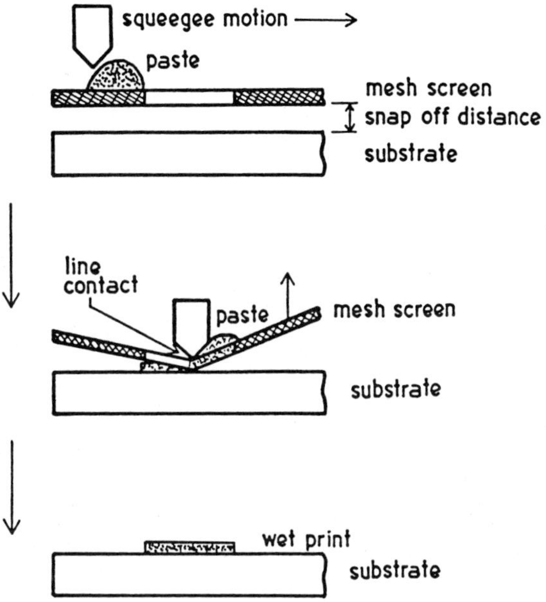

Figure 2-19 Schematic of printing operation.

During printing, a sequence of motions as illustrated in Figure 2-19 is created; these are:

The squeegee makes a line contact with the screen and a close contact with a given amount of paste.

While the squeegee is in motion, the paste moves along in front of the squeegee.

The paste is released through the openings onto the substrate behind the squeegee.

The screen behind the squeegee "instantaneously" snaps back after the squeegee passes over.

There are a number of printing variables that can be grouped into squeegee system, substrate, screen system, paste, and snap-off parameter.

Squeegee system

Blade material type	Squeegee size
Blade hardness	Squeegee speed
Blade edge angle	Downstop pressure
Blade edge flatness	External pressure
Blade configuration	

Substrate

Material type	Surface flatness
Surface condition	Substrate size

Screen systems

Mesh screen	Print pattern
Wire material type	Screen size
Wire diameter	Metal mask stencil
Mesh tension	Material type
Mesh count	Flatness
Mesh orientation	Thickness
Emulsion type	Print pattern
Emulsion thickness	Stencil size

Paste

Paste consistency	Chemical compatibility with the screen and squeegee
Paste rheology	
Paste stability	Compatible thickness with printing pattern
Paste tackiness	

Snap off

After the equipment is selected and the design of assembly is finalized, there are four major operational parameters to be monitored. They are the squeegee speed, squeegee downstop, squeegee pressure, and snap-off. In time of paste performance, these parameters are interrelated; changing one will change the effects of the others in a nonproportional manner. Nonetheless, with an understanding of the general relationship among these parameters and their effects on printing results, in conjunction with the systematic testing on the specific system (printer, board, paste), a set of optimized parameters can be established.

For a printing operation, the balance between the two primary mechanisms, paste transfer and screen release, dominates the printing results. Practically, these two mechanisms require different sets of parameters. In other words, the maximization of paste transfer efficiency may hinder the screen release. Figure 2-20 illustrates the relationship between the paste transfer (emptying of screen) as described by Equation 2-1 and screen release (no screen sticking) as defined by Equation 2-2.

$$\eta V \leqslant \frac{D^2 \cdot (S - B) \cdot \mathrm{Pa}}{128 \cdot d\omega \cdot a} \tag{2.1}$$

$$\eta V \leqslant \frac{\pi \cdot a \cdot m \cdot \sigma_s}{256 \cdot L_{sr}} - \frac{L_{sr} \cdot (m + d_R) \cdot \text{Pa}}{a} \qquad (2.2)$$

where η is viscosity, V is the squeegee speed, D is the diameter of mesh opening, d is the wire diameter of mesh, S is the screen size, B is the length of printer pattern, a is the snap-off distance, Pa is the atmosphere pressure, m is the wire pitch, $d_R = (m - d\omega)$, and L_{sr} is the distance from squeegee edge to screen frame.[5]

Current surface mount printed circuit boards mostly consist of a mixture of standard land pattern (50 mil pitch) and fine pitch land pattern. To accommodate both patterns, using a stencil with a differential thickness is one approach to deposit paste on the board with one-time printing. The metal mask stencil with a given foil thickness is recessed in areas for fine pitch patterns. The recessed areas with reduced foil thickness provide thinner print as required for fine pitch patterns. When design requires thicker and thinner print patterns close together (typically $\leqslant$ 200 mil), a two- or multiple-foil process is needed. The difference in thickness of paste print is obtained by joining foils together through a lamination technique.

Another important factor to be considered is ambient environment. Solder pastes, although to different extents, are responsive to temperature as represented by the viscosity versus temperature data as shown in Figure 2-21. Maintaining a reasonably constant temperature under which the printing operation is carried

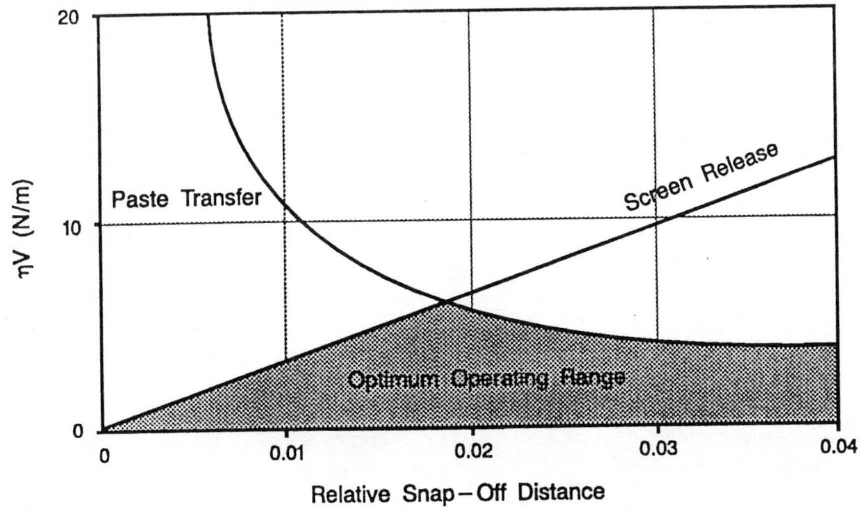

Figure 2-20 Balance of paste transfer and screen release in relation to snap-off distance.

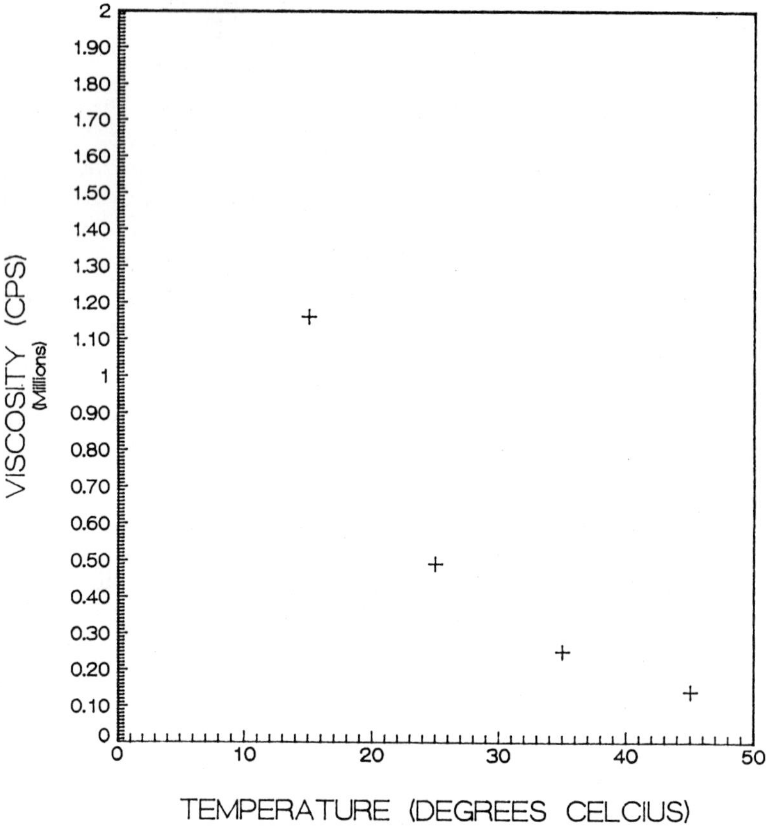

Figure 2-21 Variation of viscosity with temperature.

out is a recommended practice. The humidity is also a factor; its effect on paste performance depends on the properties of the paste. Controlled humidity, again, is a necessary condition to achieve desired quality.

2.10 DISPENSING TECHNIQUE

For the assemblies requiring solder connection on an irregular surface or on hard-to-reach areas, dispensing technique becomes a viable approach. The paste is delivered on the designated solder pads by means of needles with different length and inner diameter through pneumatic or volume displacement mechanistics. The dispensing needle gauge and corresponding inner diameters[1] are listed in Table 2-6.

Table 2-6 Dispensing Needle Gauges and Inner Diameters

	Inner Diameter	
Needle Gauge	in.	mm
15	0.054	1.37
18	0.033	0.84
20	0.023	0.58
21	0.020	0.51
22	0.016	0.41
23	0.013	0.33
25	0.010	0.25
30	0.006	0.15

Each increment in gauge number can affect paste dispensability drastically. The following is a list of some practical points:

In the dispensing system, set up the shortest distance possible that the paste needs to pass through

Use a size of paste syringe only large enough to fit production run (avoid excess size of syringe).

Select the right needle gauge that matches the targeted amount and size of paste deposit (avoid too small a needle size).

When the system is new, flush it with a compatible lubricant to wet the inner wall of passage that paste will be in contact with.

When the system will have extended downtime, flush the system thoroughly with a compatible lubricant.

Select a proper dispensing paste.

For a syringe stored at low temperature, bring it to ambient temperature prior to starting the dispensing operation.

Make sure no foreign particles fall in the system.

Keep the applied pressure at a minimum level as needed, usually 15 to 40 psi.

Maintain a correct and consistent gap between the needle and the substrate because the gap affects the size and shape of the paste deposit.

Do not leave paste in the system for an extended downtime.

Check the chemical compatibility of the paste with the materials used in the dispensing system.

2.11 SOLDERING PRINCIPLE

Soldering is the process of making solder joints by applying heat to the solder material and to the substrates to be joined at a temperature normally below

400°C. When the solder material is a solder paste, a series of reactions and interactions in sequence or in parallel are expected to occur in response to the applied heat. These may be represented in different stages as follows, where temperature increases as arrows indicate:

Preheat the substrates and solder paste	→	Volatile evaporation	→	Volatile evaporation Flux concentration Flux activation	→	Chemical pyrolysis Fluxing substrates Fluxing solder powder	→
(I)		(II)		(III)		(IV)	

→	Solder melting Solder wetting substrate Flux or vehicle protecting substrates and breaking out of molten solder Chemical pyrolysis	→	Flux or vehicle breaking out of molten solder Chemical pyrolysis	→	Solder solidification
	(V)		(VI)		(VII)

The extent of each of the above phenomena in a given stage and between stages varies with the specific makeup of the solder paste. For example, paste containing more low boiling volatiles may need an extended stage II at moderate temperature (100 to 130°C) in order to mildly expel the volatiles without causing violent paste spattering; the pyrolysis process in stages IV through VI can be controlled through paste makeup and by the temperature profile (temperature versus time) so that excessive pyrolysis will not occur and cleaning difficulty will be avoided. For no-clean paste, stages II through VI may demand a different temperature profile to achieve minimum residue. With these expected phenomena in mind, the optimum temperature profile and compatible solder paste can be established.

2.12 SOLDERABILITY

The solderability of solder paste is the ability to achieve good wetting of molten solder on the surface of substrates without forming undesirable solder balls outside the main solder pool so that sound solder joints are formed. The substrates to be dealt with in soldering field include:

Gold (Au)
Silver (Ag)
Palladium-silver (Pd-Ag)
Platinum-silver (Pt-Ag)
Palladium-platinum-silver (Pd-Pt-Ag)
Palladium-gold (Pd-Au)
Platinum-gold (Pt-Au)
Tin-lead (Sn-Pb)
Tin (Sn)
Copper (Cu)
Nickel (Ni)

In order to achieve good solderability, the demand on the fluxing activity depends on the intrinsic wettability by molten solder, the compatibility of the flux chemistry with the substrate, and the surface condition of the substrate, which may change with the time and exposure condition. In terms of intrinsic wettability, the substrates are ranked in descending order as tin, tin-lead > copper > palladium-silver, platinum-silver, palladium-platinum-silver > nickel. The surface of component leads is normally coated with tin/lead composition (100% Sn → 5% Sn-95% Pb) by using either electroplating or molten solder dipping. The coating generally is more uniform but more porous from the electroplating process than from dipping, where a thicker and denser coating is obtained.

The tin-lead coating is relatively stable under ambient environment and its oxides can be easily handled through fluxing. However, its long shelf time and heat excursion may cause solderability deterioration through other mechanisms, such as copper-lead intermetallics formation when leads are made of copper, and lead material corrosion when leads are made of alloy 42 or Kovar. To assure consistent and good solderability of components, the shelf time and heat exposure must be minimized.

With respect to solderability measurement, the edge dip test in reference to ANSI/IPC-S-804A is commonly used for solder pads and the MIL-STD-202 method for the component leads. A more quantitative test, using a wetting balance

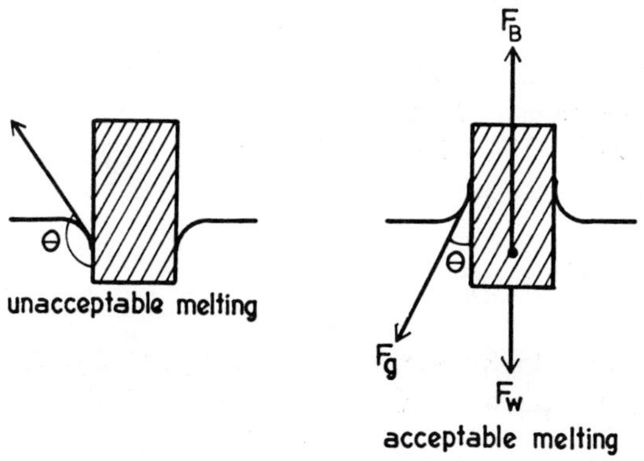

Figure 2-22 Wetting phenomena of substrate dipping into molten solder.

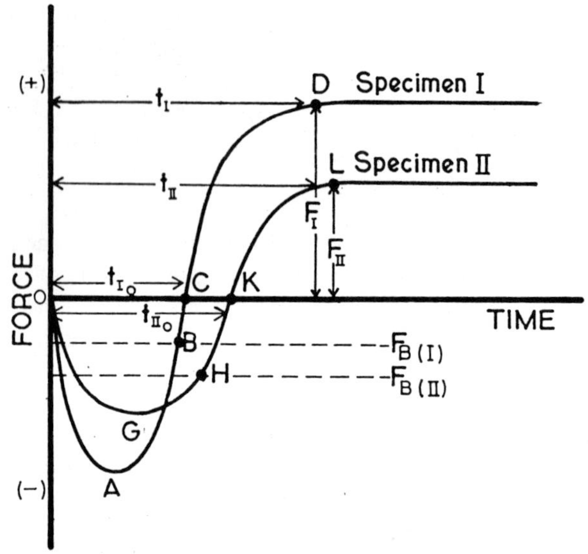

Figure 2-23 Wetting force versus time.

method, is being developed by IPC. The wetting balance measures force versus time. When a specimen is in contact with the molten solder, as shown in Figure 2-22, the forces involved are related as:

$$F_w \cong F_g \cos\theta - F_B = F_g \cos\theta - P_g V$$

Where F_w is the resultant wetting force, F_g is the interfacial force between specimen and molten solder, F_B is the buoyancy force, θ is the contact angle, P is the density of molten solder, "g" is the gravitational acceleration constant, and V is the immersed volume of specimen. Several characteristic parameters of the wetting curve, as shown in Figure 2-23, are considered as the measurement criteria. These are its minimum force, the time for zero net force corrected for buoyancy, the time for two-thirds of the wetting force, the time for 200 μN/mm, the maximum force at 2 sec, and theoretical maximum force. Specific testing conditions are also being developed.

2.13 SOLDERING METHODS

Soldering can be achieved through different heat transfer modes, including conduction, convection, infrared radiation, infrared-convection, vapor condensation, hot gas, resistance, induction, and laser beam. Each solder method has its uniqueness and beneficial features in cost, performance, and operational efficiency. Table 2-7 summarizes the strengths and limitations of each method.[1]

Table 2-7 Reflow Methods—Feature Comparison

Reflow Method	Benefits	Limitations
Conduction	Low equipment capital Rapid temperature changeover Visibility during reflow	Planar surface and single-side attachment requirement Limited surface area
Infrared	High throughput Versatile temperature profiling and processing parameter	Mass, geometry dependence
Vapor phase condensation	Uniform temperature Geometry independence High throughput	Difficult to change temperature Temperature limitation Relatively high operation cost
Hot gas	Low cost Fast heating rate Localized heating	Temperature control Low throughput
Convection Induction	High throughput Fast heating rate High temperature capability	Slow heating Applicability to nonmagnetic metal parts only
Laser	Localized heating with high intensity Short reflow time Superior solder joint Package crack prevention	High equipment capital Specialized paste requirement Limit in mass soldering

2.14 CONTROLLED ATMOSPHERE SOLDERING[6]

The control of heating rate, temperature uniformity across the mass under heating, and temperature reproducibility, as well as the need for the compatibility of heating profile with the characteristics of the solder paste, are important to the ultimate yield of the soldering process and to the properties of the resulting solder joint, which have been well covered in the literature.[1,7,8,9,10] The atmosphere under which the soldering is performed is, however, equally important and has not been studied in the soldering field.

The atmosphere, based on its function, can be considered as being of two types: protective and reactive. A protective atmosphere normally is inert toward the specific soldering material under the specific conditions, and a reactive at-

Table 2-8 Composition of Atmospheric Gases

Atmosphere	Composition (vol %)						
	Carbon Dioxide (CO_2)	Oxygen (O_2)	Carbon Monoxide (CO)	Hydrogen (H_2)	Methane (CH_4)	Nitrogen (N_2)	Trace
Air	—	21.0	—	—	—	78.1	0.9
Nitrogen	—	—	—	—	—	99.8–100.0	0–0.2
Hydrogen	—	—	—	99.8–100.0	—	—	0–0.2
Dissociated methanol	—	—	33.3	66.7	—	—	—
Dissociated ammonia	—	—	—	75.0	—	25.0	—
Exothermic gas (air/gas = 6/1)	5.0	—	10.0	14.0	1.0	70.0	—
Endothermic gas (air/gas = 2.4/1)	—	—	20.0	38.0	0.5	41.5	—

mosphere may involve oxidizing, reducing, or dissolving functions in the subject system.

Broadly, the chemicals that can provide oxidizing or reducing potential in relation to the specific metal and metal oxide system and that can generate significant vapor pressure at an operating temperature are expected to contribute to the functional nature of the total atmosphere. The following list shows some commonly used atmospheres, and their corresponding nominal compositions are summarized in Table 2-8.

Dry air
Nitrogen
Hydrogen
Nitrogen-hydrogen blends at different ratios
Dissociated methanol-nitrogen blends at different ratios
Dissociated ammonia
Exothermic gas
Nitrogen and dopants at different concentrations

Thermal cracking of methanol essentially yields hydrogen and carbon monoxide at high temperature, as represented by the chemical equation:

$$CH_3OH \rightarrow 2H_2 + CO$$

At low temperature (below 800°C), side reactions may occur, leading to the formation of H_2O, CH_4, CO_2, and C_5. Figure 2-24 shows the resulting components of methanol decomposition over a range of temperature.[11] The composition of exothermic and endothermic gases also varies with the air-gas ratio.

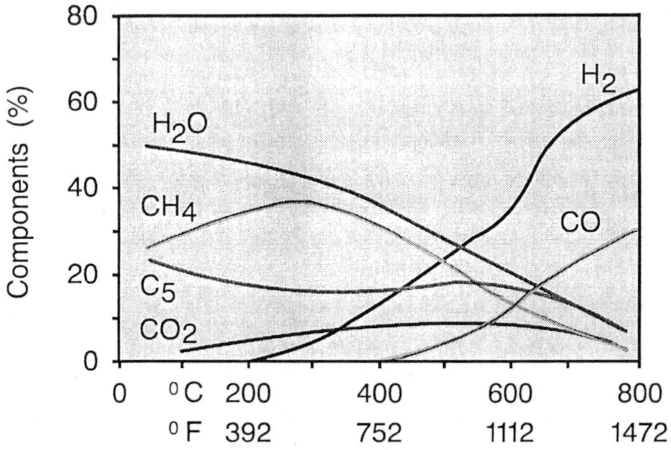

Figure 2-24 Thermodynamic equilibrium composition of methanol decomposition.

Each component of atmosphere gases may function as an oxidant or reducing agent depending on the temperature and its oxidation-reduction potential relative to that of the materials involved. Among the components of common atmosphere gases, oxygen, water vapor, and carbon dioxide normally serve as oxidant to most metals and metal oxides, and hydrogen and carbon monoxide serve as reducing agents. The ratio of oxidant content to reducing agent content in relation to that ratio at equilibrium indicates whether the resulting atmosphere is oxidizing or reducing.

Since water vapor is oxidizing in nature, its partial pressure in the furnace contributes to the overall nature of the atmosphere. The partial pressure of water vapor in an atmosphere gas is conveniently expressed as the dew point, that is, the temperature at which condensation of water vapor in the air takes place. The dew point can be measured by a hygrometer or dewpointer by means of a fog chamber, chilled mirror, or the aluminum oxide technique. The relationship of dew point to the vapor pressure is shown in Figures 2-25 and 2-26. The relative humidity (RH) is related to the actual vapor pressure of water (or represented by dew point) (P_w) and the saturated vapor pressure at the prevailing ambient temperature (P_s) as follows:

$$\mathrm{RH} = \frac{P_w}{P_s}$$

During soldering, the reactions and interactions of chemicals in solder paste and those between chemicals and metal surface can be quite complex. In simple

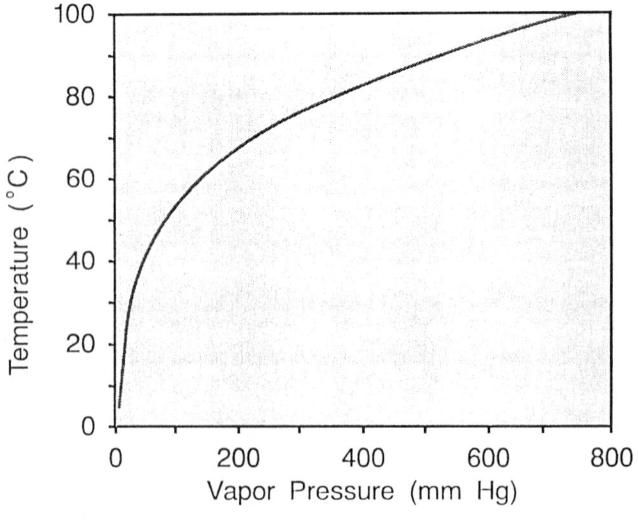

Figure 2-25 Relationship of vapor pressure and dew point from 0 to 100°C.

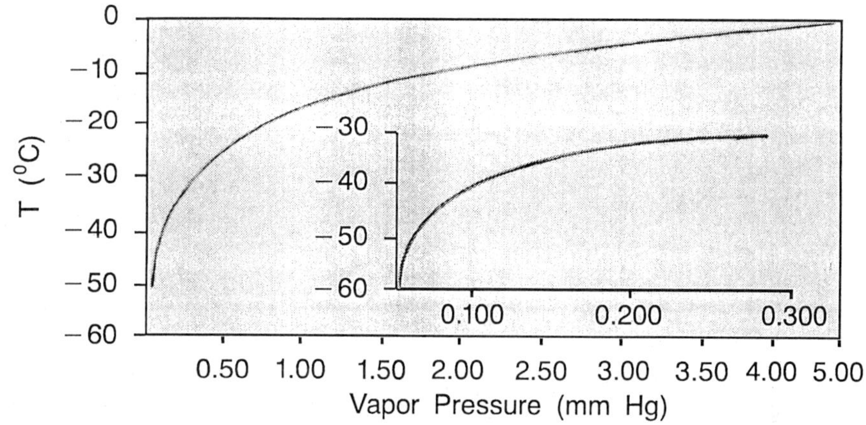

Figure 2-26 Relationship of vapor pressure and dew point from −60 to 0°C.

terms, the mechanisms may include evaporation, pyrolysis, oxidation, and reduction. The generalized oxidation and reduction reactions can be expressed as:

$$\chi M + y\, O_2 \rightleftarrows M_x\, O_{2y}$$

$$M_xO_y + y\, H_2 \rightleftarrows \chi M + y\, H_2O$$

$$\chi M + y\, CO_2 \rightleftarrows M_xO_y + y\, CO$$

To obtain the thermodynamic equilibrium constant K for each of the preceding reactions, proceed as follows:

$$\chi M\ (s) + y\, O_2(g) \rightleftarrows M_xO_{2y}\ (s) \tag{2.3}$$

$$K_1 = \frac{a_{M_xO_{2y}}}{a_M^x\, a_{O_2}^y}$$

$$M_xO_y\ (s) + y\, H_2\ (g) \rightleftarrows \chi M\ (s) + y\, H_2O\ (g) \tag{2.4}$$

$$K_2 = \frac{a_M^x\, a_{H_2O}^y}{a_{M_xO_y}\, a_{H_2}^y}$$

$$\chi M(s) + y\, CO_2\ (g) \rightleftarrows M_x\, O_y(s) + y\, CO\ (g) \tag{2.5}$$

$$K_3 = \frac{a_{M_xO_y} a_{CO}^y}{a_M^x\, a_{CO_2}^y}$$

Assuming the compositions of solids remain constant and the gases behave ideally, we have:

$$K_1 = \frac{1}{P^y_{O_2}}$$

$$K_2 = \frac{P^y_{H_2O}}{P^y_{H_2}}$$

$$K_3 = \frac{P^y_{CO}}{P^y_{CO_2}}$$

Where "a" represents the activity of each reactant and product of reactions (2-3), (2-4), and (2-5), and P_{H_2O}, P_{H_2}, P_{CO}, P_{CO_2} represent the partial pressure of H_2O, H_2, CO, and CO_2, respectively.

By introducing the relationship between free energy G^0 and the equilibrium constant, namely,

$$G^0 = -RT \ln K$$

it is shown that the reactions (2-3), (2-4), and (2-5) can proceed in either the forward or reverse direction, depending on the temperature and the ratio of P_{CO}/P_{CO_2}, P_{H_2O}/P_{H_2}, and P_{O_2}.

Figure 2-27 shows the standard free energy of formation for some metal and metal oxide systems and CO/CO_2, H_2O/H_2, CO_2/C atmospheres as a function of temperature. Assuming it is under equilibrium condition and at soldering temperature of 250°C, lead oxide and copper oxide can be reduced by hydrogen; however, hydrogen is not effective on tin oxide until the temperature reaches 600°C. Equation (2-4) also indicates that the presence of too much water vapor in the furnace atmosphere will cause oxidation to certain metals. The partial pressure of water vapor should therefore be maintained at a constant and defined value.

At a given atmosphere composition and dew point, the gas flow rate and the flow pattern of the exhaust system in the furnace are also important factors in soldering performance. A typical flow pattern in a furnace chamber is shown in Figure 2-28. The gas flow rate should be high enough to avoid localized atmosphere buildup as a result of local reactions. In order to achieve the best performance and cost results, the required flow rate is determined by the characteristics of solder paste being used, furnace belt speed, loading pattern, belt width, and other furnace parameters. The exhaust efficiency and its flow pattern in combination with the flow rate dominate the removal of volatile components

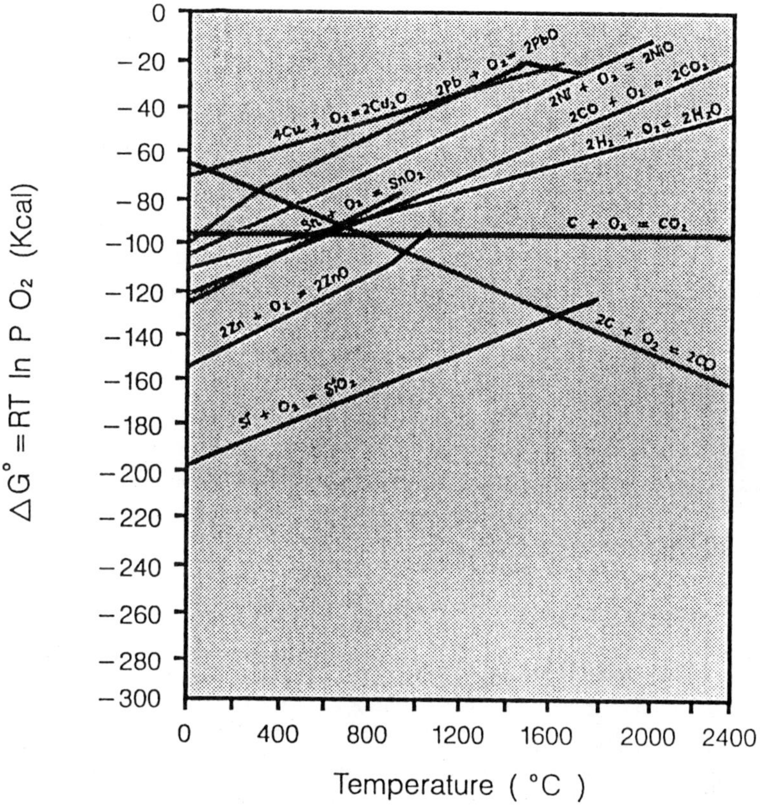

Figure 2-27 Standard free energy of formation of selected metal and metal oxides.

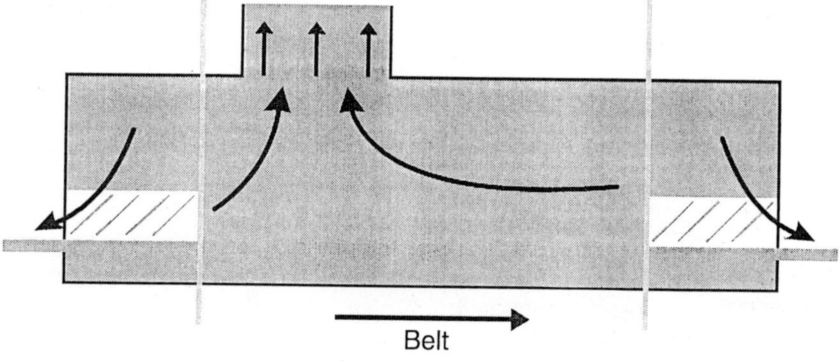

Figure 2-28 Typical flow pattern of conveyorized furnace.

generated from the pyrolysis and evaporation of chemicals in the solder material, which in turn affect solderability.

It should be noted that a complete burnout process is normally infeasible at the eutectic temperature of tin-lead or the eutectic tin-lead-silver solders, when using a solder paste.

In order to obtain good solderability and quality solder joints, a metallic continuity at the interface(s) between the solder and the substrate must be formed during soldering. When using a solder paste, a complete coalescence has to occur in synchronization with the development of metallic continuity at the interface(s). At soldering temperature, the atmosphere surrounding the workpiece protects or interacts with the surface of the substrates, with the solder alloys, and with the chemical ingredients in the flux or vehicle system. These interactions affect the resulting chemical and physical phenomena in terms of volatilization, thermal decomposition, and surface-interfacial tension.

A controlled atmosphere is expected to deliver more consistent chemical and physical interactions; therefore, a more consistent soldering process results.

In addition to consistency, the inert or reactive atmospheres possess further merits, which include:

Solderability enhancement
Solderability uniformity
Irregular residue charring prevention
Polymer-based board discoloration prevention
Wider process window
Overall quality and yield improvement

The inert and reactive atmospheres are expected to facilitate conventional fluxing efficiency during soldering. It should be noted, however, that the performance results highly rely upon the specific atmospheric composition and its compatibility with the solder material, substrates, and chemicals incorporated in the system, which must be also compatible with the soldering temperature profile.

2.15 SOLVENT CLEANING

Until recently, the solvents used for cleaning the after-reflow residue were primarily the azeotropes or blends of 1,1,2-trichlorofluoroethane with alcohol, methylene chloride, or accetone or blends of 1,1,1-trichloroethane with one or several alcohols. Those solvents have delivered successful results from a cleaning point of view. However, recent scientific data further substantiate the earlier findings that the chemicals under the group of chlorofluorocarbon (CFC) are the main contributors to the depletion of ozone (O_3) in the stratosphere, which results in biological and agricultural hardship from the transmission of excess ultraviolet radiation onto the earth. It is theorized that the stratospheric chlorine free radial (Cl·) is an active species depleting ozone through a catalytic chlorine oxide cycle, as shown in the following reactions:[12]

$$Cl\cdot + O_3 \rightarrow ClO + O_2$$

$$Cl\,O + O\cdot \rightarrow Cl\cdot + O_2$$

In addition to natural ozone removal mechanisms occurring in the stratosphere, the chlorine radical can be derived from sources such as CFC. This is because of the extraordinary stability of CFC compounds that are released to the troposphere and are not broken down by tropospheric processes. The accumulated CFC in the troposphere can then diffuse into the stratosphere where CFCs are photodissociated by ultraviolet radiation, generating chlorine.

These findings give rise to global concern about the use of CFCs and lead to the restriction of CFC production with a designated schedule set by the Montreal Protocol. Many countries and regions have actually started implementing more stringent regulations limiting the use of CFC chemicals.

Under today's limited availability and a likely future unavailability of CFC solvents, different alternatives to CFC solvent cleaning of electronic assemblies will become available through the efforts of solvent manufacturers, related material suppliers, and electronic packaging manufacturers. The alternatives and the corresponding status as of this writing are:

Cleaning Medium	Status
Hydrogenated CFC solvent	Available, in field testing
Water plus saponifier	In use
Water	In use
Terpene plus water	Available, in field testing
Other CFC substitutes	In development

The hydrogenated CFC solvents essentially possess similar properties in terms of cleaning as do the CFC solvents and perform effectively to clean conventional rosin-based residue. Table 2-9 shows some preliminary testing results of a hydrogenerated CFC solvent, HCFC-141b (CH_3, CCl_2 F), for cleaning RA-type fluxes and RMA-type solder paste.[13] The terpene solvent is able to clean rosin-based residue because of its natural compatibility with rosin. However, terpene hydrocarbon solvent, having inherent low vapor pressure, has to be used in conjunction with subsequent aqueous rinsing in order to remove the residual solvent through emulsification, and it obviously lacks vapor-degreasing capability. Aqueous cleaning is discussed in the next section.

Table 2-9 Cleaning Ability of HCFC in Relation to CFC[12]

Remaining Contaminate	Solvent: CFC-113/CH_3 OH/acetone isohexane/CH_3NO_2	Solvent: HCFC-141b/CH_3OH CH_3NO_2
Flux 1 (ugNaCl/in.^{2})*	7.2	5.2
Flux 2 (ugNaCl/in.^{2})*	21.2	17.4
Flux 3 (ugNaCl/in.^{2})*	7.0	6.7
Paste (ug/in^2)**	1198.0	1105.0

*In reference to MIL-P-28809 A 4.8.2.3.
**By high performance liquid chromatography.

Despite today's dynamic nature of cleaning solvent and technique selection and electronic assembly design, the principle of cleaning will not alter. The knowledge and information to answer the following questions are essential to perform an effective cleaning process:

What are the characteristics of soldering residue to be cleaned?
What are the properties of cleaning agent?
What are the features of cleaning equipment?
What is the compatibility among residues, cleaning agent, and equipment?
How is the cleanliness measured?
What are the criteria of cleanliness?

During cleaning, the action steps involve wetting, spreading, softening, solubilizing, and carrying away, which operate in parallel or in sequence. The physical properties of the cleaning agent that affect its efficiency are boiling point, density, viscosity, surface tension, solvency, latent heat of vaporization, vapor pressure, and solubility parameters. The relationship between each property and cleaning performance follows the general trends listed below:

Increasing temperature increases cleaning efficiency.
Increasing density increases wetting ability.
Increasing viscosity decreases wetting ability.
Increasing surface tension decreases wetting ability but increases capillary ability.
Increasing latent heat of vaporization decreases dryability.
Increasing vapor pressure increases dryability.
Increasing solvency increases solubilizing efficiency.
Closer solubility parameter between cleaning agent and gross residue facilitates cleaning.

Table 2-10 Solubility Parameters of Selected Materials*

Material	Solubility Parameter ($cal^{1/2}\ cm^{-3/2}$)			
	δ_d	δ_p	δ_h	δ
Water	6.4	15.4	16.8	23.8
Acetone	7.6	5.0	3.4	9.8
Methylene chloride	8.9	3.1	3.0	9.9
Methanol	7.4	6.0	10.9	14.5
Ethanol	7.7	4.3	9.5	12.9
2-Propanol	7.8	3.0	8.0	11.5
1-Butanol	7.8	2.8	7.7	11.3
2-Butanol	7.7	2.8	7.1	10.8
n-Hexane	7.3	0	0	7.3
n-Heptane	7.5	0	0	7.5
n-Octane	7.6	0	0	7.6
n-Nonane	7.7	0	0	7.7
n-Decane	7.7	0	0	7.7
n-Hexadecane	8.3	0	0	8.0
Cyclohexane	8.2	0	0.1	8.2
Benzene	9.2	0	1.0	9.1
Toluene	8.8	0.7	1.0	8.9
Biphenyl	10.5	0.5	1.0	10.6
1,1,1-Trichloroethane	8.3	2.1	1.0	8.7
1,1,2-Trichlorotrifluoroethane	7.2	0.8	0	7.2
Prelete	—	—	—	8.6
Freon TMS	—	—	—	7.4
Freon SMT	—	—	—	7.9
Genesolv DMS	—	—	—	7.4
Freon TA	—	—	—	7.4
Genesolv DA	—	—	—	7.4
Freon TE	—	—	—	7.4
Genesolve DE	—	—	—	7.4
Di-*n*-butyl phthalate	8.7	4.2	2.0	9.9
N, N-Dimethyl formamide	8.5	6.7	5.5	12.1
Ethylene glycol	8.3	5.4	12.7	16.1
Glycerol	8.5	6.0	14.3	17.6
Propylene glycol	8.2	4.6	11.4	14.8

*Courtesy of Allan F. M. Barton, *Handbook of Solubility Parameters and Other Cohesion Parameters* (CRC Press, Inc., 1983).

The solubility parameter is a cohesive property of materials, also known as the Hildebrand parameter, named after Joel H. Hildebrand for his pioneering fundamental concept on the solubility of nonelectrolytes. Assuming the cohesive energy (E) between molecules is a linear combination of three contributors—

dispersion interaction (E_d), polar interaction (E_p), and hydrogen bonding (E_h)—then

$$-E = (E_d + E_p + E_h)$$

Since cohesive pressure p is defined as cohesive energy per unit volume V,

$$P = \frac{-E}{V}$$

and solubility parameter (δ) is related to the cohesive pressure by

$$\delta = P^{1/2}$$

thus,

$$\delta^2 = \delta^2 d + \delta^2 p + \delta^2 h$$

Table 2-10 lists the solubility parameters with three contributing portions for the selected chemicals shown.

With respect to cleaning technique and equipment, the techniques used include vapor degreasing, liquid spray, immersion, high pressure spray (more than 50 psi), and immersion with ultrasonic aid. The single step of vapor degreasing is the mildest cleaning process and is considered inadequate to clean highly dense surface mount assemblies. A combination of steps, such as

vapor → spray → immersion-ultrasonic → spray → vapor or vapor → spray → vapor-immersion-high pressure spray → spray → vapor is generally required.

2.16 AQUEOUS CLEANING AND AQUEOUS CLEANING PASTE

Two types of aqueous cleaning prevail, namely, water saponification and water solubilization. For water-saponification cleaning, the water medium contains a saponifier that changes the water-insoluble rosin into a soluble salt as represented by the following reactions:

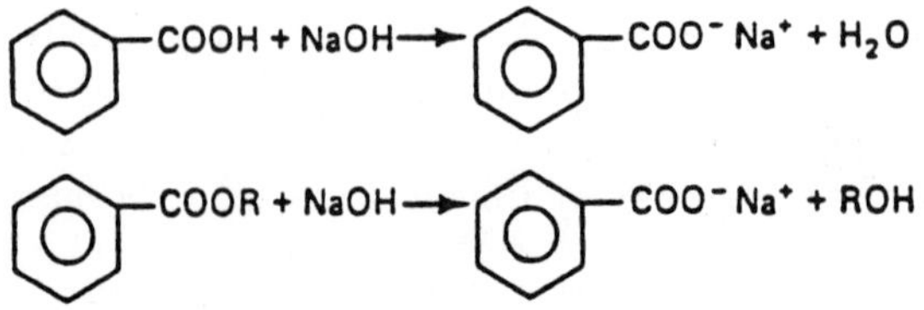

And water-solubilization cleaning refers to pure water as the cleaning agent, without incorporation of any additives.

In addition to saponifer or saponifiction, the additives and reactions involved in different chemical nature can be considered. The specific additive obviously depends on the chemical makeup of the paste.

For cleaning with water and saponifer, the analog of conventional RMA-type paste, which generally contains rosin or rosin derivatives and essentially maintains the characteristics of the RMA-type pastes, meets the requirement. The cleaning effectiveness highly relies on control of the quality and quantity of the saponifier used during the cleaning process. A water-soluble product, however, requires a different chemical makeup. The conventional OA-type paste is not considered a proper candidate. Technically, the system can be designed to deliver adequate performance on the production floor; however, tighter process control and smoother operation flow are generally needed.

It should be recognized that cleaning under the tight clearances between large components and boards, as well as being able to nondestructively measure cleanliness, has been and will continue to be a challenge, regardless of whether solvent or aqueous cleaning is used. Water-soluble products are designed to be compatible with plain water for easy removal. In addition to the chemical makeup of the water-soluble product, cleaning efficiency depends on the design and layout as well as the execution and control of the cleaning process. Furthermore, the temperature profile being used during soldering needs to be compatible with the paste characteristics.

Aqueous cleaning eliminates the use of CFC solvents; but, waste water management and disposal or recycling come into play. The waste water should not be discharged directly into public sewage systems without testing and treatment. Ion exchange and chemical treatment have been used to precipitate out heavy metals. Federal, state, and local environmental regulations should be observed for implementing proper waste disposal.

2.17 NO-CLEAN PASTE[14]

To combat the potential occurrence of "imperfect" cleaning, using the mildest amount of activity in the flux or vehicle system (nonsolder powder portion in solder paste) possible is a logical strategy. This leads to a discussion of no-clean pastes, which are expected to solve the CFC issue as well as to alleviate cleanability and cleanliness-measurement problems.

No-clean paste, therefore, is poised as an elegant and well-rounded approach. Nevertheless, the no-clean route is not a simple drop-in replacement—it not only places new demands on the paste design but also on the reflow process. To achieve the desired results, particularly the minimal residue, controlled atmosphere soldering constitutes an integral part of the system. Controlled atmosphere soldering is discussed in an earlier section. Figure 2-29 depicts an example of

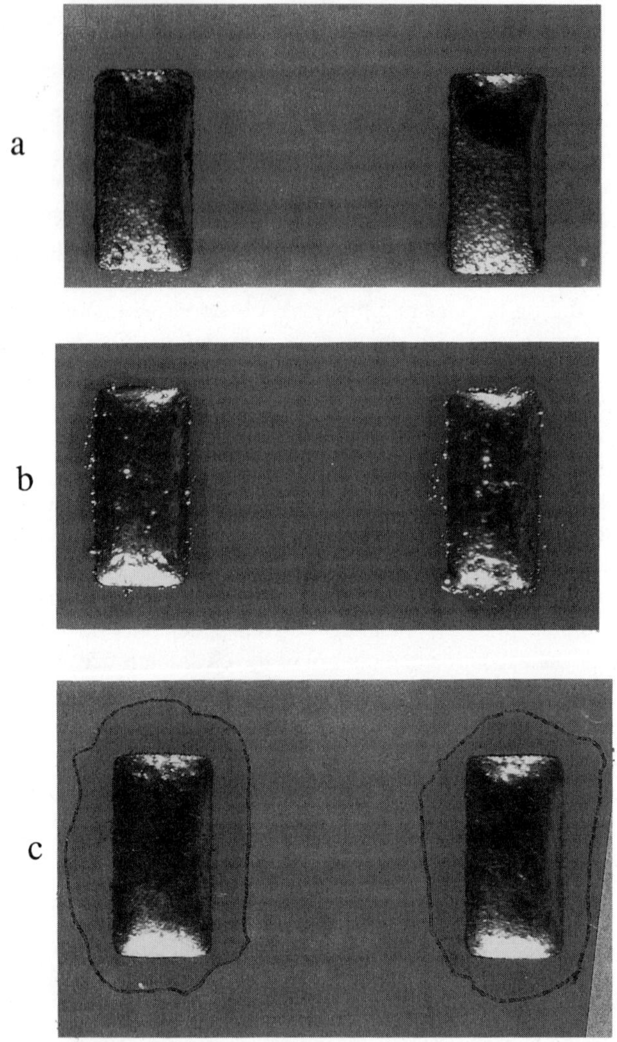

Figure 2-29 Solder joint and reflow residue of (*a*) No-clean paste after N_2 reflow; (*b*) no-clean paste after ambient air reflow; (*c*) RMA-type paste after N_2 reflow.

the effect of controlled atmosphere soldering on solder joints prepared from a standard RMA-paste and a no-clean paste after being reflowed under an N_2 atmosphere. It also illustrates the solderability enhancement of an N_2 reflow, exhibiting good wetting characteristics, minimal and inactive residue, and no solder balls, while the joint reflowed in ambient atmosphere shows solder balls and excess residue.

2.18 FINE PITCH PASTE

As the pin count of surface mount components increases, the pitch between pins becomes finer. As of this writing, a fine pitch pattern of 0.025 in. (0.63 mm) is considered the state of the art. When the pitch becomes smaller than 0.025 in. (0.63 mm), the processes and materials must be modified. What is needed in order to meet the demands of finer pitches covers several aspects of performance. Paste flow control in response to temperature becomes more important. The paste needs to retain its shape as temperature rises from ambient to near reflow, in order to avoid pad bridging. In Figure 2-30, paste 1 shows the desirable flow control and paste 2 imparts significant hot slump. The size of the solder powder incorporated in the paste will shrink from the 44- to 74-μm range to the 5- to 45-μm range. This requires practical powder production technology coupled with affective sizing technique. Solder ball-free solder paste also becomes more important for fine pitch as demonstrated in the solder ball test of Figure 2-31. In addition, the effects of the metal content, tackiness, and viscosity will become more pronounced. Overall, precision and control are key factors in the success of applying fine pitch solder paste. The following proposes some parameters in the paste and process suitable for three ranges of component pitches:

25 mil $\leq$ pitch $<$ 50 mil
- Standard solder paste
- Powder particle size 44 to 74 μm
- Selected printing parameters
- Slower squeegee speed, 0.5 to 2.0 in./sec
- Harder squeegee durometer, $\geq$80
- Thinner stencil, 6 $\pm$ 2 mil
- Precision and control

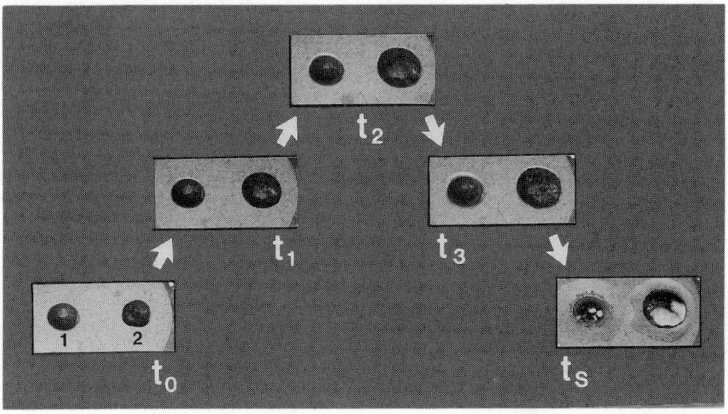

Figure 2-30 Flow dynamics of solder paste: (1) high flow restrictivity; (2) low flow restrictivity.

a

b

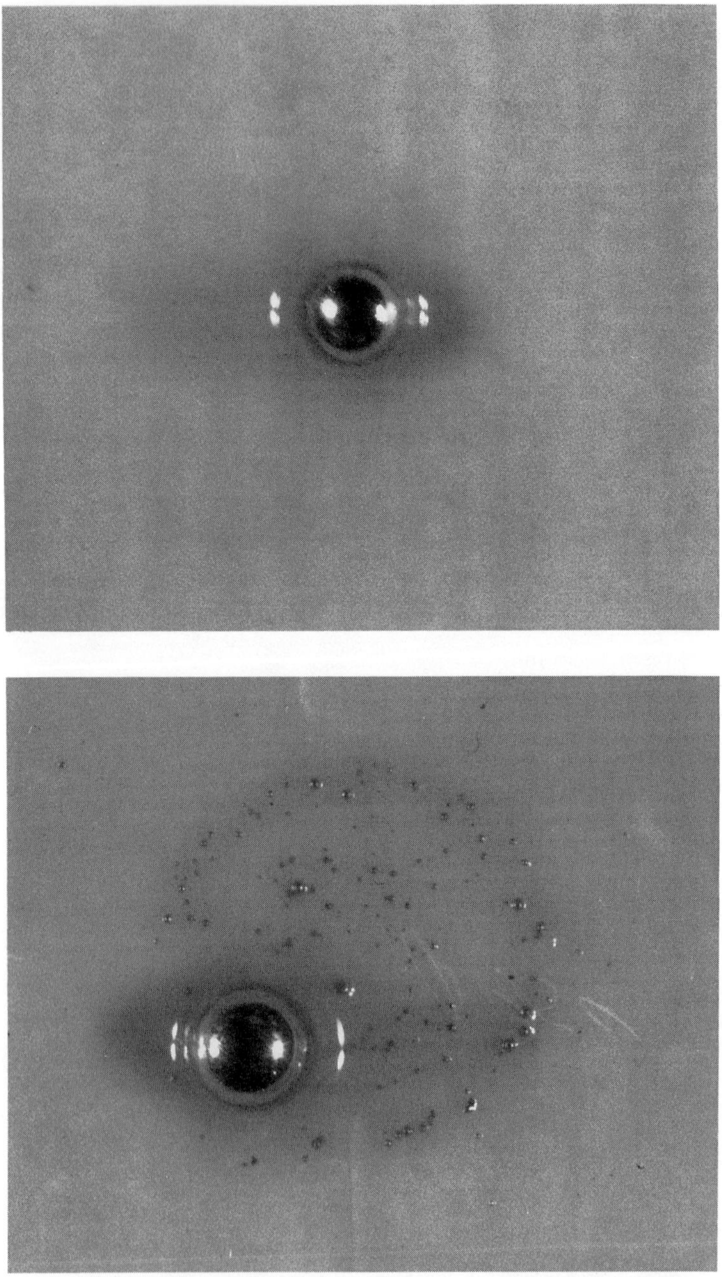

Figure 2-31 Solder ball test on alumina substrate (*a*) solder ball free; (*b*) excessive solder ball.

Table 2-11 Summary of Quality Assurance Tests for Solder Paste

Part	Tests
Paste	1. Appearance
	2. Metal content and flux and vehicle percentage
	3. Density
	4. Viscosity
	5. Viscosity versus shear rate
	6. Cold slump
	7. Hot slump/open time
	8. Molten flow
	9. Tack time
	10. Dryability
	11. Dispensability
	12. Printability
	13. Shelf stability
	14. Storage, handling, and safety
Flux and vehicle	15. Water extract resistivity
	16. Copper mirror corrosion
	17. Chloride and bromide
	18. Acid number
	19. Infrared spectrum fingerprint and other spectroscopies
Solder powder	20. Alloy composition
	21. Particle size—sieve
	22. Particle size distribution—SediGraph
	23. Particle shape
	24. Particle surface condition
	25. Dross
	26. Melting range
Reflow	27. Solder ball
	28. Solderability
	29. Exposure time
	30. Soldering dynamics
Post-reflow	31. Cleanliness—resistivity of solvent extract
	32. Surface insulation resistance, before and after cleaning
	33. Solder joint appearance
	34. Solder voids
	35. Joint strength
	36. Power cycling
	37. Temperature cycling
	38. Vibration test
	39. Simulated aging
	40. Thermal shock

10 mil < pitch < 25 mil
  Solder paste with powder particles 5 to 44 μm
  Stencil thickness 4 ±2 mil
  Designed paste deposition pattern
Pitch 10 mil
  Fine mesh screen, 200 mesh or finer
  Fine needle dispensing ≤ 25 gauge
  Designed paste deposition pattern
  Prebumping
  Solder paste with fine powder particles, 5 to 25 μm

2.19 QUALITY

With quality defined as continuous improvement, a major effort needed today in the solder paste area is to continue to advance and perfect the performance and the consistency of solder paste. Table 2-11 summarizes the parameters that affect the quality and performance of a solder paste. The detailed test procedures of each parameter are illustrated in the reference.[1]

2.20 CONCLUSION

In view of new trends and new demands, the question "can solder paste technology coupled with soldering technology continue to provide reliable interconnections for board-level packaging in the next decade and beyond?" has to be asked and answered. The answer relies on how promptly and how well the concerned scientists and engineers continue to advance and perfect the design-material-process and how committed the fund decision makers are to the advanced technology development. This is not to imply that a technology has no limits, but there are still advancements and innovations in solder paste yet to come. This has to be implemented by effectively providing a system solution involving design, material, equipment, and process.

REFERENCES

1. Hwang, J. S., *Solder Paste in Electronics Packaging—Technology and Applications in Surface Mount, Hybrid Circuits, and Component Assembly,* Van Nostrand Reinhold, New York, 1989.
2. March, J., *Advanced Organic Chemistry, Reactions, Mechanisms, and Structure,* McGraw-Hill, New York, 1968.
3. Hwang, J. S., "Solder Paste Rheology—Principle and Practice," *Technical Proceedings, Expo SMT 88,* pp. 171–177.
4. Ferry, J. D., *Viscoelastic Properties of Polymers,* Wiley, New York, 1970.

5. Riemer, D. E., "Analytical Engineering Model of the Screen Printing Process," *Solid State Technology,* September 1988, p. 85.
6. Hwang, J. S., "Controlled Atmosphere Soldering—Principle and Practice," *Technical Proceedings, NEPCON-West,* 1990.
7. Vandervoort, S., "Lamp IR Reflow Provides Repeatability Between Units," *Electronic Packaging and Production,* May 1989, p. 58.
8. Martel, M. L., "New Wrinkles in Reflow," *Circuits Manufacturing,* June 1989, p. 33.
9. Dow, S., "Reflow Soldering Survey," *Assembly Engineering,* December 1988, p. 18.
10. Glynn, M., "Full Spectrum IR Reflow through Energy Management," *Printed Circuits Assembly,* February 1989, p. 22.
11. McGinn, K. S., "Effects of Furnace Atmosphere Chemistry in Sintering," *Industrial Heating,* May 1985.
12. Stolarski, R. S., and R. J. Cicerone, *Can. J. Chem., 52,* 1974, p. 161.
13. Bonner, J. K., "Searching for a Substitute," *Circuits Manufacturing,* July 1989, p. 42.
14. Hwang, J. S., "Soldering and Solder Paste Prospects," *Surface Mount Technology,* October 1989, p. 56.

3

Technical Considerations in Vapor Phase and Infrared Solder Reflow Processes

Charles L. Hutchins

3.1 INTRODUCTION TO SURFACE MOUNT REFLOW SOLDERING

Based on the method of soldering technology that is used, there are three general types of assemblies. Some publications list other type designations, but the convention used in this chapter is the one that is most common.[1] Since the manufacturing or soldering technology is the basis for the designation, the types given below are unambiguous and clearly understandable.

The two basic methods for soldering are reflow, using either presoldered components, solder preforms, or solder paste, and wave or flow soldering, commonly used for through-hole assembly. Once the type of process is chosen, the types of components that can be used are determined. These two soldering techniques are used for the general types that are discussed below:

Type I. Surface mounted components only using reflow soldering, either single sided or double sided.

Type II. Both surface mounted components, using reflow soldering, and through-hole components, using wave or flow soldering, that are mounted on top of the printed circuit board (PCB). By using a modified or dual wave, small surface mounted components may be mounted on the bottom with adhesives and soldered at the same time as the through-hole components (see type III below).

Type III. Surface mounted components are mounted on the bottom with adhesives, through-hole components are inserted from the top, and both are soldered with a single pass through a modified or dual wave.

The details of the type I (and the first part of the type II) process are discussed in the following section because of their applicability to this chapter.

3.2 TYPE I

This process uses only surface mount components, which are attached with a reflow solder process. These include actives, such as small outline transistors (SOTs), small outline integrated circuits (SOICs), ceramic chip carriers (CCCs), and plastic leaded chip carriers (PLCCs), as well as a wide assortment of passive components. Included in this group are chip capacitors and resistors, various forms of larger filter capacitors, sockets, connectors, coils, switches, and other special hardware to complete the electronic function of the surface mount assembly. These can be mounted on either one side or on both sides.

A typical single sided type I assembly is shown in Figure 3-1. It uses several different types of surface mount packages. Surface mount assemblies are found in many different sizes and different substrate materials, from large epoxy-glass computer "mother boards" to small and medium ceramic substrates for military and avionics uses.

The basic process flow is shown in Figure 3-2. It shows the four major steps for the reflow process. The first step is the placement of solder paste on the component pads or lands on the PCB. This is typically done with either a syringe dispenser or a screen printing procedure. The amount of paste can be easily controlled and registration to ±2 mils is usually achievable with standard equipment.

Figure 3-1 Typical single-sided type I surface mount assembly. (Photo by author)

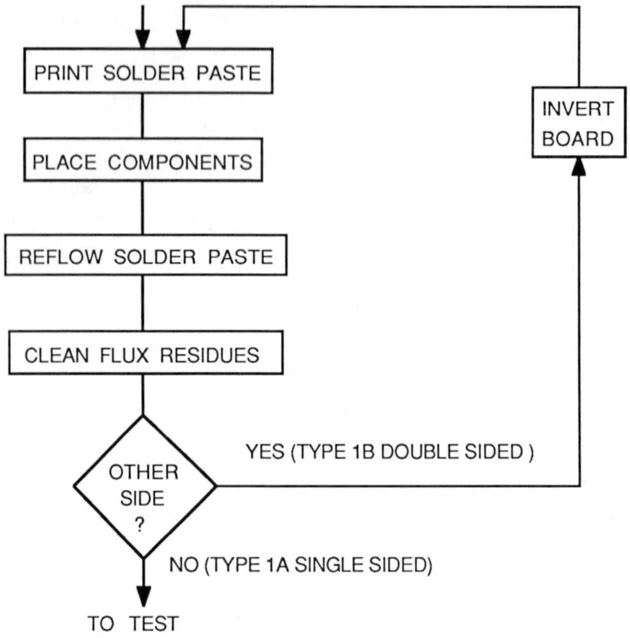

Figure 3-2 Process flow for type I A and B.

The next step is the placement of components in the paste, with particular care to align them to the PCB lands. The paste has a sticky consistency, holding the components in place during careful handling. Although during the reflow step the surface tension of the molten solder will provide some self-alignment of smaller components, minor yield problems can result from component misregistration.

The third step is the reflow of the solder paste. This can be achieved in several types of hot ambients, the most popular of which are the vapor phase (or condensation heating) and infrared (or infrared-convection) reflow systems. These latter two steps will be discussed extensively in this chapter. The final step is a clean-up operation to remove flux residues and other contamination.

3.3 SOLDERING REQUIREMENTS FOR SURFACE MOUNT TECHNOLOGY

The two major requirements for soldering SMT assemblies are coplanarity and solderability. The leads of the component and the land area of the substrate or PCB must be coplanar and wettable with the solder used in their joining. As these have a significant impact upon the solder joint strength, yield, and reliability, it is important that we understand the effect that the reflow soldering step has on both of them.

There are three major requirements for the solder alloy and the method used in the soldering step. First, the reflow temperature of the solder alloy must be compatible with the metallization used on the components, the substrate, and the reflow system. Generally, this means that the solder alloy melts at approximately the same temperature as the solder coatings on the component leads and substrate lands. Also, the constituents of all three must not form undesirable metallic compounds, such as intermetallics with brittle or weak mechanical properties or high electrical resistance.

Next, the surface tension of the molten solder should be sufficiently strong to provide a continuous surface between the component lead and the land on the PCB during the reflow process, resulting in fillet formation upon cooling. This attribute has not received much attention in previous solder research but will find increased attention as SMT implementation increases. A solder meniscograph or wetting balance can be used as a quantitative test instrument.

Surface tension of the molten solder is what makes surface mounting a manufacturable process. It is extremely difficult and costly to form component leads into an absolutely planar arrangement, so a soldering system (alloy, flux, solvent, and reflow technique) that can provide a measure of compensation for nonplanarity is very desirable and valuable. It will prevent yield losses from open solder joints and, thus, the costly manufacturing rework.

Thirdly, the strength of the alloy formed after the soldering step must be strong enough to provide reliable joints. In some cases a "flexible," or compliant, joint is desired, while in others a strong rigid connection is required. Obviously, this property is dependent upon the choice of the alloy constituents as discussed above with respect to the components and substrate, but it also can be an independent choice. Further, it can be affected by the time and temperature history, reflow operation plus any rework or repair of the total assembly. The alloys and intermetallics can continue to change if the conditions are favorable, such as extremely hot leads of high power dissipating components. In general, this is not a serious problem, but considerations for this potential must be constantly maintained.

3.4 REFLOW PROCESS PHASES

The primary purpose of the reflow process is to melt the solder particles, wet the surfaces to be joined, and then solidify the solder into a strong metallurgical bond. However, prior to this melting phase, several other phases occur. The sequence of these actions is diagrammed in Figure 3-3.

During the five phases shown, there are several local actions, followed by a global action, and then a strong metallurgical bond and fillet formation. The local actions start with the solvent evaporation. If this is too violent, there will be "mini-explosions," giving rise to the formation of small solder balls being scattered all over the board. This reduces the amount of solder volume available

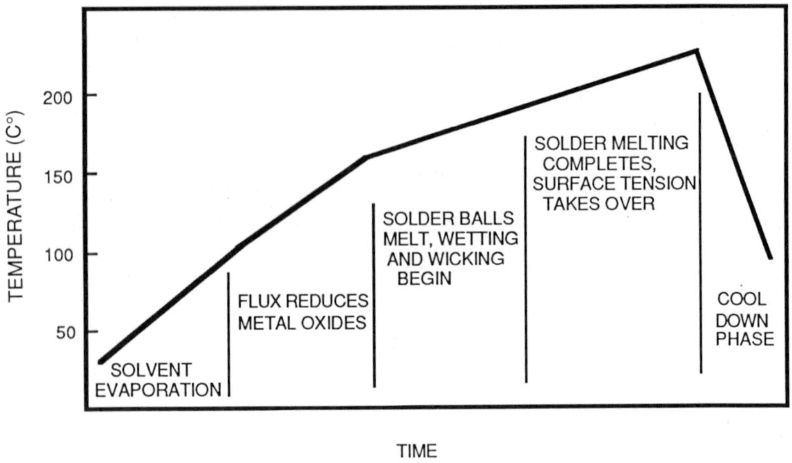

Figure 3-3 Solder reflow phases.

and can result in weak or open solder joints, or the solder balls can cause extraneous shorts if they happen to end up between two metal conductors.

Next the flux melts and begins the chemical reaction with the metal oxides, preparing a "clean" metal surface for the solder. As this metal oxide reduction action of the flux is proceeding, the solder spheres in the paste begin to melt at the hottest spots, replacing the flux and wetting the clean metal surfaces. This wetting action results in the wicking of the solder onto the hottest and cleanest metal surfaces. The speed of the action depends strongly upon the fluxing action, the local temperature, and the metal alloys of the component lead, solder paste, and substrate. Under most conditions, temperature nonuniformity is the major factor. High temperature helps to reduce surface tension and can promote significant wetting and wicking action.

The heat source should be 25 to 40°C greater than the melting, or liquidus, temperature of the selected alloy to assure complete melting of the solder and good metallurgical bond formation. The actual time for the solder to completely melt and wet the metals to be joined depends upon the localized temperature of the assembly, and at the typical reflow temperature of 215°C (Sn63 or Sn62) it will be 1 to 2 sec,[2] as shown in Figure 3-4. The time that the assembly is at the maximum temperature of the reflow system should be limited to approximately 30 to 50 sec, depending upon the thermal mass of the assembly. This will allow reflow to take place but will minimize the formation of undesirable intermetallics or damage to the board. This will also allow time for the trapped flux and residual solvents to escape, minimizing void formation as the solder cools.

After all of the solder spheres have melted, a liquid solder "volume" forms. The global action of the surface tension of the molten solder then prevents further wicking action. Thus, the surface tension action, which minimizes surface area,

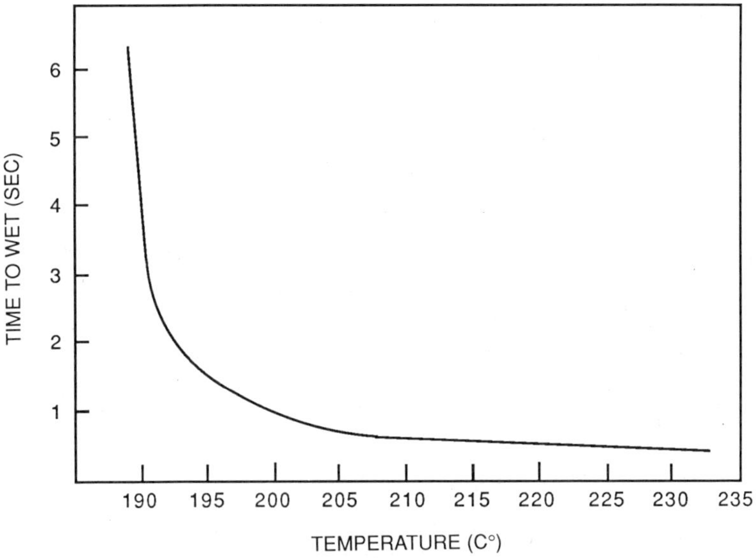

Figure 3-4 Solder wetting time versus temperature.

is beneficial in that it helps hold the solder in place and bridge a gap between the component lead and the substrate, forming a fillet upon cool down. The volume of solder, the metal alloys being joined, and the physical geometry of the metal (lead, land, via, etc., shape, and spacings) determine the resultant final shape, fillet formation, and strength of the solder joint.

If all the factors of the process are sufficiently matched, there is good solder wetting and fillet formation between the component lead and the land on the substrate. If this is not the case, several things can happen at the solder joint. There can be an open joint due to an out of specification coplanarity gap or due to a contamination-wettability problem. There could be a marginal (insufficient) solder joint formed for several reasons—volume loss due to solder ball formation, a small (not out of spec.) coplanarity gap, insufficient solder paste print, or a wicking away of the solder from the desired joint (either vertically or horizontally).

Thus, the matching of the solder paste properties and the reflow environment must comprehend several actions. These are the solvent evaporation, flux action, solder wetting and wicking, and surface tension of the molten solder. Further, minimizing temperature nonuniformity and its resultant effects on the solder is also very important. To fully optimize the process, one must control each of the actions listed above with the raw material introduced into the operation and the time-temperature profile.

The cool down after reflow should take place within 30 to 50 sec to enhance small grain formation but should be no greater than 3°C/sec to prevent thermal shock damage to some components. Larger grain formation in solder usually promotes voids and joints that are slightly weaker.

3.5 REFLOW EQUIPMENT

The major types of equipment used in reflow soldering include infrared furnace, vapor phase system, convection oven, hot platen or conductive belt, and laser. The order given is also the approximate order of popularity or frequency of use. Each is an acceptable method in principle, with the differences being primarily in the type of assembly being produced and in yield and cost trade-offs. Each of these will be discussed in the following sections, with major emphasis on infrared (IR) and vapor phase soldering (VPS).

3.5.1 Infrared

The IR process primarily uses the transfer of heat by absorption of radiant energy. Infrared radiation is energy transported by electromagnetic waves. It is the portion of the electromagnetic spectrum with wavelengths between 0.72 and 1000 μm. Radiation wavelengths are a function of the temperature of the emitter source and possibly of the material used as a "reflector," which really absorbs and reradiates the original energy.

All bodies radiate electromagnetic energy as a function of their absolute temperature. This relationship is described by the Stefan Boltzmann law:

$$W = KT^4$$

where W = total emissive power
K = Boltzmann's constant
T = absolute temperature

Thus as the temperature of the emitter changes, the radiated power changes by the fourth power. This is very useful for quick response to changes to loading of an IR furnace but can also make the control of the desired temperature very difficult.

The useful range for heating occurs between the wavelengths of 1 to 8 μm. They are generally classified as short or near wave (0.7 to 3 μm) and medium to far wave (3 to 8 μm). The temperature of the radiating element determines the wavelength. Thus, the shorter wavelengths usually are obtained from lamp emitters that have a tungsten filament, while longer wavelengths come from panels that use resistive heating elements.

Heating takes place indirectly as the IR energy is absorbed within the molecular structure and causes an increase in local temperatures. Thus it is a "volume" heating phenomenon. IR energy is absorbed primarily in organic materials such as the PCB, solder paste flux and solvents, and plastic components. The depth at which it is converted to heat is dependent upon the wavelength, with shorter wavelengths penetrating deeper. Metal reflects most of the IR energy but does act as a conductor to carry heat from hot to cool spots.

Since most of the energy is absorbed in the PCB, it gets hot first and then transfers this heat to the lands. The flux and solvent also absorb a small amount of the energy. Thus, the land gets hot first, the solder melts and wets the land, then wets the lead. This reduces the amount of solder that wicks away, allowing a given volume of solder to span a greater coplanarity gap if one exists, reducing the probability of an open solder joint. However, since the board and conductors are the hottest, solder can wick laterally away from a joint to a large conductor or via if the solder is not constrained by proper PCB layout or construction.

The major difficulty with IR reflow is that it is not an "equilibrium" method of heating, (i.e., the heating object will never reach the same temperature as the heating source). With IR sources of quartz lamps or resistance panel heaters, the source "temperatures" range from 400°C for panels to 1000°C for quartz lamps. To reach a maximum safe reflow temperature of 220°C, one must depend on accurate pulse operation of the heaters and the belt speed to be adjusted such that the heated object exits the heating chamber at the correct time-temperature. For a given mass, too slow means that there will be higher energy absorption and too high a temperature, and vice versa.

Also, since IR heat results from unequal energy absorption, variations in absorptivity or mass, either across the board or from assembly to assembly, could produce significant temperature variations. For boards of large size or large mass variations, temperature differences in excess of 20°C can exist between the corner (low mass area, typically the hottest) and the center (or high mass area). This is generalized in the typical IR reflow profile shown in Figure 3-5. This curve shows a five-zone system with appropriate temperature plateaus to optimally achieve the actions shown in Figure 3-3.

Note that this curve is also dependent upon the spacing, or loading, of the assemblies going through the heating zones. Different spacings between assemblies could result in significantly different temperature profiles. Excessive temperature, with respect to the individual components or PCB material that are used, can cause low yield and/or reliability problems. Inadequate temperature could produce unreflowed solder joints.

A typical IR system will have three or more heating zones, and some will have independent top and bottom heat source control. Other features will include microprocessor control, with a memory to store profile information, and automatic setup from memory. This latter feature is useful for production of SMAs that are different in size and mass. A typical system that provides these features is shown in Figure 3-6.

With temperatures up to 1000°C being used to generate the radiant energy, more care must be exercised. The geometry, mass, absorptivity, and thermal conductivity of the material being soldered must be considered. With a short

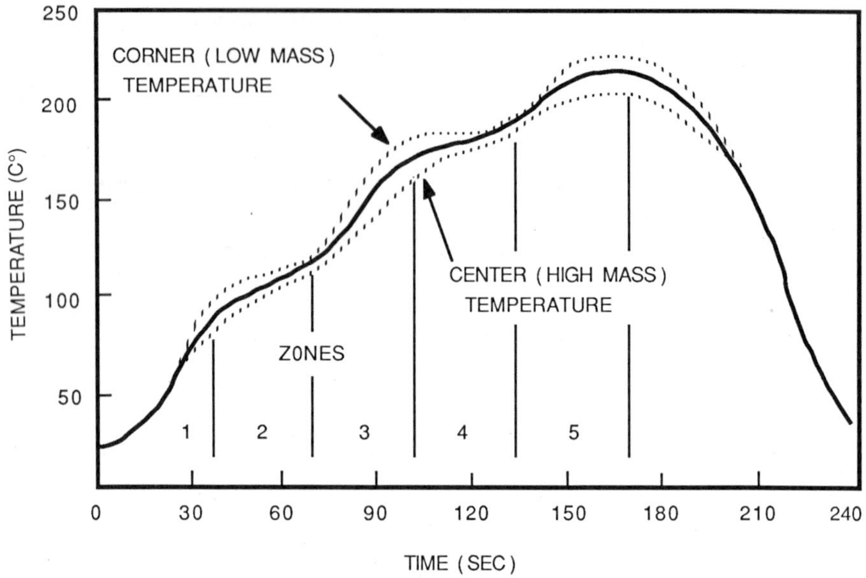

Figure 3-5 Typical five-zone infrared profile.

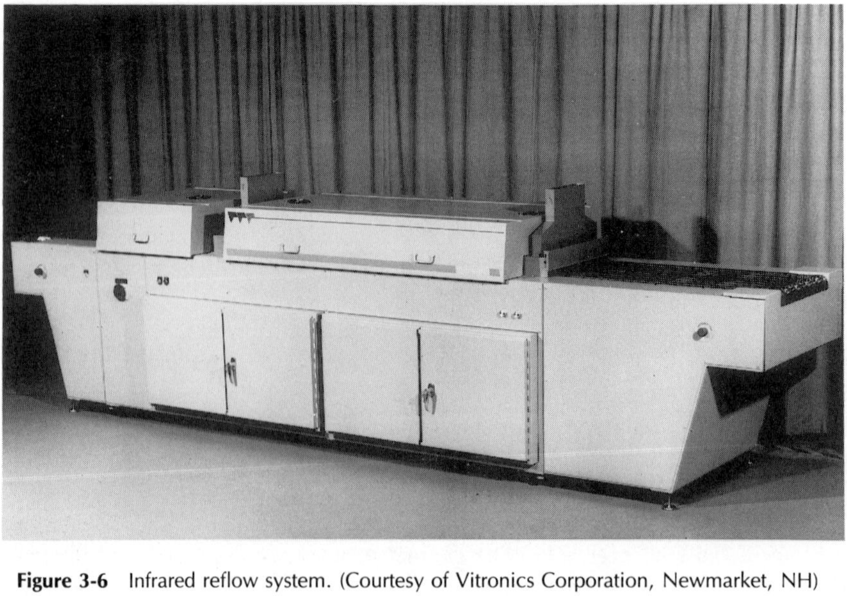

Figure 3-6 Infrared reflow system. (Courtesy of Vitronics Corporation, Newmarket, NH)

wave (lamp) furnace, ceramics, organics, and glasses are all transparent to some degree; thus the energy is more penetrating. With a longer wave (panel) system, there is more surface heating. In both types there is usually some portion of heat energy that is transferred to the assembly by convection. With lamp systems it is in the 20 to 40 percent range, while with a panel system it can be greater than 50 percent. With both systems, slower belt speed and lower power settings will increase the ratio of convection heat.

Achieving a time-temperature profile for a safe and complete reflow is an individual effort for each and every assembly, and to a lesser degree the profile must be adjusted for the solder paste being used. Some surface mount assemblies have a nonuniformity of mass due to component placements or cutouts in the board for mechanical components. These make it difficult to find the correct power settings and belt speed, and resultant zone temperatures, that will produce acceptable temperature uniformity across the PCB to prevent problems with the day-to-day range of manufacturing variables.

Since IR ovens for reflow soldering generally use three or more heating zones, they provide a wide range of heating profiles for reflow of surface mount assemblies. As the assembly passes through the oven on the belt, the temperature profile is established by the speed of the belt, the energy levels of the sources, the distance of the assembly from the sources, and the absorptive characteristics of the components and the substrate.

The IR system provides an opportunity to tailor a specific heat profile to provide optimum yields but requires more attention to initial profile setup and maintenance of the equipment. It must also have a "low mass" characteristic in order to respond quickly to changes in mass loading, such as would happen when the output of a pick-and-place changes due to component feeder problems.

3.5.2 Vapor Phase

Vapor phase soldering (VPS) was developed by the Western Electric Company in the mid-1970s as a technique to solder large backplane assemblies that were impractical to wave solder. In its simplest form, the process is carried out in a deep container that contains boiling liquid in the bottom and a condensing coil near the top. The part is introduced into the vapor just above the liquid, and the vapor condenses on its cooler surfaces, giving up its latent heat of vaporization. This is a very efficient process that transfers heat from the vapor concentration gradient as well as a temperature gradient (convection heat) within the reflow chamber.

The fluid used most commonly today is an inert perfluorinated liquid, a derivative of compounds originally developed to provide vaporization cooling of power transformers. These fluids have a high boiling point, are thermally and chemically stable and compatible with most materials of the soldering process, and have a vapor that is significantly heavier than air. The most commonly used fluids boil at approximately 215°C. There are other fluids available for higher and lower temperatures for noneutectic tin-lead or low temperature solder alloys.

VPS is the easiest reflow technique to use. It is defined as an equilibrium heating system. The object being heated eventually attains the same temperature as the heat source. This means that the vapor phase is also a very safe process in that the assembly will never see excess temperatures.

Since the temperature of a saturated vapor is the same as the boiling point of its fluid, the soldering temperature is predetermined by the fluid chosen. For any given fluid, this temperature is constant within a few degrees, resulting in a process both precisely and repeatedly maintained. This provides adaptability of the process to a variety of applications, independent of product geometry or uniformity of thermal mass of the SMT assembly.

VPS is achieved by transferring the latent heat of the vapor to the surface upon which it condenses. Thus, it is a "surface" heating phenomena. Objects in the reflow environment are heated on their surfaces first. If an object has a higher surface-to-volume ratio than its neighbor, it will heat up faster. Also, if it has a higher thermal conductance value, it will reach higher internal temperatures faster and conduct heat away from hot spots to cooler areas. This is important where there is considerable metal, such as a "J" or "I" lead. The leads will get hot much faster than the metallization on the substrate, which is thermally constrained by the PCB. This is shown in the typical VPS reflow time-temperature profile in Figure 3-7. Note that both eventually achieve the same maximum temperature, a key safety feature of VPS.

Thus, with no solderability problems, the solder will wet to the lead first. Since the lead is hotter away from the solder paste, there will be a tendency for

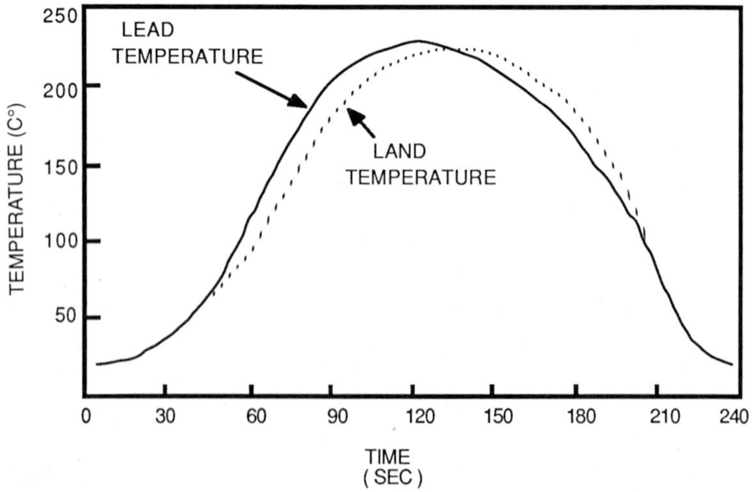

Figure 3-7 Typical in-line vapor phase profile.

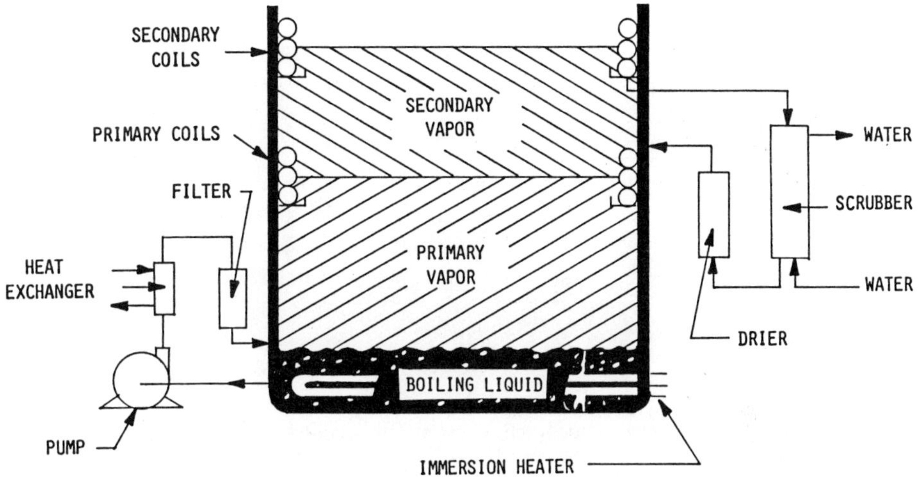

Figure 3-8 Cross section of a typical dual vapor batch reflow system.

the solder to wick up the lead. Eventually after the solder is completely molten, the surface tension effects will take over, bridging between the lead and the land and allowing fillet formation on cool down. Any disruption in the surface tension due to solvent splatter or a coplanarity gap between component lead and land greater than the solder can bridge with the surface tension will cause the solder to separate. The fillet will not form and an open solder joint will be the result. There is a relationship between the surface tension of the solder alloy, the volume of solder paste, the temperature uniformity, and the maximum allowable gap between the lead and land for successfully forming solder joints that can only be determined empirically.

The most common version of the vapor phase system uses a secondary vapor blanket to control loss of the primary fluid. A cross section of a typical batch reflow system is shown in Figure 3-8. The secondary vapor, formed by the heat of the primary condensing coil and controlled by the secondary condensing coil, forms a cap just above the primary fluid. Another type of batch system does not use the secondary vapor but provides vapor containment by the system design. A photo of a this type of batch system is shown in Figure 3-9.

Traditionally, the secondary fluid was a mixture containing primarily chlorofluorocarbon 113 (CFC113). Since the interface contains superheated vapors of this fluid, it is an inherent acid generator (HF and HCl), requiring considerable maintenance to minimize the effects on the system. Newer lower boiling temperature perflourinated compounds are now available that were developed to reduce this acid generation problem.

Other systems have been designed to operate in an in-line fashion for automated manufacturing lines. The reflow time for modules of different mass is adjustable

Figure 3-9 Typical single vapor batch reflow system. (Courtesy of Centech Corporation, Minneapolis, MN)

by changing the belt speed. Systems are available in both single and dual vapor models. Dual vapor systems are generally more expensive to purchase but use less primary fluid and are therefore somewhat cheaper to operate. They do require regular maintenance to control the acid generation problem.

Newer in-line vapor phase systems have built-in preheaters in the entrance area to dry the solvents in the solder paste and provide a more gradual heating of the assembly. A picture of a typical system is shown in Figure 3-10. This

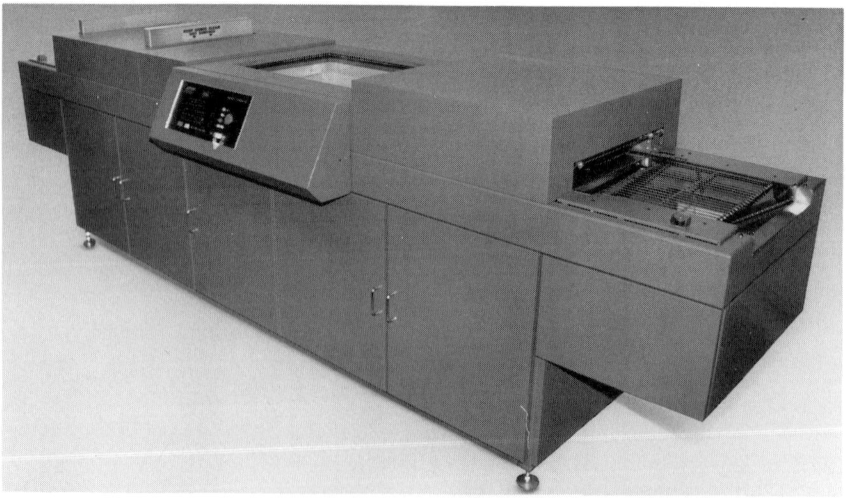

Figure 3-10 In-line vapor phase solder reflow system. (Courtesy of Centech Corporation, Minneapolis, MN)

system has the additional feature of a special "conveyor," rather than a belt, that supports the assembly in a horizontal position during reflow and cool down.

3.5.3 Convection

This process exposes the surface mount assembly to a hot oven and primarily uses the convection method to transfer heat to the solder paste for reflow. The temperature of the atmosphere in the oven and the speed of the belt establishes the temperature profile. Secondary heating occurs as a result of radiation and conduction from the hot surfaces within the furnace. The temperature-time profile will be similar to the IR oven profile shown in Figure 3-7.

Typical convection furnaces employ carefully designed entry and exit baffles to provide close control of the air temperature and flow and thus the solder reflow ambient. They can have several heat zones and possibly a cooling zone. Similar to IR ovens, inert or reducing atmospheres can be used to prevent oxidation. These systems generally operate at much slower conveyor speeds per unit length of heat tunnel than IR systems.

One of the potential benefits of the convection-based reflow system is its ability to accommodate several different board sizes, masses, and materials with one time-temperature profile. One reported test used the same profile for two assembly weights that varied by more than a 10 to 1 ratio. The temperature variation between the two was less than 25°C, more than desirable but demonstrating the versatility of the equipment.

3.5.4 Conductive Belt

This technique uses a series of temperature controlled platens, with a heat conductive belt transporting the assembly. The heat is transmitted through the belt to the assembly to accomplish solder reflow. Soldering temperature profiles are controlled by varying the temperature of the platens and the speed of the belt. This is the only reflow process that allows a continuous visual inspection of the process. However, to prevent oxidation, an inert gas hood is required. This process is widely used for soldering ceramic assemblies and generally has limited use in epoxy or glass PCB reflow applications.

3.5.5 Laser Soldering

Reflow soldering with a laser is achieved by focusing the beam on a desired solder joint, pulsing the beam, then moving the assembly, and refocusing and pulsing the beam on the next solder joint. This movement and focusing requires computer control and detail programming for each solder joint. This is costly in both time and money for use as a general production reflow but could be cost effective in some situations.

Potential applications include specialty soldering of unusual geometries or heat sensitive components that could not withstand the reflow process or for repair, soldering individual leads using computer control. A manually controlled beam locator could also be a practical solution for a repair operation.

3.6 PREREFLOW SOLDER PASTE BAKE

Baking the solder paste before reflow removes the solvents and absorbed water, reducing the tendency for splatter and solder ball formation during reflow. A typical drying operation would be approximately 30 min at 80°C. Other recommendations of different times (20 to 60 min), and different temperatures (75 to 100°C) have been found useful, with the variations generally not critical to yield.

Solder ball formation can be a significant problem if the balls get trapped in the smaller spaces between ceramic chip carriers and substrates or possibly between the leads of small pitch leaded ICs. The solder balls also reduce the solder volume available for the fillet formation, and the action of splatter interrupts the holding power of the surface tension. Both of these have minor deleterious results on the reflow and connection process and hence on yield.

Baking before vapor phase reflow has been observed to increase yield by reducing the incidence of solder joint opens on J lead components.[3] The first heating zone in an IR reflow system can provide a short, but effective, solvent drying step. This can be particularly important on large assemblies with over 1000 solder joints. Conversely, it can be a marginally beneficial and costly process for in-line vapor phase reflow production of small modules of less than a few hundred solder joints. The cost of equipment and handling for a separate in-line bake could be greater than the yield improvement achieved. IR preheaters, built into vapor phase systems, offer the most promise and are built into most currently sold systems.

3.7 MAXIMIZING SOLDER JOINT YIELD

One subtle but important difference between reflow and wave soldering is the management of the surface tension of the molten solder. In reflow soldering, it is necessary to maximize surface tension, while most wave soldering systems use temperatures well above the liquidus point of the solder and oil intermixes to reduce surface tension. This helps to prevent solder bridging, the major yield loss problem with wave soldering. Conversely, the major problem in SMT reflow soldering is open solder joints. Thus, as the reflow solder process is developed, it is necessary to look for ways to increase surface tension to maximize yields.

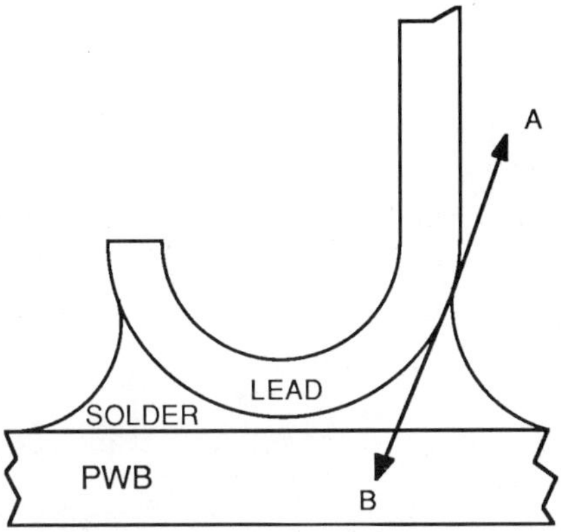

Figure 3-11 Solder joint showing representative force vectors during reflow.

To further understand the necessity for maximum surface tension, let's use the simplified vector drawing in Figure 3-11. The paste thickness will be approximately as shown. When the vapor phase reflow process starts to take place, the solder will typically melt first at the interface with the lead. The vectors shown represent the forces on the solder as the reflow starts to take place.

If there is a positive thermal gradient up the lead, vector **A** represents the force of the solder wicking up the lead. Vector **B** represents the force of the surface tension attempting to keep the solder intact between the component lead and the PCB land. To maximize the probability of a fillet formation upon cool down, reflow conditions should be such that thermal gradients are minimized and surface tension is maximized.[4] Techniques to accomplish this will be covered later in this chapter.

The wicking action of the solder can be better visualized by examining the picture in Figures 3-12 through 3-14. The first picture shows typical solder joint, two good ones in the background and two open joints in the foreground. Note the good fillet formation on the good joints and the extra solder on the shoulders of the bad joints.

The next two pictures show cross sections of similar joints. Again, note the fillet formation and the contact between the lead and land on the good joint. On the bad joint, note the excess solder on the shoulder and inside the lead. Also note the space between the lead and the land, which is approximately 4 mils. For the solder paste and reflow conditions used, the surface tension could not offset the wicking action. The wicking action is enhanced by temperature nonuniformities (i.e., the lead is hotter than the land during the initial reflow action). Both joint exhibit good wetting, as evidenced by the small angle between the wetted surfaces and by minimum formation of voids.

With infrared reflow processing, the wicking action is different. Since IR energy is absorbed primarily in the PCB and is reflected by metal, the PCB gets

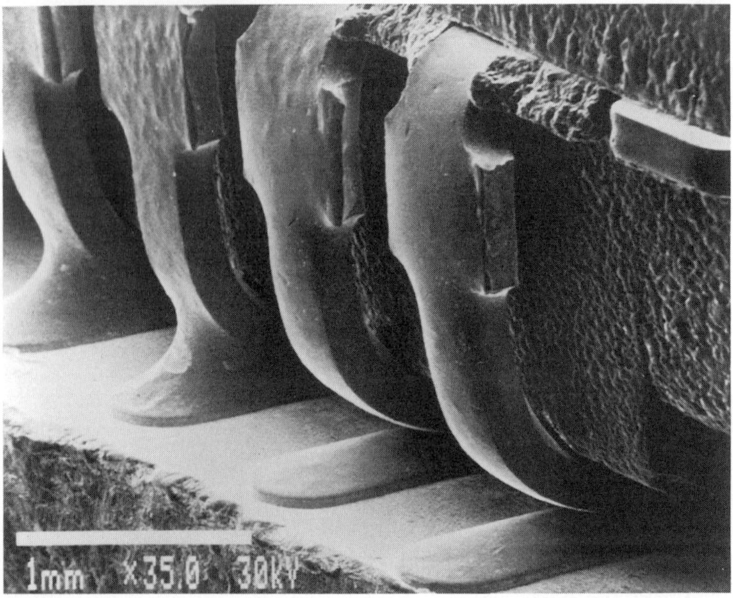

Figure 3-12 Scanning electron microscope photo of good and bad solder joints. (Courtesy of Texas Instruments, Austin, TX)

Figure 3-13 Cross section of a good solder joint. (Courtesy of Texas Instruments, Austin, TX)

Figure 3-14 Cross section of a bad solder joint. (Courtesy of Texas Instruments, Austin, TX)

hot first. This in turn heats the lands first, melts the solder paste, which then wets the leads. The temperature nonuniformity is favorable to reduce the wicking of the solder up the lead, but there can be significant nonuniformity in the plane of the PCB. This can cause wicking laterally along any solder coated metallization. This is controlled somewhat by the design technique.[1]

In summary, the key process attributes to maximize yield are coplanarity between components and the substrate, high surface tension solder, minimal "boiling" action of the solvent and flux in the solder paste, temperature uniformity between the component lead and the land, and wetting action of the solder that favors the land first, then the lead.

3.8 REFLOW PROCESSING

The solder reflow process can be achieved by using almost any available practical heat source. Hot air guns, hot plates, small soldering irons, and toaster ovens (and even a kitchen stove oven but only once for safety sake) have been used to build small experimental or prototype surface mount assemblies. The maximum temperature and time are controlled manually. While this is useful to get an initial familiarization, it will not provide the quality and reliability needed for a high volume commercial product. To do this will require a repeatable, controllable, and cost effective heat source.

It is important that the reflow time-temperature profile have a sufficiently slow warm-up to evaporate the solvent safely without a massive solder ball problem. This is usually handled with a bake or a preheat step prior to VPS reflow or with the first plateau near 100°C with IR. Higher viscosity solder paste (higher metal content) will have less solvent to evaporate, making this step less critical.

The flux activation level of the solder paste should be as mild as possible to minimize clean-up problems. Thus, potential contamination problems should be solved at the source, not at the soldering step. With this in mind, the flux activation step of the reflow profile should be appropriate for the amount and type used in the solder paste. Some large assemblies require a long convection/IR warm-up step to minimize the min-max temperature differences on the PCB during reflow, such that the flux can be activated and used up before the solder starts to melt. When this happens, there is no wetting action and the solder does not flow together to form a joint.

Excessive temperature, with respect to the individual components used can cause low yield and/or reliability problems. Inadequate temperature could produce unreflowed solder joints. Achieving a time-temperature profile for safe and complete reflow is an individual effort for each assembly and solder paste combination. Each component is sensitive to some degree to the temperature ramp-up and to the maximum temperature. The nonuniformity of absorption and mass make it difficult to find the correct power settings and belt speed in an IR system that will produce acceptable maximum temperature and uniformity control across the PCB to prevent problems with reliability and solder paste.

3.8.1 Vapor Phase

Vapor phase reflow, as discussed previously, is perhaps the safest and easiest to use. The formulation of the fluid controls the reflow temperature and, if the equipment is properly adjusted and maintained, the process is extremely repeatable. Two types of equipment exist, a top loaded system for batch processing and an in-line system for automated manufacturing. The batch system is an excellent choice for prototype development and low volume production of various sizes and shapes of SMT assemblies. The in-line system will work well for all high volume production requirements, regardless of size, shape, mass, or choice of components.

The time-temperature profile of an in-line vapor phase system easily meets most reflow requirements. There is generally a smooth transition from room temperature to reflow temperature and back to room ambient. By adjusting the preheater and belt speed, this transition can be easily controlled. The belt speed should be adjusted to provide adequate time for reflow to occur for the largest assembly to be processed. Because of the broad process range provided by vapor phase, this is easy to achieve.

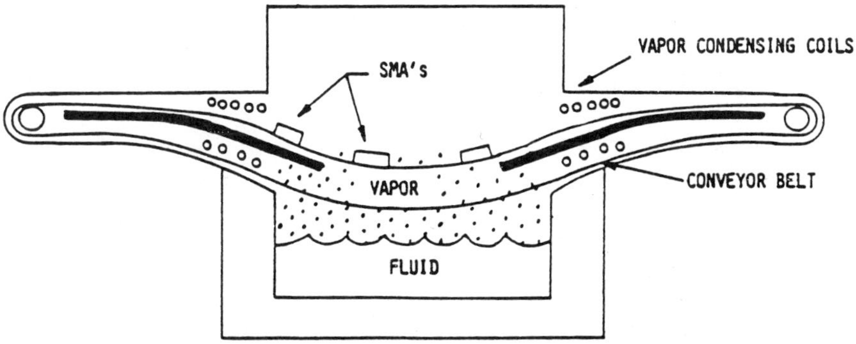

Figure 3-15 Cross section of an in-line vapor phase system.

A cross section of a typical continuous belt system is shown in Figure 3-15. Note the small slope required on both the entrance and exit to control the vapor region. This can cause large components to shift slightly while exiting the reflow region, depending upon the amount of surface tension of the solder joints and the weight of the component. This will generally only be a minor visual concern for a 68 lead PLCC.

A common complaint regarding the operation of vapor phase systems is the cost of the perfluorinated fluid. The initial cost, plus replacement cost of fluid lost to evaporation and surface adsorption on the assemblies as they move through the reflow area can become a significant, and easily identifiable, part of the operational cost of the manufacturing area.

Vapor recovery systems have recently been developed to control the exhaust vapors and to reduce this fluid loss. To obtain the advantages of the inherent control and flexibility of the vapor phase process, one must exercise care to select the proper size and type of equipment and to maintain it according to the manufacturer's specifications.

The simplest process to control, and hence to optimize, is VPS. Even if not fully optimized, the worst that generally happens is a few more open solder joints that can be easily touched up. Table 3-1 gives the steps that can be taken to minimize the solder wicking and open joint problem. If the maximum reflow

Table 3-1 Vapor Phase Soldering Optimization

Action	Benefit
Prereflow bake	Reduce boiling and splatter of solvent before reflow
Preheat	Partial bake, reduces temperature nonequilibrium
Sn62 (2% Ag)	Silver additive increases surface tension of solder
Two-part alloy paste	Produces a delay in melting and wetting of solder
SMOBC/HASL* PCBs	Contamination free solder, wets at lower temperature

*Solder mask over bare copper/hot air solder level.

temperature is matched to the solder alloy, generally 215°C for Sn62 or Sn63, and the time is sufficiently long, approximately 60 sec, very few reflow related problems occur. Most defects will be the result of component or pick-and-place problems.

3.8.2 Infrared

While the vapor phase provides inherent control based upon the formulation of the fluid, the necessary set-up and control of the IR reflow system must be provided manually or by machine. The flexibility for individual control of the top and bottom heat source, and thus the zone temperature, also provides an opportunity to provide too little heat and under-reflow the solder causing a weak joint or to overheat the assembly causing heat related damage to the PCB or components. If properly controlled, however, the yield can be slightly higher, under worst-case conditions, due to a reduction in open solder joints.

The typical IR reflow profile can easily meet the reflow requirements given earlier. First there is a rapid preheat zone where the board is heated to approximately 100°C. The second zone provides more convection heat than IR, causing an effective "plateau" in the profile. This allows the temperature differential to partially equalize (Fig. 3-5) and evaporate the solvent. The third zone heats the board to just below reflow temperature, activating the flux. The fourth zone is another equalization step to minimize temperature differentials. The fifth zone heats the assembly to the reflow temperature, followed by a cool down step. This profile has been observed to give the best yield and reliability by reducing the temperature nonuniformity between the hottest and the coolest points.

By reducing the belt speed, and correspondingly the source power to achieve the correct zone temperature, the heat retained in the surrounding walls and the belt of the IR system will provide a small amount of convection heating. This will reduce temperature variations between different components due to mass variations and differential absorption of the range of wavelengths of the IR radiation.

Profile setup and adjustment can be achieved by several techniques. Most commonly, a thermocouple and a chart recorder are used along with an assembly that represents the production unit. Thermocouples should be placed at the corners of the board as well as in the middle. Maximum temperature should not rise above 220°C, and the variation over the board should be less than 10°C. A silicon diode in an IC on the test assembly can be used to sense the temperature inside the package or as an alternative to a thermocouple. This temperature will typically be a few degrees less than the area near the IC package lead.

Type I B assemblies can be reflowed if a proper mechanical support is provided during the second reflow step. This support should prevent any physical contact to the components on the bottom of the assembly. It is not necessary to use two different solder alloys (different temperatures) since the belt in an IR oven runs

Table 3-2 Infrared Reflow Soldering Optimization

Action	Benefit
Uniform mass on PCB	Minimize temperature differential across PCB
Large component on corner	Absorb more heat from hotter area
Metal border	Reflect IR, spread heat from hotter edge area
Low temperature and slow belt speed	Reduce rate of IR energy absorption
Increased convection heat	Surface heat to minimize temperature differentials

flat and relatively smoothly and does not cause any vibration problems. The same time-temperature profile can be used to reflow both sides, although the heat source may need to be increased (or the belt speed reduced) for the second reflow to compensate for the additional component mass.

The techniques of optimization for an IR reflow process are shown in Table 3–2. These are primarily related to the design phase but will allow more latitude once the manufacturing process has started.

3.8.3 Cost Comparison

Sometimes IR, when compared to vapor phase, is given too much credit for cost reduction of the manufacturing operation. With vapor phase the cost is direct and identifiable. With IR, the cost is indirect and possibly not identifiable with IR operation. If the assembly production is typically a low volume and high mix operation, there will be some lost production time for profile changeover and checkout. This lost time and labor cost is usually not tracked as an IR cost. If the setup is not properly performed, there will be poor solder joint formation or damaged boards and components, both a latent reliability hazard. These could result in very large expenses, much greater than a year's supply of vapor phase fluid. In summary, if vapor phase is not fully optimized, there will be a few open solder joints that can be easily reworked. If IR is not fully optimized and controlled, whole production lots of boards must be scrapped.

If proper controls are maintained, the cost of the two operations (at the bottom line) are approximately the same for a low volume-high mix operation.[5] For a high volume-low mix, where the profile is setup within the 10°C min-max requirement, is rarely changed, and is monitored frequently, IR can be more cost effective.

3.9 SMT RELIABILITY

There are three areas of concern with respect to SMT reflow solder processing and the interaction with reliability. First there is the ramp-up of temperature versus time. Next there is the "integral" of the temperature and time effects. Thirdly, there is the impact of maximum temperature of the process on the components. Each has its potential detrimental effects and will be discussed in turn.

The surface mount solder reflow process takes the board and components from room temperature (25°C) to soldering temperature (215°C) in approximately 1 min (Figures 3-5 and 3-7). Then the board is held at reflow temperature for 20 to 50 sec, depending on the mass of the assembly. The thermal cycle stress and the maximum temperature are certainly higher than normal operation, but they are within the capabilities of the basic structure. For example, consider the thermal stress of the temperature ramp upon entering the reflow system. The temperature rise is given by:

$$\frac{215°\text{C} - 25°\text{C}}{60 \text{ sec}} = \frac{190°\text{C}}{60 \text{ sec}} = 3.2°\text{C/sec}$$

The vapor phase profile curve has an "S" shape, so the worst-case slope will be midway between the extremes. That value is typically 6 to 8°C/sec, with a desired upper limit for some ceramic-based components of less than 4°C/sec. This can be quite detrimental for multilayer ceramic chip capacitors.

For ICs, it is not as critical. This thermal cycle stress can be compared to that given in the IC tests reported in typical reliability testing. There the components are transferred from "hot ambient to cold ambient" in approximately 3 to 4 sec. Thus the temperature cycle gradient is:

$$\frac{150°\text{C} - (-65°\text{C})}{4 \text{ sec}} = \frac{215°\text{C}}{4 \text{ sec}} = 54°\text{C/sec}$$

Thus during IC qualification testing, the value is an order of magnitude worse and is repeated 1000 times with a component failure rate of less than 1 percent defective. Thus the inherent strength of the plastic SMT IC package is sufficiently reliable to withstand the thermal ramp.

The next concern is the amount of time that the IC components spend at the soldering temperature. As stated above, this can be as much as 50 sec for large assemblies. Further, there may be several cycles for the double-sided board and a couple of repair cycles.

The IC experiences several time-temperature variations during its lifetime. The plastic package is molded at approximately 300°C for a few seconds. Afterward, it is cured for several hours at approximately 175°C. If the electronic assembly were subjected to a fire, the flame retardant constituents of the mold compound would start to break down at 300°C in a few seconds. This would extinguish the fire but would be very detrimental to the IC.

Solder dipping of component leads, which sometimes also includes immersing the body in molten solder, has been done for many years with no observed significant component degradation (assuming corrosive flux is not used). This takes a few seconds at 260°C. Further, type III surface mount assemblies have been successfully produced for several years, which also take several seconds of immersion in the solder wave.

We know from long experiences with electronics that there is a temperature that will allow ICs to last a long time. Further, if the temperature is too high, there will be an excessively high failure rate. All of the information above, plus the simplistic (and safe) assumption of a decaying parabolic (or exponential) relationship between time and temperature, is used to construct the generalized plastic component degradation curve shown in Figure 3-16.

Note that the total elapsed time for type I processing fits safely under the curve, even for many reflow cycles. Thus, in general, the plastic IC package suffers no significant damage during the surface mount reflow processing from either thermal ramp or cumulative temperature exposure.

One of the more important problems in the SMT industry lately has to do with the maximum temperature achieved.[6] The plastic body of the IC has a TCE of approximately 16 ppm/°C below the glass transition point (approximately 160°C) and about 50 ppm/°C. This generates very large stresses inside the package at the maximum soldering temperature and depending on how these stresses are managed, internal delamination or plastic package cracking can result.

The major factors in determining the internal stress and its resultant action are plastic mold compound material, mold compound curing, mechanical design of the package, silicon chip size, and absorbed moisture. A search for a low

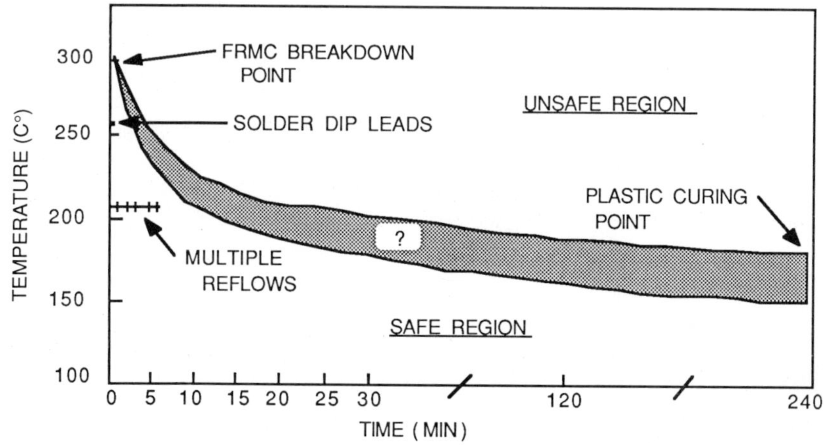

Figure 3-16 Generalized plastic component degradation curve.

stress mold compound and its proper time-temperature cycle for curing to minimize internal stress buildup during SMT reflow temperatures has been the subject of an intense study over the last few years. Improvements have been made, but there appears to be a practical physical limit. Internal package design should be such that stress build up is minimized or controlled. Perforations or indentations of the silicon chip mount pad can help control the plastic expansion. The silicon chip size is limited by the electronic function desired and generally cannot be reduced just for the benefit of package reliability.

Moisture absorption in plastic packages has received a lot of attention over the last couple of years. It is really a minor contributor to this problem, but, unfortunately, it is the only variable that the user can control. By using a drying step just before reflow, a marginally bad package can be improved into a marginally good package. Since this is the only alternative to the user, it can be used for its pragmatic benefit until sufficient improvements can be made in the mold compound and mechanical design. Baking or drying is generally beneficial with the larger PLCC/PQFP (1 in. or larger) packages with silicon chips larger than 0.300 in. on a side. Using single IC reflow soldering with a "hot bar" or "hot collet," rather than mass reflow soldering, is another method to minimize package damage.

REFERENCES

1. Hutchins, C. *Surface Mount Technology—How to get started.* Sugar Land, TX: C. Hutchins & Associates, 1989.
2. McLellan, N., and W. Schroen. 1987. "TI Solves Wicking Problem." *Circuits Manufacturing,* September 1987.
3. Hutchins, C., and J. Wallace. 1985. "Surface Mount Soldering Yield Improvement Test." *Fifth International Electronic Packaging Conference Proceedings,* October 1985.
4. Klein Wassink, R. J., and J. A. H. Gerven. "Displacement of Components and Solder during Reflow Soldering," *Soldering and Surface Mount Technology,* February 1989.
5. Hutchins, C., and P. Marcoux. 1988. "Optimizing the Vapor Phase and IR Reflow Processes." *Electronic Packaging and Production,* February 1988.
6. McKenna, R. "Surface Mount Device Package Cracking: An Overview." *Proceedings of the Technical Program, Surface Mount '89,* August 1989.

4

Optimizing the Wave Soldering Process

Janet R. Sterritt

Wave soldering is a very effective technique for joining components to printed circuit boards (PCBs) in large quantities. A printed circuit assembly (PCA) consists of the printed circuit board, connectors, and all components. As in all metallic joining applications, there are three basic steps to the soldering process. The first step prepares the surface for soldering. The next step applies molten solder to the bottom side of the assembly. The final step consists of allowing the joint to solidify by cooling. Defects that result in joint reliability issues can result from all of these stages. The goal of wave soldering process optimization is the reduction of joint defects. Several wave soldering process hardware enhancements are available to reduce certain types of defects. The process parameters should also be optimized to truly capitalize on the machine investment. Some examples of process parameters are solder temperature, solder wave height, conveyor speed, and preheat temperature setpoints. The optimal wave soldering machine and wave soldering process depend closely on the PCA design intended for production. Both wave soldering process parameter identification and process hardware selection must be examined to optimize the total process and reduce solder joint defects.

4.1 BASIC WAVE SOLDERING PROCESS OVERVIEW

The first step, surface preparation, consists of two phases: cleaning all of the metallic surfaces that have been designated for soldering and bringing these surfaces to the proper temperature to accept solder. Cleaning is generally accomplished by chemically etching the surface with an acidic compound such as flux. This step removes any oxidation or dirt that may have collected on the board prior to the wave soldering process.

The next phase in surface preparation heats the board. Heating accomplishes two tasks: flux activation and a reduction in temperature between the printed circuit assembly and the solder wave. Most fluxes require heat for thermal activation to break down the oxides. Manufacturers' directions specify the amount of heat and time needed for activation. The second role in preheating the board and components is that of bringing the PCA temperature to near solder temperature. Solder is typically used in wave soldering applications at a temperature of 500°F (260°C). If a board and components at room temperature enter the solder wave that is at a much higher temperature, the instantaneous thermal increase could cause differential thermal expansion resulting from the thermal expansion rates of the differing materials used inside components. Surface mount technology (SMT) devices are more susceptible to this expansion differential than leaded through-hole (LTH) devices, although both types of components can exhibit the related defect. Differential expansion related damage can cause the component to randomly fail in the in-line test stage or later as the PCB is being used. Poor solder joints can also be created if the thermal imbalance between the board temperature and solder wave is too great. In this case the solder thermal energy is used to bring the board to the wave temperature rather than to ensure that good wetting is occurring. The poorly soldered defects in this case look like cold joints or joints with rough or dull surfaces. These joints have a higher tendency to crack when the board is being used in the field since hairline cracks in a rough surface can later develop into cracked joints. Differential heat expansion between metallic surfaces and the PCA create a randomized soldering process that is easily overcome by a well-controlled preheating section.

The next step in a basic wave soldering process is the application of the solder to the PCA via the solder wave. The PCA is transported through the wave. This fully immerses the component leads or surface mount component and pads on the bottom side of the board in solder. The cleaned metal surfaces form a metallic bond with the solder, creating the soldered joint. As it exits the wave, the solder joint is in molten stage. Excess solder flows off the joint back into the wave.

The final step is joint solidification, which is achieved by allowing the joint to cool. Any vibration to a nonsolidified joint can cause surface roughness, which can result in cracks that cause reliability problems. Therefore, a wave soldering machine conveyor often extends a good distance beyond the solder wave to ensure that the solder joint temperature is below the liquid to solid solder transition temperature of 361°F (183°C), prior to removing the board from the conveyor.

4.2 WAVE SOLDERING PROCESS HARDWARE

The soldering process is actually performed by four subsystems. Board preparation is accomplished in two separate subsystems, flux application and preheat. Soldering is performed in a solder delivery system. Joint solidification in a basic wave soldering machine is performed by allowing the board to cool naturally while still in the conveyor. Therefore, the conveyor is in essence the cooling system. A complete wave soldering machine is shown in Figure 4-1.

4.2.1 Fluxing

Flux is formulated as active solid particles suspended in a liquid alcohol base. Key issues germane to fluxing performance are ensuring bottom surface board coverage and cleaning issues. Flux is used to remove the oxides from the metal surfaces so that the solder can form a better bond with the metal surfaces under the joint. There is no need or desire to apply flux to the top side of the board since top side flux residue will only result in a cleaning issue. The primary reason to clean a printed circuit board is to remove flux residue. Certain types of flux residues can continue to etch the board surface after the board is in use. The residues have also been found to act as a host for the growth of mold in humid climates. Both can result in PCB field failures. Therefore, the objective of fluxing is to apply an even coating to the bottom of the board with minimal excess flux.

4.2.2 Fluxers

There are three commonly used types of fluxers: foam, wave, and spray. A foam fluxer operates by pushing air through a diffuser to create bubbles that then rise through a chimney to form a flux head. The diffuser is typically a porous plastic or pumice stone. Foam fluxing is dependent on surfactants that are added to the flux enabling it to foam. The flux head is capable of being adjusted over a range of depths. The process parameter that can be used to control a foam fluxer is air flow. Air flow or pressure adjusts the amount of flux aeration as it passes over the flux stone, causing the depth of the flux head to vary. Air pressure can also cause the size of the flux bubbles to change. Higher pressures force the air through the flux at a more rapid rate, resulting in smaller bubbles. Conversely, lower air pressure can yield larger bubbles. Large bubbles imply that there is a noticeable air space in the center of each bubble. The air space results in an area of the printed circuit assembly being exposed to air and not flux; which in turn results in an uneven flux deposition. Small bubbles are the preferred formation for a good foam fluxer. Figure 4-2 shows a typical foam fluxer in position

Figure 4-1 Wave soldering machine. (Courtesy of Hollis Automation)

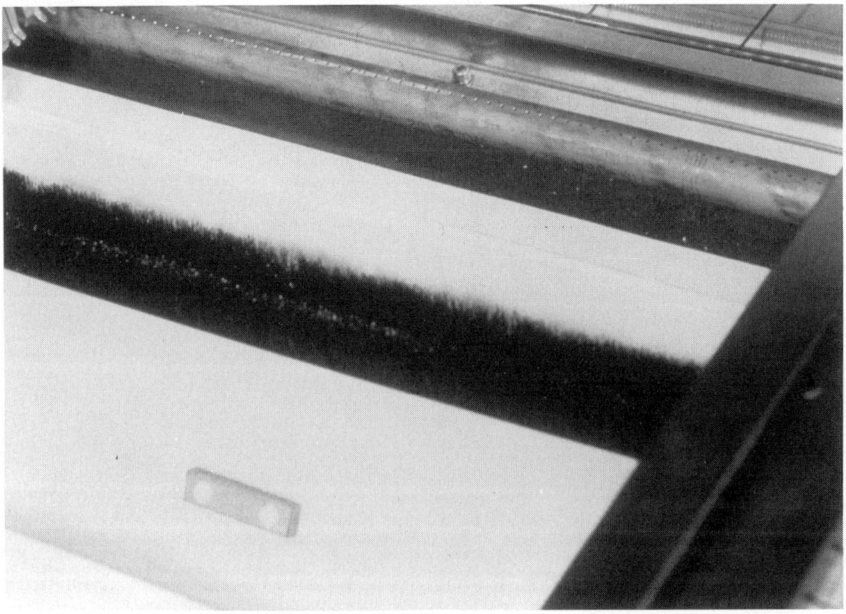

Figure 4-2 Foam fluxer. (Courtesy of Hollis Automation)

in a soldering machine. This particular fluxer has black brushes in the photograph. The brushes enable better foam head support.

A wave fluxer propels a large amount of flux up a chimney to create a wave of flux. A flux wave can be adjusted over a wide range of depths. Since the bottom side of the printed circuit assembly is submerged in flux, wave fluxers tend to be capable of applying more flux to the board than foam fluxers. Flux used in a wave fluxer does not require the addition of a foaming agent, which can be an advantage when using certain types of flux. Some fluxes containing a heavy foaming agent will leave additional residue on the board after soldering. This residue can cause additional cleaning issues. The process parameter that controls wave flux application is pump speed, which is used to adjust wave height. A wave fluxer is shown in Figure 4-3.

A spray fluxer, as shown in Figure 4-4, is used to thinly coat the printed circuit assembly with flux. An oscillator is used to inject flux into an air stream, which is then diverted to form a spray pattern onto the PCA. This method applies the least amount of flux onto a board surface. The process parameter in spray fluxing adjusts the amount of flow from the sprayer. Specific gravity is checked when first loading the flux reservoir but not required for routine checking. A

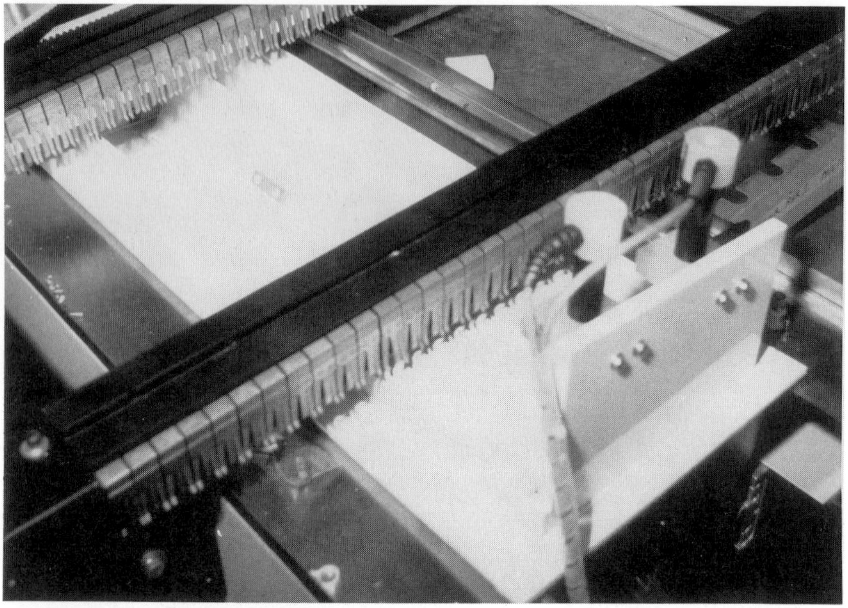

Figure 4-3 Wave fluxer. (Courtesy of Hollis Automation)

spray fluxer draws from an enclosed tank; flux that is not used does not recirculate back to the flux reservoir. Unlike the open tank used in wave and foam fluxers, the closed reservoir greatly reduces the evaporation rate.

Additional process enhancements are available for the form and wave fluxing subsystems. A fluxer air knife uses compressed air at ambient temperature to wipe excess flux off the printed circuit assembly after it has been coated by the fluxer. Some of the excess flux can drip from the board onto the preheat module as it travels through preheat on the conveyor. This increases preheat maintenance time since it requires cleaning. Most of the excess flux is baked on the board in preheat and not removed in the solder wave. The baked-on flux remains on the board as a gummy residue that is difficult to clean. Generally, flux containing higher solids will yield a flux residue by-product that should be cleaned. Low solids or no-clean fluxes yield significantly less residue issues. Fluxer air knife angle and air pressure become two more process parameters that need to be included in the optimization process.

Another foam or wave fluxer feature is closed-loop specific gravity control. Since both foam and wave fluxers reside in an open flux reservoir, the flux in the reservoir is subject to evaporation. As the alcohol in the flux evaporates, the solids content in the flux changes. The change in solids content can be monitored

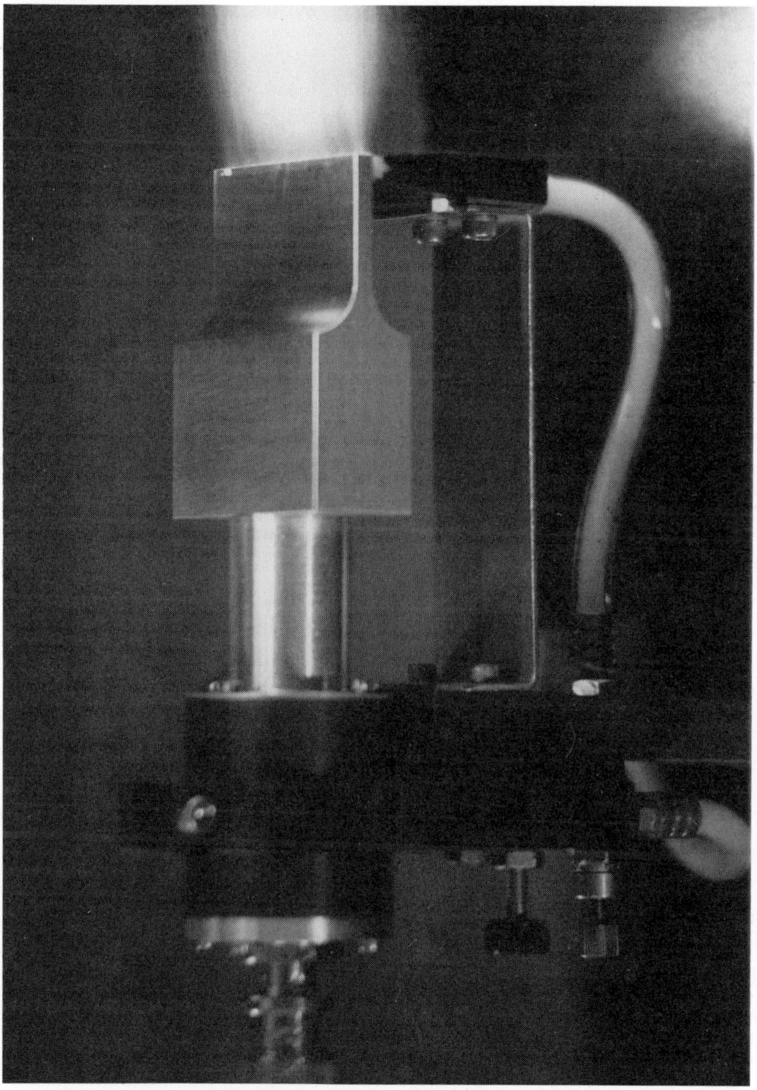

Figure 4-4 Spray fluxer. (Courtesy of Ultrasonic Systems Engineering)

through measuring specific gravity. As the specific gravity increases, thinner needs to be added to lower the solids content. Closed-loop control consists of using pumps to add flux or flux thinner to the flux reservoir as the specific gravity changes over time. Closed-loop control enables specific gravity to be used effectively in a parameter optimization.

4.2.3 Fluxer Measurement Parameters

The parameters crucial to establishing good fluxing characteristics are specific gravity, air pressure for foam fluxers, and pump speed for wave fluxers. Specific gravity is most commonly measured by taking a sample of flux, placing it in a graduated cylinder, and placing a hygrometer in the cylinder. The hygrometer is a buoy that floats to the level determined by the specific gravity. An automated system consists of bobs that float in a flux tank. The bobs display a relative displacement, which is translated into specific gravity. The displacement is read by a sensor and displayed. Closed-loop control systems can add either flux or thinner to adjust concentration and specific gravity to the desired number. The desired level is dependent on the type of flux used, varying according to the manufacturer's specification. The optimal flux specific gravity is controlled to maintain the value that the manufacturer has specified.

Foam fluxer air pressure regulates the size of the bubbles in the flux head and the height of the flux head. A sensor and pressure regulator mounted in the incoming air line enable measurement and adjustment, respectively. The air pressure is adjusted by placing a setup board in the conveyor with the lead edge immediately over the flux head. The conveyor is then stopped. The air pressure is increased until the foam head extends up to a level that is two-thirds of the way up the thickness of the board. The air pressure level that corresponds to this point is the setpoint. As a further aid in setting up the process, a glass board can be substituted for a regular board. The glass lends itself to estimating flux head bubble size for fine air pressure adjustment as a visual verification of flux bubble size when in contact with the board surface.

Wave fluxer height is measured as a function of pump speed. In a manual mode, this is achieved by using a strobe to determine the revolutions per minute (rpm) of the pump. In an automated system, a tachometer is placed on the pump and the value is displayed. The wave fluxer height is found in a method similar to the one used for foam fluxers. A setup board is placed in the conveyor with the lead edge immediately over the peak wave height position. The conveyor is stopped. Pump motor speed is increased until the wave extends up to a level that is two-thirds of the way up the thickness of the board. The pump speed that corresponds to this point is the setpoint.

4.2.4 Fluxer Optimization

The greatest challenge in fluxing is ensuring that all metallic surfaces are covered while using a minimum amount of flux. Excess flux results in cleaning issues after the wave soldering. The type of PCA product being manufactured determines the fluxer requirements. Key board descriptors that influence this decision are board depth, complexity, and line throughput rate. Board depth is the distance

Table 4-1 Fluxer Hardware Optimization

Printed Circuit Board Factors	Optimal Wave Soldering Process
Maximum depth: short Board complexity: high or low Line speed: low to medium	Foam fluxer
Maximum depth: short or long Board complexity: low Line speed: low to medium	Spray fluxer
Maximum depth: long or short Board complexity: high or low Line speed: low to high	Wave fluxer

from the bottom board surface to the longest or highest lead or component, respectively. Board complexity is a metric used to describe the combination of shadowed components and solder joint density. Highly complex boards have large areas that are shadowed by other components and/or very closely spaced leads and pads.

Wave fluxers are used most often with longer leaded devices since the wave depth can be extended to a deeper range than a foam fluxer. Wave fluxers are also capable of depositing a good layer of flux at very high throughput speeds (i.e., in the range around 10 ft/min.). A foam flux head can maintain an adequate depth for most applications. Wave and spray fluxers are used when flux additives are a concern. The spray fluxer provides a very thin coating of flux, but shadowing can occur with some board layouts. A spray fluxer does not gain any benefit from closed-loop control since the reservoir is sealed and little or no evaporation takes place. Since the spray fluxer deposits very little flux on the board, a fluxer air knife is not required either. Some excess spray residue can collect in the machine exhaust and other wave soldering machine parts over time, increasing maintenance. All of these factors need to be considered when selecting fluxer hardware. A quick reference for making these trade-offs is provided in Table 4-1.

4.2.5 Preheating

Preheat brings the printed circuit assembly temperature to a level near wave solder temperature. The key to obtaining good preheat characteristics is raising board temperature with minimal variation across the board or over the length of time for a production run. Heating issues include thermal shock, color sensitivity, and thermal variation. Thermal shock is the rapid absorption of a large quantity of heat in a very short space of time. This can cause differential expansion within components, which can result in stretching internal component connections and thus lead to component failures in the field. Color sensitivity is exhibited when a component encased in a black covering heats more rapidly than one encased in white. Thermal variation is the heating envelope provided by the wave soldering preheating section, shown as hot and cool areas on the same board. This results in a board that has various temperatures over the surface of the board as it is immersed in the solder wave and causes variations in the surface appearance of solder joints since the wave is used in a differing manner across the surface of the board. In cooler areas, the wave is actually being used as a final preheat station, while in areas that are too hot, the flux may have been evaporated off so that some oxidation can occur prior to entry into the solder wave.

4.2.6 Preheaters

Preheating can be accomplished by several different techniques. Infrared heat and convection are the most common techniques. Infrared devices are divided into three types: short, medium, and long wave length. Short wave length is defined as 1 to 3 μm. Medium wave length extends from 3 to 5 μm, and long wave length is defined as above 5 μm. Printed circuit assemblies heated with short wave length radiation tend to heat components unevenly in a color-sensitive pattern. Color sensitivity is not characteristic in medium- and long-range wave length units. Long-range wave length sources tend to require more time to stabilize at temperature, yielding longer process setup time. Medium-range wave length infrared sources are the most commonly used preheat method. A typical infrared preheater module with medium wavelength quartz lamps is shown in Figure 4-5.

Strip heaters are black body radiators and fall into the long wave length infrared heater category, most commonly used in strips, rods, or serpentine shapes. When used as an exposed rod or serpentine shape, this method can give a preheat pattern that has noticeably hot and cold spots. The spotting is caused by the heat distribution from the source. The amount of heat delivered is reduced by the distance away from the source. In a serpentine-shaped heater, for example, the region near the curved ends has a higher concentration of heat and is therefore hotter than the straight rods that connect them. A black body radiation source that is implemented in contiguous pieces would not exhibit this problem. Black body radiator preheaters are also assembled into self-contained units with a quartz or glass top. In this configuration, the preheating module is referred to as a panel heater. The cover provides a thermal barrier that makes the resulting heat pattern more uniform across the module since the barrier acts as a method of capturing and controlling convective heat from the sources along with infrared heat. Ceramic plates that have heating coils laced through the plate also fall into the

Figure 4-5 Infrared preheat module. (Courtesy of Hollis Automation)

panel heater category. Strip heaters and panel heaters require time to warm up and stabilize prior to usage.

Convective preheat is achieved by heating the air surrounding the board, which in turn heats the PCA. This method gives very even and thorough preheating; however, convective heating alone is a very slow process. Convection occurs in combination with all of the previously described process techniques. Preheaters not only heat the printed circuit assembly but also heat the air in the region surrounding the module. Some wave soldering equipment capitalizes on this effect by surrounding the preheat subsystem with a tunnel as a means of combining convective heat with infrared heat. Convective heating used in combination with other methods helps to make the preheat process more uniform.

In the actual use of all preheat technologies, individual preheaters are assembled into larger modules that are in turn controlled as one set. The modules can then be configured in different patterns, where one module represents one zone of preheat. Generally, each separate pattern defines a distinct wave soldering machine. The most common configurations used consist of mounting varying lengths of modules on the bottom side of the conveyor only or mounting preheater modules to both the top and the bottom sides.

The most significant preheat process additions are closed-loop control and zone adjustable preheat. Closed-loop control is achieved by placing a thermocouple in each preheat zone. Thermocouple placement depends on the preheating method: above the zone for convective preheaters or combination convective and infrared preheaters, in an infrared bulb, or in a panel. The preheat module is designed to provide heat to the board over a length of time. In practice, the production line throughput rate specifies the amount of time. Therefore for different mass boards varying lengths of preheat are needed to be able to fully heat the board within the time constraint. The length is further divided into separate preheat modules that are individually controlled preheat zones. Preheat modules may be configured both on the top and the bottom sides of the conveyor to give more even heating to both sides of the board.

4.2.7 Preheat Measurement Parameters

Preheat process parameters are zonal temperature setpoint, temperature uniformity, heating rate, preheat exit temperature, and delta T. The difference in temperature between preheat exit and the solder wave temperature that the printed circuit assembly experiences as a rapid change is the change in temperature or delta T. The preheat exit temperature is the average board temperature at preheat exit, which is the maximum board temperature prior to entering the solder wave. The heating rate is the slope of the thermal heating rate curve that the board experiences. Temperature uniformity is the maximum allowable variation in temperatures that occur over the surface of a PCA. The zonal setpoint temperatures are the setpoints for each individual preheat zone.

A thermal profile is shown in Figure 4-6. The profile is achieved by attaching a thermocouple to the bottom surface of a board. The profile is a graph of temperature over the length of time in the conveyor. The profile shows the gradual rise as the board passes over preheat. Immediately upon preheat exit, the board enters the first wave. Then it cools as it passes in the space between waves, rapidly rising again as it enters the laminar wave. After exiting the wave, the board displays a normal natural cooling effect. The profile is superimposed over a block diagram of the machine subsystems. This machine has 10 zones of preheat between the fluxer and solder waves. The profile also shows a further warming level at the air knife position. In the example, the temperature ramp rate is the slope of the line plotted on the graph. Delta T is the difference between the solder temperature of 500°F in this case and the maximum temperature at the exit of preheat.

Temperature uniformity is found by using a multiplicity of thermocouples and observing the collective results. Some key locations are top side surface temperature, inside bottom side fillets, under components, and in or near heat sinks. Thermocouple placement is determined by board layout and component selection issues for each individual board. The result of using several thermocouples simultaneously is shown in Figure 4-7. This uniformity profile is measured by

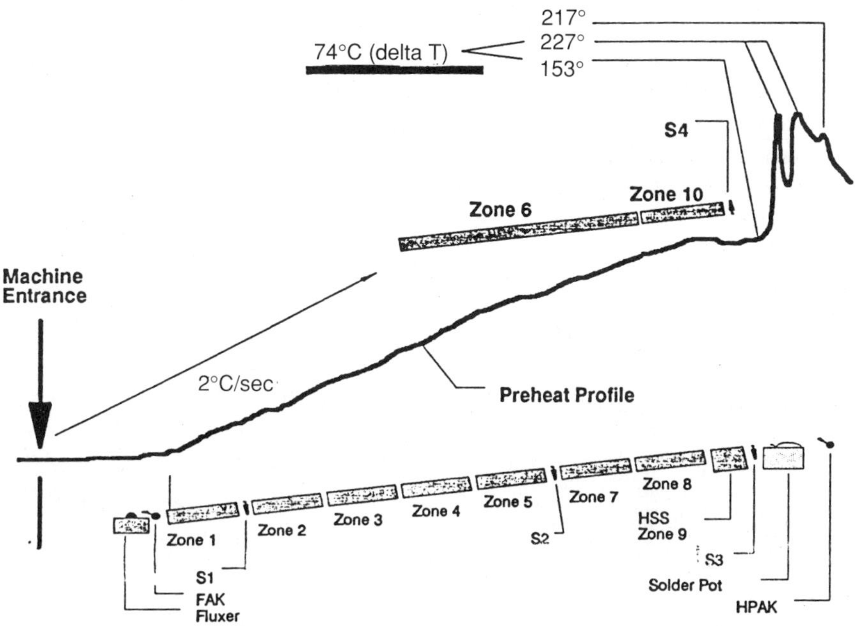

Figure 4-6 Preheat profile in relationship to subsystems. (Courtesy of Hollis Automation)

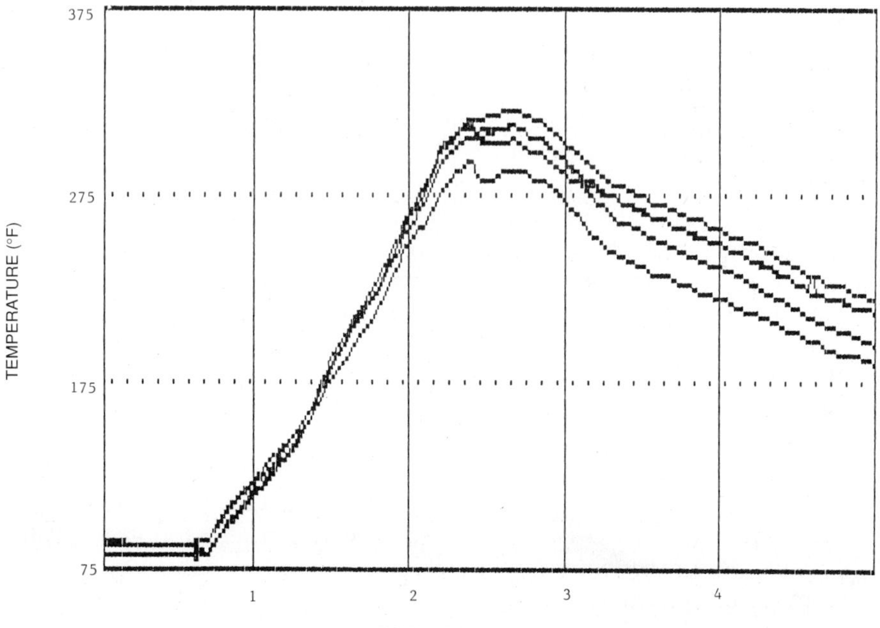

Figure 4-7 Uniformity profile. (Courtesy of Hollis Automation)

placing five thermocouples in a row perpendicular to the direction of motion. All are reading bottom side surface temperature. The uniformity range is the difference between maximum and minimum temperature.

The preheat setpoints are established by using a setup board that has thermocouples attached in the desired locations. The board is run at the desired conveyor speed and the thermal profile recorded. When the optimal profile is obtained, the temperature setpoints for each zone are noted. These are the setpoints for preheat.

4.2.8 Preheat Optimization

When considering the many preheat options, the key printed circuit product descriptions that determine the preheat choice are product size, mass, and number of board layers. The maximum amount of preheat required is a function of the size and mass of the board. Thin, small boards require less heat to raise the temperature to the desired level than heavy, thick boards. An overall metric used to determine this level in an optimization is the number of layers in the board. Multilayer PCBs require more heat to achieve the same thermal characteristics as single-layer boards. A good overall cutoff is four or more layers being considered to require more preheat. Production throughput goals result in line speed requirements that also affect overall preheat length. Printed circuit assemblies that require short preheat lengths at relatively low line speed rates of 3 to 4 ft/min can require twice the length to satisfy the same demands at 6- to 8-ft/min rates.

All preheat varieties meet the overall preheat goals as long as they are capable of providing even, non-color-sensitive heat. Single-layer or multilayer boards of up to four layers with leaded through-hole components at relatively low conveyor speeds are prime candidates for short length, bottom side preheaters. A short length preheater is considered to be from 1 to 3 ft since the boards in this category can be properly warmed in this distance at standard conveyor speeds (i.e., 3 to 4 ft/min). Higher mass, multilayer boards that are small in overall surface area and contain leaded through-hole components are matched to short lower preheat with a top side preheater module to assist in overall heating. Large multilayer, high mass boards, with leaded through-hole components are matched to longer preheaters with both top and bottom side preheating units. Long preheating length is considered to be over 3 ft. Any board containing SMT devices is matched to long preheat length on both the top and the bottom side. Since LTH devices are less susceptible to thermal shock, there is less emphasis on reducing slope. The longer-length preheat can be used to gradually increase board and component temperature. If higher conveyor speeds are desired, even longer preheater subsystems are required. If a large, massive SMT assembly is the identified product and a throughput requirement leads to line speeds in the

Table 4-2 Preheater Hardware Optimization

Printed Circuit Board Factors	Optimal Wave Soldering Process
Thin LTH	Short bottom side preheat
Multilayer LTH	Long bottom side preheat or short bottom side with top preheat
SMT	Long bottom side preheat and top side preheat

7- to 10-ft/min region, as much as 10 ft of top and bottom side preheat is needed to meet these requirements.

The greatest key to optimizing performance is understanding thermal profiling. The thermal profiling concept encompasses the preheating and soldering subsystems in the sense that the solder temperature defines the peak process temperature. Reducing the peak temperature changes solder viscosity. Even though lowering solder temperature reduces operating latitude from the eutectic ratio, tight control of the tin-to-lead solder content ratio is not required since there still remains ample operating room between the liquidus limits.[1]

The longer-length machine accommodates all types of printed circuits being manufactured; however, cost limitations in both soldering machine price and floor space may force the use of short-length machines. Optimal profiling can still be achieved by running at slower throughput speeds. The preheat trade-off optimization is shown in Table 4-2.

4.2.9 Wave Soldering

The wave soldering subsystem applies solder to the bottom surface of a PCA. It is a mass of solder that is pumped to a height above the level of a reservoir and then cascades back into the tank, creating a moving wave of molten solder. The expanded definition of a wave soldering subsystem refers to all hardware that is intended to shape solder joints while still in the molten state.

4.2.10 Solder Waves

There are many forms of solder waves. The basic solder wave provides a wide laminar flow of solder that moves in a direction opposite to the conveyor motion at the entrance to the wave and with the conveyor direction at the exit of the board from the solder wave. The type of solder flow lends its name to describe each wave. A bidirectional wave parts in the center with half of the solder flowing to the entrance and the other half of the solder flowing toward the exit. A high volume laminar wave, which is optimized to run product at faster through-

Figure 4-8 Dual solder wave. (Courtesy of Hollis Automation)

put rates, is a bidirectional-type wave but has a disproportionately larger mass of solder flowing toward the entrance than toward the exit. The laminar wave gives excellent results on most leaded through-hole technology and simplistic bottom side surface mount technology. However, component shadowing as seen in some complex designs and deep leaded through-hole barrels prevalent on multilayer boards can result in lack of penetration, skips, and misses.

In some product applications, a narrow highly turbulent wave precedes the laminar wave. The turbulent wave has a high vertical velocity that can drive up long barrels or fill in behind a large surface mount component. Used in combination with the laminar wave, the dual wave is more robust in that it can easily solder more difficult leaded through-hole technology as well as bottom side surface mount devices. Figure 4-8 depicts a typical dual wave system with a turbulent wave on the left and a laminar wave on the right.

A unique tool that helps to sculpt solder joints while they are still molten is the wave soldering air knife. The air knife produces a low pressure warm air mass, which is directed toward the bottom surface of the board upon immediate exit from the wave. This removes excess solder from molten joints, reducing bridges, excess solder, and icycling. Figure 4-9 shows an air knife in a soldering machine. It is the square tube positioned immediately after the wave in the picture.

Figure 4-9 Air knife. (Courtesy of Hollis Automation)

Another wave soldering process addition is mixing oil in with solder in the wave. Oil intermix reduces the solder alloy surface tension. This changes the soldering entrance and exit velocity latitude, making these items less critical. Opening the solder wave setup latitude enables greater ease of setup in the production environment. The oil has proven not to cause difficulty in solder joint formation since the oil does not impede solder application to wetted surfaces but rather reduces the surface tension on the solder wave. Oil intermix reduces bridging, excess solder, and icycling. Waste oil concerns offer the printed circuit manufacturer a trade-off between improved process latitude and waste removal.

4.2.11 Solder Wave Measurement Parameters

Wave soldering process parameters are solder temperature, solder wave height, and wave height uniformity across the process width. Solder temperature refers to the temperature of the solder in the wave. Since solder is a metal and the mass of solder is continually overturned as it is pumped through the wave, wave temperature rarely differs from solder reservoir temperature by more than a degree. The time in solder is actually monitored by finding the contact length of the printed circuit assembly and matching it with conveyor speed. The relative velocity between the solder wave and the PCA as it enters and exits the wave shapes the joint. Surface roughness on the smooth wave is limited so that areas of the board are not missed.

Solder wave height can either be measured directly via a wave height sensor or empirically by using solder pump speed. A wave height sensor is mounted on the wave and uses closed-loop control to monitor and control solder wave height. Pump speed measurements and setup are performed in a manner similar to the wave fluxer setup and measurement.

A very thin layer of dross can form on top of the wave. It appears to break, falling over both edges. The fine top dross layer actually floats on top of the wave. Since it is floating, it does not have the same properties as the solder in the wave. Therefore, it is not a precise indicator.

If an air knife is used, the key parameters are distance from the solder wave exit, distance to the board, angle to the board, air temperature, and air pressure. The air knife should be positioned as close to the wave as product depth limitations enable. The knife should also be set 1/16 in. lower than the component with the maximum depth on the assembly. The angle varies as a function of product type and board layout. The lower angle ranges from 38° to 50° are used for leaded through-hole boards in general. Higher angles, ranging from 50° to 90°, are generally used in SMT soldering. Extremely tight LTH components require higher air knife angles and can extend to 90°. Air knife temperature is optimized when set to any value in the range from 500°F (260°C) to 800°F (426°C) as the air exits the orifice. The high temperatures are used when excessive board mass can deplete air temperature. The effects of air pressure are determined by soldering some sample boards and examining the results. An air gauge is placed in line to measure air pressure at the knife body. The air knife is setup by repetitively setting pressure and finding performance by running a board. A trade-off between bridging and excess solder defects and blow holes must be made. In general, higher pressure is used with SMT devices.

The oil intermix process control parameter is the amount of oil being fed into the wave. This is gauged through a metering pump. When set up properly, the oil should break the wave in an even array of very small oil droplets.

4.2.12 Wave Soldering Optimization

Soldering requirements leads to determining the need for a single laminar wave, dual wave, or air knife addition. The factors used to determine the hardware trade-off for the solder wave are joint type, density, and number of board layers. The joint type is either LTH or SMT. Density refers to the spacing between joints. As joints are more closely located, the tendency to create bridges is increased. Single-layer boards or relatively few layer boards with leaded through-hole technology at a low level of complexity (30 joints per square inch or less) can be soldered effectively on a single laminar wave in a non-SMT application. SMT devices by their design present a large surface to the wave. The leading edge of the device shadows the trailing edge; the vertical velocity component in the turbulent wave fills in the regions that would be shadowed in a laminar wave. Leaded through-hole technology assemblies with long barrels for multilayer boards require a mechanism to force solder up into the barrels. This results in

Table 4-3 Solder Wave Optimization

Printed Circuit Board *Factors*	Optimal Wave Soldering *Process*
Low density, few layer LTH	Single laminar wave
High density, few layer LTH	Single laminar wave and air knife
Low density SMT	Dual wave
Low density, multilayer LTH	
High density SMT	Dual wave and air knife
High density, multilayer LTH	

the addition of a turbulent wave prior to the laminar wave. If the complexity is further increased on a leaded through-hole board with a level of joints per square inch higher than 30, the addition of a hot air knife is matched to the process.

All boards containing SMT devices benefit from a dual wave system with the addition of a hot air knife; however, cost limitations force a trade-off to be made. The solder wave trade-off optimization is described in Table 4-3.

4.2.13 Solidification

In a basic wave soldering process, joint solidification is accomplished by allowing the PCA to cool naturally. The solder joint is molten as it exits the soldering subsystem. The joint must not be disturbed as it cools, or surface roughness can result. The hardware that accomplishes this task is the conveyor, by virtue of the length after exiting the solder wave.

4.2.14 Conveyors

A conveyor can be set up for use directly with the PCA or for use with a pallet system. A finger conveyor grips the exterior of the board to carry it through fluxing, preheat, and soldering. The fingers contact the materials in each of these subsystems, which results in a gradual build-up over a period of time. If the finger residue is not removed, board jams or conveyor vibrations can result. Conveyor vibrations can cause the conveyor to shake, which results in movement in the solder joint as the solder cools. This can result in defective joints. The fingers can be cleaned through routine preventive maintenance or the implementation of a finger cleaner, which scrubs the fingers after the solder wave.

Another process enhancement that can be added to this subsystem is an active cooling module such as fans or vortex tubes. A massive PCA can take a great deal of time to cool. If this is further complicated by a desire to run at high throughput speeds, a very long conveyor would be required. A slight amount of cooling using ambient air cooling fans can reduce the overall length of conveyor needed to allow the solder to solidify without damaging solder joints. This enables a more favorable trade-off between overall machine length, board mass, and conveyor speed.

4.3 WAVE SOLDERING PROCESS PARAMETER OPTIMIZATION

Once the optimal wave soldering hardware set has been identified, the process parameters must be optimized. The subsystem parameters described in the previous sections must be set up prior to a system optimization. The system optimization is primarily focused on the thermal management throughout the wave soldering machine in preheat, wave soldering, and through the air knife.

The surfaces intended for soldering wet more easily if they are all near the solder temperature. In a nonoptimized wave soldering process, it is possible that the components and PCB are hundreds of degrees lower in temperature than the solder wave. When a cooler board enters the wave, it is actually brought up to temperature by the wave itself. This very fast thermal transition is referred to as thermal shock. An optimized wave soldering process increases preheat exit temperature to minimize thermal shock.

Another system interactive factor contributing to good wetting is solder condition. Alloy purity strongly affects joint quality and determines the acceptable range of fluxing required. An interesting aspect of solder condition is its temperature. Typical recommended wave solder temperatures range from 500°F (260°C) to 525°F (274°C). These temperatures are well above the melting point of solder and are the accepted industry standards because they offer a wide process operating range over a wide latitude of solder purity levels. The minimum is limited to 500°F (260°C) to ensure that the solder is well above the liquidus and solidus temperature range. However, for alloy composition near 37 percent lead and 63 percent in (eutectic) and near eutectic, 500°F (260°C) is 139°F (77°C) over the minimum melting point. Therefore, if solder alloy quality is maintained to a level near eutectic, the minimum solder temperature can easily be lowered by as much as 75°F (41°C) without jeopardizing good wetting conditions. Lowering the soldering temperature reduces the peak temperature to which the board is subjected. Overall, this results in less thermal stress on the assembly components.

Optimal system profiles are produced by limiting three factors: peak temperature, delta T, and temperature ramp rate or slope. The main impetus for these limits stems from SMT component manufacturers. For example, one recommendation limits the slope to not more than 3.6°F (2°C) per second. Another manufacturer recommends limiting thermal shock to less than 212°F (100°C).[2,3] Both manufacturers' limitations are based on packaging material expansion rates. As the technology matures, recommendations may be relaxed. Even though recommendations may be relaxed, process optimization from the standpoint of improved solderability will continue to enforce these concepts.

It is important to note that all of the temperatures controlled by this optimization are measured on the bottom surface of the board. One popular method of quantifying thermal performance in wide use today is the overall temperature equivalent for the top side. Typically this rule limits top side surface temperature to a range from 180°F (82°C) to 250°F (121°C). While this is an effective overall metric, it is not capable of delineating bottom side temperature performance. On an average board the top side is 30°F (16°C) to 40°F (22°C) cooler than the bottom side; however, this is greatly affected by the mass of the board, the number and type of ground planes, the presence of heat sinks, and the soak time in preheat. The longer the preheat section the less temperature differential between top and bottom side surfaces. SMT components can also shadow pad areas so that the surface to be soldered is not at the desired temperature, even though the average bottom side surface temperature may be within the desired limits. An actual bottom side surface measurement is necessary to properly optimize the profile. The ideal system thermal profile minimizes all temperature gradients throughout the process.

4.3.1 Optimization Procedure Test Study

To evaluate these theories, a test was run to compare the results of running mixed technology product on two machines, one with hardware optimized for LTH and the other with hardware optimized for SMT. One other more subtle difference existed between the machines. The LTH machine was not run with optimal thermal management while the SMT machine was. The boards chosen for this test were both mixed technology, enabling the test to produce leaded through-hole and surface mount results with fewer boards. However, one board is primarily SMT while the other is primarily LTH.

Using the hardware optimization guide, the standard LTH machine consisted of foam fluxer, short preheat, and dual wave soldering system. The hardware optimization process recommends lengthening the preheat and adding an air knife to better optimize the machine for mixed technology boards. A wave soldering machine matching the optimal description was produced and used in a comparison test to verify the optimization process.

The test was run using commercial boards in a real production environment. This type of real manufacturing production facility test is perfect for verifying hypotheses and enabling comparisons. In the full production process the boards were top side populated, reflowed, and cleaned. Then the boards were populated by pick-and-place for bottom side SMT into epoxy. The epoxy was cured and then populated with leaded components by pick-and-place equipment. Finally, off-sized parts were hand inserted. Printed circuit assemblies were built and

Table 4-4 Test Machine Configurations

Parameter	Standard Machine	Optimized Machine
Conveyor	Finger	Finger
	4 ft/min	4 ft/min
Fluxer	Foam	Foam
	SMC grade OA	SMC grade OA
Preheater	I/R	I/R
	4 ft bottom	5 ft bottom
	0 ft top	2 ft top
Solder	Dual wave	Dual wave
	500°F (260°C)	470°F (243°C)
	SMC grade solder	SMC grade solder
Air knife	None	1.5 in. to wave
		450°F (232°C) at PCA
Finger cleaner	Yes	Yes

processed in batch mode. Both machines used the same batch of flux and solder, enabling a wave soldering process machine test as opposed to a soldering materials test. Boards were run simultaneously on both machines from the same batches. The only differences between machines were the inherent system temperature profile and hot air knife.

Machine configuration is described in Table 4-4. Both machines were configured similarly. Both ran at 4 ft/min, with foam fluxers and dual solder waves. The main differences were preheater length, solder temperature, and the addition of the hot air knife in the SMT optimized unit. In order to meet the 3.6°F (2°C) per second slope and 212°F (100°C) delta T rule at 4 ft/min conveyor speed, the bottom side preheater was lengthened to 5 ft under the conveyor and 2 ft were added to the top of the conveyor. The top side preheater was directly over the bottom side preheater section at the edge nearest to the solder pot. The optimized machine soldered in a "low" temperature mode at 470°F (243°C). The air knife was set at the same temperature for both boards, but angle and pressure were further optimized for each. The LTH board used a 45° angle at high pressure, 12 psi. The SMT board used a high angle of 85° and low pressure of 6 psi. The pressure differential between boards types was mainly driven by the maximum board depth. The LTH board was 1/8 in. deeper than the SMT board.

The standard LTH machine profile is shown in Figure 4-10 and Figure 4-11 is the optimized profile. Figure 4-11 illustrates the elongated preheat thermal profile capable of being achieved on an SMT-type wave soldering machine. Both profiles are generated by using the same thermocouple placed on the bottom surface of the board. Both figures show bottom side surface temperature with solder waves on. The thermocouple was attached with high temperature solder, effectively resoldering it in a hollowed out joint. The optimized machine meets

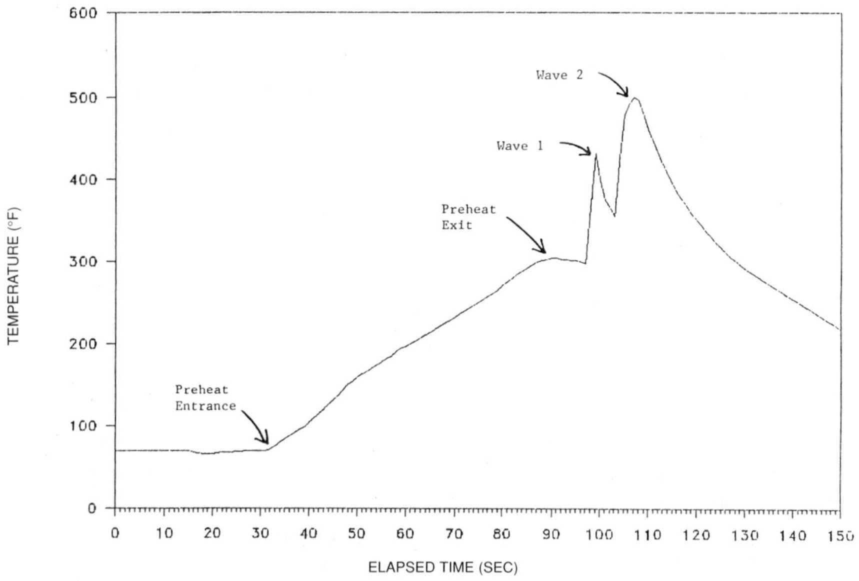

Figure 4-10 Standard machine profile. (Courtesy of Hollis Automation)

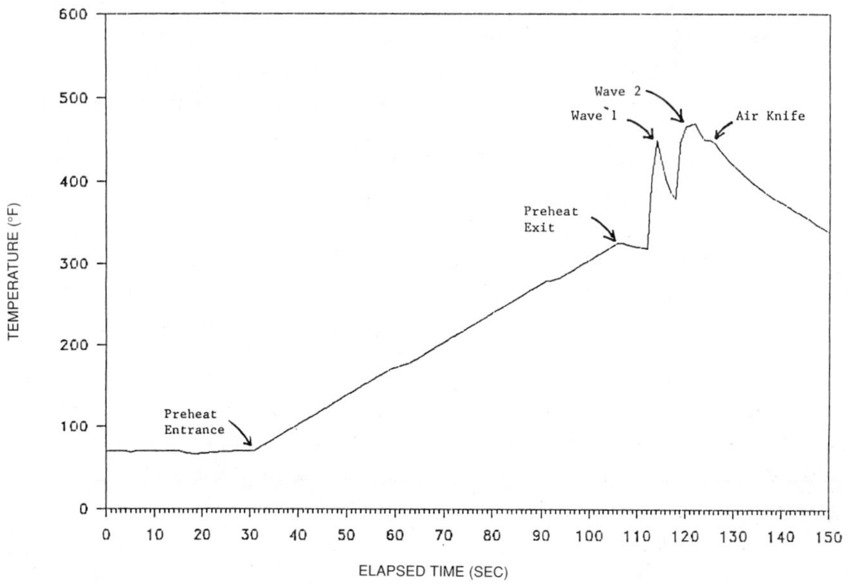

Figure 4-11 Optimized machine profile. (Courtesy of Hollis Automation)

the 3.6°F (2°C) per second limit as well as the thermal shock limit of 212°F (100°C). The actual optimized rise rate is a maximum of 3.4°F (1.9°C) per second and a maximum of 153°F (85°C) shock temperature, delta T. This compares to the standard machine maximum temperature rise rate of 5.4°F (3°C) per second and maximum thermal shock temperature of 198°F (110°C). The standard machine was intentionally set to the best profile it could achieve on these boards.

The soldered boards were inspected by the same team. The boards were batch coded and mixed so that the inspectors were not aware of the machine designation. Visual inspection focused on identifying bridges, excess solder webbing, poor solder, no solder, blow holes, and solder balls. A bridge was defined as a short or solder-filled space between any two leads or pads to leads in the case of SMT joints. If, for example, two spaces between three leads were filled, two bridges were recorded. Excess solder pertained to leaded and leadless SMT devices as well as leaded through-hole components. For leaded components of both varieties, excess solder was defined as a joint in which the lead could not be clearly discerned. For leadless SMT devices, excess solder was determined to exist when the knee of the solder fillet protruded more than half way up the edge of the device. Webbing was defined as a solder streamer condition that formed very fine joints between leads and/or pads. Poor solder was defined on SMT components only as the condition when the knee of the solder fillet occurred below the halfway mark on the edge of the device. While it was clear that some solder may have wetted to the solder joint in a poor solder condition, the main portion of the lead was exposed on leaded devices. No solder referred to a complete lack of solder. A blow hole was defined as a void in the joint. Solder balls were defined as coherently formed globules that were not attached to either lead or pad but did attach to the surface of the board.

The test used two boards in an effort to establish the effectiveness of wave soldering optimization on a wide range of product shown in Table 4-5. The first board was a mixed technology with top side reflow surface mount, bottom side wave soldered surface mount, and leaded through-hole components. It measured 2 by 8 by 0.128 in. and contained 347 bottom side joints. The bottom side surface mount were leadless capacitors including some ceramic C-size ones, small outline transistors (SOTs), and small outline integrated circuits (SOICs). A total of 201 joints were surface mount and 146 were leaded through-hole. Board 1 represented a very complicated SMT application.

Table 4-5 PCB Descriptions

Product	LTH Joints	SMC Joints	Total Joints	Sample Size No. of Boards
1	146	201	347	200
2	1980	22	2002	500

Table 4-6 Results of Comparative Processes

Product	Standard Machine	Optimized Machine	Defect Ratio
1	1	0.35	3X
2	1	0.07	14X

The second board was primarily leaded through-hole representing the other side of the spectrum. It mainly consisted of leaded through-hole DIPs with a few bottom side small chip capacitors. It measured 8 by 10 by 0.128 in. with a total of 2002 joints, all on the bottom side. These were divided into 1680 leaded and 22 SMT joints.

4.4 RESULTS

The results for both boards are shown in Table 4-6. The defect per million (ppm) rate has been calculated by averaging total defects and converting to a per million joint unit basis. Since both products were assembled and run at the same time under identical conditions, except for wave soldering, all defects not relating to wave soldering can be assumed to be the same for all boards, making direct comparisons possible. The ppm rates are shown normalized to the standard machine. The normalization process enables us to see the effect of the optimization without showing the absolute value and exposing a private manufacturer's industry secret. The ratio between the machines remains the same regardless of the reference point.

Board type 1 showed an improvement of 3 times fewer defects on the optimal machine versus the standard machine. Board type 2, the primarily leaded through-hole board, showed 14 times fewer defects than the nonoptimal machine. This study shows the enormous effect that optimizing process profile can have on reducing joint defects.

4.5 CONCLUSION

Joint specifications place stringent solderability requirements on the process physics. Surface mount components continue the trend toward finer pitches. This increases the joint density and the likelihood of developing solder joint defects, such as bridges. The bottom side surface mount components themselves are also subjected to harsh environmental changes. Many SMT components cannot tolerate rapid dramatic temperature changes. Leaded through-hole components also benefit from these thermal profiling trends. Basically, any surface that is brought closest to the soldering temperature benefits by not being subjected to a large thermal shock.

The important goal in optimizing the process is to make the best hardware selection with respect to the expected product type first and then further optimize the process by choosing the correct process setpoints. Wave soldering hardware is dictated by board design and production throughput rate. These factors are derived from two items; the first is describing the most difficult board that is expected to be run and the second is identifying the maximum expected throughput rate or line speed. The wave soldering subsystem hardware is selected. Then the process parameters are optimized to achieve the lowest peak thermal profile with the least thermal rise rate. This augments the wicking action as the board is introduced to the solder in the wave. The study shows that wave soldering process optimization increases yield for SMT devices and excels at increasing performance for LTH technology. The study further shows that optimizing both hardware and process parameters for a technology level higher than the lower cases also benefits the lower technology. If there were no cost limitations on purchasing hardware, it would always be more beneficial to optimize to a more difficult product type. The effect of longer preheat, dual waves, and air knife benefits all product types.

REFERENCES

1. Manko, H. H., *Soldering Handbook for Printed Circuits and Surface Mounting,* Van Nostrand Reinhold Company, Inc., New York, 1986.
2. AVX product literature.
3. Murata Erie product literature.
4. Woodgate, R., *Handbook of Machine Soldering,* John Wiley & Sons, Inc., New York, 1983.
5. Leonida, G., *Handbook of Printed Circuit Design, Manufacture, Components, & Assembly,* Electromechanical Publications, Ltd., 1981, pp. 331–463.
6. Bahr, K. E., "Solderability: A Critical Issue for Fabricators and Assemblers," *IPC Proceedings,* January 1990.

5

Post-Solder Cleaning Considerations

Leslie A. Guth

5.1 PURPOSE AND CHAPTER DESCRIPTION

The improper selection of either the soldering flux or the subsequent cleaning materials, processes, or equipment can greatly affect circuit pack reliability. In fact, it is necessary to choose the appropriate cleaning strategy by taking into account the activity and the composition of the soldering flux used. Residue remaining from either the flux or the cleaning agent or a combination of the two can affect the reliability of the circuit board.

This chapter begins with a section on the existing specifications for soldering fluxes and the associated test procedures for qualifying them. Following are sections on flux composition, flux application methods, and soldering process issues, all of which are directed toward wave soldering. The last two sections discuss nonliquid fluxes and future trends. Throughout this chapter, the characteristics of soldering fluxes and their use are described, with the underlying theme relating to the importance or the lack thereof for postsolder cleaning.

Due to the ever-increasing environmental concerns, an emphasis is placed in this chapter on those fluxes and associated soldering processes that require no cleaning, especially low solids fluxes (LSFs). Again, flux selection should be made realizing that this decision will directly affect the possible subsequent cleaning materials and processes and their associated environmental concerns.

5.2 ENVIRONMENTAL CONCERNS

Chlorofluorocarbons (CFCs), widely used in electronics industry for cleaning, have been implicated in the depletion of stratospheric ozone. In 1987, an international agreement, The Montreal Protocol, was signed; it went into effect in 1989 and called for a 50 percent reduction in CFC usage by 1998. In June 1990, this document was amended based on a technology assessment and now includes provisions for total phaseouts of CFCs and other chlorinated solvents by 2000 (methyl chloroform's phaseout date is 2005). Because of this agreement, cleaning trends are toward nonhalogenated solvents (e.g., terpenes, aqueous saponifiers, and water only) or no cleaning at all. The no-clean option is preferred since it eliminates other environmental concerns such as waste disposal and energy consumption, a contributor to the greenhouse effect.

5.3 DEFINITION OF SOLDERING FLUX

Soldering flux is defined in the flux specification of the Institute for Interconnecting and Packaging Electronic Circuits (IPC), ANSI/IPC-SF-818, "General Requirements for Electronic Soldering Fluxes," as:

> A chemically and physically active formula which promotes wetting of a metal surface by molten solder, by removing the oxide or other surface films from the base metals and the solder. The flux also protects the surfaces from reoxidation during soldering and alters the surface tension of the molten solder and the base metal.

That is, flux prepares the metal surfaces to be soldered by:

- Removing oxides prior to soldering
- Assisting in uniform heating of the metal surfaces during preheat
- Maintaining an oxide-free surface during the soldering process
- Lowering the surface tension at the metal-solder interface

The vast majority of soldering is done with the aid of fluxes; choosing the appropriate flux for the soldering process dictates the necessary postsoldering cleaning process.

5.4 SPECIFICATIONS

Before examining the different specifications in detail, several test methods that are cited in two or more specifications will be discussed in a general sense. These methods are the copper mirror, silver chromate paper (halide), the surface insulation resistance, British corrosion, and ionic conductivity tests.

5.4.1 Test Methods

The methods discussed below have specific purposes, advantages, and disadvantages. These, along, with their procedures, will be described.

Copper Mirror Test

The copper mirror test is referenced in a number of specifications[1,2] and is a measure of a flux's corrosivity to copper. This corrosivity is an indication of the activity level of a flux. Both the speed and relative ease with which this test can be performed make its use quite common. However, it also has several drawbacks.

The copper mirror test involves placing two drops of the flux in question on a copper mirror, a glass microscope slide on which 300 to 500 Å of copper has been vapor deposited. Typically, a flux made of pure water white gum rosin and isopropanol acts as a control and a drop of it is also placed on the mirror. The mirror is then placed in a controlled environment of 23 ± 2°C and 50 ± 5 percent relative humidity for 24 hours. Then the mirror is rinsed in isopropanol and examined for any spots or areas where the copper has been removed by the flux. Figure 5-1 illustrates two passing copper mirrors on the left and three failing copper mirrors on the right.

Results of the copper mirror test indicate the potential corrosivity of the raw or unheated flux. Although the corrosivity of unheated flux can be a concern, it is, unfortunately, not always indicative of the corrosivity of a partially heated flux (e.g., flux present on the top, or component side, of a circuit pack during

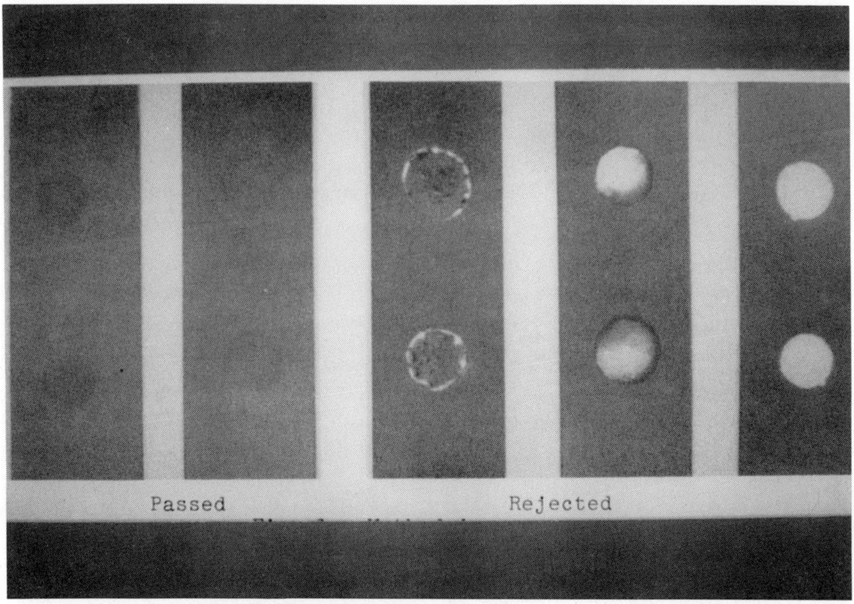

Figure 5-1 Results of copper mirror test.

wave soldering) or a fully heated flux (e.g., flux present on the bottom, or wiring side, of a circuit pack during wave soldering). Also, the results can be difficult to interpret sometimes since they are subjective and there is no middle ground between pass or fail.

Because of these drawbacks, critics have questioned the relevance of this test altogether. However, due to its speed, the relative ease of performing this test, and the evidence that these results do correlate well with those of the more lengthy and involved surface insulation resistance test,[3] this test is certainly worthwhile and justified. It is especially useful as a prescreening test when testing a new formulation or as a routine incoming materials check.

Silver Chromate Paper (Halide) Test

The silver chromate paper test is another fairly simple test to perform. Its purpose is to detect significant amounts of ionizable chlorides and bromides in a flux. Any flux that leaves the silver chromate paper unchanged has either no Cl^- or Br^-, or at least, no detectable quantities.

The test procedure involves placing a drop of flux on silver chromatic paper. After 15 s, the paper is rinsed or immersed in isopropanol and then dried for 10 min and examined for any color change. A color change means that halides are

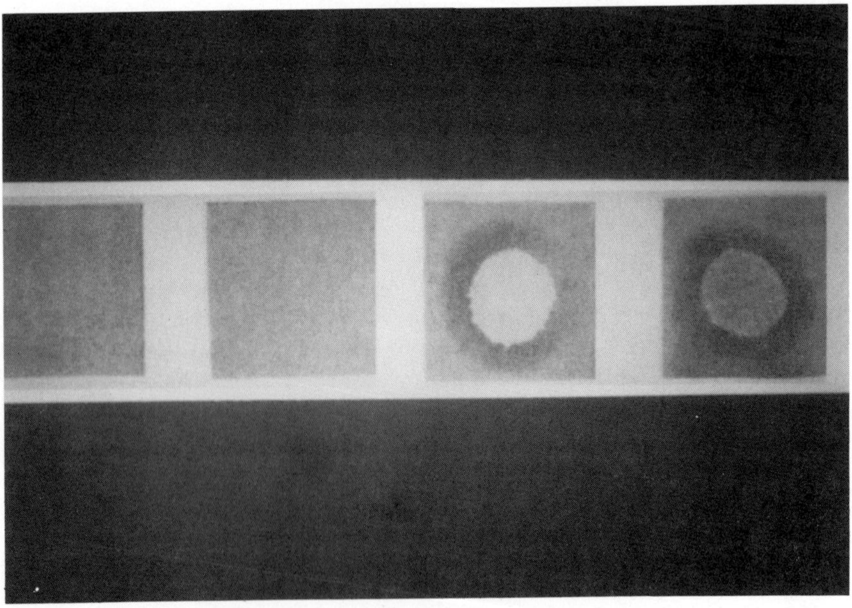

Figure 5-2 Results of silver chromate paper test.

present in detectable quantities. Examples of silver chromate paper results are shown in Figure 5-2.

Drawbacks of this test include possible subjectivity in interpretation of results, lack of quantitative data, and interference from other nonhalide materials such as amines, cyanides, or isocyanates.[1] Some chemicals can cause color change, as can fluxes that are unusually acidic ($pH < 3$). Thus, if halides are not suspected and a color change is detected, other tests should be performed to verify the results.

Another problem with this test is that the results are pass/fail only and are not quantitative. A proposal in a telecommunications specification had called for a more quantitative assessment of the amount of chlorides and bromides in a flux based on a color changes seen in the silver chromate paper.[4] Still yet another problem is that the silver chromate paper only detects chlorides and bromides, not fluorides.

Even with these shortcomings, though, this test is beneficial for prescreening or incoming material inspection for fluxes that should be noncorrosive.

Surface Insulation Resistance Test

Surface insulation resistance (SIR), also known as insulation resistance (IR) and moisture and insulation resistance (M&IR), is a lengthy, but worthwhile, flux evaluation test. SIR testing is also used to evaluate printed circuit fabrication processes. Of all test methods, this one best simulates the relevant cause and effect relationship between soldering fluxes and circuit boards. Even though SIR coupons are not identical to actual circuit boards and SIR measurements are not exactly like actual circuitry, SIR testing is better than other tests since (1) the soldering process is part of the sample preparation, (2) the sample substrate is itself a circuit board, and (3) the response variable, leakage current, is a measured attribute of circuitry. Even so, this test method does have some shortcomings that will be discussed along with the details about the types of test patterns, sample preparation, temperature and humidity selection, necessary equipment, and types of requirements.

The result of SIR testing is an electrical measurement of the effect of flux and/or cleaning residues on a printed circuit board. The change in the SIR due to the presence of some residue is measured assuming that it is much greater than the bulk resistance of the printed circuit board. To maximize the measuring sensitivity of this test, the leakage current passing between long parallel lines placed at different electrical potentials is measured. Instead of long narrow circuits, interdigitated combs that are electrically identical to two long lines are typically used. Examples of two comb patterns are shown in Figure 5-3, the telecommunications comb pattern with 0.025-in. lines and 0.050-in. spacings and the IPC 'B' comb pattern with 0.0125-in. lines and spacings.

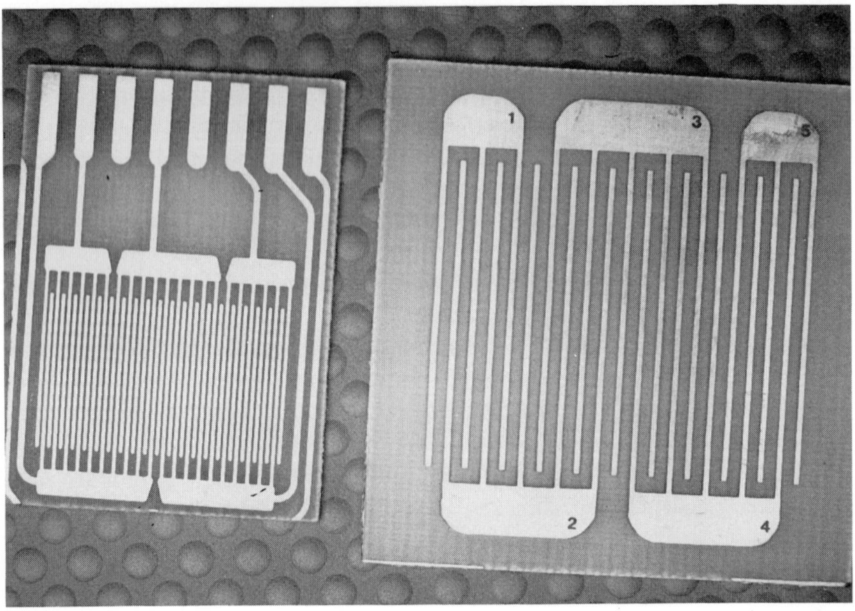

Figure 5-3 Two SIR comb patterns: telecommunications and IPC; the smaller pattern is from IPC.

Traditionally, data are reported as insulation resistance (in ohms and megohms) rather than as leakage current. The conversion is simple via Ohm's law, $V = IR$, where V is the test voltage and I is the leakage current. Diagrams in Figure 5-4 show the typical circuitry used to apply a 45 to 50-V dc bias voltage on the left and to measure the resistance with a reverse polarity 100-V test voltage on the right. SIR data are a function of a number of parameters:

1. The width of the spacing between the traces
2. The length of the traces
3. The testing voltage
4. The test voltage electrification time
5. The bias voltage, if any
6. The temperature
7. The relative humidity
8. The dielectric material of the test substrate
9. The specific solder resist coating, if any
10. The quality of the test leads, the electrical instruments, and the noise shielding

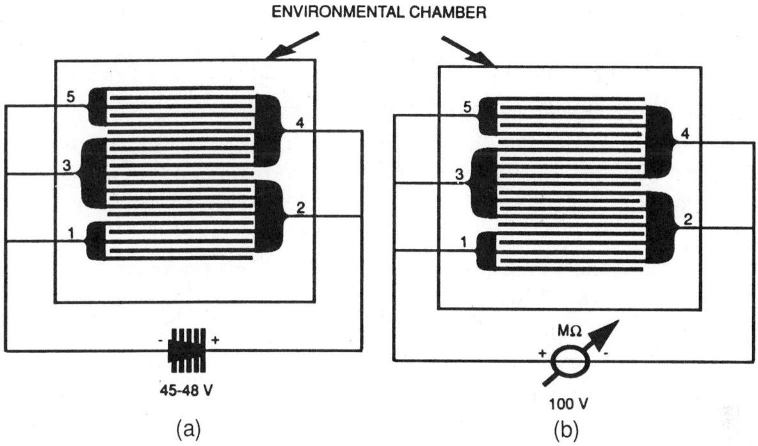

Figure 5-4 Circuit diagrams for biasing and measuring SIR comb patterns (*a*) Test, (*b*) Measurement.

Although SIR is affected by both the total length and width between the traces, it is a moot point when testing fluxes if the same pattern is always used. The absolute values are only of interest if the test coupon's circuitry is identical in both length and spacing as is the circuitry of the actual electronic product, which is rarely, if ever, the case. However, as mentioned previously, there are many comb pattern shapes and sizes. Theoretically, the data from all comb patterns can be correlated by taking into account the number of squares that make up a comb. A square is a unitless value equivalent to the total length of the comb divided by the width of the spacing between two lines. So, a comb pattern that has more lines will be longer and have more squares; additionally, a comb pattern that has narrower spacing will have more squares. Several papers that have discussed theory and concepts behind the use of squares are recommended for further reading.[5,6] More recently, a study has been documented showing that correlation is good within a range of squares of not more than one order of magnitude.[7]

The test voltage as well as the electrification time can also influence the ultimate SIR measured. Moreover, the bias voltage and its value relative to the test voltage can affect the results. These results have been documented.[8] Whether to use a bias voltage and whether this bias voltage should be the reverse polarity of the test voltage is an ongoing debate. Both the telecommunications and the IPC specifications require a 100-V dc test voltage and a 45 to 50-V dc bias of reverse polarity. Although these electrical parameters may influence the SIR results, it is important to look at them in the context of the whole picture. That is, if one simply maintains particular settings for these parameters, the changes in the SIR resulting from changes in the electrical parameters can be disregarded. Established SIR test methods specify these settings so the relative differences between different fluxes can be observed without worrying about electrical parameter effects.

Both temperature and humidity can be used to accelerate the ageing of samples. Depending on the setting, however, these factors can also be specified to simulate worst-case conditions. Figure 5-5 shows several typical temperature and humidity settings. Also shown is the envelope in which circuitry typically must operate relative to an example of a performance test condition (e.g., 35°C at 90%RH)

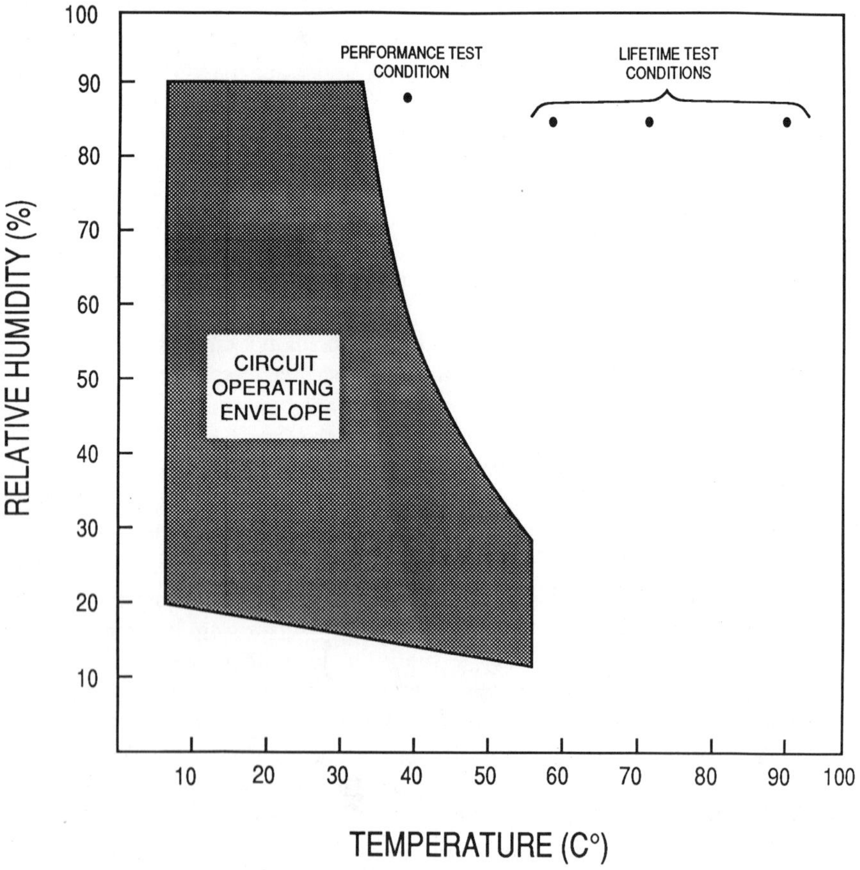

Figure 5-5 Typical temperature and humidity conditions for SIR tests.

and several lifetime or accelerated test conditions (e.g., 55, 65, and 85°C, all at 80%RH). Other conditions are also used such as 85°C/85%RH for Class III SIR testing per IPC-SF-818. As desirable as accelerated conditions may seem, it is important to realize that an extreme environment may cause some flux materials to fail by mechanisms that are nonexistent at normal operating conditions.

Sample preparation simulates the assembly process, including fluxing the comb pattern and then soldering it. Recently, the Bellcore specification[2] has added an additional test in which component side flux residues are tested by floating the comb pattern face up on a solder pot or wave, rather than face down. This additional test is important since some flux residues can potentially be more harmful when exposed to less heat.

Some specifications require minimum absolute SIR values as a pass/fail criterion whereas others require a minimum SIR relative to the SIR value of a control. A relative value is preferred from a designer's point of view since it is simply allowing a minimum degradation due to the presence of flux residues. Obviously, some guideline for an absolute value for the control is a good idea. In contrast, an absolute value for an SIR requirement is implying a knowledge of the electrical needs for a particular product, a difficult if not impossible task for widely used industry specifications. This is true since one requirement on one particular comb pattern cannot simulate any and all use environments and circuitry designs.

British Corrosion Test

Another test that is specified by a number of groups is a corrosion test, also referred to as the British corrosion test.[1] It involves reflowing solder in the dimple of a copper panel in the presence of flux solids and then ageing the panel at 40°C and 93%RH for a number of days, depending on the specification and the type of flux being tested. A flux is said to pass if no blue-green corrosion can be observed after ageing. Figure 5-6 shows both passing and failing results for the corrosion test.

This test has both benefits and drawbacks. On the plus side, the results are independent of a flux's solids content since this quantity is held constant. Because a concentrated flux material is used, the test is certainly a worst-case condition.

However, the test lacks an electrical bias, which is necessary to initiate some corrosion mechanisms (e.g., electrolytic). Also, it creates an accentuated copper-solder-flux interface. Lastly, the interpretation of results is difficult, subjective, and qualitative only. An evaluation of "how green is green" and the extent of green or blue-green coloration is difficult to quantify. As with the copper mirror test, there is no middle ground between pass and fail.

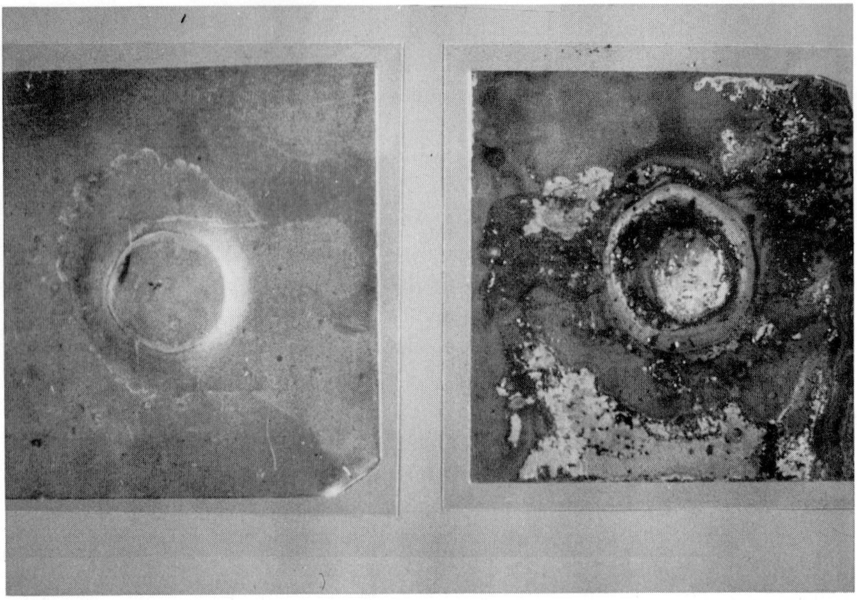

Figure 5-6 Results of corrosion test.

The difference between this test and the SIR test lies in their two distinct purposes. The corrosion test evaluates the potential for an "open" failure, whereas the SIR test determines the potential for an electrical "short."

Ionic Conductivity Test

The ionic conductivity test[2] involves extracting those species on the circuit board that are soluble in a water-alcohol solution. Those species that are ionizable are measured either by their effect on the resistivity or conductivity of the extracting solution. A number of different systems are commercially available including one that constantly "cleans" the extracting solution, a "dynamic" measurement, and several that measure the decreasing resistivity in a static situation, "static" measurements. Some contain room temperature extracting solutions whereas others both heat and spray the extracting solution to improve extraction efficiency. Data are typically provided in $\mu g/cm^2$ or $\mu g/in^2$ NaCl equivalents. This test is useful for monitoring of the cleaning process and yields an average contamination level of ionizable species on the board. Obviously, those species that are not soluble or ionizable are not going to be counted, and information on localized concentrations of residue cannot be collected. The ionic conductivity test is often used to monitor water cleaning.

5.4.2 Institute for Interconnecting and Packaging Electronic Circuits (IPC)

The IPC flux specification, IPC-SF-818, classifies fluxes in terms of activity levels rather than in terms of material types. All electronic grade fluxes should fit in one of three major categories of L, M, and H, denoting low, medium, and high activity, respectively. Because of the further groupings for the SIR testing, ultimately a flux would be categorized with L, M, or H, followed by the numeral 1, 2, or 3 (designating class of assembly) and by C, N, or CN depending on whether a flux meets the SIR after being cleaned (C), not cleaned (N), or both (CN).

Because of the classification system itself, virtually all fluxes can fit into at least one category. It is up to the user to determine what product class is being soldered and whether postsolder cleaning is part of the assembly process. After these decisions, the activity level necessary or sufficient for the assembly process must be chosen, remembering that with higher activity level, there is usually increased soldering efficiency along with increased potential for corrosion if proper cleaning procedures are not maintained.

The IPC specification requires (1) copper mirror test, (2) silver chromate (halide) test, (3) British corrosion test, and (4) surface insulation resistance test. The SIR comb pattern is a part of the IPC-B-25 test board and has 0.0125-in. lines and 0.0125-in. spacings (see Fig. 5-3). The SIR requirement is an absolute value of $\geqslant 1 \times 10^8\ \Omega$.

5.4.3 U.S. Military

The U.S. military specification, MIL-F-14256E, is limited to only rosin-based fluxes. There are three categories allowed, R, RMA, and RA designating rosin, mildly activated rosin, and activated rosin, respectively. These are equivalent to the R, RMA, and RA categories in QQ-S-571, the U.S. federal government solder specification.

Like many other specifications, MIL-F-14256 requires a copper mirror test (for R and RMA) and a silver chromate test (for R, RMA, and RA). A solder spread factor test is also required for RMA and RA fluxes.

However, what sets the military specification apart from other specifications is that it not only allows just rosin fluxes but also demands that a minimum of 51 percent of the solids content is rosin and that the rosin itself must meet a certain minimum acid number. Another unusual requirement is a minimum resistivity of water extract. For the first time, MIL-F-14256 has incorporated a surface insulation resistance test in Issue E, June 1989. It requires a $\geqslant$ 100-MΩ reading under 85°C and 85%RH conditions using the IPC B-25 'B' comb pattern (same as IPC-SF-818 Class 3 test coupon, conditions, and requirement).

5.4.4 Telecommunications

The telecommunications industry has its own specifications and requirements. After the divestiture of AT&T in 1984, the court set up a research group, Bellcore, to work for the interests of the seven Bell operating companies. The Bellcore specifications grew out of the original AT&T specifications and have in recent years dictated U.S. telecommunication industry requirements.

The most recent Bellcore assembly document that specifies flux materials is TR-TSY-000078, Issue 2.[2] It divides allowed fluxes into only two categories, compliant and noncompliant.

Compliant fluxes are "noncorrosive fluxes" (formerly called "rosin noncorrosive fluxes"). Requirements that must be met include copper mirror, silver chromate, fluoride, pH, and SIR tests. The SIR test specifies a comb pattern with 0.025-in. lines and 0.050-in. spacings (see Fig. 5-3) and an environment of 35°C and 90%RH. Originally, Bellcore required that a comb pattern soldered with a compliant flux be at least 75 percent of the SIR of the control (like the AT&T specification MS-58556). Even though Bellcore still requires an SIR value relative to the control, the requirement is really the equivalent resistance of two parallel resistors, one being the SIR for the control and the other, an ideal resistor with a value of $10^5 M\Omega$. Thus, the requirement is a sliding scale and the gap between it and the control narrows as the control value decreases. This requirement must be met for three different types of comb pattern groups: (1) fluxed, soldered, and cleaned, (2) fluxed, soldered, and not cleaned, and (3) fluxed, soldered with comb face up, and not cleaned. The reason for the third group is to simulate the exposure of the top or component side of a circuit board to flux. This is necessary since some fluxes have been found to be more detrimental when they are not heated to as high a temperature, a situation experienced by flux present on the top side of a circuit board. There is a misconception that a telecommunications noncorrosive rosin or compliant rosin flux is the same as an RMA flux. This is not necessarily true since the requirements for each are different.

The other category for telecommunication fluxes is noncompliant, which is simply a classification for aggressive fluxes. That is, any flux that cannot meet the requirements for compliant fluxes is then tested for the noncompliant requirements. Aggressive fluxes are tested per these two requirements: (1) surface insulation resistance of $\geq 3 \times 10^9 \Omega$ on a telecommunications comb pattern with stripes of solder mask perpendicular to the comb's metal traces exposed to the flux, solder, and an appropriate cleaning process and (2) ionic extraction value of $\leq 1.0\ \mu g/cm^2$ ($6.45\ \mu g/in.^2$) NaCl equivalent for a bare copper comb pattern exposed to flux, solder, and the cleaning process. Clearly, not all aggressive fluxes can necessarily meet these requirements. Since the residues of these more active fluxes are designed to be removed by cleaning, testing uncleaned boards is not required.

The inclusion of solder mask stripes on the SIR coupon has been debated as have possible alternative patterns. Some people fear that the solder mask stripes make the test vehicle unduly stringent and even unrealistic compared to actual circuit designs. Effort to modify this geometry is ongoing. The ionic extraction test is generally more well accepted by the industry as a cleaning efficiency check for aggressive fluxes since it is easier; however, nonionizable but hygroscopic and harmful species are undetected by ionic extraction even though their effect is measurable by SIR. The extraction medium can be either 50:50 or 75:25.

5.5 FLUX MATERIALS AND ASSOCIATED CLEANING

Fluxes are composed of a number of different materials including:

- Solvent: the liquid carrier for the flux ingredients
- Vehicle: a thermally stable material that acts as the high temperature solvent
- Activator: ingredients that create a wettable surface for solder by removing oxides and possibly other contaminants
- Surfactant: an ingredient that reduces the surface tension at the metal-solder interface to further promote solder wetting
- Antioxidant: a material that prevents reoxidation of the metal surfaces

In this section, flux characteristics and the corresponding cleaning process will be discussed in more detail and will be examined in terms of the following flux categories: rosin, water soluble (WS), synthetic activated (SA), and low solids (LS).

Before looking at each flux group individually, it is worthwhile to compare the makeup of these fluxes in terms of generic constituents and proportions, only looking at the three major components, solvent, vehicle, and activator. As shown graphically in Figure 5-7, these four groups of fluxes consist of different proportions of vehicle, activator, and solvent. Those rosin fluxes that pass military RMA or telecommunication specifications typically have a much lower activator to vehicle ratio than the other flux types shown. Water soluble and synthetic activated fluxes may have about the same proportion of activator relative to the amount of vehicle. All of these groups typically have 15 wt% to 35 wt% solids or nonvolatiles. In comparison, LSFs have much less solids, and the amount of activator is typically the same or more than the amount of vehicle.

Figure 5-8 shows the effect of flux selection on cleaning material choice. Not all cleaning materials can remove all flux residue types, nor do all flux residues need to be removed.

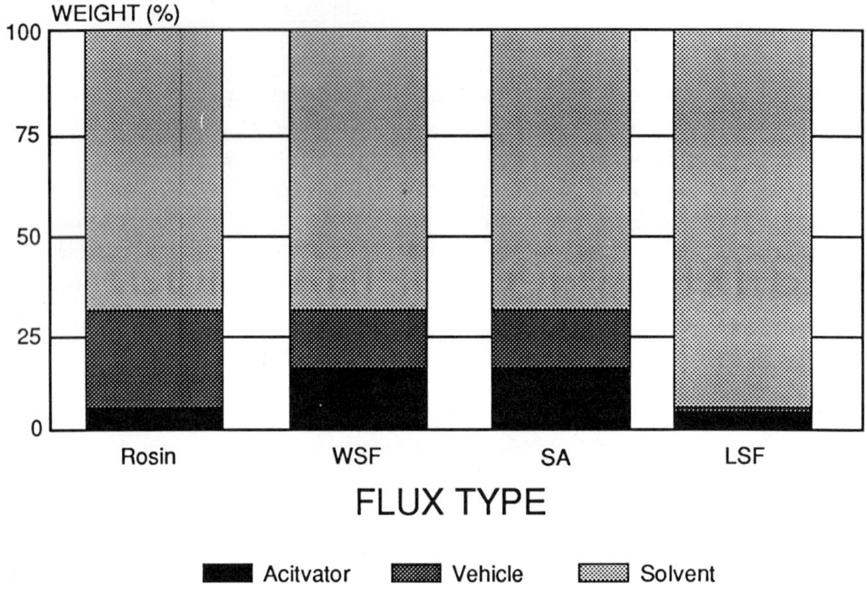

Figure 5-7 Comparison of major component proportions for four groups of fluxes.

CLEANING MEDIUM	FLUX TYPE				
	Rosin (benign)	Rosin (aggressive)	WSF	SA	LSF
None (no clean)	X	-	-	-	X
Chlorinated solvents or CFC-113 based	X	X	-	X	-
Aqueous (saponifier)	X	X	X	-	-
Semi-aqueous (terpenes)	X	X	X	X	-

Figure 5-8 Cleaning material choices as a function of flux type.

5.5.1 Rosin

The vehicle in rosin fluxes is rosin or colophony, a naturally derived material from pine trees. Rosin, consisting of a number of isomers, mostly abietic acid, which are fairly large molecules and include one carboxyl group (molecular formula: $C_{19}H_{29}COOH$), acts as a flux vehicle and is also a mild activator at soldering temperatures. Another unique property of rosin is that it is an extremely good insulator at room temperature; in fact, its bulk resistivity is at least an order of magnitude higher than the epoxy-glass printed wiring board material's bulk resistivity.[9] Because rosin hardens on cooling to room temperature, it serves as an excellent encapsulant, prohibiting the movement of other flux residue ingredients such as those from activators.

Depending on the final activity of a rosin flux, there are a number of different types of activators typically used ranging from organic acids to organo-halides. Again, depending on the quantities added, rosin fluxes may or may not pass copper mirror, silver chromate, surface insulation resistance, and corrosion tests. "L" or "RMA" fluxes should pass these tests, whereas "RA," "M," or "H" will not. Work has been carried out to maintain the noncorrosive nature of rosin fluxes, yet maximize the soldering efficiency achievable with them.[9]

Because of the insulating properties of rosin flux residue, it is not necessarily cleaned off to avoid corrosion. In fact, many companies have left rosin flux residue unremoved for years without problems. Nevertheless, there are a number of reasons for removing rosin flux residue. Often rosin residue is sticky and tacky. So, for aesthetic reasons or if the printed circuit boards may be exposed to excessive dust and dirt,[10] it may be necessary to clean. Another reason for cleaning is to avoid problems during bed-of-nails testing or in-circuit testing; rosin flux residue can cause false opens when the test pins are unable to push through insulating residue covering a test pad. Sometimes it is necessary to remove rosin residue because of its effect on the subsequent addition of conformal coating to a printed circuit board. It is unfortunate when it is necessary to remove rosin residue since its complete removal, although possible, is usually quite difficult.

Cleaning processes for rosin residue removal include semiaqueous (e.g., terpenes, other hydrocarbons) ones, degreaser types using CFCs and HCFCs (partially halogenated chlorofluorocarbons), and those using aqueous detergents or saponifiers. For environmental reasons, CFCs and other chlorinated solvents should be avoided. Aqueous detergents and saponifiers have been used for years in both underbrush and total immersion spray cleaning operations. Since these saponifiers are highly caustic and therefore corrosive, the wash stage must be followed by a rinse stage in which the saponifier solution is rinsed off the boards sufficiently. Semiaqueous processes, such as those with terpenes, have been shown to be excellent cleaners.[11] Component compatibility and environmental impact should always be addressed when selecting a cleaning process. Rosin residue removal can be accomplished with batch cleaners, underbrush in-line cleaners, or total immersion spray impingement cleaners.

5.5.2 Water Soluble

Water soluble fluxes (WSFs) are typically very active and aggressive, and the residue can be corrosive if not removed properly. However, since they are active and designed to be removed in aqueous solutions, they offer a number of advantages over less active and more difficult to remove fluxes.[12] In fact, the process window for soldering is relatively large, and extremely low solder defect levels (less than 50 ppm) are attainable. Figure 5-9 illustrates the wider process window and the lower solder defect level attainable with a WSF compared to some rosin fluxes. With the improved soldering efficiency, some companies have been able to eliminate the need for postsolder touch-up operations.

Often WSFs are referred to as organic acid, or OA, fluxes. Although it is true that there are organic acids in many WSFs, it is also true that many other fluxes including rosin, SA, and low solids include organic acids. Thus, the name OA for water soluble fluxes is a misnomer and should be avoided.

The vehicle for WSFs is typically a polyethyleneoxide or polypropyleneoxide, commonly referred to as polyglycols. These materials are nonionic but are hy-

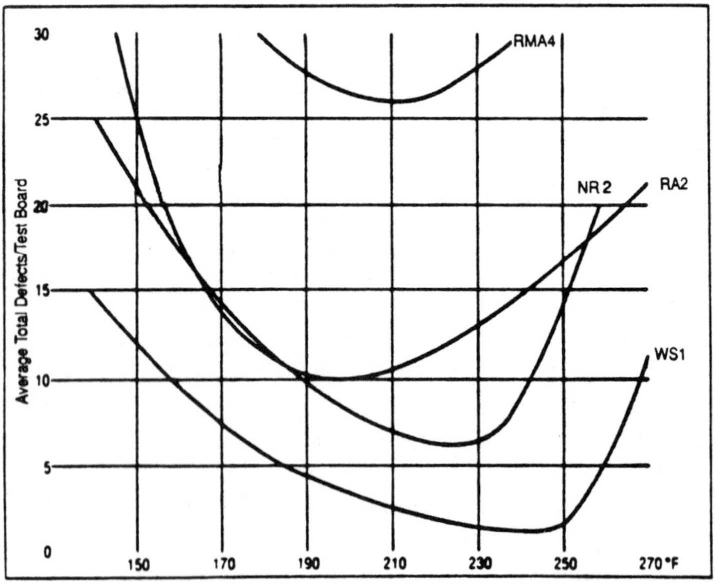

Figure 5-9 Solder defect levels as a function of preheat temperature for rosin fluxes RA2, NR2, and RMA4 compared to water soluble flux WS1.[12]

groscopic, so a deleterious, but synergistic effect is observed, when these polyglycol residues are not removed along with the ionic residues from WSFs.[13] That is, polyglycols are not going to encapsulate activator residues like rosin.

Usually the activity level is quite high in WSFs, meaning that fairly strong organic acids and halide-bearing compounds are present or that relatively large quantities are present. Some WSFs are halide-free and promoted as less corrosive, yet their postsolder residue still needs to be removed. Also, there are some WSFs that have a neutral pH since it is believed that these would be less corrosive than those that are more acidic.

Because of their high activity level, most WSFs will fail the copper mirror and silver chromate paper tests. If proper cleaning is employed, a number of WSFs can pass SIR and corrosion tests. Both WSFs and SA (see next section) fluxes must meet the requirements for noncompliant fluxes for the telecommunications industry, and neither are presently allowed by the U.S. military since they do not contain rosin.

WSFs, like other aggressive fluxes, must be removed in a spray impingement cleaning system, sometimes called "total immersion cleaning." For WSFs, the cleaning medium is either water only or aqueous detergents. In-line spray systems look much like car washes, with spray nozzles directed at the printed circuit boards from both the top and the bottom. Because of the need to thoroughly remove all residues and because of the type of cleaning used, the compatibility of components to immersion in water must be addressed. Also, it is important to ensure that components are irrigable. The use of WSF and aqueous cleaning for surface mount or mixed assemblies has been debated for a number of years. However, a number of companies have demonstrated their feasibility.[14–16,11]

5.5.3 Synthetic Activated

In the 1980s, Du Pont developed a family of high activity synthetic activated (SA) fluxes, analogous to WSFs and designed so that its residues were easily soluble in the CFC-113/methanol azeotrope cleaning medium. SA fluxes are based on a mono and di-isooctyl phosphate (IOP) chemistry and include compatible activator and solvent materials.[17] This technology was offered to flux manufacturers who subsequently marketed these SA fluxes.

Due to their high activity, these fluxes perform much like WSFs in that they will not pass the copper mirror or silver chromate paper tests. Also, SA flux residues must be removed in a spray impingement or total immersion cleaning system. Usually the cleaning medium is CFC based; however, it has been shown that terpenes can adequately remove SA flux residues also.[18] Thus, component compatibility with the cleaning medium is an issue that must be addressed. Unfortunately, a component that is compatible with water may not be compatible with CFCs, and vice versa. If the cleaning process is sufficient, the requirements for SIR and corrosion tests can be met with SA fluxes.

Like WSFs, SA fluxes typically yield lower solder defects than rosin fluxes and their residues are more easily removable. Nevertheless, even with these benefits, due to environmental concerns with CFCs, a decline in SA flux usage is already being seen and will probably continue.

5.5.4 Low Solids (No-Clean)

Unlike the fluxes discussed in the previous sections, LSFs are formulated to leave minimal or no postsolder residue. As the name implies, this category of fluxes includes those that have a much lower solids content, typically from 2 to 5 wt%, compared to traditional fluxes.

Because the intent is to avoid cleaning, these LSFs should meet the LN (preferably) or at least MN activity designations specified in IPC-SF-818. They also should be able to meet the noncorrosive or compliant flux requirements in the telecommunication specification. LSFs are not presently allowed by the U.S. military.

The elimination of postsolder cleaning eliminates machine cost and maintenance and reduces materials and operating costs. Along with cost reductions, space savings and process simplification are also realized. Additionally, the need for any postsolder assembly steps is eliminated in cases where components, incompatible with the cleaning process, must be manually inserted and soldered (this postsolder assembly of components is often necessary when aggressive fluxes like WSF and SA, which require immersion cleaning, are used). Moreover, the associated temporary solder mask used as a solder resist for plated through-holes for postsolder assembly is not needed.

Concerns over EPA regulations, international restrictions, component qualification, and waste disposal are precluded by the elimination of cleaning. The environmental concerns are becoming increasingly important in making process decisions as more is discovered about the effect of these processes on the ecosystems and atmosphere.

An unexpected benefit has sometimes been realized in terms of improved soldering yields with LSFs relative to those with rosin fluxes.[19] In fact, as much as an order of magnitude drop in defect levels has been observed.[20] These defect level changes are also dependent on the product code and process parameters. Because of the lower viscosity and freer flow compared to rosin fluxes, LSFs can more easily wet all surfaces of the wiring side of a board, which is especially important when surface mount devices are wave soldered.[21]

Ideally, a flux that does not need postsolder cleaning should be noncorrosive, leave a nontacky, noncorrosive, colorless and/or minimal amount of residue, have sufficient activity to yield acceptable soldering, and be compatible with existing fluxing equipment. Unfortunately, it is difficult, if not impossible, to find a LSF that meets all of these characteristics. Instead, it might be necessary to modify the flux application process to realize all of the above.

By maintaining a low solids content and ensuring sufficient flux activity, the activator often outweighs the vehicle. Because of this unusual ratio, one cannot rely on the vehicle being able to encapsulate activator residues.

The solvent used in LSFs is typically isopropanol, the same one used in the other fluxes, the only difference being that there is a lot more solvent in a LSF than in a traditional solids flux. The vehicle in LSFs is a synthetic resin material, rosin, or modified rosin. Those that contain rosin are amber in color, rather than clear, and can, like higher solids content rosin fluxes, leave residues that impair bed-of-nails testing. One flux manufacturer developed a number of LSFs based on a vehicle that is more thermally stable than rosin, pentaerythritol tetrabenzoate (PETB).[22] Besides being the flux vehicle, PETB supposedly acts as an antioxidant also. Most LSFs are halide-free and contain only organic acid type activators. Usually, the ones that have halides also include rosin. Preferably, a LSF will be halide- and rosin-free. Many LSFs do not naturally foam due to their low solids content and the characteristics of the particular constituents themselves. LSFs that are foamable usually contain foaming agents that are either nonvolatile or volatile. Some nonvolatile foaming agents are glycols that remain along with the other flux residue after soldering. Like the glycols in WSFs, these can be hygroscopic, creating an environment for potential corrosion. Some LSFs include, along with isopropanol, small quantities of other solvents that promote foaming but volatilize during heating. In general, it has been seen that foaming agents may have a deleterious effect on copper mirror, SIR, and corrosion test results.

It has been noted that LSFs may be relatively difficult to remove with traditional CFC azeotropes; in fact, white residues can form so caution is necessary.[21] Consequently, it should be emphasized that there are certainly some cases when cleaning does not necessarily improve a situation and can, in fact, do harm.

It has been shown that the SIR decreases as a function of the original quantity of flux applied, inferring that excessive postsolder residue can cause electrical problems.[20] Results indicated that excessive LSF residues are detrimental to circuit integrity. In hindsight, it seems obvious that flux residues that are not removed after soldering and do not contain a large proportion of highly insulating water white rosin could cause problems. To better understand this phenomenon and confirm this hypothesis, SIR studies were modified such that the quantity of flux applied to comb patterns was carefully controlled. SIR results supported the accelerated ageing results and demonstrated that heavy deposits of flux residue cause high surface leakage currents. Figure 5-10 shows the dependence of SIR on flux quantity for eight different LSFs. It is clear that, although all show the same trend, the SIR for some fluxes are more strongly dependent on flux quantity than others.

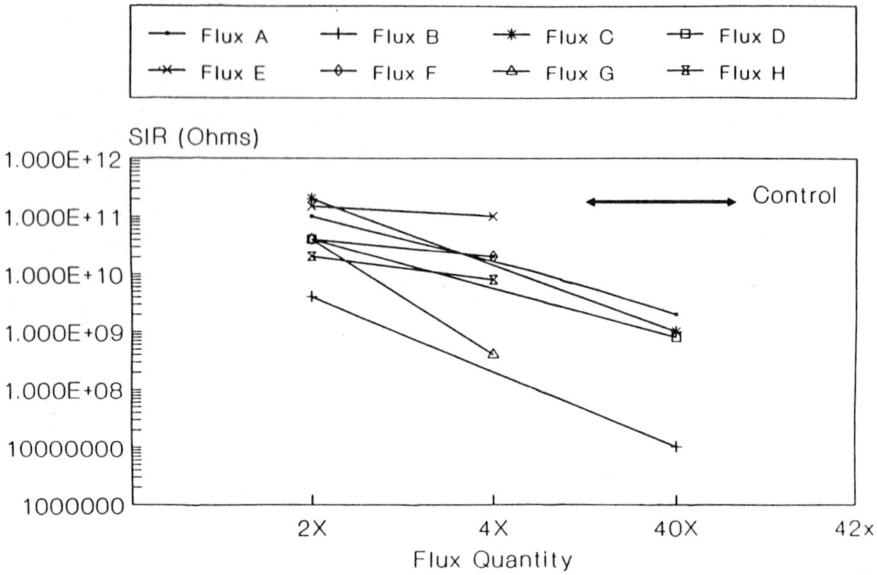

Figure 5-10 Average SIR data for comb patterns soldered with eight LSFs as a function of quantity of flux applied.[23]

The drawbacks of using LSFs compared to higher solids content fluxes are summarized below:

- Quantity is critical to ensure minimal residue on the one hand and sufficient fluxing activity on the other.
- Specific gravity is difficult to control.
- The effect of water absorption can be more pronounced.
- Foaming is difficult.
- The process window may be narrower.

Because of these drawbacks with LSF, the selection of application equipment and solder process parameters is important. However, if these are chosen wisely, the use of LSFs can yield enormous benefits.

5.5.5 Controlled Atmosphere Soldering

Controlled atmosphere soldering has also been called inert gas or fluxless soldering. Inert gas soldering or fluxless soldering are misnomers since sometimes more than inert gases are used, either reducing gases or vacuum atmospheres. Furthermore, liquids consisting of a flux solvent and activator are commonly used. Sometimes, for solder paste reflow, totally inert nitrogen atmospheres are used (see Chap. 2)

Recently, several companies have developed wave soldering systems that purportedly do not require the use of fluxes and therefore do not require postsolder cleaning. Another advantage that is cited is the decreased amount of solder dross. Dross is sometimes treated as a hazardous material. However, dross also can be sold to recycling companies that refine it for customers desiring lower grade solder; in this case, the amount of dross formed is not as problematic except that it must be removed more often.

One system, in particular, uses a combination of nitrogen and formic acid vapor to create an oxygen-free environment. However, to ensure proper through-hole soldering, a liquid consisting of an organic acid and an alcohol is applied by an ultrasonic atomizer. Although this material is not called a flux, it is, by all definitions, a flux and should be treated as such in terms of its impact on reliability. Another controlled atmosphere soldering system relies on vacuum pumps and chambers to eliminate oxygen in the wave soldering module. Although these new processes are promising, further evaluations of the reliability and compatibility impact of the gases and liquids that the PC boards encounter need to be performed.

5.6 FLUX APPLICATION METHODS

There are a number of ways to apply fluxes to printed circuit boards for wave soldering, the most common being wave, foaming, and spraying.

5.6.1 Wave

A wave fluxer is similar to the solder wave itself; the flux is pumped continuously through a wide nozzle located in an open reservoir or pool of flux. It is quite simple to operate and maintain. Another advantage is that virtually any liquid flux can be applied this way. Because a wave fluxer can apply excessive quantities of flux, it is sometimes followed by brushes or an air knife to remove excess.

With a wave fluxer, maintaining the original specific gravity requires adding solvent since the flux is in an open reservoir and exposed to the ambient. With most fluxes, the specific gravity can be controlled adequately with an automatic density controller that monitors the specific gravity and adds additional solvent or flux when the specific gravity gets out of range. However, with LSFs fluxes, this issue of changing specific gravity can be detrimental since small changes in the flux solvent content can greatly change the composition due to the small amount of solids (see the section "Application Issues for LSFs"). Water absorption is also an issue with wave fluxers due to open reservoir.

5.6.2 Foam

A foam fluxer consists of an open reservoir of flux containing a submersed porous stone through which air is pumped. With the right combination of air pressure, nozzle size and shape, and surface tension of the flux, air bubbles are created in the pool of flux and pushed through a nozzle, creating a foam of flux at the surface. Maintaining a level and uniform foam head with fine bubbles can sometimes be difficult since it depends on a number of factors. In recent years, modifications have been made to foam fluxers to make it easier to foam low solid fluxes; one manufacturer has designed a foamer with a dual aerator stone and an enlarged nozzle opening.

Like a wave fluxer, a foam fluxer causes changes in the specific gravity of the flux since the flux is not only exposed to the ambient but is aerated. So, either a density controller is necessary or, in the case of LSFs, caution is suggested. Another issue, especially important with foam fluxes, is the possibility of water absorption by the isopropanol or other hygroscopic flux constituents. The addition of water can increase solder defects, cause spattering and solder balls, and give a false confidence in specific gravity maintenance. That is, the addition of water can lead one to believe that the specific gravity is correct when, in fact, it is only in the right range due to the presence of an impurity (water).

Because of the nature of both wave and foam fluxers, fluxes applied with these methods can easily become contaminated with water, dirt from the printed circuit boards, conveyor fingers, or fixtures.

5.6.3 Spray

There are several different spray applications methods, high velocity spray, ultrasonic spray, and rotating drum spray, all with advantages and disadvantages, which are discussed in detail below.

A high velocity spray fluxer, which can have a nozzle similar to one on a spray cleaner, can propel excessive flux onto the top side of the circuit board. Moreover, units with large spray patterns can coat the inside of solder machines and can create maintenance problems. Typical nozzles create spray patterns with relatively higher concentrations in the center, making uniform application difficult. Also, because of the nozzle design and the evaporating solvent, typical spray nozzles are easily clogged. Sometimes, a solvent after-spray is discharged after the flux spray to prevent or, at least, minimize clogging.[24]

Because of the importance of controlling the deposition of LSFs, a novel means of applying them was developed and patented.[25] The theory behind ultrasonic atomization has been documented.[26] While the flow rate controls the amount of flux that will be dispensed, the power into the nozzle controls the amplitude of the sound waves into the nozzle, affecting the droplet size. A low velocity air stream directs the spray mist upward. A drawing of the flux mist being generated and subsequently shaped in the mixing chamber and directed upward is shown in Figure 5-11.

The ultrasonically generated spray fluxer offers several advantages over other fluxers, especially pertinent for LSFs.[20] First, flux quantity can be controlled by

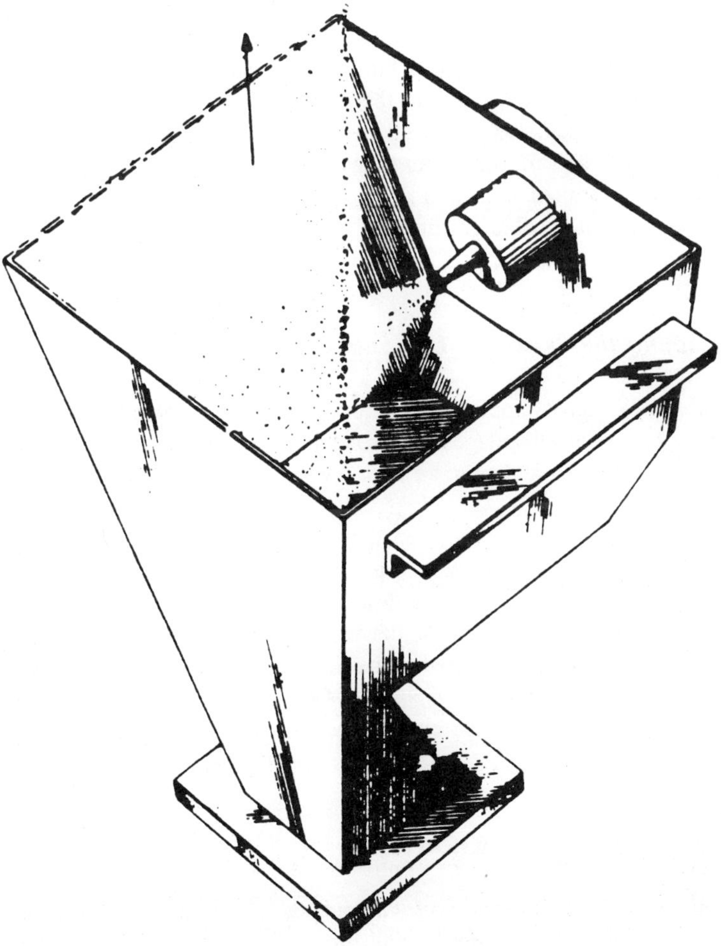

Figure 5-11 Ultrasonically atomized flux being shaped and directed in mixing chamber.[25]

adjusting the flux flow rate, ensuring product quality and reliability. Second, this is a closed system, preventing the evaporation of the flux solvent and thereby assuring constant flux composition, especially important for LSFs. Moreover, the closed system prevents the possibility of water absorption. Because of the nature of the low velocity flux mist, a minimal amount of flux is deposited on the top side of the board, important for LSFs that can degrade the SIR more severely when not heated fully to soldering temperature.

5.6.4 Rotating Drum Spray

Another spray method that has been in existence for a number of years has recently been touted as an application method for LSFs. It consists of a partially immersed mesh drum rotating in a flux tank. Located on its center axis is an air knife, which, in combination with the rotating drum, creates a fine flux spray. The quantity of flux that is sprayed depends on the mesh size, the rotation speed, conveyor speed, and the air velocity of the air knife. Therefore, the quantity of flux can be controlled with this application method. However, the issues regarding control of specific gravity and the possibility of water absorption exist with this method since there is an open flux tank.

5.6.5 Application Issues for Low Solids Fluxes

A number of different application techniques are commercially available, each with its own list of advantages and disadvantages. From a flux quantity standpoint, a wave fluxer applies more than a foam fluxer, which applies more than a spray fluxer.

Already discussed is the fact that the amount of LSF applied may affect the SIR and ultimately the reliability of a printed circuit board. Consequently, it is important to control the quantity of flux applied. Normally, a board's cleanliness is controlled by the efficiency of the cleaning material and cleaning process; here, the control must be directed at the flux application process step since there is no postsolder cleaning. As important is the amount of flux that is deposited onto the top side of the printed circuit board since partially heated LSF residues may be corrosive.

5.7 PROCESS ISSUES ASSOCIATED WITH RELIABILITY

There are a number of process issues associated with the type of flux used. These issues include flux residue, solder ball formation, top-side fillet formation, and conformal coating compatibility, all of which are discussed below.

5.7.1 Flux Residue

Depending on the flux type, flux residue may or may not be a reliability concern. The more aggressive fluxes leave behind more corrosive residue and should therefore be removed properly. Less active fluxes or those that meet certain tests requirements without removing their postsolder residue may be removed for aesthetic reasons or to avoid problems with testing, solder balls, or conformal coating.

5.7.2 Solder Ball Formation

Material incompatibilities can cause excessive solder balls and/or other solder-type residue. The number of solder balls is a function of the type of flux used, the type of solder mask on the printed circuit board, and the soldering conditions. The presence of solder balls becomes a real issue when there is no postsolder cleaning (e.g., some rosin fluxes; LSFs, controlled atmosphere soldering) that could remove the balls along with the other residue. In comparisons of a number of solder mask types and flux types (rosin, SA, WSF, and LSF), the industry experience is that the number of solder balls is greater with dry film solder masks and with low LSFs. Work has been done showing that the following factors can affect solder ball formation: dry film solder mask type, printed wiring board cleanliness, surface roughness, and flux type.[27] Khutoretsky's study included dry film solder masks and both water soluble and rosin fluxes; the results indicated that both the flux residue and the solder mask polymer can retain solder balls on the board surface. Photomicrographs in Khutoretsky's paper illustrate this "connecting chain" phenomenon.

5.7.3 Top-Side Fillet Formation

Top-side fillets are required in some specifications, the definition of which varies amongst specifications. Some specifications simply require positive wetting on both the component leads and the barrel and a certain amount of barrel fill, whereas other specifications might require a fillet on the top side that looks almost identical to that on the bottom side. The exact amount of solder necessary to create a reliable joint is debatable and obviously is affected by the conditions under which a printed circuit board will ultimately be exposed. That is, an electronic product that is made for military aircraft will be exposed to much harsher temperature and vibration conditions than a household consumer product.

The formation of sufficient fillets on the component side of the printed circuit board is affected by the board and component temperatures through the solder wave, their solderability, board thickness, the component-to-hole aspect ratio, the solder wave height, the conveyor speed, the flux application method, the activity of the flux, and the solids content of the flux. If the circuit board and components are not adequately heated, the solder will not rise properly in the holes. Moreover, if solderability is poor or the activity of the flux is not adequate, the solder will not readily wet. Differences in top-side fillets have been observed between a 37 percent solids rosin flux and a 15 percent solids rosin flux; additionally, other parameters also affected top-side fillet formation in this study including conveyor speed, height of flux foam, and the preheat profile.[28] For LSFs, problems with top-side fillet formation can be more pronounced.

In summary, the solder parameters, the board design, the solderability, and the type of flux can all affect top-side fillet formation.

5.7.4 Conformal Coating Compatibility

Conformal coating is commonly applied after soldering to printed circuit boards that will be exposed to harsh or extreme conditions. These include military applications and automobile electronics. Some flux residues underneath these coatings can cause blistering, peeling, or measling. Any moisture or hygroscopic residues underneath the coating can cause these problems. Obviously then, coating compatibility with a flux and its residue should be taken into account. Although rosin fluxes are usually associated with product requiring conformal coating, reports by members of the electronics industry indicate that coatings exist that are compatible with other flux types including WSF and LSF; cleaning before the coating process is usually necessary and depends on the type of flux used.

5.8 NON-LIQUID FLUXES

5.8.1 Core Solder Material

Instead of liquid fluxes that are used for wave soldering, hand soldering is usually performed with solder wire that contains a core or several cores of flux material. This material is solid or semisolid at room temperature, but it flows and is active at soldering temperatures to promote adequate wetting of the solder. Rosin-based core solder has been the standard in the electronics industry but in the last few years there has been an increase in water soluble core flux and, more recently, LSF core solders. Activators may be organic acids and/or organo-halides.

Core materials can be tested just like liquid fluxes by obtaining the core material from the manufacturer or extracting it from the solder wire. (Since using the core material is easier than extracting it, this is preferred.) However, although some specifications state that the core material may be removed by melting the solder wire and then testing, this practice is emphatically not recommended. By melting the solder, the core material is also heated and can thus react with the solder or simply undergo changes by itself. As a consequence, the material tested will have undergone one more heating step than allowed or required in tests such as copper mirror or SIR. It is best to use the unheated core material, apply it to the appropriate substrates, and perform the tests as described earlier.

5.8.2 Solder Paste Material

Solder pastes are often used for attaching surface mount devices. The paste is stenciled or screen-printed on pads on the printed circuit board, the components are placed onto the pads, and the whole assembly is heated so that the solder paste is reflowed. The solder paste is a combination of solder powder, rheology modifiers, and normal flux constituents such as high temperature solvent, a vehicle, and activators. It is difficult to find a good solder paste because of the number of requirements imposed upon these materials. That is, along with acceptable flux and solder materials, a paste must also be formulated to include other properties such as tackiness, printability, and minimal slumping. Therefore, solder pastes are usually tested like other fluxes for chemical reliability purposes and then undergo further tests to ensure the presence of certain physical properties. The commercial electronics industry specification for solder pastes is IPC-SP-819.

Solder pastes based on any of the four major flux types exist. However, most solder paste used today is rosin-based with minimal or no halide activators due to cleaning concerns and question regarding need for higher activity. Since paste is placed exactly where it is needed and the type of joint that is made is so different for surface mount than for through-hole, the activity of a solder paste is not as important as it might be for a liquid wave soldering flux. Because of the additional materials in solder pastes and the localized concentrations of paste residue compared to wave soldering flux residue, cleaning is more difficult and more of a concern than cleaning after wave soldering.

Removal of solder paste residue is sometimes, but not always, necessary for the same reasons as for liquid fluxes. However, with the increasing problems with cleaning, lower solids content solder pastes are being developed. Water soluble solder pastes are available for applications such as mixed technology where water soluble liquid flux and solder paste residue will be removed in a water cleaning system. The concern is that the residue be removable after the heat of soldering and that the electrical properties (SIR) of the assembly have not been excessively degraded. Even though the flux-type ingredients may be water soluble, the rheological agents are not necessarily so. Hwang[29] offers an extensive discussion about solder pastes in general (also, see Chapter 2).

For liquid fluxes, core solders, or solder pastes, less flux solids means less residue. An effective way to reduce flux residue when soldering with cored wire solder is to use smaller core sizes. Typically the standard 3 percent flux (by weight) can be reduced to 2, 1, or 0.5 percent. With less flux being used, solderability and heat control become more important to ensure that reliable solder connections are made. For solder pastes, it is difficult to truly make a lower solids content paste and also maintain the proper physical properties. That is, the less nonmetal material that exists, the more solder powder that is present; obviously, the printability and tackiness properties will then be affected. Thus, it seems a paste that leaves less or no residue should contain the usual proportion of solder powder (approximately 90 wt %) but include other materials that decompose during reflow. Low solids pastes reflowed in controlled atmospheres are being investigated.[30]

5.9 TRENDS

In 1990, the major push is away from CFC and, if possible, toward the elimination of cleaning altogether. There are several possible ways to avoid cleaning including the use of noncorrosive rosin fluxes, the use of LSFs, and the use of controlled atmosphere soldering. According to Szymanowski,[31] controlled atmosphere (inert gas) soldering and low LSFs offer the most attractive alternatives to CFCs. In those cases where cleaning is still necessary, HCFCs, terpenes, or aqueous cleaning are the choices. However, HCFCs are an interim solution only, since their use may be restricted in the future. Presently, there are still no soldering methods that are truly "fluxless" and there may never be any that do not use some kind of fluxing material. But with LSFs and controlled atmosphere soldering, at least, we are getting closer to that ideal.

In the future, improvements in wave soldering LSFs will be seen as well as improvements in low solids solder core wire and low solids solder pastes.

Other types of fluxes can and will still be used with appropriate non-CFC cleaning media. For rosin flux residue removal, semiaqueous cleaners are recommended over CFCs and saponifiers for environmental reasons. If a completely aqueous system is preferred, water soluble flux can be used if the assembly allows complete removal of the flux residue with water. However, the addition of a cleaning process brings additional expense, maintenance, capital equipment, and floor space. With the ever-increasing restrictions on the wastes that can be expelled from factories, minimizing or eliminating cleaning altogether is certainly the preferred direction and the trend.

REFERENCES

1. IPC, ANSI/IPC-SF-818, "General Requirements for Electronic Soldering Fluxes," Lincolnwood, IL: Institute for Interconnecting and Packaging Electronic Circuits (IPC), 1988.

2. Bellcore, TR-TSY-000078, Issue 2. "Generic Physical Design Requirements for Telecommunications Products and Equipment," Red Bank, NJ: Bell Communications Research, Inc., 1988.
3. Guth, L. A., "Low Solids Flux Material Characterization Studies," IPC Technical Paper IPC-TP-735, 1988.
4. Bellcore, TA-TSY-000078, Issue 2. "Proposed Revision of TR-TSY-000078: Generic Physical Design Requirements for Telecommunications Products and Equipment," Red Bank, NJ: Bell Communications Research, Inc., 1987.
5. Zado, F. M., "Electrical/Electronic Reliability Considerations in Modern PWB Manufacturing and Assembly Operations," *Printed Circuit World Convention III,* WCIII-70, 1984.
6. Gorondy, E. J., "Surface Insulation Resistance—Part I: The Development of an Automated SIR Measurement Technique," IPC Technical Paper IPC-TP-518, 1984.
7. Gorondy, E. J., "Surface/Moisture Insulation Resistance (SIR/MIR)—Part III: Analysis of the Effects of the Test Parameters and Environmental Conditions on Test Results," IPC Technical Paper, 1988.
8. Chan, A. S. L. , and T. A. Shankoff, "Interrelating Surface Insulation Resistance Test Patterns," *Circuit World,* **15**(4), 1989, pp. 34–38.
9. Zado, F. M., "Increasing the Soldering Efficiency of Noncorrosive Rosin Fluxes," *The Western Electric Engineer* (1), 1983, pp. 22–29.
10. Reagor, B. T., and C. A. Russell, "A Survey of Problems in Telecommunication Equipment Resulting from Chemical Contamination," *IEEE Transactions on Components, Hybrids, and Manufacturing Technology,* **CHMT-9**(2), 1986, pp. 209–214.
11. Dickinson, D. A., L. A. Guth, and G. M. Wenger, "Advances in Cleaning of Electronic Assemblies," *Nepcon West 1989,* 1989.
12. Chung, B. C., M. P. Cassidy, and G. W. Graham, "Evaluation of Flux Performance, Cleaning and Reliability," *The Western Electric Engineer,* (1), 1989, pp. 30–39.
13. Zado, F. M., "Effects of Non-Ionic Water Soluble Flux Residues," *The Western Electric Engineer,* (1), 1983, pp. 40–48.
14. Wargotz, W. B., L. A. Guth, and C. V. Stroud, "Quantification of Cleanliness Beneath Surface Mounted Discretes Assembled by Wave Soldering," *Printed Circuit World Convention IV,* WCIV-72, 1987.
15. Danford, A., and P. Gallagher, "SMD Cleanliness in an Aqueous Cleaning Process," *Nepcon East 1987,* 1987, pp. 246–255.
16. Aspandiar, R., A. Piyarali, and P. Prasad, "Is OA OK?" *Circuits Manufacturing,* April 1986, pp. 29–36.
17. Kenyon, W. G., "Synthetic Activated (SA) Flux Technology: Development, Commercialization, Benefits and Future Applications," *Proceedings of Internepcon Japan '86,* 1986.
18. Hayes, M., and C. Hood, "Alternative Cleaning Methods: One Possibility," interview in *Printed Circuit Assembly,* September 1989, pp. 33–35.
19. Sholley, C., "New Fluxes Eliminate Cleaning of PCB's," *Electronic Manufacturing,* March 1989, pp. 32–33.
20. Guth, L. A., "Low Solids Flux Technology for Solder Assembly of Circuit Packs," *Proceedings of the Electronic Components Conference,* 1989, pp. 748–753.
21. Toubin, A., "Low Solids Content Fluxes," *Circuit World,* **15**(2), 1989, pp. 17–18.

22. Rubin, W., "Some Recent Advances in Flux Technology," *Brazing and Soldering,* (2), Spring 1982, pp. 24–28.
23. Guth, L. A., "Flux Selection and Application to Negate Cleaning Needs," presentation at Nepcon East, Boston, MA, June 14, 1989.
24. Chung, B. C., W. A. Jancuk, and V. A. Corsaro, "Aqueous Detergent for Removing Rosin Fluxes," *The Western Engineer,* (1), 1983, pp. 62–68.
25. Fisher, J. R., L. A. Guth, and J. A. Mahler, "Method and Apparatus for Applying Flux to a Substrate," U.S. Patent 4,821,948, April 18, 1989. Assigned to AT&T.
26. Berger, H., "Ultrasonic Nozzles Atomize Without Air," *Machine Design,* July 21, 1988, pp. 58–62.
27. Khutoretsky, M., "An Evaluation of Solder Balls on Printed Wiring Boards with Dry Film Solder Resist," *Proceedings of AMS's Third Conference on Electronic Packaging: Materials and Processes & Corrosion in Microelectronics,* 1987, pp. 225–233.
28. Wittmer, P., Conversation with the author. Unpublished Magnavox data, 1989.
29. Hwang, J. S. *Solder Paste in Electronics Packaging,* New York: Van Nostrand Reinhold, 1989.
30. Morris, J. R., "No Clean Alternatives for Solder Paste Reflow," *Proceedings of SMART VI Conference,* (2), 1990, pp. 97–106.
31. Szymanowski, R. A., "Fluxing Options for CFC Elimination," *IPC Technical Review,* April-May 1989, pp. 19–23.

ADDITIONAL READINGS

Deram, B., "Considerations for Use of No-Clean Fluxes in Soldering PCB's," *Electronic Manufacturing,* February 1989, pp. 32–34.

Elliott, D., "How to Avoid Problems with Wave Soldering and IR Reflow," *Surface Mount Technology,* October 1989, pp. 47–54.

Johnson, R. O., "Solderability Issues for Hand Soldering," *Printed Circuit Assembly,* June 1989, pp. 30–32.

Klein Wassink, R. J., *Soldering in Electronics,* Ayr, Scotland: Electrochemical Publications, Ltd., 1984.

Wade, R. L., "A View of Low-Solids Fluxes," *Printed Circuit Assembly,* March 1989, pp. 31–34.

6

Scanning Electron Microscopy and Energy Dispersive X-ray (SEM/EDX) Characterization of Solder—Solderability and Reliability

James L. Marshall

6.1 SCANNING ELECTRON MICROSCOPY/ENERGY DISPERSIVE X-RAY ANALYSIS

SEM/EDX is a powerful way to characterize solder in resolving solderability and reliability problems. SEM/EDX is a combination tool, with two instruments working in partnership. These two instruments operate simultaneously to complement each other's data acquisition, guided by the instrumental operator. Scanning electron microscopy (SEM) provides an image (i.e., morphological information or surface features) on a magnified scale—X10 to X100,000, although the usual range is perhaps X50 to X5000.[1] Energy dispersive x-ray (EDX) determines the elemental composition of an area, with a sensitivity of perhaps 0.1 to 1 percent composition and with a spatial resolution of 1 μm.[2] EDX is commonly used for elements with atomic number ⩾11 (sodium), but thin window EDX systems can also detect elements with atomic number ⩾5 (boron). With these two instruments operating together, the instrumentalist can scan areas of potential interest, zoom in with higher magnification, and determine elemental compositions in selected areas of interest.

6.2 OTHER METHODS—WDX

A less commonly used method of x-ray analysis is wavelength dispersive x-ray (WDX). One advantage of WDX is a much higher resolution, permitting the clean separation of poorly resolved signals in EDX, such as indium and tin. This higher resolution for WDX permits a higher sensitivity and the detection of impurities that otherwise might be masked by major constituent peaks. Another advantage of WDX is the facile detection of elements with low atomic number

($\geqslant$5). Instrumentally, however, WDX involves a more complex mechanical operation to select appropriate wavelengths (EDX uses a solid-state detector).[3] WDX also involves longer analysis times since the spectrometer is mechanically scanned throughout its wavelength range—consuming minutes or even hours versus seconds or minutes for EDX.[2] EDX therefore is more useful for rapid qualitative analysis, particularly for unknown specimens. Since EDX is more commonly used than WDX in SEM, discussion in this chapter will be limited to SEM/EDX.

6.3 DETECTION MODES

Two detection "modes" are available for SEM—secondary electron emission and backscattered electron emission.[1] The secondary image generates high resolution images and is used for morphological studies. Hence, this mode of detection is appropriate for minimally prepared samples for rapid analysis. For example, a solder joint clipped from a circuit board can be quickly mounted and inserted into a spectrometer and within minutes can supply important information such as:

Where is the fracture?
Through the bulk solder?
Or is the solder peeling from the component lead?
What is the composition of the solder?
Of the component lead?

In contrast, the backscattered mode of detection emphasizes elemental identification and presents higher visual contrast for compositional differences—dark images for low atomic number, light images for high atomic number. The resolution for backscattered images is lower because the sampling volume of detected electrons is much greater, perhaps 0.5 μm.[1] Although the backscattered mode may be used for rapid-mount samples, it is more commonly reserved for carefully prepared specimens that present a flat, polished surface. Backscattered images are particularly useful in distinguishing different phases in metallurgical samples, for example, the differentiation among various phases of copper-tin elements and intermetallic compounds.[1] In this chapter, both secondary and backscattered images will be presented to illustrate the suitability of each method in different situations.

6.4 SAMPLE PREPARATION

For morphological images of solder specimens, a quick mount can customarily be used for sample preparation because solder is conductive. For a cross section, a flat polished specimen must be prepared, using the steps of potting, cutting, grinding, and polishing.[2] The specimen is then gold- or carbon-coated to render the sample conductive. Care must be taken in all these steps to prevent chemical artifacts. For example, chemical etching can change the composition of the surface and must be avoided. Smearing must be carefully avoided to prevent misinterpretation of results; smearing can occur during the grinding or polishing steps. If materials of different hardness exist in the sample (such as solder on Al_2O_3), the solder can be preferentially removed in a lapping operation, producing a specimen with roughened solder recessed beside the durable ceramic. To prevent such useless specimens, low pressure grinding and polishing of short duration should be used. It may be useful to the instrumentalist to compare secondary images with backscattered images or EDX to verify agreement of observed structures, remembering that secondary images are more superficial (which would include smearing), whereas backscattered images or EDX disclose data from deeper portions. The instrumentalist may have to perfect diverse techniques for different situations. For techniques of sample preparation, the reader is referred to excellent literature distributed by Buehler, Ltd.[4–8]

6.5 DIFFERENT PHASES IN ALLOYS

Different compositional phases, identified in phase diagrams,[9] can be differentiated in SEM/EDX. Backscattered images, which emphasize composition, are particularly useful to distinguish phases. EDX data are sufficiently quantitative to determine compositions and thereby identify phases that differ only in stoichiometry (e.g., the copper-tin phases Cu_3Sn and Cu_6Sn_5 or the gold-tin phases $AuSn_4$, $AuSn_2$, AuSn). Phases in alloys generally have specific stoichiometries, although the composition of a given phase may have a wide range. These phases are sometimes designated by Greek letters. However, binary compositions of different combinations of elements may use the same Greek designation, and ambiguity may result (e.g., two different α phases may exist in the same sample). To avoid confusion, the description of elemental and compound phases will be done by formula (e.g., Ag_3Sn, C_6Sn_5, Sn) with the understanding that stoichiometry is rarely exact.[10]

6.6 INTERMETALLICS

The term "intermetallics" in this chapter has the special meaning generally attributed in solder technology (i.e., an alloy formed between a solder and the base metal during the soldering operation[11]). This intermetallic forms the bond between the solder and the substrate. Intermetallics are generally more brittle than the base metal or the solder. Sometimes the intermetallic can grow needles or prisms into the solder and can even fracture and drift into the solder.

6.7 SCOPE OF THE CHAPTER

This chapter will present typical SEM/EDX examples as an interpretative guide for solving problems for the instrumentalist. We will assume an instrumental system commonly used in solder work, a SEM/EDX with thick window (atomic number ⩾11). For more detailed discussions of SEM/EDX instrumentation and techniques, the reader is referred to specific treatises.[12,13,1–3]

6.8 SEM/EDX CHARACTERIZATION—GENERAL

6.8.1 *Tin-Lead Solders*

In a cross section of a typical 60% Sn-40% Pb solder joint to copper (Fig. 6-1), the darker tin and the lighter lead domains are easily visible.[11,14] The intermetallic[19] is not so easily seen and it takes more effort to clearly distinguish this Cu_6Sn_5 compound (Fig. 6-2). For a quick-mount sample (external appearance), these tin and lead domains can be easily seen in the secondary mode of detection if the surface of the solder is clean and not heavily oxidized (Fig. 6-3).

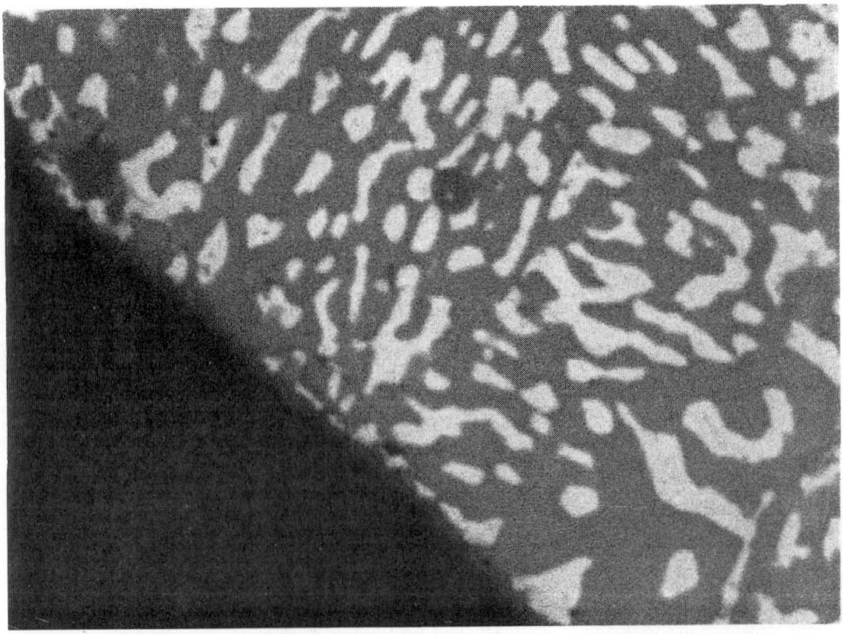

Figure 6-1 Cross section of 60% Sn-40% Pb solder joint. Lower left, dark region: copper substrate. Upper right: darker domains, tin; lighter domains, lead. X2000, backscattered.

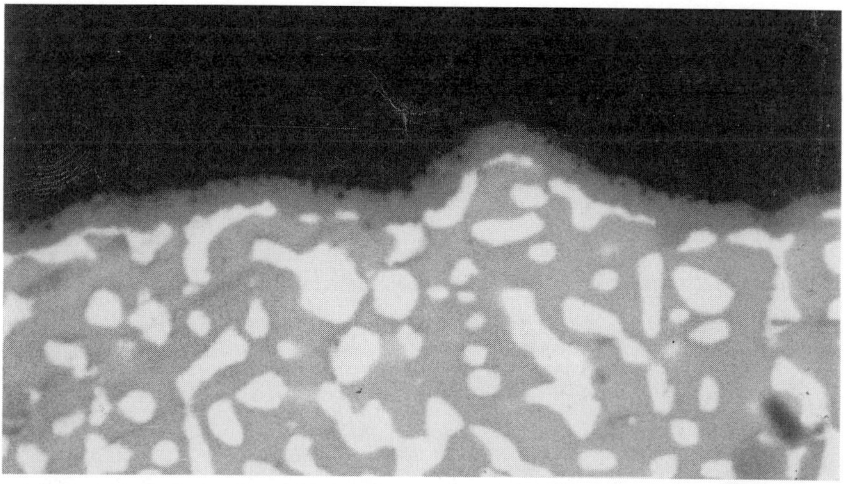

Figure 6-2 Cross section of 60% Sn-40% Pb solder joint, with careful polishing and careful adjusting of the instrumentation, to manifest the copper-tin Cu_6Sn_5 intermetallic, about 1 μm thick. Top half, dark region: copper substrate. Bottom half: darker domains, tin; lighter domains, lead. Intermediate region, Cu_6Sn_5 intermetallic. X2000, backscattered.

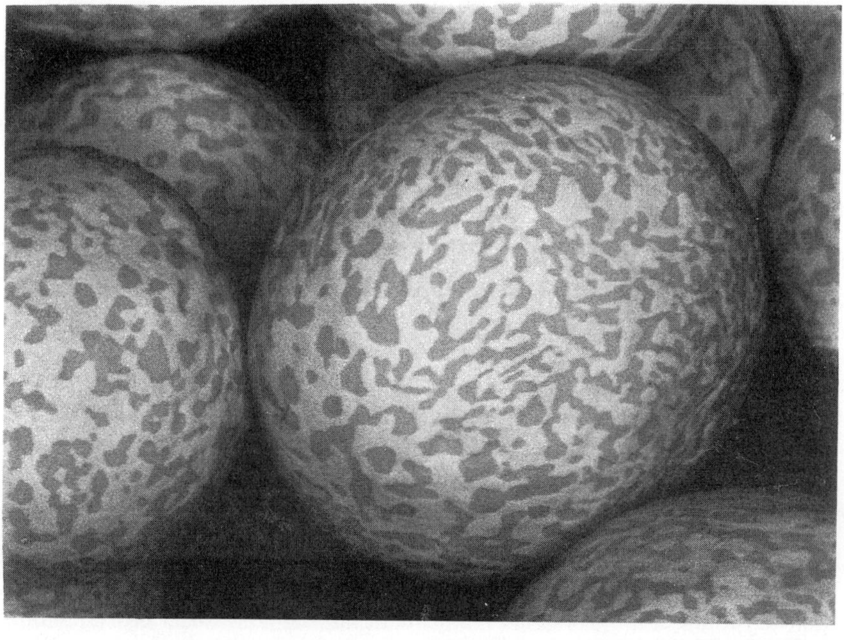

Figure 6-3 Cleaned solder paste, 60% SN-40% Pb. X200, secondary.

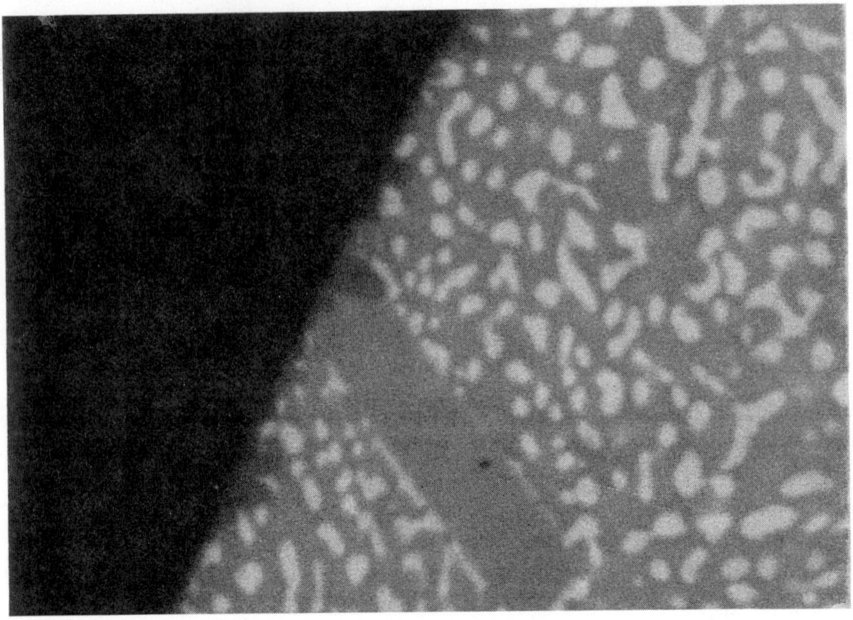

Figure 6-4 Cross section of 62% Sn-36% Pb-2% Ag solder joint. Left, dark region: copper substrate. Right: darker domains, tin; lighter domains, lead. The crystal dominating the view is Ag_3Sn, growing from the intermetallic. Because of extended heat treatment, some Cu_6Sn_5 is growing from the interface into the solder (darker small areas, e.g., top of connection of Ag_3Sn crystal to substrate interface). X2000, backscattered.

6.8.2 Two Percent Silver Solder

Ordinarily, the appearance of 62% Sn-36% Pb-2% Ag solder is virtually identical to 60% Sn-40% Pb or 63% Sn-37% Pb solder, but sometimes one may observe a silver-rich crystal[9] (Ag_3Sn) growing from the solder joint interface (Fig. 6-4). Chunks or crystals of binary alloys growing into the solder are indications of extended heat treatment (also, see Figs. 6-28 through 6-30).

6.8.3 Gold- and Silver-Based Solders

Like tin-lead, silver solder (95% Sn-5% Ag) possesses two phases.[9] With a different ratio of phases, however, the appearance is quite unlike that of tin-lead (Fig. 6-5). Gold can form many phases with tin, but gold solder possesses only one phase with the composition 80% Au-20% Sn (Fig. 6-6).[9]

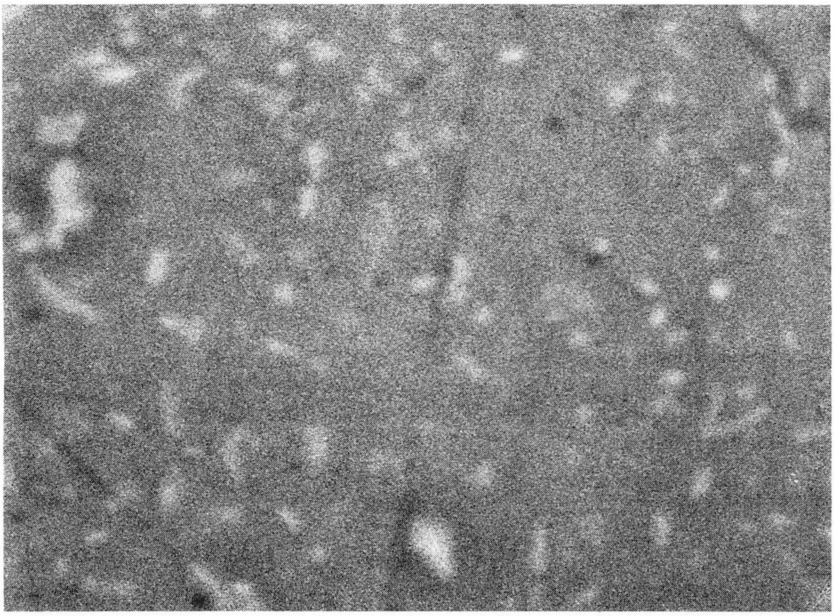

Figure 6-5 Cross section of 95% Sn-5% Ag solder. The light domains are silver-rich phases. X7800, backscattered.

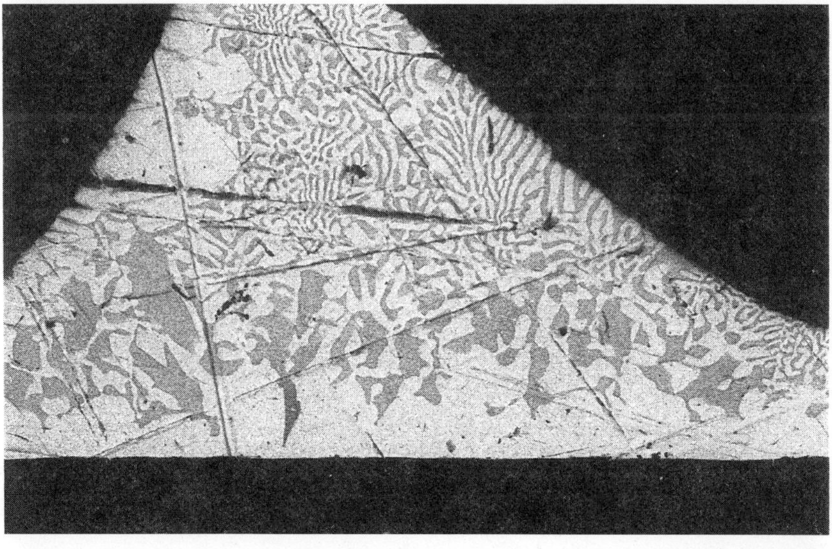

Figure 6-6 Cross section of gold solder joint. Upper left, dark region: silver wire. Bottom, dark region: copper substrate. Solder: lighter regions, gold; darker regions, 80% Au-20% Sn solder. X100, backscattered.

6.8.4 Indium Solders

The common solders of indium include lead and tin alloys. Indium-lead solder of a given composition (e.g., 50% In-50% Pb) possesses only one phase[9] and a cross section of this solder is featureless. For indium-tin (Fig. 6-7), it is very difficult to distinguish between the indium-rich and the lead-rich phases[9] because these two elements have almost identical atomic numbers. Hence, the intensities of the two phases are virtually identical in SEM micrographs, even in the back-scattered mode of detection. Obtaining a dot map, therefore, is laborious (Fig. 6-8). Figures 6-7 and 6-8 point out the caution that must be exercised during interpretation of SEM micrographs: artifacts in Figure 6-7 could have been misidentified as a different phase.

Although the SEM of indium solders is not very informative compositionally, the indium-copper or indium-tin-copper intermetallic[9] can be easily seen (Figs. 6-9 and 6-10).

The resoldering of an indium solder joint with tin-lead solder can result in a weakened joint (Fig. 6-11). It may appear superficially (i.e., optical examination)

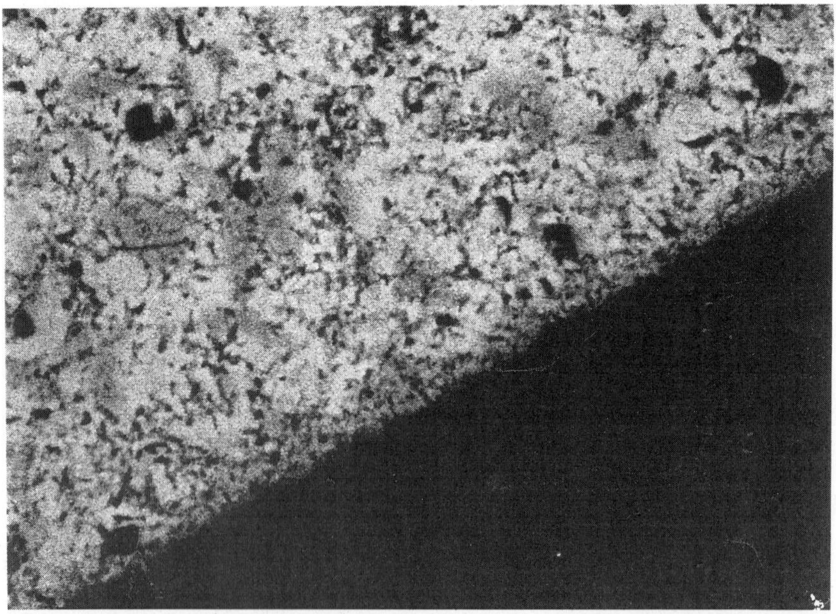

Figure 6-7 Cross section of 50% In-50% Sn solder joint. Lower right, dark region: copper substrate. Upper left: solder. The darker and lighter regions of the solder do not identify the two different phases of the indium-tin solder but are artifacts. X2000, backscattered.

Figure 6-8 Tin dot map of Figure 6-7. Darker regions of solder: tin-rich phases. Lighter regions of solder: indium-rich phases.

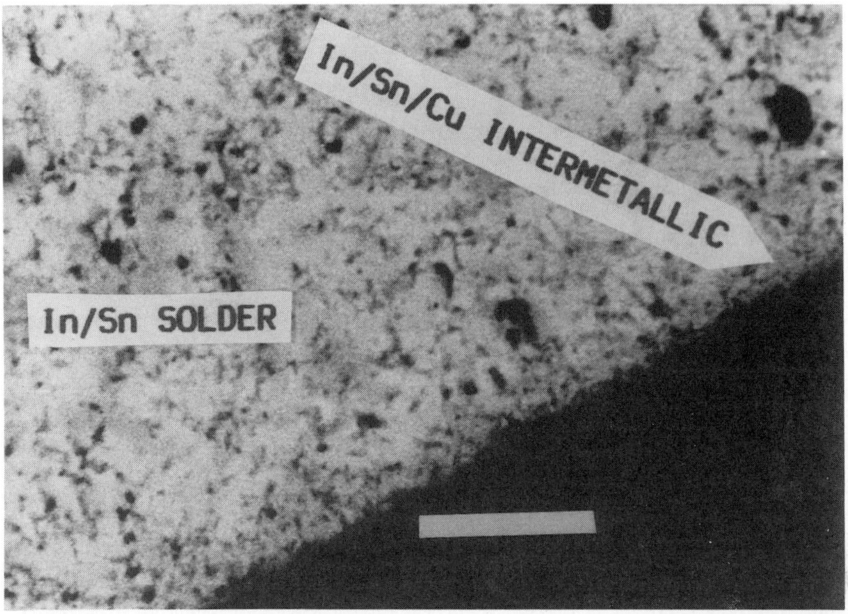

Figure 6-9 Expansion of Figure 6-7. Lower right, dark region: copper substrate. Between the solder and substrate lies the indium-tin-copper intermetallic, about 2 μm thick. X4000, backscattered.

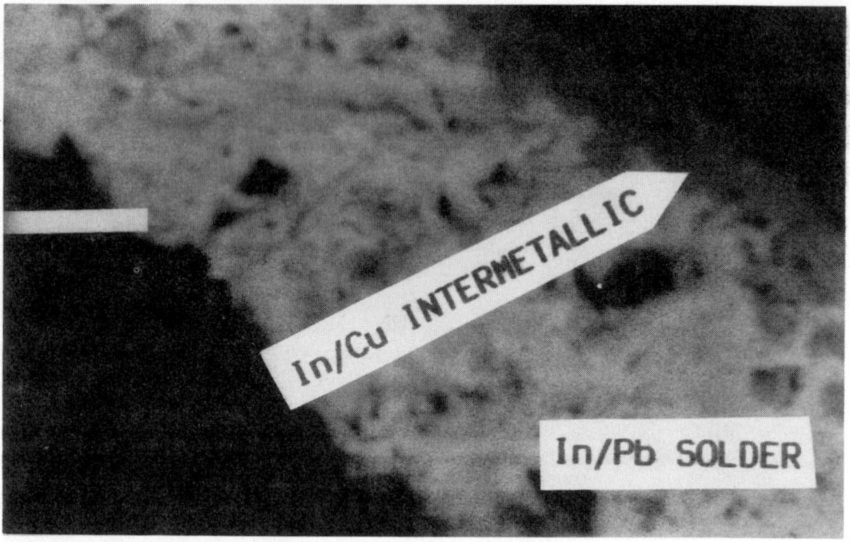

Figure 6-10 Cross section of 50% In-50% Pb solder joint. Upper right, dark region: copper substrate. Center region: 50% In-50% Pb solder joint. Inbetween is the indium-copper intermetallic, about 2 μm thick. X4000, backscattered.

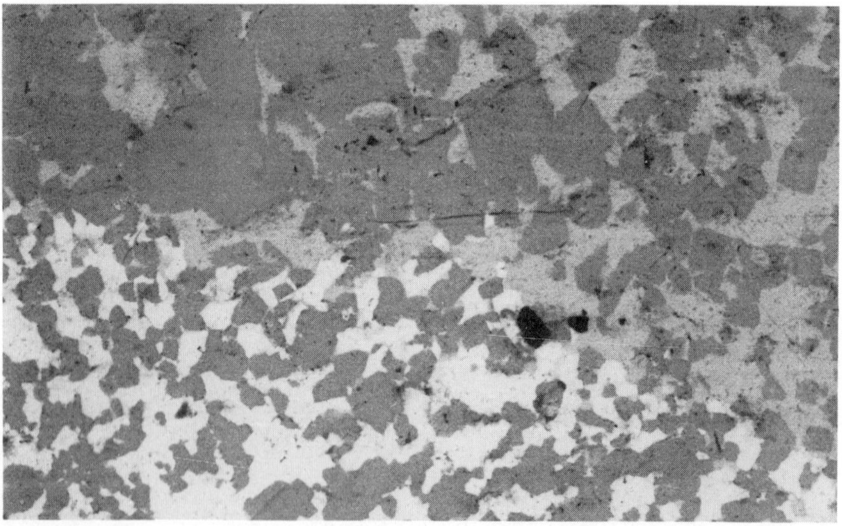

Figure 6-11 Cross section of 50% In-50% Sn solder joint resoldered with 60% Sn-40% Pb. The original indium-tin solder joint was positioned at the top, and the 60% Sn-40% Pb solder intrusion was from the bottom. Lightest phase, at bottom: lead-indium, high lead concentration. Darkest phase, scattered throughout view: indium-tin. Gray phase, mostly at top: lead-indium, low indium concentration. Commonly, only the darkest and lightest phases are observed in a resoldered joint. X1000, backscattered.

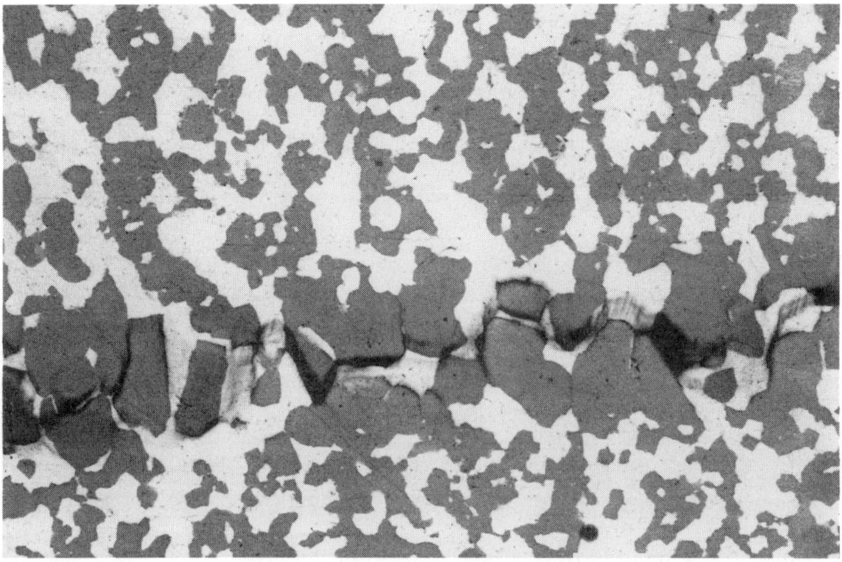

Figure 6-12 Cross section of failing 50% In-50% Sn solder joint resoldered with 60% Sn-40% Pb. Lightest phase: lead-indium. Dark phase: indium-tin. X1000, backscattered.

that the solder joint is normal, but SEM shows an unusual combination of phases that later may promote failure. In Figure 6-11, three different phases are shown, but more commonly only two phases of indium-tin and indium-lead are observed. A fracture develops between these two domains and propagates through the sample (Fig. 6-12). The mechanism of this failure is similar to that for tin-lead (see below)—a fracture develops mainly at the interface between different domains—but the process is accelerated for indium-tin-lead, which are compositions of inferior fatigue resistance.

6.8.5 Bismuth Solders

The appearance of 43% Sn-43% Pb-14% Bi (Fig. 6-13) is very similar to that of tin-lead solders, with two phases—tin and lead-bismuth.[9] This solder has two very attractive properties that ordinarily are mutually exclusive—a low melting point (ca. 144 to 163°) and good fatigue resistance properties.[15] Other bismuth solders include tin-lead-bismuth (16%-32%-52%), melting at 96° and 42% Sn-58% Bi melting at 138°. Although these latter two solders previously had a good reputation for reasonable shear strengths and fatigue properties,[16,17] they suffer from the disadvantage that the addition of tin-lead (a common occurrence during soldering to tin or tin-lead coated substrates) can actually lower the melting point (by approaching the tin-lead-bismuth ternary eutectic) and cause rapid failure.[15]

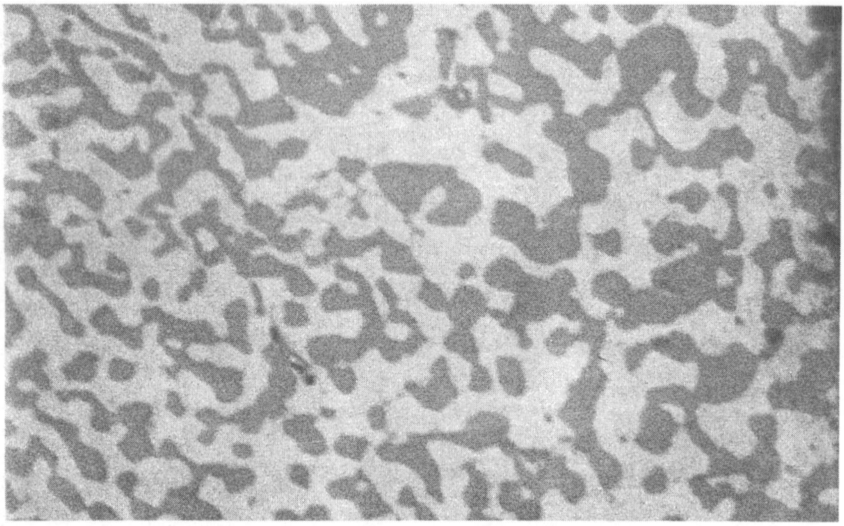

Figure 6-13 Cross section of 43% Sn-43% Pb-14%Bi. Darker domains: tin. Lighter domains: lead-bismuth. X2000, backscattered.

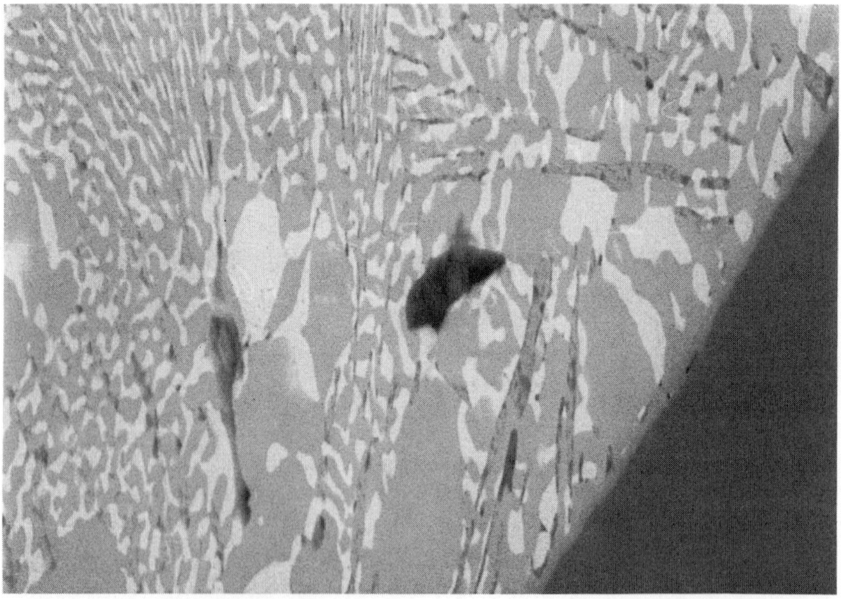

Figure 6-14 Cross section of solder joint to cadmium-plated brass. Lower right, dark region: brass. Solder region: tin and lead domains, heavily admixtured with cadmium. Interface between substrate and solder: copper-cadmium. Cadmium needles are reaching from the interface into the bulk of the solder. X600, backscattered.

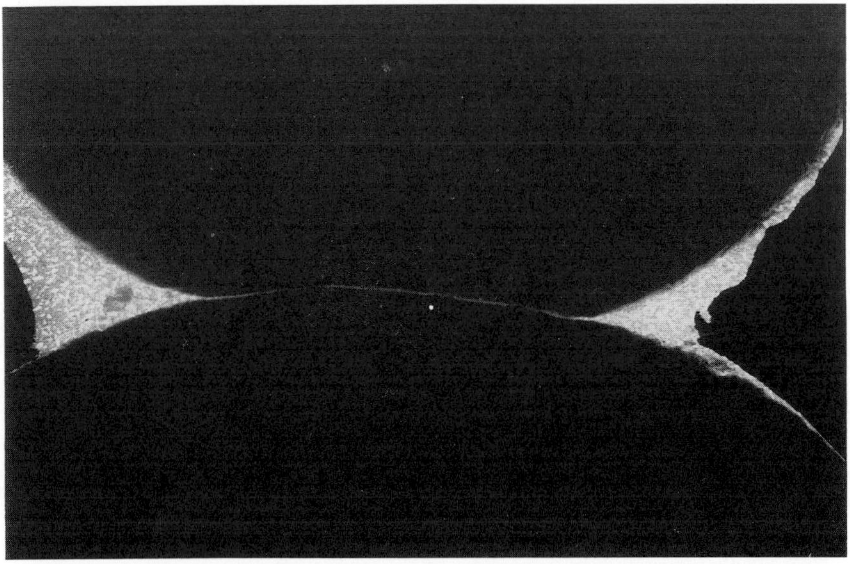

Figure 6-15 Cross section of "weld" of two tinned copper wires. X360, backscattered.

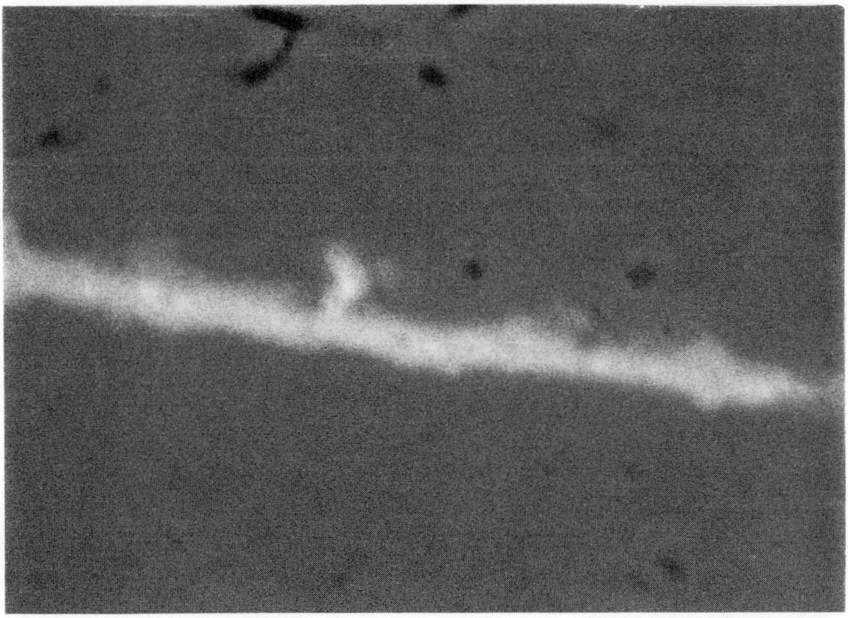

Figure 6-16 Expansion of Figure 6-15. The boundary between the two copper wires is a 90%-10% composition of copper-tin and hence is not an ordinary solder. X6000, backscattered.

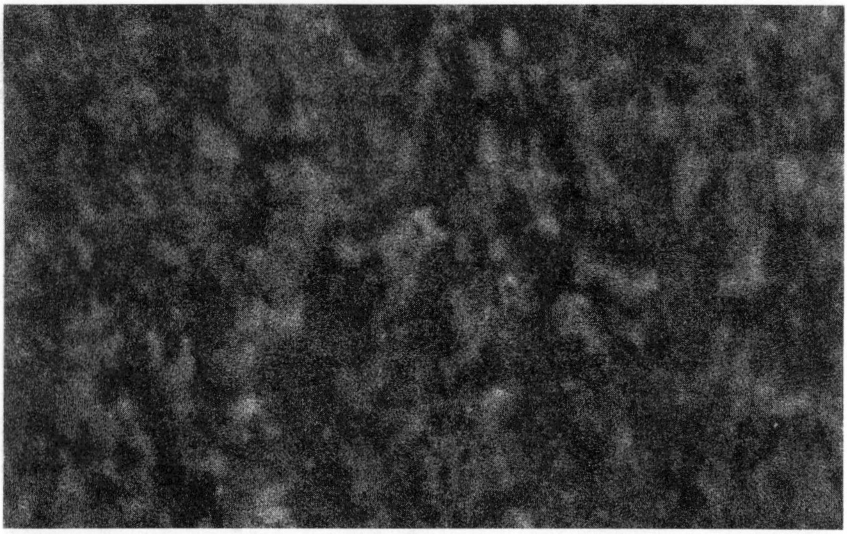

Figure 6-17 Cross section of alloy J. The lighter regions are silver-rich. X10,000, back-scattered.

Miscellaneous

Soldering to cadmium plate is not as common as in the past since it is now realized that this creates new compounds and intermetallics (Fig. 6-14). Tin-lead solders under special situations can actually create a new composition of "solder" (Figs. 6-15 and 6-16)—with tin-plated copper wires, high temperatures (1000°C) produced by resistance welding technique can produce an intermediate alloy with a composition of 90% Cu-10% Sn.[9] Alloy J, a high-melting rapid solidification (RS) alloy,[18] has a unique appearance with small domains (Fig. 6-17).

SOLDERABILITY ISSUES

Maintaining Solderability

Wetting of a copper substrate can clearly be prevented if the surface is oxidized[11,14] (Fig. 6-18). However, a tinned surface can also experience loss of solderability if copper from the substrate migrates to the surface with subsequent oxidation[19] (Fig. 6-19). For brass (Figs. 6-20 through 6-22), both copper and zinc migrate.[20,21] Copper migrates via the growth of intermetallic Cu_6Sn_5,[19] and zinc moves more rapidly via diffusion.[20,21] The migration of copper can be prevented by an adequate thickness of tin or solder (perhaps 400 μ in.), and the diffusion of zinc can be prevented by a copper barrier. Alternatively, migration of copper and zinc can be prevented by a nickel barrier.

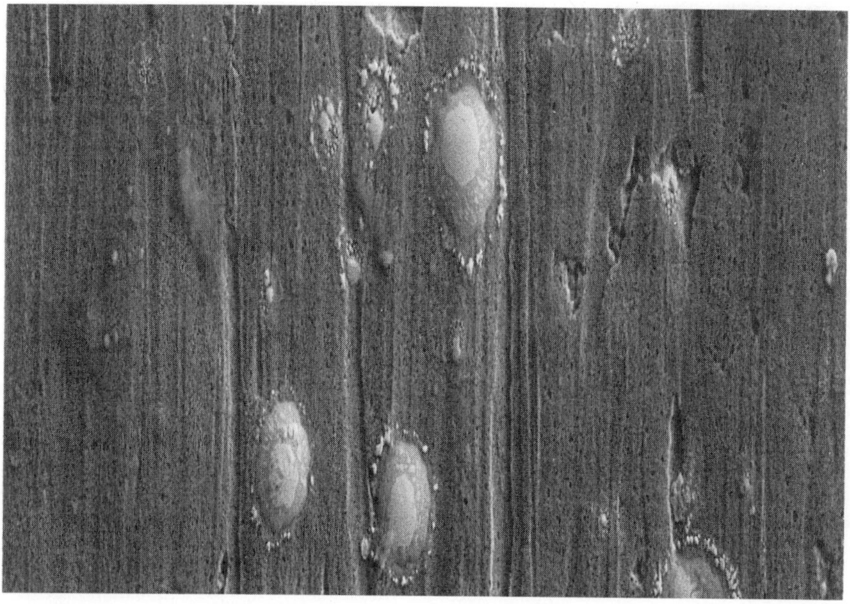

Figure 6-18 An attempt to solder a heavily oxidized copper surface. X300, secondary.

Figure 6-19 This tinned copper substrate, over a period of a year, experiences migration of copper from the substrate in the form of Cu_6Sn_5. These crystals of Cu_6Sn_5 are sometimes masked by extensive oxidation (darker regions) of the copper. X2000, secondary.

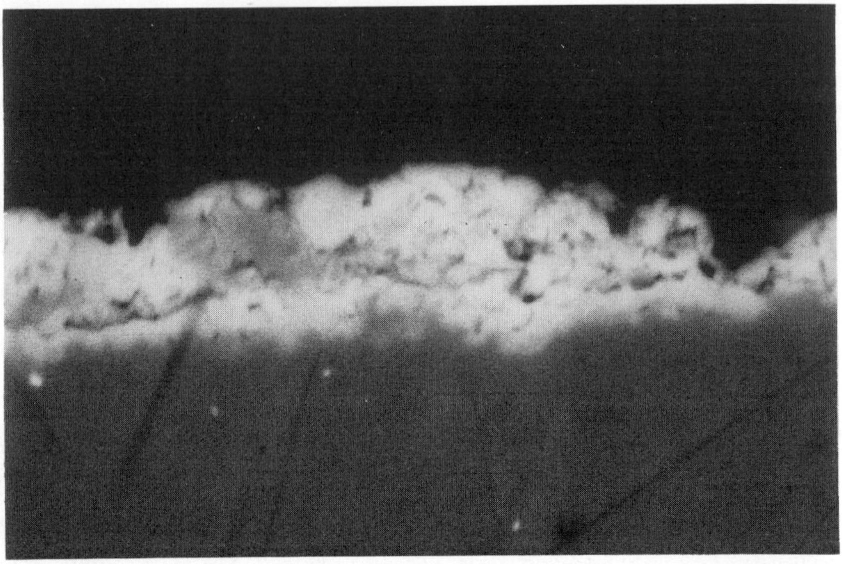

Figure 6-20 Cross section of tinned substrate with extensive base metal migration. Below: brass substrate. Middle portion: tinned layer. X4000, backscattered.

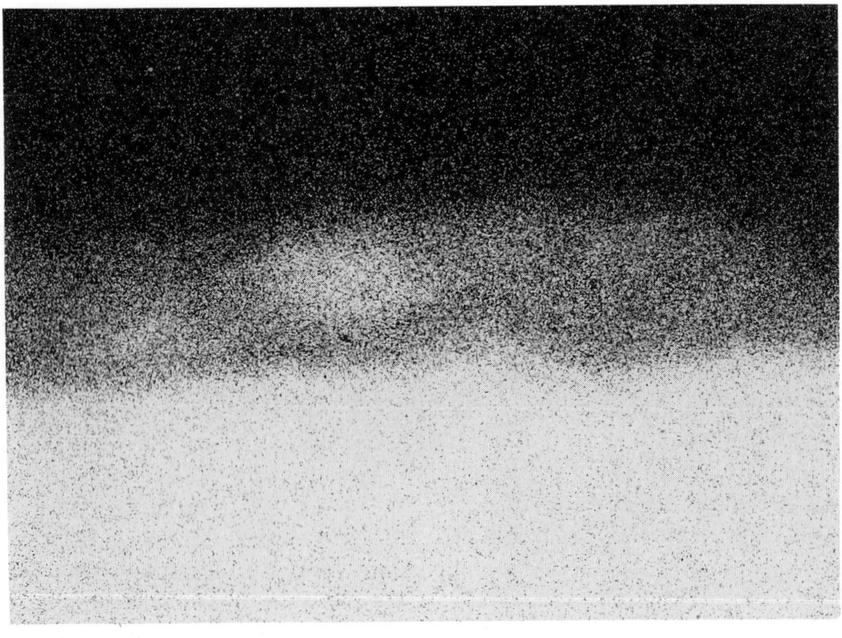

Figure 6-21 Copper dot map of Figure 6-20. Cu_6Sn_5 has migrated extensively through the tinned layer.

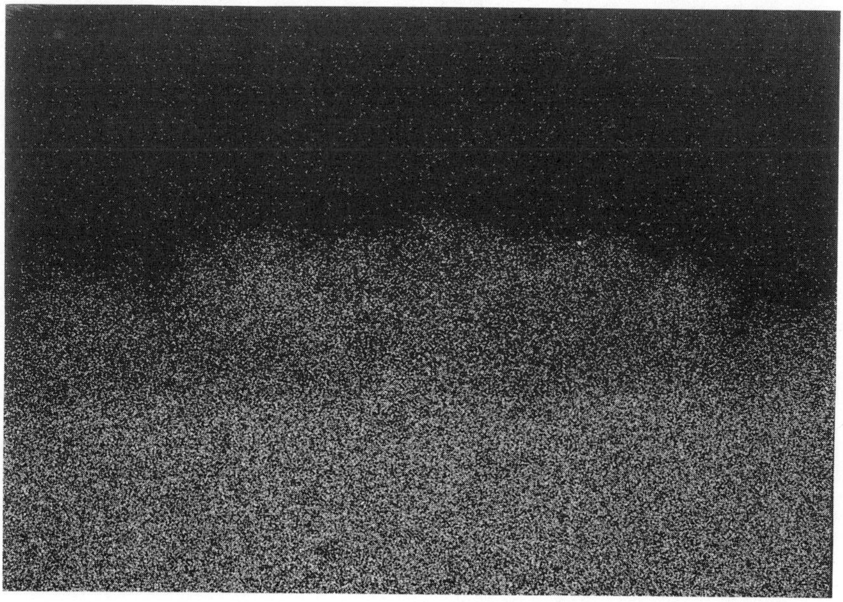

Figure 6-22 Zinc dot map of Figure 6-20. Zinc has diffused rapidly through the tinned layer.

6.9.2 Inadequate Tin Protective Coatings

Figures 6-19 through 6-22 above exemplify the dangers of an inadequate tin protective barrier for copper or brass since the base metal can migrate to the surface and oxidize. Nevertheless, a freshly tinned surface will solder satisfactorily, even with minimum thickness. However, its appearance is unusual. Figures 6-23 through 6-27 show such a surface with a tin thickness gradient.

6.9.3 The Dangers of "Forcing" Poor Solderability

Occasionally a situation of poor solderability is "solved" by excessive heating, to "force" the soldering process. This excess operation may be performed, for example, by extended periods in direct contact with a soldering iron whose tip temperature may far exceed usual wavesoldering, vapor phase, or infrared reflow operations. Evidence for this extended heat treatment is conspicuous: Cu_6Sn_5 chunks (Fig. 6-28) suspended in the solder. Nickel, also, can form intermetallics; upon extended heat treatment, nickel-tin intermetallic can grow in the form of prisms from a nickel barrier and can fracture and drift in the solder. If a silver-containing solder is used, Ag_3Sn crystals sometimes develop (Figs. 6-29 and 6-30).

Figure 6-23 Inadequately tinned copper surface (fresh). EDX indicates only the presence of tin-copper bimetallic. X1000, secondary.

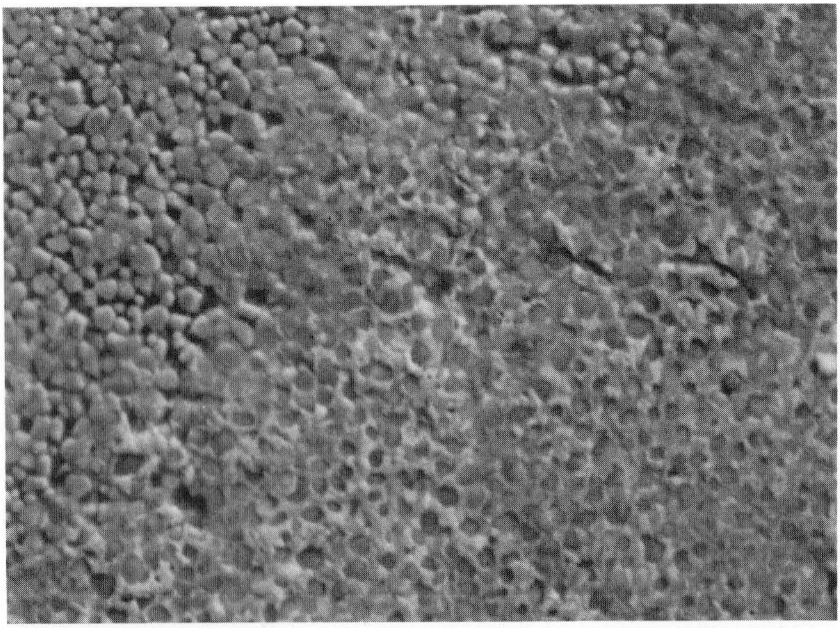

Figure 6-24 Similar situation to Figure 6-23, with slightly more tin. The upper left-hand corner indicates incomplete fusion. X1000, secondary.

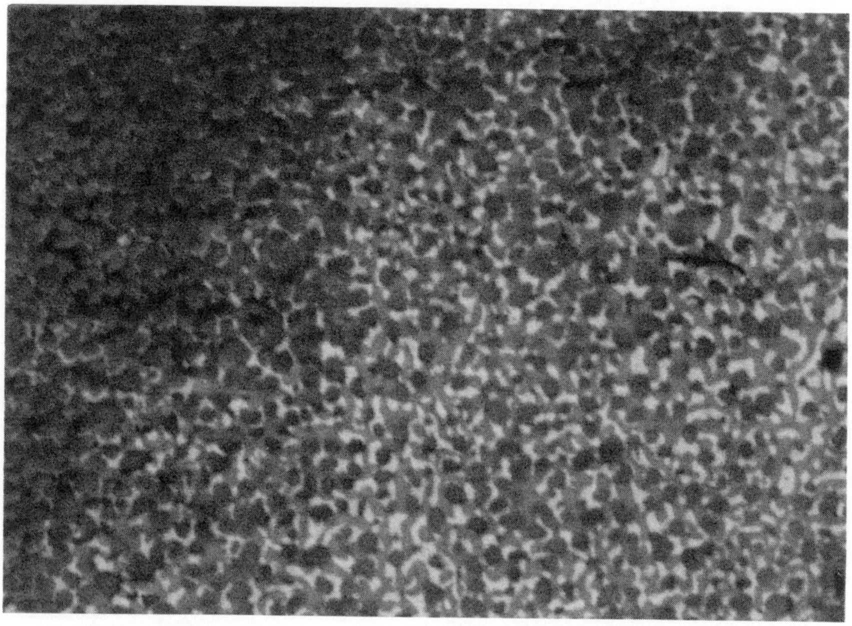

Figure 6-25 Backscattered view of Figure 6-24. Upper left corner, dark: tin-copper. Remainder: dark regions, tin; light regions, lead. X1000, backscattered.

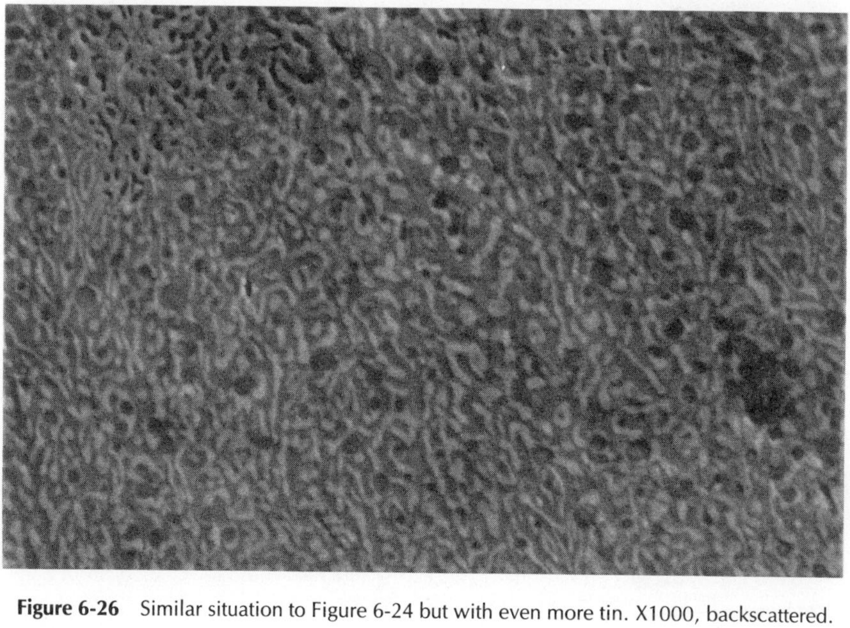

Figure 6-26 Similar situation to Figure 6-24 but with even more tin. X1000, backscattered.

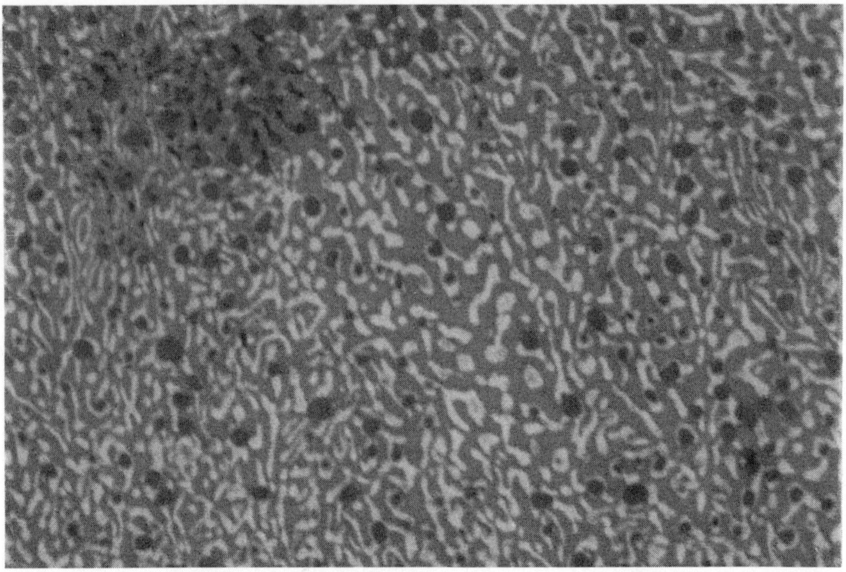

Figure 6-27 Backscattered view of Figure 6-26. The tin-copper bimetallic regions are now confined to isolated areas. X1000, backscattered.

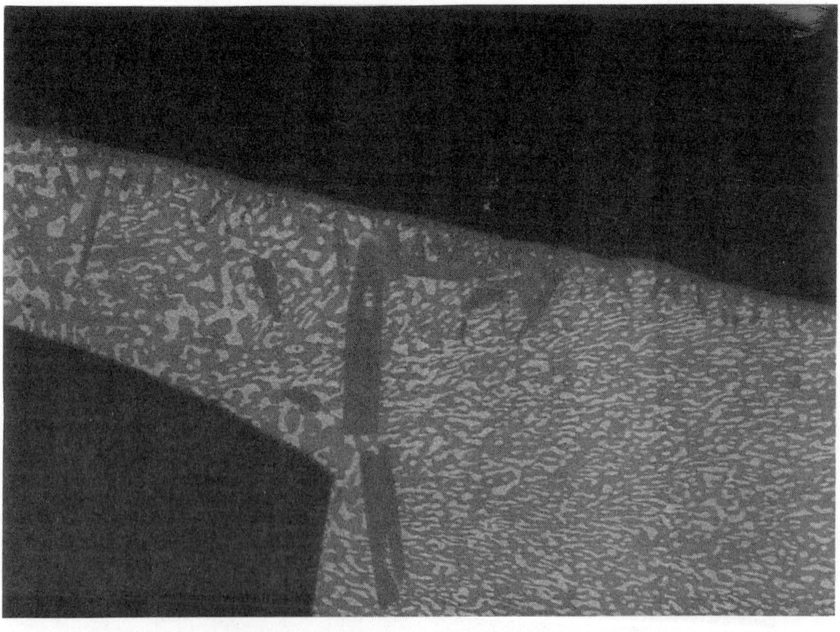

Figure 6-28 Cross section of solder joint, excessively heated to "solve" problem of non-solderability. The large chunks are Cu_6Sn_5. X500, backscattered.

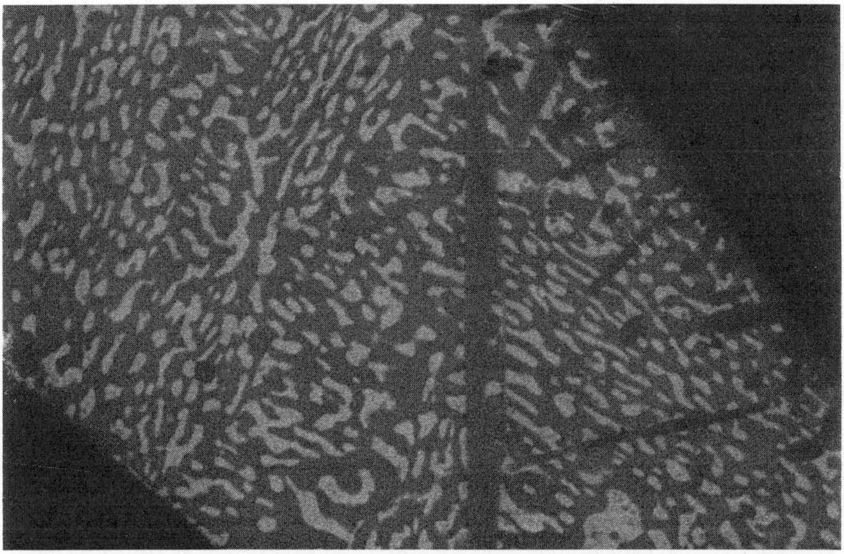

Figure 6-29 Expansion of Figure 6-28. The large vertical needle is Ag_3Sn. X1000, backscattered.

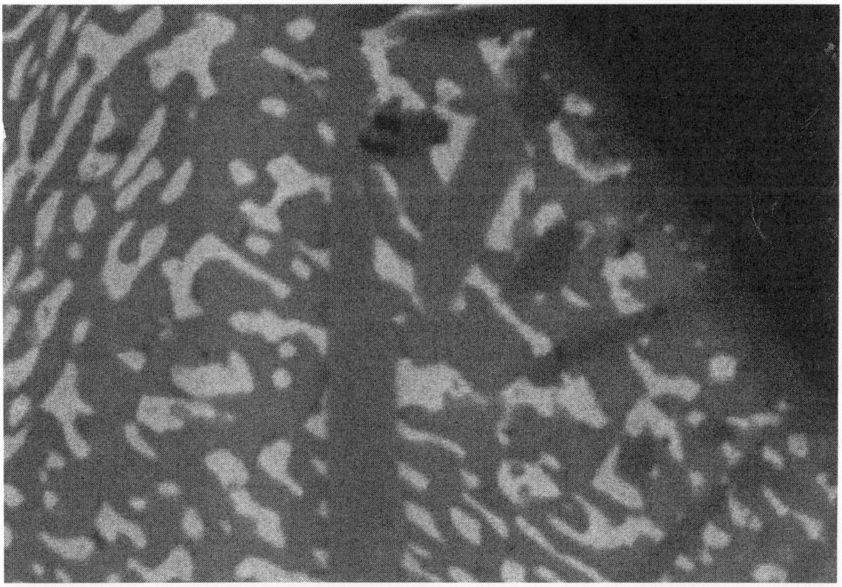

Figure 6-30 Expansion of Figure 6-29. Upper right corner, dark region: copper substrate. Careful inspection can differentiate the lead (light regions) from the darker regions—the tin (mosaic) and the Ag_3Sn (needles) and the Cu_6Sn_5 (smaller, darker needles, generally emanating from the substrate interface. X2000, backscattered.

6.10 RELIABILITY ISSUES—LEACHING OF SUBSTRATE

Silver is commonly plated onto ceramic to permit soldering to this substrate. Scavenging of this silver is possible when excess heating is experienced during the soldering process, and failure of the silver-ceramic bond can occur. If the component is originally defective, tin intrusion has already occurred into the silver metallization and failure is difficult to avoid. Effective inspection may be difficult because the silver metallization may superficially appear to be sound (Figs. 6-31 and 6-32), and a silver dot map may show minimal scavenging of the silver into the solder (Fig. 6-33). However, tin intrusion may be sufficiently severe to scavenge the silver from the substrate (Fig. 6-34). Push tests of this component easily shear the metallization away from the ceramic substrate—the component failure rate is high. An acceptable metallization layer (Fig. 35) displays an intact silver layer with no tin intrusion. Push tests of this component demonstrate a strong bond between the metallization and the ceramic, and the component passes quality control inspection. After reflow, an intact silver metallization layer remains (Fig. 6-36). An example of extended heat treatment during reflow shows complete scavenging of the silver (Fig. 6-37).

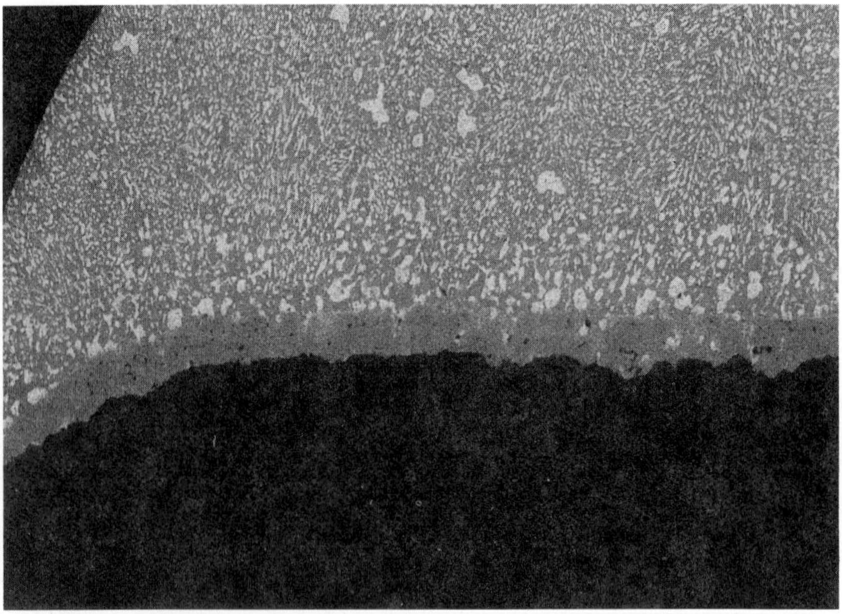

Figure 6-31 Cross section of a failing ferrite inductor. Below, dark region: ferrite. Above, 62% Sn-36% Pb-2% Ag solder. Between the ferrite and the solder lies the silver metallization, apparently intact but nevertheless defective. X480, backscattered.

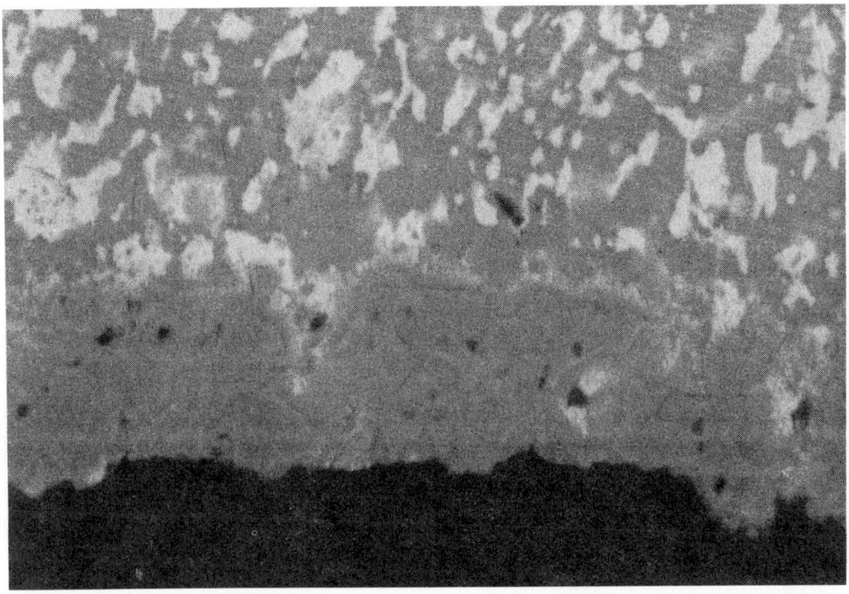

Figure 6-32 Expansion of Figure 6-31. The silver metallization appears to be sound. However, see Figure 6-34. X2000, backscattered.

Figure 6-33 Silver dot map of Figure 6-32. The silver layer is not extensively leaching away into the solder. However, see Figure 6-34. X2000.

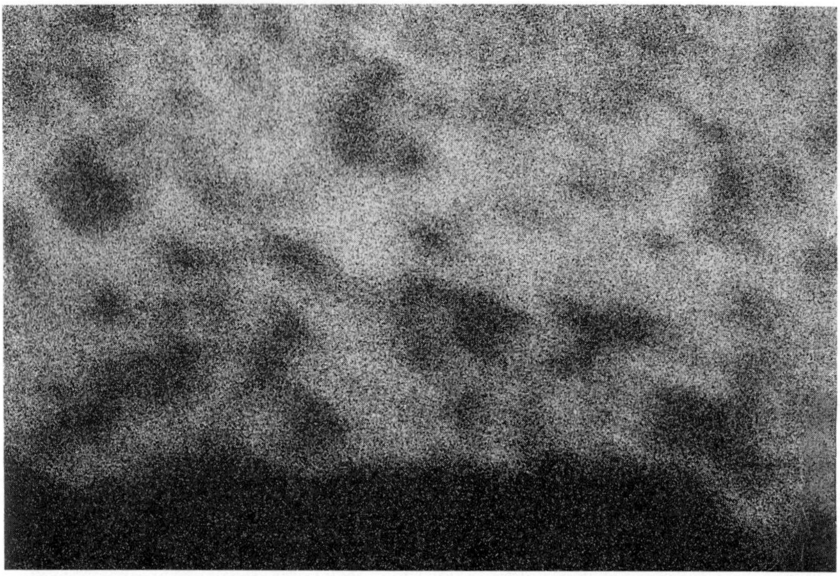

Figure 6-34 Tin dot map of Figure 6-32. The tin of the solder has extensively intruded into the silver metallization.

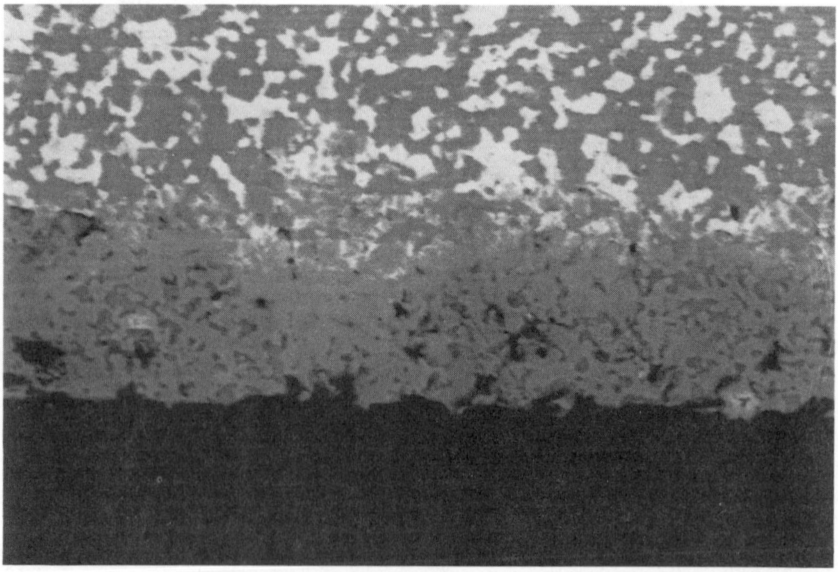

Figure 6-35 Cross section of acceptable ferrite inductor. Superficially, this view may even appear to be inferior to the faulty inductor because the metallization layer is thinner. However, tin dot maps demonstrate tin intrusion is minimal. X2000, backscattered.

Figure 6-36 Cross section of specimen of Figure 6-35 after reflowing. Minimal temperature excursion was experienced, and the silver metallization is still intact.

Figure 6-37 Cross section of ferrite inductor with extended heat treatment. The silver has been largely leached away into the solder, developing Ag_3Sn needles. This component quickly fails. X480, backscattered.

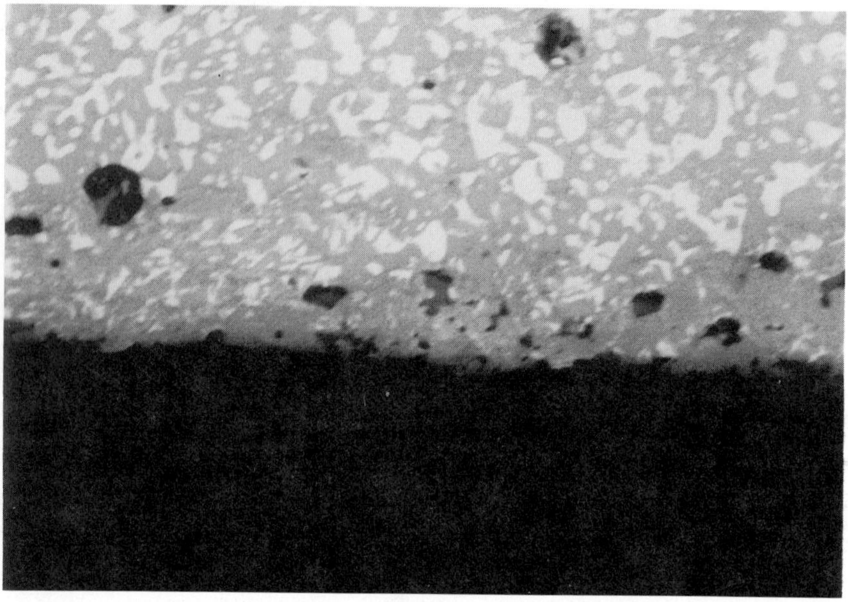

Figure 6-38 Cross section of a faulty solder joint. Below, dark region: alumina ceramic. Above, 62% Sn-36% Pb-2% Ag solder. Between the ceramic and solder is the silver-palladium metallization. X1000, backscattered.

In other cases, the silver may be partially scavenged, and yet the component is not defective. The intrusion of tin into the metallization is again the key. In a defective reflowed solder joint (Fig. 6-38), a silver metallization appears to remain between the solder and the alumina ceramic. However, a silver dot map (Fig. 6-39) demonstrates that the silver is dispersing into the solder, and a tin dot map shows complete tin intrusion to the ceramic substrate (Fig. 6-40). A lead dot map (Fig. 6-41) shows no unusual distribution of this element. Push tests of this component fail.

An acceptable solder joint (Fig. 6-42) superficially appears similar to the faulty component, with a silver dot map (Fig. 6-43) demonstrating some silver mixing with the tin. Nevertheless, a distinct layer of original conductor remains next to the ceramic. The tin dot map shows that tin does not intrude into this original conductor layer (Fig. 6-44). In this satisfactory solder joint, the lead distributions (Fig. 6-45) are again not unusual. Push tests of this component pass because a silver layer to the ceramic is intact.

A common solution to the silver scavenging problem is a nickel barrier. A series of micrographs show how the integrity of this system is preserved in a chip capacitor soldered to a metallized ceramic substrate (see Figs. 6-46 through 6-50).

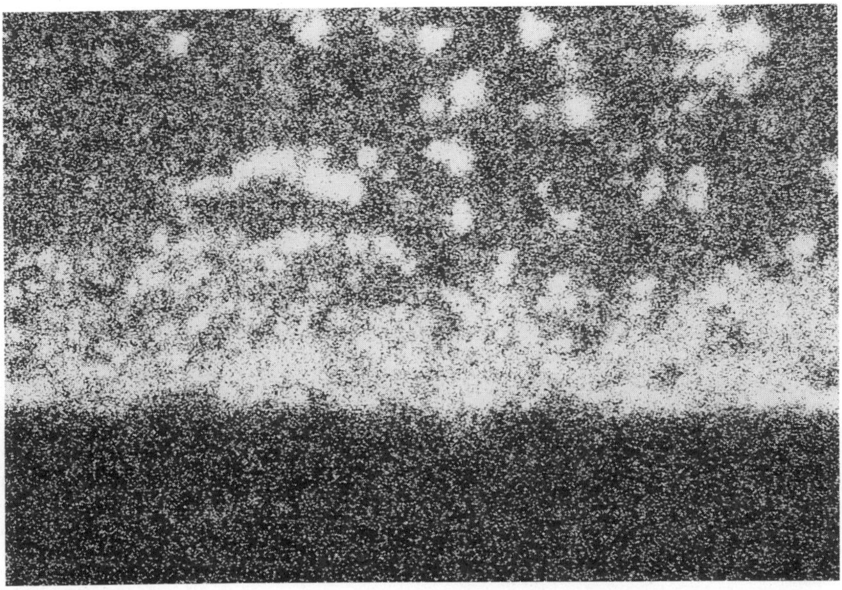

Figure 6-39 Silver dot map of Figure 6-38. Extensive scavenging of the silver into the solder is occurring. X1000.

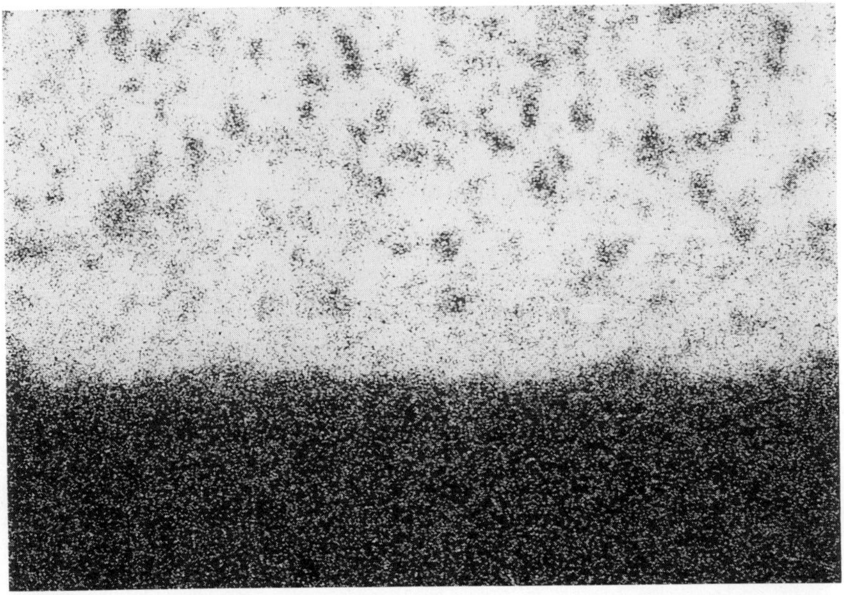

Figure 6-40 Tin dot map of Figure 6-38. Tin intrusion is complete to the alumina ceramic. X1000.

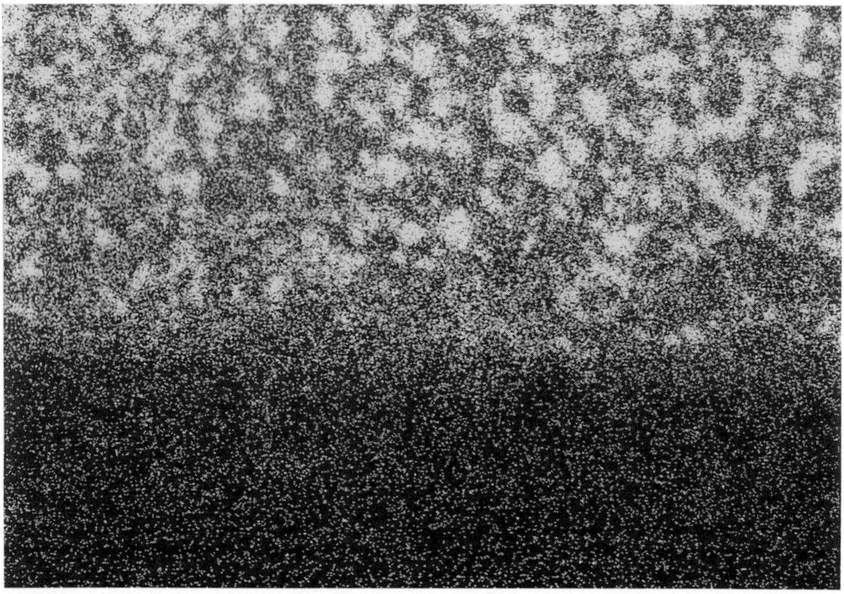

Figure 6-41 Lead dot map of Figure 6-38. X1000.

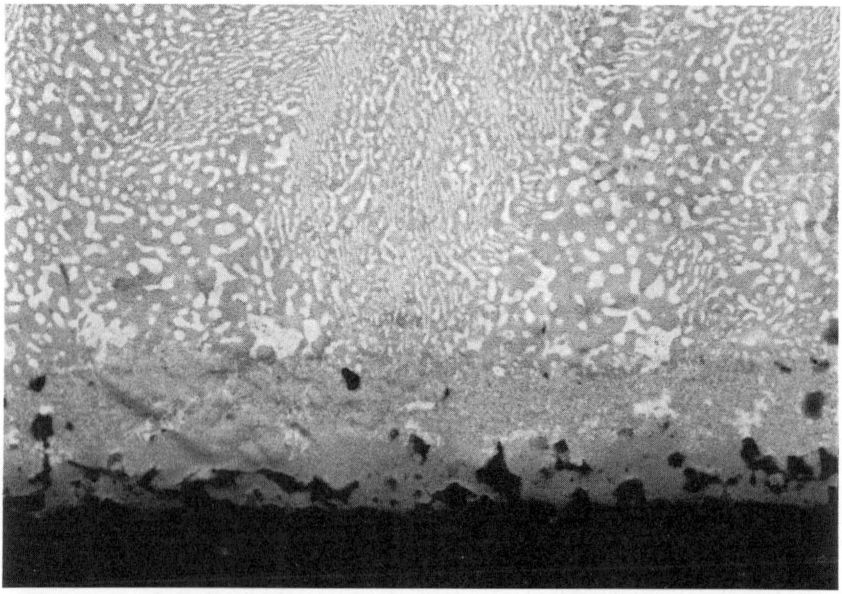

Figure 6-42 Cross section of satisfactory solder joint. Below, dark region: alumina ceramic. Above, 62% Sn-36% Pb-2% Ag solder. Between the ceramic and solder is the silver-palladium metallization. X1000, backscattered.

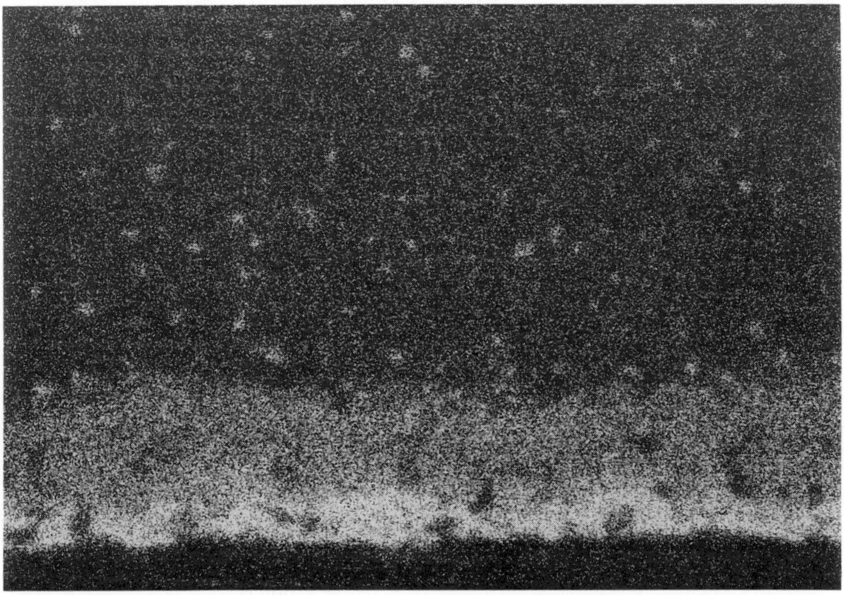

Figure 6-43 Silver dot map of Figure 6-42. Although the silver mixes with the tin, a distinct layer of original conductor remains next to the ceramic. X1000.

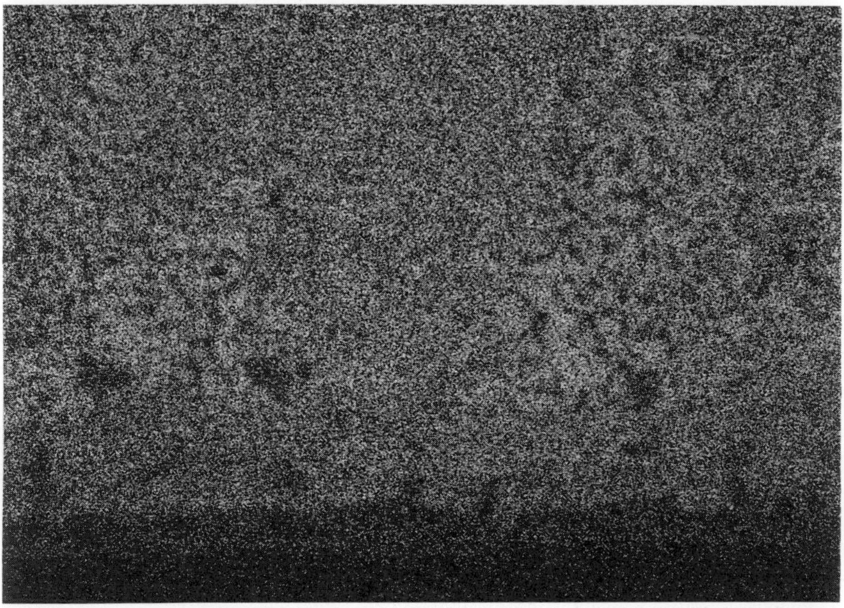

Figure 6-44 Tin dot map of Figure 6-42. The tin does not penetrate to the ceramic. X1000.

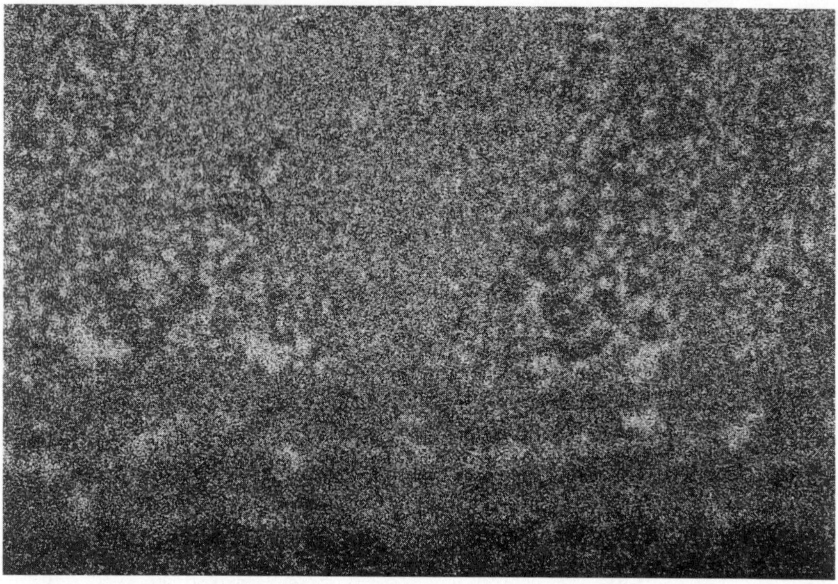

Figure 6-45 Lead dot map of Figure 6-42.

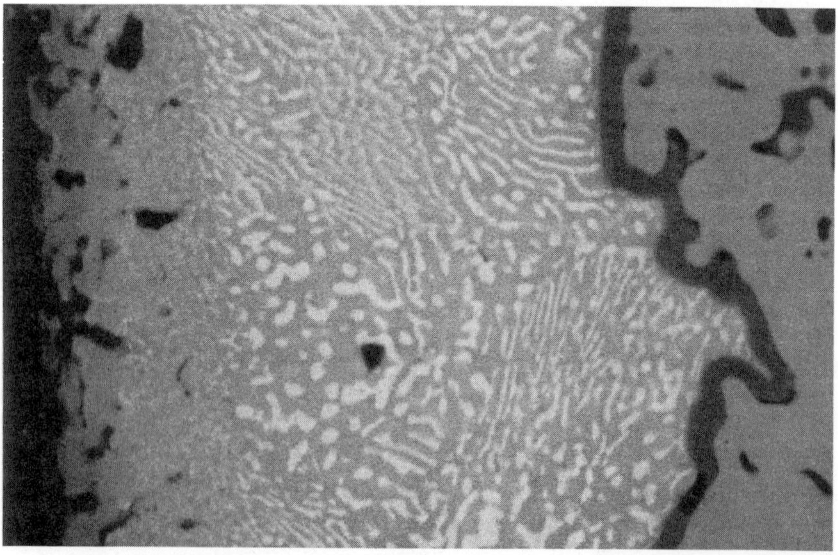

Figure 6-46 Cross section of chip capacitor to metallized ceramic. Far left, dark region: alumina ceramic substrate. Far right, gray region: silver metallization to chip capacitor (not shown). Middle of micrograph: 62% Sn-36% Pb-2% Ag solder. Between the solder and the alumina lies the silver-palladium metallization. Between the chip metallization and the solder lies the nickel barrier. X1300, backscattered.

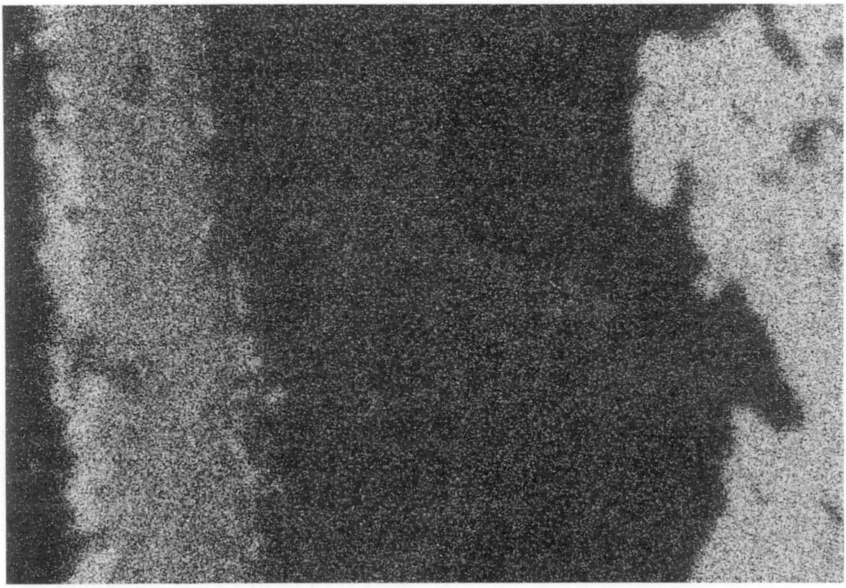

Figure 6-47 Silver dot map of Figure 6-46. The silver metallization to the alumina ceramic is intact. X1300.

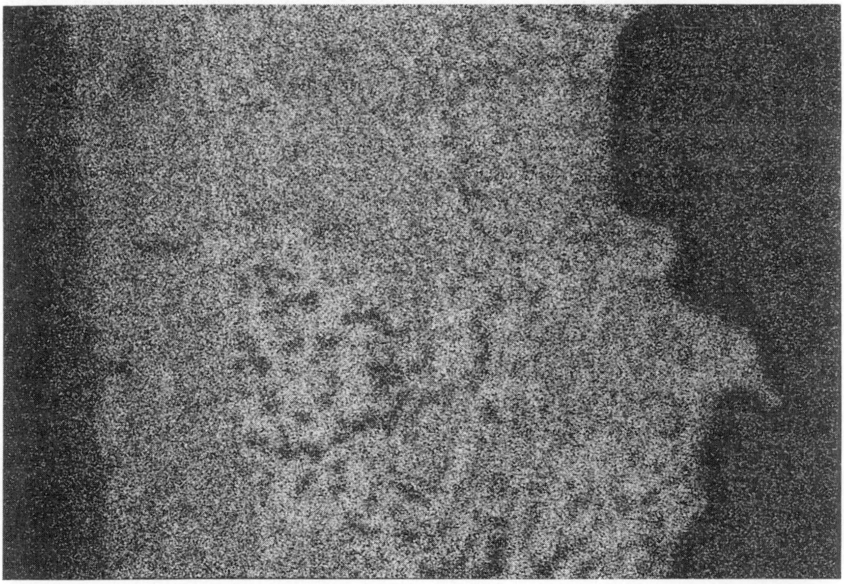

Figure 6-48 Tin dot map of Figure 6-46. The tin does not intrude to the silver metallization to the left. X1300.

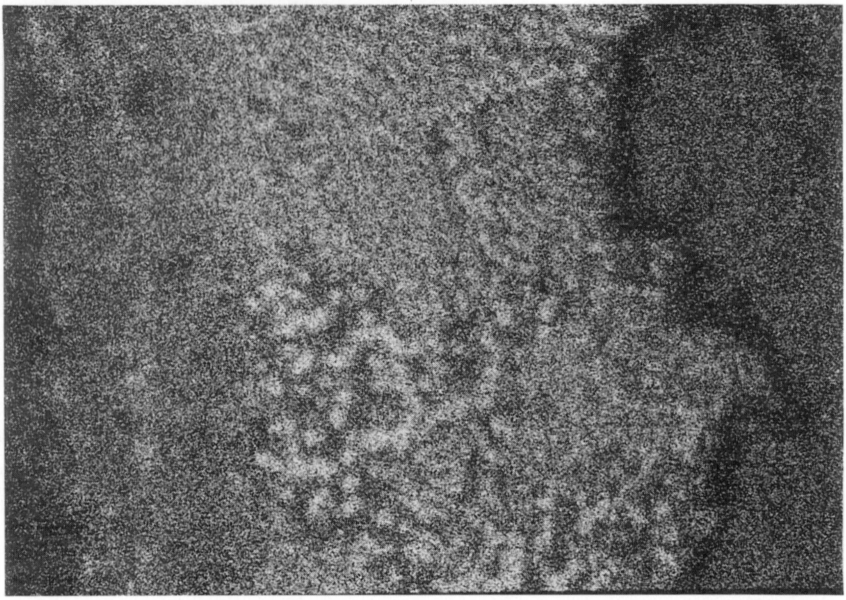

Figure 6-49 Lead dot map of Figure 6-46. X1300.

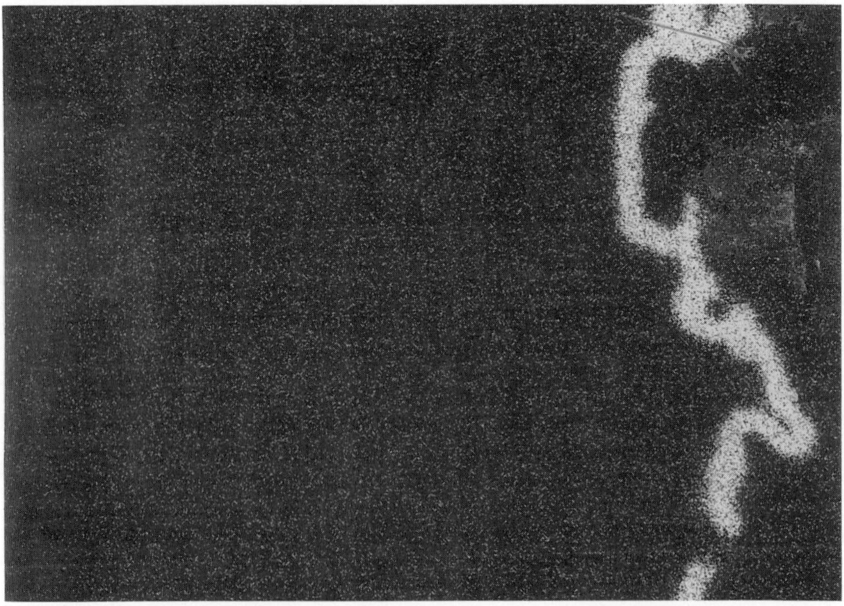

Figure 6-50 Nickel dot map of Figure 6-46. This is the nickel barrier between the solder and silver end cap of the chip capacitor. X1300.

6.11 RELIABILITY ISSUES—GOLD EMBRITTLEMENT

It is well understood that contamination can frequently cause solderability or reliability problems, but it is not universally understood that gold—a common protective coating to preserve solderability[22]—can cause reliability problems.[23] Figures 6-51 through 6-57 display the morphological characterization of soldering to gold-plated substrates. Superficially the solder joint has a monotonic composition (Fig. 6-51), but a backscattered micrograph (Fig. 6-52) manifests a great deal of elemental redistribution, where tin from the solder is scavenged by the gold underneath to produce a lead-rich area. Further magnification shows that the interface between the solder and gold possesses gold-tin intermetallics (Figs. 6-53 and 6-54). In some regions the scavenging of the tin is sufficient to leave only lead-rich metal, which does not wet well with the gold-plated substrate (Fig. 6-55). Further up the solder slope, gold occasionally migrates to the surface to produce $AuSn_4$ needles (Fig. 6-56). A backscattered view of this region displays the redistribution of $AuSn_4$ not only in the form of needles but also in the form of large aggregates among the lead spheroids (Fig. 6-57).

A cross-sectional characterization of soldering to gold-plated substrates is shown in Figures 6-58 through 6-60. If the gold layer is thick (perhaps above

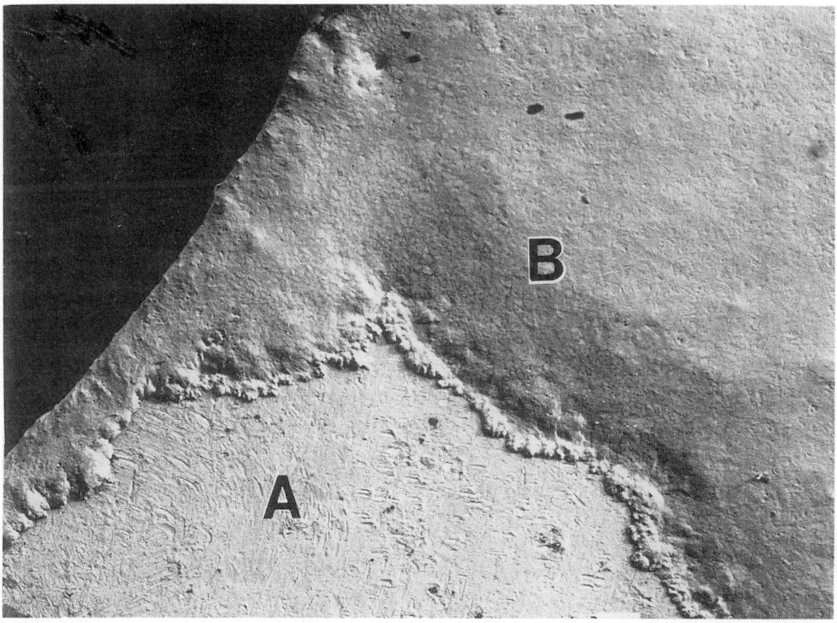

Figure 6-51 Solder joint to gold-plated substrate. Region A: gold-plated nickel substrate. Region B: 62% Sn-36% Pb-2% Ag solder. X60, secondary.

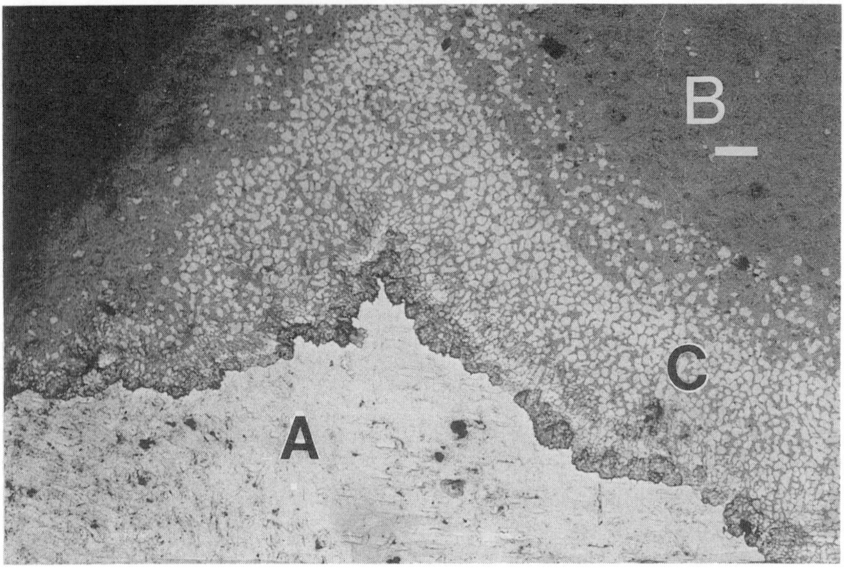

Figure 6-52 Backscattered view of Figure 6-51. The solder region is divided into the "normal" region (B) and the tin-deficient region (C). Region C consists mainly of lead domains. X100, backscattered.

Figure 6-53 Expansion of Figure 6-51. Top region (to the right): lead spheroids. Bottom region (to the left): gold-plated substrate. Intermediate region: gold-tin intermetallic region. X1000, secondary.

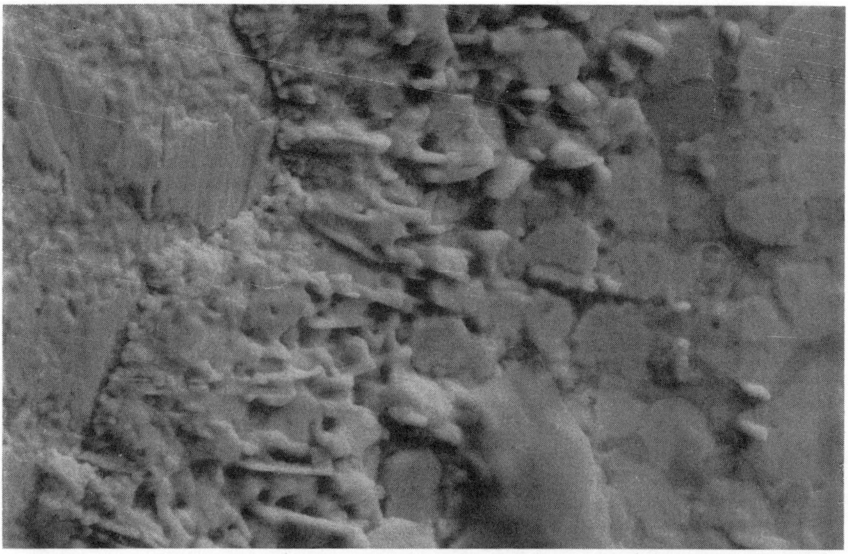

Figure 6-54 Expansion of Figure 6-53. Top region (to the right): lead spheroids. Bottom region (to the left): gold-plated substrate. In the intermediate region, the gold-tin intermetallics are now clearly visible in the form of $AuSn_4$ needles. X3000, secondary.

Figure 6-55 Interface area where tin has been thoroughly scavenged, leaving only lead-rich solder, which does not wet well with the gold-plated surface. X1200, secondary.

Figure 6-56 Area in solder where gold has migrated to the surface, producing $AuSn_4$ needles. Left: lead-rich spheroids. X540, secondary.

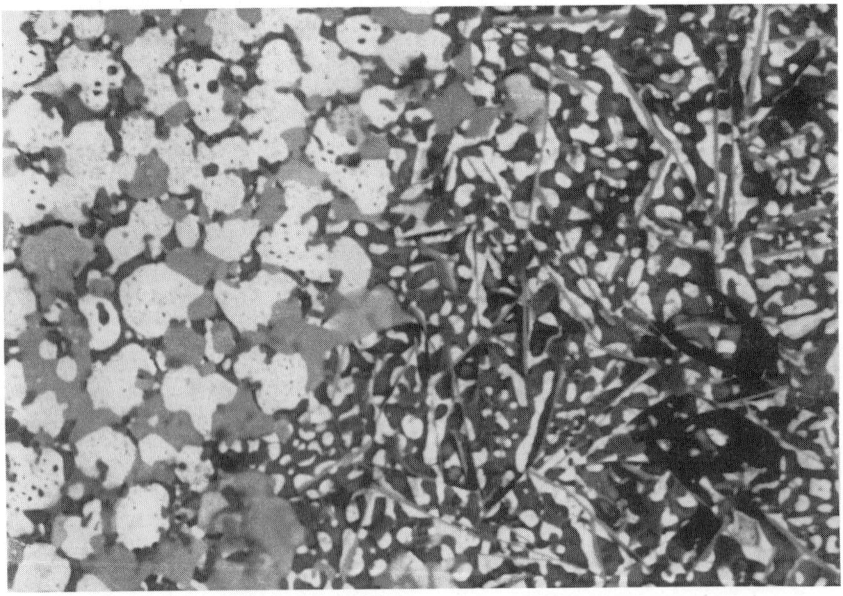

Figure 6-57 Backscattered view of Figure 6-56. Left: light regions, lead; gray regions, $AuSn_4$. Right: dark regions, tin; light areas, lead; gray needles, $AuSn_4$. X1000, backscattered.

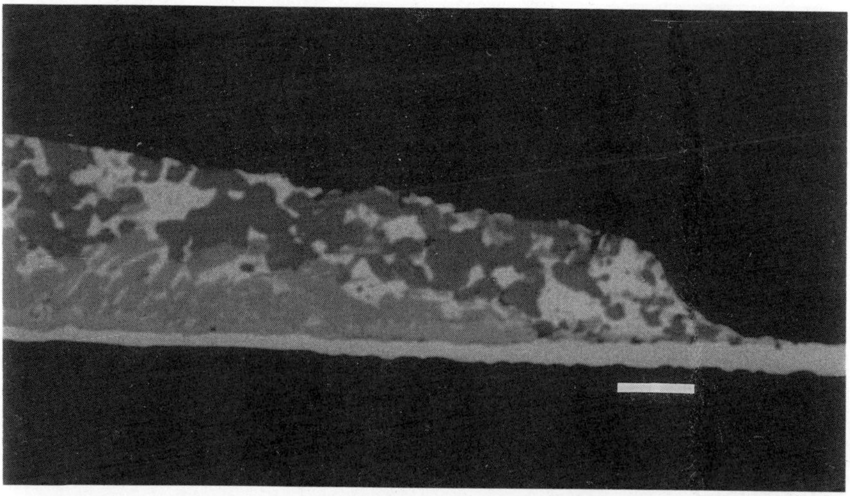

Figure 6-58 Cross section of solder joint with excess original thickness of gold. The original gold layer was 5 μm thick, and this solder joint has been reduced to 3 μm. Bottom, dark region: nickel substrate. White horizontal strip: gold layer. Solder: dark regions, tin; light regions, lead; gray regions, $AuSn_4$. X1000, Backscattered.

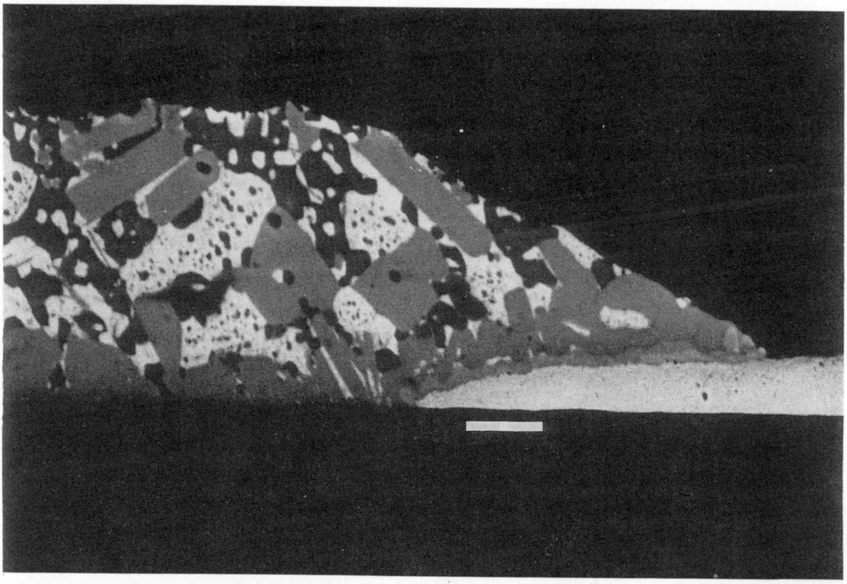

Figure 6-59 Cross section of solder joint with excess original thickness of gold, and reflowed with excess solder in an attempt to remove gold-tin intermetallics from the substrate. However, large amounts of $AuSn_4$ still adhere to the substrate. Bottom, dark region: nickel substrate. White horizontal tongue, leading from right: gold layer. Solder: dark regions, tin; light regions, lead; gray regions, $AuSn_4$. X1000, backscattered.

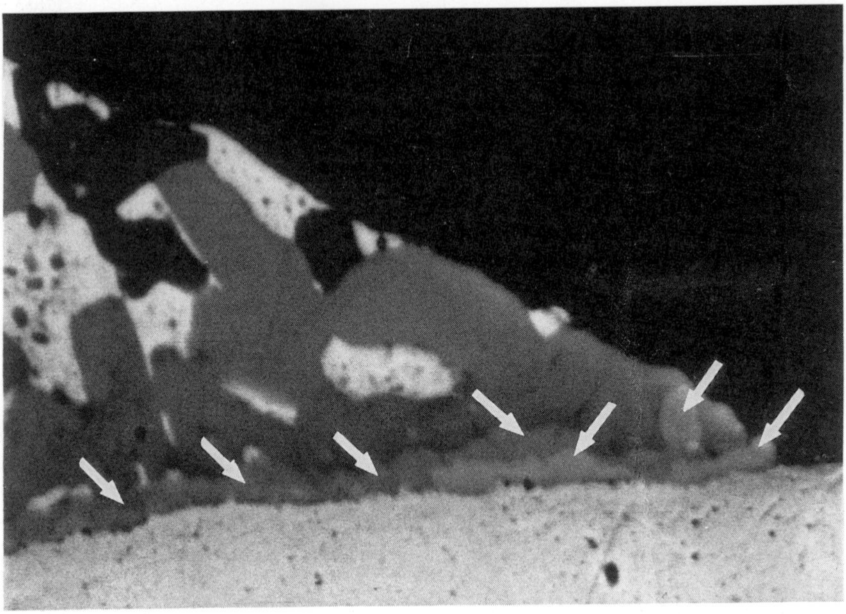

Figure 6-60 Expansion of Figure 6-59 at the interface between the solder and gold. Bottom white region: gold. Solder: light regions, lead; dark regions, tin; gray areas, $AuSn_4$. In addition, $AuSn_2$ (arrows pointing right) and AuSn (arrows pointing left) appear in very minor amounts. X2700, backscattered.

3 μm), a residual layer can remain on the substrate (Fig. 6-58), with extensive gold-tin intermetallics. Even if excess solder is used so that all the gold is redistributed (Fig. 6-59), considerable amounts of gold-tin intermetallics still may remain. Most of these intermetallics are $AuSn_4$, although at the interface of the gold and solder itself sometimes $AuSn_2$ and AuSn are found (Fig. 6-60).

Reliability issues arise with soldered gold-plated substrates when gold-tin intermetallics still adhere to the soldered substrate (Fig. 6-61). Since the gold-tin intermetallics are brittle, they can crack through normal temperature excursions and cause failure (Fig. 6-62). The examples given here are for gross failures; however, deterioration in performance can be experienced even before this cracking can be observed by SEM. The solution to the problem is to ensure that no gold-tin intermetallics remain at the substrate. If the soldering process guarantees that the gold-tin intermetallics drift away from the substrate (Fig. 6-63) so that the solder interface is substrate-tin (Fig. 6-64), gold embrittlement should not be a concern.

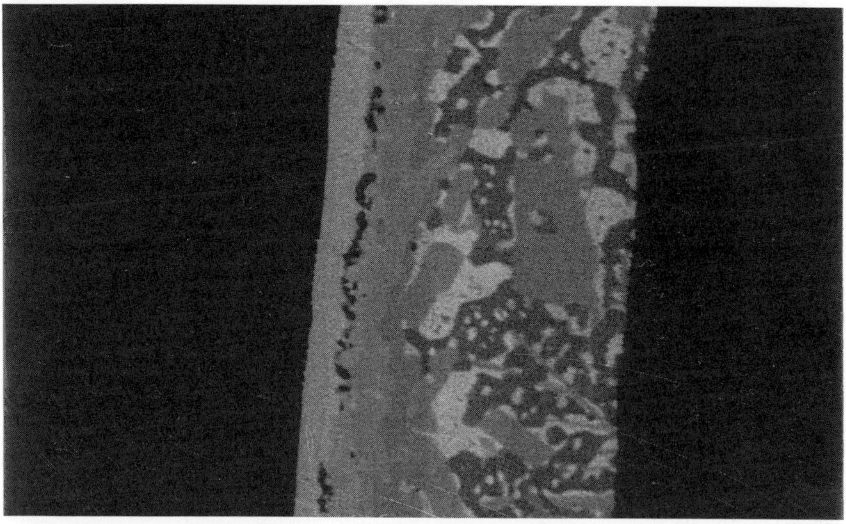

Figure 6-61 Cross section of gold-embrittled, failing solder joint. The cracks are developing at the gold-tin intermetallic. Left, dark region: nickel substrate. Next to the substrate, white vertical strip: gold. Solder: dark regions, tin; light regions, lead; gray regions, gold-tin intermetallics. X1000, backscattered.

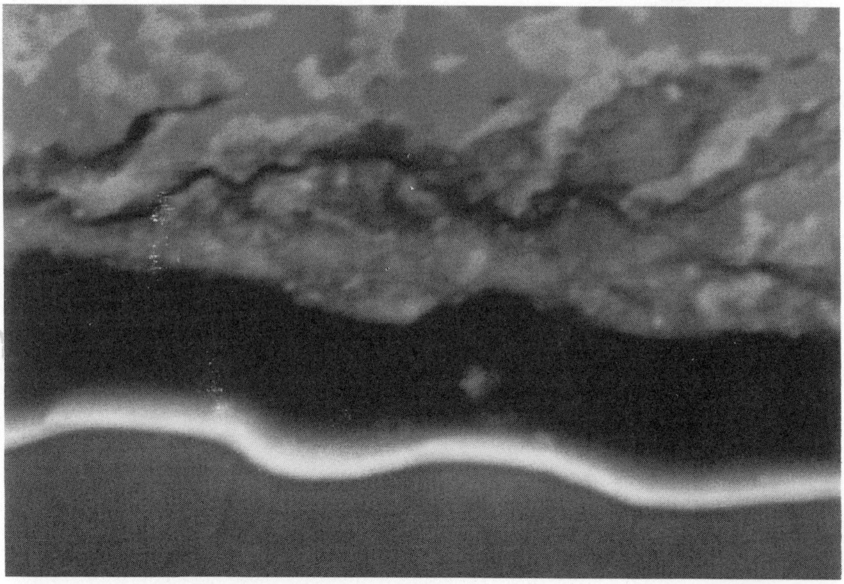

Figure 6-62 Cross section of gold-embrittled, failed solder joint. The horizontal crack between the gold and solder is complete. Bottom: copper substrate. Next to the substrate, white horizontal strip: gold. Top: solder. X3000, secondary.

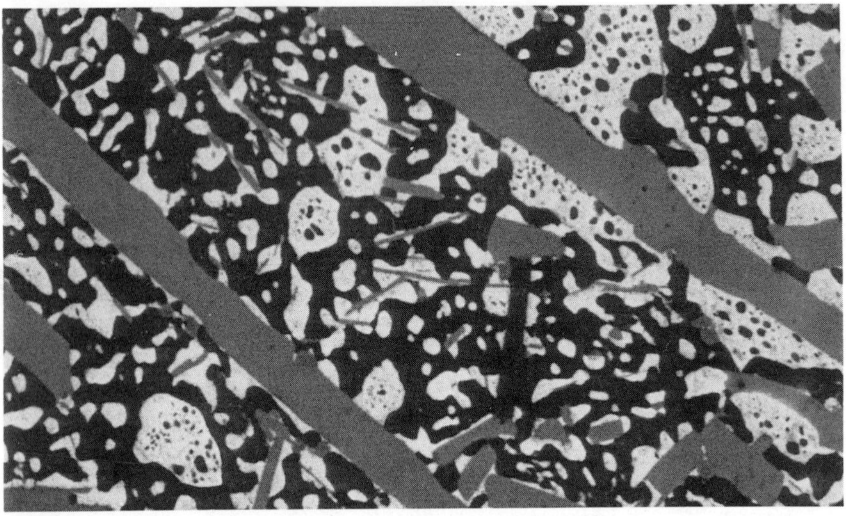

Figure 6-63 Cross section of well-soldered joint to gold-plated substrate. The $AuSn_4$ lamellae have drifted well into the bulk of the solder. Dark regions: tin. Light regions: lead. Gray regions: $AuSn_4$. X1000, backscattered.

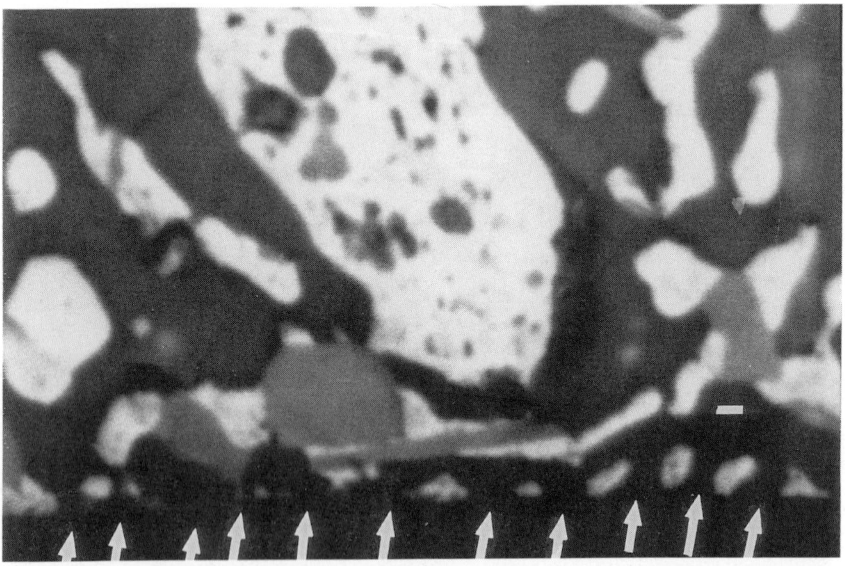

Figure 6-64 Cross section of well-soldered joint to gold-plated substrate at the solder-substrate interface. The gold-tin intermetallics have drifted far from the interface, leaving the tin to form the normal solder interface (arrows). Bottom, horizontal dark area: nickel substrate with nickel-tin intermetallic. Solder: dark areas, tin; light areas, lead; gray areas, $AuSn_4$. X3900, backscattered.

6.12 RELIABILITY ISSUES—FATIGUE

Solder fatigue is a process whereby mismatch of the thermal coefficients (TCE) of different materials,[24] coupled with an inherent design deficiency, causes stress and/or strain cycling in a solder joint, ultimately resulting in failure.[25,26] This fatigue can occur either isothermally[16,17,27,28] or during thermal cycling.[29] The isothermal process can be more easily characterized,[16,17] but the latter process more closely describes the true field situation.[30] The duration of the stress is important in determining the rate of fatigue.[31,32] The sequence of fatigue (Figs. 6-65 through 6-69) progresses from a pristine solder joint, through crazing (development of a matted surface) and pitting (development of small cavities), to cracking and failure. Hard solders can experience more catastrophic failure (Fig. 6-70) (catastrophic failure is the rapid development of failure after a long, quiescent induction period).

Surface SEMs (i.e., external appearance) during this fatiguing process have different appearances for various solders, principally because of different ductilities of these solders.[11] Tin-lead solders show the extensive working of the material, leading to the crazing phenomenon (Figs. 6-71 and 6-72). Other soft solders (Figs. 6-73 and 6-74) exhibit similar appearances, with a higher degree

Figure 6-65 Pristine solder joint (60% Sn-40% Pb). X40, secondary.

Figure 6-66 Fatiguing solder joint (60% Sn-40% Pb) progressing through the initial stage of crazing. X40, secondary.

Figure 6-67 Fatiguing solder joint (60% Sn-40% Pb) progressing through the stage of pitting. X40, secondary.

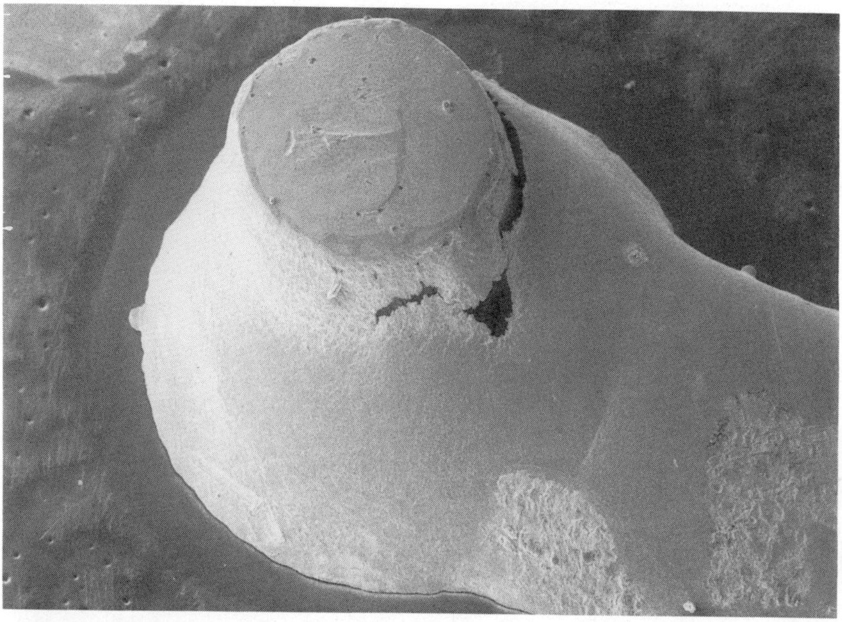

Figure 6-68 Fatiguing solder joint (60% Sn-40% Pb) beginning to crack. X40, secondary.

Figure 6-69 Complete failure of fatiguing solder joint (60% Sn-40% Pb). X40, secondary.

Figure 6-70 Complete failure of 95% Sn-5% Ag solder joint. The superficial evidence of fatiguing (crazing, pitting) observed with tin-lead solders does not exist here; instead, catastrophic failure occurs. X40, secondary.

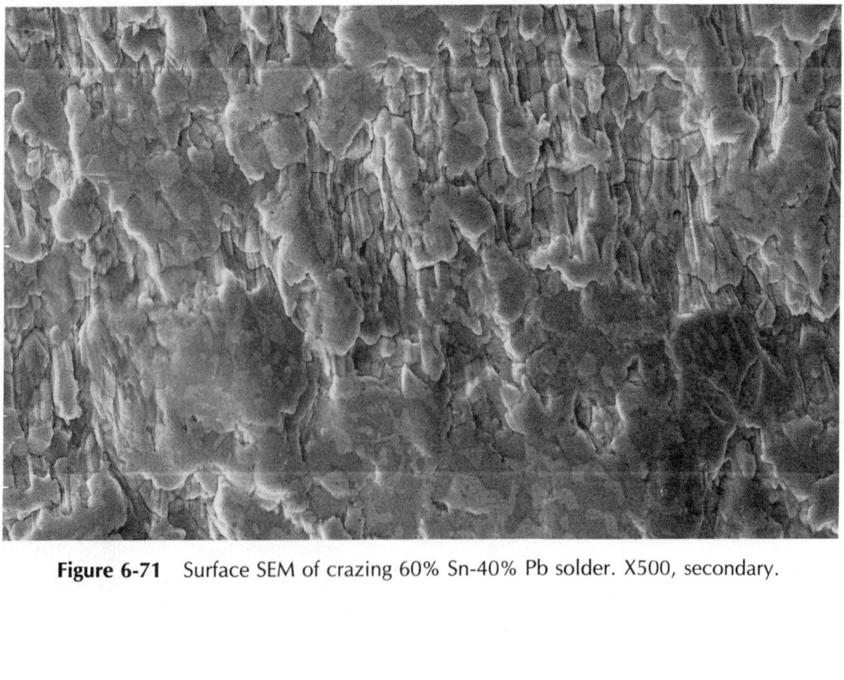

Figure 6-71 Surface SEM of crazing 60% Sn-40% Pb solder. X500, secondary.

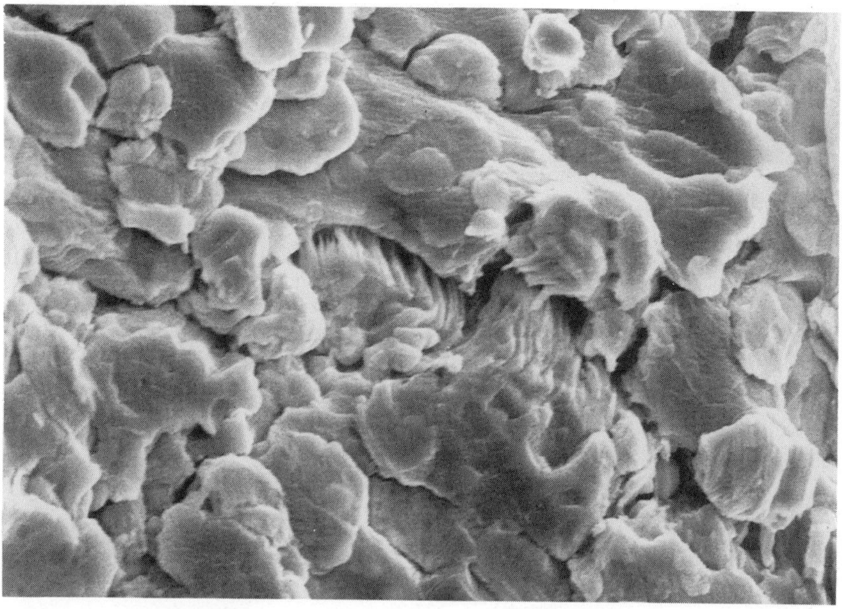

Figure 6-72 Expansion of Figure 6-71. X4000, secondary.

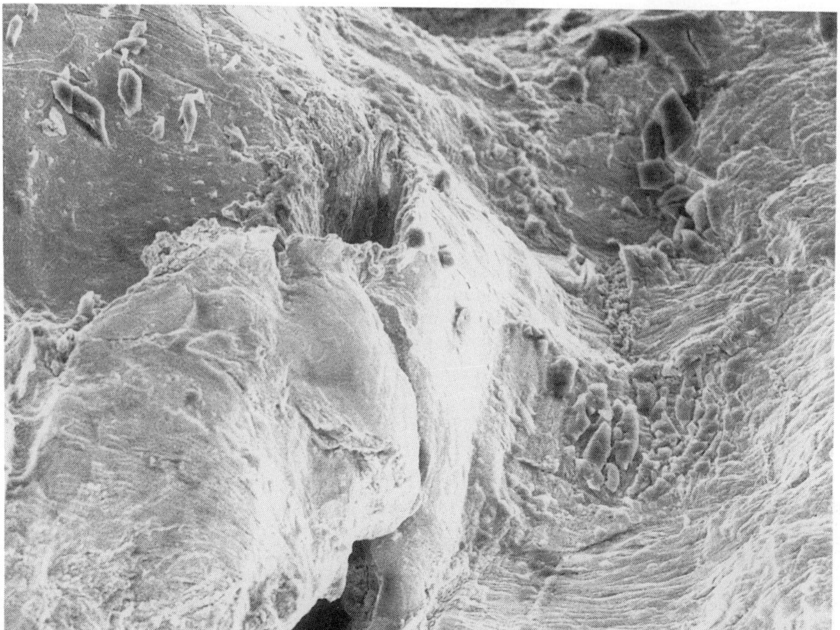

Figure 6-73 Surface SEM of 50% In-50% Pb solder joint before fatiguing. X500, secondary.

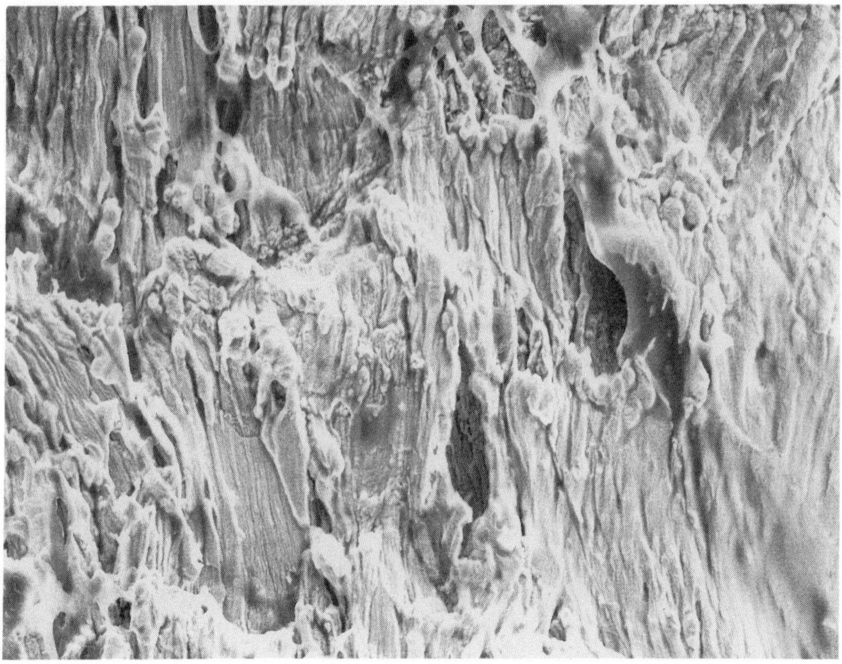

Figure 6-74 Surface SEM of 50% In-50% Pb solder joint after failure. X2000, secondary.

of creep. The behavior of hard solders (Figs. 6-75 and 6-76) reflect the catastrophic behavior of this class of alloys: failure along a definite slip, with the remainder of the surface relatively unchanged.

Cross-sectional analysis of tin-lead solders shows a well-defined and reproducible fatiguing mechanism.[33,34,35] Failures develop along the area of mechanical activity (movement) that enlarges the tin and lead domains, weakening the structural material (Figs. 6-77 and 6-78). Long before failure occurs, this developing site of domain enlargement can be observed (Fig. 6-79).

The mechanism for tin-lead-silver solders (62% Sn-36% Pb-2% Ag) appears the same as for 60% Sn-40% Pb or 63% Sn-37% Pb, with one additional feature—silver concentrates into larger domains during the developing fatigue (Fig. 6-80). Tin-lead-silver solders are more durable than tin-lead solders,[30] probably because of their greater shear and creep strengths.[11]

For soft solders, cross-sectional analysis does not show this mechanism of domain enlargement. The higher ductility of these solders can significantly affect their proper application.[36] For hard solders, failure can frequently occur at the intermetallic[30] (Fig. 6-81).

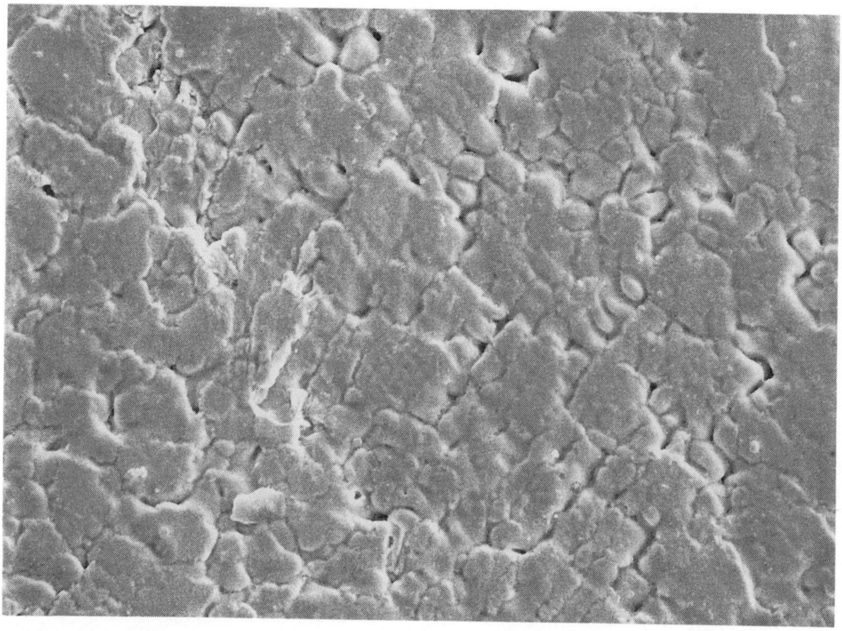

Figure 6-75 Surface SEM of 95% Sn-5% Ag solder joint during fatiguing. X1000, secondary.

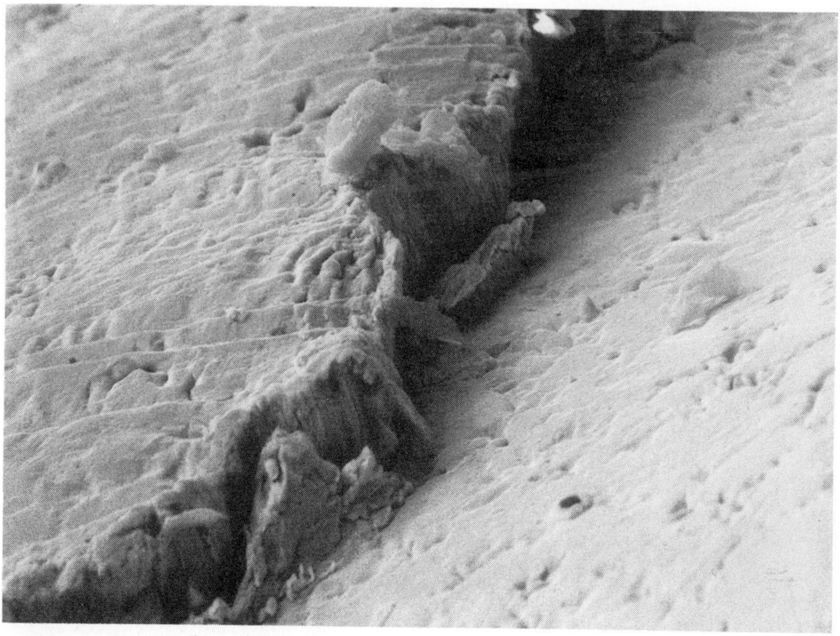

Figure 6-76 Surface SEM of 95% Sn-5% Ag solder joint after failure. X1000, secondary.

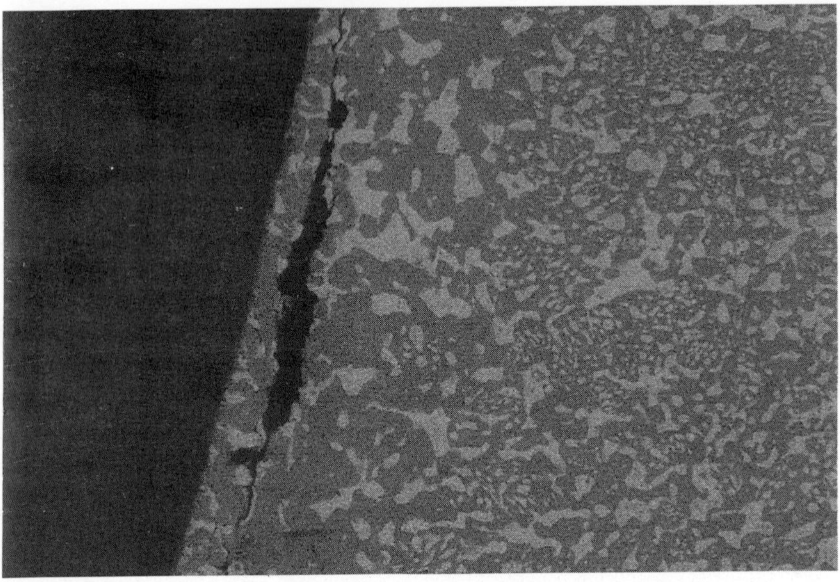

Figure 6-77 Cross section of fatiguing 60% Sn-40% Pb solder joint. Left, dark region: copper pin. Solder: dark domains, tin; light domains, lead. The area of mechanical activity in this example is occurring parallel to the solder joint, resulting in domain enlargement and eventual failure of the solder material. X470, backscattered.

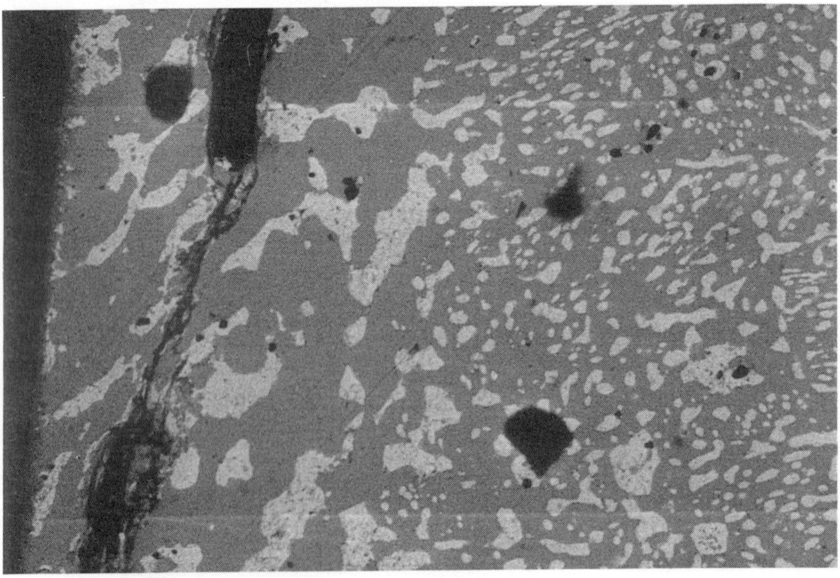

Figure 6-78 Enlargement of Figure 6-77. X780, backscattered.

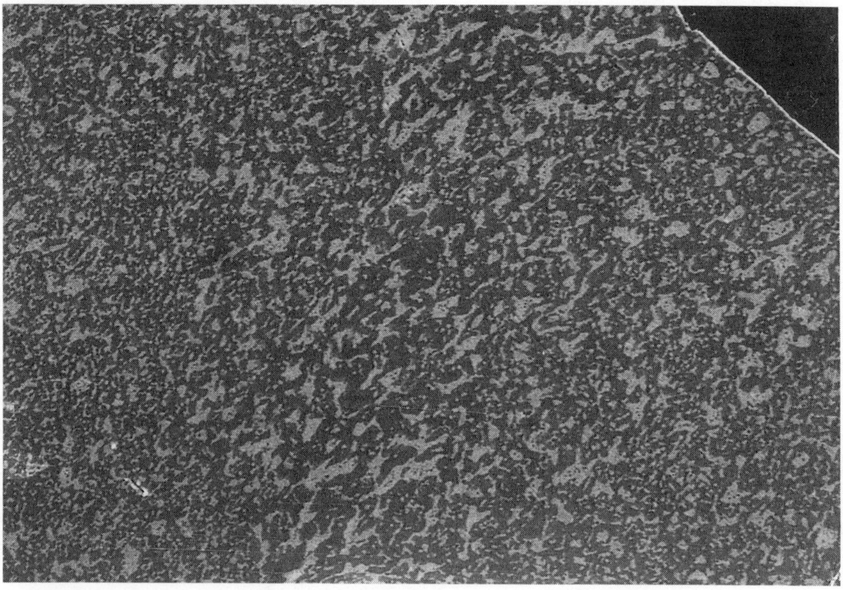

Figure 6-79 Cross section of solder joint at initial stages of fatiguing. Careful inspection discloses an area of mechanical activity (middle of view, vertical zone). X300, backscattered.

Figure 6-80 Silver dot map of cross section of fatiguing 62% Sn-36% Pb-2% Ag solder. The silver is aggregating into domains. This domain enlargement occurs not only at the fatiguing site but throughout the solder. X1000.

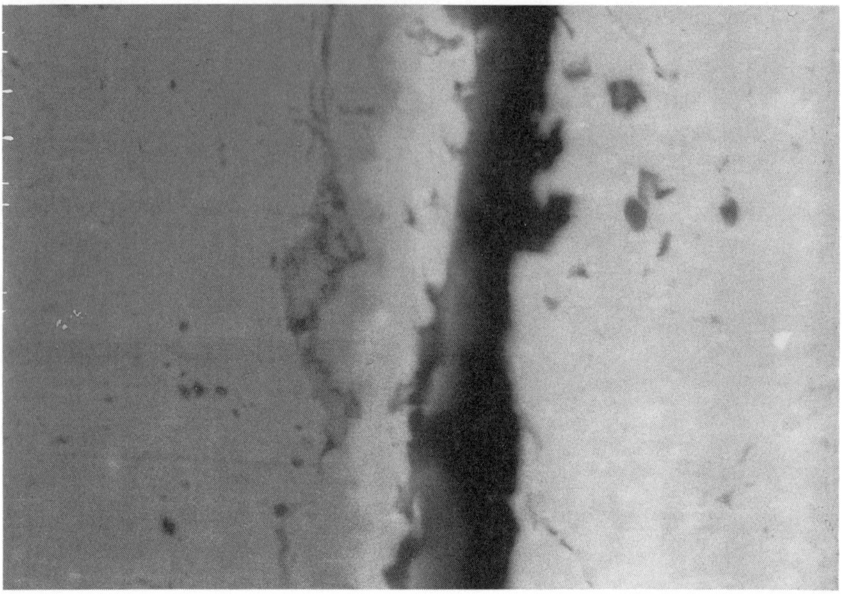

Figure 6-81 Cross section of failed 95% Sn-5% Ag solder joint. A major portion of failure occurs at the intermetallic itself. Left: copper pin. Right: solder. X2000, backscattered.

REFERENCES

1. Verhoeven, J. D., "Scanning Electron Microscopy," *Metals Handbook Ninth Edition Volume 10 Materials Characterization,* American Society for Metals, Metals Park, OH, 1986, pp. 491–515.
2. Heinrich, F. J., and D. E. Newbury, "Electron Probe X-Ray Microanalysis," *Metals Handbook Ninth Edition Volume 10 Materials Characterization,* American Society for Metals, Metals Park, OH, 1986, pp. 516–535.
3. Leyden, D. E., "X-Ray Spectrometry," *Metals Handbook Ninth Edition Volume 10 Materials Characterization,* American Society for Metals, Metals Park, OH, 1986, pp. 82–101.
4. Buehler Digest, vol. 25, no. 1. "Cross-Sectioning Techniques for Analysis of Microelectronic Components." Lake Bluff, IL: Buehler, Ltd., 1988.
5. ———, vol. 25, no. 2. "Cross-sectional Analysis of Printed Circuit Boards." Lake Bluff, IL: Buehler, Ltd., 1988.
6. ———, vol. 25, no. 3. "Cross-sectional Analysis of Soldered Connections, Connectors, and Electrical Components." Lake Bluff, IL: Buehler, Ltd., 1988.
7. Nelson, J. A., "Basic Steps for Cross Sectioning PCB's," *Insulation/Circuits,* Lake Publishing Corporation, Libertyville, IL, May 1977.
8. Nelson, J. A., "How Microstructural Analysis of Printed Circuit Boards Provides a Versatile Quality Control Method," *Electrionics,* Lake Publishing Corporation, Libertyville, IL, July 1984.

9. Hansen, M., and K. Anderko, *Constitution of Binary Alloys,* McGraw-Hill, New York, 1958.
10. Lyman, Ed., "Metallography, Structures, and Phase Diagrams," *Metals Handbook,* vol. 8, American Society for Metals, Metals Park, OH, 1973.
11. Wassink, R. J. K., *Soldering in Electronics*. Electrochemical Publications Ltd., Ayr, Scotland, 1984.
12. Goldstein, J. *Practical Scanning Electron Microscopy,* Plenum Press, New York, 1975.
13. Goldstein, J. *Scanning Electron Microscopy and X-ray Microanalysis: A Text for Biologists, Materials Scientists, and Geologists,* Plenum Press, New York, 1981.
14. Manko, H. H., *Solders and Soldering,* 2nd ed., McGraw-Hill, New York, 1979.
15. Marshall, J. L., "Scanning Electron Microscopy Characterization of Solder Failure—Standard and Nonstandard Joints," *Proceedings NEPCON '90,* Anaheim, CA, February 1990, pp. 1361–1370.
16. Wild, R. N., "Some Fatigue Properties of Solders and Solder Joints," IBM No. 73Z000421, IBM Publications, Owego, NY, January 1973.
17. ———, October 1975. "Some Fatigue Properties of Solders and Solder Joints," IBM No. 74Z000481, IBM Publications, Owego, NY, October 1975.
18. Olsen, D. R., and K. G. Spanjer, *Solid State Tech.,* September 1961, pp. 121–125.
19. Kay, P. K., and C. A. MacKay, "The Growth of Intermetallic Compounds on Common Basis Materials Coated with Tin and Tin-Lead Alloys," Publication 517, International Tin Research Institute, Columbus, OH.
20. Carter, G. F., *Principles of Physical and Chemical Metallurgy,* American Society for Metals, Metals Park, OH, 1979.
21. Askill, J., *Tracer Diffusion Data for Metals, Alloys, and Simple Oxides,* pp. 45–54. New York: IFI/Plenum Press, 1970.
22. Thwaites, C. J., "Soldering Gold Surfaces," *Electroplating & Metal Finishing,* no. 26, 1973, pp. 21–26.
23. Duckett, R., and M. L. Ackroy, "The Influence of Solder Composition on the Embrittlement of Soft-soldered Joints on Gold Coatings," Publication 520. International Tin Research Institute, Columbus, OH.
24. Wolverton, W. M., *Brazing and Soldering,* Vol. 13, Autumn 1987, pp. 33–38.
25. Manson, S. S., *Thermal Stress and Low-Cycle Fatigue,* McGraw-Hill, New York, 1966.
26. Coffin, L. F., Jr., "Introduction to High-temperature Low-cycle Fatigue," *Experimental Mechanics,* May 1968, pp. 218–224.
27. Solomon, H. D., "Fatigue of 60/40 Solder," *Proceedings 1986 36th Electronics Components Conference,* 1986, p. 622.
28. Stone, D., H. Wilson, R. Subrahamanyan and C.-Y. Li, "Mechanisms of Damage Accumulation in Solder during Thermal Fatigue," *Proceedings 1986 36th Electronics Components Conference,* 1986, p. 630.
29. Shine, M. C., L. R. Fox, and J. W. Sofia, "A Strain Range Partitioning Procedure for Solder Fatigue," *Brazing and Soldering,* Vol. 11, no. 9, 1985, p. 11.
30. Marshall, J. L., and S. R. Walter, *Int. J. Hybrid Microel.,* Vol. 10, 1987, pp. 11–17.
31. Engelmaier, W. "Functional Cycles and Surface Mounting Attachment Reliability," *Surface Mount Technology,* 1984, p. 87.

32. Coffin, L. F., Jr., "Low-Cycle Fatigue: A Review," *Applied Materials Research,* vol. 1, October 1962, pp. 129–141.
33. Lichtenberg, L. R., "Comparison of Environmental Thermal Cycle Tests on Reflow Soldered Assemblies," *ISHM 1984 Proceedings,* The International Society for Hybrid Microelectronics, Reston, VA, 1984, p. 65.
34. Wright, E. A., and W. M. Wolverton, "The Effect of the Solder Reflow Method and Joint Design on the Thermal Fatigue Life of Leadless Chip Carrier Joints," *Proceedings 1984 34th Electronics Components Conference,* 1984, p. 149.
35. Smeby, J. M., "Solder Joint Behavior in HCC/PWB Interconnects," *Proceedings 1984 34th Electronics Components Conference,* 1984, p. 117.
36. Coombs, V. D., "Fatigue Properties of Pure-Metal Solders," *ISHM 1976 Proceedings,* The International Society for Hybrid Microelectronics, Reston, VA, 1976, p. 67.

7

The Role of Microstructure in Thermal Fatigue of Pb-Sn Solder Joints

J. W. Morris, Jr., D. Tribula, T. S. E. Summers, and D. Grivas

The Pb-Sn eutectic alloy is widely used as a joining material in the electronics industry. In this application the solder acts as both an electrical and mechanical connection within and among the different packaging levels in an electronic device. Advances in packaging technologies driven by the desire for miniaturization and increased circuit speed result in severe operating conditions for the solder joint and thus solder joint reliability problems. Specifically, the mismatched thermal expansion characteristics of the materials joined by the solder and the cyclic temperature fluctuations normally encountered during service constitute a condition of thermal fatigue for the constrained solder. This is especially a problem in the surface mounting of leadless components where a shearing of the solder joint occurs with each temperature excursion. Repeated temperature cycling, such as that associated with Joule heating as the device is turned on and off, and environmental temperature changes fatigue the solder joint ultimately causing its failure. Increased solder joint reliability and the development of new more fatigue-resistant solder alloys require a fundamental understanding of the metallurgical mechanisms operative during the fatigue cycle and, hence, an understanding of the solder microstructure.

From the metallurgical perspective the thermal fatigue problem in solder contacts is at least as complex as any fatigue problem that has been researched. Since the solder is mechanically soft and is used at a high homologous temperature, deformation is introduced by plasticity and creep throughout the strain cycle. In addition, the cyclic deformation is affected by the fact that the solder contact is a thin layer bonded to relatively rigid materials with a complex intermetallic layer at the bonding plane. An understanding of the deformation is especially complicated by the inherent complexity of the as-solidified eutectic solder microstructure and its instability as it is cycled and aged. The problem of predicting the rate of thermal fatigue in a solder joint is formidable; there is no reliable predictive theory for thermal fatigue under creep conditions for even simple, single phase, monolithic materials. Nonetheless, mechanistic research on the fatigue and related deformation modes of solder joints can provide valuable information to help guide the design of accelerated tests and the development of improved solders.

The work that is reviewed here concentrates on the role of the solder microstructure and its association with the mechanisms of thermal fatigue. Particular emphasis is placed on a fundamental understanding of the microstructure-property relationship in Pb-Sn alloys and its dependence on processing history. It has been observed that in all cases where shear deformation at elevated temperatures is a factor, the solder fails by the same basic mechanism—inhomogeneous deformation of the solder microstructure resulting in inhomogeneous recrystallization and softening and thus joint failure. The metallurgical causes of this mechanism are outlined here. Possible paths toward the defeat of this failure mechanism in the design of superior, more fatigue-resistant solder alloys are discussed and evaluated.

7.1 EXPERIMENTAL DETAILS

This research on the microstructural mechanisms of fatigue in solder has included the design of samples that permit controlled deformation of well-characterized solder joints, the development of appropriate metallographic procedures, and the design and construction of equipment to test joints under various loading conditions including high temperature creep in shear, isothermal shear fatigue, and thermal fatigue. The specimen design and mechanical testing procedures are such that they are capable of producing the loading conditions and geometries of real solder joints; since thermal fatigue failures are usually associated with shear loading, these have been emphasized. The mechanical tests conducted as part of this work include tests to failure in shear, isothermal and thermal fatigue testing, and isothermal creep testing in shear. The specimens designed for these experiments are such that they will produce loading in a well-defined stress state under simple shear, for shear to failure, creep, and isothermal and thermal fatigue conditions. Figures 7-1 through 7-3 show the specimens used in this work. Metallographic preparation techniques were developed[1–3] that ensure flat solder joint surfaces for detailed examination in both optical and SEM microscopes.

The specimen in Figure 7-1 is that used in shear to failure, isothermal shear fatigue, and creep in shear experiments. The design is such that, under loading in tension, the solder joints that correspond to the shaded regions in Figure 7-1 are loaded in nearly simple shear. Manufacture of the specimen consists of assembling polished and fluxed copper plates with appropriate spacers to obtain

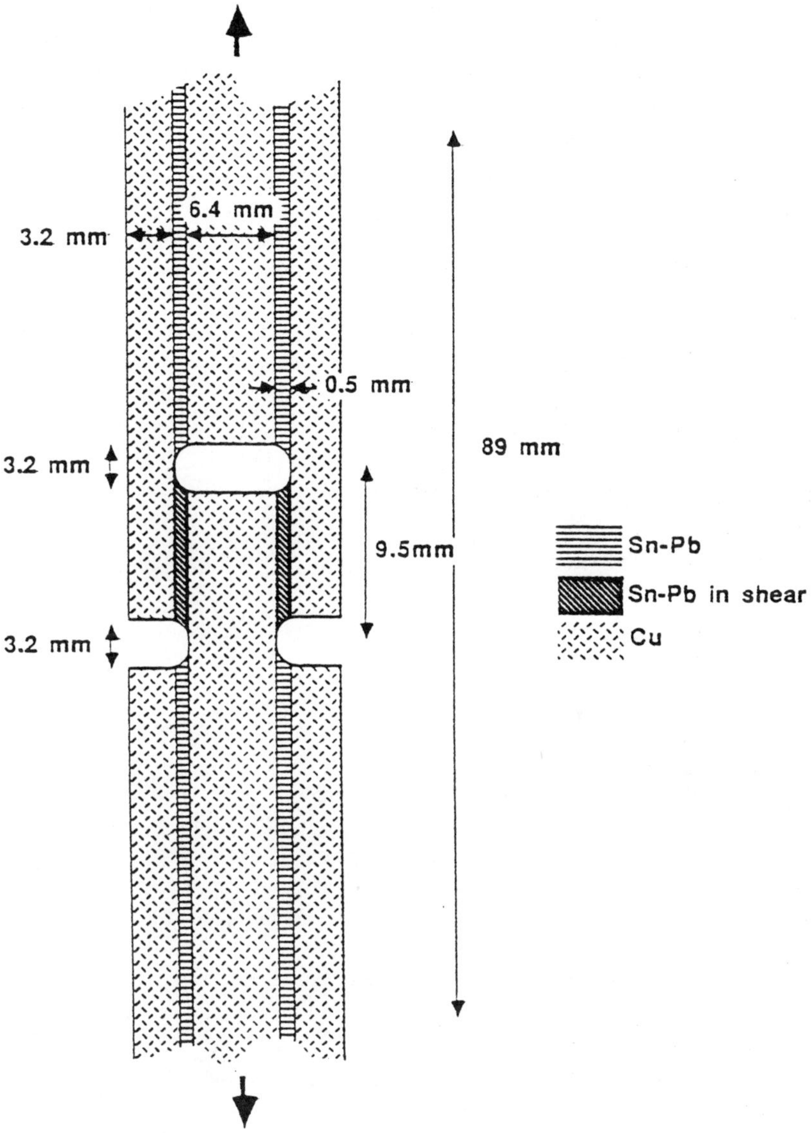

Figure 7-1 The specimen configuration used for shear strength, isothermal shear fatigue, and creep testing. The specimen is a multilayer copper-solder sandwich. On axial loading, the two solder joints, shaded regions in figure, are loaded in nearly simple shear. Dimensions are in millimeters.

desired solder joint width. The assembly is then immersed in a molten solder bath. After cooling, the excess solder is removed and the block is sliced to yield several specimens such as that shown in Figure 7-1. Figure 7-2 shows the second specimen configuration used in the creep tests. Unlike the specimen in Figure 7-1, this geometry consists of 18 individual and separate solder joints within each specimen (9 on either side of the central plate). Again, under loading in tension, the solder contacts are deformed in simple shear. To make these specimens, polymeric circuit boards with plated Cu pads are assembled together and immersed in molten solder. These two specimen configurations differ most importantly in the solder volume per joint and thus the cooling rate of the solder material.

The thermal fatigue specimen appears in Figure 7-3. Specimen manufacture is similar to the specimen in Figure 7-1, In this case, the central Cu plate is

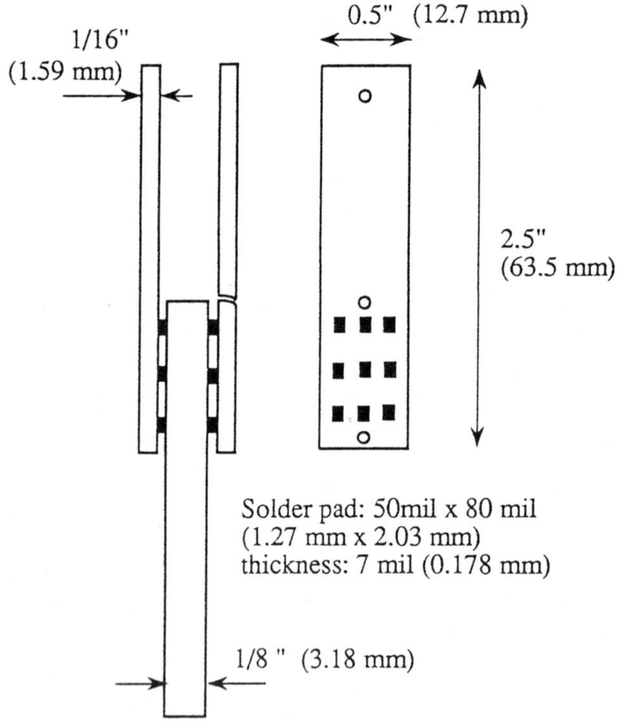

Figure 7-2 Specimen configuration used for creep testing of rapidly solidified solder joints. The specimen consists of two sets of nine individual solder contacts (on Cu) on a polymeric circuit board. On loading, an appropriate spacer is inserted between the two boards on top to ensure symmetric loading in shear of the solder contacts under axial loading of the specimen. The small solder volume per joint results in faster cooling rates of the solder material. (All dimensions are in inches. Pad size 0.03 by 0.07.)

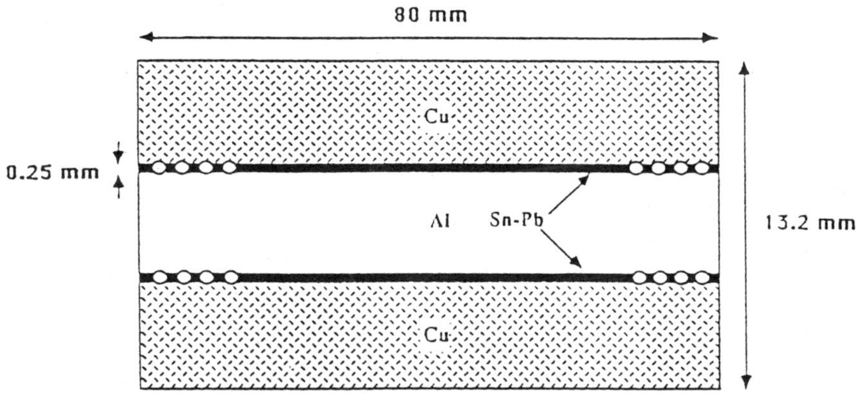

Figure 7-3 Specimen used for thermal fatigue experiments. The dark central regions are the solder joints. The mismatched thermal expansion characteristics of the constraining copper and aluminum strain the solder joints in shear with each temperature excursion. Since the strain magnitude varies linearly from the center of the specimen to a maximum strain at the edges, the holes along the length of the solder joints serve to isolate individual solder regions at a particular strain level.

replaced with a plate of Al (Ni and Cu plated to allow wetting by the solder). After polishing, fluxing, and assembly with an appropriate spacer, the assembly is immersed in molten solder. Slicing the block yields the individual specimens. A final drilling procedure results in isolated solder joints along the length of the specimens. The thermal expansion difference between the Cu and Al layers imposes a shear strain on the solder; the magnitude of this shear strain increases linearly from the center to the ends of the specimen. The holes along the length of the specimen serve to isolate individual solder regions at a particular shear strain.

7.2 EUTECTIC MICROSTRUCTURES

A central principle of materials science states that a material's properties originate not only from its atomic composition but also from the atomic arrangements within that material—the material's microstructure. However, for a material of a given chemical composition, the material's microstructure is not constant but can vary greatly with its processing history. It follows that if one is to understand and control the mechanical behavior of solder joints, one must begin not only with an understanding of the solder microstructure but also with an understanding of its origin in terms of various processing parameters. In the case of solder alloys, the application dictates use in the as-cast state, and hence, the cooling rate of the alloy is the only variable processing parameter. The local cooling rates of a solder contact depend on many factors, including the soldering process, the specific solder joint thickness and geometry, the composition and configuration of the carrier board, and the neighboring components. To understand the effect of cooling rate on solder joint microstructure and thus mechanical properties, it is instructive to first consider more simple solidification scenarios and the resultant solder microstructures.

7.2.1 Lamellar Eutectics

In the case of preferentially oriented plane-front steady-state growth, the eutectic Pb-Sn material grows as alternating lamellae of the two constituent phases parallel to the direction of growth. Eutectic solidification is a cooperative growth process since the solute rejected ahead of one phase region becomes immediately incorporated as the solvent phase in the adjacent region, and the plates thus grow at the same rate. Many lamellae constitute a lamellar grain and will continue to grow in such a coupled manner until contact with a mold wall or a similarly growing grain. A distinct boundary delineates two adjacent eutectic grains; an example appears in Figure 7-4. The lamellar spacing within the eutectic grain is determined by the cooling rate. A faster cooling rate results in fine lamellae,

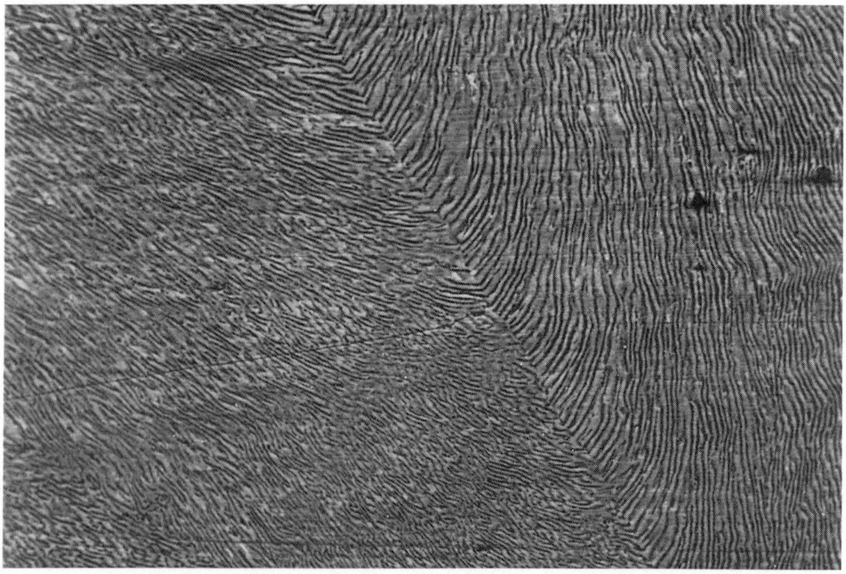

Figure 7-4 SEM micrograph of a eutectic Pb-Sn solder slowly cooled from 250°C. Slow cooling results in a highly lamellar structure as shown. In the figure, two eutectic grains are seen. Each grain is distinguished by a specific direction of lamellar arrangement, and a distinct boundary delineates each grain. The light regions are the Pb-rich phase; the dark regions are the Sn-rich phase.

and, in the case of a multigrained structure, the eutectic grains will be smaller and the lamellar spacing within each grain will be finer.[4–6] The microstructural features that make up such a structure are the eutectic grain size and the finer phase structure contained within.

It is important to note that within such a morphology, many individual phase regions (e.g., the lamellae) constitute a single eutectic grain. Within these eutectic grains all phase regions of the same kind exhibit a single and constant crystallography, and there is a unique and preferred crystallographic interface plane between the two constituent phases.[6–11] The lamellar grain just described is, in fact, two single crystallites of the two constituent phases.[10–14] The crystallographic relationships within and between the two phases are those that minimize growth and interfacial energies for a given set of solidification conditions. The stability of the liquid solid interface determines the inter- and intraphase crystallographic and morphological perfection.[7–10]

7.2.2 Degenerate Eutectics

A breakdown of the lamellar structure results from instabilities of the advancing liquid-solid interface. The resultant morphology may be wavy, rod-like, or globular (degenerate lamellar) and lacks the long-range perfection of the regular lamellar structures.[9] Interface instabilities may arise from growth along unfavorably oriented growth directions, local supercooling effects, and rapid cooling rates.[8–10] Solidification under such conditions results in discontinuities and faults of the individual phase regions and phase interfaces that disrupt the continued growth of the aligned lamellar structure. Instead, only a short-range phase alignment is maintained and a colony substructure develops within the individual eutectic grain. The colony is a subset of the eutectic grain and is characterized as a short-range phase alignment over a region of the eutectic grain separated

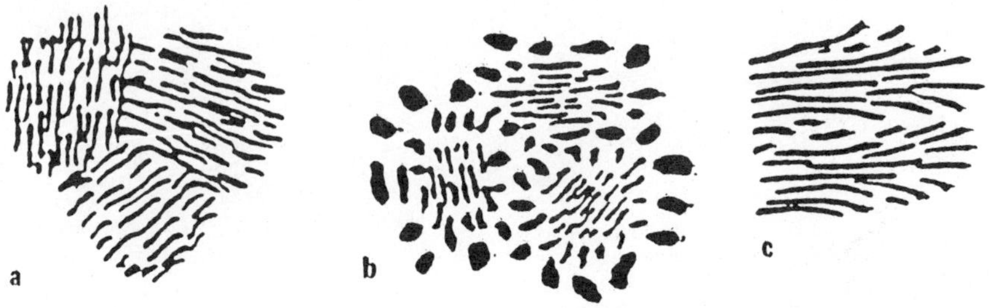

Figure 7-5 Schematic representations of several features of a eutectic microstructure. (*a*) Several lamellar grains showing distinct grain boundaries, (*b*) several lamellar grains showing irregular degenerate material at the grain boundaries, (*c*) a single eutectic grain showing a colony substructure.

from other eutectic colonies by a small crystallographic mismatch. Figure 7-5 schematically illustrates several features of a eutectic microstructure including the colony structure within the eutectic grain.[15] These degenerate structures lack the structural perfection of the lamellar eutectics, especially at the colony boundaries. However, they do maintain a special and unique crystallographic alignment with the growth direction and, as much as possible, between the two phases.

A dramatically different microstructure is obtained for quenching of the Pb-Sn eutectic material from the melt.[13] In this case, the solidification occurs so quickly that the eutectic colony formation is suppressed, and instead the microstructure appears as a dispersion of the Pb-rich phase in a Sn-rich matrix. The quenched microstructure, unlike the slower cooled eutectic colony structure, is uniform in phase size and shape distribution throughout the entirety of the specimen. And, unlike the eutectic grain structure described above, there is no apparent crystallographic relationship between the two phases; each individual phase region appears to be an individual phase grain. Figure 7-6 shows a bulk Pb-Sn eutectic specimen water quenched from 230°C. In contrast to a more

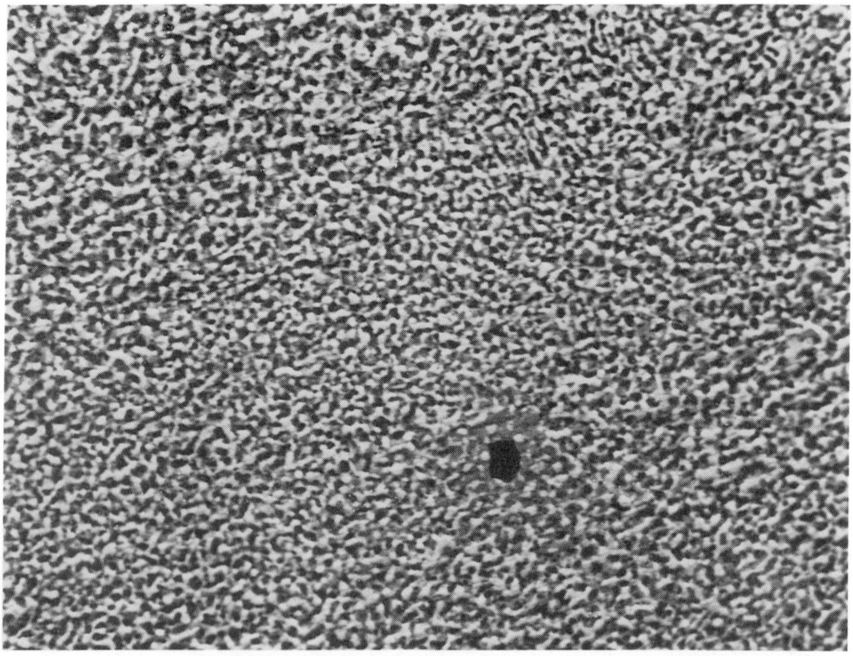

Figure 7-6 SEM micrograph of a eutectic Pb-Sn alloy water quenched from 250°C. The microstructure lacks the eutectic colony features characteristic of slower cooled specimens and appears instead as a homogeneous dispersion of the Pb-rich phase in a Sn-rich matrix. The light regions are the Pb-rich phase; the dark regions are the Sn-rich phase.

moderately cooled specimen the quenched microstructure is much finer and lacks the morphological heterogeneity that characterizes the colony structure.

The description of a eutectic microstructure is not straightforward. Unlike a single-phase material, a grain size is not readily apparent. Much of the published work on solder microstructures uses the phase diameter (i.e., the average diameter of a phase region in a two dimensional section) as the grain size. The phase diameter may not be a true grain at all. By definition, a grain refers to an element of a material within which a single crystallography exists. In a eutectic structure, many individual phase regions may, in fact, constitute a single eutectic grain (e.g., the two lamellar grains in Figure 7-4). The terminologies that apply to eutectic microstructures, as established by Weart and Mack,[14] include the eutectic grain, the eutectic colony, and the individual phase regions. The eutectic grain refers to that portion of the structure that nucleated at a single site and that shares a specific and unique crystallography. The eutectic colony may be a subset of a eutectic grain and refers to a region within the microstructure where the phase particles have a characteristic arrangement.

7.2.3 Solder Joint Microstructures

In a casting, such as a solder joint, solidification initiates from many different nucleation events. A favorably oriented seed may not be present and the cooling rate is such that there is insufficient time to select preferred growth directions. The final microstructure will thus contain both lamellar and degenerate eutectic features within a eutectic colony/grain arrangement. The distribution and size of these will vary with global and local cooling rates, larger lamellar eutectic colonies/grains dominating at slower cooling rates. Such a microstructure is characterized by a local homogeneity—within a eutectic colony/grain and a global inhomogeneity as the phase features differ from eutectic grain to eutectic grain and, especially, at the eutectic colony/grain boundary. Faster cooling rates yield progressively less distinct eutectic colony/grain structures and an increasingly more degenerate and globally uniform structure.

In the solder microstructures that are described in this work no distinction is made between a eutectic grain and eutectic colony. The term "eutectic colony" here refers to a region of the microstructure that exhibits a specific phase particle size and shape arrangement. For convenience, the term "colony" is being used to distinguish it from the grain size referred to in the superplastic and highly degenerate structures where an individual phase region constitutes in fact an individual crystallographic grain. It should be noted that polarized light microscopy of the colony or grain structures studied here reveals constant contrast within a colony feature and contrast changes between these eutectic structural features, thus indicating that the colonies are indeed individual and distinct crystallographic grains.

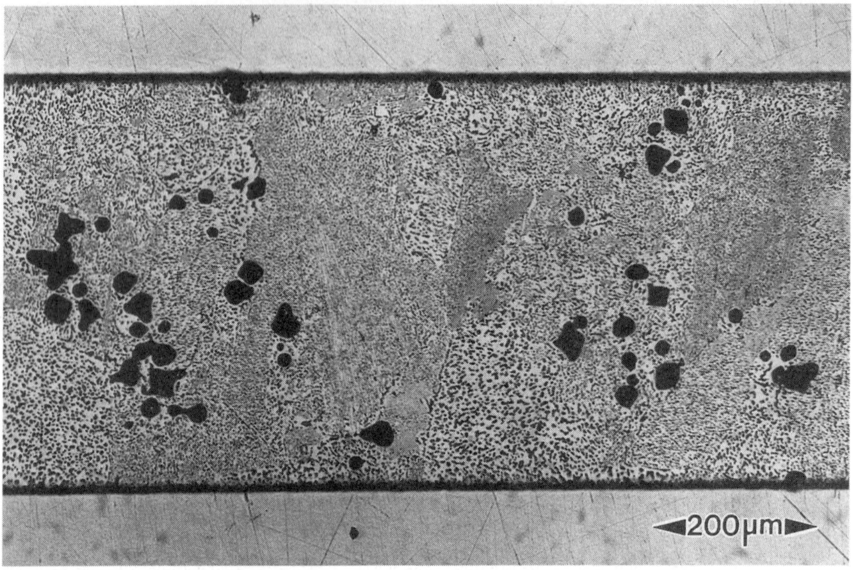

Figure 7-7 Optical micrograph of a typical 60% Sn-40% Pb solder joint. The dark regions are the Pb-rich phase. Note the phase distribution into grainlike features (i.e., colonies). This solder microstructure is representative of the solder microstructures in the specimen configuration shown in Figure 7-1. The light regions are the Sn-rich phase; the dark regions are the Pb-rich phase.

The eutectic Pb-Sn solder joint microstructure shown in Figure 7-7 is representative of the experimental joints used in this work. The bulk joint material consists of lamellar and degenerate lamellar features distinguishable as eutectic colonies with distinct boundaries separating the individual colonies. Note that the eutectic colony size is far larger than the individual phase regions and is almost on the order of the actual joint thickness. The complex and irregular nature of the solidification is consistent with the many different microstructural features observed. The microstructure in Figure 7-8 corresponds to a joint from the specimen configuration in Figure 7-2. The solder volume in this joint is quite small, and thus the solder experiences a much faster cooling rate. This is evidenced by the smaller occurrence of distinct colony features and the greater overall degeneracy of the eutectic structure. Figure 7-9 shows a cross section through a "real" solder joint. This particular joint attaches a resistor to a polymeric circuit board and was made by the wave soldering method.[16] The microstructure consists of individual and distinct eutectic grains that contain phase morphologies of both lamellar and degenerate character. Solder joint microstructures vary considerably with soldering processes and joint geometries. The regular lamellar

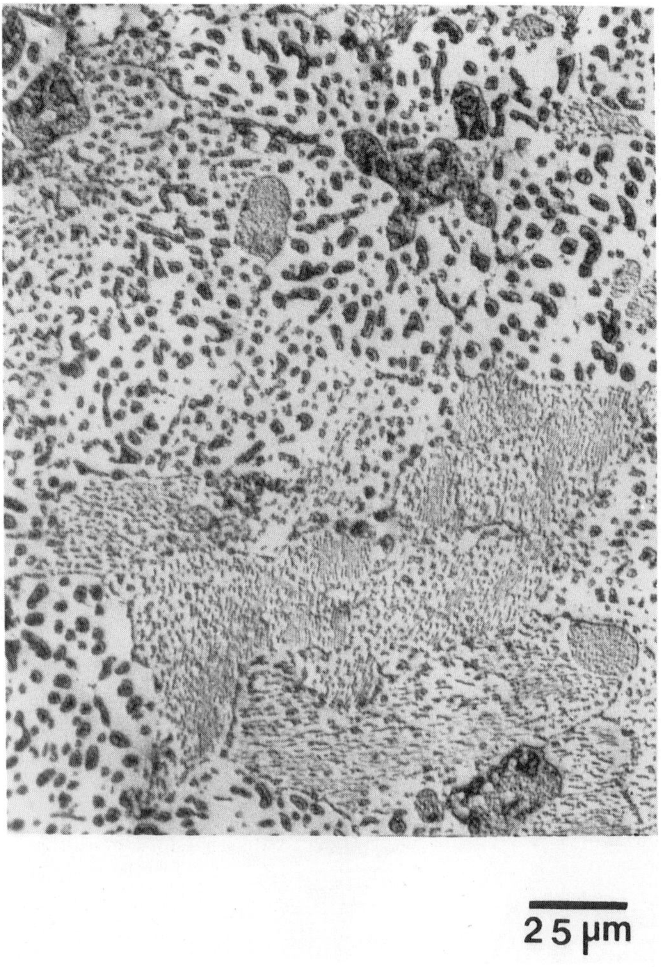

Figure 7-8 Optical micrograph of a 60% Sn-40% Pb solder joint. This solder microstructure is representative of the solder microstructures in the specimen configuration shown in Figure 7-2. The smaller joint solder volume results in a faster solder cooling rate. This is evidenced by the slightly smaller scale of the colony features and a greater degeneracy of the eutectic structure. The light regions are the Sn-rich phase; the dark regions are the Pb-rich phase.

colony features dominate with moderately cooled solder joints of reasonable thickness, and the more degenerate dispersed structures are characteristic of faster cooling rates and very thin joints. An understanding of the microstructural mechanisms of thermal fatigue should make clear the role of the initial microstructure and its importance to fatigue life.

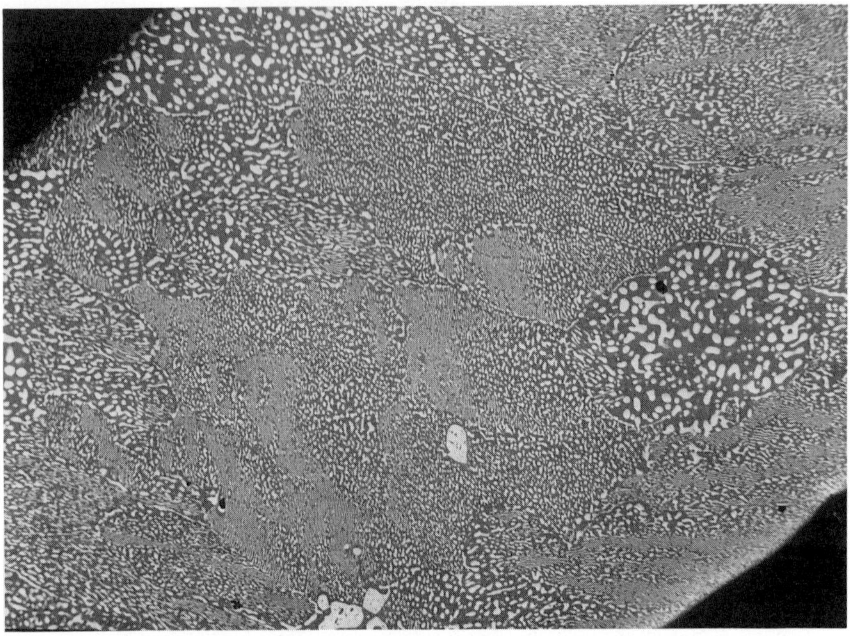

Figure 7-9 SEM cross section of a "real" solder joint. This particular joint was made by the wave soldering method. The microstructure contains both distinct lamellar colony features and regions of more degenerate character. The light regions are the Pb-rich phase; the dark regions are the Sn-rich phase.

Intermetallics

Since a Pb-Sn solder joint most frequently occurs on a Cu substrate, another microstructural component of the Pb-Sn solder joint is the intermetallic layer at the Cu-solder interface. The intermetallics form when the molten solder comes into contact with the clean Cu surface. In low Sn solders, the intermetallic is Cu_3Sn which grows as densely packed globular particles on the Cu surface. When the Sn content is higher, such as in the near-eutectic compositions, the intermetallic phase is a double layer with Cu_3Sn on the Cu substrate and Cu_6Sn_5 in contact with the molten solder. The Cu_6Sn_5 grows as hexagonal rods that protrude into the molten solder. Turbulence in the melt causes these spikes to break off and they appear as intermetallic particles in the bulk of the joint material.[17] Figure 7-10 shows detail of the interfacial and bulk intermetallic phase in a 60% Sn-40% Pb solder joint. It has been shown that the intermetallics have little influence on the solder mechanical properties reviewed in this work.[9,18–20]

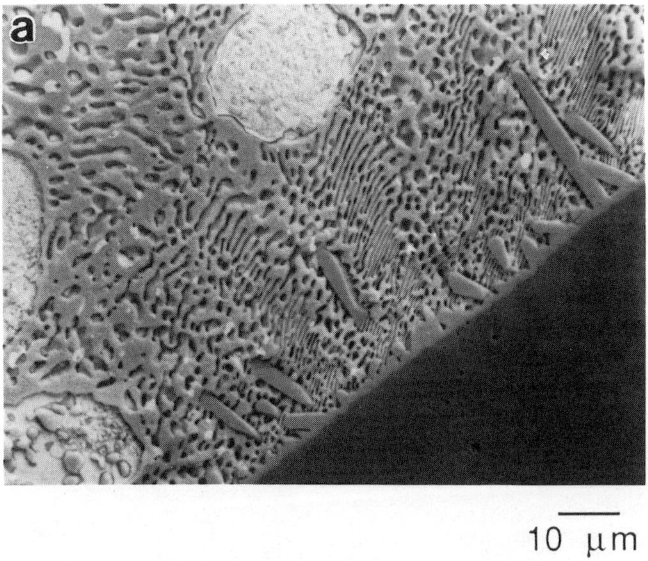

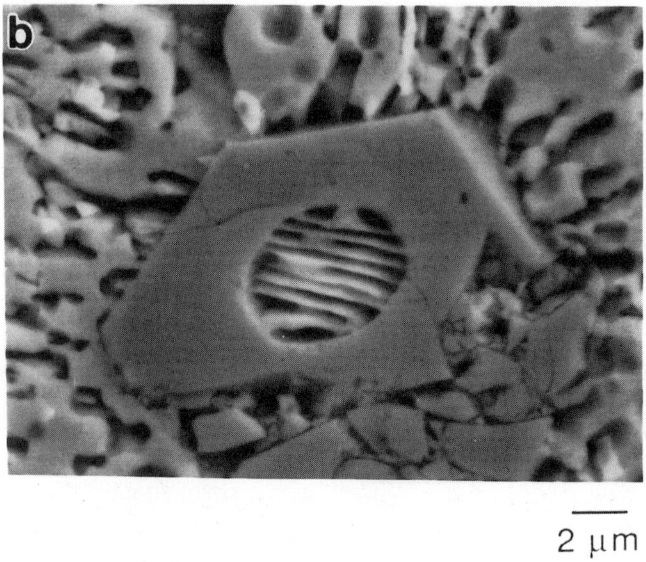

Figure 7-10 SEM micrograph of a polished 60% Sn-40% Pb surface etched to reveal details of the microstructure. (*a*) Interfacial intermetallics are clearly seen at the Cu-solder interface. The interfacial intermetallic phase is in fact a bilayer consisting of a Cu_3Sn layer and a layer of Cu_6Sn_5 adjacent to the solder. (*b*) Detail of a Cu_6Sn_5 intermetallic particle found in the bulk solder joint material. These intermetallics originate at the Cu-solder interface. Turbulence during solidification causes them to break off and locate within the bulk of the solder material. The light regions are the Pb-rich phase; the dark regions are the Sn-rich phase.

7.2.4 Effects of Composition

Other frequently used solder compositions are the off-eutectic, Pb-rich alloys. The Pb-rich solders exhibit microstructures that are very similar to those of the near-eutectic alloys and are thus also of interest to this work since a characteristic failure mechanism operates as long as the microstructure contains a sufficient volume fraction of the eutectic structure. The Pb-rich microstructures differ only in the appearance of the primary Pb-rich phase, as seen in Figure 7-11 for several Pb-rich alloys. The presence of this feature can be understood in terms of the

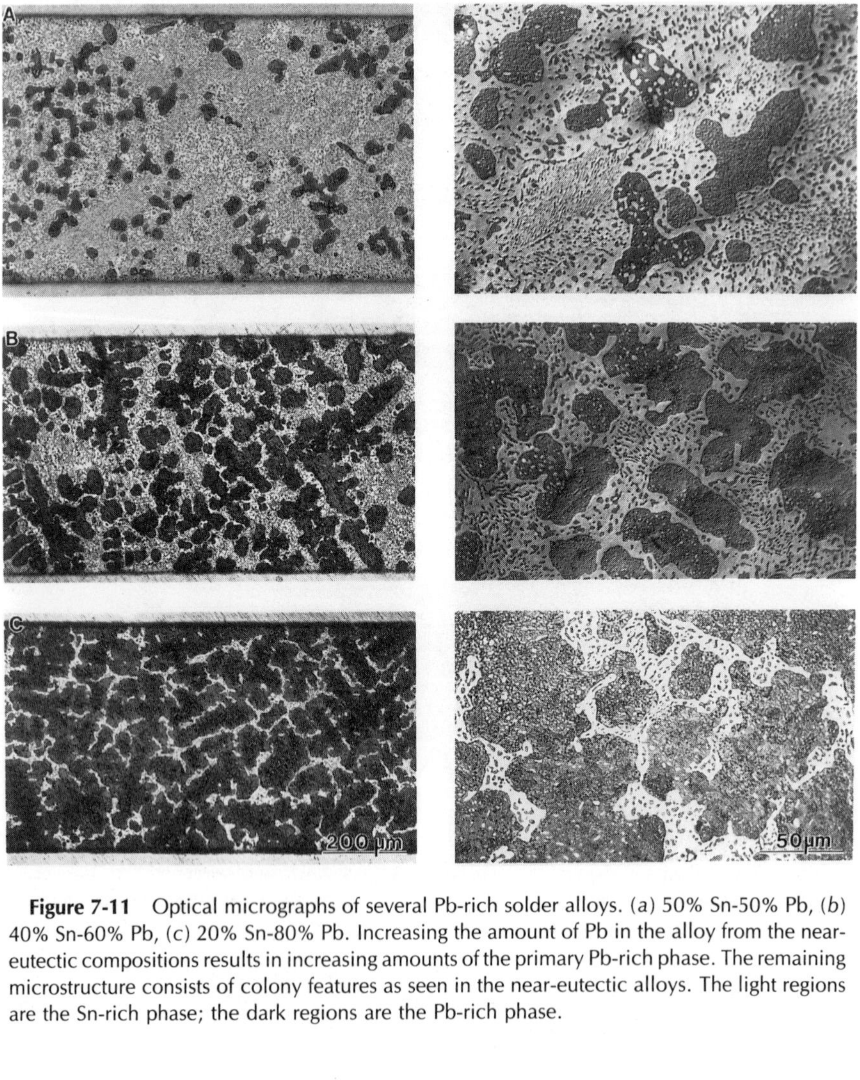

Figure 7-11 Optical micrographs of several Pb-rich solder alloys. (*a*) 50% Sn-50% Pb, (*b*) 40% Sn-60% Pb, (*c*) 20% Sn-80% Pb. Increasing the amount of Pb in the alloy from the near-eutectic compositions results in increasing amounts of the primary Pb-rich phase. The remaining microstructure consists of colony features as seen in the near-eutectic alloys. The light regions are the Sn-rich phase; the dark regions are the Pb-rich phase.

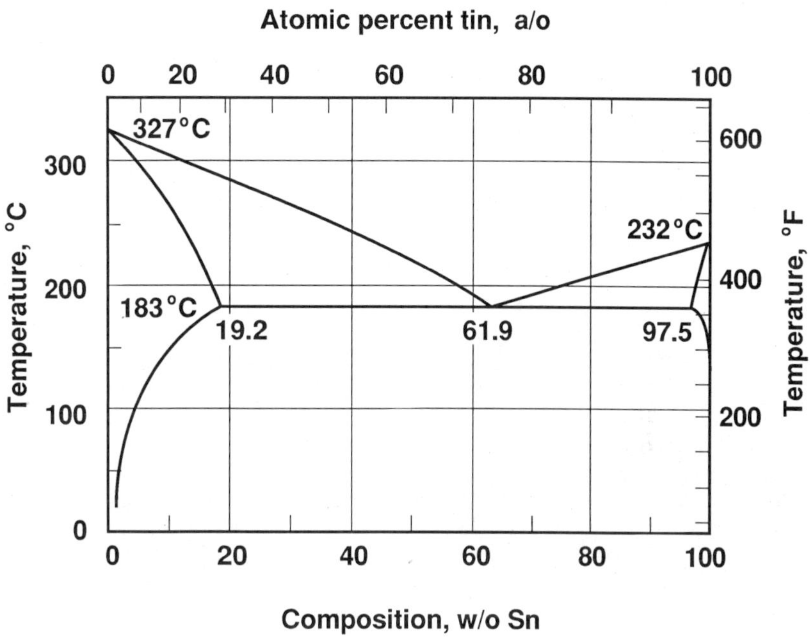

Figure 7-12 The Pb-Sn phase diagram. The eutectic composition occurs at 61.9% Sn-38.1% Pb.

binary Pb-Sn phase diagram of Figure 7-12. On cooling from the melt, the off-eutectic compositions pass through the Pb-rich α plus liquid two-phase region. At the temperatures that fall inside this phase field, the Pb-rich phase begins to solidify in a dendritic morphology. Upon reaching the eutectic temperature of 183°C, the remaining liquid solidifies into the classic eutectic microstructure. The final microstructure thus consists of primary Pb-rich dendrites in the eutectic structure. The volume fraction of the primary Pb-rich dendrites is determined by the initial composition and cooling rate.

A relaxation of the regularity and perfection of the eutectic lamellar structure is also possible with a small addition of a third element.[21] During solidification, the third element in the eutectic Pb-Sn alloy, acting analogously to an impurity in a single component system, will be rejected from the solidifying phases into the remaining liquid. (This analogy is valid when the partition coefficient is <1, as is expected here based on an analysis of the binary diagrams of the impurity and constituent phases.)[4–6] The resultant compositional changes result in an area of constitutional supercooling directly ahead of the advancing solid-liquid interface and allow for the perturbation and breakdown of the solidification front into a colony-(cellular) structure. The coarser and irregular features at the colony boundaries result from the slower growth rates encountered at the curved colony interfaces.[6,14,15,22,23]

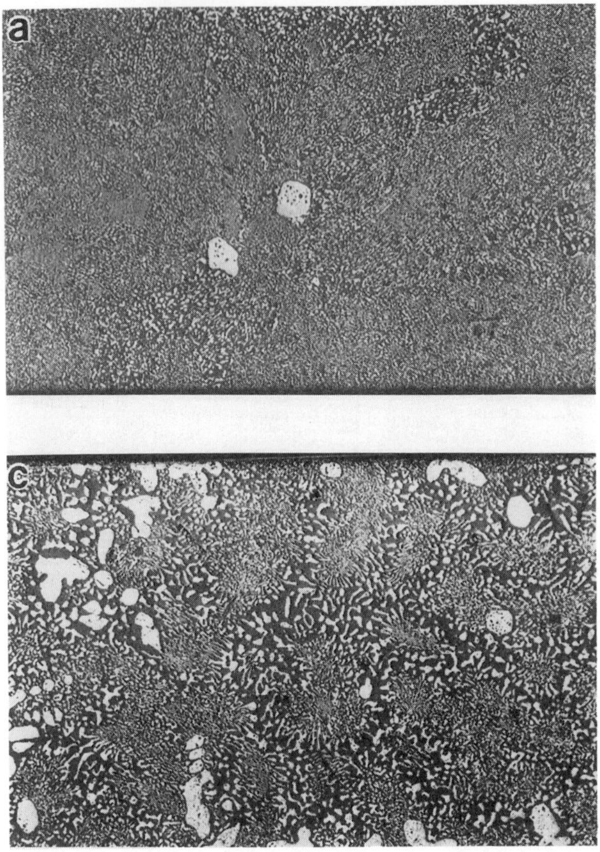

Figure 7-13 Optical micrographs of several ternary alloys. (*a*) 58% Sn-40% Pb-2% Sb, (*b*) 58% Sn-40% Pb-2% Bi, (*c*) 58% Sn-40% Pb-2% In, 58% Sn-40% Pb-2% Cd. All microstructures are the result of identical processing histories; hence, the microstructural changes are due to compositional changes only. Note the appearance of diffuse colony boundaries and the associated increase in the amount of degenerate material with the In and Cd additions. The light regions are the Sn-rich phase; the dark regions are the Pb-rich phase.

Figure 7-13 shows microstructures of several ternary alloys, 58% Sn-40% Pb-2%X where X is Sb, Bi, Cd, and In. All samples were prepared using equal additions of the third element and identical processing procedures; hence the microstructural changes are due to compositional changes only. The addition of Sb does not dramatically alter the solder microstructure. However, the addition of Bi, In, and Cd changes both the colony macrostructure and the eutectic microstructure within. The boundary area between individual colonies is much wider and consists of coarse irregular phase particles. Additionally, the phase-

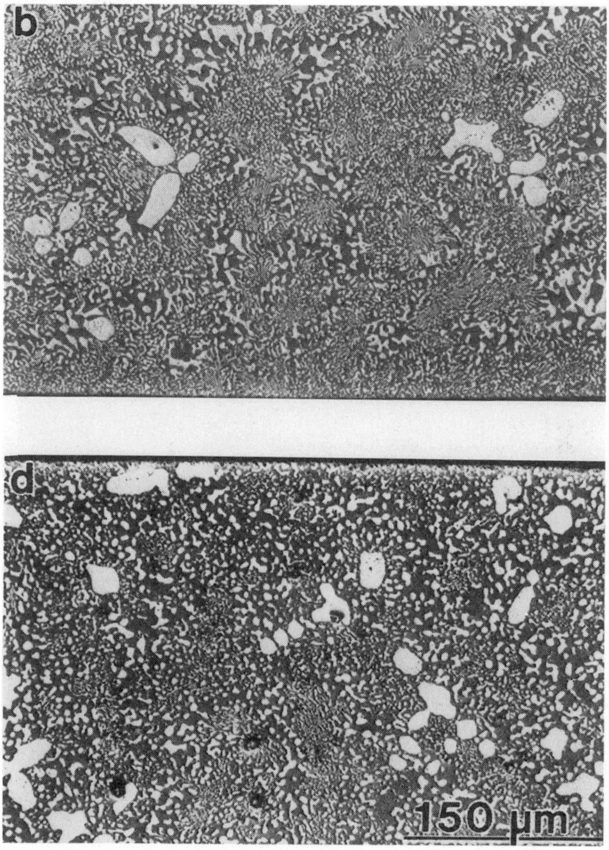

Figure 7-13 *(continued)*

to-phase arrangement within the colonies is more faulted and the areal extent of the lamellar eutectic arrangement is reduced. Of the alloys studied here, the Cd alloyed solder exhibits the least distinct colony structure (i.e., this solder possesses the most degenerate, most globally uniform microstructure).

7.2.5 Recrystallized Pb-Sn Microstructure

A very different Pb-Sn microstructure is achieved after a mechanical working of the as-solidified Pb-Sn material. The plastic deformation introduced with, for example, rolling, swaging, or extrusion of the as-cast bulk solder introduces a very high dislocation density into the Pb-Sn microstructure that in turn initiates a recrystallization of the deformed material. The recrystallization process in-

Figure 7-14 Optical micrograph showing a worked and annealed 60% Sn-40% Pb microstructure. The thermomechanical treatment results in a recrystallized microstructure in which the colony features characteristic of the as-solidified structure have been completely replaced by a uniform distribution of equiaxed Pb-rich and Sn-rich grains. The light regions are the Sn-rich phase; the dark regions are the Pb-rich phase.

volves the nucleation of new strain-free grains in the deformed material that grow until the latter is consumed. Thermal energy above some threshold is necessary to initiate this process; however, since solder is already at a high homologous temperature at 25°C, recrystallization may occur dynamically as the material is being deformed. The recrystallized microstructure consists of equiaxed, randomly oriented grains of the two constituent phases uniformly distributed throughout the material. As shown in Figure 7-14, this type of Pb-Sn microstructure bears little resemblance to the eutectic colony structure that is characteristic of the normally cooled, as-solidified material. The ease of recrystallization and the dramatic difference in appearance of the recrystallized material are well documented for this alloy.[24–27]

7.2.6 Coarsening Behavior

The coarsening behavior of as-cast eutectics varies significantly with the various possible morphologies. The perfect lamellar structures exhibit exceptionally high morphological stability at elevated temperatures.[7,27–29] This contrasts greatly with the rapid spheroidization and coarsening of the more faulted degenerate and dispersed structures and the irregularly shaped material comprising the colony boundaries. Coarsening always initiates in these less regular regions and very slowly consumes the more perfect material. The difference is attributed to the higher interfacial energies of the less favorably aligned structures and the enhanced diffusion afforded by the greater fault and grain boundary areas.[30–32]

7.3 MECHANICAL PROPERTIES

Just as the processing history determines a material's microstructure, the microstructure in turn determines the mechanical properties. Since the Pb-Sn eutectic structure exhibits dramatically different morphologies with different processing histories, it is not surprising that these different microstructures exhibit significantly different mechanical properties.

7.3.1 Eutectic Structures

The regular lamellar microstructures characteristic of the as-cast Pb-Sn eutectic material possess the highest yield strengths and creep resistances and the lowest elongations and strain rate sensitivities within this alloy system.[24,25,33–37] The exceptional strengths are attributed to the large-scale, regular arrangement of the lamellar structure and the associated difficulty of dislocation motion through the aligned two-phase structure.[12] Deformation of the multigrained material is most easily accommodated at the eutectic colony or grain boundaries where these strengthening mechanisms are less effective. For a multicolony structure, such as that observed in the joints in Figures 7-7, 7-8, and 7-9, the deformation character is thus highly inhomogeneous. Deformation is more easily accommodated in the softer intercolony areas than in the stronger regularly aligned phase arrangement inside the colony. This is readily apparent on observing the patterns of surface relief that develop on a flat polished eutectic colony structure as it is deformed. Figure 7-15 shows such a pattern. Relatively little deformation occurs in the colony interior; rather, the surface relief indicates that the strain is accommodated by the degenerate material at the colony boundaries. Furthermore, microhardness traces across such a eutectic structure, as shown in Figure 7-16, illustrate that the irregular phase arrangement at the colony boundary is softer than the regular phase arrangement within the eutectic colonies. Processing changes that reduce the colony size and thus increase the amount of soft boundary material predictably affect the observed mechanical properties. Castings of bulk specimens show a dramatic decline in the yield strength[34,38] and creep resistance and an increase in strain rate sensitivity and elongation[35] with faster cooling rates. Faster cooling rates increase the amount of degenerate

Figure 7-15 Optical micrograph of a 60% Sn-40% Pb solder joint showing the surface relief that develops as the joint is deformed. The deformation pattern shows that the response of the solder microstructure is highly inhomogeneous; deformation occurs predominantly at the colony boundaries where the strengthening mechanisms that characterize the interior of a eutectic colony are lacking. The light regions are the Sn-rich phase; the dark regions are the Pb-rich phase.

irregular material through reduced colony size and structural perfection and thus lead to an overall softening of the material.

The shear strengths of the ternary alloys appear to follow a similar trend. The alloys that exhibit the greatest volume of lamellar aligned eutectic material also exhibit higher shear strengths. The Cd and In alloyed solders, those microstructures possessing larger fractions of irregular degenerate phase particles, exhibit slightly lower yield strengths. Table 7-1 lists the shear strength data for the various alloys. All data refer to the specimen configuration in Figure 7-1.

A softening also occurs within the simple binary system when the amount of eutectic structure is decreased by increasing the amount of Pb in the alloy. In the Pb-rich alloys, the primary Pb-rich phase replaces the classic lamellar eutectic structure. Since the Pb-rich phase is relatively soft, its presence in significant amounts leads to a decreased shear strength in the Pb-rich alloys.[39]

An even greater difference in mechanical behavior is exhibited by the recrystallized Pb-Sn material. Within this homogeneous, fine-grained microstructure the strengthening mechanisms associated with the composite nature of the lamellar eutectic are lost and the deformation character is greatly altered. The fine equiaxed grain size and associated large grain boundary area contribute to the exceptional softness and ductility of the recrystallized microstructure.[24,40–42]

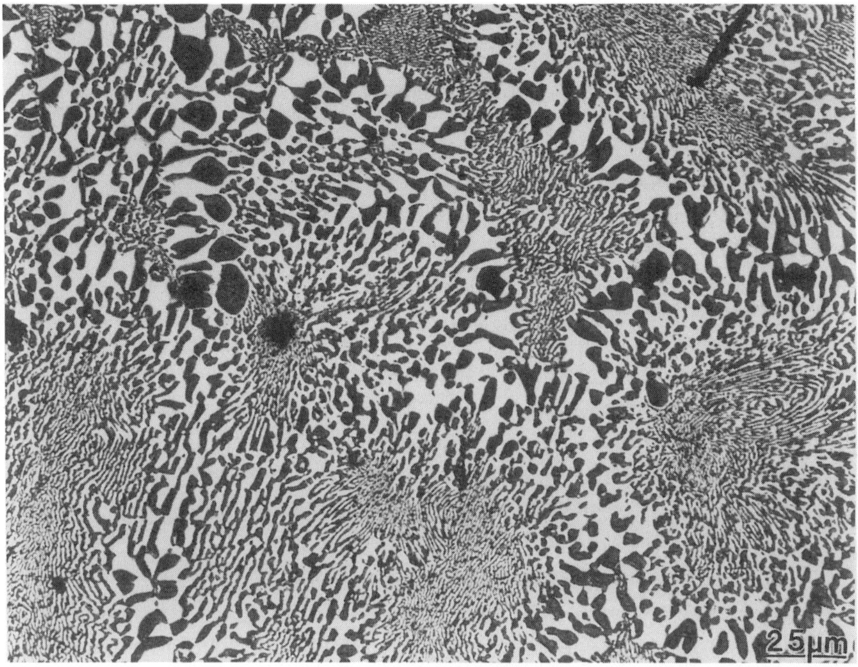

Figure 7-16 Optical micrograph of microhardness traces across a multicolony eutectic Pb-Sn microstructure. The hardness values illustrate the difference in mechanical response between the regular phase arrangement inside a colony, relatively hard material (17 Vicker's microhardness number), and the degenerate features at the colony boundary, relatively soft material (11 Vicker's microhardness number). The light regions are the Sn-rich phase; the dark regions are the Pb-rich phase.

Table 7-1

	Stress Exponent*	Shear Strength (MPa)
60% Sn-40% Pb	6.0	22
61.9% Sn-40% Pb	6.3	23
58% Sn-40% Pb-2% Bi	5.5	25
58% Sn-40% Pb-2% Sb	5.6	26
58% Sn-40% Pb-2% In	3.7	19
58% Sn-40% Pb-2% Cd	3.3	18

*Stress exponent refers to Equation (7-1).

Note: Right column: Shear strengths of several Pb-Sn alloys. These data were obtained on the specimen configuation shown in Figure 7-1. Left column: Stress exponents, as determined from creep experiments according to Equation (7-1) of several ternary alloys. The simple binary Sb-alloyed and Bi-alloyed solders exhibit stress exponents characteristic of conventional plasticity. The In- and Cd-alloyed solders exhibit slightly lower stress exponents. This division is also consistent with the extent of lamellar eutectic nature of the solder microstructure. The higher stress exponents are associated with the greatest amount of lamellar eutectic material.

In fact, under the appropriate deformation conditions, this type of microstructure is superplastic, exhibiting elongations on the order of hundreds of percent.[24,43,44]

The strength of the eutectic Pb-Sn material arises from the regular lamellar eutectic arrangement of the individual phase regions. A reduction in the areal extent and perfection of this classic eutectic structure also reduces the material's strength. Faster cooling rates and appropriate compositional changes increase the amount of degenerate material and thus soften the material. The microstructure that exhibits no trace of the colony structure (i.e., the recrystallized microstructure) is by far the softest. In the microstructures that exhibit a colony structure, the colony size is an important microstructural parameter since this determines the ratio of hard to soft material. This microstructural heterogeneity also accounts for the highly inhomogeneous deformation that is observed in the colony microstructures. In the degenerate and recrystallized material, deformation is more uniformly distributed since the area over which the deformation can be easily accommodated, the grain boundary regions, is larger.

7.3.2 Deformation Mechanisms

Solder service conditions entail operation at temperatures that represent high solder homologous temperatures; creep deformation is thus an important deformation mode, and the study of these low strain-rate high temperature deformation mechanisms is thus warranted.

To investigate the predominant deformation mechanisms the steady-state creep data of an alloy system are represented by the equation

$$\frac{d\gamma}{dt} = A\tau^n e^{-E_a/kT} \tag{7-1}$$

where $d\gamma/dt$ is the steady-state shear strain rate and τ is the shear stress.[27,43–51] In this representation, it is believed that, within a wide strain rate-stress regime, there exist regions of constant n and E_a values within each of which a different deformation mechanism dominates. For example, superplastic deformation of fine equiaxed microstructures is characterized by a stress exponent in the range of 2 and an apparent activation energy equal to the activation energy of grain boundary diffusion.

Superplastic deformation of the eutectic Pb-Sn system has been extensively studied and the values of these constants are well known. The microstructures that exhibit superplasticity are not those of the as-cast material but rather those microstructures that result after a mechanical working of the as-cast alloy (i.e., a recrystallized microstructure that is uniformly fine grained and equiaxed).[26,27,44,51–60] Superplasticity of the recrystallized Pb-Sn microstructure has been observed for strain rates between 10^{-7} to 10^{-2} sec^{-1} with a stress exponent ~ 2[26,27,44,51–59] and activation energy on the order of 12 kcal/mol (a favorable agreement with the activation energy for grain boundary diffusion).[27,52,53,55,57]

Other deformation mechanisms, such as those operative in nonsuperplastic microstructures or in different strain rate regions, are characterized by different values of these constants. For example, conventional dislocation climb controlled plasticity is associated with much higher stress exponents, 4 to 10, and activation energies on the order of those for self-diffusion.[38] Within the Pb-Sn system, conventional plasticity (e.g., deformation of nonsuperplastic microstructures), is associated with a stress exponent of ~6 to 7 and an activation energy of ~20 kcal/mol.[27,44,49,61,62]

The fine equiaxed microstructures necessary for superplastic deformation are not characteristic of as-solidified joint solder microstructures, such as those shown in Figure 7-7. Hence, these joints are not expected to deform superplastically. Results of creep experiments on these solder joints are consistent with this expectation and reveal deformation constants representative of conventional plasticity. Figure 7-17 shows a plot of strain rate versus stress for these solder joints. The slope of this curve, according to the representation of Equation 7-1,

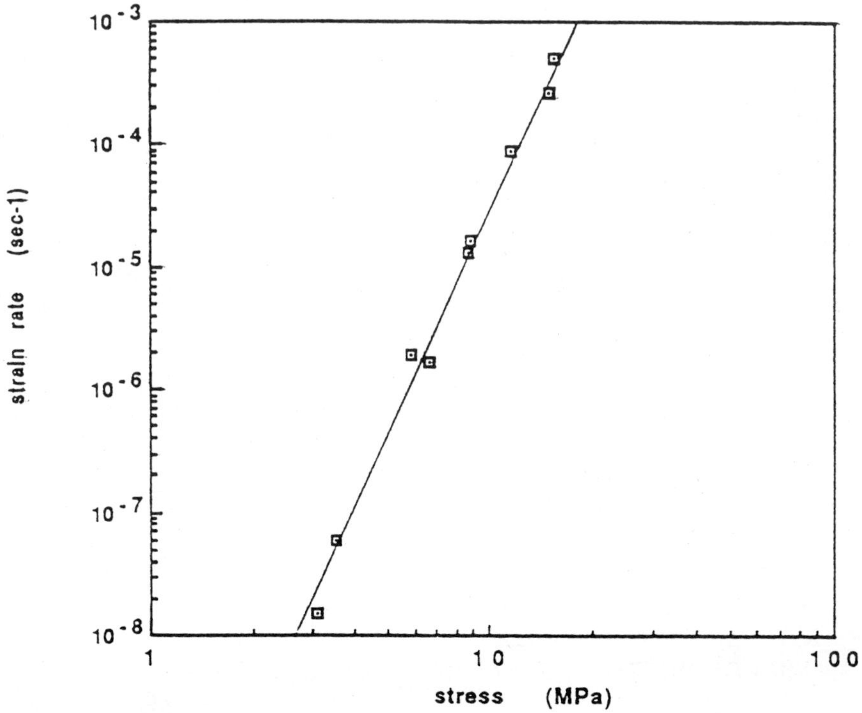

Figure 7-17 A plot of steady-state strain rate versus stress, at constant temperature, for the eutectic Pb-Sn alloy. The slope of the line is equal to 6.0, indicating conventional plasticity.

is the stress exponent; in this case, the stress exponent has a value of 6, indicating conventional plasticity.

Creep experiments on the more rapidly cooled solder joints, where the smaller colony size and decreased amount of lamellar eutectic material exhibit slightly different, lower, stress exponents, suggest different deformation mechanisms. The smaller joint volumes of the joints in the specimen configuration of Figure 7-8 result in faster cooling rates and thus a more degenerate, less "eutectic" structure. Preliminary measurement of creep deformation of this solder microstructure reveals a lower stress exponent, approaching 2—the value normally associated with superplastic deformation.[63] Superplasticity in an unworked unrecrystallized rapidly solidified eutectic solder has not been unambiguously demonstrated; the possibility has recently been suggested by Shine and Fox.[64] However, a different microstructure may indeed imply a different deformation mechanism. Furthermore, it seems reasonable that as a microstructure acquires more and more degenerate eutectic material (i.e., the more closely it resembles a superplastic microstructure), the deformation character will too acquire a superplastic component and thus exhibit a lower stress exponent. This area is currently under investigation.

A survey of the work on the creep deformation of near eutectic Pn-Sn solder in a joint configuration reveals stress exponents in the range ~ 2 to 6.[64–67] These data are not always accompanied by microstructural information; however, the different joint geometries and joint manufacturing techniques used in these experiments most probably result in a variety of joint solder microstructures. As with the data presented earlier, the scatter may be due to the varying degree of classic lamellar eutectic structure that occurs with the different joint manufacturing techniques.

A similar observation is made for the ternary alloys. The stress exponents measured for solder joints of the configuration shown in Figure 7-1 appear in Table 7-1. It is possible to separate these solders into two groups. The Cd and In alloys exhibit a stress exponent between 3 and 4, and the Sb, Bi, and simple binary alloys exhibit stress exponents of ~ 6. This classification is also consistent with the observed microstructures. The Cd and In alloys, though solidified under similar conditions, possess the more degenerate microstructures. The remaining three possess the most lamellar structures. As with the binary alloys, it is possible that the reduced stress exponent of the Cd and In is the result of the more degenerate nature of the joint solder microstructure and an associated superplastic component in the deformation.

7.4 MICROSTRUCTURAL EVOLUTION UNDER THERMAL FATIGUE

Thermal fatigue of solder joints is driven by the difference in the thermal expansion coefficients of the surfaces that bind the solder. Figure 7-18 schematically

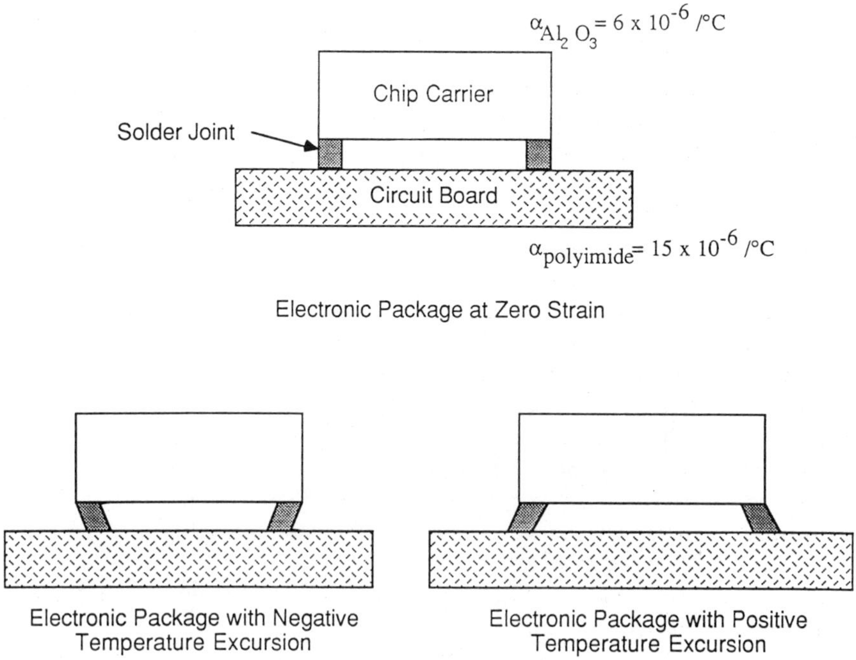

Figure 7-18 Schematic illustration of the shear that develops in a solder joint between chip and substrate with high and low temperature excursions.

illustrates a typical joint geometry and the origin of the thermal fatigue problem. As shown in the illustration, for a surface mounted leadless component, the dominant load on the joint is in shear; even for more complex joint geometries, initial failure of solder joints in service often occurs in that portion of the joint that experiences shear loading. For these reasons thermal fatigue under shear loading and associated microstructural evolution is of particular interest.

7.4.1 Thermal Fatigue in Shear

Figure 7-19 illustrates the microstructural changes that occur in a 60% Sn-40% Pb solder joint as it is cycled from −55 to 125°C. The specimen configuration is shown in Figure 7-3. The uppermost image is the initial microstructure; subsequent images represent the same spot in the specimen with an increasing number of thermal cycles. The initial microstructure is the classic eutectic structure described earlier consisting of eutectic colonies made up of alternating phase regions of the Sn- and Pb-rich phases. As the cycling proceeds, the shear deformation concentrates into bands that are clearly marked by an associated rapid

Figure 7-19 A series of optical micrographs showing the microstructural evolution as a joint is cycled between −55 and 125°C. Relatively early on in the cycling, a band of coarsened microstructure appears near the substrate-solder interface. This band becomes more pronounced with increased thermal cycles, and eventually cracks initiate within this coarse microstructure causing the joint's failure. The light regions are the Sn-rich phase; the dark regions are the Pb-rich phase. The top micrograph was taken prior to thermal cycling while the bottom one after 1700 cycles.

microstructural coarsening. The band of coarsened material traverses the initial microstructure parallel to the Cu-solder interface, along the direction of maximum shear. After 1700 cycles, cracking of the solder joint is clearly visible within this coarsened band. Higher magnification SEM images of the coarsened band, Figure 7-20, clearly show its definition along with the equiaxed and coarsened individual Sn- and Pb-rich phase grains. As shown in Figure 7-19, failure occurs exclusively through this coarsened material. An examination of published microstructural studies of thermally fatigued and failed solder contacts shows that this fatigue behavior has been widely observed.[19,20,68–72]

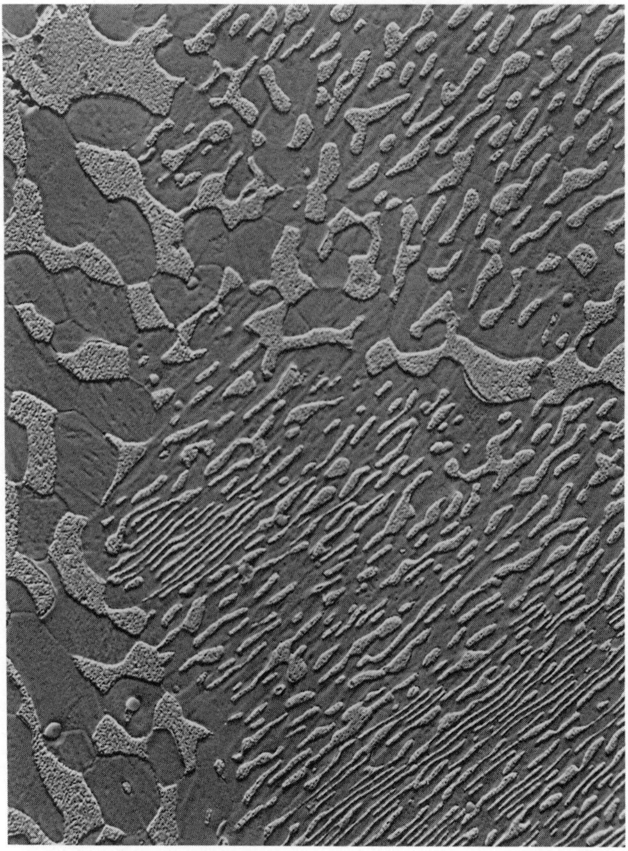

Figure 7-20 High magnification SEM micrograph showing the definition and detail of the coarsened band during thermal fatigue. Within the coarsened band, individual Pb-rich and Sn-rich phase grains are readily discernible. The light regions are the Pb-rich phase; the dark regions are the Sn-rich phase. The uppermost micrograph corresponds to the initial microstructure after 1700 cycles.

7.4.2 Microstructural Mechanisms of Thermal Fatigue

The fatigue failure of the solder joint is directly related to the formation of a planar band of microstructurally coarsened material. The formation and growth of the coarsened band are intimately related to the inhomogeneous deformation character of the as-solidified solder. This fatigue failure behavior is also characteristic of solder joint failures under thermal fatigue,[19,20,66–72] isothermal fatigue,[65,69,73] or high temperature creep,[20,74] where the solder joint has reasonable

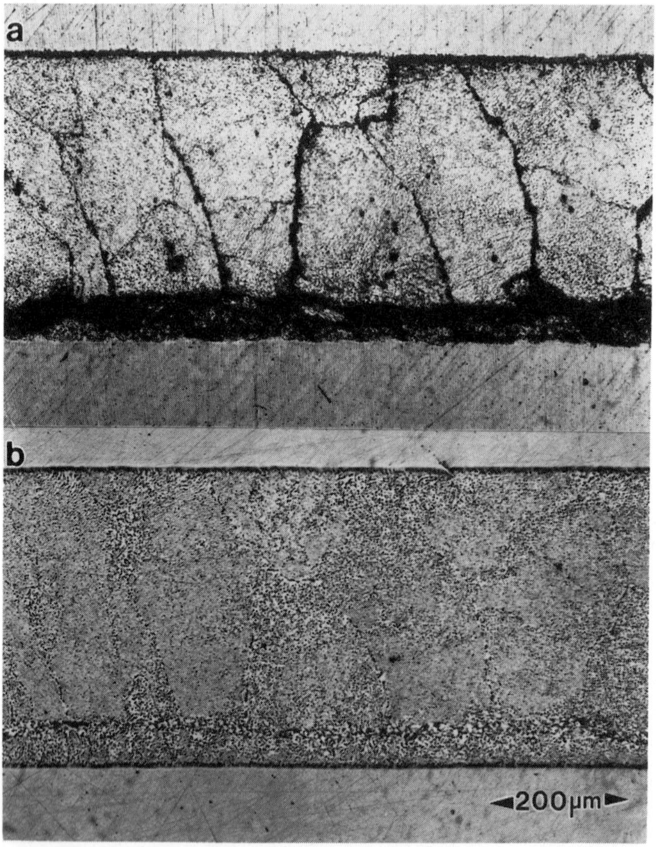

Figure 7-21 Optical micrographs of a polished solder joint deformed in shear. (*a*) The surface relief that develops with deformation. The deformation is localized to a thin band near the Cu-solder interface and along colony boundaries elsewhere in the joint. (*b*) The same specimen as in (*a*) but polished to remove the surface relief. There is an exact correspondence between the location of the major deformation band and a band of recrystallized and coarsened microstructure. The deformation elsewhere in the joint corresponds to the irregular material at the colony boundaries. The light regions are the Sn-rich phase; the dark regions are the Pb-rich phase.

thickness and the lamellar eutectic structure predominates. A closer study of the sequence of events that initiate and govern this failure mechanism is readily possible with appropriate creep experiments. The sequence of metallurgical changes that lead to fatigue failures can be described as follows.

The microstructural event that initiates the pattern of failure is the inhomogeneity of the plastic deformation of the as-solidified eutectic microstructure as it is deformed in shear. Figure 7-21 shows the deformation pattern that forms when the solder joint diagrammed in Figure 7-1 is deformed slowly in shear. The deformation pattern is easily observed by polishing the surface of the solder joint prior to testing and observing the surface relief that appears as the joint deforms. The most common pattern is that shown in Figure 7-21 and consists of a nearly straight band of deformation that runs slightly displaced and parallel to the Cu-solder interface along the direction of maximum shear and more irregular deformation features that lie approximately perpendicular to this direction. A comparison of the deformation pattern and the corresponding polished solder joint surface reveals the direct correspondence between the deformation and the location of microstructurally coarsened material. The definition of the coarsened band and the microstructure within are seen in greater detail in Figure 7-22. The material within the coarsened band consists of equiaxed, coarsened

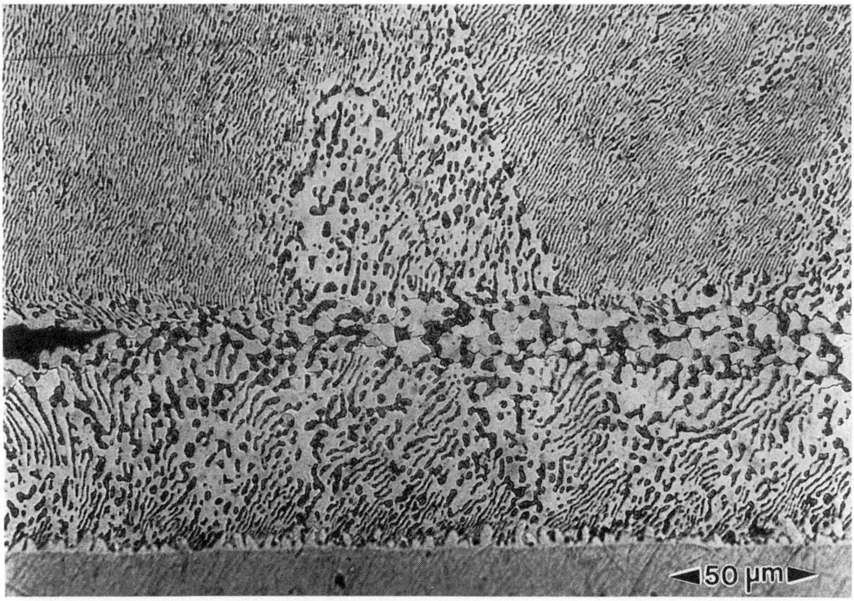

Figure 7-22 Detail of the coarsened band seen in Figure 6-21 (optical) showing its definition and the presence of equiaxed Pb- and Sn-rich phase grains apparently formed by recrystallization. The light regions are the Sn-rich phase; the dark regions are the Pb-rich phase.

grains of the Sn- and Pb-rich phases with no trace of the initial as-solidified colony structure visible. Closer inspection shows that the parallel band in Figure 7-21 traverses eutectic colonies, while the perpendicular coarsening occurs at colony boundaries. In all cases, the parallel deformation originates at a lateral free surface of the joint, a point that corresponds to the principal stress concentration in this sample geometry (see Fig. 7-23).[75]

The initiation of the coarsened band seems straightforward. The joint solder microstructure deforms inhomogeneously; deformation occurs primarily along colony boundaries where the softest material is found. The inhomogeneity of the shear strain is further enhanced in the vicinity of a stress concentration. It is well known that the Pb-Sn is unstable with respect to recrystallization if it is sufficiently worked at a sufficiently high homologous temperature or exposed to high temperature after deformation. The inhomogeneous shear strain that develops during deformation creates regions of concentrated plastic deformation. This concentration is greatest near the principal stress concentration within the joint configuration, and the eutectic recrystallizes there to relieve the accumulated deformation. (Figure 7-23 shows the location of the stress concentration in this joint geometry, the four lateral free surfaces.)

The mechanism of growth of this material into the recrystallized bands observed also seems straightforward. Once a recrystallized region has formed, it grows through the progressive recrystallization of the deformed material. Since

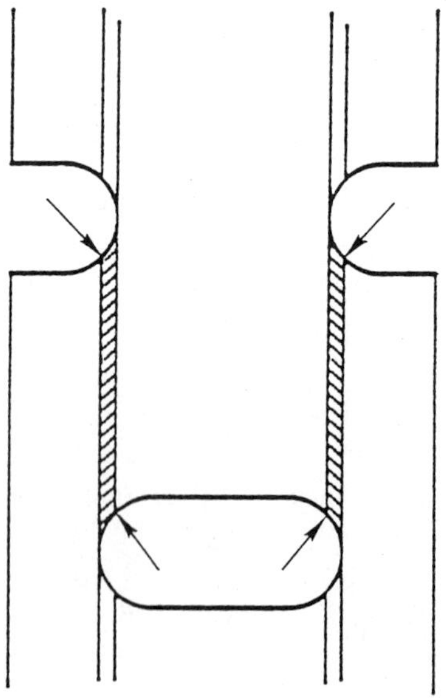

Figure 7-23 Schematic of the double shear specimen showing the principal stress concentration of this joint geometry.

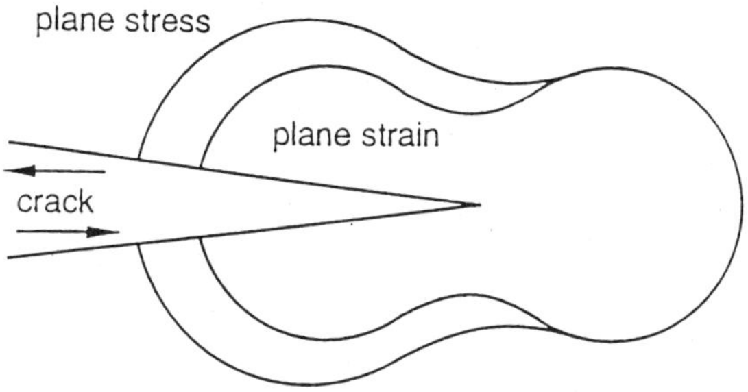

Figure 7-24 The plastic deformation field ahead of a mode II crack in a Von Mises material (after McClintock et al.[76]).

the recrystallized material is soft compared to the material surrounding it, it behaves mechanically like a narrow crack. The applied stress is a simple shear, and the associated strain field is characteristic of a mode II crack. The plastic strain field for a mode II crack has been solved for a Von Mises material by McClintock, et al.,[76] and is diagrammed in Figure 7-24. Note that the deformation field is narrow and tends to confine the deformation to the plane of the crack. It follows that recrystallization and coarsening will always occur at the periphery of the recrystallized region in the direction of the applied shear since this is the most highly deformed material. Thus the recrystallized band tends to remain planar and narrow while it extends itself along a shear plane (parallel to the Cu-solder interface), as is observed.

Inhomogeneous shear deformation, resulting from the inhomogeneity of the as-solidified microstructure, triggers inhomogeneous coarsening of the eutectic microstructure that is concentrated in bands of maximum shear. Since the microstructure softens as it coarsens, shear deformation concentrates in the coarsened regions, which therefore coarsen more rapidly. Eventually, the accumulated strain leads to cracks within the coarsened material, and the joint ultimately fails.

7.4.3 Other Microstructures

The proposed mechanism and its dependence on the mechanical inhomogeneity and instability of the solder microstructure suggest that it may be possible to develop solders with improved fatigue properties. The guidelines for solder alloy design are relatively clear. Since the coarsened band that precipitates failure is the result of an initiation and growth process, steps should be taken to inhibit its initiation and/or inhibit its propagation. Both approaches have been explored and the results are consistent with our current understanding of the fatigue failure mechanism.

Ternary Alloys

High temperature creep deformation for certain ternary alloys is associated with the development of relatively uniform deformation patterns. Surface relief observations, such as those in Figure 7-25, show that the deformation in the degenerate microstructures of the In- and Cd-containing alloys is accommodated over a much greater area than in the case of the simple normally cooled eutectic structures. The diffuse deformation behavior of the degenerate microstructures

Figure 7-25 The surface relief that is observed after creep deformation of the (*a*) In- and (*b*) Cd-alloyed solders. The deformation patterns reveal significantly more uniform deformation throughout the entire joint than was observed for the simple binary alloy. (Optical micrographs: the light regions are the Sn-rich phase; the dark regions are the Pb-rich phase.)

can be rationalized through the greater volume of soft material. As with the classic eutectic structures, deformation will always concentrate in the softest material (i.e., the degenerate material at the colony boundaries). However, since the areal extent of the degenerate material is greater in the In and Cd containing alloys, the distribution of the deformation is also extended. Furthermore, the coarsened band formation associated with creep in the lamellar eutectic structures is not present in these degenerate, uniformly deforming morphologies. Figure 7-26 shows a Cd-containing specimen crept under the same conditions as the simple binary alloy in Figures 7-21 and 7-22. As seen in Figure 7-26, the creep deformation in the Cd alloy is not associated with coarsened band formation. Since the coarsened band formation is intimately related to the thermal fatigue failure of solder joints, these results indicate improved fatigue resistance.

In fact, in terms of inhibiting the coarsened band formation, the addition of the third element to the binary alloy is potentially beneficial in two ways. First, due to the creation of a more mechanically uniform microstructure, local plastic strain levels are lower. The microstructural instability that leads to the coarsened band formation may thus be suppressed since local strain concentrations may

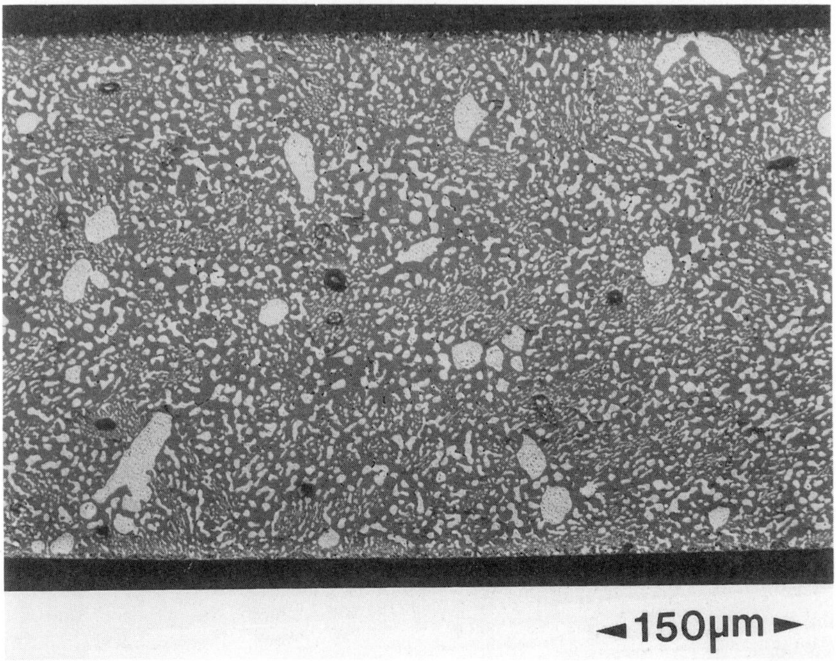

Figure 7-26 The polished microstructure of the deformed Cd-alloyed specimen seen in Figure 7-25. The deformation is not associated with microstructural changes, as was the case with the binary alloys. (SEM micrographs: the light regions are the Pb-rich phase; the dark regions are the Sn-rich phase.)

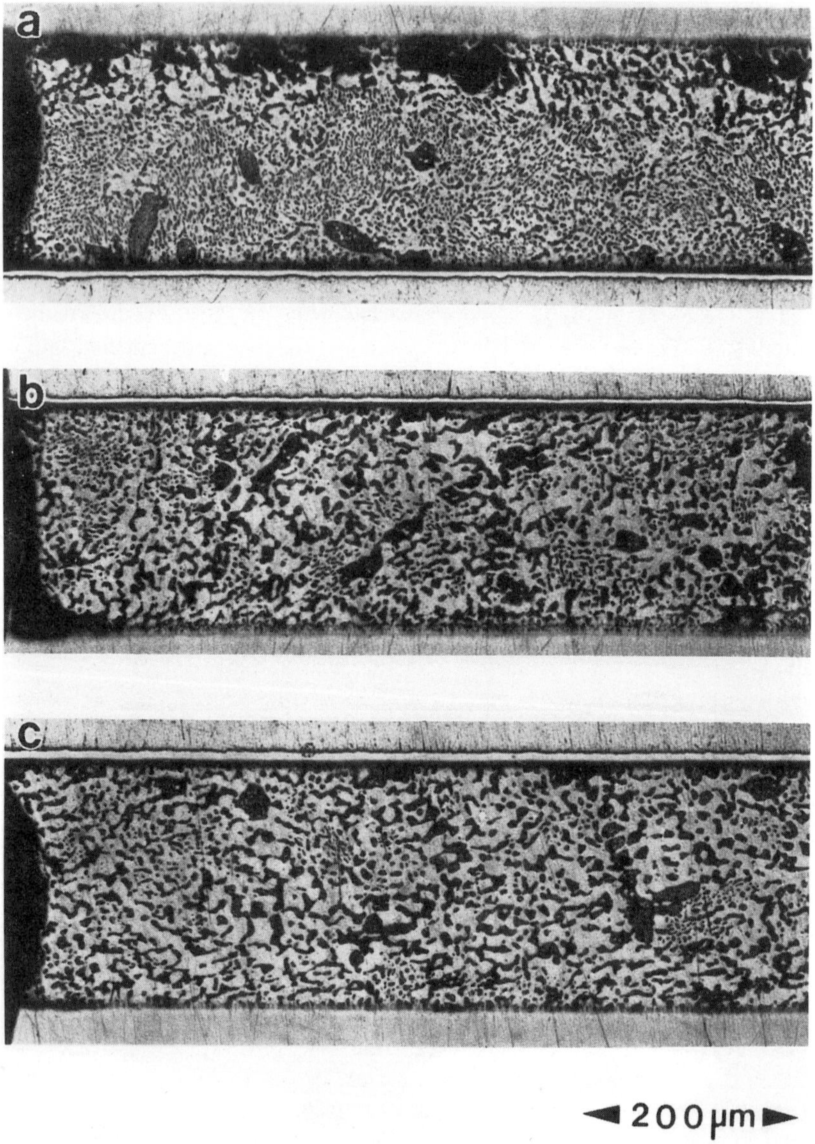

Figure 7-27 Optical micrographs of several thermally fatigued solder joints, 3000 cycles between −55 and 125°C. (*a*) A 60% Sn-40% Pb joint showing the definition and equiaxed grain structure of the coarsened band material. Joint cracking has occurred exclusively through this coarser microstructure. (*b*) 58% Sn-40% Pb-2% In and (*c*) 58% Sn-40% Pb-2% Cd solder joints after the same thermal treatment. Both these solder joints are still intact. These alloyed solder joint microstructures have coarsened with the thermomechanical exposure; however, the coarsening is global and not confined to a band near the solder substrate interface. (Optical micrographs: the light regions are the Sn-rich phase; the dark regions are the Pb-rich phase.)

not exceed the critical level needed to initiate the recrystallization. Furthermore, the addition of certain elements, including In and Cd, are known to inhibit recrystallization and creep deformation in the Pb and Sn systems.[77–80] Thus independent contribution of these effects to the inhibition of coarsened band formation in the crept ternary alloys is not yet clear.

Thermal fatigue of these ternary alloys reveals similar deformation patterns. Surface relief examinations of thermally fatigued ternary solder joints reproduce the relatively uniform deformation patterns seen under creep loading conditions. A thermal fatigue life comparison between the simple binary eutectic structures and the degenerate ternary alloys reveals the enhanced fatigue resistance of the latter alloys. Figure 7-27 shows a comparison of the three different compositions after 3000 thermal cycles. The 60% Sn-40% Pb joint is clearly cracked and both the In and Cd alloys appear intact. The cracking in the 60% Sn-40% Pb solder is, as expected, confined to the coarsened band, and the coarsened band is clearly visible and clearly microstructurally distinct from the remaining joint material. However, unlike the creep deformation, thermal fatigue deformation does result in a coarsening of the solder microstructures. But, in contrast to the 60% Sn-40% Pb alloy, the coarsening is global and not confined to a band near a joint interface. The addition of the third element has diffused the initial microstructures and thus the deformation patterns. The increased fatigue resistance of the ternary alloys appears to be the result of a postponement of the fatigue recrystallization and softening failure mechanism rather than a complete inhibition. The more uniform microstructures result in less localization of strain and thus defer the initiation and propagation of the recrystallized band.

Pb-rich Solders

The isothermal fatigue behavior of several compositions within the Pb-Sn system has recently been studied.[73] In these high Pb alloys, islands of the relatively soft Pb-rich phase displace the harder eutectic material and cause the surrounding eutectic to become more degenerate. It was found, however, that as long as the joint microstructure contained a significant fraction of the classic lamellar eutectic structure, failure occurred by the same mechanism as seen previously in fatigue and creep of near-eutectic alloys—inhomogeneous deformation resulting in the formation of a recrystallized and coarsened band of material. Figure 7-28 shows joints fatigued with a 10 percent strain amplitude at 75°C for several Pb-rich alloys. For Sn concentrations between about 40 and 63 wt %, formation of the coarsened band was not prevented or postponed with respect to its formation in the near-eutectic alloys. Its propagation, however, was inhibited by the presence of the Pb-rich proeutectic phase, with the 50% Sn-50% Pb alloy showing the most resistance to fatigue. Since the 40% Sn-60% Pb alloy has the most proeutectic phase and the least Sn of the alloys exhibiting

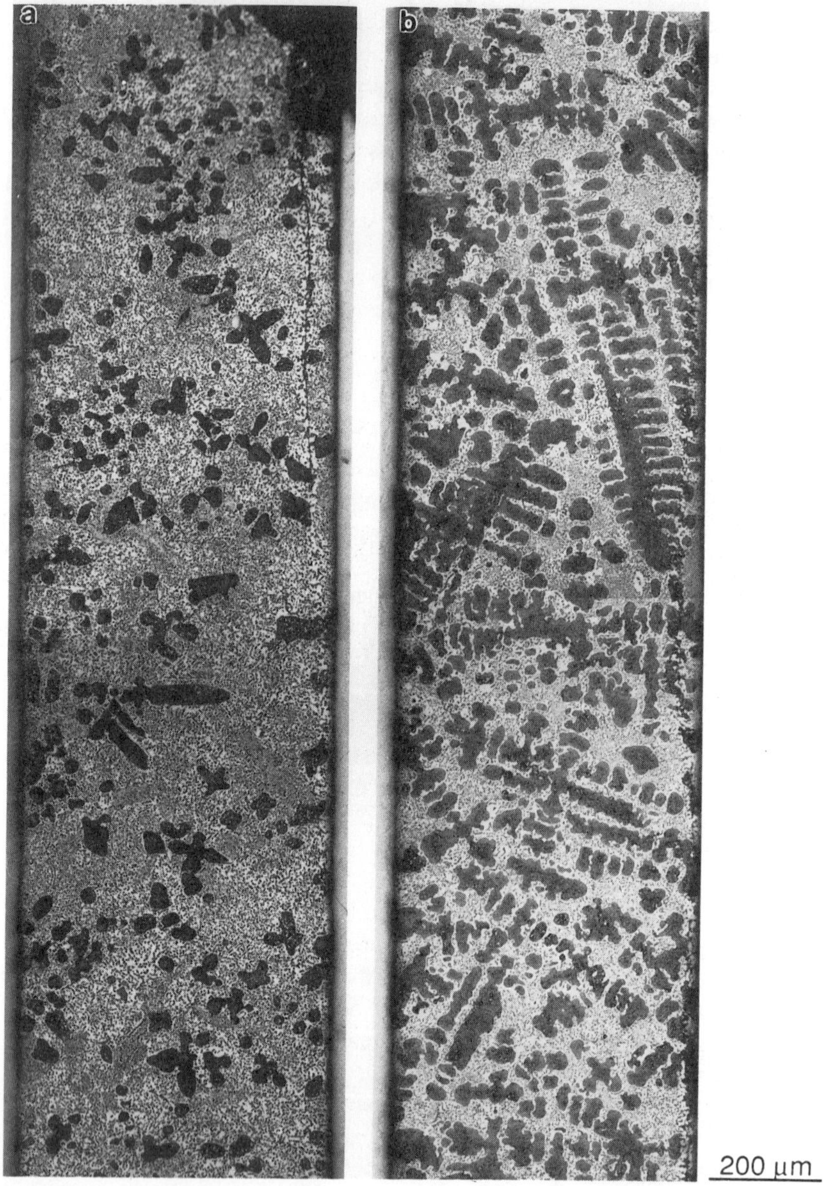

Figure 7-28 Optical micrographs of several Pb-rich alloys isothermally fatigued with a 10 percent strain amplitude at 75°C. The (*a*) 50% Sn-50% Pb and (*b*) 40% Sn-60% Pb joints both exhibit the formation of a coarsened band that eventually becomes the site of joint failure. The increased fatigue life of the Pb-rich alloys is due to the more difficult propagation of the coarsened band through a microstructure that contains many primary Pb-rich islands. (Optical micrographs: the light regions are the Sn-rich phase; the dark regions are the Pb-rich phase.)

coarsening, it is believed that the Pb-rich phase acts as a barrier to the propagation of the coarsened band, but its effectiveness is determined by the amount of Sn present. When a significant amount of the joint material consists of the primary Pb-rich phase, its fatigue resistance relative to the eutectic-containing alloys depends greatly on test conditions and it fails by a different mechanism.

Rapidly Solidified Joints

More rapidly solidified solder joints, such as those shown in Figure 7-8, possess a smaller colony size and a larger fraction of degenerate eutectic material. These microstructures also fail through the formation of coarsened bands that concentrate the strain and thus lead to failure. However, recent creep experiments indicate that the failure in these more degenerate structures is delayed. The presence of a larger amount of degenerate microstructure, as in the ternary solders, is expected to diffuse the deformation and thus postpone the microstructural instability that leads to failure. The possibility that this postponement is actually due to a superplastic deformation component is currently under investigation. The idea was suggested by Shine and Fox,[64] who used it to formulate a quantitative model of creep fatigue deformation. Since superplastic deformation occurs without accumulating damage, it should postpone the plastic strain intensities that initiate failure.

7.5 CONCLUSION

The problem of solder fatigue is a complex one; however, an understanding of the metallurgical mechanisms that lead to joint failure clarifies the role of the solder microstructure and thus indicates paths toward improved solder alloy development. The complexity of the problem is, at least partly, attributable to the complex nature of the two phase eutectic microstructure. An understanding of the microstructure and its relationship to process parameters and influence on properties is an important first step. The work outlined here reviews the eutectic Pb-Sn microstructure, its microstructure-property relationship, and, finally, its impact on the fatigue behavior.

As in any other material, the mechanical properties depend on the alloy's microstructure. A Pb-Sn alloy of a given composition may exist with many different microstructures depending on the processing history. Each microstructure has its own characteristic mechanical properties and the differences in mechanical behaviors may be significant. The exceptional strength and stability of the lamellar eutectic microstructures arises from the long-range regularity and alignment of the two-phase crystal structure. These strengthening mechanisms are lost in more degenerate, less regular structures, and such microstructures deform more easily.

The solder joint microstructure contains both lamellar eutectic and degenerate features and thus a tendency toward inhomogeneous deformation. In the eutectic colony structures routinely observed in solder joints, deformation is most readily accommodated at the colony boundaries where the long-range lamellar phase arrangement terminates and a more faulted phase structure exists. The inhomogeneous deformation character results in strain concentrations at the colony boundaries as the joint is deformed. Since the Pb-Sn eutectic is unstable with respect to recrystallization and coarsening under mechanical working, it will recrystallize in these highly strained regions. The recrystallized material is very soft relative to the remaining joint material and will further concentrate deformation and eventually will lead to joint failure.

The metallurgical mechanism of fatigue involves two processes: shear deformation develops in an inhomogeneous pattern that triggers recrystallization, which forms soft, coarsened bands. It follows that a microstructure that deforms more uniformly or one that is inherently more resistant to the initiation and propagation of the recrystallized bands should show enhanced fatigue resistance. It is possible to alter the amount of lamellar eutectic material within a solder joint with appropriate compositional and processing changes and thus make the joint more uniform in its mechanical response. Chemical additives that are known to enhance the microstructural stability may also be used. In this work, several such metallurgical techniques have been proposed and evaluated. The replacement of the lamellar eutectic microstructure with a more equiaxed degenerate one results in a microstructure that naturally deforms more uniformly and possibly superplastically. Such microstructures have shown enhanced creep and thermal fatigue resistances. In Pb-rich solders, the presence of the islands of proeutectic Pb-rich phase results in a difficulty in propagating these coarsened bands and thus an improved fatigue life.

7.6 ACKNOWLEDGMENTS

This work is jointly supported by the director, office of Basic Energy Science, Materials Science Division of the U.S. Department of Energy under contract No. DE-ACO3-76SF00098.

REFERENCES

1. Grivas, D., D. Frear, L. K. Quan, and J. W. Morris, Jr., *J. Electr. Mat.*, **15,** 1986, p. 355.
2. Frear, D., D. Grivas, L.K. Quan, and J. W. Morris, Jr., *MRS Symposium,* **72,** 1986, p. 181.
3. Frear, D., D. Grivas, and J.W. Morris, Jr., *J. Electr. Mat.*, **17,** 1988, p. 171.
4. McLean, M., *Directionally Solidified Eutectics,* The Metals Society, London, 1983.

5. Shewmon, P. G., *Transformations in Metals,* J. Williams Book Co., OK, 1983.
6. Elliott, R., *Eutectic Solidification Processing,* Butterworths, London, 1983.
7. Verhoeven, J. D., D. P. Mourer, and E. D. Gibson, *Met Trans.* A, **8A,** 1977, pp. 1239–1247.
8. Hunt, J. D., *J. Inst. Met.* **94,** 1964, pp. 125–129.
9. Hopkins, R.H., and R. W. Kraft, *Trans. AIME,* **242,** 1968, pp. 1627–1633.
10. Labulle, B., and C. Petipas, *J. Cryst. Growth,* **28,** 1975, pp. 279–287.
11. Blanchin, M. G., A. Guinier, C. Petipas, and G. Sauvage, *Acta Met.,* **20,** 1972, pp. 1251–1258.
12. Cagnon, M., M. Suery, A. Eberhardt, and B. Baudelet, *Acta Met.* **25,** 1977, pp. 71–75.
13. Chilton, J. P., and W. C. Winegard, *J. Inst. Met.,* **89,** 1960, pp. 162–164.
14. Weart, H. W., and D. J. Mack, *Trans. Met. Soc.* AIME, **10,** 1958, pp. 664–670.
15. Quan, L. K., "Tensile and Shear Behavior of Alloyed Sn-Pb Solder Joints," M.S. Thesis, University of California, Berkeley, 1988.
16. Specimen courtesy of J. Maki, EMPF, China Lake, CA.
17. Frear, D., D. Grivas, J. W. Morris, Jr., *J. Elect. Mat.,* **16,** 1987, p. 181.
18. ———*J. Elect. Mat.,* **17,** 1988, p. 171.
19. ———*J. Met.,* **40,** 1988, p. 18.
20. Tribula, D., D. Grivas, D. Frear, and J. W. Morris, Jr., *Weld. J. Res. Supp.,* **68,** 1989, p. 404s.
21. Eady, J. A., and W. C. Winegard, *Can. Met. Quart.,* **10**(3), 1971, pp. 213–214.
22. Kraft, R. W., D. L. Albright, J. A. Ford, *Trans. Met. Soc.* AIME, **227,** 1962, pp. 540–552.
23. Kraft, R. W., *J. Met.,* **2,** 1966, pp. 192–200.
24. Cline, H. E., and T. H. Alden, *Trans. AIME,* **239,** 1967, pp. 710–714.
25. Avery, D. H., and W. A. Backhofen, *Trans. ASM,* **58,** 1965, pp. 551–562.
26. Morrison, W. B., *Trans. AIME,* **242,** 1968, 2221–2227.
27. Kashyap, B. P., and G. S. Murty, *Mat. Sci. and Eng.* **50,** 1981, pp. 205–213.
28. Frebel, M., and B. Otte, *Scripta Met.,* **9,** 1975, p. 1317.
29. Graham, L. D., and R. W. Kraft, *Trans. AIME,* **236,** 1966, p. 94.
30. Frebel, M., G. Duddek, W. Graf, M. N. Faridani, and B. Otte, *B. Bunsenges Phys. Chem.,* **82,** 1978, pp. 259–265.
31. Livingston, J. D., and J. W. Cahn, *Acta Met.,* **22,** 1974, p. 495.
32. Frebel, M., and G. Duddek, *Mat. Sci. Eng.,* **32,** 1974, p. 125.
33. Kashyap, B. P., and G. S. Murty, *Met. Trans. A.,* **13A,** 1982, pp. 53–58.
34. Fine, M., and R. L. Dowdell, *Trans. ASM,* **37,** 1946, pp. 245–277.
35. Baker, W. A., *J. Inst. Met.,* **65,** 9 1939, pp. 277–300.
36. Pearson, C. E., *J. Inst. Met.,* **54,** 1934, pp. 110–124.
37. Jenkins, C. H. M., *J. Inst. Met.,* **40,** 1928, pp. 21–39.
38. Rack, H. J., and J. K. Maurin, *J. Test. and Eval.,* **2**(5), 1974, pp. 351–353.
39. Inoue, H., Y. Kurihara, and H. Hachino, *IEEE Trans. CHMT-12,* 1986, p. 190.
40. Hargreaves, F., *J. Inst. Met.,* **40,** 1928, p. 41.
41. ———, *J. Inst. Met.,* **39,** 1928, p. 301.
42. ——— and J. Hills, *J. Inst. Met.,* **41,** 1929, p. 257.
43. Bird, J. E., A. K. Mukherjee, and J. F. Dorn, *Quantitative Relations Between Properties and Microstructure,* Israel University Press, 1969.

44. Zehr, S. W., and W. A. Backhofen, *Trans. ASM,* **61,** 1968, pp. 300–313.
45. Mukherjee, A.K., J. E. Bird, and J. F. Dorn, *Trans. ASM* **62,** 1969, p. 155.
46. Weertman, J., and J. R. Weertman *Physical Metallurgy,* R. W. Cahn, P. Haasen, Eds., Elsevier Science Publishers, 1983, p. 1310.
47. Vaidya, M. L., K. L. Murty, and J. E. Dorn, *Acta. Met.,* **21,** 1973, p. 1616.
48. Mohamed, F. A., K. L. Murty, and J. W. Morris, Jr., *Met. Trans.,* **4,** 1973, p. 935.
49. Grivas, D., "Steady State Creep in Pb-Sn Eutectic Alloys," M.S. thesis, University of California, Berkeley, CA, 1974.
50. Murty, G. S., *J. Mat. Sci.,* **8,** 1973, p. 611.
51. Grivas, D., "Deformation of Superplastic Alloys of Relatively Low Strain Rates," Ph.D. Thesis, University of California, Berkeley, CA, 1979.
52. Schneibel, J. H., and P. M. Hazzledine, *J. Mat. Sci.,* **18,** 1983, pp. 562–570.
53. Geckinli, A. E., and C. R. Barrett, *J. Mat. Sci.,* **11,** 1976, pp. 510–521.
54. Murty, G. S., and B. P. Kashyap, *Trans. Ind. Inst. Met.,* **37**(5), 1984, pp. 509–513.
55. Mohamed, F. A., and T. G. Langdon, *Phil. Mag.,* 1975, pp. 697–709.
56. Yadava, R. K., and K. A. Padmanabhan, *J. Mat. Sci.,* **17,** 1982, pp. 2435–2440.
57. Baudelet, B., and M. Suery, *J. Mat. Sci.,* **7,** 1972, pp. 512–516.
58. Kashyap, B. P., and G. S. Murty, *Met. Trans. A.,* **12A,** 1981, pp. 1923–1925.
59. Yadava, R. K., and K. A. Padmanabhan, *Mat. Sci. and Eng.,* **37,** 1979, 127–136.
60. B. P. Kashyap, A. Arieli, and A. K. Mukherjee, *J. Mat. Sci.,* **20,** 1985, pp. 2662–2686.
61. Rohde, R. W., and J. C. Swearengen, *Trans ASME, J. Eng. Mat. and Tech.,* **102,** 1980, p. 207.
62. Weinbell, R. C., J. K. Tien, R. A. Pollack, and S. K. Kang, *J. Mat. Sci.,* 1987, p. 3901.
63. Mei, Z., M. C. Shine, and J. W. Morris, Jr., submitted to *J. Electr. Mat.*
64. Shine, M. C., and L. R. Fox, *Low Cycle Fatigue,* ASTM STP 942, H.D. Solomon, Ed., American Society for Testing and Materials, Philadelphia, 1988, p. 588.
65. Solomon, H. D., *Braz. and Sold.,* **11,** 1986, p. 68.
66. Hall, P. M., *IEEE Trans., CHMT-12,* 1987, p. 556.
67. Enke, N. F., and B. I. Sandor, ASME publication 88-WA/EEP-12, 1988.
68. Wild, R. N. "Some Properties of Solder and Solder Joints," IBM Report no. 74z000481, Oct. 1975.
69. Bangs, E. R., and R. E. Beal, *Weld. Res. Supp. Weld. J.,* **54,** 1978, p. 377.
70. Wolverton, W. M., *Braz. and Sold.,* **13,** 1987, p. 33.
71. ———, "Know Your Enemy: How to Stop Solder Joints from Cracking," *Circ. Manuf.,* Nov. 1972, p. 10.
72. Frear, D., D. Grivas, M. McCormack, D. Tribula, and J. W. Morris, Jr., *Effect of Load and Thermal Histories of Mechanical Behavior* TMS, 1987, p. 113.
73. Summers, T. S. E., and J. W. Morris, Jr., Trans. of the ASME, *Journal of Electronic Packaging,* **112** (June 1990) pp. 94–99.
74. Tribula, D., D. Grivas, D. Frear, and J. W. Morris, Jr., *J. Elect. Pack.,* 1989, p. 83.

75. Chen, F. K. "Applications of the Finite Element Method to the Plastic Deformation Analysis of Perforated Sheet Metals and the Process Simulation of Shape Rolling," Ph.D. Thesis, University of California, Berkeley, 1989.
76. McKlintock, F. A., and G. R. Irwin, "Plasticity Aspects of Fracture," ASTM STP 381 (1965) p. 84 as referenced in *Elementary Engineering Mechanics,* D. Broek, Ed., Martinus Nijhoff Pub., 1983, p. 99.
77. Hanson, D., and E. J. Sandford, *Trans AIME,* **54,** 1935, p. 159.
78. Sully, A. H., *Metallic Creep and Creep Resistant Alloys,* Butterworths, London, 1949.
79. Hopkins, L. M. T., and C. W. Thwaites, *J. Inst. Met.,* **54,** 1953, p. 181.
80. Jenkins, C. H., *J. Inst. Met.,* **40,** 1921, p. 21.

8

Microstructure and Mechanical Properties of Solder Alloys

Harold J. Frost

The reliability of solder joints often depends directly on the mechanical properties of the solder alloy. The tensile strength of the alloy determines the resistance of the joint to mechanical rupture under short-term direct loading, as might occur during impact testing. If the solder joint is subjected to cyclic stresses and strains generated by thermal expansions during thermal cycling, the reliability depends on resistance to fatigue. Because the temperatures of application of solder joints are typically a substantial fraction of the melting temperature of the alloy, the low frequency fatigue behavior is related to stress relaxation and creep behavior.

It is commonly recognized that the mechanical properties of metals and alloys depend on their microstructures. Solder alloys are no exception to this general pattern. Solder alloys are often a special case, however, because unlike most structural alloys, they are used at such high temperature (relative to the melting temperature) that the microstructures can change and evolve during service. What is more, application conditions that involve thermal cycling may (or may not) produce cyclic microstructural evolution, depending on the temperatures and frequencies involved. Finally, cyclic mechanical stresses or deformations may influence or promote microstructural changes.

The elements of microstructure that are typically important are the grain size and shape, the size, shape, and spacing of second phase particles, and the dislocation density. In eutectic solders the grain size and the second-phase size and spacing are directly related because (at least) two phases are always present. There are some solders that are essentially single-phase solid solutions in which the precipitation of a second phase is not important (for example, Pb-5 wt % In). Finally there are some solder alloys that are predominately one phase, but in which an important second phase may precipitate (for example, Pb-5 wt % Sn).

8.1 THERMAL CYCLING FATIGUE

The standard mechanical reliability concern for solder joints is the measurement of resistance to failure by low cycle fatigue due to thermal cycling. If the service conditions involve any thermal cycles, strains due to thermal expansion mismatch are very difficult to avoid. Even with matched thermal expansion coefficients there is likely to be some strain imposed on the solder because of differential rates of heating and cooling in different regions. In many designs the imposed strains exceed the elastic limit of the solder and produce plastic deformation. In these conditions we may expect eventual failure by low cycle fatigue. The duration of the fatigue lifetime that is required in service typically far exceeds the time available for reliability testing. This means that service reliability must be estimated using some sort of extrapolation. The situation becomes complex because of the relatively high temperatures involved. Low cycle fatigue at more than half the melting point of the solder will involve creep fatigue interactions. Under these conditions, the number of cycles to failure depends on the cycle frequency and on the stress cycle shape (that is, on hold times at load).

An understanding of the thermal fatigue of a solder joint requires that we know the stresses and strains experienced by the solder. In the typical case, the solder joint is much less stiff and more compliant than the other components that the solder joint joins together. In this case, a specified thermal cycle results in a specified strain cycle on the solder. The specified strain is made up of both elastic and plastic strains. The stresses reached in such a cycle depend on the elastic and plastic compliances of the solder. Plastic compliance results from both very rapid plastic deformation at high stresses and much slower plastic creep at lower stresses. During long hold times, plastic creep deformation replaces elastic deformation during stress relaxation. If the strain cycle history is known, and if the relationship between creep rate and stress is known, it is possible to predict the stress cycle history. This sort of modeling has recently been reported by Wilcox, Subrahmanyan, and Li and Subrahmanyan, Wilcox, and Li.[2] In their work it is necessary to include the temperature dependence of the plastic creep behavior. Creep relaxation is much faster at high temperatures. The magnitude of the stresses reached upon heating is therefore lower than the magnitude reached upon cooling.

The eventual failure of the solder results from the initiation and propagation of fatigue cracks. Crack initiation probably results from the accumulation of creep cavitation damage, the rate of which is determined by the time spent at stress. (The rate of damage accumulation may also depend on temperature.) Propagation of the cracks after initiation may also depend on the time spent at stress. The fact that the stress relaxation is different in the heating and cooling parts of the cycle means that the accumulation of damage will also be different. Although the situation is complicated, it can be modeled if we have two relationships: one for the rate of plastic deformation as a function of stress and temperature and another for the rate of damage accumulation as a function of stress and temperature. In the general case these two relationships would also depend on the strain and the temperature histories because of the effects of work hardening and annealing. In some cases it may be reasonably accurate to neglect such history effects. In the case where these two relationships are valid, it is possible to determine the relationships from experiments at one set of conditions (frequency, temperatures, and strain levels) and use the relationships to predict behavior (fatigue lifetime) at another set of conditions. That is, the long-term service reliability could be extrapolated from short-term accelerated tests.

8.2 PRECIPITATION AND DISSOLUTION IN PB-SN ALLOYS

The equilibrium phase diagram for binary Pb-Sn alloys is shown in Figure 8-1. For alloys with less than 19 wt% Sn it is possible to achieve a single-phase solid solution of Sn in Pb by annealing at a sufficiently high temperature. At lower temperatures (below the solvus temperature) the equilibrium structure has two phases: precipitates of the tin phase within a matrix of the lead phase. For a range of low tin compositions the solvus temperature is within the temperature range that may be expected during service or during reliability testing (Fig. 8-2). For these alloys, temperature cycles may produce cycles of precipitation and

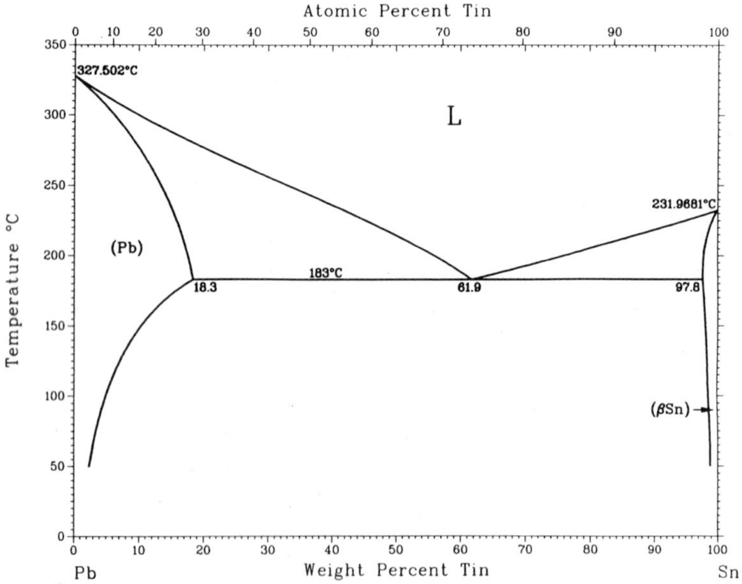

Figure 8-1 Equilibrium phase diagram for binary lead-tin alloys.[3]

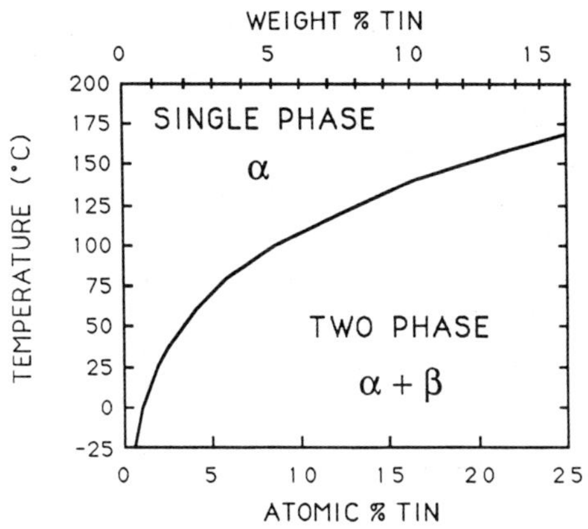

Figure 8-2 Limit of solid solubility (solvus) of tin in lead as a function of temperature, from Tu and Turnbull.[4]

dissolution, depending on the frequency. If the thermal cycle frequency is very low, the precipitation process will go to completion at the low temperature and the dissolution process will go to completion at the high temperature. The microstructure will therefore be different in different portions of the thermal cycle. If, on the other hand, the thermal cycle frequency is very high, neither the precipitation nor the dissolution will be able to occur during a cycle. In this case the microstructure will achieve some approximate steady state that may differ from the equilibrium structure at either temperature.

The precipitation reaction usually proceeds as cellular precipitation, in which cells of lamellae of the tin phase nucleate at the grain boundaries and grow into the matrix. This process, also known as discontinuous precipitation, has been well studied in Pb-Sn alloys by Turnbull and Treaftis[5,6] and Tu and Turnbull.[4,7–9] More recent studies include Afanas'yev and Yelsukova,[10,11] Varkey and Padmini,[12] Osorio,[13] and Hackney and Biancaniello.[14] An example of the progressive growth of cells is shown in Figure 8-3.[15] The fraction transformed f as a function of time t may be expressed by the Avrami equation for a given temperature:

$$f = 1 - \exp(-Kt^n)$$

where the constant K depends on temperature, and n is independent of temperature. The exponent n is between 1 and 4 for most transformations; Lutender found $n \approx 2$. Figure 8-4 shows the results as time-temperature-transformation curves for 50 percent completion, as calculated from fits to the Avrami equation. Also shown in Figure 8-4 are curves for 6 and 9 wt% Sn taken from Borelius, Larris, and Ohlsson.[16] The temperature for the maximum transformation rate increases with increasing tin content. For 3 to 5 wt% Sn, the maximum rate is

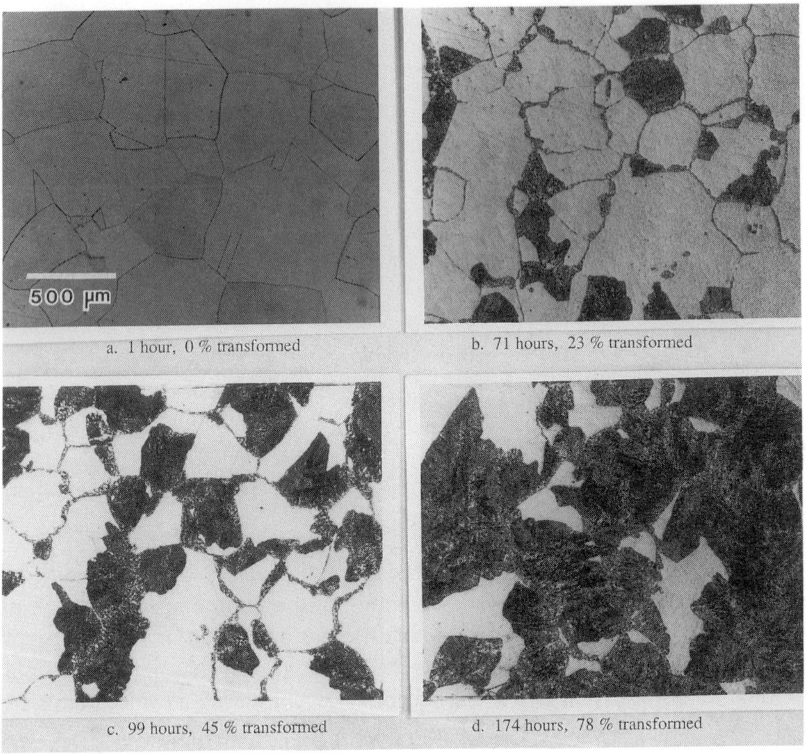

Figure 8-3 Micrographs of 97% Pb-3% Sn aged at 23°C for various times after solution annealing. (*a*) 1 hr, 0 percent transformed, (*b*) 71 hr, 23 percent transformed, (*c*) 99 hr, 45 percent transformed, (*d*) 174 hr, 78 percent transformed.

near room temperature; for 2 wt% Sn, it is at a lower temperature. For practical purposes, transformation in the 2 wt% Sn alloy does not become significant for many days. Conversely, transformation in the 5 wt% Sn alloy near room temperature is completed within 1 day.

An alternative mode of the precipitation reaction has recently been described by Frear, Posthill, and Morris.[17] They found that the alloy Pb-5 wt% Sn showed cellular precipitation if it was slowly cooled from a high temperature but underwent a homogeneous precipitation reaction when it was quenched to room temperature. Cellular precipitation is usually the preferred reaction because heterogeneous nucleation at grain boundaries is easier than homogeneous nucleation within the lead matrix and because the cell boundaries provide fast diffusion paths for the tin to agglomerate. The homogeneous nucleation observed by Frear et al. probably occurred at vacancies clusters formed during the quench. Unfortunately the quenching rate that is required to produce the homogeneous nucleation is not known.

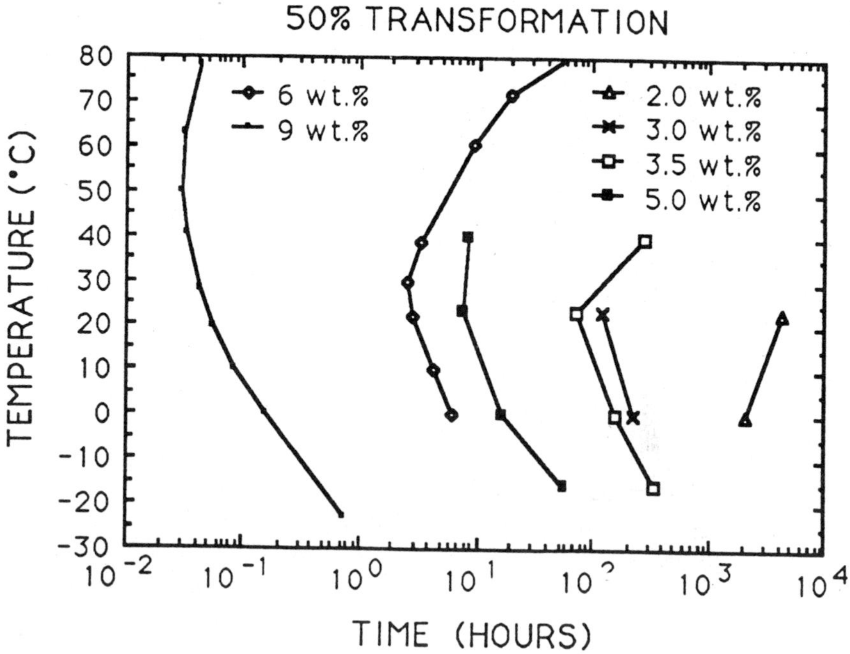

Figure 8-4 Time required for a 50 percent transformation for the precipitation reaction in Pb-Sn alloys of various compositions. The curves for 6 and 9 wt % Sn are taken from the resistivity measurements of Borelius, Laris and Ohlsson.[16]

Early studies of the cellular precipitate structure (e.g., Tu and Turnbull[4]) discovered that the average spacing between tin lamellae decreased with increasing tin concentration and decreasing precipitation temperature. Frost et al.[18,20] confirmed these findings, as shown in Figure 8-5. This behavior is certainly the expected behavior for diffusion controlled phase transformation. It becomes important for our consideration here because the flow stress of the alloy depends on interlamellar spacing. Figure 8-6 shows examples of stress-strain behavior for different microstructures. When the interlamellar spacing is smaller than about 1 μm, the precipitate structure provides significant strengthening.

When the temperature is raised above the solvus temperature (Figure 8-2), the precipitate structure begins to dissolve. This may be a complicated process. One possible mechanism is the direct diffusion of the tin into the lead phase. Another mechanism is the retraction of the cell boundaries as a reversal of the cellular precipitation process. The grain boundaries here provide a fast diffusion path for the dispersal of the tin into the lead phase. This process is dominant at the relatively lower temperatures that are only slightly above the solvus. Higher temperatures result in faster dissolution as shown in Figure 8-7. This temperature

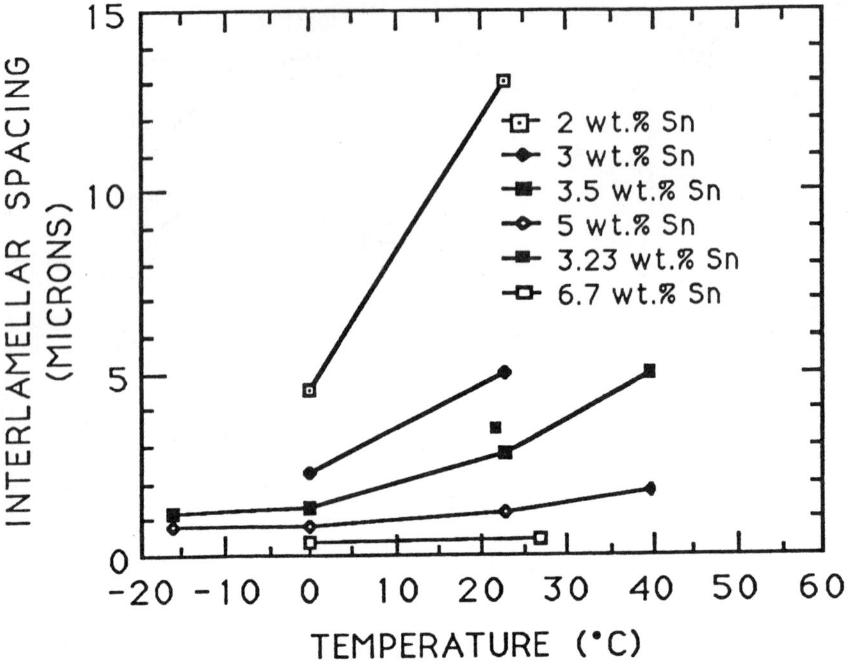

Figure 8-5 Average interlamellar spacings as a function of transformation temperature, for various tin concentrations.

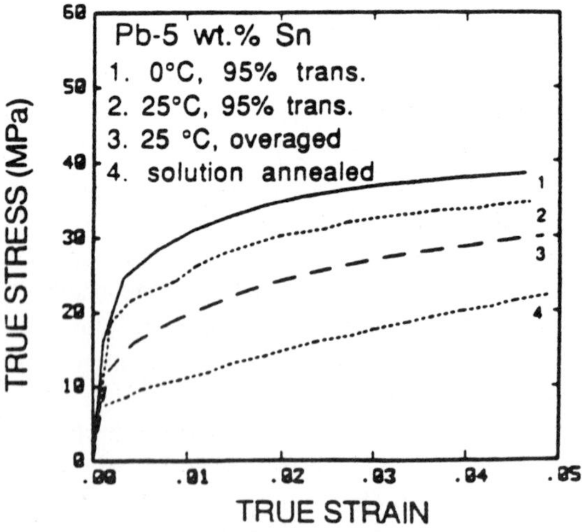

Figure 8-6 True stress and true strain curves for tensile tests on the 95% Pb-5% Sn alloy, tested at 25°C, with an initial tensile strain rate of 0.1/sec.

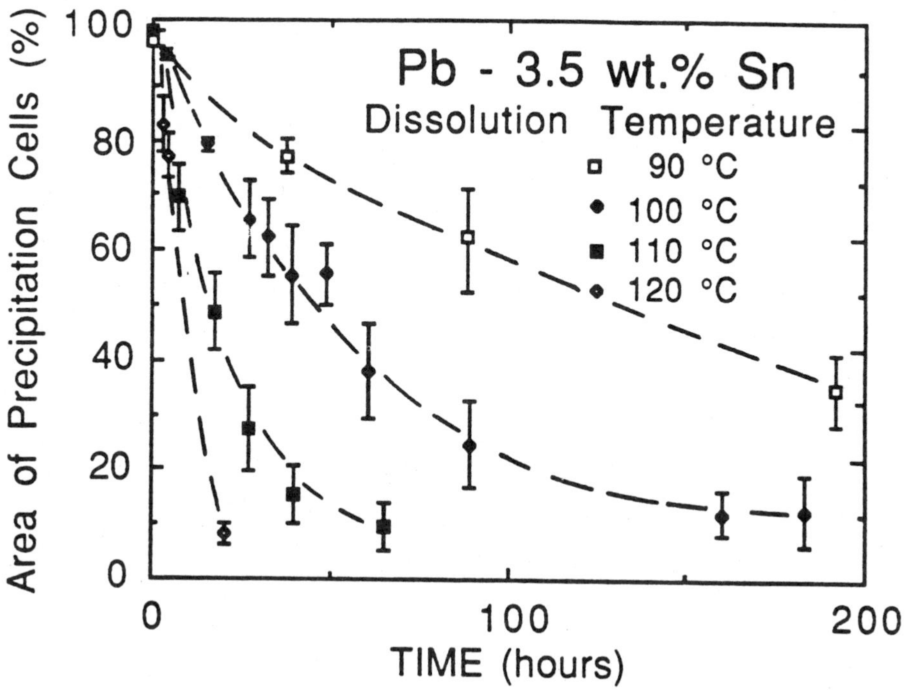

Figure 8-7 Dissolution behavior of the cellular precipitate structure at various annealing temperatures, as measured by the area fraction containing the cellular precipitates as a function of time.

effect results from both temperature dependence of diffusion and the fact that the thermodynamic driving force for dissolution also varies with temperature. The apparent activation energy for the process is therefore not directly comparable to the activation energy of the rate-limiting diffusion process.

Tu and Turnbull[4] observed that dissolution by cell boundary retraction would occur even at temperatures slightly below the solvus. (The effect extends to about 10°C below the solvus.) One driving force for this retraction is the surface energy of the interfaces between the lead and tin phases. Frost, Howard, and Stone[19] have reported that the rate of dissolution by cell boundary retraction increases with smaller interlamellar spacings. This effect may be the result both of shorter diffusion distances required to disperse the tin and of the additional driving force for retraction provided by the interphase interfaces. This means that the dissolution rate depends on the temperature at which the precipitate structure initially formed, as shown in Figure 8-8.

During the process of dissolution by cell boundary retraction, there is also an evolution in the structure of tin lamellae within the cells. The tin phase actually

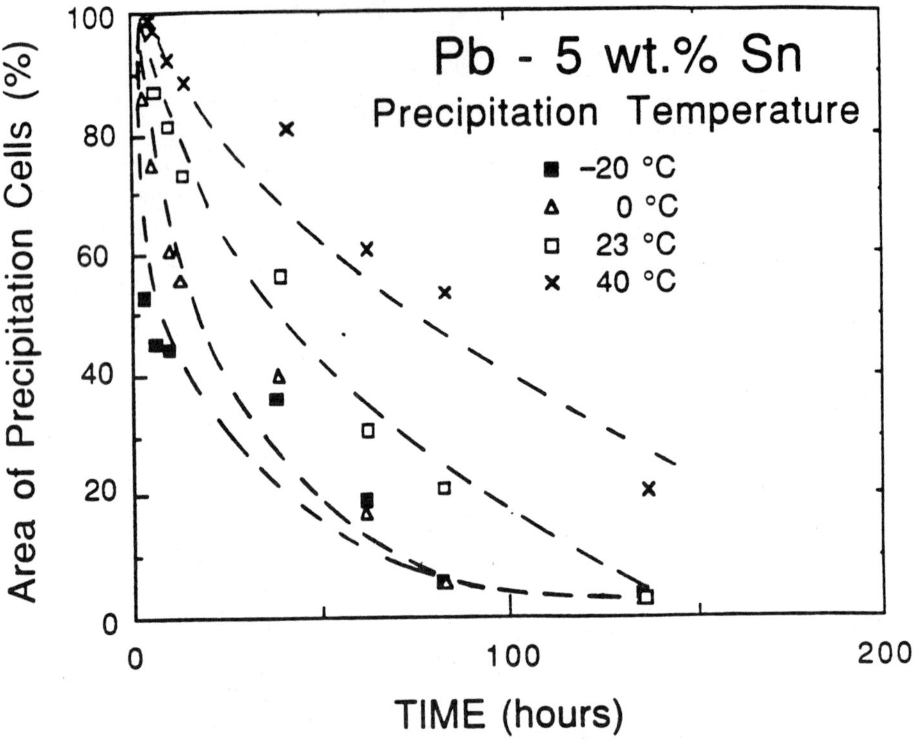

Figure 8-8 Dissolution behavior of the cellular precipitate structure in the 95% Pb-5% Sn alloy at the annealing temperature of 100°C for four different ageing temperatures. The different aging temperatures that correspond to different interlamellar spacings, as shown in Figure 8-5, result in remarkably different dissolution rates.

occurs as blades with numerous rifts and holes instead of complete plates (Tu and Turnbull[4,9]). They are therefore subject to instabilities due to capillarity effects that lead to break-up of the lamellae and spheroidization. This pheroidization was observed by optical metallography,[1,4] and is also evident in Figure 8-5. The extent of spheroidization is also shown by a comparison of Figures 8-9 and 8-10. Figure 8-9 shows a high magnification view of the early stage of dissolution by cell boundary retraction. The tin phase remains predominantly in the form of well-defined lamellae. Figure 8-10 shows a much later stage of dissolution in which a region of tin precipitates has spheroidized.

During the spheroidization process there is also some dissolution of the tin phase by lattice diffusion directly into the lead matrix. For temperatures near the solvus, however, the lattice diffusion is so slow that direct dissolution by lattice diffusion is not completely accomplished before the cellular dissolution is nearly complete. Cellular dissolution by cell boundary migration continues after the spheroidization has become prominent, but the boundaries appear to slow down. Spheroidization should reduce two components of the driving force for the cell boundary migration. It reduces the area of lamella-matrix interface, and it also disperses some tin into the lead matrix, thereby reducing the chemical driving force for cell boundary migration.

Just as the precipitate microstructure is unstable to coarsening and spheroidization during dissolution above the solvus, it is also unstable below the solvus. The initial cellular precipitation process cannot produce a microstructure that is completely at equilibrium. There must be some free energy of the interfaces between the two phases, and the compositions of the two phases will deviate slightly from their equilibrium values. These factors provide a driving force for evolution toward a coarser microstructure. If the tin were distributed in small spherical particles, then those particles should coarsen in the usual manner. Frear et al.[17] report a decrease in hardness during ageing that is directly attributable to such coarsening. If the tin were distributed in planar lamellae, without imperfections, then coarsening might occur in a discontinuous fashion by the passage of interfaces that consumed the fine lamellae and left behind coarse lamellae.

Figure 8-9 Cell boundary retraction in the 95% Pb-3.5% Sn alloys. The precipitate structure formed at 23°C. The dissolution anneal was for 39.7 h at 90°C. The migrating grain boundary is visible on one side of the denuded zone.

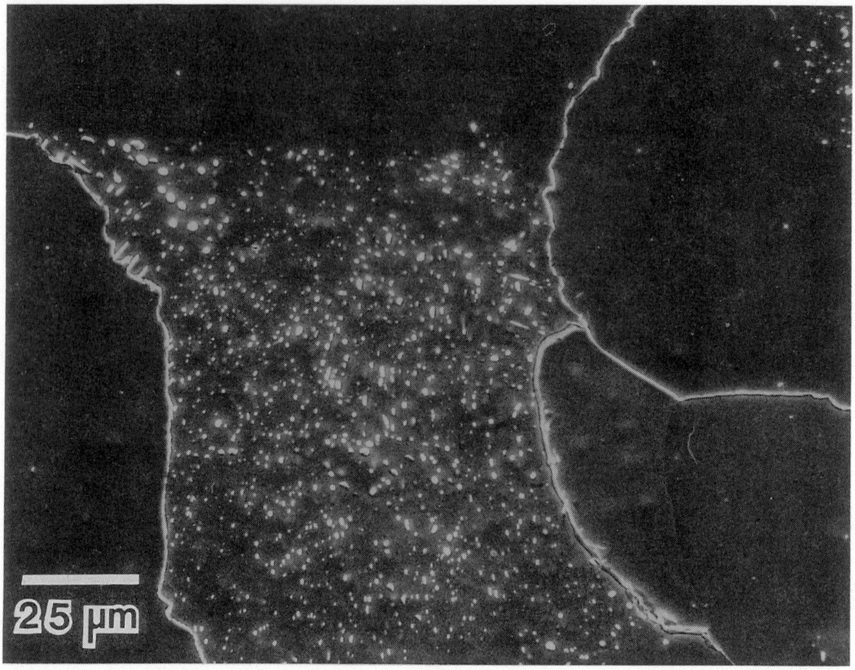

Figure 8-10 Microstructure of 95% Pb-5% Sn after precipitation at room temperature, followed by annealing for 17 h at 110°C. Most of the precipitate structure has undergone cellular dissolution. In the region shown the lamellar structure of the tin precipitates has undergone spheroidization.

Although the process of discontinuous or cellular coarsening has long been recognized (e.g., Livingston and Cahn[21]), it has not yet been quantitatively studied in Pb-Sn alloys.

8.3 DISCUSSION

The complexity of the dissolution and coarsening processes means that it is somewhat difficult to predict exactly what microstructure would result from an arbitrary cyclic thermal history. We may expect precipitation to proceed at the lower temperatures and dissolution to proceed at the higher temperatures. If the cycle time is long compared to the phase transition time, the microstructure should cycle in a manner consistent with the simple experiments. The rate of precipitation from the solid solution is fairly well known, and although the dissolution rate depends on the precipitate structure, it may also be estimated. If the cycle time is short, neither precipitation nor dissolution will be complete in one cycle, and the microstructure will drift toward a structure determined by whichever process is dominant.

The importance of microstructural evolution to the thermal fatigue problem involves the boundary between very low frequency behavior (cyclic microstructure) and high frequency behavior (steady-state microstructure). If the cyclic strain is specified, the cyclic stresses will depend on the microstructure. If the initiation and propagation of creep-fatigue damage depends on stress, the fatigue lifetime will also depend on the microstructure. It may therefore be inaccurate to extrapolate across the frequency boundary between high and low frequency behavior. Furthermore, for either high or low frequency cycling, the strength of the alloy should depend on the temperature limits. This means that an extrapolation from thermal cycling from one temperature range will involve more than just the extrapolation from one level of thermal strains to another level of thermal strains.

From the discussion above it is clear that the effect of changing cycle frequency is intertwined with the effect of change in the cyclic temperature limits. Because both the precipitation and dissolution rates depend on temperature, we expect that the boundary between low and high frequency depends on the temperature limits involved. The situation is complicated by the fact that the dissolution rate depends not just on alloy composition and dissolution temperature but also on the temperature at which the precipitate structure was formed. A proper estimate of the frequency boundary must take this into account. In conclusion, a proper extrapolation of the mechanical behavior from one set of conditions to another must take the microstructural evolution into account.

REFERENCES

1. Wilcox, J. R., R. Subrahmanyan, and Che-Yu Li, "Thermal Stress Cycles and Inelastic Deformation in Solder Alloys," *Proceedings of 1989 ASM Electronic Packaging Conference—Materials and Processes*. ASM International, 24–28 April 1989, Philadelphia, PA., pp. 203–211.
2. Subrahmanyan, R., J. R. Wilcox, and Che-Yu Li. "A Damage Integral Approach to Solder Joint Fatigue," *Proceedings of 1989 ASM Electronic Packaging Conference—Materials and Processes*. ASM International, 24–28 April 1989, Philadelphia, PA, pp. 213–221; also "A Damage Integral Approach to Thermal Fatigue of Solder Joints." *Proceedings of 39th Electronics Components Conference,* IEEE, 22–24 May 1989, Houston, TX.
3. Karakaya, I., and W. T. Thompson, "The Pb-Sn (Lead-Tin) System," *Bulletin of Alloy Phase Diagrams,* **9**(2), 1988, pp. 144–152.
4. ———, "Morphology and Kinetics of Cellular Dissolution of the Pb-Sn Alloy," *Metallurgical Transactions,* **2,** 1971, pp. 2509–2515.
5. Turnbull, D., and H. N. Treaftis, "Kinetics of Precipitation of Tin from Lead-Tin Solid Solutions," *Acta Metallurgica,* **3,** 1955, pp. 43–54.

6. ______, "Micrographic Investigation of Precipitation in Pb-Sn Alloys," *Trans AIME* **212,** 1958, pp. 33–42.
7. Tu, K. N., and D. Turnbull, "Morphology of Cellular Precipitation of Tin from Lead-Tin Bicrystals," *Acta Metallurgica,* **15,** 1967, pp. 369–376.
8. ______, "Morphology of Cellular Precipitation of Tin from Lead-Tin Bicrystals—II," *Acta Metallurgica,* **15,** 1967, pp. 1317–1323.
9. ______, "Morphology and Structure of Tin Lamellae Formed by Cellular Precipitation," *Acta Metallurgica,* **17,** 1969, pp. 1263–1279.
10. Afanas'yev, N. I., and T. F. Yelsukova, "Ageing Peculiarities of Lead-Tin Alloys and Experimental Testing of Continuous Decomposition Theories," *Phys. Met. Metall.,* **53,** 1982, pp. 129–134.
11. ______, "The Influence of Impurities on the Rate of Discontinuous Decomposition of Lead-Tin Alloys," *Phys. Met. Metall.,* **57,** 1984, pp. 86–91.
12. Varkey, P. A., and A. R. K. L. Padmini, "Precipitation Studies in Lead-Tin Alloys," *Acustica* **52,** 1983, pp. 290–295.
13. Osorio, F. S., "Morphology of the Cellular Precipitation Reaction in a Pb-3.0 wt% Sn Alloy," Thesis, University of Missouri-Rolla, 1973.
14. Hackney, S. A., and F. S. Biancaniello, "Experimental Indications of a Low Undercooling Instability in the Cellular Mechanism of Discontinuous Precipitation," *Scripta Metallurgica,* **21,** 1987, pp. 371–376.
15. Lutender, S. D., "Micrographic Study of the Precipitation and Dissolution Kinetics of Selected Pb-Sn Alloys," M.E. Thesis, Dartmouth College, 1986.
16. Borelius, G., F. Larris, and E. Ohlsson, "Kinetics of Precipitation in Pb-Sn Alloys," *Arkiv För Matematik, Astronomi och Fysik,* **31A**(10), 1944, pp. 1–19.
17. Frear, D. R., J. B. Posthill, and J. W. Morris, Jr., "The Microstructural Details of β-Sn Precipitation in a 5Sn-95Pb Solder Alloy," *Metallurical Transactions,* **20A** 1989, pp. 1325–1333.
18. Frost, H. J., R. T. Howard, P. R. Lavery, and S. D. Lutender, "Creep and Tensile Behavior of Lead-Rich Lead-Tin Alloys," *IEEE Trans. on Components, Hybrids and Manufact. Tech.,* **11,** 1988, pp. 371–379.
19. Frost, H. J., R. T. Howard, and G. J. Stone, "The Effects of Thermal History on the Microstructure and Mechanical Properties of Solder Alloys," *Microelectronic Packaging Technology, Materials and Processes.* W. T. Shieh, Ed. ASM Internation, 1989, pp. 121–127.
20. Frost, H. J., P. R. Lavery, and S. D. Lutender, "Microstructure and Mechanical Properties of Lead-Tin Alloys," in *Electronic Packaging and Corrosion in Microelectronics,* M. E. Nicholson, Ed. ASM International, 1987, pp. 259–267.
21. Livingston, J. D., and J. W. Cahn, "Discontinuous Coarsening of Aligned Eutectoids," *Acta Metallurgica* 22, 1974, pp. 495–503.

9

The Interaction of Creep and Fatigue in Lead-Tin Solders

John K. Tien, Bryan C. Hendrix, and Abbas I. Attarwala

For 2000 to 3000 years, Pb-Sn solders have found extensive use in plumbing, and more recently, in electrical connections because of their good wetting properties and low melting temperature. But solder has been avoided for mechanical connections because of its low strength at ambient operating temperatures, which is above one-half of the absolute melting temperature ($0.5T_m$) of Pb-Sn solder. However, during the past 15 years, the packaging of electronic components has moved from technology that used solder predominantly as an electrical connection (plated through-hole) to technology that uses solder as both a mechanical and electrical connection (surface mount technology, SMT). In addition, electronic components are expected to withstand increasingly harsh temperature cycles. Accordingly, thermal fatigue and creep of solder joints has emerged as a source of failure of SMT components,[1] where temperatures play a role not only in inducing the stresses but also in thermally activated deformation and rupture.

9.1 CURRENT APPROACHES TO ACCELERATED TESTING

Thermal testing is used to measure electronic component reliability and to estimate component life. Testing for the thermal fatigue or cyclic creep life of components with realistic temperature cycles would take as long as the design life of the component. For this reason, components are often tested by accelerating the thermal cycles (i.e., increasing the heating rate and/or decreasing the cycle hold time).

In order to predict lives from accelerated tests, the damage stored during a realistic thermal cycle must be related to the damage stored in the accelerated thermal cycle. Damage in solder joints will accumulate from thermally activated, time dependent mechanisms (creep), from athermal cyclic mechanisms (fatigue), and from microstructural changes. These damage mechanisms are expected to interact with one another and to have different relative magnitudes as a test is accelerated. *Ab initio,* one would expect that for any thermal cycle where the solder does not completely relax during the time at maximum temperature, the creep damage inflicted will be hold time dependent. For subzero minimum temperatures, the time dependent damage should be small compared to the athermal damage and so hold time would not be expected to be important. By the same reasoning, the heating or cooling rate could be accelerated in the cold regions of the cycle but not in the hot parts without resulting in time or cycle dependent complications. As will be discussed below, these complications include not only creep but also anelastic relaxation.

Coffin-Manson-type relationships have been modified to account for frequency and temperature effects.[2–6] These approaches are empirical methods for keeping track of how much damage is stored in each cycle and relating this damage to the life of the material. This empirical approach is effective as long as the model is developed carefully within a particular temperature range and electronic package type. It can be argued that the most trustworthy models fit data from accelerated thermal tests of components,[2,5] while others use bulk specimens[3,4] and isothermal tests.[3,4] One of the problems with this empirical approach is that it takes too much time to develop good tests for a particular application and then tests need to be redeveloped for other applications.

The cyclic nature of the stress history of electronic packages implies at least some sort of fatigue behavior, if not creep/fatigue behavior, and has led to the current approach of a modified Coffin-Manson equation. This may still be inadequate because of the high homologous temperatures involved with solder applications since, even at room temperature, creep mechanisms will also contribute significantly to the damage. In order to improve the way in which models are developed and applied, the fundamental behavior of the solder needs to be examined at realistic temperatures and with the application of cyclic stresses.

9.2 DAMAGE BY FATIGUE AND CREEP MECHANISMS

The cyclic nature of the temperature variation causes the strain experienced by the solder joints to be cyclic and, therefore, the damage in the solder to be a function of the number of thermal cycles. By this definition, SMT solder joints experience not only fatigue but thermal fatigue. At low temperatures (below $T/T_m = 0.5$), fatigue can cause damage by the initiation and propagation of microcracks. In the above equation, T is temperature and T_m is melting temperature. The usual equations used to describe fatigue life N_f are the Coffin-Manson relation:[7]

$$N_f = \left(\frac{\Delta\varepsilon_p}{\varepsilon_f}\right)^{-A} + \left(\frac{\Delta\sigma}{\sigma_o}\right)^{-B} \tag{9-1}$$

This is an empirical equation that predicts fatigue life. Low cycle fatigue is controlled by the rate of crack growth. In general, low cycle fatigue controls life up to around 10^4 cycles and high cycle fatigue controls life above 10^6 cycles. The first term describes low cycle fatigue where the damage stored in each cycle is related to the plastic strain range $\Delta\varepsilon_p$ raised to the negative A power. The second term describes high cycle fatigue where most of the life of the component is spent in the initiation of cracks. Here the damage stored in each cycle is best described by the stress range $\Delta\sigma$ raised to the negative B power. In equation (9-1), A is the low cycle fatigue Coffin-Manson exponent, B is the high cycle fatigue Coffin-Manson exponent, ε_f is the fracture strain, and σ_o is the characteristic stress for high cycle fatigue.

In general, surface mounted components see thermal cycling that can range from $0.4T_m$ to $0.75T_m$. Accordingly, it can be argued that in most microelectronic applications, the life of a component is spent at temperatures above $0.5 = T/T_m$, which is about room temperature for the highest melting point Pb-Sn solders. As a rule of thumb, creep is an operative deformation mechanism whenever the temperature is above $0.5T_m$. Creep strain can result from any of a number of thermally activated, time dependent mechanisms. These mechanisms can obey a constitutive relation of the form:

$$\dot{\varepsilon} = C\left(\frac{1}{d}\right)^a \frac{1}{kT}\left(\frac{\sigma - \sigma_b}{E(T)}\right)^n \exp\frac{E_a}{kT} \tag{9-2}$$

C is a constant, d refers to grain size and a to grain size sensitivity, σ is the applied stress with σ_b being the back stress and $E(T)$ the modulus as a function of temperature, making n the stress exponent. The thermal activation of creep is characterized by an activation energy E_a and k is Boltzman's constant.

At high stresses, creep is controlled by dislocation motion. When dislocation entanglement and recovery reach an impasse where the rate of hardening is equal to the rate of recovery, a quasi-steady state is defined that obeys Equation (9-2). This creep rate is controlled by the rate at which edge dislocations can climb out of their slip planes. In this case the grain size exponent a will be negligible, the stress exponent n will be between 3 and 6, and the activation energy E_a will be the energy for vacancies to travel to climb sites at dislocations[8–10] whether by bulk diffusion or some short circuit path.

At lower stresses, creep strain results directly from the motion of vacancies from grain boundary to grain boundary, which results in the elongation of grains. Here, the grain size sensitivity a becomes important. At the highest temperatures, the vacancies move by lattice diffusion and the creep mechanism is referred to as Nabarro-Herring creep. Nabarro-Herring creep is characterized by grain size exponent $a = 2$, a stress exponent $n = 1$, and an activation energy E_a of the lattice diffusion energy.[11] At lower temperatures, the vacancies move by grain boundary diffusion and the creep mechanism is referred to as Coble creep. Coble creep is characterized by a grain size exponent $a = 3$, a stress exponent $n = 1$, and an activation energy E_a of grain boundary diffusion.[12]

In some cases at intermediate stresses, materials can undergo superplastic deformation. Although as-cast solders do not typically exhibit superplastic behavior, solder alloys in the rolled and annealed condition are often superplastic, and their superplastic deformation can provide some insight into the deformation of as-cast solder. Superplastic deformation is characterized by extremely large strains caused by steady stresses. In practice, strains of 500 to 1000 percent are often necessary before failure can occur. Stress exponents are usually less than 3 regardless of the particular mechanism since higher stress exponents make specimens sensitive to necking.

The most common type of superplastic flow involves grain boundary sliding and migration in fine grained materials. This mechanism of superplastic flow[13,14] has the same grain size, stress, and temperature dependence of Nabarro-Herring or Coble creep, depending on the temperature. However, the strain rate is a couple orders of magnitude higher than Nabarro-Herring or Coble creep because the shape change of the grains merely allows grains to slide past one another and reorganize. No permanent elongation of the grains is required.

The damage that is stored during creep deformation can be of three basic types: creep cracks, void nucleation and growth, and microstructural degradation. Creep cracks are formed and grow at lower temperatures where flow mismatch at grain boundary triple points or persistent slip bands cannot be accommodated by the flow of the adjacent grains. Void nucleation and growth at grain boundaries occur throughout creep under most conditions even when voids have little to do with the final failure. There are many models for void nucleation,[15,16] most of which link nucleation to grain boundary sliding and impurities or particles at the grain boundaries. As voids grow and link, the supporting cross section is decreased and the stress increases. Microstructural degradation is often the most serious damage in alloys that depend on the phase morphology for creep resistance. Stresses at high temperatures allow strengthening precipitates to coarsen and change shape, weakening the alloy. Recrystallization can also occur as dislocation tangles recover and grow. It should be noted that the localized weakening by recrystallization can localize the creep strain, causing necking and failure.

9.3 ASSESSING ACTUAL JOINT DAMAGE

The above reviews damage storage leading to failure in bulk specimens. The next issue is the more realistic or at least the perceived realities of the damage in actual joints. During a thermomechanical cycle, the SMT component, the solder, and the wiring board expand or contract at different rates. Under this action, damage is stored in the solder. At the lower temperature reaches of the

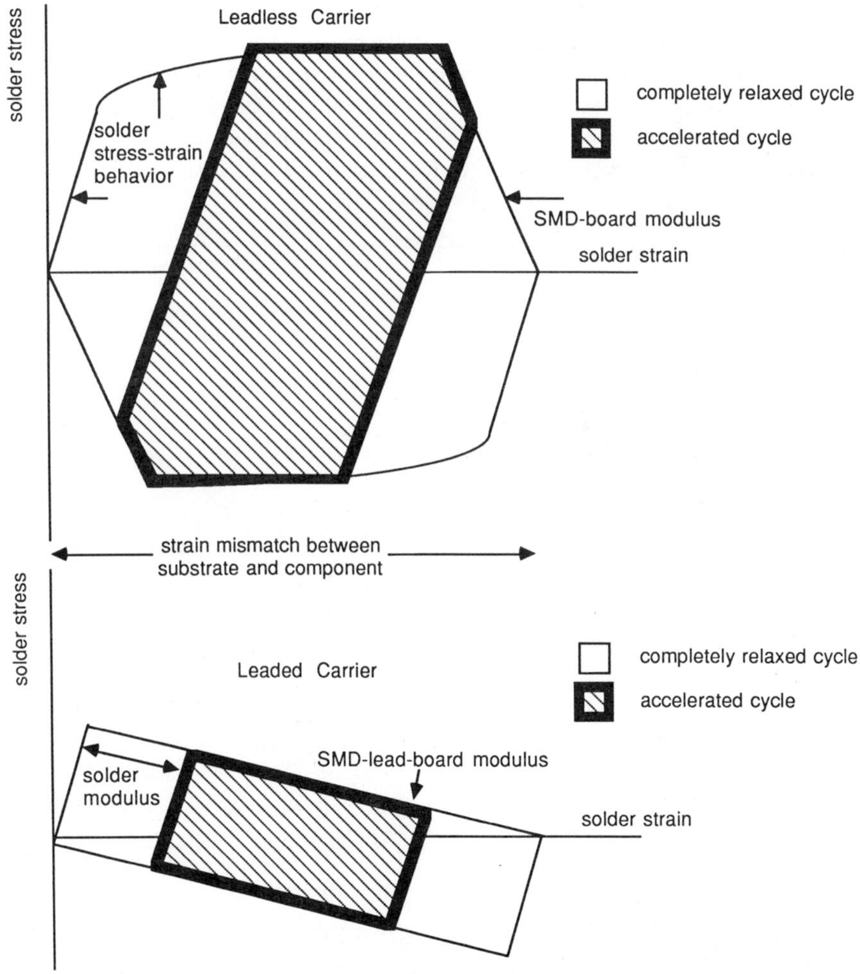

Figure 9-1 Stress-strain hysteresis in the solder are a measure of the amount of damage stored. The figure compares accelerated hold times to long hold times.[2]

thermal cycle, at least some of the damage will be athermal dislocation entanglement. During the higher temperature parts of the cycle, nearly all of the damage is stored in a time and temperature dependent manner by one of the creep mechanisms.

The integration of the solder stress versus solder strain to give the energy stored in the solder over a thermal cycle is shown in Figure 9-1 for generic leaded and leadless packages.[2] The slope and plastic yield point of the on-load ramp is a function of the temperature rate and the solder response. The constitutive relation of the solder at that strain rate and each temperature determines the shape of the stress-strain relation. In this way, any change in temperature rate can change the entire shape of the stress-strain relationship. The shape of the stress-strain relation during the relaxation of the solder during the upper temperature hold time (see Fig. 9-1) is controlled by the bending compliance of the SMT component-wiring board or SMT component-lead-wiring board system, depending on whether the package is leadless or leaded, respectively. If the

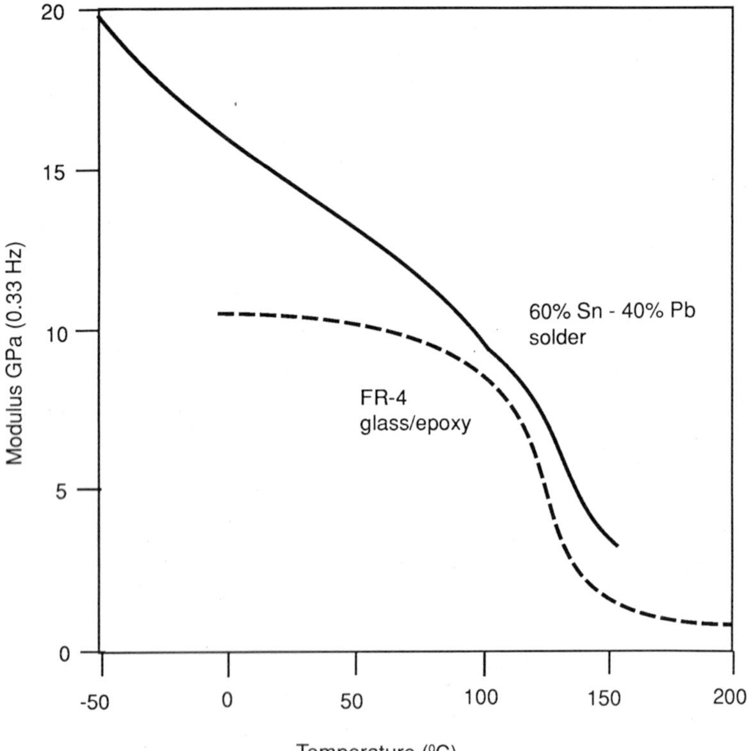

Figure 9-2 Schematic modulus as a function of temperature for solder and FR-4 glass or epoxy board.

package system is very compliant, large amounts of mismatch can be accommodated at low stresses that store less damage in the solder joint. High compliance can be achieved by using long, thin leads or thin, compliant boards. It is important to note that the compliances of both solder and polymeric board materials are quite temperature dependent as shown in Figure 9-2.

The stress relaxes toward zero during the upper temperature hold at a rate depending on the temperature through a yet to be determined constitutive equation for the solder joint. One problem with the development of a solder constitutive equation is that it must incorporate both deviatoric and hydrostatic stresses. An even more difficult factor is that the microstructure of a solder joint can change during the thermal cycle. These changes can include coarsening of the Pb and Sn phases, recrystallization, and grain growth.

The damage stored in an accelerated test can be compared with the damage stored in a realistic thermal cycle by comparing the stress-strain hysteresis experienced by the solder. The accelerated test can be valid if the dominant damage storage mechanism is the same as in the real thermal cycle (i.e., the mechanism that controls failure during the application must be the active mechanism in the test.) In addition, if the accelerated and realistic cycles are well defined, some estimation of life might be made.

9.3.1 In-service Testing

There is always a need to measure the damage in a component after it has experienced some amount of in-service use in order to determine the residual life left in the component. There have been a number of proposed in-service tests, some of which destroy the component in question and some which attempt to probe the damage without destroying the joints.

The nondestructive methods focus upon measuring cracks that have developed in the joints. Ultrasonic microscopy detects the discontinuity of the modulus in the solder at the crack surface and can map out the cracks in all the joints.[17] Electrical conductivity techniques measure as a function of time the loss of joint cross section by crack or void growth. Conductivity is also a function of the growth of brittle intermetallic layers and microstructural changes in the solder. Likewise, the thermal conductivity of joints can be examined by infrared mapping of the temperature distribution during operation. These nondestructive techniques are in the stage of initial development.

Another sort of nondestructive test that is being examined is the measurement of the hysteresis in the temperature versus bending strain of the wiring board. The bending strain is usually measured with a strain gauge mounted on the wiring board surface opposite the SMT component. It has been found that the area of the hysteresis decreases according to a log-linear relationship with the number of cycles that a component has seen up to failure.[18]

The destructive examination of joints is being proposed for measuring microstructural damage in the joints.[19–22] A more recent development is the monitoring of the voids on grain boundaries.[23] Grain boundaries can be exposed for examination by embrittling the boundaries with liquid gallium. There is a correlation between the void distribution on the grain boundary surfaces and the amount of creep-fatigue damage that has been stored. The details are still under investigation.[24,25]

9.4 UNDERSTANDING THE DAMAGE MECHANISMS

It should be evident that in order to understand and implement accurate and dependable accelerated testing and in-service testing, we need to more fully understand solder damage storage behavior as a function of time, temperature, stress, and microstructure. Conversely, as mentioned, a priority for mechanical property understanding is the determination of what the dominant modes of damage storage are in realistic temperature and strain rate regimes.

In what follows, a self-consistent set of experiments and their results are presented. These experiments are designed to separate the different types of strain in isothermal cyclic creep tests in order to understand the damaging strain from benign strain. Bulk cast material was used in this development of the form of constitutive equations for joints. Cyclic creep tests were performed on two alloys of nominal composition 63% Sn-37% Pb (eutectic) and 81% Pb-19% Sn,

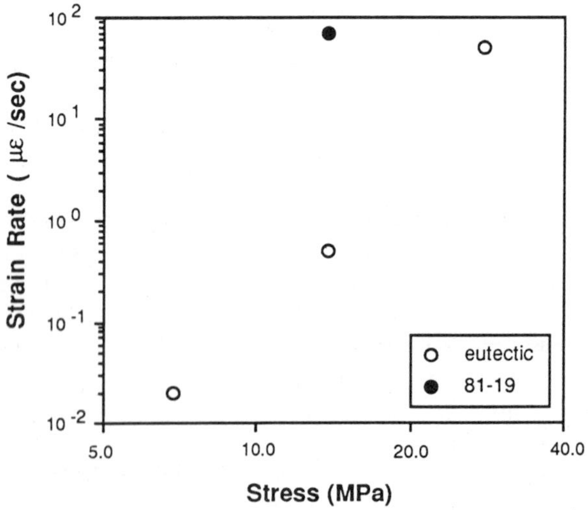

Figure 9-3 Minimum creep rate as a function of stress for bulk eutectic and 19% Sn-81% Pb solders.

both of which have their solidus at the eutectic temperature (183°C). Experimental details can be found elsewhere.[24–29]

9.4.1 Creep and Tensile Test Results

The creep results shown in Figure 9-3 indicate that the minimum creep rate can be described by a power law Equation (9-2) with a stress exponent n equal to 5 to 6 in the range of stresses tested at ambient temperature. Cyclic tests were

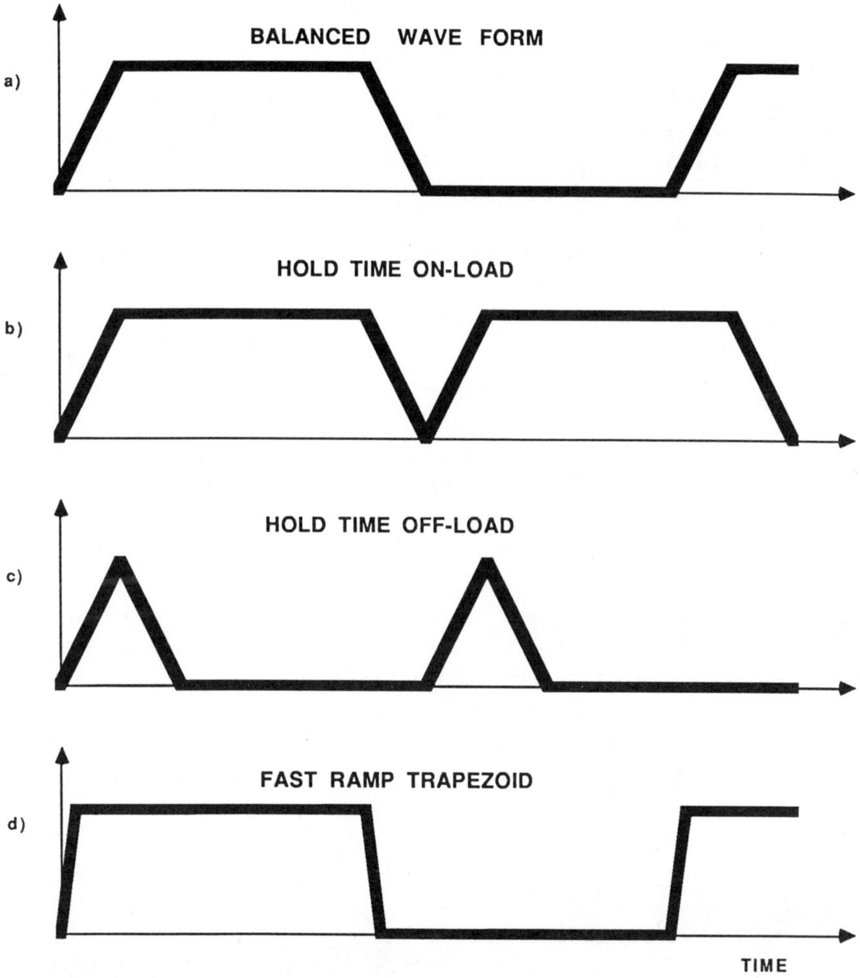

Figure 9-4 Waveforms used during cyclic testing. (*a*) Balanced wave form. (*b*) Hold time on-load. (*c*) Hold time off-load. (*d*) Fast ramp trapezoid.

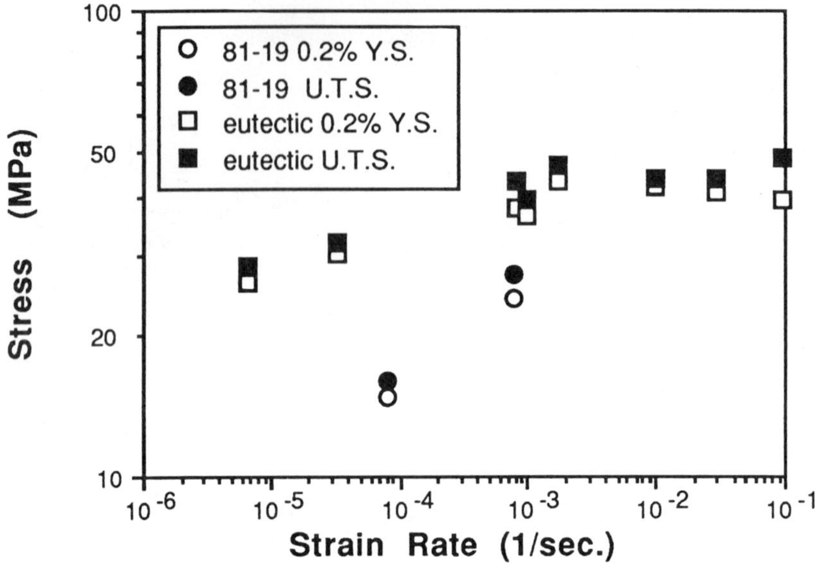

Figure 9-5 Ultimate tensile and 0.2 percent yield stresses for bulk eutectic and 19% Sn-81% Pb solders.

designed with the waveforms shown in Figure 9-4 and stresses such that the corresponding static tests had lives greater than 3600 sec (1 h). Initial creep tests were performed on joint specimens at 27.6 and 55.2 MPa. The 27.6-MPa test elongated immeasurably after 60 h and so 55.2 was chosen as the stress at which to test the frequency dependence of the joint configuration. The 19% Sn-82% Pb alloy was tested at 0 to 13.8 MPa since this was below the 0.2 percent yield stress down to 8×10^{-5} sec^{-1}. The tensile results in Figure 9-5 indicate a power law material at lower strain rates and a less strain rate dependent behavior at higher strain rates. The fast ramping trapezoidal waveform (Figure 9-4d) was designed to have a strain rate around 10^{-2} sec^{-1}. This served to minimize the amount of time dependent strain being stored during the on-load ramp and provides for the separation of strain storage mechanisms. Strain controlled cyclic creep tests were designed at 200 $\mu\varepsilon$ for comparison to the 27-MPa stress controlled tests. Higher alternating strains were used to find the effects of plastic increments on life and stress relaxation.

9.4.2 Cyclic Creep

The rate of damage storage in a stress controlled cyclic creep test is the envelope strain rate shown schematically in Figure 9-6. Envelope strain is the measure of the nonrecoverable strain stored in each cycle. For each test, an envelope strain curve was obtained by plotting the point of maximum strain in each cycle as a function of time. The minimum slope of the resulting curve gives the minimum

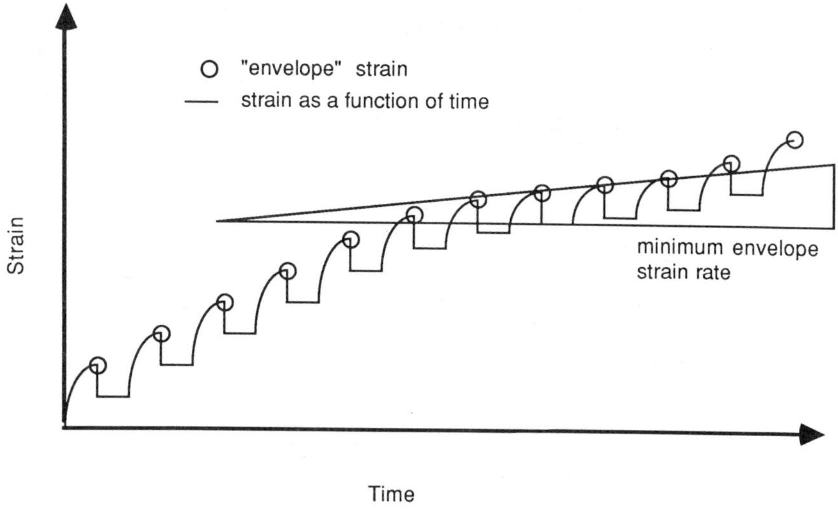

Figure 9-6 Schematic envelope strain curve. ○ = envelope strain; ——— = strain as a function of time.

envelope strain rate that can be fit to a power law such as equation (9-2). The envelope strain rate in a cyclic test can be compared with the strain rate of a static creep test. In the case of zero minimum stress, the creep rate of the cyclic test should be measured as a function of the time spent on-load rather than total time, which includes time spent off-load. In the case of nonzero minimum stress, a method is developed below for comparison of test results.

The cyclic creep results for the eutectic alloy specimens are shown in Figure 9-7 as a function of stress range and in Figure 9-8 as a function of mean stress. A comparison of the two methods of plotting indicates that the rate of storing permanent damage is a weak function of stress range but a strong function of mean stress. This underscores the importance of the creep component as opposed to a strictly fatigue component. A factor of 2 in stress range can make a difference of perhaps a factor of 5 in damage storage rate or life; a factor of 2 in mean stress produces a change of around 2 orders of magnitude in damage storage rate.

If the stress exponent n in Equation (9-2) is other than 1, the effective mean stress during a cyclic creep test should be defined as the average of each stress raised to the n power.

$$\varepsilon(\text{avg}) = C\,(\sigma_{\text{eff}})^n \tag{9-3}$$

where

$$\sigma_{\text{eff}}{}^n = \frac{\sigma_{\max}{}^n + \sigma_{\min}{}^n}{2} \tag{9-4}$$

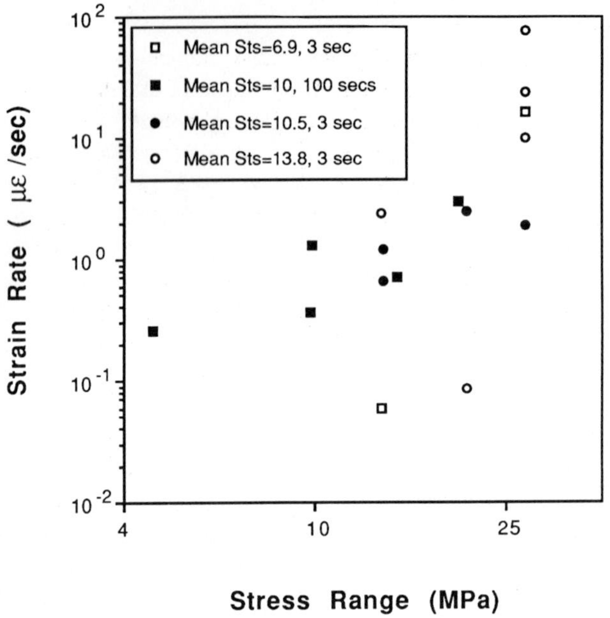

Figure 9-7 Minimum envelope strain rate as a function of stress range.

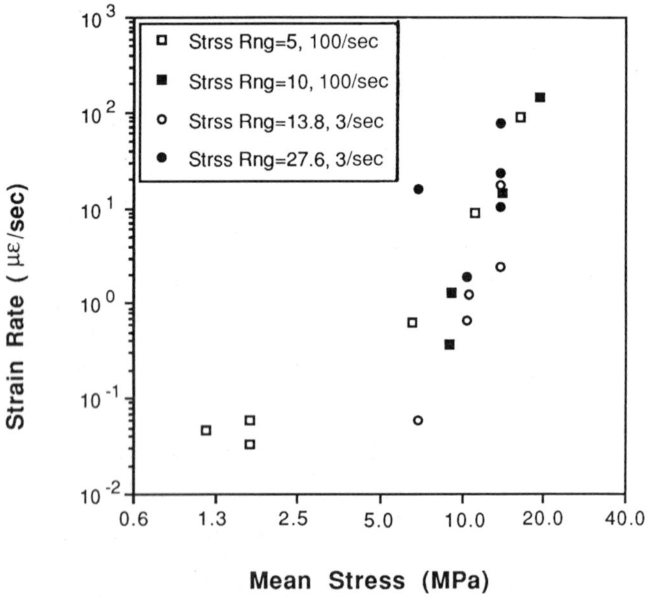

Figure 9-8 Minimum envelope strain rate as a function of mean stress.

In the above equations, σ_{eff} is the effective average creep stress, σ_{max} is the maximum stress, and σ_{min} is the minimum stress. Hence, the cyclic creep data from Figures 9-7 and 9-8 are replotted as a function of effective mean stress in Figure 9-9 with *n* taken equal to 6. In these plots, the use of the effective mean stress allows us to correlate the information of the specimen tested at a constant mean stress (E110) to that of the specimens tested at constant stress range (E103 and E105). Effective mean stress also takes into account the "time on load" for specimens with zero minimum stress. The plot of log strain rate versus log effective stress shows that the data points from the same specimens are approximately parallel to each other. This behavior with respect to effective mean stress indicates that even the small stress range effect shown in Figure 9-7 is, in fact, a result of creep damage and not an indication of some small amount of athermal fatigue damage.

The 100-sec hold time data is separated according to specimen so that the scatter caused by different specimens is obvious. The 100-sec hold time tests were performed on only three specimens. Multiple tests were performed on single specimens in order to minimize the effects of microstructural variations in the initial castings. Tests were interrupted after they had reached a minimum strain rate and then the specimen was used again for tests at other stresses. The influence

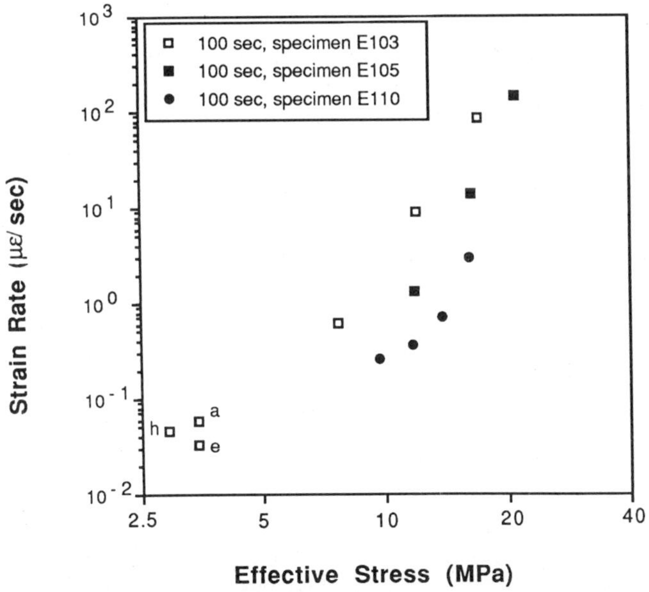

Figure 9-9 Minimum envelope strain rate as a function of effective mean stress defined by Equation (9-4). E103 and E105 had constant stress ranges of 5 MPa and 10 MPa, respectively; E110 had a constant mean stress of 10 MPa.

of the prior stress history on the strain rate is demonstrated by comparing the points *a, e,* and *h* in the lower left-hand corner of Figure 9-9 from specimen E103. These three tests were performed at the beginning, middle, and end of specimen E103's test history. Comparing the scatter between points *a, e,* and *h* to the scatter between specimens indicates that the scatter introduced by the evolving microstructure is less than the scatter introduced as a result of the initial microstructural variations between the different specimens. It should be noted that the casting conditions for bulk specimens in the lab are quite well controlled as compared to typical solder joint reflow conditions.

The temperature dependence of the envelope strain rate is shown in Figure 9-10. The 27.6-MPa 3-sec hold time data show an apparent activation energy of 0.95 eV. The 10-MPa 100-sec hold time data show an apparent activation energy of 0.86 eV. These activation energies are in line with the bulk diffusion energies for Pb or Sn[30,31] and with other creep studies on lamellar eutectic Pb-Sn.[32] In practical terms, this implies that for 1°C variation in test temperature around room temperature, the rate of nonrecoverable strain storage will change by about 12 percent.

In a strain controlled test, the strain range dependence of solder life obeys a Coffin-Manson-type relationship when all other variables remain constant, as

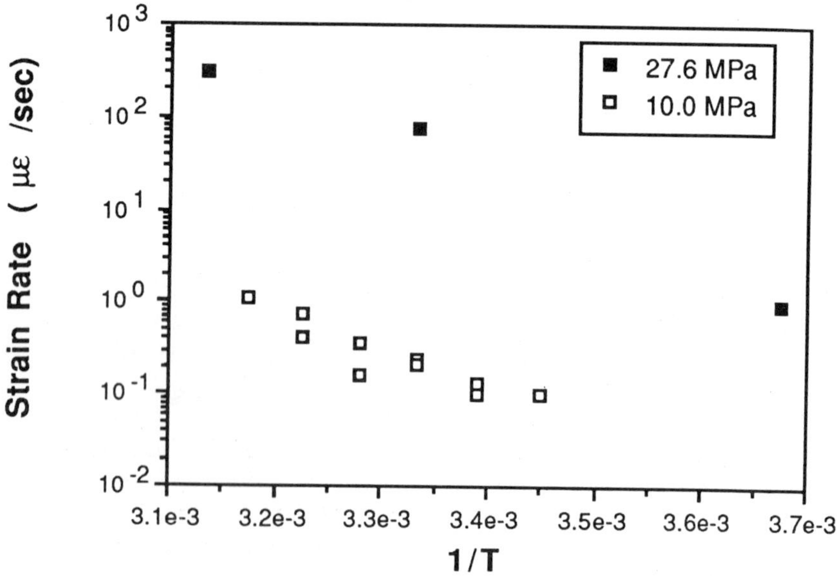

Figure 9-10 Minimum envelope strain rate as a function of temperature for 100-sec hold time at 10 MPa and 3-sec hold time at 27.6 MPa.

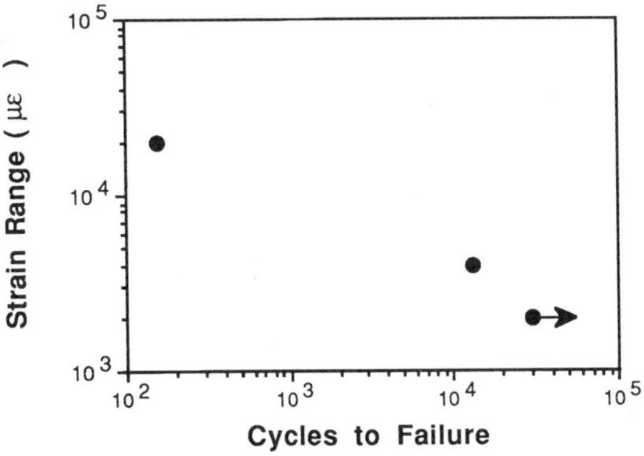

Figure 9-11 Cycles to failure N_f as a function of strain range for 3-sec hold times.

shown in Figure 9-11. The failure of a specimen was declared when a measurable drop in the apparent modulus resulted from crack growth reducing the cross section of the specimen. All tests were cycled from zero strain to the maximum strain of the test. The minimum stress started at zero but relaxed to a negative value during a shakedown period until the mean stress approached zero. The shakedown of the stress is a measure of permanent damage in the specimen, but some of the creep strain stored on the positive side of the cycle is macroscopically recovered on the negative side of the cycle even though it could still be quite damaging. Likewise, damage is stored on the negative side of the cycle, though the resulting strain is erased on the positive side of the cycle. The larger strain range tests stored a significant amount of plastic strain on the ramp-up; therefore the mean stress is essentially zero after the first cycle. In these large strain range tests both athermal and thermal components of the nonrecoverable strain contribute to the stress relaxation in the shakedown process.

9.4.3 Hold Time Effects

The acceleration of a fatigue or thermal fatigue test pivots about the cyclic hold time and the ramp rate. The ramp rates determine the relative amounts of time dependent and time independent strains that are stored during the ramp. The hold time controls the amount of time dependent strain stored during the hold. These tests were designed to help understand the effect of accelerating a fatigue test. The earliest of these tests kept a constant wave shape (Figure 9-4*a*) and varied the frequency. Once it became apparent that the damage controlling mechanisms were time dependent, the ramp rate was kept constant and only the hold time was varied (Fig. 9-4*d*).

The effect of hold time on the envelope creep is shown in Figure 9-12. As the hold time decreases (i.e., as the test is accelerated), the creep rate decreases

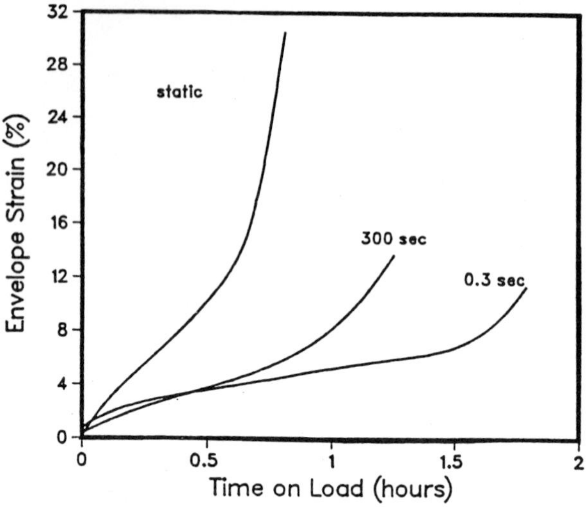

Figure 9-12 Envelope strain as a function of time for a static test, a 300-sec hold time test, and a 0.3-sec hold time test.

and the life becomes much longer. This increase in life is shown as a function of hold time for different stresses, temperatures, and alloys in Figure 9-13. All cases show a decrease in damage storage rate by a factor of 5 to 10 as the test is accelerated from static creep into the short hold time regime.

A similar increase in life with decreased hold time is also observed in the joint specimens, as shown in Figure 9-14. In this case, life, rather than strain rate, is plotted because the gauge length of a joint is too short to allow accurate measurement of strain. Since the life of a joint increases significantly over the same range of frequencies as the bulk solder, the effect should be examined further in the bulk solders and applied to joint testing.

The detailed strain per cycle data is shown in Figure 9-15 for a 3-sec hold time test. Quite a significant strain recovery is observed upon each unloading. This strain recovery was found to range from 10 to 40 percent of the elastic strain depending upon the temperature and stress range. This strain recovery resembles an anelastic effect in that it is a time dependent recovery of strain. Strictly, an anelastic effect should be linear with stress (i.e., the elastic strain).[33] Confirmation or denial of the linearity of this strain storage with the stress requires more exact control of the stress than was available at the time of testing. For the lack of a better term, the effect will be referred to as anelastic in this work.

In order to determine whether this anelastic strain recovery or some cyclic hardening effect is responsible for the decrease in strain rate with decreasing hold time, tests were run that eliminated either the off-load hold time or the on-load hold time, as shown in Figures 9-4*b* and 4*c*. All three tests had 3-sec hold

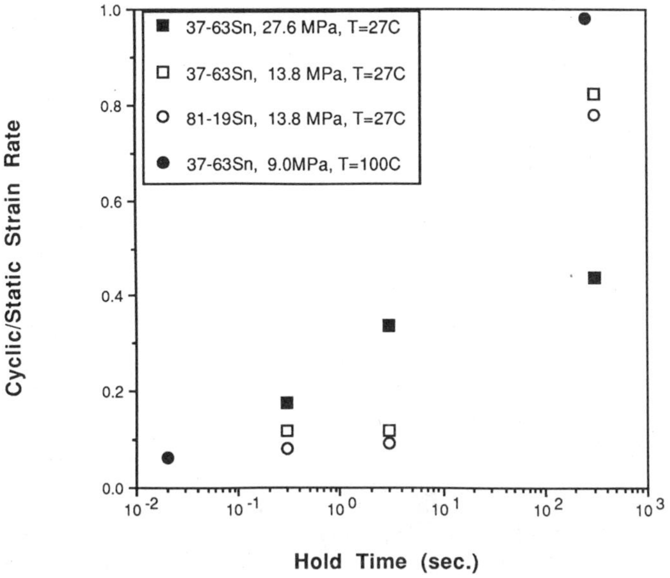

Figure 9-13 Normalized minimum strain rate (i.e., minimum strain rate divided by the minimum strain rate for the static creep test) as a function of hold time for eutectic and 19% Sn-81% Pb solders.

times and identical ramp rates. The time on load, which is defined here as the time spent above the mean stress, is used rather than total test time or number of cycles to compare the severity of the tests. The envelope strain curves for these tests are shown in Figure 9-16. The rate of damage storage is much faster

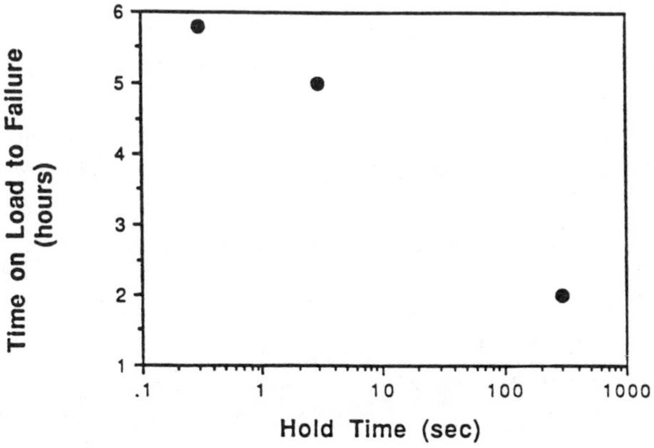

Figure 9-14 Joint life as a function of cyclic frequency.

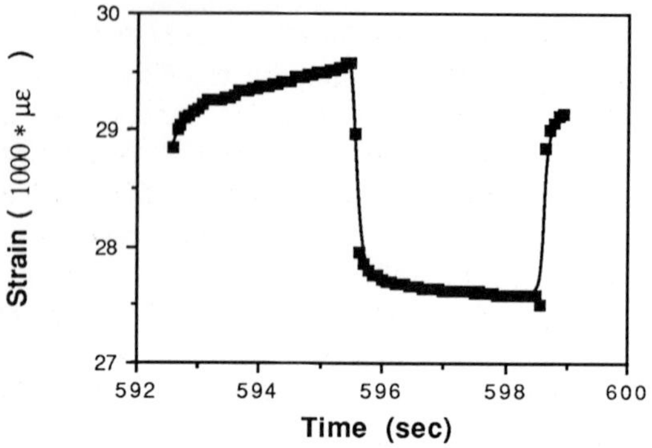

Figure 9-15 Strain as a function of time for a single cycle showing the large strain recovery during the off-load hold time.

when there is relatively little time spent relaxing off-load. In fact, the on-load hold time test behaved almost identically to the static creep test. This indicates that the increased life with decreased hold time is the result of recovered anelastic strain.

If the decrease in creep rate had been the result of cyclic hardening, the on-load hold and the balanced hold time tests would have had the same strain rates.

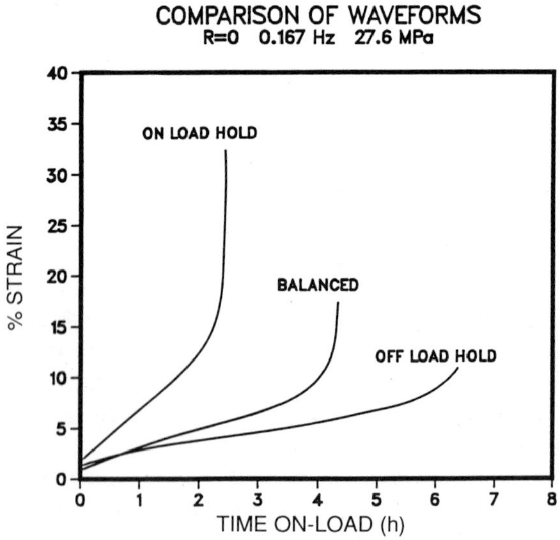

Figure 9-16 Envelope strain as a function of time for different waveforms shown in Figure 9-4. Comparison of waveforms: $R = 0$; 0.167 Hz; 27.6 MPa.

Both tests see the same number of cycles for the amount of time they spend on-load. In the case where no time is spent recovering the anelastic strain during an off-load hold, the anelastic storage is unable to prevent the storage of non-recoverable strain during the on-load hold. And so, nonrecoverable strain is being stored right from the beginning of each on-load hold. The test with only an off-load hold spends very little time storing time dependent strain and spends much more time recovering anelastic strain. In this case all of the anelastic strain that was stored can be recovered during the off-load hold time and therefore all of the time during the on-load is spent storing recoverable strain.

As permanent creep damage is retarded by increasing the off-load hold time, the reduction of creep rate with decreasing hold time in the balanced cyclic tests must be caused by the recovery of the initial portion of the time dependent strain that is stored during each on-load hold time. Therefore, the creep strain and the anelastic strain are serial processes during the on-load hold. In other words, the storage of anelastic strain at the beginning of the on-load hold retards the storage of creep strain. However, as the anelastic process becomes saturated, creep strain will be stored. Accordingly, for short hold time tests, almost all of the time dependent strain that has been stored during the on-load hold is anelastic and can be recovered during the off-load hold. Conversely, for longer hold time tests, the reverse will be the case.

For the purpose of developing constitutive equations, the time dependent strain stored in each cycle may be separated into the recoverable anelastic strain and the nonrecoverable creep strain. The strain storage and recovery during the steady-state stage of the creep tests can be analyzed using data from individual cycles as shown schematically in Figure 9-17. The storage of strain can be described by

$$\varepsilon(\sigma,t) = \varepsilon_{EL}(\sigma) + \varepsilon_{PL}(\sigma) + \varepsilon_{AN}(\sigma,t) + \varepsilon_{CR}(\sigma,t) \qquad (9\text{-}5)$$

The recovery of strain can be described by

$$\varepsilon(\sigma,t) = \varepsilon_{EL}(\sigma) + \varepsilon_{AN}(\sigma,t) \qquad (9\text{-}6)$$

In the above equations ε_{EL} is the elastic strain, ε_{CR} is the creep strain, and ε_{AN} is the anelastic strain. Each cycle then has two time independent strain components, one of which is recoverable and the other of which is nonrecoverable; therefore they can be separated by comparing the strain stored during the on-load ramp to the strain recovered during the off-load ramp. Likewise, the two time dependent strain components, anelastic strain and creep strain, can be separated by comparing the strain stored during the on-load hold time to the strain recovered during the off-load hold time.

The temperature dependence of the anelastic relaxation is small, and small drifts in the minimum stress during the off-load hold prevented accurate mea-

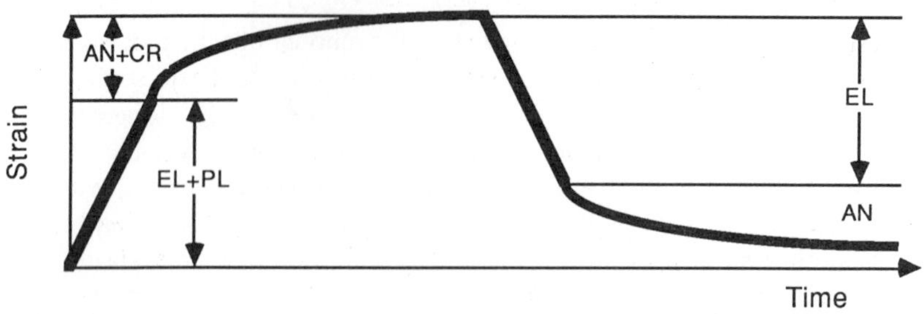

Figure 9-17 Schematic of the separation of the different type of strain in a single cycle. The time independent strains, elastic EL, and plastic PL, are separated by comparing the strain stored during loading to the strain recovered during unloading. Likewise, the time dependent strains, creep CR and anelastic AN, are separated by comparing the strain stored during the on-load hold time to the strain recovered during the off-load hold time.

surement of the relaxation rate. Relaxation times of 1 to 10 sec were measured using a simple exponential decay model. These were measured over the temperature range of 17 to 42°C. There was no systematic variation in the relaxation time over this temperature range. Assuming that the change in relaxation time is hidden in the scatter, the activation energy for the relaxation must be less than 0.73 eV.

Many joint loadings can be described as strain or displacement controlled. Stress-strain hysteresis loops are shown in Figure 9-18 for hold times of 0.3 and 300 sec with a 2000 $\mu\varepsilon$ strain range. In all cases, although the minimum stress started at 0, it became negative as nonrecoverable strain was stored. This shakedown continued until the mean stress approached zero and, after cracks became significant, even dropped below zero. At short hold times this relaxation takes much longer than it does with long hold times. In the 300-sec hold time test, the average stress is relaxed to close to 0 in the first cycle and relaxes little more over the next 9000 sec. In the 0.3-sec hold time test, even after 9000 sec very little relaxation has occurred by comparison. The rate of relaxation is plotted as a function of time for three hold times in Figure 9-19.

A problem with strain controlled testing is in defining an easily measurable rate of damage accumulation before cracks become measurable. The rate of stress relaxation measures the storage of unrecovered plastic strain, but it is unclear as to whether strain storage of this sort is damaging in terms of the life of the specimen. The strain stored in the specimen during the on- or off-load holds of the strain controlled test can be compared to the stress controlled test, equations (9-5) and (9-6).

$$\varepsilon_o = \varepsilon_{EL}(\sigma(t)) + \varepsilon_{PL}(\varepsilon_o) + \varepsilon_{AN}(\sigma(t)) + \varepsilon_{CR}(\sigma(t)) \qquad (9\text{-}7)$$

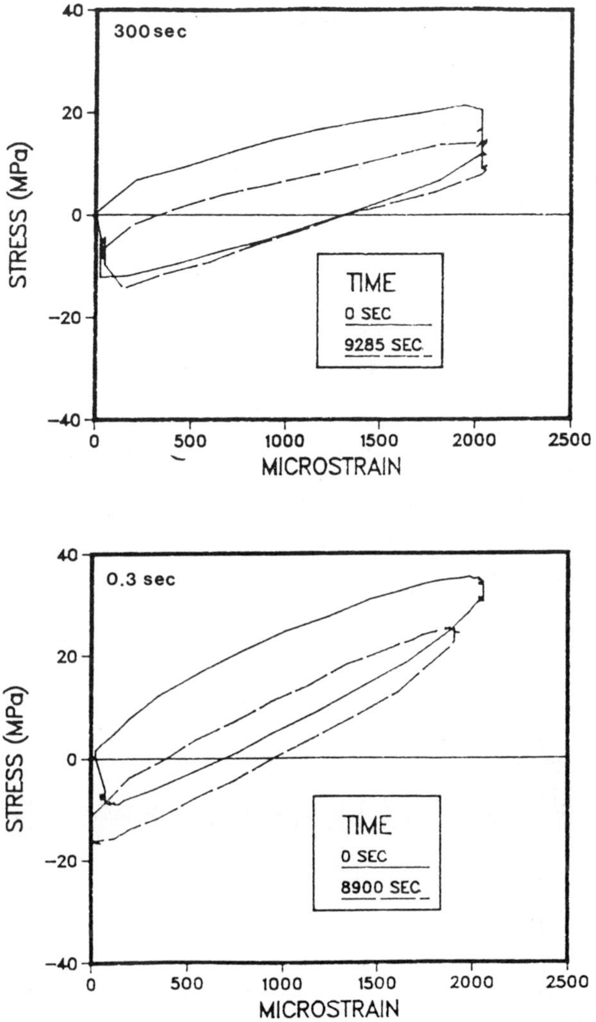

Figure 9-18 Stress-strain hysteresis for a strain range of 2000 $\mu\varepsilon$ comparing the first cycle of a test to a cycle after approximately 9000 sec of testing for 300-sec and 0.3-sec hold times.

The anelastic and creep strains cannot be separated in this case, and so there is no way to determine at this time how much time dependent damage has been stored in the material and how much of the strain was benign. But if hold times are long enough, the anelastic portion of the strain will be insignificant, and a simple comparison of the hysteresis areas between an accelerated cycle and a real cycle could give a reasonable acceleration factor. In equation (9-7), ε_o is the total strain range.

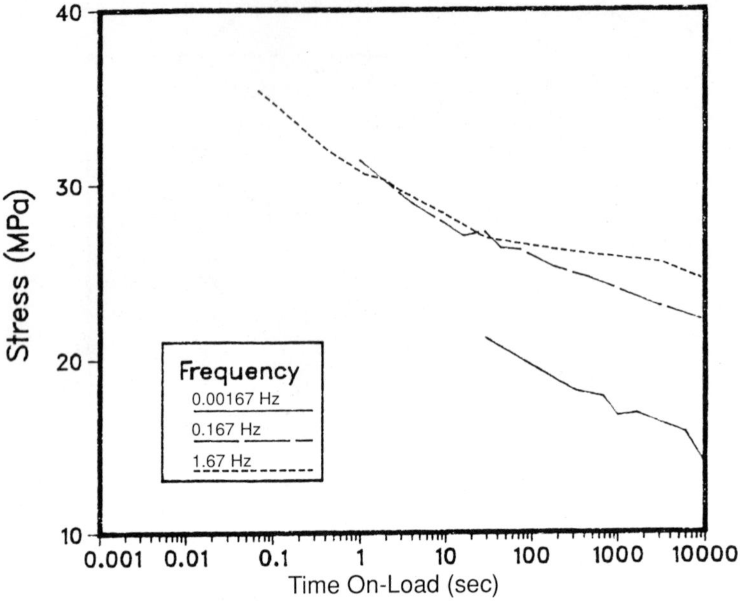

Figure 9-19 Maximum stress in a cycle as a function of time comparing different hold time tests all with strain ranges of 2000 $\mu\varepsilon$.

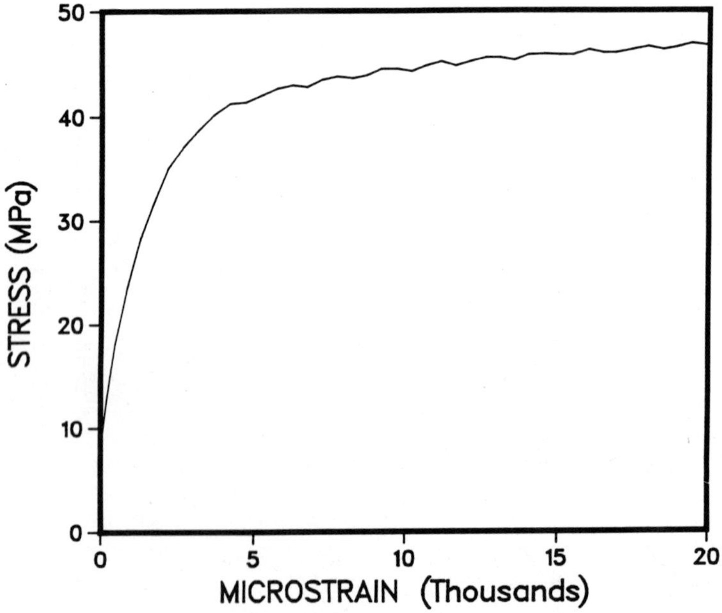

Figure 9-20 Stress as a function of strain for a tensile test at 10^{-1} sec^{-1}. Note the large amount of plastic strain stored at low stresses.

The effect of decreasing the hold time is the increase in the average stress in a half cycle. This is a natural consequence of allowing less time for relaxation in each cycle, both recoverable and nonrecoverable. At the shortest hold times, the time dependent strain that is being stored during the hold should not be storing any damage since it is all recovered during the hold of opposite sign by the anelastic mechanism. But the shorter hold times also have higher ramp rates that disallow the storage of time dependent strain during the ramp (these tests were performed with a constant wave shape) and thus cause some greater amount of athermal plastic flow to occur during the ramp. The storage of athermal plastic strain occurs at stresses much lower than the 0.2 percent offset yield point even at high strain rates, as can be seen in Figure 9-20.

9.5 INTERPRETATION FOR PACKAGING APPLICATIONS

9.5.1 Deformation

Creep deformation and anelastic recovery have been shown to be the dominant deformation and recovery mechanisms through whose interaction damage is stored near and above ambient temperature. As mentioned, the anelastic deformation acts as the initial time dependent deformation during an on-load and acts to recover the last bit of time dependent deformation that was stored prior to an off-load. The relaxation time for the anelastic mechanism is less than 10 sec over the range of tempertures tested, -1 to 100°C. Whenever the hold time of a cycle is on the order of or less than the relaxation time, the anelastic strain acts to prevent the creep deformation from creating damage in the solder. Also, the parallel nature of the creep and anelastic strains implies that they are mechanistically one type of event.

The eutectic and the 19% Sn-81% Pb alloys show nearly identical behavior with respect to accelerating the cyclic test, as was shown in Figure 9-13. This is despite the fact that the eutectic alloy has a matrix of Sn with lamellar Pb phase and the 19% Sn-81% Pb has a Pb matrix with Sn precipitates. The similarity of the behavior implies that the anelastic effect is independent of the matrix phase and matrix grain boundaries. Rather, it is probably the result of the interaction between the Pb and the Sn phases, which must persist through the change in microstructural morphology from one alloy to the other. This is consistent with the large anelastic effects that have been identified in the superplastic microstructure of eutectic Pb-Sn[34,35] that is a random mixture of equiaxed Sn and Pb phase grains. The anelastic and creep deformation behavior can be mechanically modeled as two nonlinear viscoelastic phases, Pb and Sn, connected by a viscoelastic phase boundary, as shown in Figure 9-21. This model of the strain in Equation (9-5) is the beginning of constitutive equations that will be life predictive.

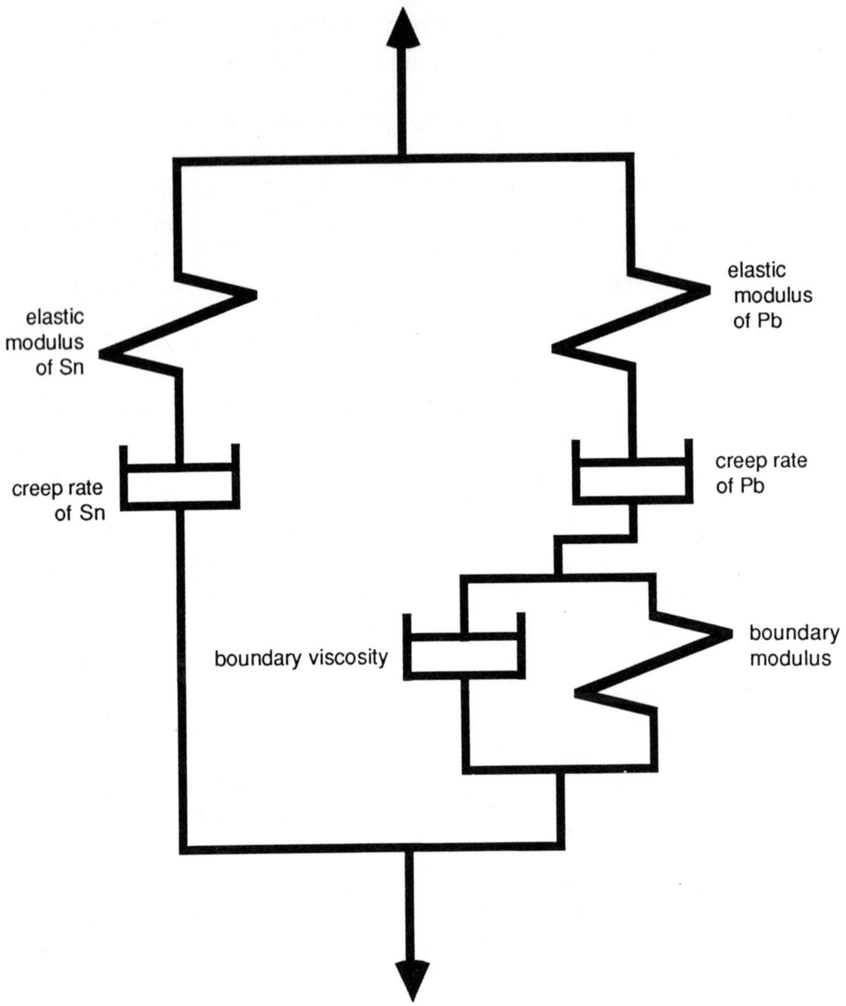

Figure 9-21 A mechanical model that can account for solder's deformation behavior.

9.5.2 Thermomechanical Test Guidance

In order to rank designs, an accelerated test must store damage by the same mechanisms that a realistic thermal cycle stores damage. Also, in order to predict component life, the damage stored during an accelerated test cycle must be related to the amount of damage stored in a realistic cycle.

The effects of hold time and ramp rate on the amount and types of damage stored during the cycle are illustrated in Figure 9-1. In either the stress or strain controlled case, the ramp rate controls the relative amounts of plastic and time dependent strain. Faster ramp rates cause the accumulation of more plastic damage and less anelastic strain and creep damage. Slower ramp rates minimize the storage of athermal plastic strain. If parts of the thermal cycle are in a temperature region where little or no time dependent damage is stored, acceleration of this part of the cycle requires no modification of damage stored per cycle. Short hold times of accelerated tests decrease the amount of relaxation and thereby decrease the amount of damage stored, as shown in Figure 9-1. This change in damage storage per cycle can be accounted for by considering the change in the energy stored per cycle. But, if hold times are decreased below, say, around 3 to 10 sec, the storage and recovery of anelastic strain will retard the storage of creep damage. The energy of the hysteresis caused by anelastic strain storage and recovery is not damaging to the solder and so the calculation of a damage storage rate from the hysteresis becomes meaningless. Hold times should be designed long enough to cause the storage of creep damage to be dominant (i.e., to make the anelastic strain that is stored insignificant compared to the creep strain). By this analysis, the practical effect of anelastic strain recovery is to define a limit to the acceleration of a thermomechanical test such that hold times should be significantly larger than the anelastic relaxation time of 3 to 10 sec.

9.6 CONCLUDING REMARKS

Accelerated thermomechanical test cycles and life or residual life predictive equations should be carefully designed with the consideration of the type of damage being stored at each temperature. At ambient temperature and above, the damage will be controlled by cyclic creep. Acceleration of the ramping rate can certainly be made in parts of the cycle where athermal damage is dominant. Acceleration at high homologous temperatures should take into account the time dependent nature of the creep damage storage and the anelastic recovery phenomena. But a thermomechanical test should be accelerated only with caution to the point where the anelastic storage and recovery of strain prevents the storage of creep damage. Around ambient temperature, hold times should be kept above 3 to 10 sec in order to keep the anelastic effect from becoming a significant part of the strain storage.

Further study should be aimed at the modes of deformation and fracture at higher temperatures as a function of acceleration of the cycle. As temperature increases, the frequency at which anelastic deformation suppresses creep damage may change enough to allow more acceleration of the test than the limit advised here of 3- to 10-sec hold times. The effects of acceleration on higher Pb content solders (e.g., 90% Pb-10% Sn, 95% Pb-5% Sn, or 98% Pb-2% Sn) should also be tested. As the amount of Sn phase decreases, the anelastic relaxation should become less pronounced, and it may be practical to accelerate tests much more in these cases. Acceleration of the strain controlled tests showed a clear dependence of the relaxation rate on hold time, but further work needs to be carried out on the effects of acceleration on life in strain controlled tests.

Small volume specimens should be developed and examined in order to understand how the details of solder behavior may change in going from bulk solder to micrometer dimensioned solder joints. These details could affect the limit of acceleration of a cycle.

REFERENCES

1. Engelmaier, W., "Surface Mount Solder Joint Long-Term Reliability: Design, Testing, Prediction," *Soldering and Surface Mount Technology,* **1**(1) Feb. 1989, p. 14.
2. "Functional Cycles and Surface Mounting Attachment Reliability," Circuit World, *J. Inst. Cir. Techn.,* **11**(3), 1985.
3. Solomon, H. D., "Low-Frequency, High-Temperature, Low-Cycle Fatigue of 60/40 Solder," *Brazing and Soldering,* No. 11 Aug. 1986, pp. 68–75.
4. Vaynman, S., M. E. Fine, and D. A. Jeanotte, "Isothermal Fatigue of Low Sn Pb Based Solders," in *Load and Thermal Histories on Mechanical Behavior of Materials,* P. K. Liaw, Ed. Warrendale, PA, TMS, 1987.
5. Fox, L. R., J. W. Sofia, and M. C. Shine, "Investigation of Solder Fatigue Acceleration Factors," *IEEE Transactions on Components, Hybrids, and Manufac. Tech.,* **CHMT-8**(2), June 1988.
6. Sinclair, N. A., "Thermal Cycle Fatigue Life of LSI Solder Interconnections," *Proc. Intl. Electrical and Electronics Conf. and Expo,* Toronto, 1982, pp. 56–57.
7. Manson, S. S., *Thermal Stress and Low Cycle Fatigue,* New York, McGraw-Hill, 1966.
8. Weertman, J., *J. Appl. Phys.,* **26,** 1955, pp. 1213–1217.
9. Friedel, J., *Dislocations,* Ch. 11, London and Oxford, Pergamon Press, 1967.
10. Frost, H. J., and M. F. Ashby, in *Fundamental Aspects of Structural Alloy Design,* R. I. Jaffe and B. E. Wilcox, Eds., New York and London, Plenum Press, 1977, pp. 27–58.
11. Herring, C., "Diffusional Viscosity of a Polycrystalline Solid," *J. Appl. Phys.,* **21,** 1950, p. 437.
12. Coble, R. L., "A Model for Boundary Diffusion Controlled Creep in Polycrystalline Materials," *J. Appl. Phys.,* **34,** (6), 1963, p. 1679.
13. Ashby, M. F., and R. A. Verrall, "Diffusion Accommodated Flow and Super plasticity," *Acta Met.,* **21,** 1973, p. 149.
14. Sherby, O. D., R. D. Caligiuri, E. S. Kayali, and R. A. White, "Fundamentals of Superplasticity and Its Application," in *Advances in Metal Processing,* J. J. Burke, R. McHrabian, and V. Weiss, Eds., New York, Plenum Press, 1978.
15. Goods, S. H., and L. M. Brown, "The Nucleation of Cavities by Plastic Deformation," *Acta Met.,* **27,** 1979, pp. 1–15.
16. Riedel, H., *Fracture at High Temperatures,* Chs. 6–9, New York, Springer-Verlag, 1987.

17. Burgess, J. F., R. O. Carlson, H. H. Glascock, C. A. Neugebauer, and H. F. Webster, "Solder Fatigue Problems in Power Packages," *IEEE Transactions on Components, Hybrids, and Manufac. Tech.* **CHMT-7**(4), 1984, pp. 405–410.
18. Honke, D., private communication.
19. Zubelewicz, A., R. Berriche, L. M. Keer, and M. E. Fine, "Lifetime Prediction of Solder Materials," presented at the ASME Annual Meeting, Chicago, IL. Dec. 1988.
20. Frear, D., D. Grivas, M. McCormack, D. Tribula, and J. W. Morris, "Fatigue and Thermal Fatigue of Pb-Sn Solder Joints," *ASM Conf. Electronic Packaging,* 1987.
21. Frear, D. R., D. Grivas, and J. W. Morris, "Thermal Fatigue in Solder Joints," *J. of Metals,* June 1988, pp. 18–22.
22. O'Clock, G. D., M. S. Peters, J. R. Pater, G. A. Kleese, and R. V. Martini, "Pb-Sn Alloy Microstructure: Potential Reliability Indicator for Interconnects," *IEEE Transactions on Components, Hybrids, and Manufac. Tech.,* **CHMT-10**(1), 1987, pp. 82–88.
23. Greenwood, P. J., T. C. Reiley, V. Raman, and J. K. Tien, *Scripta Met,* **22,** 1988, pp. 1465–1468.
24. Hendrix, B. C., Ph.D. Thesis, Columbia University, 1989.
25. Hendrix, B. C., A. I. Attarwala, and J. K. Tien, unpublished work.
26. Weinbel, R. C., J. K. Tien, R. A. Pollak, and S. K. Kang, *J. Mat Sci,* **22,** 1987, pp. 3901–3906.
27. Tien, J. K., B. C. Hendrix, and A. I. Attarwala, *IEEE Transactions on Components, Hybrids, and Manufac. Tech.* **CHMT-12,** 1989, pp. 502–505.
28. Hendrix, B. C., J. K. Tien, S. K. Kang, and T. Reiley, in *Effects of Load and Thermal Histories,* P. K. Liaw, Ed. Warrendale, PA. TMS, 1987, p. 139.
29. Weinbel, R. C., E. A. Schwarzkopf, and J. K. Tien, *Scripta Met.,* **21,** 1987, pp. 1165–1168.
30. Oberschmidt, J., K. K. Kim, and D. Gupta, *J. Appl. Phys.,* **53,** 1982, p. 5672.
31. Askill, J., *Tracer Diffusion Data for Metals, Alloys, and Simple Oxides,* New York, Plenum Publishing Corporation, 1970, pp. 35–39.
32. Cagnon, M., M. Suery, A. Eberhardt, and B. Baudelet, "High Temperature Deformation of the Pb-Sn Eutectic," *Acta Met.,* **25,** 177, pp. 71–75.
33. Nowick, A. S., and B. S. Berry, *Anelastic Relaxation in Crystalline Solids,* New York, Academic Press, 1972.
34. Homer, C., and B. Baudelet, "Anelastic Behavior of the Pb Sn Eutectic Alloy," *Scripta Met.,* **11,** 1977, pp. 185–189.
35. Schneibel, J. H., and P. M . Hazzledine, "The Influence of Anelasticity on the Stress Relaxation Behavior of Superplastic Pb-Sn Eutectic," *Scripta Met.,* **11,** 1977, pp. 953–956.

10

Creep and Stress Relaxation in Solder Joints

Peter M. Hall

Stress-strain analysis of solder joints is a wonderful field. Stress and strain both have six components, and there are thirty-six elastic "constants" (which of course are not constant at all). All these depend on the position in the joint, the composition, purity, history, temperature, and geometrical configuration of the solder, as well as the surrounding structure and its properties and temperature distribution. In addition, they are functions of time, with time "constants" ranging from milliseconds to centuries. All together, there are enough variables to keep the journals supplied with papers for years to come. (A bibliography compiled a few years ago[1] already contained some 372 references.) This provides opportunities for us to exercise our skills in materials science, mechanical engineering, physics, chemistry, and metallurgy. The game of course, is to focus on one or two independent variables and one or two dependent variables and study their interaction. The hope is that the others can be kept more or less under control or that they don't matter too much, and if we don't oversimplify something important, perhaps some insights can be gained. The problem is that the reliability of these joints is a deadly serious matter. They are created by the billions, and all kinds of important (and unimportant) devices rely on their performance.

This chapter is a summary of some thinking and some experiments that we have done over the last few years from a fundamental point of view. It is hoped that the insights gained nay be helpful to those who design, test, and specify these joints in production and those who use them in the field. The focus here is on surface mount leadless solder joints and on the stresses caused by thermal expansion mismatches.

The objective of this study is to provide some information for understanding and interpreting thermocyclic testing of components involving solder joints. It is well known that solder creep and stress relaxation are important in such cycling, but information on these processes is difficult to come by, so it is usually ignored, by assuming either that there is no relaxation or that there is complete relaxation. Depending on the temperature range, either assumption can be quite misleading. It is important to understand the stress relaxation kinetics during any temperature cycle and especially during the "hold times."

For years we have complained to makers of printed wiring boards (PWBs) and other substrates that most of our problems arise because they provide us with substrates that don't have the optimum coefficient of linear thermal expansion. (This is often acronymized as CTE, or as some prefer, TCE. My own preference is for the term "expansivity," or the symbol α.) Indeed, the number of thermal cycles to failure is a strong function of the expansivity of the substrate (e.g., Sherry and Hall,[2] Fig. 17). Recently, however, the substrate makers have called our bluff and offered to provide us with whatever expansivity we feel is optimum. This has forced us to reconsider the question, "Given a surface mount component with a given expansivity, what is the ideal expansivity for the substrate?" In the next section, we deal with this question, as always, by simplifying it down to its fundamentals.

10.1 IDEAL EXPANSIVITY OF A SUBSTRATE

We treat this subject in various degrees of sophistication, looking in turn at six of the effects that may be involved. In each case, we consider how much strain would be imposed on the solder if it were completely compliant or fully relaxed and how we could adjust the expansivity of the substrate to minimize that strain.

10.1.1 No Temperature Gradients, No Transients

Given a component surface mount soldered to a substrate, let its temperature be changed from T to T_0 as in Figure 10-1*a*. Let there be neither temperature gradients nor transients, and assume that the solder is fully compliant, or relaxes fully. The temperature change will then result in an overall shear strain in the joint of

$$\gamma = \frac{L}{H}(T - T_0)(\alpha_S - \alpha_C) \tag{10-1}$$

where γ is the shear strain, $2L$ is the in-plane dimension of the component to be surface soldered. (In a chip carrier, we take $2L$ as the diagonal, to provide a worst case.) H is the height of the solder, T is the temperature, T_0 is the temperature at which the shear strain is zero, α_C is the (given) expansivity of

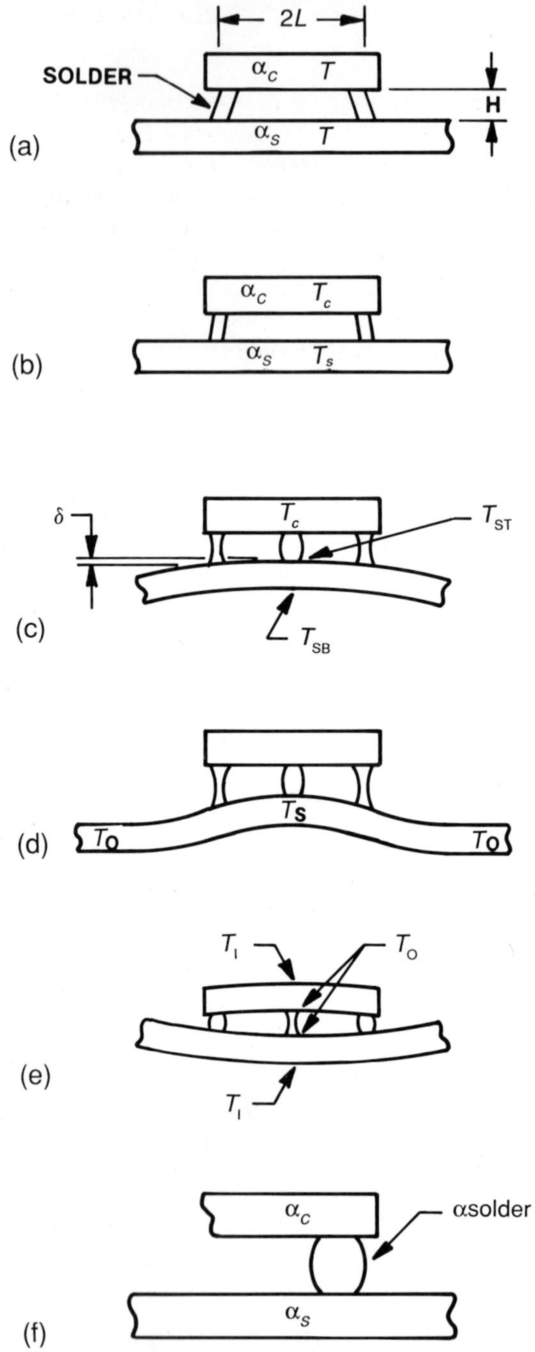

Figure 10-1 Idealized thermal strains. (*a*) Negligible temperature gradients. (*b*) Component hotter than substrate. (*c*) Through-board temperature gradient. (*d*) In-plane temperature gradient. (*e*) Thermal shock. (*f*) Solder-substrate mismatch.

the component, and α_S is the expansivity of the substrate. Many authors have quoted this equation. Here we have ignored nonuniformity of strain within the joint as well as from one joint to the other (for a multijoint component). Nevertheless, for this case it makes good sense to have the expansivity of the substrate equal to that of the component so as to minimize the shear (strain). So the answer to our question for this case is simply $\alpha_S(\text{ideal}) = \alpha_C$.

10.1.2 Power Dissipation in the Component

If the component is dissipating heat, there will be temperature gradients in the system. A first order analysis assumes that the component is at a uniform temperature T_C, whereas the substrate is at some other uniform temperature, T_S, and generally, $T_C > T_S$ (Fig. 10-1*b*). Again, we let the solder relax completely, and here we find that the shear strain is

$$\gamma = \frac{L}{H}[(T_C - T_0)\alpha_C - (T_S - T_0)\alpha_S] \tag{10-2}$$

In this case, to minimize the shear strain, we would set the substrate expansivity equal to

$$\alpha_S(\text{ideal}) = \alpha_C\left(\frac{T_C - T_0}{T_S - T_0}\right) \tag{10-3}$$

Normally, $T_C > T_S$ so we now conclude that the ideal coefficient of the substrate is greater than that of the component. In one case in point, for a ceramic chip carrier on a PWB, the ideal α_S was 1.6 times that of the ceramic chip carrier. This concept is often referred to as "tailoring" the expansivity, and it is well documented.[3,4] It was received with enthusiasm among many substrate engineers, partly because a factor of 1.6 applied to a typical ceramic expansivity (6 ppm/K) results in 9.6 ppm/K, which is readily achieved by adding certain metal inserts into the PWB.[5] It was soon realized, however, that not all the components on a substrate would have the same power or temperature, so the best one can do is optimize the expansivity for one of the components on the substrate.[4] Already, our answer to the question posed is becoming complex. But there are other considerations.

10.1.3 Z-Gradients in the Substrate

If the ceramic chip carrier is dissipating power, as above, there will be thermal gradients in the substrate normal to the plane of the substrate (Fig. 10-1*c*). Figure

10-2 shows results of some thermocouple measurements of the temperature rise above ambient due to 0.75 W dissipated in a 0.65- by 0.65-in. (1.65- by 1.65-cm) ceramic chip carrier. The average temperature rise of the chip carrier was 24 K, and the average rise of the PWB in the vicinity was 15 K. There was a temperature difference of 10 K through the thickness of the board (under the center of the carrier). Waller, Fox, and Hannemann[6] used a finite element analysis on a similar geometry and also showed that there is a considerable temperature drop across the board. This gradient causes the board to bow, as we have seen,[7] and remain bowed even after the solder has completely relaxed. The result is that the solder joints along any one side of the carrier will have a distribution of tension and compression strains. Again we simplify the problem, assuming the bowing will result in a spherical deformation of the substrate. Let the deflection of the substrate at the corners relative to that in the center of one side be δ. If we assume that the algebraic sum of the deformations is zero, the corner joints will be stretched by $^2/_3\delta$, whereas the joint in the center of a side will be compressed by $^1/_3\delta$. The deflection difference is

$$\delta = \frac{L^2(T_{ST} - T_{SB})\alpha_S}{4h_S} \tag{10-4}$$

where h_S is the thickness of the substrate and T_{ST} is the temperature of the top of the substrate and T_{SB} is the temperature at the bottom of the substrate. So the solder tensile strain in the corner (worst case) joint is

$$\varepsilon = \frac{L^2}{6Hh_S}(T_{ST} - T_{SB})\alpha_S \tag{10-5}$$

Thus to minimize the tensile strain, we should have the substrate expansivity equal to zero. We can't do this without increasing the shear strain given by Equation 10-2, so some engineering compromise must be made. One way would be to minimize the sum of the squares of ε and γ. We assume that T_S from Equation 10-2 can be approximated by $^1/_2(T_{ST} + T_{SB})$. This results in the following "ideal" expansivity:

$$\alpha_S(\text{ideal}) = \frac{\alpha_C(T_C - T_O)(T_{ST} + T_{SB} - 2T_O)}{\frac{1}{2}(T_{ST} + T_{SB} - 2T_O)^2 + \left(\frac{L^2}{18h_S^2}\right)(T_{ST} - T_{SB})^2} \tag{10-6}$$

This will produce a result between those of minimizing the shear and minimizing the tension. In our case, the numerical result was that

$$\alpha_S(\text{ideal}) = 0.88\alpha_C \tag{10-7}$$

Now we find the ideal substrate expansivity is *less* than that of the component, an unexpected result.

10.1.4 In-Plane Gradients

In addition to through-board gradients, for the case of powered components there will be in-plane gradients (Fig. 10-2). If there is a hot spot under each carrier, there will be buckling of the substrate in those regions (Fig. 10-1*d*), as shown by Howard, Sobeck, and Sanetra.[8] The substrate might conceivably buckle in either direction if this were the only effect, but as we already have seen, the through-board gradient will cause convex bowing, so the buckling caused by the in-plane gradient will add to it. We don't have a simple formula for this buckling, but clearly it will be proportional to the expansivity of the substrate, so it causes us to lower the value of our ideal expansivity even more. We have, then, for our particular case,

$$\alpha_S(\text{ideal}) < 0.88\alpha_C \tag{10-8}$$

Depending on the circumstances, this may or may not be a large effect, but the direction of the effect is clear.

10.1.5 Temperature Shock

Another environment of interest is that of temperature shock. If the temperature is changed so rapidly that the inner parts of the system lag well behind the outer

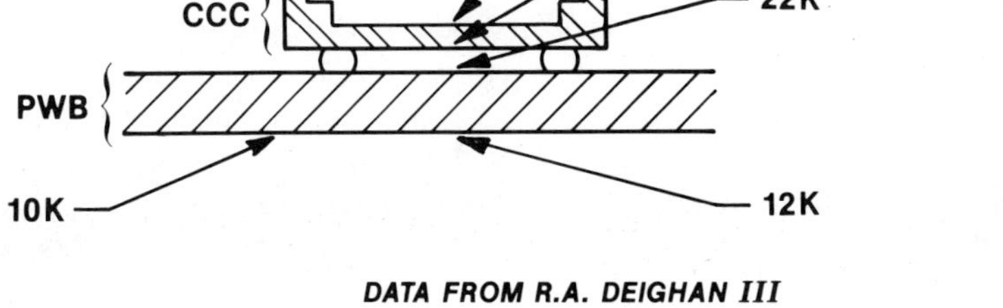

Figure 10-2 Temperature distribution with 0.75 W-dissipated.

parts, we will find opposed bowing, as suggested by Engelmaier[9] and by Figure 10-1*e*. Here both the chip carrier and the substrate have the temperatures of their "exposed" surfaces changed, but their "interior" surfaces remain at the original temperature. They will both bow but in opposite directions. The figure is for a positive temperature change and positive expansivities. As usual, it is highly simplified. We can obtain an idea of the strains involved by assuming a linear gradient through each of the materials and setting the sum of the solder joint deflections equal to zero. Here we are assuming that the solder relaxes faster than the thermal time constants of the system, which is normally not the case. Instead of the strains calculated here, there would be stresses built up. The general conclusion, however, remains the same:

$$\varepsilon = \frac{L^2}{2H}\left(\frac{T_C - T_0}{h_C}\alpha_C + \frac{T_S - T_0}{h_S}\alpha_S\right)\frac{2}{3} \tag{10-9}$$

where h_C is the thickness of the component (e.g., a chip carrier). To minimize the tension due to this effect we would have to set the substrate expansivity to less than zero. Specifically,

$$\alpha_S(\text{ideal}) = -\alpha_C\left[\frac{(T_C - T_0)h_S}{(T_S - T_0)h_C}\right] \tag{10-10}$$

where h_C and h_S are the thicknesses of the component and the substrate, respectively. For the particular case we are interested in, the factor in brackets was 1.5. Thus, if we are concerned about stresses induced by thermal shock, we should lower our overall value of ideal substrate expansivity even farther than indicated by Equation 10-8. The amount by which it should be lowered depends on the perceived hazard of temperature shock relative to that of power cycling for field conditions. As Engelmaier[9] has pointed out, temperature shock is more commonly encountered in testing than in the field.

10.1.6 Solder-Substrate Expansivity Mismatch

Another effect that should be considered here is the expansivity mismatch between the solder and the substrate (Fig. 10-1*f*). This mismatch has received little attention until recently,[10] no doubt because the relevant dimension of a solder joint is normally much less than that of the component (L). In terms of free bodies, if the solder were not connected to the pads, the in-plane displacement due to the expansivity mismatch between the component and the substrate would be given by $L(\alpha_S - \alpha_C)$. On the other hand, if the substrate and component expansivities are equal, and again the solder is not connected, the in-plane displacement due to the solder-substrate mismatch is given by $D(\alpha_S - \alpha_{SOLDER})$, where D is the half-diagonal of the solder pad. The expansivity of solders is in the range of 25 ppm/K,[11] so there can be some large mismatches at the joints even if the component and the substrate are perfectly matched (or even if there is no component, only a "blob" of solder). To minimize this effect, of course, the substrate expansivity should match that of the solder. This would normally be impractical, however. Even if it were possible, there would still be the mismatch of the solder at the component side of the joint. So optimizing the substrate expansivity would only shift the failure to the other side of the joint. Still, when solder joints are made to a low expansivity material like ceramic or silicon, large local mismatches can and do occur, and they can and do cause cracking, especially for large area joints, both during thermal stresses of soldering and temperature cycling after the joints are made.

10.1.7 Overall Judgment

Thus we find there is no one answer to the question posed by the substrate manufacturers. We are reduced to making an engineering compromise, based on our best judgment. My own judgment, based on the above concepts, is that the ideal substrate expansivity should be less than that of the component. If, for example, the component has an expansivity of 6 ppm/K, an ideal substrate expansivity might be in the range 2 to 4 ppm/K. It seems well to remember that for the world to be ideal, it is insufficient for all expansivities to be equal. The ideal world would have them all zero. And most of the time, we improve matters by moving toward this ideal, that is by lowering expansivities.

10.2 CREEP AND STRESS RELAXATION

Solder at room temperature and above is a ductile material. A wire made of 60% Sn-40% Pb can be stretched at room temperature by more than 60 percent before breaking (reference 12, pp. 9–71). Thus when stresses are applied to solder joints, they yield, and they do so in such a way as to reduce the stresses. When a sample's strain is held constant and the stress varies (decreases), we call the process "stress relaxation." When the stress is held constant and the strain changes in time, we call it "creep." The distinction between the two has been useful in interpreting stress-strain data where one of the variables is controllable. In both cases there is plastic flow, and microstructurally, the same things happen. Curves relating plastic strain rate to stress normally give identical results for creep and stress relaxation experiments.[13] In the field, solder joints experience time-dependent changes in the strains and the stresses simultaneously.

We have called the combined process "stress reduction,"[14] but a better term might be "creep-relaxation."*

When thermal stresses are set up in a soldered assembly, the members applying the stresses to the solder (i.e., the components and the substrate) are normally stressed within their elastic range. Thus, we can speak of a "stiffness" (k) of the assembly, which is the derivative of force with respect to displacement, as in a spring constant.

$$k = -\frac{dF}{dx} \tag{10-11}$$

So if the solder yields by a distance Δx, the force on the joint is reduced by an amount $k\Delta x$, where k is the stiffness of the assembly. There are six components of stress and six of strain, so in general there are thirty-six stiffness constants involved. In the following sections, we make the usual three assumptions (references 15 and 16, p. 282): (1) The main effects are shear and bending in the solder joint, (2) the configuration has axial symmetry, and (3) only radial forces and tangential moments are applied to the solder (Fig. 10-3).

Let k be the stiffness of the assembly for these shear stresses. For constant temperature we have $\Delta\gamma = \Delta x/H$, or

$$\Delta\gamma = \frac{-\Delta F}{kH} \qquad \text{for } \Delta T = 0 \tag{10-12}$$

where F is the force per joint. We also know from Equation 10-1 that if $F = 0$ (complete relaxation),

$$\Delta\gamma = \frac{(\alpha_S - \alpha_C)L}{H}\Delta T \qquad \text{for } F = 0,\ \Delta F = 0 \tag{10-13}$$

(provided we have no temperature gradients). We can combine these two equations to obtain

$$\Delta\gamma = \frac{(\alpha_S - \alpha_C)L}{H}\Delta T - \frac{\Delta F}{kH} \tag{10-14}$$

This equation is important despite its simplicity since it relates the shear strain, the shear stress, and the temperature in a linear equation. It simplifies the analysis considerably, provided the stiffness k is not a function of temperature, which

*This author at one time proposed coining the term "creelaxation." Fortunately AT&T management talked me out of it.

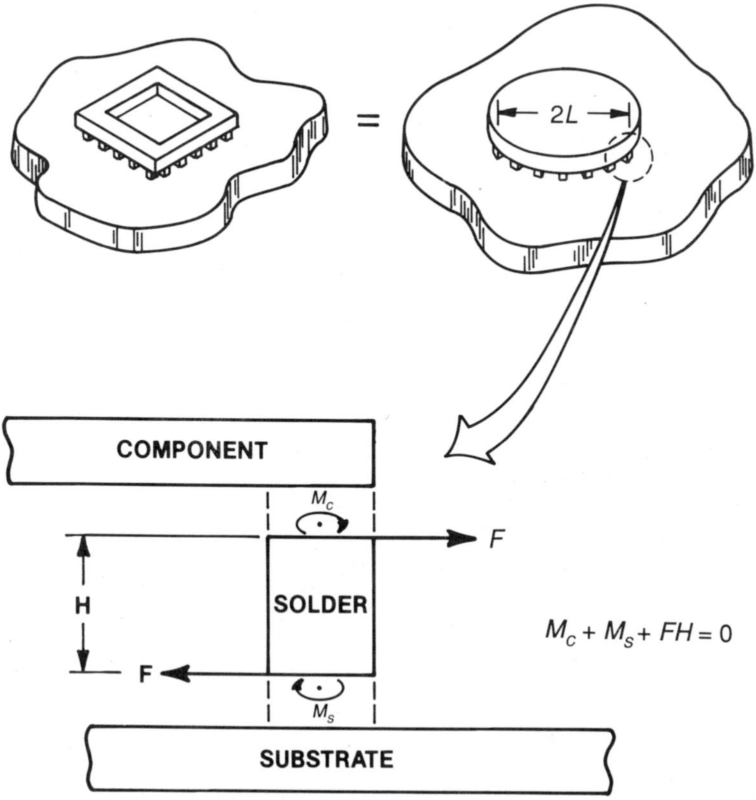

Figure 10-3 Idealized surface mount assembly.

has been shown to be a reasonable approximation for our case, at least.[14] For instance, the equation tells us that if the temperature is held constant, the force will be linearly related to the shear strain, with slope $-kH$. This is the condition of creep-relaxation mentioned above. If k is large (stiff), we have conditions approximating stress relaxation. If k is small, we are closer to creep. In any real case, we have simultaneous time dependent changes in stress and strain.

Another way to look at this equation is to realize that when the temperature is changed, the expansion mismatch (the first term on the right) causes a change in the force (the second term on the right) or a change in the shear strain (the term on the left) or some combination, depending on the properties of the solder. If the solder is highly compliant, the mismatch causes strain, as at high temperatures. If the solder is stiff, the mismatch causes mostly stresses, as at low temperatures. Equation 10-14 shows how the stresses and the strains together make up the thermal mismatch.

This equation is most useful in combination with strain measurements. Using strain gauges, we can measure the shear strain $\Delta\gamma$.[15]

$$\Delta\gamma = L\left[\frac{\Delta(\varepsilon_{ST} - \varepsilon_{CB})}{H} + \frac{\Delta(\varepsilon_{CT} - \varepsilon_{CB})}{2h_C} + \frac{\Delta(\varepsilon_{ST} - \varepsilon_{SB})}{2h_S}\right] \tag{10-15}$$

where ε_{ST} and ε_{SB} are the strains on the top and bottom of the substrate, and ε_{CT} and ε_{CB} are the strains on the top and bottom of the component. And if we know the elastic constants of the substrate and component, we can obtain a measure of the shear force by measuring changes in their bending. Their radii of bending are given by

$$\Delta\left(\frac{1}{R}\right)_C = \frac{\Delta(\varepsilon_{CB} - \varepsilon_{CT})}{h_C} \tag{10-16}$$

and

$$\Delta\left(\frac{1}{R}\right)_S = \frac{\Delta(\varepsilon_{SB} - \varepsilon_{ST})}{h_S} \tag{10-17}$$

The change in the shear force per joint ΔF is then given by Hall,[15]

$$\Delta F = \frac{P/6}{2H + h_C + h_S}\left[\frac{E_C h_C^2 \Delta(\varepsilon_{CB} - \varepsilon_{CT})}{1 - \nu_C} + \frac{2E_S h_S^2 \Delta(\varepsilon_{SB} - \varepsilon_{ST})}{1 - \nu_S^2}\right] \tag{10-18}$$

where E_C and E_S are the elastic moduli of the component and substrate, and ν_C and ν_S are their Poisson's ratios. The quantity P is the "pitch," or the center-to-center spacing of the joints around the perimeter of the component. The reason that the component and the substrate contributions appear differently in the above equation is that the substrate extends over a larger region than just under the component, so additional force is required to deform the portion of the substrate that is not under the component.

This equation explicitly provides for the component and substrate to have different radii of curvature, so it does not depend on the "bimetallic strip" model. From strain gauge measurements and using Equations 10-15 and 10-18, we can follow the shear force F and shear strain γ for any given temperature cycle as a function of time. Of course, we have to know the temperature of each gauge to correct for the apparent strain, so these equations are most useful for the special case of negligible temperature gradients. One simple thing that can be done is to measure the effective stiffness k by observing the change of shear force as the temperature is changed under constant shear strain. This can be readily done at lower temperatures, where the shear strain rate is negligible, and Equation 10-14 reduces to

$$\frac{dF}{dT} = kL(\alpha_S - \alpha_C) \qquad \text{at constant } \gamma \qquad (10\text{-}19)$$

From this equation it is easy to determine the stiffness k. For our case,[15] over the range -25 to $+40°C$, the parameter k was remarkably constant at 1.7 lb/mil, or 300 N/mm.

Figure 10-4 is a schematic plot of shear force F as a function of shear strain γ for a given constant temperature. If we start at a given value of force F_0, as the solder yields the shear strain will increase, and the force will decrease along the straight line shown. In principle, this will continue spontaneously until the force reaches the elastic limit at the temperature in question. At high enough temperatures, the elastic limit is small, and the force can be considered to go to zero at some final shear value γ_f. Similarly, if the force starts at a negative value, F_1 at the same temperature, the force will move toward zero, and the strain will decrease, to arrive ultimately at γ_f. So far, this analysis is independent of the properties of the solder, except for the assumption that it creeps-relaxes

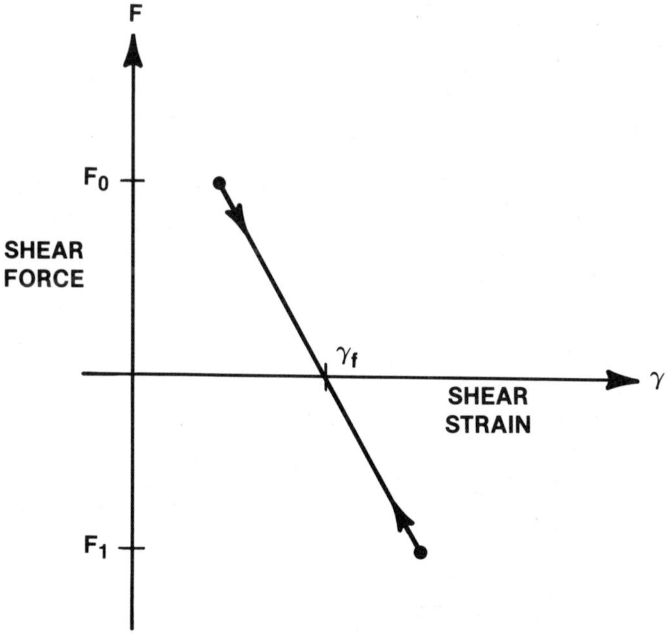

Figure 10-4 Creep-relaxation at constant temperature.

to its end point. The slope and location of the line depend on the component, the substrate, and the configuration. The solder properties only control how fast the system moves along the straight line shown. We deal with this aspect next.

10.3 SOLDER PROPERTIES

The mechanical properties of ductile materials are often described in terms of elastic and plastic (i.e., nonelastic) strain. Any change in strain is assumed to be made up of these two components. Thus,

$$\Delta\gamma = \Delta\gamma_e + \Delta\gamma_p \tag{10-20}$$

Note that when we hold γ constant (stress relaxation), the two parts are equal in magnitude but opposite in sign. The elastic part is assumed to be reversible and linear in the force. For shear, this is expressed as

$$\Delta\gamma_e = \frac{\Delta F}{GA} \tag{10-21}$$

where G is the shear modulus of the material and A is the area of the joint, usually taken as the area of the pad. Here we have assumed that the shear stress is uniform within the joint, which is certainly not the case.[17,18] We can, however, speak of some effective or average shear modulus, which makes this equation useful. Here we note that if we hold the force (or the stress) constant (creep conditions), the elastic strain is constant, and the change in plastic strain equals the change in total strain. In creep-relaxation conditions, we see from Equations 10-12 and 10-21 that both the total strain and the elastic strain are linearly related to the force, so the plastic strain must be also. Combining Equations 10-20 and 10-21 we can eliminate $\Delta\gamma_e$ and find that

$$\Delta F = GA(\Delta\gamma - \Delta\gamma_p) \tag{10-22}$$

We then substitute this into Equation 10-14 and solve for $\Delta\gamma_p$

$$\Delta\gamma_p = \left(1 + \frac{kH}{GA}\right)\Delta\gamma + \frac{kL}{GA}(\alpha_S - \alpha_C)\Delta T \tag{10-23}$$

With the strain gauges, we can measure $\Delta\gamma$ and ΔF, so if we also measure ΔT (with thermocouples, for example), we are in a position to measure the plastic shear strain $\Delta\gamma_p$ and the shear force ΔF as a function of time and temperature.

The first term in parentheses in Equation 10-23 gives us an idea of how large the stiffness must be to approach stress relaxation and how small to approach creep conditions. When $kH \ll GA$, for constant temperature, the plastic shear strain change is approximately equal to the total shear strain change, which is the condition we know as creep. When $kH \gg GA$, (again for constant temperature) the plastic shear strain change is much greater than the total shear strain change. This seems anomalous until it is remembered that in creep-relaxation, the change in elastic strain always has the opposite sign of the change in plastic strain. Thus, in this condition, the plastic and elastic shear strain changes are approximately equal and in opposite directions, which is the condition of stress relaxation. The value of G for eutectic Sn-Pb solder is a function of temperature,[12] ranging from 0.78 Mlb/in.^{2} at -28°C to 0.43 Mlb/in.^{2} at 97°C. (In metric units, this is 5.4 and 3.0 GPa). In using these values, we are near the middle of a wide range of reported results for the elastic modulus of eutectic Sn-Pb solder, ranging from 0.8 Mlb/in.^{2} (5.7 GPa)[19] to 4.5 Mlb/in.^{2} (31 GPa) (reference 11, p. 345) both at room temperature. For room temperature, we use 2 Mlb/in.^{2} (14 GPa) for the elastic modulus. The shear modulus is then calculated from the standard relation $0.5E/(1 + \nu)$ where E is the elastic modulus and ν is Poisson's ratio. Thus, for our conditions, kH/GA varied from 0.11 to 0.18 over the temperature range of the experiments. This is not small enough to be negligible, so we applied the term as given in Equation 10-23 when processing data. If the "true" value of the modulus is higher than we have assumed, the correction is even smaller than 0.18. If it is actually a factor of 2 lower than we have assumed, our values of $d\gamma_p/dt$ should be increased by 10 to 15 percent.

10.4 CONSTITUTIVE RELATIONS

The rate of plastic shear strain change is a function of the shear force and the temperature. This relation is often expressed for ductile materials in terms of a power-law constitutive relation.[20]

$$\frac{d\gamma_p}{dt} = b\left(\frac{F}{A}\right)^N \tag{10-24}$$

where t is the time, and b and N are functions of temperature but (so it is assumed) independent of time. Now as we have seen, we can measure $\Delta\gamma_p$ and ΔF, so within a constant, we can measure F and thus evaluate the above equation. This is especially easy when we hold temperature constant. We have done this for several different temperatures, by changing the temperature of the sample to a given temperature. We then hold it there while we observe the force as it decays and the strain as it changes.[14] Figure 10-5 is a log-log plot of the force as a function of time for several temperatures. These curves are similar to the stress relaxation curves of Baker.[21] We cannot readily control the force at which the creep-relaxation starts. We simply change the temperature, accept whatever

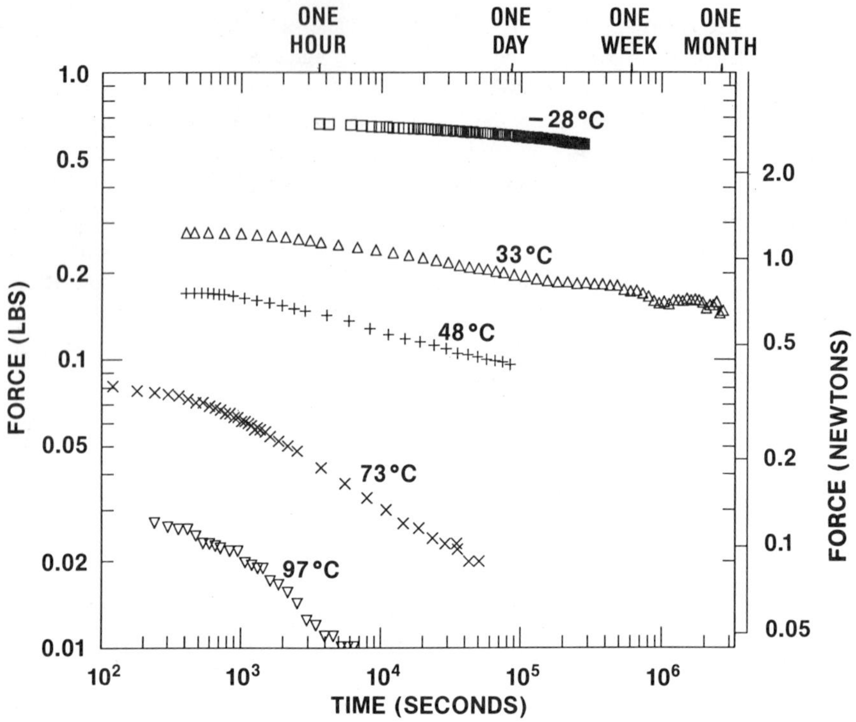

Figure 10-5 Creep-relaxation in solder joints on a chip carrier.

is there, and watch it creep-relax at constant temperature. It is clear, however, that higher temperatures generally result in smaller forces. That is expected because creep-relaxation takes place during the time it takes to increase the temperature. Previously, we derived an expression for the force as a function of time,[14] assuming a power-law constitutive relation. It predicted that on a log-log plot, the force per joint would asymptotically approach a straight line at long times and that its slope would be $-1/(N - 1)$. The curves in Figure 10-5 are consistent with that relation. We did not see any leveling off. As near as we can tell, for as far as the experiments were carried out, the creep-relaxation process was still continuing and with the same relation. Of course, in practice, the decay rate of the force eventually becomes too small to be of practical import, but that does not mean that the force has gone to zero, even after half an hour at 97°C or 24 h at 73°C. The curve at 33°C was continued to a full month, and there was still a large force left (i.e., the solder was constraining the carrier and the board in a bowed condition). The waviness at the far end of that curve is unfortunate, because it might have obscured some deviation from the straight line. The waviness had a period of 7 days, and we suspect it is associated with

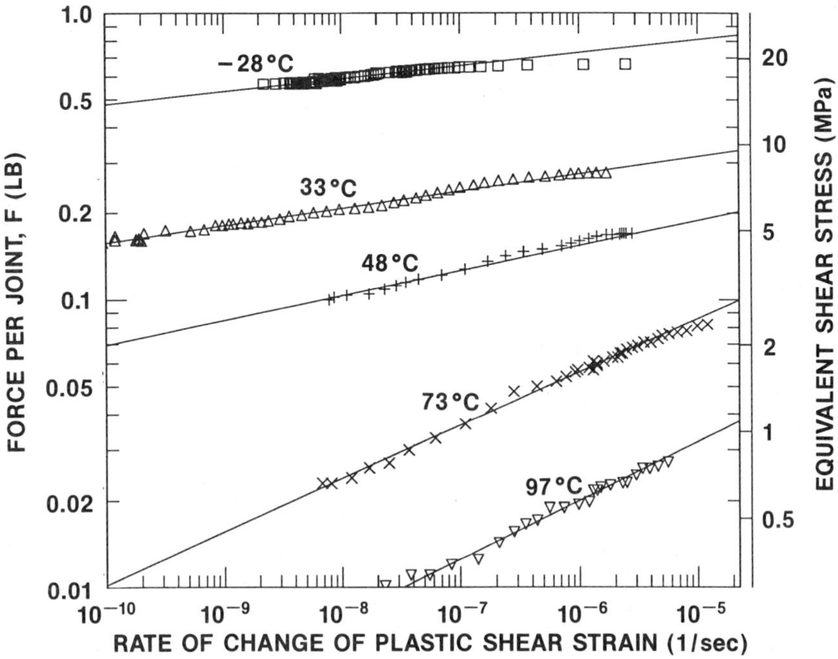

Figure 10-6 Constitutive relations for solder joints.

the weekend room temperature change (the experiment was performed in December).

Figure 10-6 is a plot from the same experiment of the force as a function of the rate of change of plastic shear strain $d\gamma_p/dt$. (These curves start at the right and proceed to the left.) In the author's opinion it would be better to plot $d\gamma_p/dt$ on the y axis, as it is normally the dependent variable. We bow to convention, however, in Figure 10-6. According to Equation 10-24, for any given temperature, the curve should be a straight line on this plot, and it should have slope $1/N$ and x-intercept $b/(A^N)$. Table 10-1 gives the results for b and N. As nearly as we can tell, the power-law constitutive relation holds for all these temperatures for the time periods of the experiments.

The form of the constitutive relation is as expected, but the constants are not. In this range of temperature and strain rate, solder is expected[22,23] to show "superplasticity," with values of N from 1.5 to 2.5 instead of the 5 to 22 shown in Table 10-1. Our values are indicative of normal plasticity, which is expected at larger values of shear strain rate. Solomon[24] found values of N from 2.6 to over 40 for strain rates from 10^{-5} to 0.1 sec^{-1} at temperatures from −50 to 150°C. Shine and Fox[25] found a transition from normal to superplastic behavior for strain rates below 10^{-5} sec^{-1}. Their results are especially interesting since

Table 10-1 Constants for Constitutive Relation[14]

Temperature (°C)	N	b sec^{-1} (MPa)$^{-N}$	Duration of Experiment (h)
−27.7	21.9	1.2×10^{-35}	80
33.2	16.5	1.6×10^{-21}	800
48.3	11.5	3.7×10^{-14}	24
72.7	5.4	7.7×10^{-8}	16
96.8	4.9	1.5×10^{-5}	1

they were performed on real solder joints on chip carriers, not on bulk samples or specially shaped test specimens. It is interesting to note that their values for stress and strain rate are not far different from ours. In fact, they overlap. Nevertheless, their experiment was quite different. They applied a mechanical stress rather than a thermal stress, and they applied it cyclically, so they used different data processing to extract the rate of change of plastic shear strain. We still do not understand why we did not find a transition to superplasticity, whereas they did. This would be a good area for future work.

10.5 TEMPERATURE CYCLING

We have been treating the special case of isothermal creep-relaxation. Similar considerations apply to temperature cycling. Their application to temperature cycling is of great practical importance because the number of cycles to failure is often related to the strain range through the "Coffin-Manson" relation,[26] strain range partitioning,[27] damage integral calculations,[28,29] or other theories. In the Coffin-Manson development, the number of mechanical strain cycles to mechanical failure is postulated to be inversely proportional to a power of the plastic strain range during the cycle. Wild[30] plotted the number of cycles to failure as a function of total strain and found the power to be of the order of magnitude of 2 but to depend on temperature and frequency. He reported nine data points for eutectic Sn-Pb solder, involving two temperatures (25 and 100°C), two frequencies (96 and 7200 cycles/day), and three shear strains (3, 8, and 20 percent). His results have been conveniently parameterized by Engelmaier[3,31] for extrapolation to other temperatures, frequencies, and strain ranges. Wild applied his strains mechanically, rather than by temperature cycling, so he knew the total strain range. In temperature cycling, we generally know the temperature range ΔT but not the strain range $\Delta\gamma$, let alone $\Delta\gamma_p$. At high temperatures, where the creep-relaxation process goes rapidly to completion, we can confidently use Equation 10-1 to obtain the plastic strain range for the Coffin-Manson relation. At lower temperatures, this is not valid. To apply the Wild or similar data to temperature cycling, we need to know (or assume) how $\Delta\gamma$ varies with temperature and frequency. Wild[30] attributes his temperature dependence to the changing ratio of $\Delta\gamma_p$ to $\Delta\gamma$. In temperature cycling, we must also consider the temperature dependence of the total strain range, a factor that was not included in Wild's experiment. In spite of this, the Engelmaier equation has been successfully used in correlating temperature cycling results.[3,4,9]

10.5.1 Small Temperature Range Cycling

When we consider temperature cycling over small temperature ranges, the elastic constants can be considered constant. Figure 10-7 shows hysteresis loops of force versus strain, measured with strain gauges, as before. The oven temperature was cycled from 42 to 52°C, with ramp time equal to hold time. A schematic loop is shown in the upper right corner. Based on our earlier concepts, if we cycle between two temperatures, we should trace out a hysteresis loop between the two creep-relaxation lines for the two temperatures. As we start the ramp from one temperature to the other, we initially change the force and shear strain with a slope *GA*. We see this from Equations 10-20 and 10-21. At the start of the temperature ramp the change in the plastic shear is initially negligible, so the

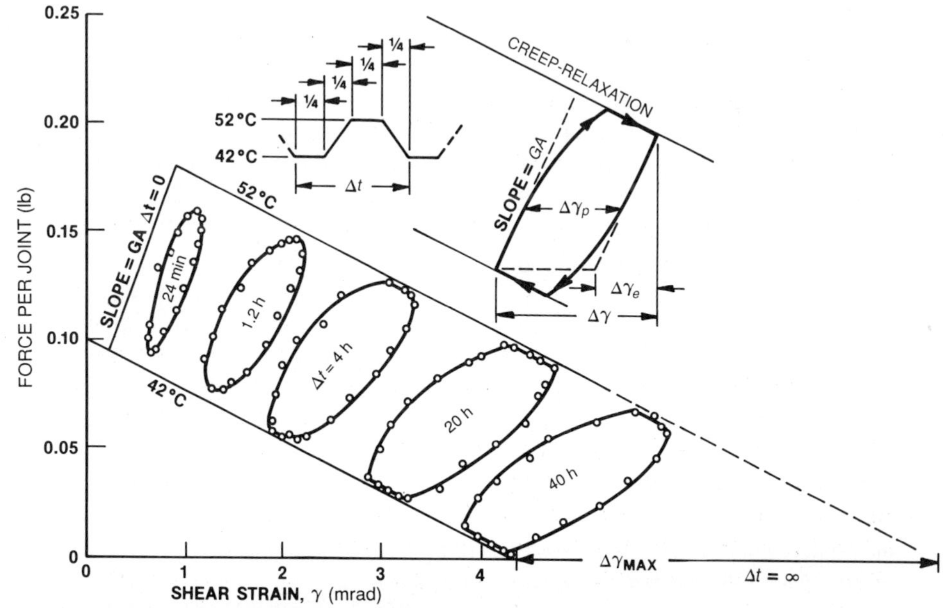

Figure 10-7 Hysteresis loops between 42 and 52°C. The loops have been arbitrarily displaced along the creep-relaxation slope to eliminate overlap.

total shear strain change is simply $\Delta F/(GA)$. Handbook data[17] for the shear modulus G at 47°C was used to draw the expected initial slope (Figure 10-7). As mentioned, above, this number is subject to considerable uncertainty. The literature contains values twice as large, as well as values half as large. If the cycle were traversed infinitely fast, so that no creep-relaxation could take place, the system would simply go up and down that line, the hysteresis loop would have no width, and there would be no plastic strain. The total strain would then be elastic and would be given by solving Equation 10-23 for $\Delta\gamma$ with $\Delta\gamma_p = 0$. The result is 0.71 mrad. This is the minimum strain range expected at this temperature for a ΔT of 10 K.

If we proceed up the ramp slowly, however, the creep-relaxation becomes appreciable, and the curve deviates from the initial slope. After arriving at the new temperature, the sample creeps or relaxes down the line until the temperature is changed again. If we make the assumption that the initial slope extends at least halfway up the ramp, we can measure the loop at its widest part and call that the plastic strain range for the loop. The figure illustrates how the total range $\Delta\gamma$ is then divided up between plastic and elastic portions. Since we are dealing here with ranges, the plastic strain range is always less than the total strain range. This is a classical method for determining plastic strain range for any hysteresis loop (e.g., Solomon[32]).

Before discussing the data, we should review some experimental considerations. These hysteresis loops are for five different frequencies: 0.6, 1.2, 6, 20, and 60 cycles/day. For clarity, we have displaced them along the creep-relaxation lines to remove overlaps. The periods are shown on the figure. The experiments involved typical frequencies for electronic equipment in the field, substantially lower than those used by Wild[30] (*viz.* 96 and 7200 cycles/day). Note that the faster cycles did not reach the creep-relaxation lines for 42 and 52°C. That is because of the temperature lag in the sample with respect to the oven air. The sample temperatures were measured with thermocouples, which indicated that in the worst case (6-min ramp and dwell times), the temperature range was only 6.9K instead of 10K. In each case, the cycle was repeated six times, and only the last cycle is represented here, as there was some "global" creep-relaxation for the first few cycles, which was convoluted with the effects of interest. One other confusing factor here is the unknown effect of temperature gradients, which play a more significant role for the faster cycles. Despite this, we can learn a lot from these results.

For example, the fastest cycle came close to approximating the elastic slope, and it had a small plastic portion (i.e., a small width). The temperature gradients mentioned above may have caused the initial slope of this loop to be (slightly) greater than the expected elastic slope. For slower cycles, the plastic part was much greater. But even at a 40-h period (at this temperature) the strain range was well below (less than half) that of a material with complete creep-relaxation (i.e., $\Delta\gamma_{max}$), which we postulate would be the case for an infinite period.

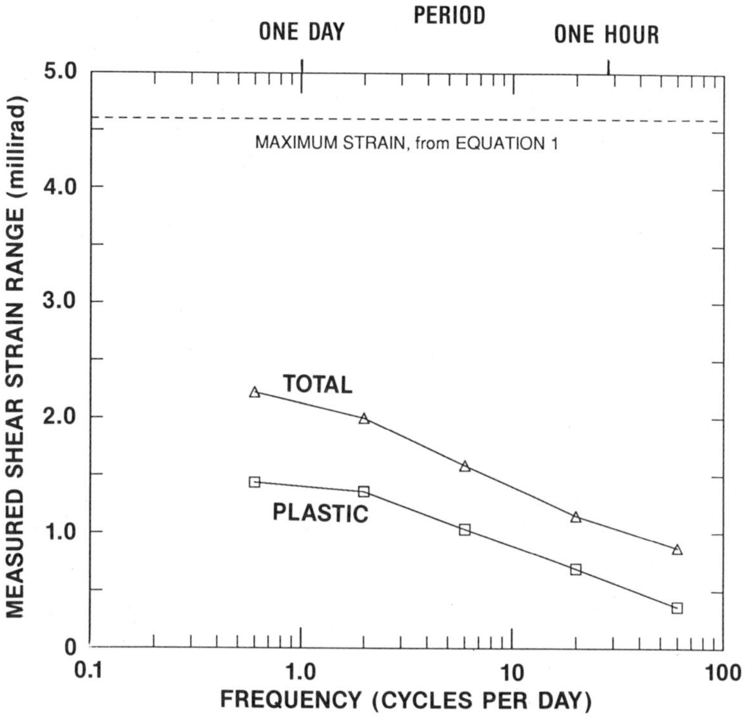

Figure 10-8 Strain ranges during cycling from 42 to 52°C.

For each of these loops, we can measure the total strain range $\Delta\gamma$ and (from the widest part of the loop) the plastic strain range $\Delta\gamma_p$. We normalized the results from the fast cycles to a ΔT of 10 K by multiplying by the correction factor of $10/\Delta T$. The results are plotted on Figure 10-8. Both the total and plastic strain range decrease as the frequency is increased. But all are well below the value given by Equation 10-1 for a material in which the creep-relaxation process has gone to completion. Note that at the lowest frequency the strain is mostly plastic, while at the highest frequency it is about half elastic. As discussed above, the total strain range at this temperature is not expected to go below 0.71 mrad. At 60 cycles/day it is already 0.8 mrad, so it must be leveling off. We cannot extend it to higher frequencies because of the temperature gradients.

Figure 10-9 shows results[14] from similar small temperature cycles where we have maintained the frequency at 6 cycles/day and varied the temperature midpoint from 0 to 100°C. Here we see how sharply the temperature affects the strain range. We found that at 80°C, the (total) strain range was 70 percent of that obtained from Equation 10-1. At 20°C, it was only 12 percent. The plastic strain range was 60 and 4 percent of that obtained from Equation 10-1 for the

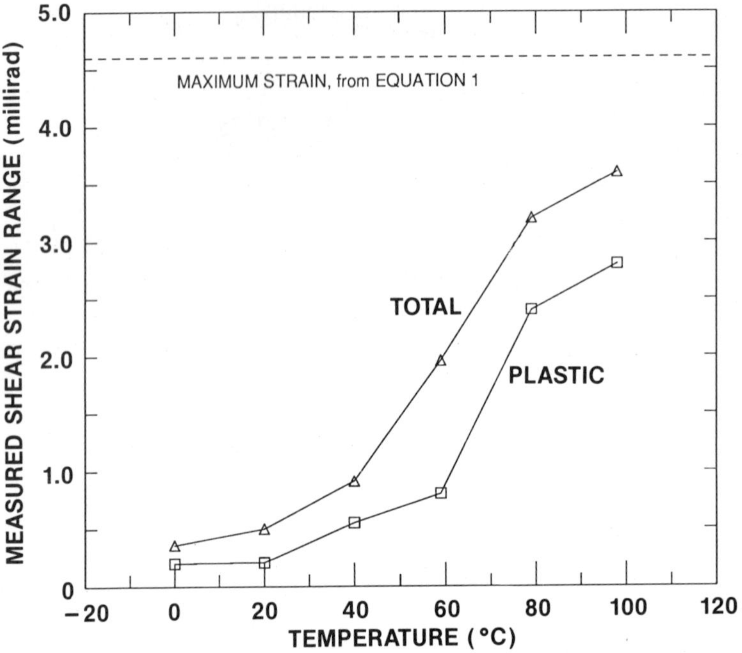

Figure 10-9 Strain ranges during cycling with $\Delta T = 10$K and frequency = 6 cycles/day.

same two temperatures. Judging from this figure, for temperature ranges above 100°C it is reasonable to assume that the strain range is proportional to ΔT, as given by Equation 10-1. If the temperature range under consideration includes temperatures below 80°C, the plastic strain range is less than half that calculated from Equation 10-1, so it would be well to realize that $\Delta\gamma_p$ is not proportional to ΔT. Since the lifetime is roughly inverse to the square of the strain range,[30] a factor of 2 in the strain range means a factor of 4 in the calculation of the number of cycles to failure.

10.6 LARGER TEMPERATURE CYCLES

The situation is considerably more complicated for larger temperature cycles, especially when they involve both temperature ranges: those where the mismatch produces mostly forces and those where it produces mostly strains. We next consider briefly such a case from our previous work.[14,15]

A leadless ceramic chip carrier without castellations was surface mount soldered to an FR-4 PWB, with strain gauges attached to the top and bottom of the middle of the carrier and also to the top and bottom of the middle of the

PWB. Experimental details are given in Hall.[15] The assembly was slowly cycled in a chamber according to the schedule shown in Figure 10-10*b*. Using Equations 10-15 and 10-18, we obtained the shear strain and shear force during the cycle. The measurements showed good repeatability after the first half cycle.[15] Therefore, we only show the second cycle here. This cycle is considered moderately severe. The assembly would begin to show electrical failure after a few hundred of these cycles.[2]

Figure 10-10*a* shows the shear strain as a function of temperature. The almost-straight line with the triangle data points is the strain expected from Equation 10-1, given the expansivities as a function of temperature as measured independently from similar strain gauges. This would be the case of complete creep-

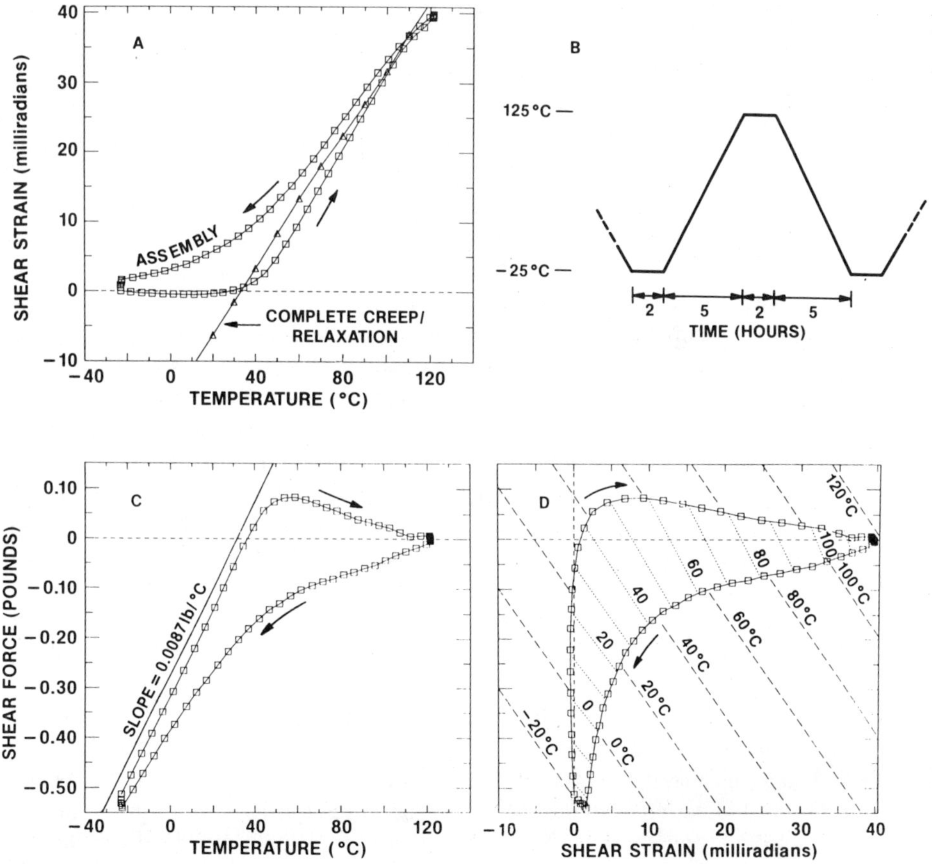

Figure 10-10 Temperature cycling with large ΔT.

relaxation as the temperature is changed. In this plot, the zero for the strain is arbitrary. For the assembly curve, we used the lowest temperature value as the zero. For the complete creep-relaxation curve, we adjusted its zero such that it would match the assembly curve at high temperatures. At high temperatures, the shear strain for the assembly follows the complete creep-relaxation curve well. That is because the creep-relaxation is fast enough to be almost complete as the temperature is changed. The deviation from the complete creep-relaxation line as the temperature is reduced is opposite to that obtained as the temperature is raised. That is because the shear force on the joints is in the opposite direction in the two cases. As the temperature comes down, the deviation from the complete creep-relaxation curve increases as the solder becomes stronger, and at −25°C the slope is close to zero, indicating high rigidity of the solder and little creep-relaxation. During the 2-h dwell at −25°C, there is little change in the shear since the creep-relaxation is extremely slow at these temperatures (despite the large forces). As the temperature is increased again, the shear strain stays almost constant until it reaches the complete creep-relaxation line, at which point the force reverses, creep-relaxation takes over, and the curve asymptotically approaches the complete creep-relaxation curve (as given by Equation 10-1).

Figure 10-10*c* shows the force measured during the same cycle, using Equation 10-18. We also have to choose a zero value for the force since a strain gauge only indicates changes from a reference state. In this case, we have a better indication of where the zero is since we know that it always relaxes toward zero and that it is small at the highest temperatures. The zero chosen for this plot is the force at the end of the 2-h dwell at 125°C.

As the temperature is decreased from 125°C, the force gradually increases in the negative direction, and its slope gradually increases also, as the solder gets "stronger." The curve traced out during decreasing temperature is similar in shape to that of the ultimate shear strength.[33] As the temperature goes below room temperature, the forces get very large indeed. This puts the largest stress on the metallization and components that are soldered together. At −25°C, the 2-h dwell causes a barely detectable decrease in the force. As the temperature is increased again, the force increases quite linearly, all the way up to 40°C. During this time, the large bend in the chip carrier and the PWB are steadily decreasing, until they cross the point of zero force (i.e., no bend). This particular part of the cycle is much like a classical bimetallic strip, where there is no shear between the two members, but the force and the bend are linearly related to the temperature. The slope of the line in this region is 0.0087 lb/K, and this is consistent with an estimate of the shear forces expected for a bimetallic strip.[14]

Thus, we can roughly divide the temperature range into two regions: (1) Above 50°C, the solder creeps-relaxes enough that Equation 10-1 is adequate and (2) below 50°C, the solder is stiff enough that the bimetallic strip analogy holds. That is, the force changes but the shear strain does not. The lesson is that temperature cycling in the high range is completely different from cycling in the low range. Thus we concur with the conclusion of Engelmaier[9] that if an assembly is not to be used at low temperatures, it may be misleading to evaluate its reliability from temperature cycles that include the low temperature region.

Figure 10-10*d* displays the same data, plotting force as a function of strain. Again the two regions are clearly shown. At high temperatures, the force is small and the strain changes are large. At low temperatures the force is large and the strain changes are small. In this plot we have added the creep-relaxation isothermal lines (large dashed lines outside the loop) as predicted by Equation 10-14. They all have the same slope, and their separation is uniform. The slope is $-kH$, where H is 0.011 in., and k was taken as 2.0 lb/mil, as calculated from a bimetallic strip model.[14] We could also have obtained the slope from Figure 10-10*c* since we know that if $\Delta\gamma$ is negligible,

$$k = \frac{1}{L\Delta\alpha}\frac{dF}{dT} \tag{10-24}$$

Using this relation, the value of k is 1.7 lb/mil, not far different from the bimetallic strip value. We have used the value 2.0 in making the creep-relaxation lines of Figure 10-10*d* since that was used before, and it fits the data a little better. The separation between the creep-relaxation lines is given directly from Equation 10-1 since it must be valid for the x-intercept (zero force) as assumed by Equation 10-1.

The dotted lines inside the loop were obtained as follows. As we cycle through the loop, we cross any given temperature value twice; once while increasing and once while decreasing the temperature. We have interpolated the raw data to obtain the force and strain at which we crossed even values of the temperature. Then we drew a dotted line connecting each pair of points thus determined. This created an experimental grid of creep-relaxation lines that can be compared with the simple linear theory indicated by the dashed lines outside the loops. The agreement is satisfactory. The slope at 40°C, for example, is just as predicted, although the location is slightly off. At −20°C, the location is slightly off, but the slope is still in good agreement. At 100°C, the slope is slightly off, but the location is as predicted. This is considered good evidence that the simple linear relation among force, strain, and temperature is valid for this type of cycling.

One can, in principle, use the constitutive relations along with the temperature profile (Fig. 10-10*b*) to numerically generate this entire loop. This has been done.[28,29,34,35] Such approaches can be used to extrapolate these results to temperature cycles of different frequencies, different ΔT's, and different geometries. That is beyond the scope of this chapter. Suffice it to say that we have obtained some insights into the behavior of these solder joints as we change the temperature. The most important result is that low temperatures are fundamentally different from high temperatures, and we should bear that in mind as we consider the reliability of these joints.

10.7 ACKNOWLEDGMENTS

I would like to acknowledge the contribution of R. A. Deighan, who wrote the programs for processing the strain gage data, and F. L. Howland, who provided stimulation, experience, and advice.

REFERENCES

1. Moore, R., *Guidelines for Surface Mounting and Interconnecting Chip Carriers,* IPC-CM-78, Evanston, IL, Inst. for Interconnecting and Packaging Electronic Circuits, 1983.
2. Sherry, W. M., and P. M. Hall, "Materials, Structures, and Mechanics of Solder-Joints for Surface-Mount Microelectronics Technology," in *Proc. Conf. Interconnection Technology in Electronics,* Fellbach, W. Germany, Deutscher Verband fur Schweisstechnik, 1986, pp. 47–61.
3. Engelmaier, W., "Functional Cycling and Surface Mounting Attachment Reliability," *ISHM Technical Monograph Series,* Int. Soc. for Hybrid Microelectronics, Silver Spring, MD, Series 6984–002, 1984, pp. 87–114.
4. ———, "Surface Mount Attachment Reliability of Clip-Leaded Ceramic Chip Carriers on FR-4 Circuit Boards," *Proc. International Electronics Packaging Conference,* 1987, pp. 104–113.
5. Dance, F. J., and J. L. Wallace, "Clad Metal Circuit Board Substrates for Direct Mounting of Ceramic Chip Carriers," *Electronic Packaging & Production,* Jan. 1982, pp. 226–237.
6. Waller, D. L., L. R. Fox, and R. J. Hannemann, "Analysis of Surface Mount Thermal and Thermal Stress Performance," *IEEE Trans. on Components, Hybrids, and Manufacturing Technology,* **CHMT-6**(3), 1983, pp. 257–266.
7. Hall, P. M., T. D. Dudderar, and J. F. Argyle, "Thermal Deformations Observed in Leadless Ceramic Chip Carriers Surface Mounted to Printed Wiring Boards," *IEEE Trans. on Components, Hybrids, and Manufacturing Technology,* **CHMT-6**(4), 1983, pp. 544–557.
8. Howard, R. T., S. W. Sobeck, and C. Sanetra, "A New Package-Related Failure Mechanism for Leadless Ceramic Chip Carriers (LC-3s) Solder Attached to Alumina Substrates," *Solid State Technology,* **26**(2), 1982, pp. 115–121.
9. Engelmaier, W., "Is Present-Day Accelerated Cycling Adequate for Surface Mount Attachment Reliability Evaluation?" IPC-TP-653, Proc. IPC. Lincolnwood, IL: Institute for Interconnecting and Packaging Electronic Circuits, 1987.
10. Clech, J.-P., "Analysis of Thermomechanical Strains in Solder Joints of Surface-Mount Assemblies," Abstract in *Proc. IEEE International Electronic Manufacturing Symposium,* 1987, p. 89.
11. King, J. A., *Materials Handbook for Hybrid Microelectronics,* Boston, Artech House, 1988.
12. Harper, C. A., *Handbook of Materials and Processes for Electronics,* New York, McGraw-Hill, 1968.

13. Murty, G. S., "Stress Relaxation in Superplastic Materials," *J. Materials Sci.*, **8,** 1973, pp. 611–614.
14. Hall, P. M., "Creep and Stress Relaxation in Solder Joints of Surface-Mounted Chip Carriers," *IEEE Trans. on Components, Hybrids, and Manufacturing Technology,* **CHMT-12**(4), 1987, pp. 556–565.
15. ———, "Forces, Moments, and Displacements during Thermal Chamber Cycling of Leadless Ceramic Chip Carriers Soldered to Printed Boards," *IEEE Trans. on Components, Hybrids, and Manufacturing Technology,* **CHMT-7**(4), 1984, pp. 314–327.
16. Tummala, R. R., and E. J. Rymaszewski, Eds., *Microelectronics Packaging Handbook,* New York, Van Nostrand Reinhold, 1989.
17. Sherry, W. M., J. S. Erich, M. K. Bartschat, and F. B. Prinz, "The Effect of Joint Design on the Thermal Fatigue Life of Leadless Chip Carrier Solder Joints," *IEEE Trans. on Components, Hybrids, and Manufacturing Technology,* **CHMT-8**(6), 1985, pp. 417–428.
18. Lau, J. H., D. W. Rice, and P. A. Avery, "Nonlinear Analysis of Surface Mount Solder Joint Fatigue," *Proc. Int. Electronic Manufacturing Techn. Symp.*, New York, IEEE-CHMT, 1986, pp. 173–184.
19. Payne, R. K., and B. Livesay, *Task I Report, General Approach to SMD Fault Identification,* submitted to Summit Technologies, Inc., and U.S. Army Laboratory Command, Georgia Institute of Technology, 1989.
20. Jones, J. W., "Theory of Creep Deformation," in *Metals Handbook,* Volume 8, *Mechanical Testing,* Metals Park, OH, American Society for Metals, 1985, p. 309.
21. Baker, E., "Stress Relaxation in Tin-Lead Solders," *Materials Sci. and Engineering,* **38**, 1979, pp. 241–244.
22. Kashyap, B. P., and G. S. Murty, "Experimental Constitutive Relations for the High-Temperature Deformation of a Pb-Sn Eutectic Alloy," *Mater. Sci. Eng.*, **50**, 1981, pp. 205–213.
23. Lam, S. T., A. Arieli, and K. Mukherjee, "Superplastic Behavior of Pb-Sn Eutectic Alloy," *Materials Sci. And Engineering,* 1979, pp. 40, 73–79.
24. Solomon, H. D., "Fatigue of 60/40 Solder," *Proc. Electronic Components Conference,* 1986, pp. 622–629.
25. Shine, M. C., and L. R. Fox, "Fatigue of Solder Joints in Surface Mount Devices," in *Low Cycle Fatigue,* ASTM Special Tech. Publication 942, H. D. Solomon, G. R. Halford, L. R. Kasand, and B. N. Leis, Eds., Philadelphia, American Soc. for Testing and Materials, 1987, pp. 588–610.
26. Manson, S. S., *Thermal Stress and Low Cycle Fatigue,* New York, McGraw-Hill, 1966.
27. Halford, G. R., and S. S. Manson, "Life Prediction of Thermal-Mechanical Fatigue Using Strain Range Partitioning," in *Thermal Fatigue of Materials and Components* Spera and Mowbray, Eds., ASTM Special Technical Publication 612, 1975, pp. 239–254.
28. Subrahmanyan, R., J. R. Wilcox, and C.-Y. Li, "A Damage Integral Approach to Thermal Fatigue of Solder Joints," *Proc. Electronic Components Conference,* 1989, pp. 240–252.

29. ______, "A Damage Integral Approach to Solder Joint Fatigue," *Proc. ASM Electronic Packaging Conference—Materials and Processes, ASM International, 1989,* pp. 213–221.
30. Wild, R. N., "Some Fatigue Properties of Solders and Solder Joints," in *Nat. Electronic Packaging Conf.,* 1974, pp. 409–414.
31. Engelmaier, W., "Accelerated Solder Joint Reliability Testing and Result Evaluation," Chapter 16 of this book, 1991.
32. Solomon, H. D., "Low Cycle Fatigue of 60/40 Solder- Plastic Strain Limited *vs.* Displacement Limited Testing," *Proc. ASM Electronic Packaging: Materials and Processes Conference,* 1985, pp. 29–47.
33. Shamash, M. B., *Development of Highly Reliable Soldered Joints for Printed Circuit Boards,* NASA Contract NAS 8-21233, Final Report, 1968, pp. 4–60.
34. Wilcox, J. R., R. Subrahmanyan, and C.-Y. Li, "Thermal Stresses and Inelastic Deformation in Solder Joints," *Proc. ASM Electronic Packaging Conference—Materials and Processes,* ASM International, 1989, pp. 203–211.
35. Clech, J.-P., and J. A. Augis, "Engineering Analysis of Thermal Cycling Accelerated Tests for Surface-Mount Attachment Reliability Evaluation," *Proc. 7th International Electronics Packaging Conference,* 1987, pp. 385–410.

11

Effects of Strain Range, Ramp Time, Hold Time, and Temperature on Isothermal Fatigue Life of Tin-Lead Solder Alloys

Semyon Vaynman and Moris E. Fine

An understanding of the mechanical and fatigue properties of solder alloys used in electronics is required for better design of solder joints, accelerated test development, and development of improved solders.

Solders of different compositions are used in electronic devices, and while there are some similarities in fatigue behavior of different solders, they are actually different materials with their specific microstructural, mechanical, and fatigue properties. Therefore, application of data on solder alloys found in the literature to a specific device must consider solder composition, microstructure, specimen design, testing conditions, fatigue life definition, and other factors to avoid erroneous assessment of fatigue reliability of the device.

Strain range, ramp time, hold time, and temperature are the most important factors affecting isothermal fatigue of materials in the given atmosphere. The effect of these parameters on the isothermal fatigue life of solder materials in the air will be discussed in this chapter.

11.1 DEFINITION OF FAILURE, SPECIMEN DESIGN, AND MODE OF LOADING

Fatigue data discussed here have been obtained for specimens of different design, ranging from a solder joint in a real device[1–3] and two plates joined by solder[4–8] to a bulk specimen[9–13] tested under different loading conditions (tension,[10–12] shear,[2–8] bending,[14] and torsion[1,9]). Different definitions of end of fatigue life have been used at different laboratories: visible cracking proportioned from total fracture,[14] a predetermined drop in load,[1–5,9,15] start of the drop of the maximum tensile stress,[10–13] or of the tensile stress-compressive stress ratio[13] and predetermined increase in electrical resistance,[2,16] etc. Of coarse, a different definition of failure leads to a different fatigue life. For example, when end of fatigue life in near-eutectic solder was defined as the number of cycles to reduce the tensile stress to 75, 50, or 10 percent of its initial value, the different definitions of fatigue life led to different slopes of Coffin-Manson plots.[15] Thus, it is very often difficult, if not impossible, to compare data developed at different laboratories and correlate these data to the fatigue life of a solder joint in the device.

11.2 EFFECT OF STRAIN RANGE ON FATIGUE LIFE

Increasing the strain amplitude leads to a decrease in the number of cycles to failure. Low temperature (< 0.3 absolute melting point, T_m), low cycle fatigue data for many metals may be fitted to the empirical Coffin-Manson relation:[17,18]

$$N_f^{\beta} \varepsilon_p = C \tag{11-1}$$

where N_f = number of cycles to failure

ε_p = plastic strain range

β, C = constants

For most metals the constant β is equal to approximately 0.5.

The Coffin-Manson relation was found to be valid for 60% Sn-40% Pb solder[4,5] tested in shear in the 1 to 10 percent plastic strain range at 35 and 150°C (Fig. 11-1). The exponent β was approximately 0.5 at temperatures below 150°C and decreased to 0.37 at 150°C. An intergranular mode of failure was observed at all testing conditions.

Limited isothermal experimental bending fatigue data for 95% Pb-5% Sn solder[14] was found to obey the Coffin-Manson relation. The exponent β was found to be equal to 0.43 to 0.46 in the strain range of 0.1 to 5.0 percent and the temperature range of 25 to 85°C. For temperatures equal to or higher than 120°C at strain ranges higher than 1 percent this exponent became about 1. 95% Pb-5% Sn solder also was found to approximately obey the Coffin-Manson relation when tested in torsion from 0.05 to 2.0 percent (Fig. 11-2).[9] In these tests at 1.0 percent total strain at room temperature transgranular cracking was observed. At 150°C the solder failed in the grain boundaries.

The Coffin-Manson relation was shown to describe approximately the fatigue life of the Indalloy 151 solder (92.5% Pb-5% Sn-2.5% Ag) tested in shear (0.5

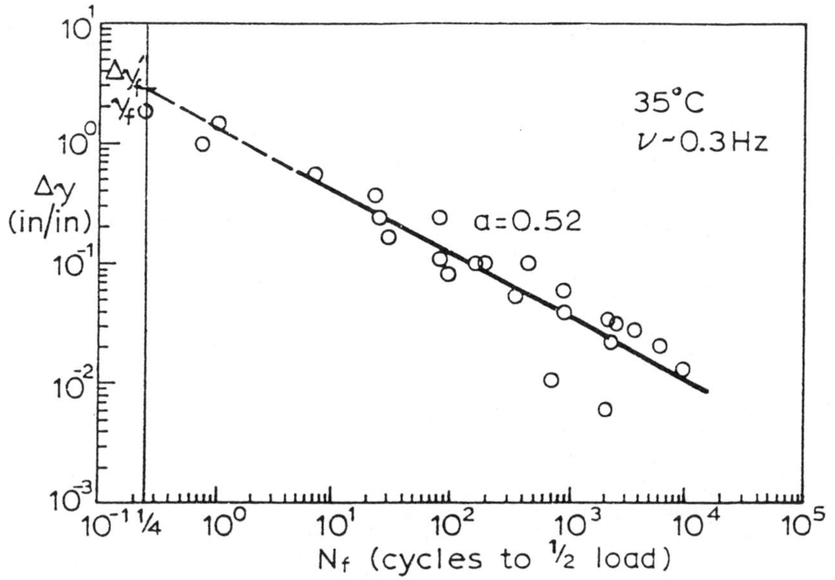

Figure 11-1 Number of cycles to failure versus plastic shear strain for 60% Sn-40% Pb solder. Tests without hold time. (After H. D. Solomon[4])

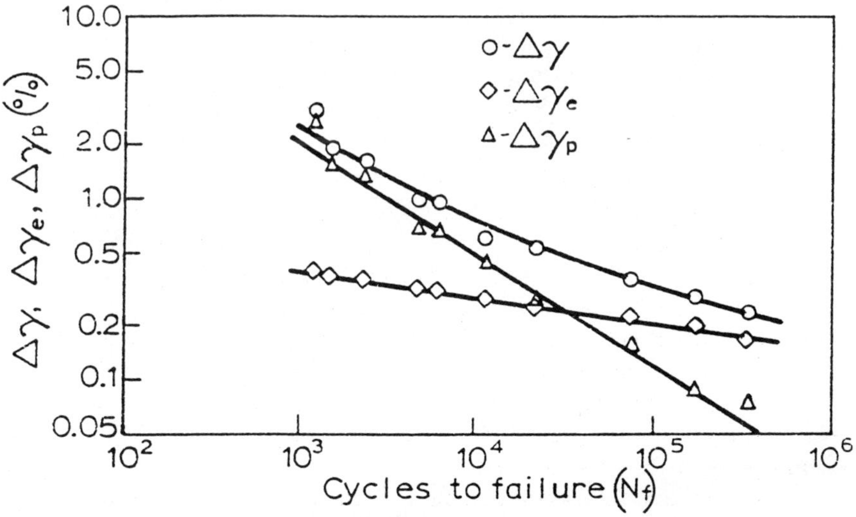

Figure 11-2 Number of cycles to failure versus plastic strain for 95% Pb-5% Sn solder. Tests without hold time. (After M. Kitano, T. Shimizu, and T. Kumazava[9])

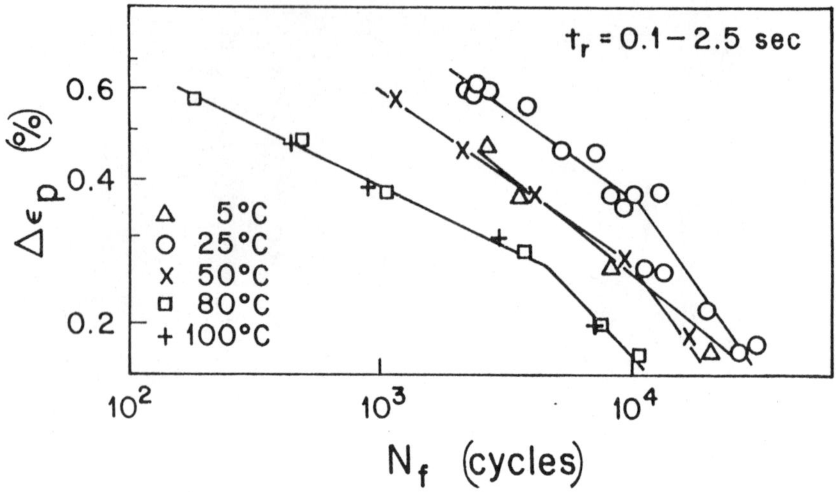

Figure 11-3 Number of cycles to failure versus plastic strain for 96.5% Pb-3.5% Sn solder. Tests without hold time. (After S. Vaynman, M. E. Fine, and D. A. Jeannotte[10])

to 100 percent plastic strain) with a cycling frequency of 0.3 Hz.[15] Ductile fracture was detected at all testing conditions.

However, for 96.5% Pb-3.5% Sn solder[10] tested at low strain ranges in tension-tension it is evident that the data for each temperature cannot be well fit to a single log-log straight line (Figure 11-3). The data are much better represented by two straight lines with a break point at approximately 0.3% plastic strain. The best fit values of the constants β and *C* from Coffin-Manson equations for this solder at different temperatures are summarized in Table 11-1.

In general, cracks in this low-tin lead-based solder could be observed under all testing conditions very early in the fatigue life. Figure 11-4 shows that at 0.75 total strain range intergranular fatigue cracks as well as transgranular cracks along slip bands are formed. At low strain ranges (e.g., 0.3 percent total strain range) coarse slip lines and the transgranular cracks associated with them are absent (Fig. 11-5). Thus, the break in the lines in the Coffin-Manson plot (Fig.

Table 11-1 Values for Constants in Coffin-Manson Relation for 96.5% Pb-3.5% Sn Solder[11]

Temperature	Low Strains		High Strains	
(°C)	β	C	β	C
5	0.45	0.15	0.45	0.15
25	0.62	1.07	0.39	0.14
50	0.56	0.43	0.37	0.08
80	0.47	0.13	0.24	0.02

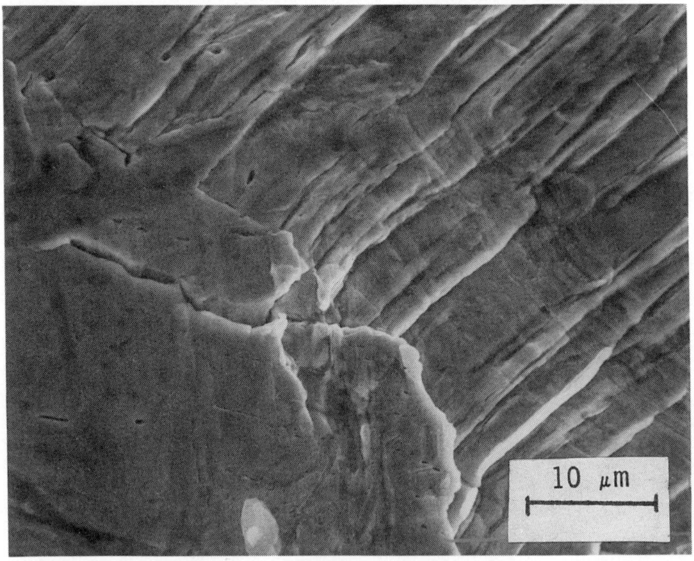

Figure 11-4 Scanning electron micrograph of surface of failed specimen. Total strain range 0.75 percent, 25°C. Ramp time 2.5 sec. No hold. 2550 cycles. N_f = 2500 cycles. (After S. Vaynman, M. E. Fine, and D. A. Jeannotte[10])

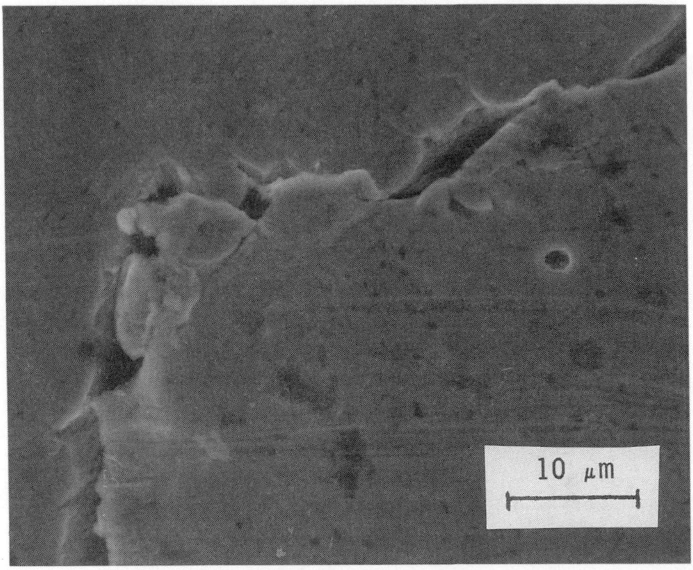

Figure 11-5 Scanning electron micrograph of surface replica and failed specimen. Total strain range 0.30 percent, 25°C. Ramp time 2.5 sec. No hold. 30,000 cycles. N_f = 29,000 cycles. (After S. Vaynman, M. E. Fine, and D. A. Jeannotte[10])

11-3) has been attributed to the change in the fracture mode from intergranular at low strains to mixed transgranular-intergranular at high strains.[10]

11.3 EFFECT OF FREQUENCY ON THE FATIGUE LIFE

At temperatures well below one-half of the absolute melting point, frequency has little effect on the fatigue life of most metals. When the temperature is increased to over half of the absolute melting point, a reduction in frequency decreases the number of cycles to failure for many metals including solders.[4,5, 10–12, 19–30] Failure tends toward a dependence on the total time of stress application rather than on the number of strain cycles.[31] This behavior arises because at high temperatures creep failure such as void formation and environmental attack, which are time dependent processes, play important roles in damage accumulation.

A general trend in the effect of frequency on the number of cycles to failure for a number of metals[23,24] is observed (Fig. 11-6); there is almost no frequency effect at very high and very low frequencies but a large effect at intermediate frequencies. Of course, the frequency effect depends on the material and on testing conditions.

J. F. Eckel[19] studied the fatigue behavior of lead in a rotating bending test at the strain ranges of 0.27 to 0.65 percent and found a marked influence of frequency on fatigue life. The time to failure t_f and the number of cycles to

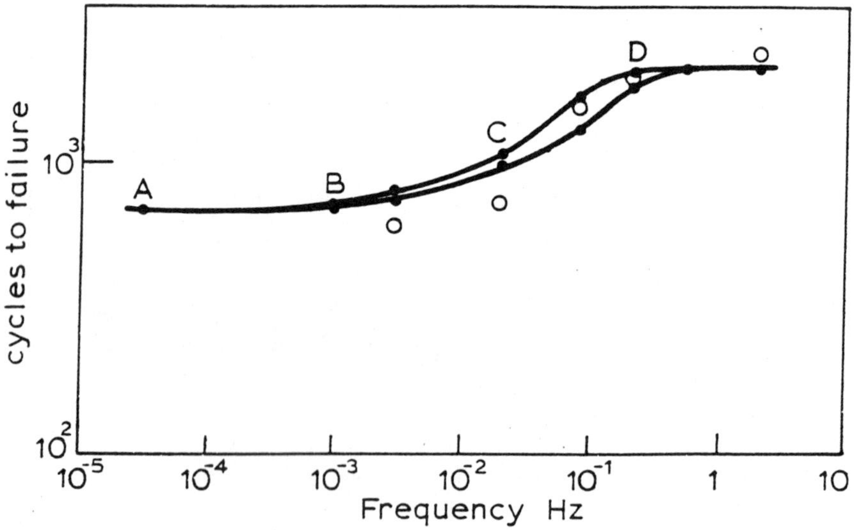

Figure 11-6 Schematic representation of number of cycles to failure versus cycling frequency. (After S. S. Manson[23])

failure were related to frequency f by the following empirical relations:

$$t_f = bf^{-m} \tag{11-2}$$

$$N_f = bf^{1-m} \tag{11-3}$$

where b and m are constants

The frequency exponent m for lead was found to be equal to 0.7 at 43°C and at all strains tested.[20]

Equations (11-2) and (11-3) were also found to be valid for chemical grade lead and lead with 1 percent Sb[29] tested at room temperature in reverse bending (zero mean strain) in the strain range from 0.1 to 1.0 percent. However, the value of m was found to depend on the strain range and to vary from 0.4 to 0.8. The effect of frequency on fatigue life was found to be stronger at high strains. It was also found that the stiffer the lead alloy the less its fatigue life was affected by the variation in frequency.

The number of cycles to failure for 60% Sn-40% Pb solder tested in shear at 35 and 150°C was reduced when frequency of cycling was decreased below approximately 3×10^{-4} Hz at 35°C and 3×10^{-3} at 150°C.[4,5] The frequency constant was 0.42 for $\gamma_p = 10$ percent at 35°C (Fig. 11-7) and was equal to 0 at 5.0 and 10.0 percent of γ_p. At higher frequency the decrease in the number of cycles to failure with increasing time per cycle was less significant.

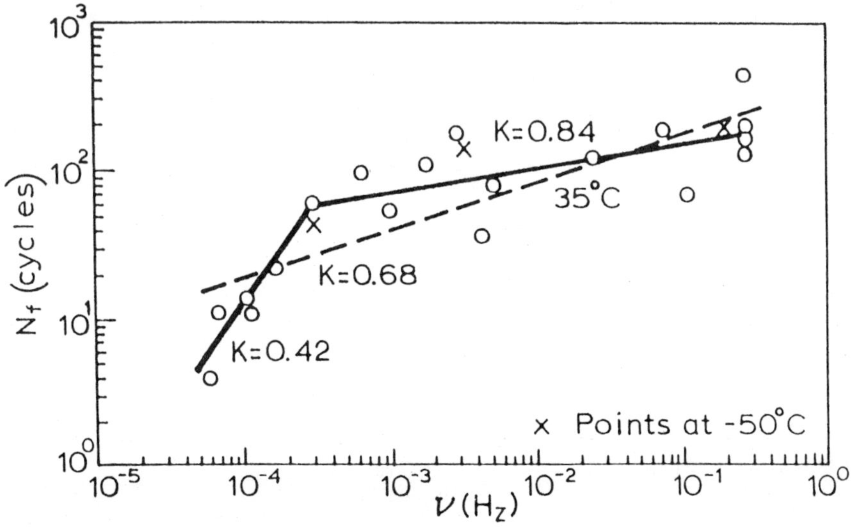

Figure 11-7 Number of cycles to failure versus cycling frequency for 60% Sn-40% Pb solder. Test run at 35°C (with some −50°C data). (After H. D. Solomon[4])

The effect of frequency on fatigue life of 95% Pb-5% Sn solder was evaluated in bending tests from 0.033 to 192 sec/cycle and in strain ranges from 0.5 to 4.0 percent.[14] Eckel's relations in Equations (11-2) and (11-3) were found to be valid with a frequency exponent *m* equal to 0.25 at all strains tested.

For 96.5% Pb-3.5% Sn solder practically no change in the number of cycles to failure was observed on varying frequency in the comparatively high frequency region.[10,11] A reduction in the number of cycles to failure with decreasing frequency below 10^{-2} Hz was found at 25, 50, and 80°C (Fig. 11-8 through 11-10) for all strain ranges tested. The behavior of this solder obeys the general trends found by Eckel. The values of the constants *m* and *b* for 96.5% Pb-3.5% Sn solder are summarized in Table 11-2. While the frequency exponent *m* at 50 and 80°C in tests with total strain of 0.60 percent are practically the same, it differs from exponents found in tests with a total strain of 0.75 percent at room temperature and in tests with a total strain of 0.50 percent at 80°C. This indicates that the frequency effect on fatigue life of this solder may be practically independent of temperature for a given strain range but is a function of the strain range. The effect of frequency on the fatigue life of this solder is related to intensification of damage in grain boundaries with increasing time per cycle during fatigue (Fig. 11-11). At these strain ranges and temperatures some transgranular cracking can be observed; however, the grain boundary cracking dominates. Thus, creep processes, such as grain boundary cavitation and grain boundary sliding, may be the main modes of failure for this solder under conditions of slow cycling. Since the failure of this solder at low strains and at high frequency is already mainly intergranular, the effect of frequency on fatigue life is less pronounced at low strains than at high ones where the mode of failure changes

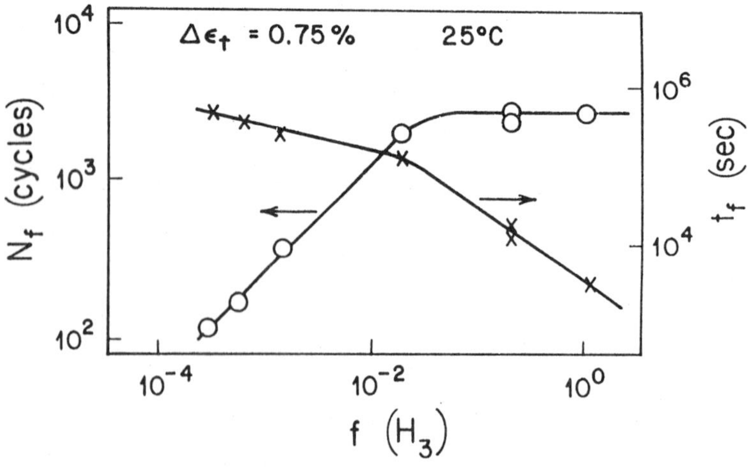

Figure 11-8 Effect of frequency on fatigue life of 96.5% Pb-3.5% Sn solder. (After S. Vaynman, M. E. Fine, and D. A. Jeannotte[10])

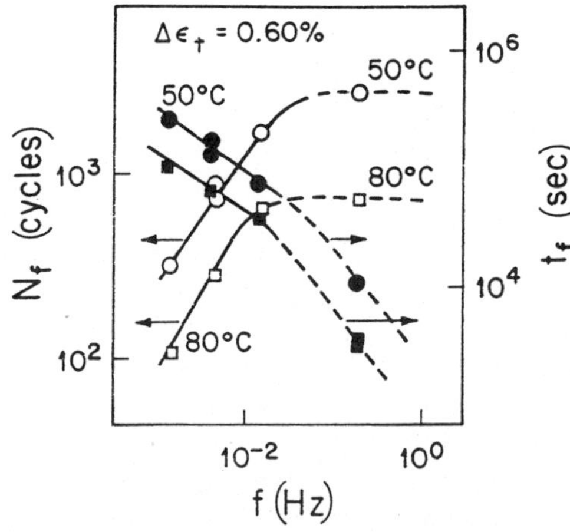

Figure 11-9 Effect of frequency on fatigue life of 96.5% Pb-3.5% Sn solder. (After S. Vaynman, M. E. Fine, and D. A. Jeannotte[10])

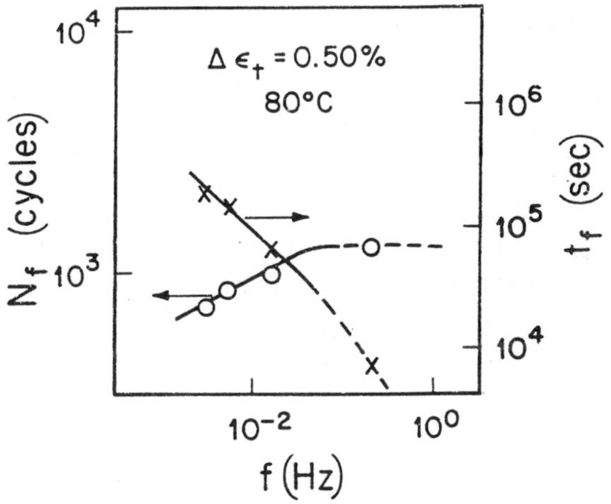

Figure 11-10 Effect of frequency on fatigue life of 96.5% Pb-3.5% Sn solder. (After S. Vaynman, M. E. Fine, and D. A. Jeannotte[10])

Table 11-2 Values of Constants *m* and *b* from Eckel's Equation for 96.5% Pb-3.5% Sn solder. Frequency below 10^{-2} Hz[10]

Total Strain Range (%)	Temperature (°C)	*m*	*b*
0.75	25	0.70	36100
0.60	50	0.56	10070
0.60	80	0.59	8090
0.50	80	0.25	2820

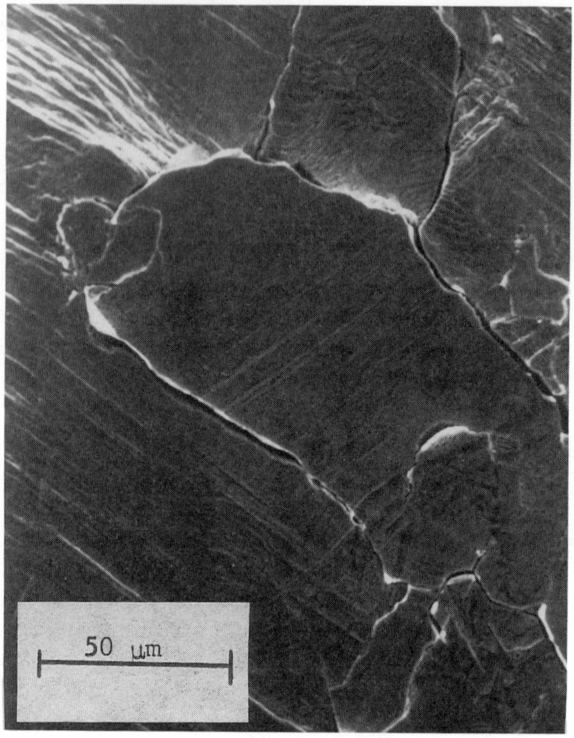

Figure 11-11 Scanning electron micrograph of surface of failed specimen. Total strain range 0.75 percent, 25°C. Ramp time 60 sec. No hold. 1300 cycles. N_f = 1230 cycles. (After S. Vaynman, M. E. Fine, and D. A. Jeannotte[10])

from mixed transgranular-intergranular at high frequency to predominantly intergranular at low frequency.

A combination of the Coffin-Manson Equation (11-1) and Eckel's Equation (11-3) leads to the frequency-modified Coffin-Manson relation. The frequency-modified Coffin-Manson relation may be used only when the parameter m is independent of the strain range. However, in case of low-tin lead-based solder tested at total strain ranges under 1 percent this relation cannot be used since the m is a function of strain range.[10,11]

11.4 EFFECT OF HOLD TIME ON FATIGUE LIFE

Hold (dwell) times introduced in the cycle at maximum and/or minimum strains, stresses, or temperatures affect the fatigue life of materials depending on the kind of material and testing conditions.

Tensile hold time is very damaging during high temperature fatigue of many materials such as steels,[22, 31–35] alloys,[27,33,36,37] and solders.[10–13,38–40] The number of cycles to failure decreases when tensile hold time is increased; however, a saturation in the number of cycles to failure with increasing hold time is detected for a number of metals.

In a few cases, however, as in high temperature isothermal fatigue of such metals as Rene 80,[41] 2.25% Cr-1% Mo alloy,[42] Udimet 700,[43,44] and nickel-base superalloy In 738,[45] tensile hold time gives greater fatigue lives due to formation of a mean compressive stress during cycling.

Often the reduction in cycles to failure of metals due to tensile hold is accompanied by a transition from the transgranular to the intergranular mode of fracture. Cavitation on grain boundaries is observed; therefore, the effect of tensile hold time on fatigue life was attributed mainly to creep (cavitation).[25–28,33,46,47]

In some cases the reduction in fatigue life due to tensile hold was attributed to the effect of environment. Oxidation was found to have a much higher effect on fatigue life of some steels at elevated temperature than creep.[24]

Hold time at maximum strain has a dramatic effect on the number of cycles to failure of low-tin lead-based and lead-tin eutectic solders[10–13] (Fig. 11-12 and 11-13); a few minutes of hold time reduce the number of cycles to failure almost

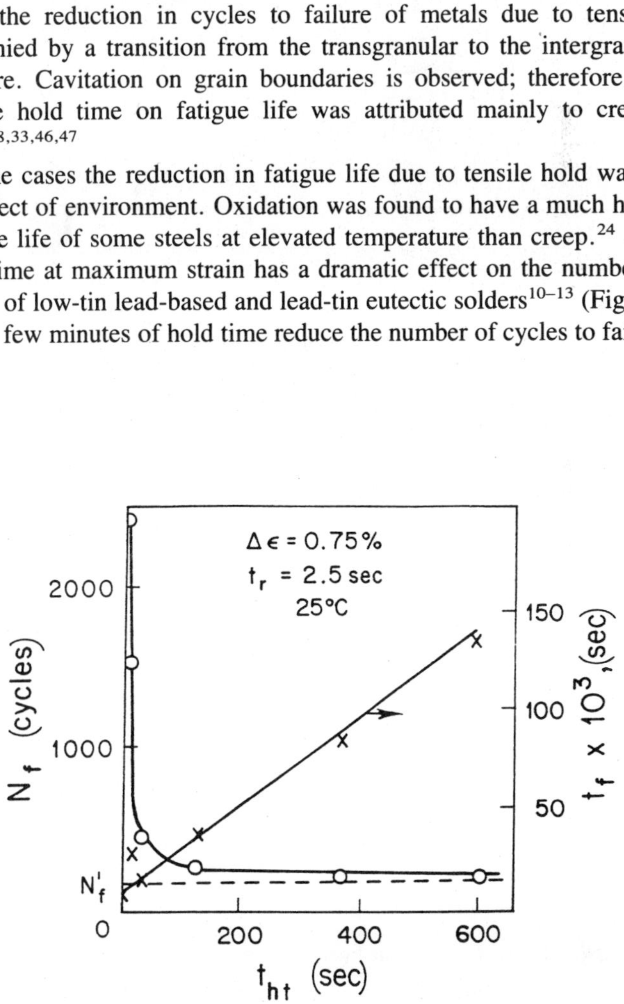

Figure 11-12 Effect of tensile hold time on fatigue life of 96.5% Pb-3.5% Sn solder. (After S. Vaynman, M. E. Fine, and D. A. Jeannotte[10])

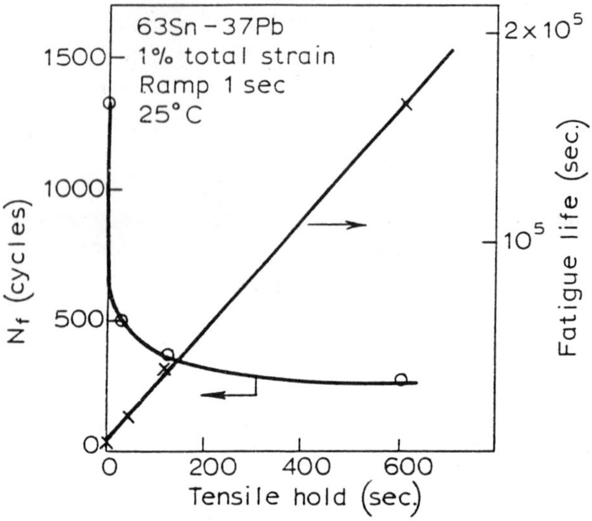

Figure 11-13 Effect of tensile hold time on fatigue life of 63% Sn-37% Pb solder. (After S. Vaynman and M. E. Fine[13])

by an order of magnitude and a further increase in hold time eventually leads to fatigue life saturation. There exists a linear relation between time to failure and tensile hold time per cycle (when hold time ≫ ramp time) (Figs. 11-12 and 11-13):

$$t_f = A + Bt_{th} \tag{11-4}$$

where t_f = time to failure

t_{th} = hold time at maximum strain (tensile hold)

Since time per cycle t_c is equal to

$$t_c = 2t_r + t_{th} \tag{11-5}$$

where t_r = ramp time

the number of cycles to failure can be found as:

$$N_f = \frac{t_f}{2t_r + t_{th}} = \frac{A + Bt_{th}}{2t_r + Bt_{th}} \tag{11-6}$$

When the ramp time is much less than hold time, as in the present study, it can be neglected and finally

$$N_f = \frac{A}{t_{\text{th}}} + B \tag{11-7}$$

For long hold times, the first term in Equation (11-7) becomes negligible and

$$N_f = B = N_f' \tag{11-8}$$

From the Equation (11-8) the limit in number of cycles to failure under a given set of conditions is equal to the slope of the time to failure versus tensile hold time per cycle line (Figs. 11-12 and 11-13). Thus, when $t_{\text{th}} \gg t_r$, solders will survive not less than B number of cycle with any tensile hold time in the cycle under a given strain range and temperature. Equations (11-7) and (11-8) were found to hold for other materials also.[48]

Using Equation (11-7) the data for a few strain ranges were extrapolated to higher hold times and approximately the same number of cycles to failure was found for tensile hold times of 1 and 10 h and this essentially gives N_f' values (Fig. 11-14).[10] It is obvious that the Coffin-Manson relation does not hold for this solder. There are three regions in this figure; at high and low strains there is dependence of the number of cycles to failure on the strain range, while at intermediate strains the dependence is very weak. Such behavior of this solder, similar to the behavior of a number of steels at high temperature,[49] indicates a strong creep-fatigue interaction in low-tin lead-based solder. Thus, the simple

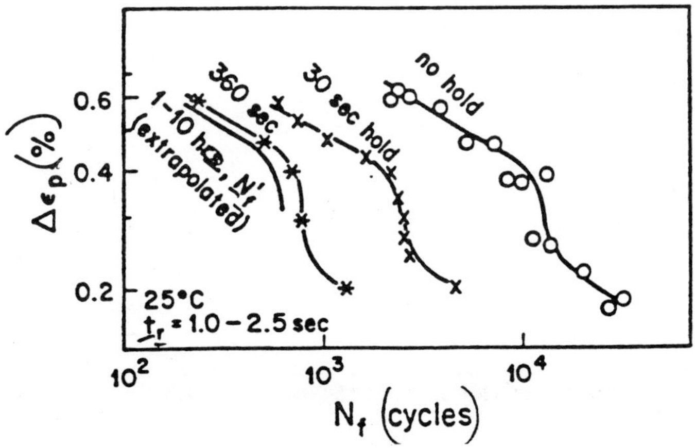

Figure 11-14 Number of cycles to failure versus plastic strain for 96.5% Pb-3.5% Sn solder. Tests with tensile hold time. (After S. Vaynman, M. E. Fine, and D. A. Jeannotte[10])

Coffin-Manson relation cannot be used for isothermal fatigue life prediction of solders at strain ranges below 1 percent.

Tests performed for low-tin lead-based solder in air and vacuum[50,51] at 0.80 percent total strain range indicate that only some of fatigue life reduction with increasing tensile hold time may be attributed to environmental attack; the number of cycles to failure saturated in vacuum as it did in air; however, fatigue life in vacuum was found to be only slightly higher than in air (Fig. 11-15). Therefore, most damage during fatigue with hold time at this strain range has to be attributed to creep and in particular to the formation and growth of voids on grain boundaries. Voids were found on grain boundaries during fatigue of 98% Pb-2% Sn[52] and 97.5% Pb-3.5% Sn[50,51] solders.

While mixed transgranular-intergranular fracture was observed in low-tin lead-based solder in no-hold time tests at higher strains (Fig. 11-4), fracture of solder along grain boundaries was characteristic for this solder in tests with tensile hold time in air at all strain ranges tested[10] (Fig. 11-16).

While tensile hold time is found to be very damaging for a number of metals, including solders, compressive hold times are usually not nearly as damaging. The processes of formation and growth of grain boundary cavities were found to be detrimental in fatigue life under tensile hold time conditions. The less damaging effect of compressive hold time is attributed to the processes of shrink-

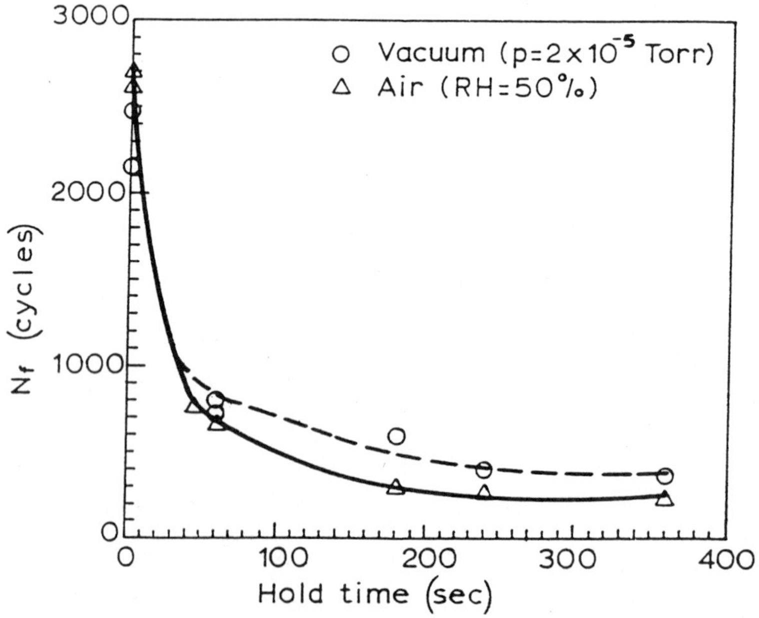

Figure 11-15 Effect of tensile hold time on fatigue life of 96.5%-3.5% Sn Pb solder in vacuum. (After R. Berriche, S. Vaynman, M. E. Fine, and D. A. Jeannotte[50])

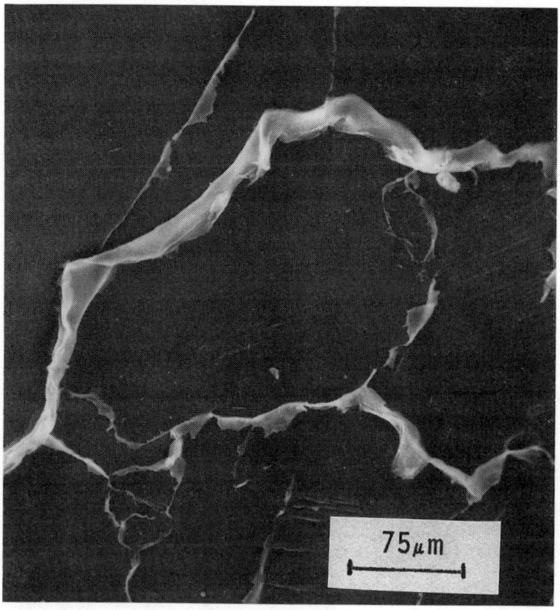

Figure 11-16 Scanning electron micrograph of surface replica. Total strain range 0.75 percent, 25°C. Ramp time 2.5 sec. Tensile hold time 90 sec. 200 cycles. N_f = 330 cycles. (After S. Vaynman, M. E. Fine, and D. A. Jeannotte[10])

age and annealing of such cavities. For example, the number of cycles to failure in tests with equal tensile and compressive hold times for Rene 80[41] and for 60% Sn-40% Pb solder[53] was found to be close to the number of cycles to failure in continuous fatigue cycling with the same period per cycle and the same plastic strain (Fig. 11-17).

Compressive hold time alone in a cycle was found to be very damaging for Rene 80 and nickel-base superalloy IN 738, both tested at 1600°F.[41,45] Since elastic strain in these alloys is much higher than the plastic strain, a significant mean stress (tensile in the case of compressive hold time in the cycle) is built in these alloys leading to enhanced crack growth. The build up of a significant mean stress is not a case for most metals, including solders.

A limited number of tension-tension fatigue tests performed with 96.5% Pb-3.5% Sn solder at 0.75 percent total strain range in air under conditions of fast ramp time (2.5 sec) and hold time at zero strain (compressive stress hold time) indicated practically no effect of such hold on fatigue life.[10] However, the compressive hold dramatically reduced the number of cycles to failure for this solder when combined with a tensile hold. Figure 11-18 shows how variable compressive hold time affects the number of cycles to failure in tests with fixed tensile hold time of 120 and 360 sec. Line 1 in this figure corresponds to tests

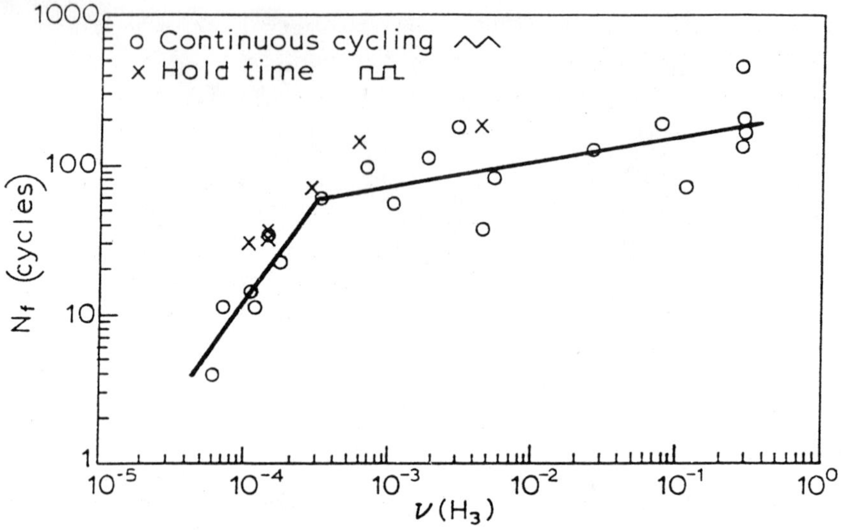

Figure 11-17 Influence of cycling frequency and hold time on fatigue life of 60% Sn-40% Pb solder. (After H. D. Solomon[53])

with only tensile hold time. By using Equation (11-8) the limit for tensile hold tests under these conditions was found to be approximately 200 cycles. However, adding compressive hold reduced the fatigue life below this value. As in tests with only tensile hold, reduction in N_f is greatest during the first few minutes of compressive hold, and the addition of more compressive hold time leads to

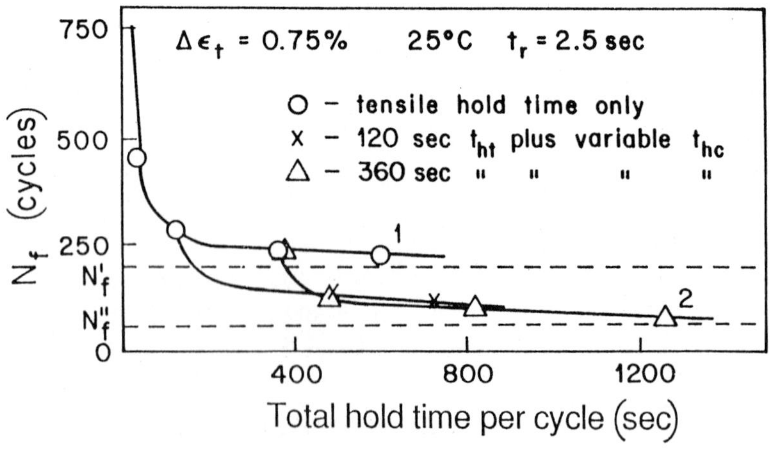

Figure 11-18 Effect of combined tensile and compressive hold time on number of cycles to failure for 96.5% Pb-3.5% Sn solder. (After S. Vaynman, M. E. Fine, and D. A. Jeannotte[10])

another limit N_f'', where N_f'' is the minimum number of cycles to failure all specimens made of this solder will survive at any combination of tensile and compressive hold times at given conditions of strain range, temperature, and ramp time.

The times to failure for tests with combined tensile and compressive hold times were plotted versus total time per cycle spent in tensile and compressive hold (Fig. 11-19). While experimental results for tests with only tensile hold time in the cycle fall on line 1, experimental results for tests with combined tensile and compressive hold time (t_{hc}) fall on line 2. It is evident from this figure that a linear relation between the time to failure and combined hold time per cycle exists provided that compressive hold time is not too small:

$$t_f = C + D(t_{ht} + t_{hc}) \tag{11-9}$$

Procedures identical to those for obtaining Equations (11-7) and (11-8) lead to Equations (11-10) and (11-11) provided total hold time per cycle is long compared to ramp time.

$$N_f = \frac{C + D(t_{ht} + t_{hc})}{2t_r + t_{ht} + t_{hc}} \tag{11-10}$$

$$N_f = D = N_f'' \tag{11-11}$$

The N_f'' is determined by D, the slope of line 2 in Figure 11-19.

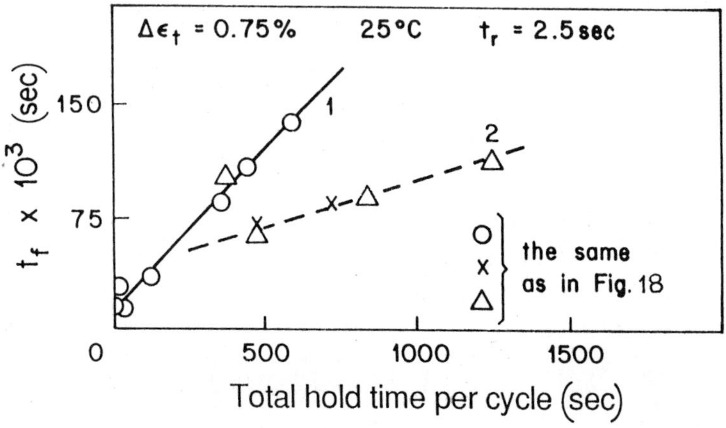

Figure 11-19 Effect of combined tensile and compressive hold time on time to failure for 96.5% Pb-3.5% Sn solder. (After S. Vaynman, M. E. Fine, and D. A. Jeannotte[10])

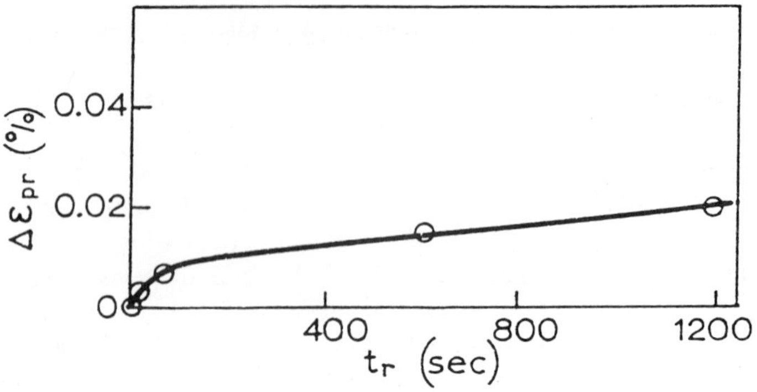

Figure 11-20 Increase in plastic strain due to increase in ramp time. 96.5% Pb-3.5% Sn solder. Total strain range 0.75 percent, 25°C. (After S. Vaynman[54])

It is generally considered that damage per cycle during cycling is proportional to the amount of plastic deformation per cycle. The small changes in plastic strain range per cycle (for example, for tests with 96.5% Pb-3.5% Sn solder at 0.75 percent total strain range discussed previously) with increasing ramp time (Figure 11-20) or hold time (Figure 11-21) are not significant enough to account for observed large, order of magnitude reductions in the number of cycles to failure. It is also well know that creep-fatigue damage is very sensitive to strain rate. As will be shown later, the strain rates in tests were varied over a few orders of magnitude; therefore, changes in fatigue life of solder with increasing ramp or hold times are attributed to the changes in strain rates.[54] The fact that hold is much more destructive than ramp was assumed to be related to the much lower strain rates operating during hold time.

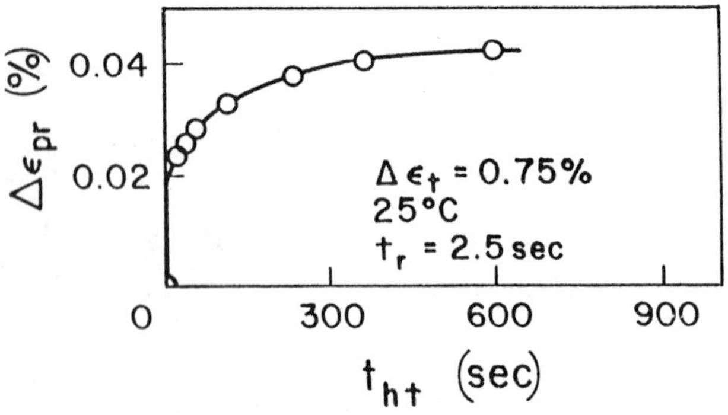

Figure 11-21 Relaxation strain per cycle versus tensile hold time. 96.5% Pb-3.5% Sn solder. Total strain range 0.75 percent, 25°C. (After S. Vaynman[54])

The strain rates during ramps in no-hold tests were calculated as:

$$\dot{\varepsilon} = \frac{\varepsilon_t}{t_r} \tag{11-12}$$

where $\dot{\varepsilon}$ = strain rate

ε_t = total strain range

The stress relaxation data for 96.5% Sn-3.5% Pb solder[54] (Fig. 11-22) were used to calculate the strain rates operating during hold times. It is well known that

$$\varepsilon_t = \varepsilon_c + \varepsilon_p + \varepsilon_{el} \tag{11-13}$$

where ε_c = creep strain range

ε_p = plastic strain range

ε_{el} = elastic strain range

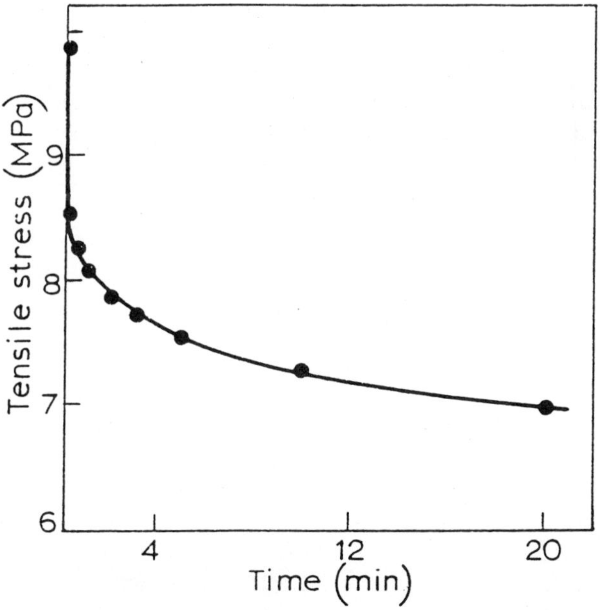

Figure 11-22 Stress relaxation in 96.5% Pb-3.5% Sn solder. Total strain range 0.75 percent, 25°C. (After S. Vaynman[54])

During stress relaxation the total strain range is constant and plastic strain is zero; therefore

$$\varepsilon_c = -\varepsilon_{el} \tag{11-14}$$

Since

$$\dot{\varepsilon}_{el} = \frac{\dot{\sigma}}{E} \tag{11-15}$$

$$\dot{\varepsilon}_c = -\frac{\dot{\sigma}}{E} \tag{11-16}$$

For each test performed, the strain rates, which are calculated by Equations (11-12) and (11-16), are plotted in Figure 11-23 together with the values of the accumulated strains to failure.

It is obvious that the damage in continuous wave tests occurs because of inelastic (plastic and creep) strains, which are a sum of plastic and creep strains, both operating at the same strain rate. Therefore, accumulated strains to failure for these tests were defined as

$$\varepsilon_f = N_f\, \varepsilon_p \tag{11-17}$$

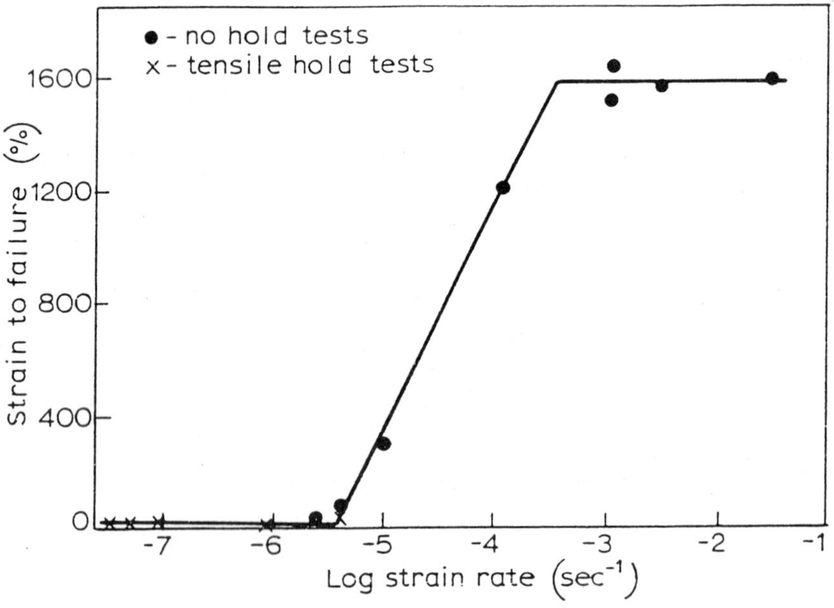

Figure 11-23 Effect of strain rate on strain to failure for 96.5% Pb-3.5% Sn solder. Total strain range 0.75 percent, 25°C. (After S. Vaynman[54])

In hold tests with very short (2.5 sec) ramp time, it was assumed that the material experiences almost all the damage during hold time because strain rates during hold are a few orders of magnitude lower than during ramp. As a result, accumulated strains to failure for hold time tests are defined as

$$\varepsilon_f = N_f \, \varepsilon_{\mathrm{pr}} \tag{11-18}$$

where $\varepsilon_{\mathrm{pr}}$ = relaxation (creep) strain per cycle (shown in Fig. 11-24).

It is evident that strain to failure is a complex function of strain rate (Fig. 11-23). At high strain rates, the accumulated strain to failure does not vary with strain rate. When strain rate is reduced below 10^{-3} sec^{-1}, strain to failure in tests without hold time starts to drop steadily. No tendency for saturation in the values of strain to failure is observed for these tests when time per cycle is increased up to 1 h. However, at longer than 1 h time/cycle, saturation in strain to failure may be expected. In hold time tests, strain to failure is very low. It drops very fast and reaches the plateau at hold times of 30 sec. As shown in Fig. 11-23, the strain rates in tests with almost any hold times are lower than those in continuous wave tests with long ramp time over the ranges usually investigated. This drastic difference in the strain rate magnitude explains why hold time has a much stronger effect on the fatigue life of solder than ramp time.

The variation of strain to failure with strain rate for low-tin lead-base solder is very similar to that found for a number of engineering alloys such as 304, 316 and 20%-25% Ni-Nb doped stainless steels at elevated temperatures[46] and depicted in general form in Figure 11-25; decreasing strain rate leads to reduction in strain to failure from a high plateau level via a transition region to a low strain to failure level at lower strain rates. This behavior is explained by considering different cavity growth (creep) mechanisms that may operate at various strain

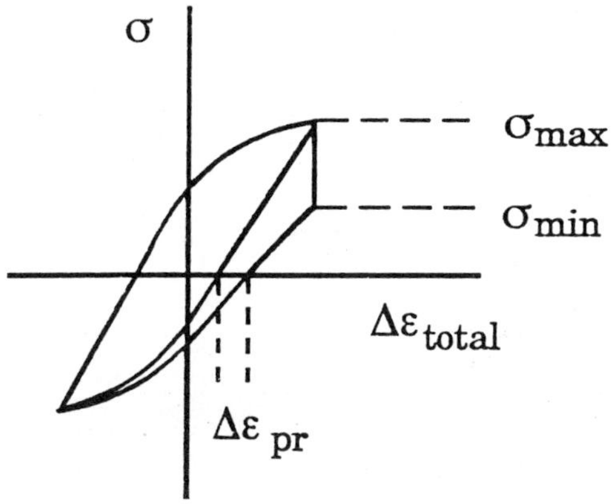

Figure 11-24 Typical hysteresis curve for tensile hold time tests.

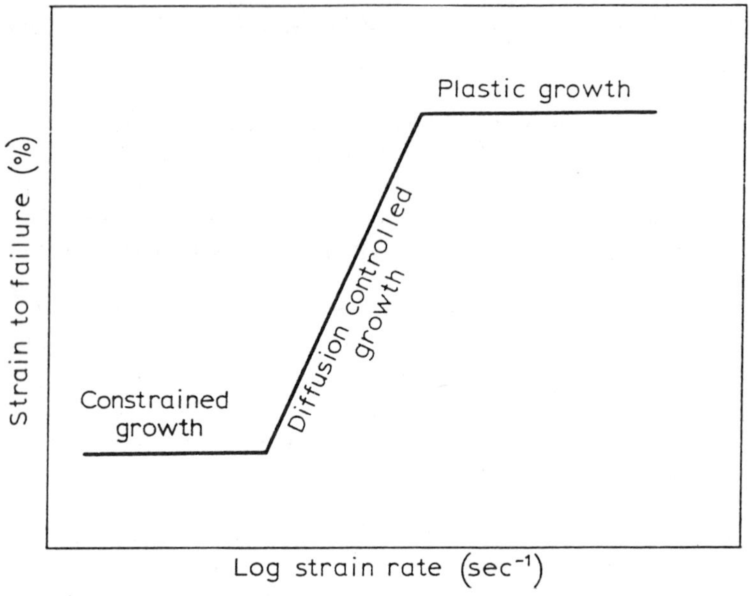

Figure 11-25 Typical strain rate-strain to failure relation for a number of commercial alloys. (After D. A. Miller, R. H. Priest, and E. G. Ellison[46])

rates.[46] At high strain rates, cavity growth occurs by deformation and at these strain rates an alternative mechanism of fracture, such as transgranular fracture, may operate. Both transgranular and intergranular modes of fracture are operational in tests with 96.5% Pb-3.5% Sn solder at high strain rates.[10,12] When the strain rate is reduced, the cavity growth mechanism changes from deformation to unconstrained (transition region in Figs. 11-23 and 11-25) and constrained cavity growth mechanisms (lower plateau in Figs. 11-23 and 11-25). Constrained cavity growth mechanism is considered to be very efficient in converting plastic strain into cavities (i.e., to be the most damaging mechanism during fatigue). Formation of voids in low-tin lead-based solders was detected during fatigue tests of low-tin lead-based solder.[51,52]

In an electronic device, the strain rates below 10^{-7} sec^{-1} operate during hold times; therefore, the most damaging constrained cavity growth mechanism is expected to be operational under these conditions. Solder joints experience strain rates of an order of 10^{4} to 10^{-5} sec^{-1} during ramping and then less damaging unconstrained cavity growth mechanism leads to solder fatigue failure. Thus, the most attention in solder fatigue research should be paid to the hold time, during which solder experiences the most damage, and therefore fatigue life prediction methods should be based on hold time tests.

11.5 EFFECT OF TEMPERATURE ON ISOTHERMAL FATIGUE OF SOLDERS

As a rule the isothermal fatigue life of metals decreases as temperature increases. However, the degree of fatigue life change depends on material and testing conditions. For near eutectic tin-lead solders (63% Sn-37% Pb[13] in the 25 to 80°C, 60% Sn-40% Pb in 35 to 150°C,[4,5] and −60 to 150°C[9] temperature ranges) the temperature effect is insignificant for all strain ranges and modes of loading investigated.

The effect of temperature on low-tin lead-based solder is much more pronounced than for near-eutectic tin-lead solders. The fatigue life of 95% Pb-5% Sn solder at strains lower than 1.0 percent was found to be much smaller at 150°C than at room temperature or −60°C.[9] The fatigue life of 95% Pb-5% Sn solder was shown to decrease when temperature was increased from room temperature to 120°C.[14] The fatigue life for this solder at 120 at 150°C was shown to be practically the same.

Fatigue life of 96.5% Pb-3.5% Sn solder at 80°C is practically the same as at 100°C (Fig. 11-26).[10] The fatigue life of this solder was found to be the lowest at these temperatures. A gradual decrease in fatigue life with increasing tem-

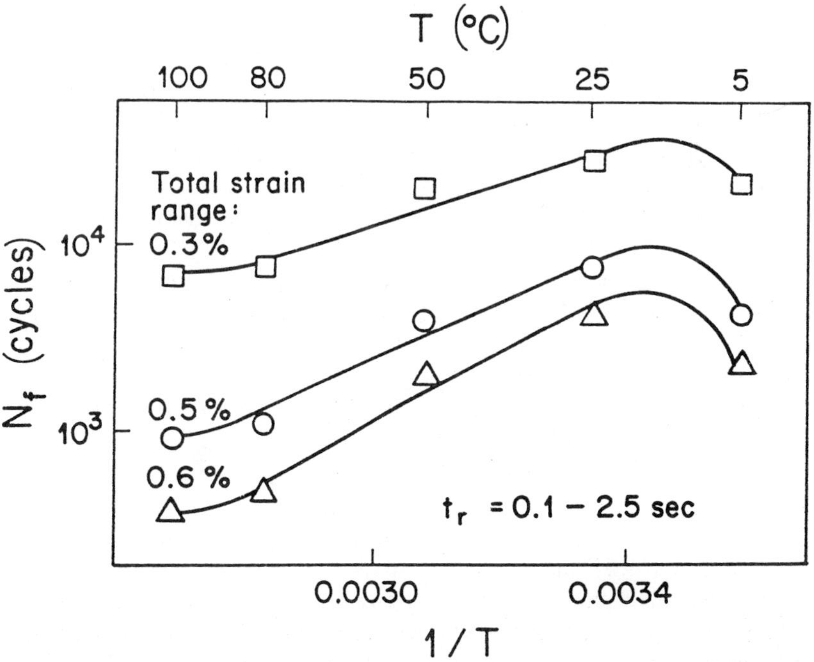

Figure 11-26 Effect of temperature on fatigue life of 96.5% Pb-3.5% Sn solder in tests without hold time. Ramp time 0.1 to 2.5 sec. (After S. Vaynman, M. E. Fine, and D. A. Jeannotte[11])

Table 11-3 Apparent Activation Energy for 96.5% Pb-3.5% Sn Solder. 25 to 80°C. 0.1-2.5-sec Ramp Time. No Hold.[11]

Total Strain Range (%)	Apparent Activation Energy, (KJ/mol)
0.30	22.7
0.40	22.7
0.50	30.1
0.60	40.8
0.75	46.0

perature was observed in the 25 to 80°C region; however, fatigue life decreased when temperature was further reduced to 5°C. The solder fatigue life dependence on temperature appears to follow approximately an Arrhenius equation only between 25 and 80°C. The apparent activation energy for this temperature range is summarized in the Table 11-3. As is evident from this table, the lowest apparent activation energy is associated with grain boundary failure at low strains (Fig. 11-5).

At higher strain (0.6 percent) the temperature effect on fatigue life of this low-tin lead-based solder is a function of frequency (Fig. 11-9) and hold time (Fig. 11-27). The change in the number of cycles to failure with temperature at the frequency less than 10^{-2} Hz is less than that at higher frequency (Fig. 11-8). This converts to an apparent activation energy of 24 to 26 KJ/mol at low frequency and 40.6KJ/mol at high frequency testing conditions.

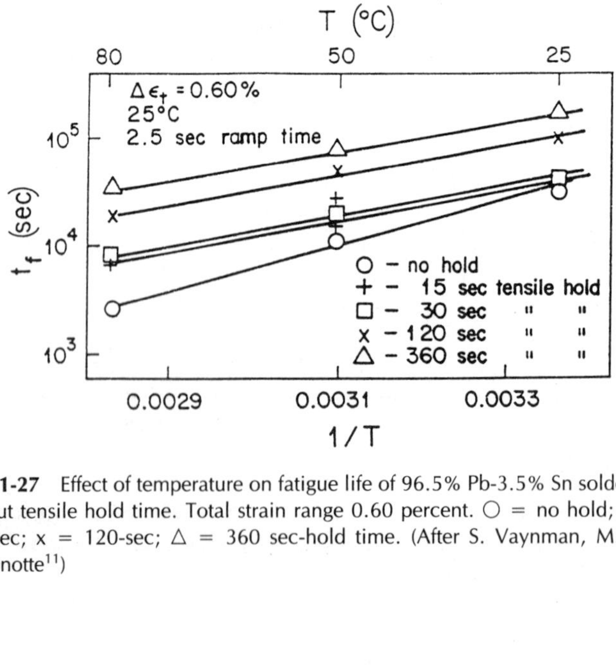

Figure 11-27 Effect of temperature on fatigue life of 96.5% Pb-3.5% Sn solder in tests with and without tensile hold time. Total strain range 0.60 percent. ○ = no hold; + = 15-sec; □ = 30-sec; x = 120-sec; Δ = 360 sec-hold time. (After S. Vaynman, M. E. Fine, and D. A. Jeannotte[11])

The temperature effect on fatigue life of low-tin lead-based solder is different in tests with hold time and without hold time(0.6 percent total strain range, 2.5-sec ramp time; Fig. 11-27); the effective activation energies are 24 KJ/mol and 40 KJ/mol, respectively.

Thus, grain boundary fracture in low-tin lead-based solder under low strain rate conditions is characterized by low activation energy.

11.6 CONCLUSION

There is no unique relationship describing the dependence of isothermal solder fatigue life on variables for all solder alloys. This is because of differences in composition and microstructure. The relationship should be determined experimentally for each solder material.

The following is the summary of the isothermal fatigue behavior of solder materials:

1. While at strain ranges over 1 percent the Coffin-Manson relation may be used for extrapolation of isothermal solder fatigue data; this relation is found to fail at lower strains due to the strong creep-fatigue interaction in solders during cycling.
2. Increasing time per cycle over approximately 60 sec in no-hold tests leads to reduction in the number of cycles to failure.
3. Tensile hold time reduces the number of cycles to failure and at longer hold times leads to the saturation in the number of cycles to failure.
4. The effect of temperature on isothermal fatigue of tin-lead near-eutectic solders is negligible. Temperature affects the fatigue life of high-lead solders. The effect of temperature on fatigue life of these solders is a function of testing conditions.

REFERENCES

1. Shine, M. C., and L. R. Fox, "Fatigue of Solder Joints in Surface Mount Devices," *Low Cycle Fatigue,* ASTM STP 942, H. D. Solomon et al., Eds. Philadelphia, ASTM, 1988, pp. 588–610.
2. Solomon, H. D., "Low Cycle Fatigue of Surface Mounted Chip Carrier/Printed Wiring Board Joints," *Proceedings of 39th Electronic Components Conference,* 1989, pp. 277–292.
3. ______, "Influence of Temperature on the Low Cycle Fatigue of Surface Mounted Chip Carrier/Printed Wiring Board Joints," *GE Information Series,* September 1987.
4. ______, "Low-Cycle Fatigue of 60/40 Solder-Plastic Strain Limited vs. Displacement Limited Testing," in *Electronic Packaging: Materials and Processes,* J. A. Sortell, Ed., ASM, 1986, pp. 29–49.

5. ———, "Low-Frequency, High-Temperature Low-Cycle Fatigue of 60Sn/40Pb Solder," *Low Cycle Fatigue,* ASTM STP-942, H. D. Solomon et al., Eds. Philadelphia, ASTM, 1988, pp. 342–369.
6. Frear, D., D. Grivas, and J. W. Morris, Jr., "Thermal Fatigue Failures in Solder Joints," *Journal of Metals,* **40,** 1988, pp. 18–22.
7. Frear, D., D. Grivas, M. McCormack, D. Tribula, and J. W. Morris, Jr., "Fatigue and Thermal Fatigue Testing of Pb-Sn Solder Joints," *Proc. 3d Ann. Electronic Packaging and Corrosion in Microelectronics Conf.,* M. E. Nicholson, Ed., ASM International, 1987, pp. 269–274.
8. ———, "Fatigue and Thermal Fatigue of Pb-Sn Solder Joints," *Proc. Effect of Load and Thermal Histories on Mechanical Behavior Symp.,* P. K. Liaw and T. Nicholas, Eds., 1987, pp. 113–126.
9. Kitano, M., T. Shimizu, and T. Kumazava, "Statistical Fatigue Life Estimation: The Influence of Temperature and Composition on Low-Cycle Fatigue of Tin-Lead Solders," *Current Japanese Materials Research,* **2,** August 1987, pp. 235–250.
10. Vaynman, S., M. E. Fine, and D. A. Jeannotte, "Isothermal Fatigue of Low Tin Lead Based Solder," *Metall. Trans.,* **19A,** 1988, pp. 1051–1056.
11. ———, "Prediction of Fatigue Life of Lead-Base Low Tin Solder," *Proceedings of 37th Electronic Components Conference,* IEEE 1987, pp. 598–603.
12. ———, "Isothermal Fatigue Failure Mechanisms in Low Tin Lead Based Solder," *Proc. Effect of Load and Thermal Histories on Mechanical Behavior Symp.,* P. K. Liaw and T. Nicholas, Eds., 1987, pp. 127–137.
13. Vaynman, S., and M. E. Fine, "Fatigue of Low-Tin Lead-Based and Tin-Lead Eutectic Solders," *Microelectronic Packaging Technology: Materials and Processes. Proc. of the 2d ASM International Electronic Materials and Processing Congress,* W. T. Shieh, Ed., ASM, 1989, pp. 255–259.
14. Rathore, H. S., R. C. Yih and A. R. Edenfeld, "Fatigue Behavior of Solders Used in Flip-Chip Technology," *J. Test. Eval.,* **1,** 1978, pp. 170–178.
15. Solomon, H. D., "Room Temperature Low Cycle Fatigue of a High Pb Solder (INDALLOY 151)," *Microelectronic Packaging Technology: Materials and Processes, Proc. of the 2d ASM International Electronic Materials and Processing Congress,* W. T. Shieh, Ed., ASM, 1989, pp. 135–146.
16. Wild, R. N., "Fatigue Properties of Solder Joints," *J. of Welding Research,* **51,** 1972, pp. 521s–526s.
17. Coffin, L. F., "A Study of the Effect of Cyclic Thermal Stress on a Ductile Metal," *Trans, ASME,* **76,** 1954, pp. 931–950.
18. Manson, S. S., "Fatigue: A Complex Subject—Some Simple Approximations," *Experimental Mechanics,* **5,**(7), 1965, pp. 193–226.
19. Eckel, J. F., "Influence of Frequency on the Repeated Bending Life of Acid Lead," *Proc. ASTM,* **51,** 1957, pp. 745–760.
20. Solomon, H. D., "Frequency Dependent Low Cycle Fatigue Crack Propagation," *Metall. Trans.,* **3,** 1972, pp. 341–347.
21. James, L. A., "Effect of Frequency upon the Fatigue Crack Growth of Type 304 Stainless Steel at 1000°F," *Stress Analysis and Growth of Cracks,* ASTM STP 513, ASTM, 1972, pp. 218–229.

22. Conway, J. R., J. T. Berling, and R. H. Stentz, "Strain Rate and Hold Time Saturation in Low Cycle Fatigue: Design Parameter Plots," *Fatigue at Elevated Temperatures,* ASTM STP 520, ASTM, 1973, pp. 637–647.
23. Manson, S. S., "The Challenge to Unify Treatment of High Temperature Fatigue—A Partisan Proposal Based on Strainrange Partitioning," *Fatigue at Elevated Tempertures,* ASTM STP 520, ASTM, 1973, pp. 744–782.
24. Sidney, D., and L. F. Coffin, "Low-Cycle Fatigue Damage Mechanisms at High Temperature," *Fatigue Mechanisms,* ASTM STP 675, ASTM, 1979, pp. 528–568.
25. Hales, R., "A Quantitative Metallographic Assessment of Structural Degradation of Type 304 Stainless Steel During Creep Fatigue," *Fat. Eng. Mater. Struct.,* **3,** 1980, pp. 339–356.
26. Baik, S., and R. Raj, "Mechanisms of Creep-Fatigue Interaction," *Metall. Trans.,* **13A,** 1982, pp. 1215–1221.
27. Tien, J. K., S. V. Nair, and V. C. Nardone, "Creep-Fatigue Interaction in Structural Materials," *Flow and Fracture at Elevated Temperatures,* Ed. R. Raj ASTM, 1985, pp. 179–214.
28. Raj, R., "Mechanisms of Creep-Fatigue Interaction," *Flow and Fracture at Elevated Temperatures,* ASTM, 1985, pp. 215–249.
29. Gohn, G. R., and W. C. Ellis, "The Fatigue Tests as Applied to Lead Cable Sheath," *Proc. ASTM,* **51,** 1951, pp. 721–740.
30. Solomon, H. D., and L. F. Coffin, "Effect of Frequency and Environment on Fatigue Crack Growth in A286 at 1100°F," *Fatigue at Elevated Temperatures,* ASTM STP 520 ASTM, 1973, pp. 112–122.
31. Forrest, P. G., *Fatigue of Metals,* New York, 1962, Pergamon Press.
32. Wareing, J., "Fatigue Crack Growth in a Type 316 Stainless Steel and a 20% Cr/25% Ni/Nb Stainless Steel at Elevated Temperature," *Metall. Trans.,* **6A,** 1975, pp. 1367–1377.
33. ______, "Creep-Fatigue Interaction in Austenitic Stainless Steel," *Metall. Trans,* **8A,** 1977, pp. 711–721.
34. Maiya, P. S., and S. Majumdar, "Elevated-Temperature Low-Cycle Fatigue Behavior of Different Heats of Type 304 Stainless Steel," *Metall. Trans.,* **8a,** 1977, 1651–1660.
35. Majumdar, S., and P. S. Maiya, "A Mechanistic Model for Time-Dependent Fatigue," *Eng. Mater. Technology,* **102,** 1980, pp. 159–167.
36. Day, M. F., and G. B. Thomas, "Microstructural Assessment of Fractional LIfe Approach to Low Cycle Fatigue at High Temperatures," *Metall. Sci.,* **13,** 1979, pp. 25–33.
37. Meleka, A. H., "Combined Creep and Fatigue Properties," *Metallurgical Reviews,* **7,** 1962, pp. 43–93.
38. Ellozy, A. R., P. M. Dixon, and R. N. Wild, "Combined Low-Cycle Fatigue and Stress Relaxation of Some Pb-In and 96/4 PbSn Alloy at Room Temperature," *Proc. of the 2d International Conference on Mechanical Behavior of Materials,* 1976, pp. 903–907.
39. Shah, H. J., and J. H. Kelly, "Effect of Dwell Time on Thermal Cycling of the Flip-Chip Joint," ISHM, 1970, pp. 3.4.1–3.4.6.

40. Agarwala, B. N., "Thermal Fatigue Damage in Pb-In Solder Interconnections," *23rd Annual Proceedings, Reliability Physics* IEEE/IRPS, 1985, pp. 198–205.
41. Lord, D. C., and L. F. Coffin, "Low Cycle Fatigue Hold Time Behavior of Cast Rene'80," *Metall. Trans.*, **4A** 1973, pp. 1647–1654.
42. Teranishi, H., and A. J. McEvily, "The Effect of Oxidation on Hold Time Fatigue Behavior of 2.25Cr-1Mo Steel," *Metall. Trans.*, **10A,** 1979, pp. 1806–1808.
43. Wells, C. H., and C. P. Sullivan, "Interaction Between Creep and Low Cycle Fatigue in Udimet 700 at 1400°F," *Fatigue at Elevated Temperatures,* ASTM STP 520, ASTM, 1973, pp. 59–74.
44. Sadananda, K., and P. Shaninian, "Hold Time Effects in High Temperature Fatigue Crack Growth in Udimet 700," *J. Mater. Sci.,* **13,** 1978, pp. 2347–2357.
45. Ostergren, W. J., "A Damage Function and Associated Failure Equations for Predicting Hold Time and Frequency Effects at Elevated Temperature, Low Cycle Fatigue," *J. of Testing and Evaluation,* JTEV, **4,** 1976, pp. 327–339.
46. Miller, D. A., R. H. Priest, and E. G. Ellison, "A Review of Material Response and Life Prediction Under Fatigue-Creep Loading Conditions," *High Temp. Mater. and Processes,* **6,** 1984, 155–194.
47. Hong, J. W., S. W. Nam, and K-T. Rie, "A Model for Life Prediction in Low-Cycle Fatigue with Hold Time," *J. Mater. Sci.,* **20,** 1985, pp. 3763–3779.
48. Vaynman, S., "On Fatigue Life Prediction Under Conditions of Tensile Hold," *Scripta Metall.,* **21,** 1987, pp. 1385–1387.
49. Skelton, R. P., *Fatigue at High Temperature,* London and New York, Applied Science Publishers, 1983.
50. Berriche, R., S. Vaynman, M. E. Fine, and D. A. Jeannotte, "The Effect of Environment on Fatigue of Low Tin Lead Base Alloys," *Proc. 3d Ann. Electronic Packaging and Corrosion in Microelectronics Conf.,* M. E. Nicholson, Ed., ASM International, 1987, pp. 169–174.
51. Berriche, R., "Environmental Effects on Time Dependent Fatigue of Low Tin-High Lead Solder," Ph.D. Thesis, Northwestern University, Evanston, Il. 1989.
52. Greenwood, P. J., T. C. Reiley, V. Raman, and J. K. Tien, "Cavitation in a Pb/Low-Sn Solder During Low Cycle Fatigue," *Scripta Metall.,* **22,** 1988, pp. 1465–1468.
53. Solomon, H. D., "The Influence of Hold Time and Fatigue Cycle Wave Shape on the Low Cycle Fatigue of 60/40 Solder," *38th Electronic Components Conference,* IEEE, 1988, pp. 7–12.
54. Vaynman, S., "Effect of Strain Rate on Fatigue of Low-Tin Lead-Base Solder," *39th Electronic Components Conference,* IEEE, 1989, pp. 273–276.

12

A Damage Integral Methodology for Thermal and Mechanical Fatigue of Solder Joints

Che-Yu Li, R. Subrahmanyan, J.R. Wilcox, and D. Stone

The differential thermal expansion of the connected members of a soldered assembly during temperature cycles produces mechanical displacements in solder joints. The resulting cyclic stresses are the driving force for damage processes that eventually cause fatigue failures. Conventional approaches to solder joint fatigue rely on correlations between fatigue life of the solder joint and some measurable or calculable parameter characteristic of the fatigue loading. These approaches often require extensive fatigue data to establish such correlations. Furthermore, their validity is often limited to the range over which the parameters in the correlation are uniquely related to the stress in the solder joint. The plastic strain range is an example of such a parameter. For a given plastic strain range, the stress in the solder joint may vary, however, depending on the displacement rate, such that the correlation between the plastic strain range and the number of cycles to failure is not unique.

A variety of damage processes such as wedge cavitation,[1] grain boundary voiding,[2] and crack growth[3,4] have been found to be operative. The accumulation of fatigue damage is manifested as a reduction in the load bearing area of the solder joint, which eventually leads to failure. The task of describing the accumulation of fatigue damage is essentially two-fold. First, the time profile of the stresses due to the applied displacement profile must be known. Second, the rate of the damage process as a function of stress must be estimated. The fatigue life of the solder joint may then be determined by the integration of the rate of damage process to some specified failure criterion. This is the basic concept of a damage-integral approach.[5,6]

The total applied displacement is accommodated by both the solder joint and other connecting members of the soldered assembly. The connecting members usually deform elastically; hence the load supported by the solder joint is influenced by their compliance. The solder joint, however, can also deform inelastically. Therefore, the stress in the solder joint can depend on the applied displacement rates and temperature. The damage processes mentioned above determine the rate of fatigue damage and its dependence on the stress state, temperature, and environment. While the rate of internal damage processes such as wedge cracking and cavitation is determined by the stress state and temperature alone, the growth of surface cracks is additionally influenced by the external environment.

As discussed previously, when the stress is a unique function of a particular parameter, such as plastic strain range, it may be possible to correlate the fatigue life of the solder joint in terms of that parameter. A good example is the Coffin-Manson approach,[7,8] which correlates isothermal mechanical fatigue life to the plastic strain range. This approach is often used to describe low cycle fatigue behavior of copper and steel at room temperature. The success of this approach to such applications can be attributed to the fact that at low homologous temperatures the strain rate dependence of stress is small; therefore the stress profile is essentially determined by the applied plastic strain range. For these conditions it is possible to show that the accumulated fatigue damage per cycle is a nearly single-valued function of the plastic strain range.[9,10]

At temperatures above roughly a third of the melting point, typical in solder joint fatigue, the stress will not only depend on the plastic strain but also on the plastic strain rate. Hence, the fatigue life becomes sensitive to the time profile of the applied displacement. In fact, it has been experimentally shown[11–13] that for such conditions the number of cycles to failure depends on the frequency and hold time at the displacement limits. For a given plastic strain range, decreasing frequency or increasing the hold time can result in a smaller number of cycles to failure.

Thermal fatigue adds another level of complexity. In addition to the expansion mismatch-induced displacements, the temperature can change dynamically as a function of time. Since both the rate of inelastic deformation and the various damage processes are temperature dependent, the calculation of the stress and the damage rates must further include the effects of temperature cycles. During such conditions, a Coffin-Manson-type relation in its original form may not be able to adequately characterize the low cycle fatigue behavior (e.g. reference[14]).

A damage integral approach, on the other hand, provides a convenient framework to account for the aforementioned complexities. The time profile of the stress and the damage rates are explicitly considered in such an approach. Since the time variation of both the stress and the damage rate for arbitrary applied displacement profiles can be complex, the use of damage-integral methods often requires numerical computations.

Damage integral approaches were first applied to isothermal "high cycle fatigue" (see e.g., references[15,16]). During high cycle fatigue, since the bulk of the specimen deforms elastically, the stress in the material is well defined. Furthermore, the failure occurs only by crack growth. The phenomenon of crack growth under mostly elastic loading has been extensively studied and is well described by a fracture mechanics approach.

The application of the same approach to low cycle fatigue, however, is not straightforward. This is to a large extent because the description of the stress at temperatures high relative to the melting point is not simple. They are a sensitive function of the specimen microstructure and composition, stress, and temperature. High temperature phenomena such as grain boundary sliding,[2,17,18] grain growth and the coarsening of dispersed phases,[19] and solute strengthening[20] can significantly influence the inelastic deformation behavior. In addition, the relative importance of individual damage processes and their interactions, particularly for nonmonotonic loading, is not well established.

In this chapter, some basic concepts involved in the application of a damage integral approach to solder joint fatigue are discussed. We shall show that a phenomenological approach to the description of the inelastic deformation properties, such as a state variable approach, can be successfully applied to the calculation of the solder joint stress profile. An equivalent approach is needed to calculate the fatigue damage. Some preliminary efforts toward this latter direction are presented here.

12.1 INELASTIC DEFORMATION AND STRESS CALCULATION

12.1.1 Governing Equation for Solder Stress

The solder joint can deform both elastically and inelastically. For any given displacement profile, the stress in the solder joint is dependent on the compliance of the assembly (including the solder joint) and the inelastic deformation properties of the solder joint. We can write the total displacement rate as the sum of the total elastic displacement rate (in the solder joint and assembly) and the displacement rate corresponding to the inelastic deformation of the solder joint. A governing equation for the load P supported by the solder joint can be written as[21]

$$\frac{d}{dt}\left(\frac{P}{S}\right) = \dot{X} - \dot{X}_i \qquad (12\text{-}1)$$

where $\dot{X}$ is the total displacement rate and $\dot{X}_i$ is the inelastic displacement rate. S, the total stiffness, relates the total elastic displacements to the load. It incorporates the stiffness of both the solder joint and the connecting members and is dependent on their geometry, moduli, and the loading mode. It may be experimentally measured or calculated.

For the case of shear loading, the solder joint height h and area A remain constant. Equation (12-1) can be rewritten in terms of the shear stress rate $\dot{\tau}$ and the inelastic shear strain rate $\dot{\gamma}$ as

$$\dot{\tau}A = S(\dot{X} - \dot{\gamma}h) \tag{12-2}$$

Consider the thermal cycling of a solder joint as shown schematically in Figure 12-1. The applied displacement rate for this case arises from the differential thermal expansion between the members of the soldered assembly; therefore $\dot{X}$ is given, for example, by $\dot{X} = \Delta\alpha\dot{T}\xi$ where $\Delta\alpha$ is the difference in thermal expansion coefficients, $\dot{T}$ is the heating or cooling rate, and ξ is the distance from the neutral point.

The calculation of the stresses in the solder joint for an arbitrary loading profile requires the integration of Equation (12-2). The dependence of stress on the inelastic strain rate and temperature must be known in the form of appropriate constitutive relations to facilitate the integration. Experiments such as displacement controlled cyclic loading, creep, and load relaxation are often used to characterize the inelastic deformation properties of solder joints to establish required constitutive relations. Inadequate constitutive relations can greatly increase the margin of error in the stresses and, therefore, the damage rates in the solder joints.

The inelastic deformation behavior of the solder joint can be significantly affected by the composition and microstructure of the solder joint. These are in turn changed by the cooling rates during solder joint reflow and the dissolution of the adjoining metallization. The thermal history of the solder joint can significantly influence the stability of the solder joint microstructure.[4,22,23] In addition, the deformation properties can depend on the contribution of multiple mechanisms. The relative contribution of each varies depending on the stress

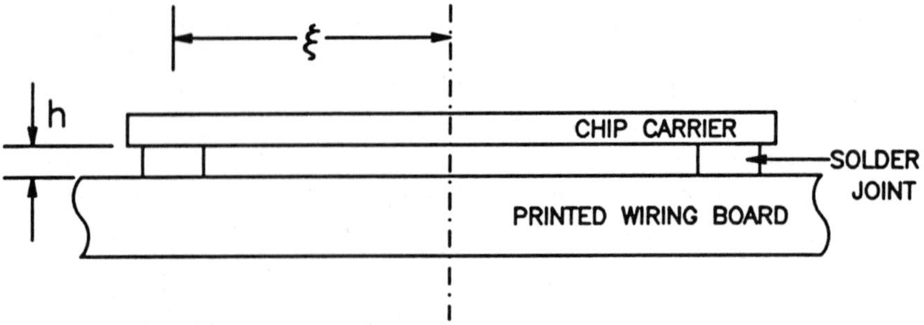

Figure 12-1 Schematic of leadless surface mount solder joint.

and temperature as well as microstructure and composition. A comprehensive theoretical basis is yet to be developed for the establishment of the required constitutive relations.

12.1.2 Inelastic Deformation Behavior and Constitutive Relations

In addition to solder joint-specific concerns discussed above, some basic issues remain to be addressed for the development of appropriate constitutive equations for inelastic deformation at high homologous temperatures. At these temperatures thermally activated structural and compositional changes usually accompany inelastic deformation. It is, therefore, necessary to account for the effect of such thermo-mechanical history on the deformation properties. On the other hand, the need for reducing experimental effort has always been of paramount importance. A phenomenological framework is therefore needed to deduce constitutive relations based on a minimum amount of experimental effort.

A state variable approach originally proposed by Hart[24] for macroplastic deformation is preferred in the present work. Hart's model was later extended by Jackson et al.[25] to describe transient deformation phenomena such as microplasticity and the Bauschinger effect. Although the initial models treated inelastic deformation due to dislocation-based processes, other deformation processes such as grain boundary sliding[26] can also be included in a state variable treatment. It has been shown[17] that a variety of test results involving inelastic deformation in solders and solder joints, obtained by load relaxation, creep, and displacement-controlled cyclic loading experiments can be represented in a unified manner using a state variable concept. This demonstrates the applicability and versatility of constitutive relations based on Hart's concept even in the presence of the various complexities discussed previously.

A key feature of Hart's formulations for inelastic deformation is the presence of a microstructural and mechanical state of the material. This state is uniquely specified by the current deformation properties, for example, by the isothermal stress-inelastic strain behavior. The deformation properties can be scaled from one mechanical state to another at a given temperature or from one temperature to another at a given mechanical state. It follows automatically that experimental effort required to determine the deformation properties over a wide range of states and temperatures can be reduced.

At high homologous temperatures deformation processes involving grain boundary phenomena will be active as will dislocation-based processes in the grain matrix. These phenomena include grain boundary sliding and migration. We have assumed that the inelastic deformation of the solder joint, represented in terms of tensor invariant, can be described by the rheological model shown in Figure 12-2. This model assumes that steady-state macroplastic deformation and transient deformation are separable. By transient deformation we refer to

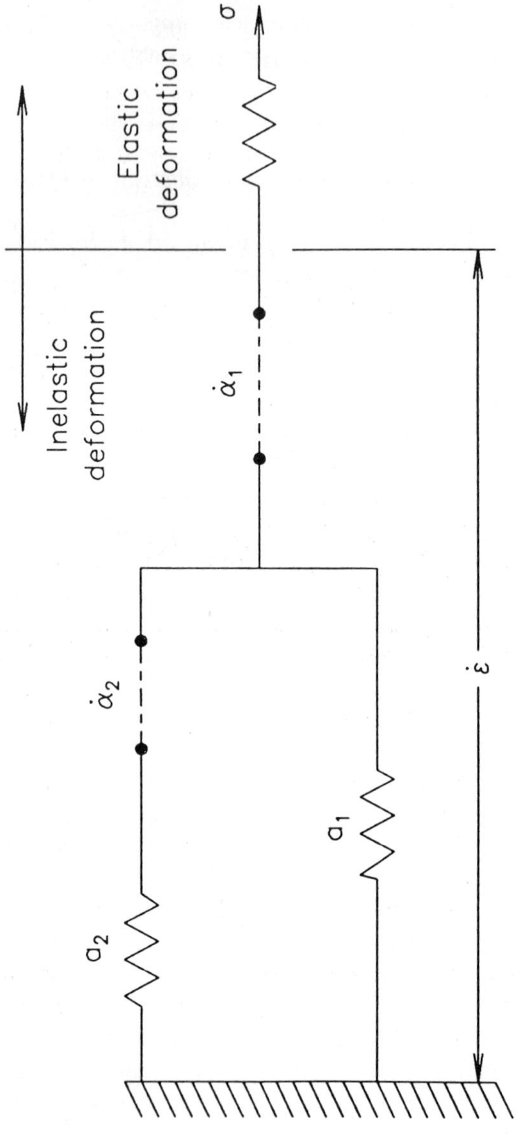

Figure 12-2 A rheological model representing the deformation behavior of solders and solder joints.

both anelastic (recoverable but time dependent) deformation and microplasticity.[25] Furthermore, for the present purposes, it is assumed that these elements represent the combined contribution of the grain boundary and grain matrix processes. The inelastic strain rate due to an applied stress σ is represented by $\dot{\varepsilon}$. The macroplastic deformation behavior is represented by the element $\dot{\alpha}_1$. The springs a_1 and a_2 represent the anelastic deformation; M_1 and M_2 are the anelastic moduli. The element $\dot{\alpha}_2$ represents the microplastic element.

The operation of the rheological model may be illustrated based on a stress-strain loop obtained during isothermal displacement controlled cycles (Figure 12-3). Upon loading at point A, only the elastic spring and the anelastic springs are active. Therefore, the initial linear slope represents the combined elastic and anelastic moduli. Initial deviation from linearity occurs when the microplastic element becomes active (at point B). Further increasing the stresses will eventually cause the solder joint to flow macroplastically (point C). A similar behavior is observed for loading in the reverse direction (along points D, E, F, and A).

Accurate and comprehensive experimental data are required to establish the constitutive relations to describe deformation behavior such as that shown in

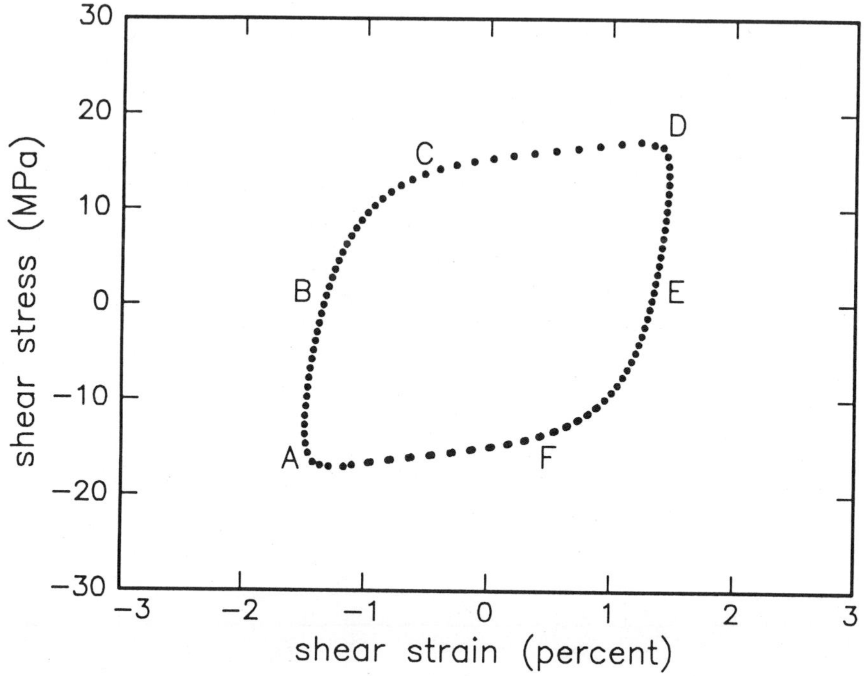

Figure 12-3 Measured room temperature stress-strain loop for surface mount type solder joints[21] during isothermal cycling. This loop generated using a triangular wave form with a cyclic period of 100 sec.

Figure 12-3. In the following, an experimental method is considered that has been useful in characterizing macroplastic deformation properties of solders and solder joints over a wide range of loading conditions. We shall show that even with the presence of several deformation mechanisms, simple and yet accurate phenomenological constitutive relations can be established. The considerations involved in the deconvolution of the individual deformation mechanisms involved are discussed in references[27,17] and are outside the scope of this chapter.

Wilcox[17] has used load relaxation experiments to determine isostructural (constant mechanical state) deformation properties of the macroplastic element. The bulk eutectic alloy was prepared using high purity Sn and Pb. A cylindrical tensile sample, with a gauge length and area of 20.5 mm and 32.2 mm^2, respectively, was used to perform load relaxation and constant load creep experiments at several temperatures. Initial tests were performed after ageing at the maximum temperature of interest (125°C) for 24 h. Subsequent tests were performed at progressively decreasing temperatures.

In each load relaxation experiment, the sample was typically loaded to an additional 2 percent strain. The cross head was then fixed. The load and the sample displacements, measured as a function of time, were used to determine

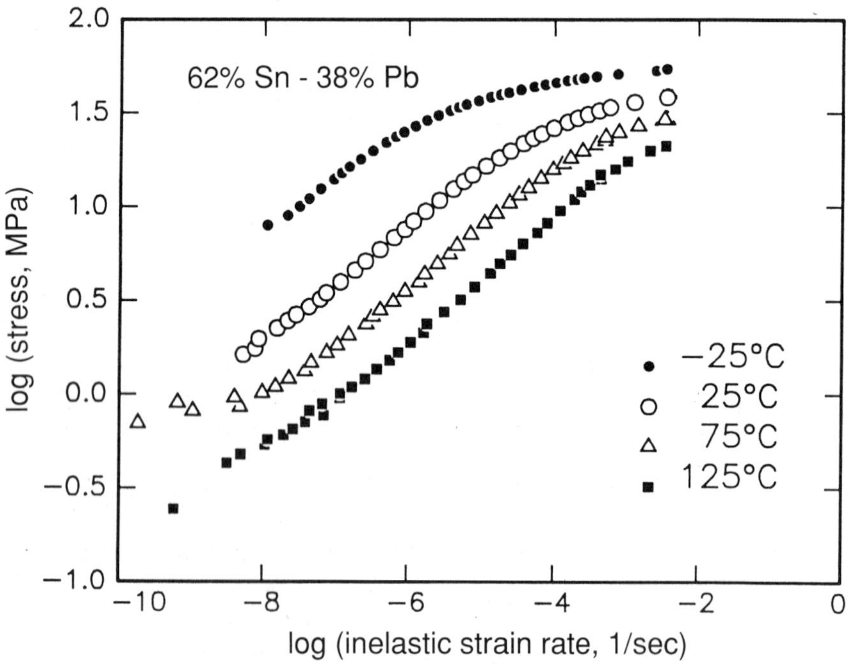

Figure 12-4 Inelastic deformation behavior of bulk eutectic tin-lead solder measured by tensile load relaxation experiments.

the stress and the inelastic strain rate of the solder specimen. The significance of stress-strain rate relations obtained from the load relaxation experiments is contingent on the stability of the microstructure throughout the relaxation run. The validity of a constant structure assumption was confirmed by reloading relaxation experiments.[28,17] Except at −25°C, the extent of work hardening during loading was found to be negligible. The measured constant structure stress-inelastic strain rate data at various temperatures are presented in Figure 12-4.

The data in Figure 12-4 represent the dependence of the macroplastic flow stress on the inelastic strain rate. It is seen that at each temperature, the slope of the stress-inelastic strain rate curves is not constant. It is small at high strain rates, then increases at intermediate strain rates, and decreases at small strain rates. The specimen surface after completion of all the relaxation experiments showed steps at the boundaries of the dispersed phase colonies and at the Pb-Sn phase boundaries, suggesting boundary sliding at these sites. It was concluded[17] that a steady-state combination of boundary migration and sliding contributes to the observed sigmoidal shape of the stress-inelastic strain rate curves. It is important to recognize that the shape of the stress-inelastic strain rate curves in Figure 12-4 is essentially the same for almost the entire temperature range. This observation is an important basis for the temperature scaling approximations made in this work and in Hart's phenomenological approach.

For the present purposes, we shall describe the macroplastic deformation behavior represented in Figure 12-4 by the macroplastic element in Figure 12-2. The macroplastic strain rate $\dot{\alpha}$ is expressed as a function of the stress σ through a constitutive relation of the form

$$\ln\left(\frac{\sigma^*}{\sigma}\right) = \left(\frac{\dot{\varepsilon}^*}{\dot{\alpha}_1}\right)^{\lambda} \tag{12-3}$$

where σ^* is the hardness parameter of the grain matrix, ε^* is the rate parameter, and λ is a shape parameter. As seen by Kashyap and Murty[29] in their creep data, different activation energies were found for different regimes of strain rate and temperature. Using an activation energy Q_d of 46 KJ/mol, the stress-strain rate curves in Figure 12-4 have been translated to 25°C to illustrate the temperature scaling described previously [$\Delta \ln(\dot{\varepsilon}) = -(Q_d/R)\,\Delta\,(1/T)$ for a given mechanical state]. The result, shown in Figure 12-5, indicates reasonable temperature scaling. This temperature scaling is facilitated by the fact that work hardening has been found to be unimportant[17] in the temperature range involved for this solder composition and microstructure. Simplification of the description of the inelastic deformation behavior over the range of measured data is thus possible although a variety of deformation mechanisms and related phenomena are involved.

It is similarly possible, using isothermal load relaxation and displacement controlled cycling experiments, to determine the constitutive relations for the

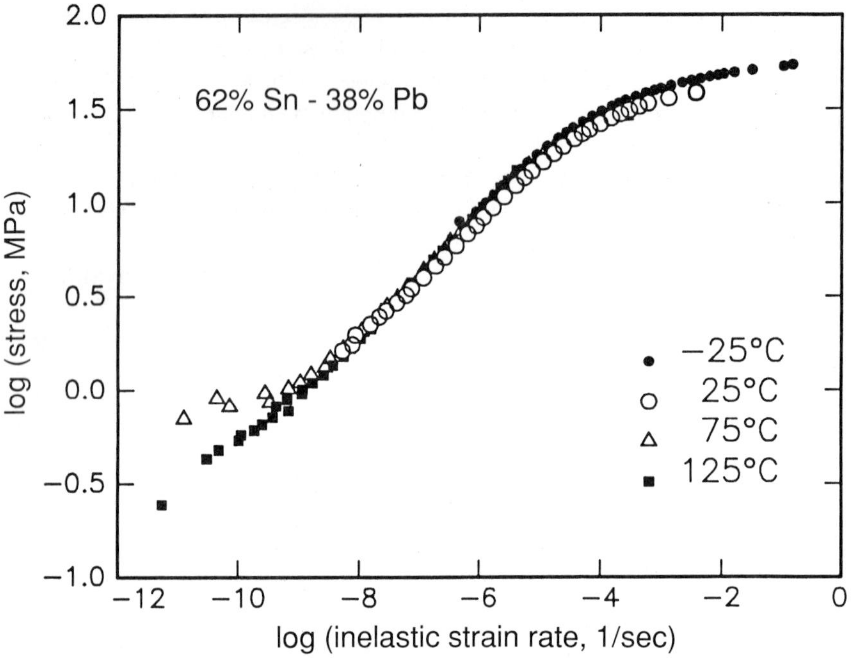

Figure 12-5 Stress inelastic strain rate data for eutectic tin-lead solder (Figure 12-4) shifted to 25°C from the actual test temperatures using a thermal activation energy of 46 KJ/mol.

anelastic spring elements and the microplastic element in Figure 12-2. A detailed description of experimental methods is given in the reference.[28]

12.1.3 Stress Calculation

The applicability of the phenomenologically based constitutive relations developed above can be demonstrated by solder joint stress calculations during isothermal displacement controlled cycling. Solomon[30,31] has reported such data for 60% Sn-40% Pb solder joints obtained under total displacement control and plastic strain limits. The solder joints, which connect brass or copper loading blocks, have a nominal area and height of 32.25 mm^2 and 0.19 mm, respectively. He has reported stress-strain loops at several temperatures and plastic strain ranges. The measured stress loops were simulated by numerical integration of Equation (12-2) and the constitutive relations described above. The shear stresses and inelastic shear strain rates are converted to their tensor invariants as $\sigma = \sqrt{3}\tau$ and $\dot{\varepsilon} = \dot{\gamma}/\sqrt{3}$. Details of the stress calculation are provided in the reference.[21] Figure 12-6 compares the results of the numerical integration to the measured data at two temperatures 150 and 35°C. An adequate description of isothermal stress cycles appears to be possible. In a separate work[21,10] it was shown that it may be possible to apply the constitutive relations similar to those measured in bulk eutectic solder to small solder joints. This result is consistent with the stress calculations described here.

It is also possible to use the same constitutive relations to calculate the stresses during a thermal cycle. To simulate the thermal stress cycles in surface mount solder joint used by Hall in Figure 12-7, his bimetallic strip approximation is used (see Chapter 10 and the reference[32]) to determine the assembly stiffness $S = 0.28$ MN/m. The loading parameters are chosen to match his experimental conditions.[34] Equation (12-2) is numerically integrated along the thermal profile including the peak temperature holds. The resulting stress simulation is that shown by the solid line in Figure 12-7. The calculated stress shows good agreement with the experimental data at all temperatures.

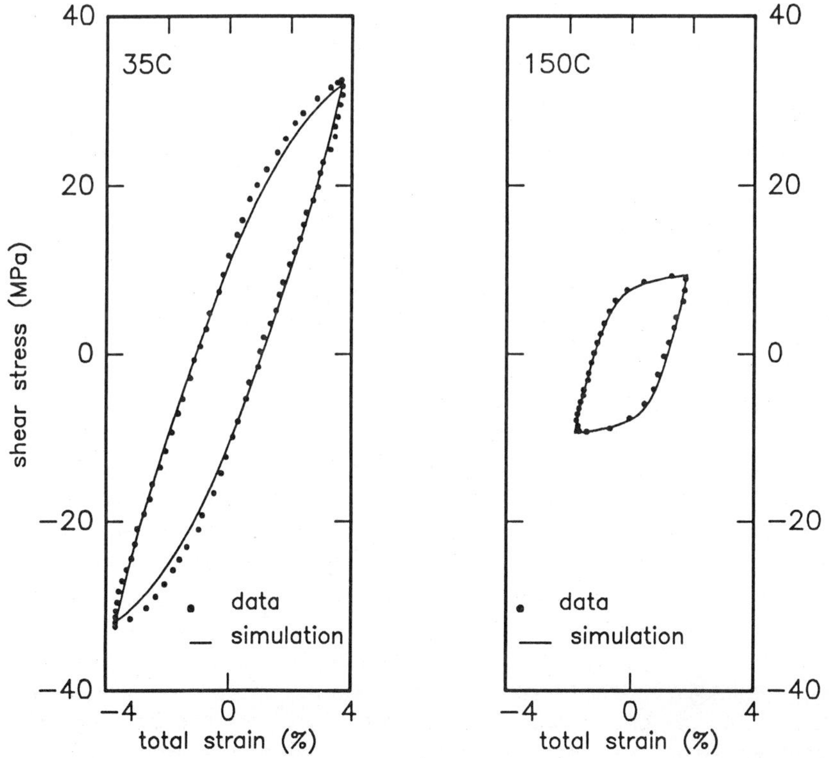

Figure 12-6 Isothermal shear stress-shear strain loops from reference [31]. The solid line is the simulation of the stresses based on the assumed flow Equation (12-3).

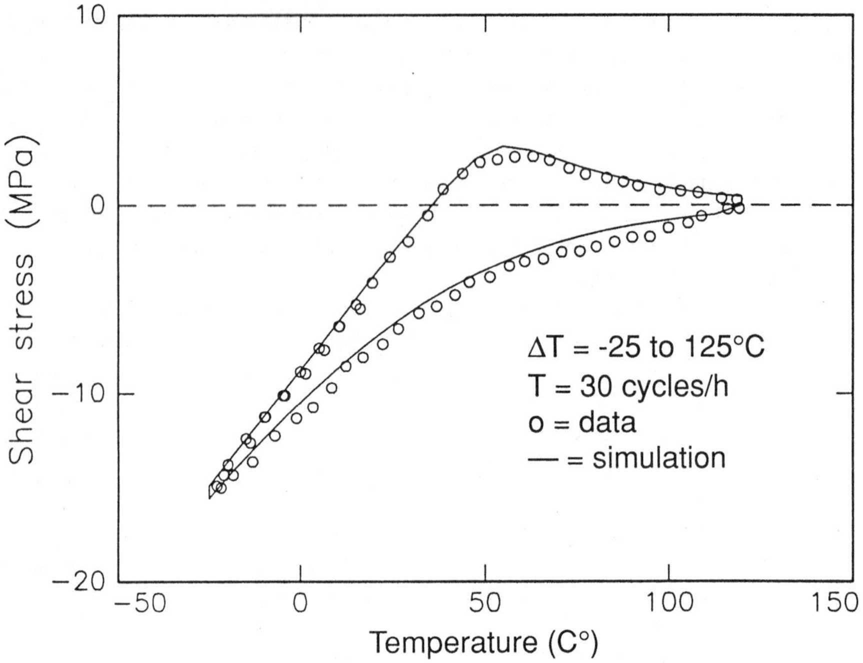

Figure 12-7 Thermal stress cycles measured by Hall.[32] The solid line is the simulation.

It must be emphasized that damage rates, such as fatigue crack growth, are approximate power functions of the applied stress, with exponents that are large. Errors in stress calculations can therefore amplify the errors in the estimated damage rates and the fatigue life. Accurate calculation of the stress during a thermal or mechanical cycle is, therefore, the first task in the development of an integral approach to low cycle fatigue of solder joints. It was shown that a phenomenological framework based on a state variable concept can be used to accurately estimate the stresses during both isothermal and thermal cycling of solder joints. The development of a similar phenomenological approach that can be applied to characterize the damage rate in the solder joint will be examined next.

12.2 DAMAGE RATE FORMULATION

12.2.1 Damage Mechanisms

Common damage mechanisms that are observed during low cycle fatigue and at high homologous temperatures are crack growth, grain boundary cavitation, and grain boundary wedge cracking. Sliding at phase and grain boundaries is operative at these temperatures. These phenomena produce stress concentrations at sites along the boundaries where flow incompatibility occurs. Cavity nucleation and wedge cracking are processes that tend to relieve such stress concentrations. The growth of wedge cracks and cavities act together with the growth of a surface crack, producing the fatigue damage that eventually leads to failure of the solder joint. Unlike inelastic deformation studies, the description of damage rates for the above processes and their interactions is not well developed, particularly for high homologous temperature fatigue.

Boundary cavitation and wedge cracking are related phenomena. Grain boundary cavitation is a diffusive process that occurs at the stress concentration sites along the grain or phase boundary.[33] Wedge cavitation also occurs because of the stress concentrations but at the grain boundary triple points. The latter process is a more important phenomenon at strain rates higher than those corresponding to boundary cavitation. Note that these damage mechanisms are driven by grain and phase boundary sliding. The deconvolution of the observed deformation behavior to deduce both the relative contributions of the various boundary phenomena and the operative stress-strain rate windows is possible. Such an exercise will, however, not be simple for solder joints because of the presence of complicated microstructures.

Since the failure of a solder joint usually involves the growth of a major crack, it may be possible to represent the combined effects of the various damage processes by the rate of growth of such a major crack. The parameters of the crack growth rate formulation will then contain the synergistic effects of all the damage processes. The description of the rate of crack growth is phenomenologically based on the applied stress intensity factors. While such an approach has been successfully used to describe the crack growth behavior under mostly elastic loading, the application of the same approach for the present interest requires extensive examination.

Even without the presence of other damage processes, the use of the fracture mechanics approach to describe crack growth under plastic loading is a controversy in itself. In the following, an attempt at a phenomenological description of fatigue damage under plastic loading will be discussed for the purpose of stimulating further work in such a direction. The nominal stress intensity factor K in the crack growth law should be considered an effective value. Although nominally in shear, there will always be a mode I contribution to this effective K, the extent of which depends on the exact loading conditions. It is further assumed that the stress contained in such stress intensity formulations is uniquely related to the operative bulk stress that is the driving force for the nucleation and growth of wedge cracks and boundary cavities. Thus the parameters in the crack growth rate formulation will reflect those for the growth of the major crack as well as the synergistic effects of all the damage processes and the environment.

12.2.2 A Phenomenological Formulation for Crack Growth Rates

The damage rate that results from the stress is calculated according to a phenomenological crack growth rate formulation.[15,16] In general, crack growth rate is a function of the stress intensity factor, the chemical environment, and the temperature. The fatigue crack propagation rate is expressed in terms of a nominal stress intensity factor K such that

$$\frac{dc}{dt} = \nu^* K^r \tag{12-4}$$

where dc/dt is the time rate of crack growth. The nominal stress intensity factor is given by $K = \tau_N \sqrt{c}Y(c)$ where c is the instantaneous crack length and $Y(c)$ is a geometric factor, τ_N is the nominal stress based on the initial loading area, and r is a crack growth exponent that can be considered constant over small ranges of K. The rate constant ν^* is presumed to have the physical significance discussed previously. It has an Arrhenius temperature dependence with an activation energy Q_c and a preexponential ν_0^*.

Extensive and good quality experimental data are needed to establish and verify the phenomenological framework. The parameters in Equation (12-4) can be deduced from isothermal fatigue data of Solomon[30,31] for 60% Sn-40% Pb solder joints. He has reported the peak load versus number of cycles N for cracking solder joints at several temperatures and plastic strain ranges. If the solder flow stress and the uncracked area remains constant and the inelastic strain is sufficiently large to achieve macroplastic flow during every cycle, the peak load directly reflects the uncracked area of the solder. Two linear crack fronts originating from opposite ends are assumed (see Solomon[30]). Then, we can calculate the crack length and crack growth per cycle from the published peak load drop versus cycles data. If the crack length change per cycle is small compared to crack length, the nominal stress and the stress intensity factor K are nearly constant in any cycle. Then, a conversion to time rates of crack growth $dc/dt = \nu(dc/dN)$ is possible ($1/\nu$ is the time period).

Since the load and crack length are now known as a function of cycles N, the nominal stress intensity factor can be estimated for any cycle. The geometry factor $Y(c)$ is assumed to be unity. When the plastic strains are large, most of the cycle is spent close to the peak stress intensity factor; hence the crack growth rate estimated approximately corresponds to the peak stress intensity factors. Thus the crack growth rate can be estimated as a function of the stress intensity factor. Repeating this procedure for isothermal load drop data for large plastic strain ranges at several different temperatures, we can deduce the relationship between dc/dt and K. These data are plotted in as the symbols in Figure 12-8.[5,6]

Figure 12-9 plots the crack growth rates versus reciprocal temperature at a constant K ($=1$ MPa$\sqrt{\text{m}}$). From this plot the activation energy for crack growth

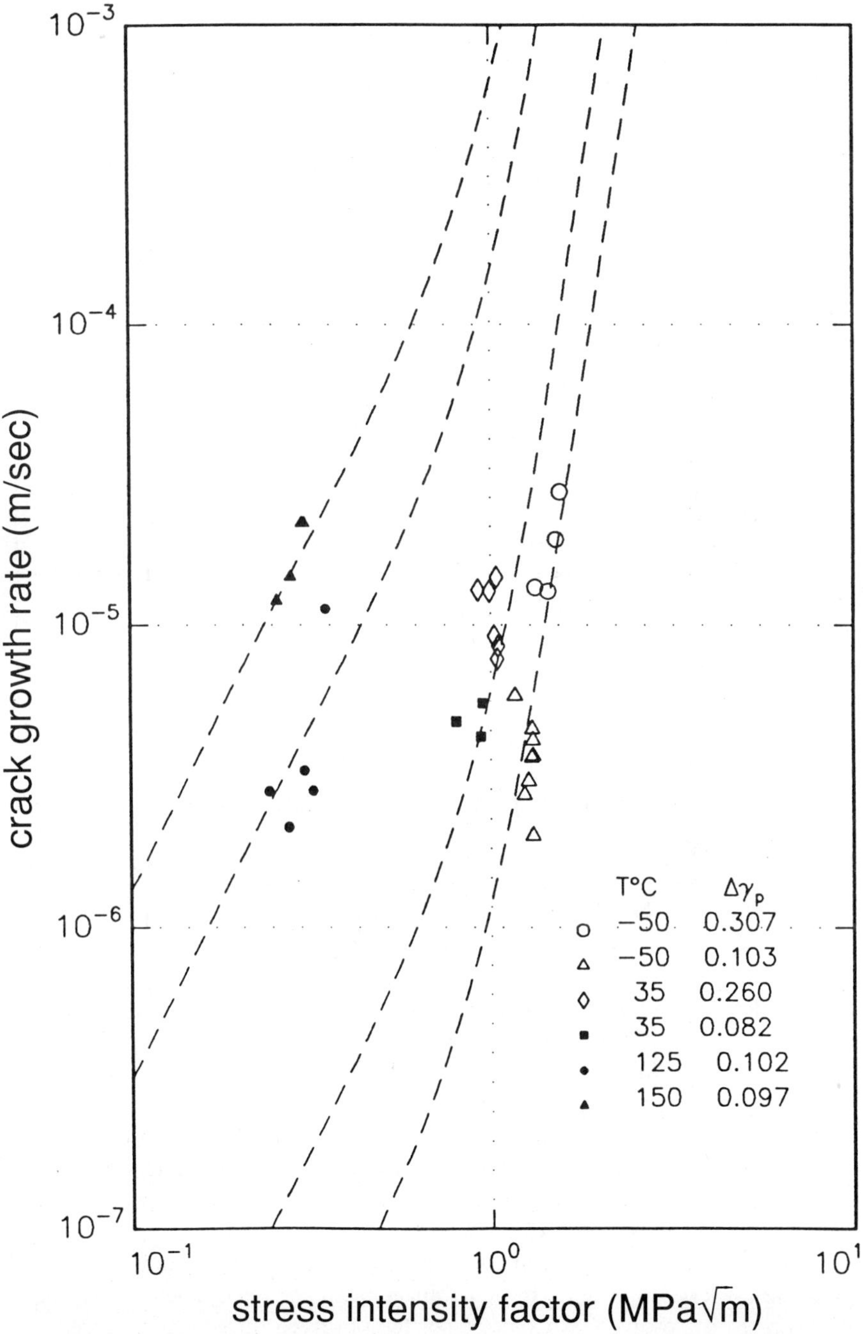

Figure 12-8 Deduced crack growth behavior as a function of temperature and nominal stress intensity factor. The dashed lines represent typical crack growth behavior. These lines are shifted vertically through an activation energy Q_c.

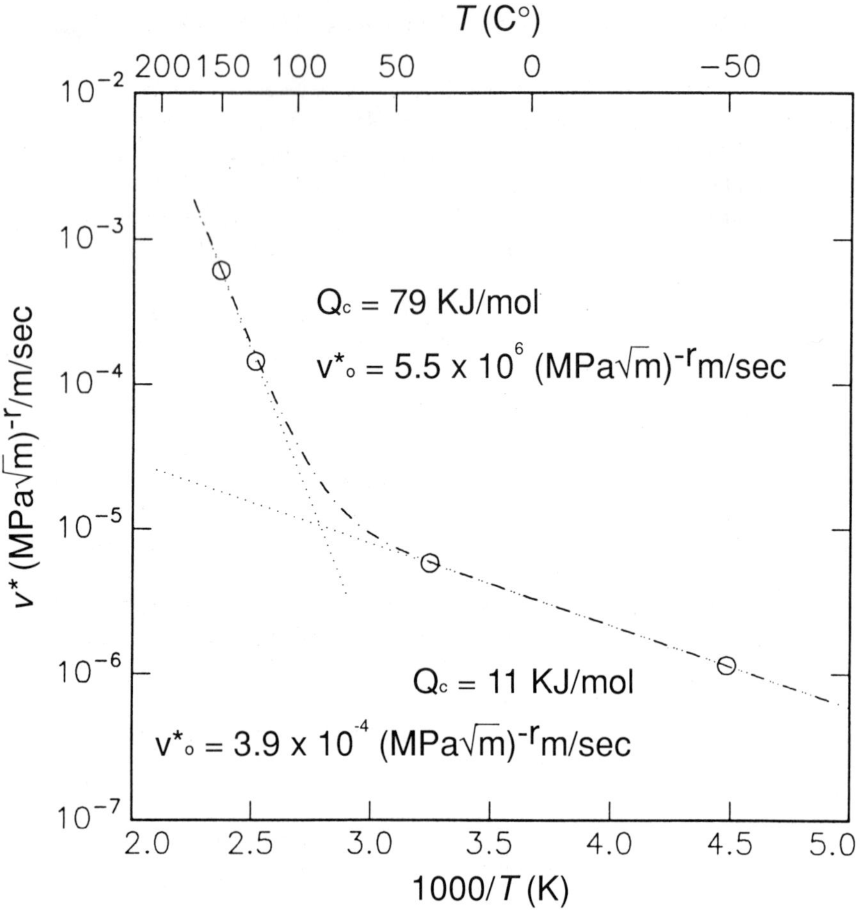

Figure 12-9 Temperature dependence of crack growth deduced at $K = 1$ MPa$\sqrt{\text{m}}$ indicating that at least two regimes of crack growth exist, one above 100°C and one below 100°C.

Q_c and the athermal rate parameter v_0^* can be determined. Different crack growth behavior is observed above and below approximately 100°C implying that at least two regions of damage behavior may be operative. For T greater than about 100°C, $Q_c = 80$ KJ/mol and $v_0^* = 5.5 \times 10^7$ ($\text{MPa}^{-2.3}$ $\text{m}^{-0.15}$ s^{-1}) while below 100°C $Q_c = 11$ KJ/mol and $v_0^* = 3.9 \times 10^{-5}$ ($\text{MPa}^{-2.3}$ $\text{m}^{-0.15}$ s^{-1}). Solomon has observed intergranular failures at high temperatures while at lower temperatures the crack path appears to be both transgranular and intergranular. However, it remains to be resolved[6] whether or not the observed difference is caused by a change in crack path due to damage mechanisms such as boundary cavitation or caused by interaction of crack growth process with the chemical environment or both.

12.3 DAMAGE INTEGRATION AND FAILURE CRITERION EFFECTS

It has been demonstrated that with good constitutive relations and accurate experimental data an adequate calculation of the stress in the solder joint is possible. A phenomenological crack growth rate relation that describes the damage rates for a range of temperatures and stresses has also been established. The application of these considerations to thermal fatigue of solder joints will now be examined.

Separating variables and integrating Equation (12-4) over the fatigue life,

$$\int_{c_i}^{c_f} \frac{dc}{(Y(c)\sqrt{c})^r} = \int_0^{t_f} dt\, \nu_0^* \exp\left(-\frac{Q_c}{\mathrm{RT}}\right) \tau_N^r \qquad (12\text{-}5)$$

where c_i and c_f are the initial and final crack length, respectively and t_f is the time to failure. Each side of Equation (12-5) is a measure of the total damage at failure, one in terms of the crack length and the other in terms of time. For a given combination of c_i and c_f, the crack length integral on the left will be a constant. Furthermore, if the exponent r is greater than 2, this integral is essentially independent of the final crack length c_f and for a given solder joint design, this integral will, in the limiting case, depend principally on the initial flaw size c_i.

For some applications[5] where the damage per cycle is nearly constant, a reasonable simplification is to consider the fatigue damage over a single cycle. The cycles to failure N_f is then given by

$$\mathcal{F} = N_f \int_0^{1/\nu} dt\, \nu_0^* \exp\left(-\frac{Q_c}{\mathrm{RT}}\right) \tau_N$$

where $\mathcal{F}$ is the crack length integral, and $1/\nu$ is the time period. This single cycle approximation is valid, for example, where the changes in nominal stress profile are small over most of the fatigue life.

12.3.1 Thermal Fatigue Life Estimation

As a numerical illustration of the previous considerations we shall perform sample calculations for a solder joint assembly at thermal cycle conditions for which the thermomechanical fatigue life has been measured. Such data is available on 60% Sn-40% Pb solder joints. Sherry and Hall[14] have measured the thermomechanical fatigue life of SMT solder joints that connect a ceramic chip carrier to an epoxy-glass printed wiring board. The assembly geometry is similar to the schematic in Figure 12-1. They have reported the cycles to failure for temperature cycles with minimum temperature $T_{min} = -20°C$ or $T_{min} = 30°C$. These tests were performed for various temperature ranges ΔT at a constant cyclic frequency of 58 cycles/day. The measured fatigue data are reproduced in Figure 12-10 as open symbols. These data vividly illustrates that the temperature range, which is a measure of total thermal displacement range, and fatigue life are not uniquely related.

The assembly details of these solder joints have been reported by Clech and Augis.[34] The thermal expansion difference $\Delta \alpha$ is 3.5 ppm/°C. The temperature gradients in the solder joint are assumed to be negligible. The solder joint height $h = 200$ μm and the area $A = 0.155$ mm^2. The elastic moduli of the ceramic and the circuit board are 255 and 12.3 GPa, respectively. Other assembly details are listed in Appendices B and C of reference.[34] The approximate thermal profiles for these tests are also given in reference.[34] The $T_{min} = -20°C$ profiles have nearly equal dwell times at the minimum and maximum temperatures while the $T_{min} = 30°C$ profiles have longer dwells at T_{min}. The temperature profiles for each simulation are chosen to match these measured profiles.

The details of the damage integrations are provided in reference.[5] The stress profiles are calculated in the same way as that for the data in Figure 12-7. The

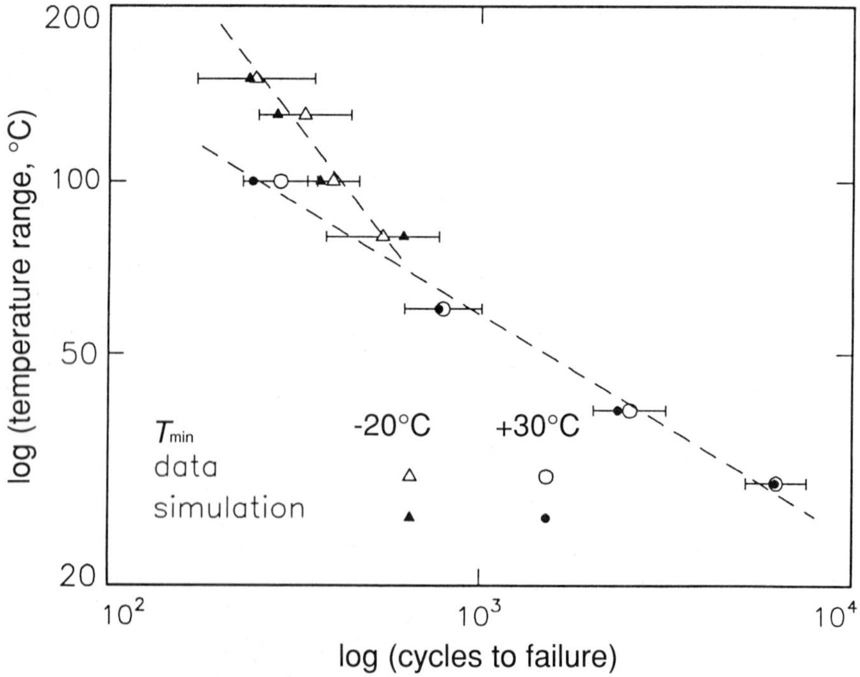

Figure 12-10 Comparison of measured thermal fatigue data for leadless surface mount solder joint with simulations. Frequency is 58 cycles/day.

results of the numerical integration performed based on Equations (12-2) and (12-6) for each of these data points are shown as the filled symbols in Figure 12-10. The agreement between simulation and experimental data is apparent. These results suggest that a phenomenological approach may be successfully applied to describe the thermal fatigue behavior under conditions where more than one damage mechanism may be operative.

12.3.2 Failure Criterion Effects

Failure of soldered interconnects is often specified in terms of damage parameters that represent the decrease in solder joint area during fatigue. For mechanical fatigue tests, the failure criterion is usually specified in terms of a load drop (e.g., see reference[30]). For thermal cycling or during service, the electrical or thermal resistance increase is often monitored to determine failure.[35,4,36]

The dependence of fatigue life on failure criterion specified in terms of various damage parameters is reflected by the profile of the parameter as a function of cycles or time. This profile can be established based on two considerations: (1) the relationship between the measured damage parameter and the solder joint area and (2) the time profile of the solder joint area during fatigue. The former is used to calculate the change in the measured damage parameter corresponding to a change in solder joint area. The time or the number of cycles required to produce a given solder joint area change can be estimated based on damage integration [Equation (12-5)].

An interest in the effect of the failure criterion on fatigue life arises from the need to correlate fatigue life data obtained from laboratory isothermal mechanical tests to thermal fatigue life. As mentioned previously, the failure criterion in each case is often specified in terms of different damage parameters. Furthermore, these criteria can correspond to different crack lengths. For such situations, the relationship between the solder joint area and the corresponding damage parameter is required to compare fatigue lives.

The dependence of load drop on the solder joint area is a function of the stiffness of the test machine or assembly relative to that of the solder joint and the magnitude of the applied displacement. When the applied displacements are small, the deformation of the solder joint is essentially linearly proportional to the load. During such conditions, the load will reflect the solder joint area only if the solder joint stiffness is comparable to or smaller than the test machine or the assembly stiffness. For an infinitely stiff machine, the stress in the solder joint is proportional to the solder joint strain through a combination of the elastic and anelastic moduli. Therefore, the load will be proportional to the solder joint area. When the applied displacements are large, the extent of plastic deformation can be large. As discussed previously, the load drop will then be proportional to the solder joint area in the absence of work hardening.

In general, when the electrical resistance is used as a damage parameter, a significant decrease in solder area is necessary for appreciable change in total resistance.[37,38] At small crack lengths the dependence of the resistance on the solder joint area is usually small. As the solder joint approaches final separation, the total resistance increase can accelerate dramatically. Correspondingly, if the failure is specified in terms of resistance increases corresponding to small crack lengths, the effect of failure criterion on the cycles to failure can be significant. On the other hand, the effect of failure criterion on fatigue life is expected to be small if a resistance increase failure criterion corresponding to the final stages of crack growth is specified.

12.4 DISCUSSION AND CONCLUSIONS

Deformation and damage during solder joint fatigue are distinct processes, and hence they can be examined separately. However, both processes can involve multiple mechanisms. A comprehensive treatment of the interactions of the various mechanisms requires a framework that can be used to design relevant experimental methods and to correlate experimental data. Phenomenological approaches for the description of deformation and fatigue damage in solder joints has been demonstrated. While for scientific knowledge such approaches provide a basis for deconvoluting individual operative mechanisms, its merits in engineering practice deserve further exploration. The development and application of phenomenological approaches to inelastic deformation of materials has received considerable attention.[39] However, significant efforts have not been made toward the use of similar approaches to damage processes. It is shown here that it may be possible, just as in the treatment of deformation phenomena, to view damage processes phenomenologically. The success of the latter will further enhance the power of integral approaches.

12.5 ACKNOWLEDGMENTS

This work has supported by IBM Corporation and by Semiconductor Research Corporation.

REFERENCES

1. Homa, T. R., "Cracking of Electronic Packaging Joints due to Creep-Fatigue," in *Proc. of ASM 3d Conf. on Electronic Packaging: Materials and Processes and Corrosion in Microelectronics,* Minneapolis, MN, ASM, 1987, pp. 209–216.
2. Raman, V., and T. C. Reiley, "Cavitation and Cracking in As-Cast and Superplastic Pb-Sn Eutectic during High-Temperature Fatigue," *Journal of Materials Science Letters,* **6**(5), 1987, pp. 549–551.

3. Agarwala, B. N., "Thermal Fatigue Damage in Pb-In Solder Interconnections," in *23rd Annual Proc. Int. Reliability Physics Society,* Orlando, Florida: IEEE, 1985, pp. 198–205.
4. Inoue, H., Y. Kurihara, and H. Hachino, "Pb-Sn Solder for Die Bonding of Silicon Chips," *IEEE Transactions on Components, Hybrids and Manufacturing Technology,* **CHMT-9**(2), 1986, pp. 190–194.
5. Subrahmanyan, R., J. R. WIlcox, and C.-Y. Li, "A Damage Integral Approach to Thermal Fatigue of Solder Joints," *IEEE Transactions on Components, Hybrids and Manufacturing Technology,* **CHMT-12**(4), 1989, pp. 480–491.
6. ———, "A Damage Integral Approach to Solder Joint Fatigue," in *Microelectronic Packaging Technology: Materials and Processes, Proc. 2d ASM Int. Electronic Materials and Processing Congress,* ASM, 1989, pp. 213–221.
7. Coffin, L. F., "Predictive Parameters and Their Application to High Temperature Low-Cycle Fatigue," in *Fracture 1969 (Proc. 2d Int. Conference on Fracture),* London, Chapman and Hall, 1969, p. 643.
8. Manson, S. S., "Interfaces between Fatigue Creep and Fracture," *Int. Journal of Fracture Mechanics,* **2**(1), 1966, pp. 327–363.
9. Majumdar, S. and P. S. Maiya, "A Unified and Mechanistic Approach to Creep Fatigue Damage," in *Proc. 2d Int. Conf. on Mechanical Behavior of Materials,* ASM, 1976, pp. 314–327.
10. Stone, D., H. Wilson, R. Subrahmanyan, and C.-Y. Li, "The Mechanisms of Damage Accumulation in Solders during Thermal Fatigue," in *Proc. 36th Electronic Components Conf.,* IEEE, 1986, pp. 630–635.
11. Solomon, H. D., "Effect of Hold Time and Fatigue Cycle Wave Shape on the Low Cycle Fatigue of 60/40 Solder," in *Proc. 38th Electronics Components Conf.,* Los Angeles, IEEE, 1988, pp. 7–12.
12. Shah, H. J., and J. H. Kelly, "Effect of Dwell Time on Thermal Cycling of Flip-Chip Joint," in *Proc. 5th Int'l. Hybrid Microelectronics Symposium,* 1970, pp. 3.4.1–3.4.6.
13. Eckel, J. F., "The Influence of Frequency on the Repeated Bending Life of Acid Lead," in *Proc. American Society of Testing Materials,* ASTM, 1951, pp. 745–760.
14. Sherry, W. M., and P. M. Hall, "Materials, Structures, and Mechanics of Solder Joints for Surface Mount Microelectronics Technology," in *Proc. 3d Int. Conf. on Interconnection Technology in Electronics,* Fellbach, West Germany, 1986, pp. 47–81.
15. Rolfe, S. T., and J. M. Barsom, *Fracture and Fatigue Control in Structures,* Englewood Cliffs, NJ, Prentice-Hall, 1977, pp. 232–267.
16. Hertzberg, R. W., *Deformation and Fracture Mechanics of Engineering Materials,* New York: John Wiley & Sons, 1983, pp. 519–598.
17. Wilcox, J. R., "Inelastic Deformation and Fatigue Damage in Metals at High Homologous Temperatures," Ph.D. Thesis, Cornell University, Ithaca, NY, 1990.
18. Lee, S.-M., and D. S. Stone, "Grain Boundary Sliding in Surface Mount Solders during Thermal Cycling," to be published in *Proc. 10th Electronic Components and Technology Conf.,* Las Vegas, Nevada, May 1990.
19. Tribula, D., D. Grivas, D. R. Frear, and J. W. Morris, "Observations on the Mechanisms of Fatigue in Eutectic Pb-Sn Solder Joints," *Journal of Electronic Packaging,* ASME, **111**(2), 1989, pp. 83–89.

20. Stone, D., H. Wilson, R. Subrahmanyan, and C.-Y. Li, "An Investigation of the Creep Fatigue Interaction in Solder Joints," in *Proc. National Electronic Packaging Conf. East,* 1986, pp. 175–180.
21. Wilcox, J. R., R. Subrahmanyan, and C.-Y. Li, "Thermal Stresses and Inelastic Deformation of Solder Joints," in *Microelectronic Packaging Technology: Materials and Processes, Proc. 2d ASM Int. Electronic Materials and Processing Congress,* ASM, 1989, pp. 203–211.
22. Frost, H. J., P. R. Lavery, and S. D. Lutender, "Microstructure and Mechanical Properties of Lead Tin Solder Alloys," in *Proc. ASM 3d Conf. on Electronic Packaging: Materials and Processes and Corrosion in Microelectronics,* 1987, pp. 259–267.
23. Frost, H. J., R. T. Howard, and G. J. Stone, "Effects of Thermal History on Microstructure and Mechanical Properties of Solder Alloys," in *Proc. ASM 2d Int. Electronic Materials and Processing Congress,* 1989, pp. 121–127.
24. Hart, E. W., "Constitutive Relations for the Nonelastic Deformation of Metals," *Journal of Engineering Materials and Technology,* **98**(3), 1976, pp. 193–202.
25. Jackson, M. S., C. W. Cho, P. Alexopoulos, and C.-Y. Li, "A Phenomenological Model for Transient Deformation Based on State Variables," *Journal of Engineering Materials and Technology,* **103**(4), 1981, pp. 314–325.
26. Li, C.-Y., "State Variable Theories for Nonelastic Deformation," in *Metallurgical Treatises,* J. K. Tien and J. F. Elliott, Eds., TMS-AIME, 1981, pp. 469–486.
27. Stone, D., S.-P. Hannula, and C.-Y. Li, "The Effects of Service and Material Variables on the Fatigue Behavior of Solder Joints during the Thermal Cycle," in *Proc. 35th Electronic Components Conf.,* IEEE, 1985, pp. 46–51.
28. Alexopoulos, P., R. L. Keusseyan, G. L. Wire, and C.-Y. Li, "Experimental Investigation of Nonelastic Deformation Emphasizing Transient Phenomena by Using a State Variable Approach," in *Mechanical Testing for Deformation Model Development,* R. W. Rohde and J. C. Swearengen, Eds., 1981, pp. 148–184.
29. Kashyap, B. P., and G. S. Murty, "Experimental Constitutive Relations for the High Temperature Deformation of a Pb-Sn Eutectic Alloy," *Materials Science and Engineering,* **50**(2), 1981, pp. 205–213.
30. Solomon, H. D., "Low Cycle Fatigue of 60/40 Solder—Plastic Strain Limited vs. Displacement Limited Testing," in *Proc. ASM's 3d Electronic Packaging: Materials and Processes Conf.,* Minneapolis, MN, ASM, 1985, pp. 29–47.
31. Solomon, H. D., "Low-Frequency, High Temperature, Low Cycle Fatigue of 60 Sn/40 Pb Solder," Tech. Rep. 85CRD238, GE Corporate Research and Development, 1985; also published, in part, in *Low Cycle Fatigue, ASTM-STP 942,* H. D. Solomon et al., Eds., ASTM, 1985, pp. 342–371.
32. Hall, P. M., "Forces, Moments, and Displacements during Thermal Chamber Cycling of Leadless Ceramic Chip Carriers Soldered to Printed Boards," *IEEE Transactions on Components, Hybrids and Manufacturing Technology,* **CHMT-7**(4), 1984, pp. 314–327.
33. Garafalo, F., *Fundamentals of Creep and Creep Rupture in Metals,* New York, Macmillan Company, 1965.
34. Clech, J. P., and J. A. Augis, "Engineering Analysis of Thermal Cycling Accelerated Tests for Surface-Mount Attachment Reliability Evaluation," in *Proc. 7th Int. Electronics Packaging Society Conference,* IEPS, 1987.

35. Wild, R. N., "Some Factors Affecting Leadless Chip Carrier Solder Joint Fatigue Life II," *Circuit World,* **14**(4), 1988, pp. 29–36, 41.
36. Solomon, H. D., "Low Cycle Fatigue of Surface-Mounted Chip-Carrier/Printed Wiring Board Joints," *IEEE Transactions on Components, Hybrids and Manufacturing Technology,* **CHMT-12**(4), 1989, pp. 473–479.
37. Solomon, H. D., "Strain Life Behavior in 60/40 Solder," *Journal of Electronic Packaging,* ASME, **111**(2), 1989, pp. 75–82.
38. Wilcox, J. R., R. Subrahmanyan, and C.-Y. Li, "Assembly Stiffness and Failure Criterion Considerations in Solder Joint Fatigue," *Journal of Electronic Packaging,* Transaction of the ASME, **112,** (2), June 1990, pp. 115–122.
39. Korhonen, M. A., S.-P. Hannula, and C.-Y. Li, "State Variable Theories based on Hart's Formulation," in *Unified Constitutive Equations for Creep and Plasticity,* A. K. Miller, Ed., London/New York, Elsevier Applied Science, 1987, pp. 89–138.

13

Modern Approaches to Fatigue Life Prediction of SMT Solder Joints

Thomas J. Kilinski, Jon R. Lesniak, and Bela I. Sandor

Surface mount technology (SMT) has been widely adopted in recent years, brought on by the need to produce small, dense electronic packaging. This technology has many advantageous characteristics:

- SMT allows for components to be mounted on both sides of the board.
- Holes through the board are only needed for communication between the surface and the circuit planes; thus the size of the holes can be decreased.
- Since less room is needed for holes, a 50 percent increase in line density from a similar through-hole board can be obtained.[1]

Unfortunately, SMT also brings with it a new era of failures. The joint is responsible for both electrical and mechanical connections. Solder does not have adequate ductility to ensure the repeated relative displacements due to the mismatch between expansion coefficients of the chip carrier and the circuit board, as it has been believed.

A major finding from mechanical tests is that solder behavior involves a creep-fatigue interaction, making it a poor material for use in mechanical connections. These considerations lead to the conclusion that it will take a combined effort in mechanics and metallurgy to control the failures of these joints. Even then, finding a technique for increasing the service life of future connections is not the total solution. A method must be developed for predicting the remaining service life of the many SMT joints already in use.

This chapter expresses our thoughts and research direction regarding SMT solder joint reliability analysis, keeping in mind fundamental and practical issues. There is no simple, complete solution to the problem at hand, but significant progress can be made toward understanding and controlling it.

13.1 MECHANICAL TESTING

13.1.1 Determination of Elastic Properties

The elastic modulus (E), Poisson's ratio (ν), and the shear modulus (G) are determined with strain gages, or mechanical extensometers on a specimen of uniform cross section in directions parallel and perpendicular to the applied loading; E and ν can be measured directly, while G is commonly computed from

$$G = \frac{E}{2\,(1 + \nu)} \qquad \text{valid for isotropic materials} \qquad (13\text{-}1)$$

Extreme caution must be taken because solders have both elastic and inelastic response at the instant of load application. If testing is performed at low strain rates, the resulting stress versus axial strain plot may appear fully linear. However, the slope of this curve *does not* provide an accurate value for the elastic modulus because inelastic as well as elastic strains are present. To obtain an accurate value, testing must be performed at low stresses so that minimal yielding occurs and at a rate high enough to eliminate any viscoplastic effects. Screw-driven testing machines cannot achieve high enough strain rates under proper control to accurately perform these tests. Closed-loop, servo-hydraulic machines may be needed to obtain accurate data. In order to know if the testing rate is sufficiently high, the stress-strain slope must be obtained for increasing loading rates. When there is no change in the measured slope with further increases in rate, it is safe to assume that the solder is undergoing purely elastic deformations. Using this technique, it was found that a ramp speed equivalent to 80 MPa/sec was sufficient to minimize viscoplastic effects in 60% tin-49% lead solder at 25°C. It is important to realize that these elastic constants may vary with temperature and are slightly dependent on microstructural properties such as eutectic grain size, provided that the microstructure is isotropic (a reasonable assumption for 60% tin-40% lead solder). Therefore, it is necessary to account for ambient temperature effects while performing any solder behavior analysis.

The experimentally determined values for E and ν are summarized in Table 13-1, where they are compared with other values found in the literature. Note that there are large differences in E obtained by various researchers. Based on our data, the average shear modulus at 25°C is $G = 14.4$ GPa.

As a further check of our results, handbook values for the elastic moduli of tin and solder were obtained. Since the solder microstructure is near-eutectic

Table 13-1 Elastic Constants for 60% Tin-40% Lead Solder at 25°C

Reference	Elastic Modulus (GPa)	Poisson's Ratio
Authors' data	38.6 ± 4	0.36 ± 0.04
3	16	—
4	15	0.40
5	30	—
6	35	—

and consists of a random mixture of lead-rich and tin-rich grains combined with a random distribution of lead-rich dendrites, a simple rule-of-mixtures calculation, used to estimate the elastic modulus of macroscopic composite materials,[1] should provide a reasonable estimate for the elastic modulus of 60% tin-40% lead solder. As seen below, this estimate is in agreement with our experimental data:[2]

$$E_{\text{tin}} = 50 \text{ GPa} \quad (\text{Ref.}^7)$$
$$E_{\text{lead}} = 14 \text{ GPa} \quad (\text{Ref.}^8)$$
$$E_{60\% \text{ tin-}40\% \text{ lead solder}} = 0.6(50) + 0.4(14) = 36 \text{ GPa}$$

13.1.2 Mechanical Properties

Other kinds of laboratory testing are needed in order to fully understand the behavior of solders during thermal-mechanical fatigue loading. Ambient temperature fluctuations, stress and strain amplitudes, superimposed mean stresses and strains, loading rates, and even the shelf life of the specimens contribute to the solder behavior causing scatter in the fatigue life prediction.

Creep and Stress Relaxation

Stress relaxation and creep properties determined in the laboratory help model the joint compliancy during thermal expansion cycles. Tests must be performed on bulk solder as well as on model SMT joints. Creep tests are somewhat unrealistic because real joints are constrained. Also creep tests may require much time. The mechanistically equivalent stress relaxation tests, on the other hand, can be performed swiftly because the primary information, stress relaxation rate or rate of load drop, occurs "immediately" upon loading. In fact, with solders, the initial strain amplitude must be applied rapidly; otherwise, the load will have significantly relaxed by the time the desired strain amplitude has been reached, resulting in an unrealistically small recorded peak load.

If the stress-time data are fitted to a power equation of the form

$$\tau = At^n \tag{13-2}$$

where τ is the nominal shear stress, t is time, and A and n are constants, the stress relaxation rates can be determined by differentiating this equation.

The stress relaxation data can also be used to obtain stress versus steady-state creep rate data by remembering that throughout a stress relaxation test the total deformation (elastic plus inelastic) remains constant. The solder strain (in this case shear strains) can be divided into its elastic and creep components,

$$\gamma = \gamma_e + \gamma_c = \frac{P}{AG} + \gamma_c \tag{13-3}$$

where P is the load acting on the solder joint and A is the cross-sectional area in shear. The steady-state creep rate is determined by substituting Equation (13-2) into Equation (13-3):

$$\gamma = \tan^{-1}\left(\frac{\delta_s}{H}\right) \approx \left(\frac{\delta_s}{H}\right) \text{ for } \gamma \ll 1 \tag{13-4}$$

where δ_s is the displacement across the solder joint and H is the height of the joint. Caution must be taken when using small angle approximations for deflec-

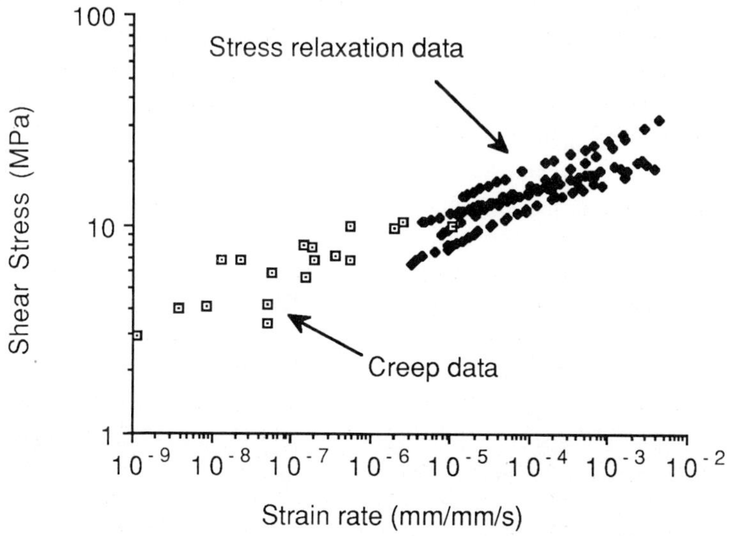

Figure 13-1 Steady-state creep rates for 60% tin-40% lead solder at 25°C.

tions. Data have been reported stating shear strains at failure exceeding $\pi/2$ rad, which is unrealistically large. The resulting equation is then differentiated with respect to time to obtain the steady-state creep rate:

$$\dot{\gamma}_c = -\left(\frac{\dot{P}}{AG}\right) \tag{13-5}$$

Figure 13-1 shows the correlation between the steady-state creep rate calculated from stress relaxation tests and the data obtained from numerous independent creep tests.

13.2 LIFE PREDICTION TECHNIQUES

This section is devoted to describing the most popular empirical models available, along with their applicability to SMT fatigue life prediction.

13.2.1 Fatigue Models

SMT solder joint fatigue failure is caused most often by a thermal expansion mismatch between the surface mounted device and the substrate material. These conditions result in the total inelastic strain range in the solder being a combination of instantaneous inelastic strains (plastic strains) and time dependent inelastic strains (creep strains). Fatigue models can be divided into several major categories. The two most prominent categories are:

Mechanistic models. The damage mechanisms involved are mathematically described. These models tend to be quite sophisticated; in addition, a given model is usually only applicable to a specific material. For example, a model developed for 60% tin-40% lead solder may be inappropriate for a tin-lead-indium solder.

Empirical models. A measurable physical quantity, usually the inelastic strain range, is used as the basis for life prediction. The particular damage mechanisms causing this physical quantity are usually not known. Despite this, empirical models have been successful for several common materials.

Coffin-Manson Equation

The Coffin-Manson equation can be written as

$$\frac{\Delta\varepsilon_p}{2} = \varepsilon_f' \, (2N_f)^c \tag{13-6}$$

where $\Delta\varepsilon_p/2$ = plastic strain amplitude (Figs. 13-2 and 13-3)

ε_f' = fatigue ductility coefficient

c = fatigue ductility exponent

$2N_f$ = reversals to failure (1 cycle = 2 reversals)

The equation was developed in the 1950s.[9,10] It represented a major breakthrough in the field of low cycle fatigue. It is therefore not surprising that this is the model most often seen in the solder fatigue literature.

It has long been recognized in the aerospace establishment that the Coffin-Manson equation is inadequate for modeling in-service thermal fatigue conditions in high temperature alloys. The primary difficulty is the inability to account for variations in waveform. Thermal fatigue involves a combination of creep and plastic strains, and each can have a different effect on the component life. The Coffin-Manson equation does not allow differentiation between these two strains, resulting in inaccurate life predictions.

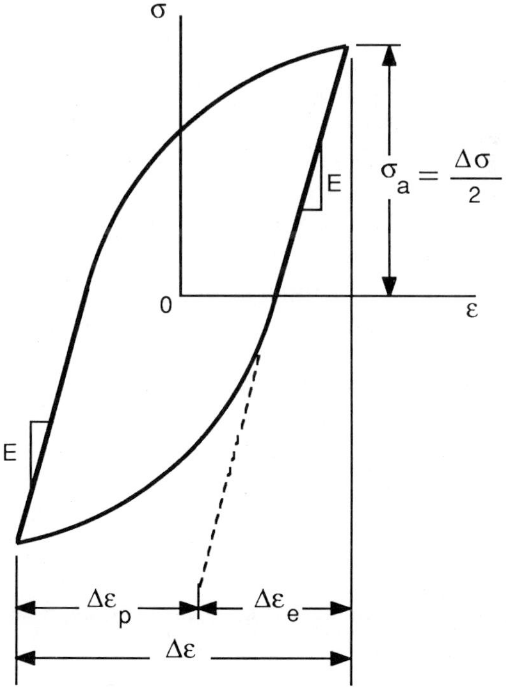

Figure 13-2 Representative stress-strain hysteresis loop for a material undergoing low cycle fatigue.

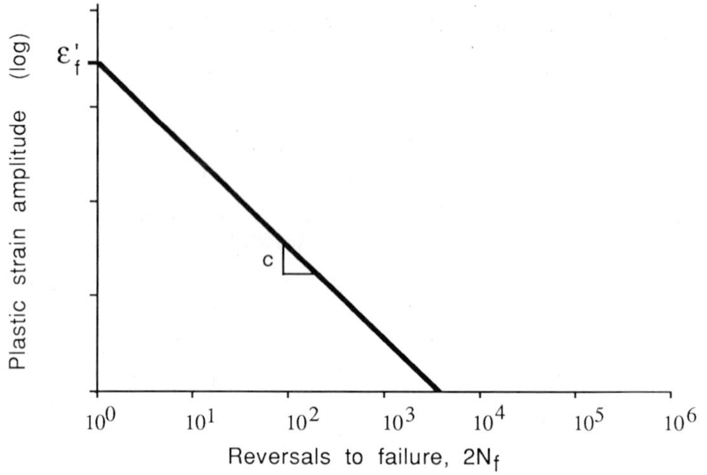

Figure 13-3 Schematic of Coffin-Manson equation.

Strain-Life Relationship (Total Strain Equation)

In this approach, the Coffin-Manson equation is combined with Basquin's equation for high cycle fatigue:

$$\sigma_a = \sigma_f'(2N_f)^b \tag{13-7}$$

where $\sigma_a = \Delta\sigma / 2$ = true stress amplitude

σ_f' = fatigue strength coefficient

b = fatigue strength exponent (Basquin's exponent)

The total strain amplitude, $\Delta\varepsilon/2$, can be written as the sum of elastic and plastic components

$$\frac{\Delta\varepsilon}{2} = \frac{\Delta\varepsilon_e}{2} + \frac{\Delta\varepsilon_p}{2} \tag{13-8}$$

Also, the elastic strain amplitude and the true stress amplitude are related by

$$\frac{\Delta\varepsilon_e}{2} = \frac{\sigma_a}{E} \tag{13-9}$$

Combining Equations (13-6), (13-7), (13-8), and (13-9) results in the following:

$$\frac{\Delta\varepsilon}{2} = \frac{\sigma_f'}{E}(2N_f)^b + \varepsilon_f'(2N_f)^c \tag{13-10}$$

This equation, displayed graphically in Figure 13-4, is often referred to as the strain-life relationship, or the total strain equation. The strain-life relationship has the advantage of combining both high and low cycle fatigue life prediction into a single equation. Note that it does not account for elastic and plastic components of damage. In high cycle fatigue, elastic strains predominate and plastic strains are too small to measure accurately; consequently, the elastic strain amplitude is used as a means to estimate long fatigue life. Although the plastic strains are much smaller than the elastic strains in high cycle fatigue, it is still the plastic strains that are causing the actual fatigue damage. It should be realized that, fundamentally, elastic strains do not cause damage.

To establish the strain-life relationship, tests performed under low cycle fatigue conditions can be used to determine the empirical constants b, c, σ_f', ε_f', and E. These constants are all that are needed to predict the fatigue response of the material over the entire range of practical lives. It would be useful to have a method that uses accelerated tests to predict in-service life. However, it is necessary to account for creep strains, which is difficult to do reliably.

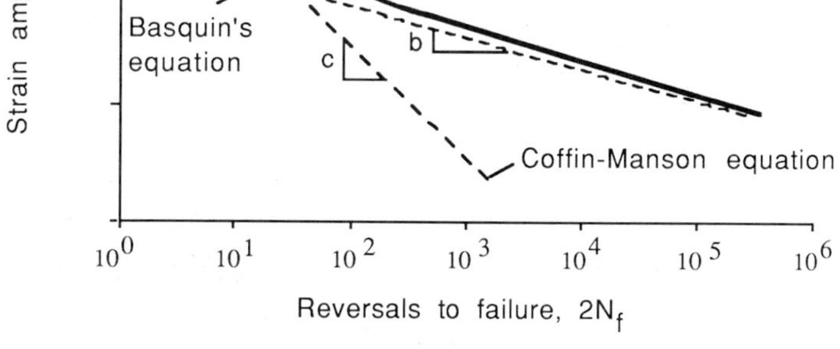

Figure 13-4 Schematic of total strain versus life equation.

Frequency-Modified Coffin-Manson Equation

The Coffin-Manson equation can be modified to include a term that accounts for the cycling frequency:

$$\frac{\Delta\varepsilon_p}{2} = \varepsilon_f'(2N_f \times \nu^{k-1})^c \tag{13-11}$$

where ν is the cycling frequency and k is a material parameter. For very low life ($2N_f < 100$), frequency is often found to have an insignificant effect on the life, and hence $k = 1$. In comparison, lowering the frequency at longer lives ($2N_f > 100$) usually results in a decreased life, and k takes on a value less than 1. A plot of total inelastic strain amplitude versus frequency-modified life ($2N_f \times \nu^{k-1}$) often reveals a bilinear relationship. More details can be found in the reference.[11]

The frequency-modification approach is good for situations in which the in-service loading is uniform for all cycles and is a sinusodial wave of constant frequency. The in-service loading waveform must have the same shape as that used in the laboratory testing. This represents a severe limitation of this approach: There is no easy way to account for variations in waveform. Consider, for example, the two hypothetical load waveforms that are assumed to result in the same total inelastic strain range per cycle, shown in Figure 13-5. Although both waveforms have the same frequency, the waveform shown in Figure 13-5*b* involves a longer hold time at the maximum stress, resulting in more creep per

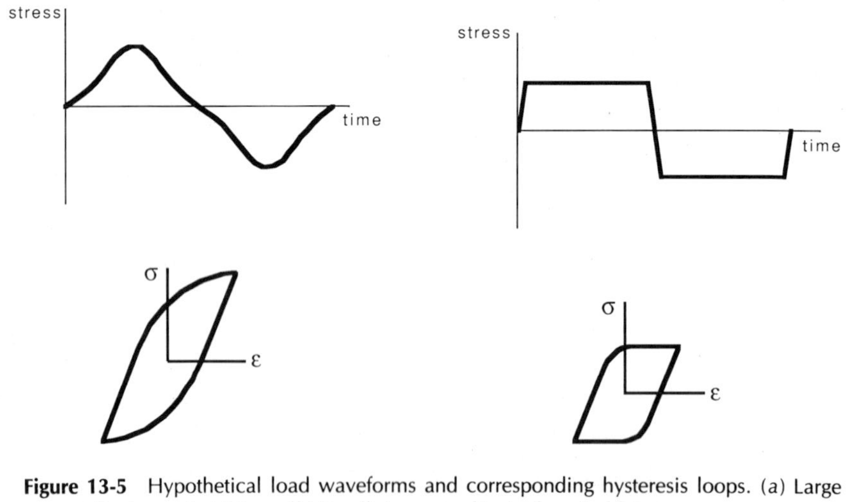

Figure 13-5 Hypothetical load waveforms and corresponding hysteresis loops. (a) Large stress amplitude; no hold time. (*b*) Small stress amplitude; long hold time.

cycle than in the waveform of Figure 13-5*a*. Since creep strains are usually more damaging than plastic strains, the waveform of Figure 13-5*b* will tend to result in a lower life than the waveform of Figure 13-5*a*. However, because the total inelastic strain range is the same in both cases, the method predicts the same lives. This equation is not recommended for solder joint fatigue analysis, except in well-defined special cases, although it has already found wide acceptance in the SMT field.

Strain Range Partitioning (SRP) Procedure

The SRP approach accounts for different types of strains causing different amounts of damage. The hysteresis loop for any waveform can be separated into four possible strain range components[12] (Fig. 13-6):

$\Delta\varepsilon_{pp}$ = plastic strain in tension and compression (PP)
$\Delta\varepsilon_{cc}$ = creep strain in tension and compression (CC)
$\Delta\varepsilon_{cp}$ = creep strain in tension; plastic strain in compression(CP)
$\Delta\varepsilon_{pc}$ = plastic strain in tension; creep strain in compression (PC)

At most, only three of the four SRP components can exist in any hysteresis loop, since there can never be both CP and PC loading in the same loop. For example, if there is an excess of plastic strain in tension over that in compression,

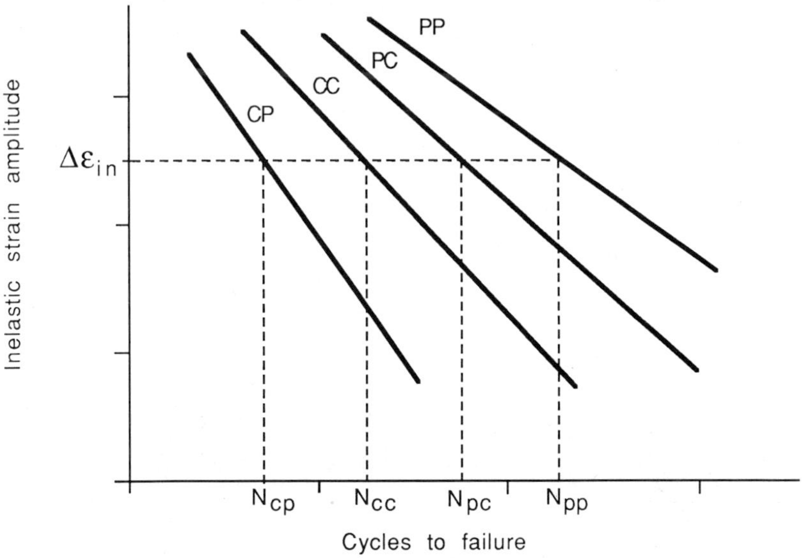

Figure 13-6 Idealized hysteresis loops for strain range components.

there will be some PC loading but no CP loading. Each of the four components is usually associated with a different microstructure, caused by creep and plastic strains.

To apply the SRP approach, the hysteresis loop corresponding to the in-service cycling conditions must be known. This complex hysteresis loop is resolved into its strain range components. Next, an interactive damage rule is applied:

$$\frac{1}{N_f} = \frac{F_{pp}}{N_{pp}} + \frac{F_{cc}}{N_{cc}} + \frac{F_{cp}}{N_{cp}} + \frac{F_{pc}}{N_{pc}} \tag{13-12}$$

where N_f = predicted cycles to failure of the material for the hysteresis loop being considered

N_{ij} = cycles to failure for the material if the total inelastic strain range, $\Delta\ \varepsilon_{in}$, of the hysteresis loop were of type ij, where ij is either PP, CC, PC, or CP (see Fig. 13-7)

F_{ij} = fraction of the total inelastic strain range of the hysteresis loop that is actually of type ij.

Note that to apply this damage rule, it is necessary to have strain range versus life relationships for the four types of strain range components. A Coffin-Manson-type relationship will usually fit the data for all four component types. In fact, the strain-life equation describing PP-type loading is precisely the Coffin-Manson equation in its commonly used form.

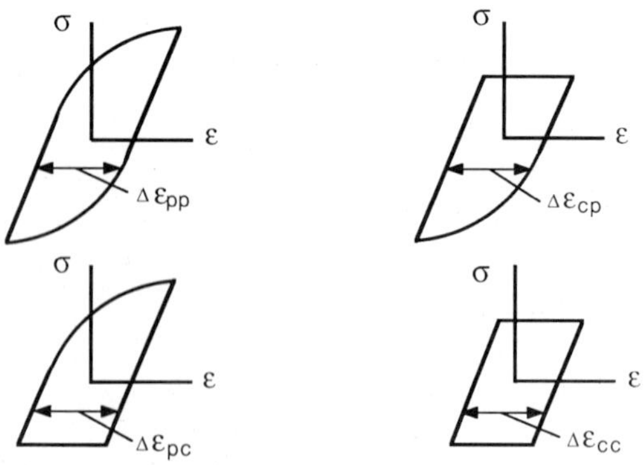

Figure 13-7 Strain-life plots for strain range components.

Total Strain Version of the SRP Procedure

A disadvantage of SRP is that it is difficult to use low cycle fatigue life results to estimate high cycle creep-fatigue lives. It often predicts in-service lives much longer than actually observed. It was this difficulty that lead to the development of the total strain version of the SRP procedure. This approach combines the advantages of low cycle SRP and high cycle strain life approaches. The technique involves estimating the elastic strain range-life relationships for the PP, CC, PC, and CP types of loading. For PP-type loading, the elastic strain range-life relationship is simply Basquin's equation. For the other three types of loading, the elastic strain range-life relationship is not as easily defined. Instead, it has been found to depend on the waveform (PC, CP, or CC), frequency, and the method by which creep strains are caused (stress hold, strain hold, or a combination thereof). Saltsman and Halford now recommend accounting for these effects by conducting cyclic stress-strain tests in order to obtain the flow characteristics of the material under the conditions of interest.[13]

Although the total strain version of SRP may help to more accurately predict high cycle creep-fatigue life, it is considerably more complicated to use than the basic version of SRP. In addition, some of the material constants required for this approach actually depend on waveform, frequency, and other effects. Thus, it is probably not practical to attempt to use the total strain version of SRP for solder joint fatigue modeling at this time. A thorough analysis of the standard SRP approach must first be tried in any case. In addition, the work by Halford and others has shown that both versions of SRP, total strain and conventional, give very similar life predictions (that is, both are usually accurate to within a factor of 2 of the actual life) when the cycles to failure is approximately 10,000 or less.[14] Since many SMT uses fall within this realm, the standard SRP procedure should prove adequate for most applications.

13.3 HYBRID LIFE PREDICTION TECHNIQUES

One goal of our program is to marry the fatigue prediction techniques with mechanistic models. It is difficult to calculate the number of cycles to failure from a purely theoretical model. However, it is meaningless to experimentally determine a life plot if the test strategy is not driven by some knowledge of the failure modes. It is imperative for those collecting life data to work closely with those analyzing the failures.

Before determining which fatigue technique to use, mechanistic damage models must be understood. It is useful to consider two types of cracks, those traversing the bulk of the solder and those traveling through the intermetallic layer between the solder and the substrate. Also, cracks can start in the interface but may continue through the bulk solder. Each type of crack is influenced by different mechanisms, and each mechanism is influenced by many parameters.[15]

The total creep rate is a combination of matrix deformation and grain boundary sliding and, in the case of near-eutectic solders, phase boundary sliding. Using polycrystalline models developed by Hart and Ghahremani, Stone[15] predicts solder to have three distinct regions of stress versus strain rate (Fig. 13-8). At high strain rates the creep stresses are dependent on matrix properties. As the strain rate decreases, grain boundary sliding becomes involved. At increasingly lower frequencies of loading, matrix properties are again the limiting factor since the matrix must deform to allow sliding at triple points (the junction of three grain boundaries).

An SRP approach lends itself to the combination of mechanistic models with life prediction functions. As discussed above, the intent of a partitioning approach is to separate the various types of strains that drive various kinds of damages. For low cycle fatigue of structural alloys at elevated tempertures this entails separating plastic strains from creep strains. In the case of solder, it is particularly important to separate creep by grain boundary sliding and matrix deformation.

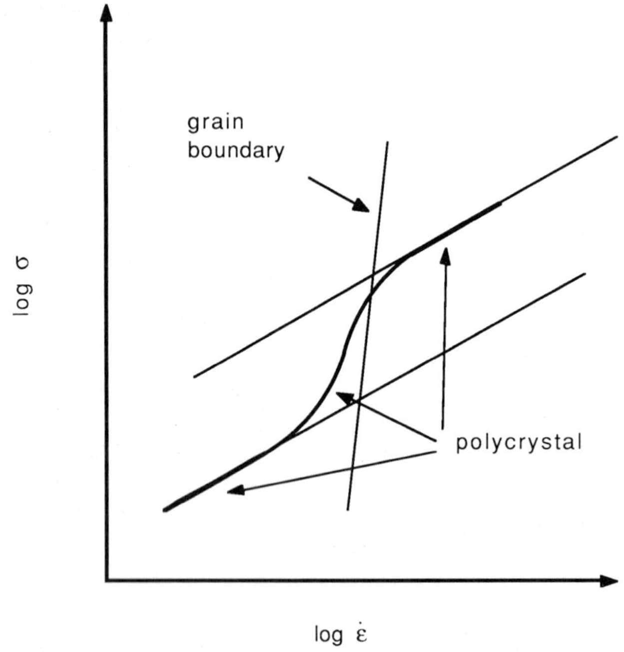

Figure 13-8 Stress versus strain rate curves that incorporate grain boundary sliding for polycrystalline material. (From D. Stone, "Creep-fatigue interactions in solders and solder joints," unpublished paper, University of Wisconsin-Madison, 1988)

13.3.1 Strain Range Partitioning Rule

Standard SRP handles the separation of different strains for a given situation well. The cumulative damage rule suggested by Hirschberg and Halford[12] works well for structural alloys at elevated temperatures. It is important to understand why this relatively simple equation works if this technique is to be extended to creep-fatigue of solder. For applying the simple damage rule, several things must hold true:

- All kinds of strains must contribute to a common damage.
- The common damage must be the unique cause of failure.
- The damage versus life rates must be similar for different strain levels and types of strain.

The first two criteria mean that if the damages caused by various types of strains are to be analyzed in this simplistic fashion, they must be the same damages. If more than one kind of damage occurs, one must predominate so that the lives for all types of strains are dependent only on that mechanism. Also, the presence of any other damage or condition should not interact with the dominant damage mechanism. The third criterion states that the damage rates need to be similar. This implies that if the pertinent damage parameter could be plotted versus fraction of life exhausted instead of reversals, the amount of damage would be identical for the same percentage of lives exhausted. It also implies that failure will occur at the same level of damage for any strain range or any type of strain. For example, consider a test where 80 percent of the life is exhausted at a type 1 strain of range $\Delta\ \varepsilon_b$, and the condition is switched to a type 2 strain of range $\Delta\ \varepsilon_a$ (Fig. 13-9). Because the same amount of damage would have accumulated by exhausting 80 percent of the life for a type 2 test, damage accumulation continues ignorant of the type of strain that caused the damage in the first part of the test. It is interesting to note that the damage versus reversal curves must be similar, which does not demand the damage be accumulated linearly. Of course, the damage function probably will not and need not be exactly known for most practical applications of SRP. If the damage versus cycles curves are not similar for all relevant types of strain or strain ranges, these nonsimilarities must be accounted for.

In light of all that is known about solder and its damage mechanisms, a simple damage accumulation rule cannot be expected to work. The nature of these mechanisms violates many of the basic criteria. There is not a unique damage occurring, several types of cracks are possible, the mechanisms interact, and they are dependent on strain rate, temperature, and time.

Shine and Fox have contributed much to these ways of thinking.[16] They carried SRP one step further by partitioning the creep strain into matrix and grain

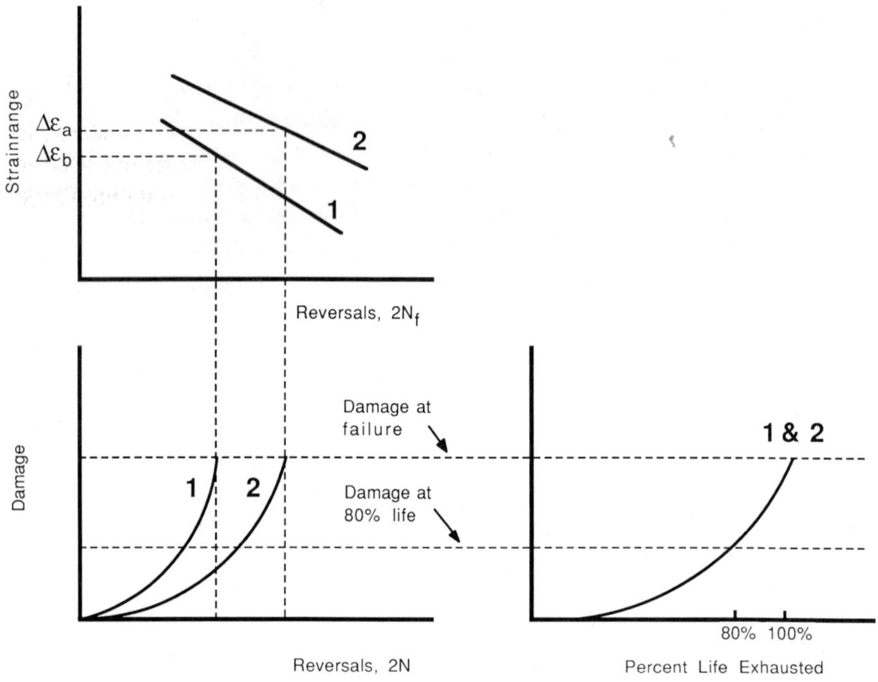

Figure 13-9 Strain range partitioning damage relationships.

boundary slip. Fatigue lives are correlated with percentages of these strains and with strain rates. The creep strain rates are separated assuming the total creep rate is the sum of matrix deformation creep rate and grain boundary sliding.

It is our intention to complement approaches such as that of Shine and Fox by developing an interactive damage rule. This rule needs to incorporate the interaction of time, temperature, strain rates, other components of strains, and any parameter that influences the damage process. This relationship will take an integral form, thus accounting for damage in each individual cycle, based on the conditions at that time.

13.4 MODEL JOINTS

13.4.1 Quality Control

It is important to control the process of manufacturing model joints to maintain a consistent representation of actual SMT connections. The process of maintaining quality control can be assured through the following techniques.

Reflow Process

There are two important aspects to control while making model SMT joints: metallization diffusion and flux burning. Metallization diffusion occurs when the specimen is held at or above the reflow temperature for too long. This allows the metallization to completely diffuse into the molten solder, weakening the bond between the solder and the substrate. Flux burning occurs when the solder does not reach reflow temperature quickly enough. The flux burns off before the solder melts, leaving flux residue within the solder.

Cooling Rates

The cooling rate affects the grain size of the solder joint. Maintaining uniform grain size is crucial for relating the fatigue life results obtained from model joints to actual SMT connections. The cooling rates have to be adjusted to account for the larger solder volume differential in the model joint. However, caution must be taken so micro- or macro-cracks are not introduced by rapid cooling.

Inspection

Optical and SEM inspection techniques should be used as a method of quality control for model joints both before (Fig. 13-10) and after testing (Fig. 13-11).

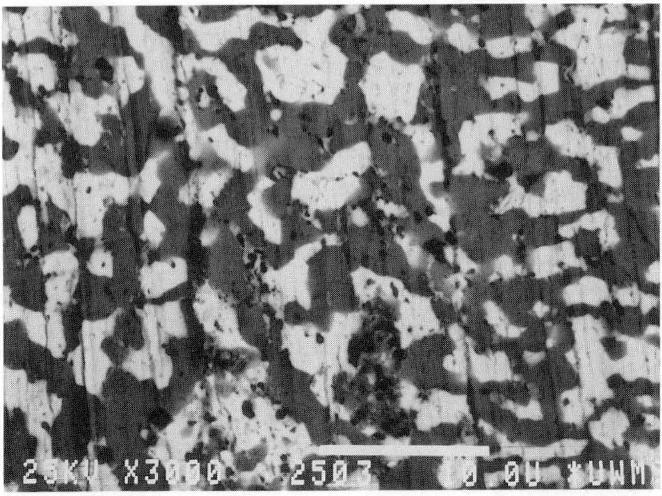

Figure 13-10 SEM backscattered electron image showing initial solder microstructure of a typical lap joint. Light regions are lead-rich grains; dark regions are tin-rich grains.

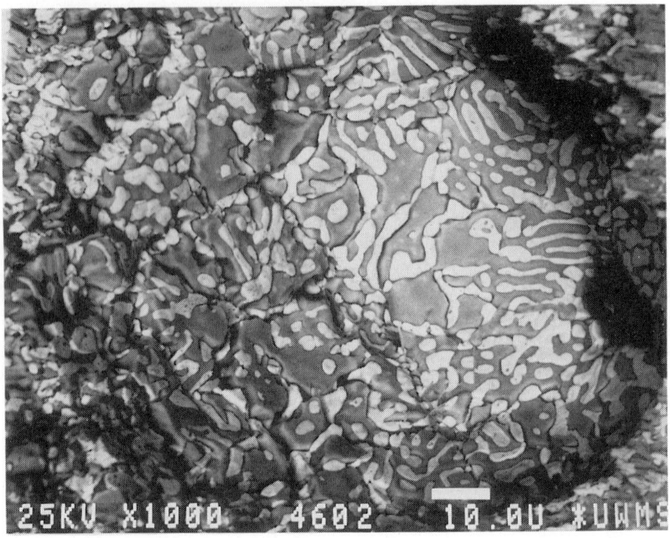

(a)

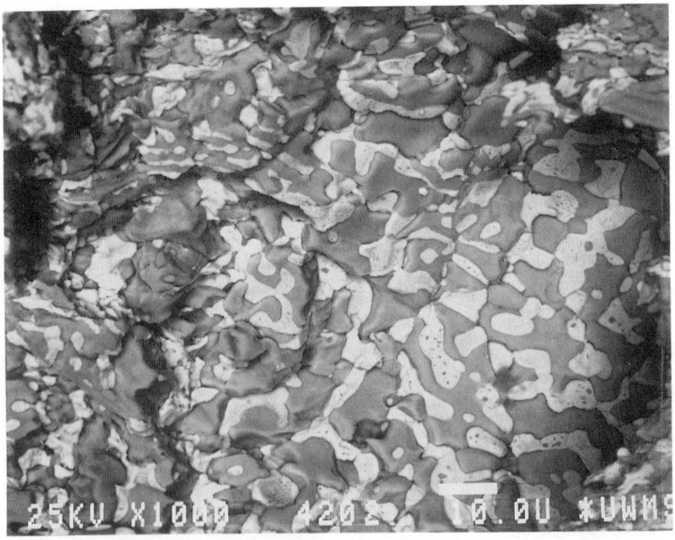

(b)

Figure 13-11 SEM backscattered electron images of fracture surfaces of several lap joint specimens tested in creep showing regions of enhanced grain growth. Light regions are lead-rich grains; dark regions are tin-rich grains. (*a*) Initial shear stress = 10.3 MPa. Time to failure = 6 h. (*b*) Initial shear stress = 7.7 MPa. Time to failure = 48 h. (*c*) Initial shear stress = 3.4 MPa. Time to failure = 45 days.

(c) 

Figure 13-11 Continued

These periodic inspections of specimens ensure consistency of grain size and void concentrations to help obtain consistent fatigue life data. Simple optical inspection using high powered microscopes is sufficient for grain size analysis.

13.4.2 Lap Joint Specimens

Single-ball solder specimens are used to isolate fatigue properties free of the geometric influences found in multilead specimens. They offer the flexibility of quickly exploring and isolating individual influences on fatigue lives. They lend themselves to the collection of basic material properties. The specimens used in our investigations have small metallic pads on an alumina substrate (Fig. 13-12). Solder balls from 0.38 to 0.89 μm in diameter are used.

13.4.3 Straddle Board Specimens

The straddle board approach is a new direction for bridging the gap between the simplest model joints and complete manufactured boards. The idea is to perform tests on real SMT connections to look for trends in life versus joint geometry.

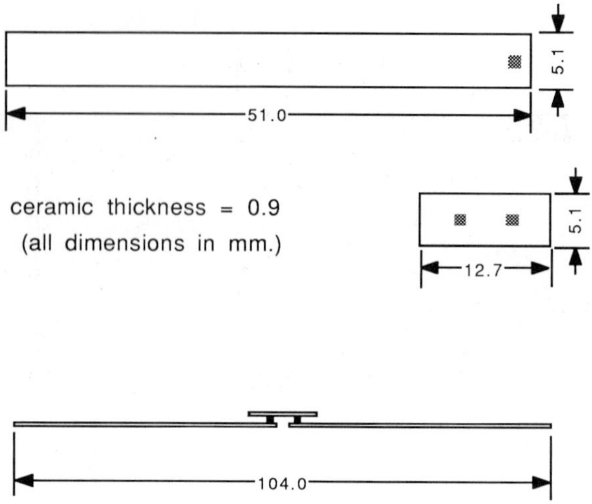

Figure 13-12 Ceramic surface mount lap joint design. Upper: top view. Lower: side view.

The simplest application for the straddle board is to enable an A to B comparison of different lead, joint, and connection systems. The straddle board is actually a fragment of a circuit board. A single chip carrier is mounted to each side of a printed wiring board (Fig. 13-13). To ensure that the joints are of the same specifications as a typical SMT joint, the boards are manufactured in a conventional manner, using the same vapor phase reflow assembly process as for regular components.

Aluminum stiffeners are attached to the straddle board before its placement into the grips. After the board has been aligned in the grip fixtures, the remaining mid-section of the board is cut free. The load and strain are zeroed, and the stiffeners removed. The straddle board is now ready for testing, with load carrying capabilities coming only from the leads.

Besides performing A to B comparisons, the straddle boards are used to obtain fatigue-life prediction data for combinations of load, deflection, and temperature. The majority of testing employs shear deformations. This type of loading most accurately resembles the thermal strains induced by the difference in thermal expansions between the board and the chip carrier. A major obstacle to overcome when using straddle board-type specimens is accounting for the extent of damage occurring in individual joints. A typical straddle board configuration has eight leads per chip, the corner leads of each side remaining connected. It is impossible to measure the loads on each individual lead given this setup. However, for correct fatigue analysis it is crucial to distinguish between a drop in the measured load caused by the failure of a single lead and the load drop caused by the general fatigue damage occurring in all the leads. An attempt has been made by

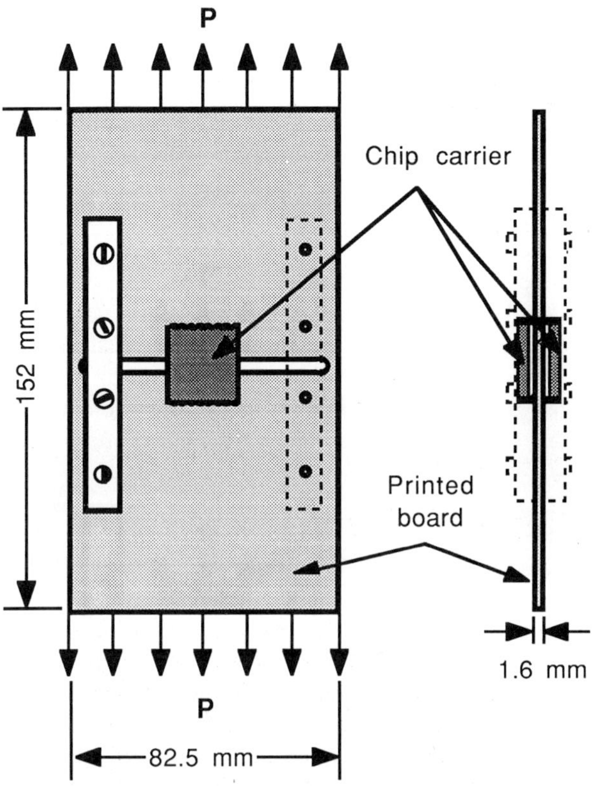

Figure 13-13 Straddle board configuration.

other researchers to correlate the amount of damage occurring in individual leads by measuring the resistance increase within that lead. The change in resistance occurs because of the change in cross-sectional area from cracks propagating within the lead. Although the concept is attractive, caution must be taken when dealing with shearing forces. Even though the connection has failed completely with respect to load carrying capabilities, electrical connection can still be made through the contact of fracture surfaces sliding past one another. Therefore, the resistance measurement will be determined by the contact area of the fracture surfaces. Arguments can be made that the electrical attachment is the only critical connection (i.e., if the joint can no longer hold a load but is still passing a signal, it is still connected). However, under real operating conditions the joint is subject to a multiaxial state of loading, not just shear loading. In this case a tensile force acting on the connection, due to shock, vibrations, or thermomechanical loading, will tend to separate the joint causing an electrical failure. Separation for even an instant is unacceptable.

13.5 EXPERT SYSTEMS

An expert system is a computer program designed to emulate an expert within a given field. The software makes use of factual knowledge and rules of thumb to perform given tasks. It consists of a knowledge base and an inference engine. In the case of SMT life prediction, the knowledge comes in the form of a large database of fatigue tests performed on actual joints and model laboratory specimens. The inference engine defines the methods by which the stored information is retrieved and used for problem solving.

The strength of the software is its capacity to store and analyze the huge quantities of life prediction data obtained through mechanical testing. This means that data from many investigators can be used in the data bank. A key aspect of a complete SMT expert system is that it could perform nonlinear finite element analysis to predict the stresses occurring in the joint connection for any given loading situation.

A major obstacle in SMT life prediction is that many fatigue life curves are needed to account for all possible combinations of joint geometries, loading conditions, and temperature cycles. The expert system has the ability to interpret much data to predict an outcome for any situation. This way, a finite number of loading combinations and lead geometries can be analyzed in the laboratory to determine their individual fatigue behaviors. The finite element method will be used to predict the stress distribution for these same constraints. The advantage of the expert system is that the geometry and loading schematic of a lead or connection system (any geometry or loading) is entered by a user. The finite element analysis predicts the stress distribution in the connection. The expert system program will then make a prediction of the service life based on an interpolation of known results.

Another advantage of the expert system is in optimizing a life prediction procedure. A limited number of tests, such as three or four, could be performed on the specimen in question and the results compared to the prediction. The program may ask for additional tests or use the available data to make a self-adjustment on its prediction. The expert system will learn from its mistakes through this method and apply this knowledge to the next interpolation.

13.6 CONCLUSIONS

Our ultimate goal is not just to establish one life rule for one particular joint but to develop and establish the most efficient test methodology to address both present and future problems. By combining the knowledge of the pertinent engineering disciplines, we are setting down guidelines by which material properties and fatigue data should be collected. Furthermore, this will be done in an expert system format that provides researchers a means of combining or comparing their findings with those of others and will also serve as a medium accessible to all electronic packaging engineers.

13.7 ACKNOWLEDGMENTS

This work has been funded by the Electronic Packaging and Interconnection Program, an industrial consortium at the University of Wisconsin-Madison.

REFERENCES

1. Seraphim, D., R. Lasky, and C. Li, *Principles of Electronic Packaging,* New York, McGraw-Hill, 1989.
2. Enke, N. F., et al., "Mechanical Behaviors of 60/40 Tin-Lead Solder," *Proceedings of the 39th Electronic Components Conference,* Houston, 1989, pp. 264–272.
3. Engelmaier, W., "Test Method Considerations for SMT Solder Joint Reliability," *IEPS,* Oct. 1984, pp. 360–369.
4. Harper, C. A., *Handbook of Materials and Processes for Electronics,* New York, McGraw-Hill, 1970.
5. Greenfield, L. T., and P. G. Forrester, "The Properties of Tin Alloys," *Int. Tin Research Institute Publication No. 155,* 1947.
6. Clatterbaugh, G. V., and H. K. Charles, Jr., "Thermo-mechanical Behavior for Soldered Interconnects for Surface Mounting: A Comparison of Theory and Experiment," *Proceedings of the 35th IEEE Electronic Components Conference,* Washington, DC, 1985, pp. 60–72.
7. Bever, M. B., (ed.), *Encyclopedia of Materials Science and Engineering,* vol. 7, Cambridge, MA, MIT Press, 1986, p. 5070.
8. Blake, A. (ed.), *Handbook of Mechanics, Materials, and Structures,* New York, John Wiley & Sons, 1985, pp. 181.
9. Coffin, L. F., Jr., *Transactions, American Society of Mechanical Engineers,* TASMA, **76,** 1954, p. 931.
10. Manson, S. S., "Behavior of Materials Under Conditions of Thermal Stress," NACA Techn. Note. 2933, 1954.
11. Coffin, L. F., Jr., "A Note on Low Cycle Fatigue Laws," *Journal of Materials* JMLSA **6,**(2), June 1971, pp. 388–402.
12. Hirschberg, M. H., and G. R. Halford, "Use of Strainrange Partitioning to Predict High-Temperature Low-Cycle Fatigue Life," NASA Technical Note D-8072, January 1976.
13. Saltsman, J. F., and G. R. Halford, "An Update of the Total-Strain Version of SRP," NASA Technical Paper 2499, October 1985.
14. Halford, G. R., and J. F. Saltsman, "Strainrange Partitioning—A Total Strain Range Version," NASA Technical Memorandum 83023, 1983.
15. Stone, D., "Creep-Fatigue Interactions in Solders and Solder Joints," Unpublished paper, University of Wisconsin-Madison, 1988.
16. Shine, M. C., and L. R. Fox, "Fatigue of Solder Joints in Surface Mount Devices," *ASTM STP 942,* Philadelphia, 1987, pp. 588–608.

14

Predicting Thermal and Mechanical Fatigue Lives from Isothermal Low Cycle Data

Harvey D. Solomon

The surface mounting of electronic components on printed wiring boards has exacerbated the problem of the fatigue of the solder joints that attach the component to the board. With surface mounting, the joints serve as the mechanical as well as the electrical connection for the component. As such, they must accommodate any relative displacement between the component and the board and support the stresses that are developed by this displacement.

Thermal displacements, resulting from the action of temperature changes on the different coefficients of expansion of the board and component, are a major problem. Cyclic temperature changes cause cyclic displacements, which can lead to thermal fatigue failures.[1] Thermal displacements are not limited to cases where the board and component have different coefficients of expansion. They can also develop due to temperature gradients, even when the board and component have the same coefficient of thermal expansion. The solder will also experience strains due to the difference in its coefficient of expansion compared to that of the component or board.

Fatigue failures are not limited to those produced by thermal displacements. Purely mechanical displacements of the component relative to the board can produce fatigue failures, even when the temperature is kept constant and there are no temperature gradients. Such displacements are developed when vibrations cause resonance or when impacts produce repetitive inertial forces. Fatigue, such as that due to mechanical displacements, is termed "isothermal fatigue" when it occurs at a constant temperature.

The long-term reliability of solder interconnections is determined by their resistance to fatigue, corrosion, creep rupture, or other such processes. This chapter will concentrate on solder fatigue but will also consider creep and environmental influences. Solder fatigue can be estimated by accelerated life testing of actual components and joints or by mechanical testing of modified laboratory specimens. The former approach has the advantage of testing the actual hardware but the disadvantage of being time consuming, with the results not easily applied to other designs or when the acceleration approach may not be appropriate. The mechanical approach allows for the generation of more generic data and the separation of the various variables that control the fatigue life. This chapter is concerned with this mechanical approach, in particular to results obtained in isothermal tests.

Isothermal, mechanical fatigue is important not only as a mode of solder joint failure, such as that due to vibrations, but also insofar as it can be used to describe and predict the more complex problem of thermal fatigue. Figure 14-1 shows the development of a displacement Δ due to the mechanical displacement of a component attached to a board by simple solder posts. Figure 14-1*a* shows the component and board at rest. Figure 14-1*b* shows the displacement of the component to the right by an amount Δ_R and FIgure 14-1*c* with a displacement to the left by Δ_L. For cycling from Δ_R to Δ_L the cyclic displacement is Δ, which is just $\Delta_R + \Delta_L$. The shear strain γ developed by this displacement is, for this very simple case, just

$$\gamma = \frac{\Delta}{h} \qquad (14\text{-}1)$$

where h is the height of the component above the board. Equation (14-1) is very simplified, using the simplest solder geometry and neglecting any constraint effects at the solder component and solder board interfaces. More detailed calculations will be discussed later. The important point to remember is that a cyclic repetition of this strain can cause a fatigue failure.

Figure 14-2 shows the same joint subjected to thermal strains. Again Figure 14-2*a* shows the at-rest case (i.e., at a state where there are no thermal strains). Figure 14-2*b* shows the case where the temperature is increased from T_0 to T_{max} and the coefficient of thermal expansion α_B is greater for the board than for that of the component α_C. This larger expansion is accommodated by shearing the solder joint. The assumption is being made that the component and board are not strained or bent. Later on we shall consider the more general situation when this is not the case. The expanded length of the chip carrier is given by $L[1 + \alpha_C(T_{max} - T_0)]$ and that of the board by $L[1 + \alpha_B(T_{max} - T_0)]$. The difference between these two lengths defines the net displacement of *both* joints. If it is assumed that both joints are equally displaced, the thermal displacement of each joint is given by $L/2(\alpha_B - \alpha_C)(T_{max} - T_0)$. Figure 14-2*c* shows the displacement

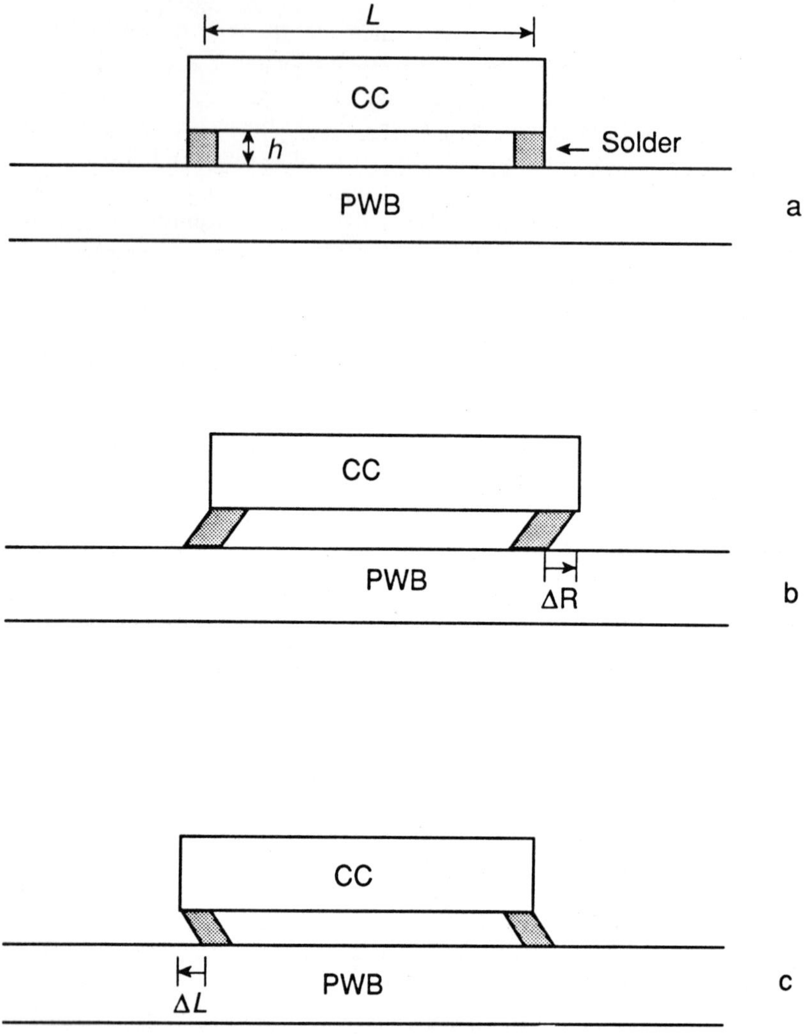

Figure 14-1 Development of mechanical shear strains.

developed by cooling from T_0 to T_{min}, with the thermal displacement of each joint being given by $L/2(\alpha_B - \alpha_C)(T_{min} - T_0)$.

The thermal displacement Δ developed by heating to T_{max} and then cooling to T_{min} is given by the difference in the displacements at T_{max} and T_{min} or by

$$\Delta = (\alpha_B - \alpha_C)(T_{max} - T_{min})\frac{L}{2} \qquad (14\text{-}2)$$

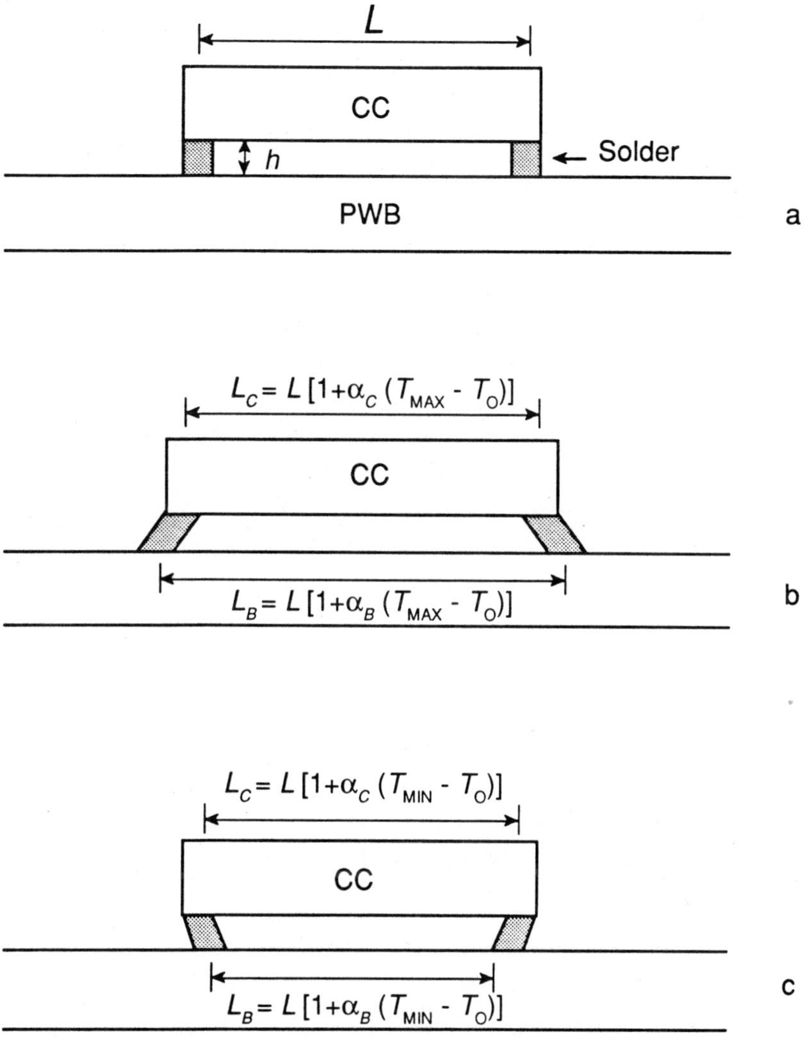

Figure 14-2 Development of thermal shear strains.

We are making the assumption that the coefficient of thermal expansion is not a function of temperature. This is not true in the general case. With $\alpha = f(T)$ Equation 14-2 becomes

$$\Delta = \frac{L}{2}\int_{T_{\min}}^{T_{\max}} (\alpha_B(T) - \alpha_C(T))dT \tag{14-3}$$

Equations 14-2 and 14-3 assume that the displacement is taken up equally by both the solder joints (on opposite sides of the component). This need not be the case. In general, the factor of 2 in Equation (14-2) or (14-3) can be replaced by a partition coefficient P that varies from 1 to 2. For $P = 2$ the displacement is shared equally by the opposite rows of joints, as is the case for Equations 14-2 and 14-3. For $P = 1$, all of the displacement is taken by a single side of the component. This can develop when the soldering is done asymmetrically, strengthening one side and causing all of the displacement to be taken up by the weaker side.

The shear strain associated with the displacement given by Equation (14-1) or (14-2) is approximated as just the displacement divided by the distance h between the component and the board, that is, for temperature independent coefficients of thermal expansion and using the partition coefficient P

$$\gamma = (\alpha_B - \alpha_C)(T_{\max} - T_{\min})\frac{L}{Ph} \tag{14-4}$$

As with Equation (14-1) the simplest geometry is assumed and more complicated questions of geometry are left for a subsequent discussion.

Equations 14-1 and 14-4 are the same for Δ of Equation (14-1) equal to

$$\Delta = (\alpha_B - \alpha_C)(T_{\max} - T_{\min})\frac{L}{P} \tag{14-5}$$

The development of a thermal strain is thus assumed to be equivalent to a mechanical displacement Δ, where Δ is given by Equation (14-5). Actually there are a number of additional factors that *might* alter this equivalence and these will be discussed later. They shall be neglected, however, for most of this discussion and the assumption shall be made that the fatigue produced by a thermal strain can be approximated by that due to a mechanical strain. This assumption, done to simplify the problem, is a customary procedure and serves as the basis for many design approaches used for thermal fatigue. For example, see the reference.[2] We shall concentrate on the factors that must be taken into account in applying isothermal test data to describe the influence of thermal strains and then consider how this approach might be altered to correct for the differences between isothermal, mechanical fatigue and thermal fatigue.

This chapter is based largely on the authors' data and represents his view of the problem. No attempt was made to synthesize all the other data available, although an attempt was made to cite it and discuss it where necessary. The reader is directed to the other chapters of this book to get an in-depth presentation of other views on this subject.

14.1 LOW CYCLE FATIGUE (LCF)

Fatigue life is related to the imposed cyclic stress or strain. Either stress or strain can be used. We shall consider only the use of strains because they are more relevant to thermal cycling. Equations (14-2) through (14-4) show that thermal cycling imposes a displacement or strain. This strain will generate a stress, so stress could be used to define the problem, but strain is used because it is the imposed variable; the stresses are only derivative. The fatigue life can be expressed in terms of the applied elastic strain, the plastic strain (including all nonelastic strains such as the anelastic strain), or by the total elastic plus plastic strain. The elastic dependence is given by the Basquin equation,[3]

$$N_f \gamma_e^{\beta} = \theta_B \tag{14-6}$$

where β, θ_B = constants

N_f = cycles to failure

γ_e = elastic shear strain

The plastic strain dependence of the fatigue life is given by the Coffin-Manson law,[4–7]

$$N_f \gamma_p^{\alpha} = \theta_{CM} \tag{14-7}$$

where α, θ_{CM} = constants

γ_p = plastic shear strain

Morrow[8] added the elastic and plastic strains to form the total strain and thus was able to define the fatigue life in terms of the total strain,

$$\gamma_T = \gamma_e + \gamma_p = \left(\frac{N_f}{\theta_B}\right)^{-1/\beta} + \left(\frac{N_f}{\theta_{CM}}\right)^{-1/\alpha} \tag{14-8}$$

Rather than use a shear strain, these expressions are often written in terms of the shear strain range, $\Delta\gamma$. This is the strain applied between the strain limits rather than just the maximum strain. With a zero minimum strain, the strain range and maximum strain are the same, but when equal positive and negative strains are employed (as is generally the case when the plastic strains are large) the strain range is twice the maximum strain.

It is important to remember that the fatigue life N_f in Equations 14-6 through 14-8 is the same for each expression. Remember that the fatigue life for a given total strain is N_f and that this strain is composed of an elastic and a plastic component, either one of which could be used to define the fatigue life. This is illustrated in Figure 14-3, which shows the fatigue life N_f correlated by the elastic, plastic, and total strains. It is generally the correlation with the plastic strain [i.e., the Coffin-Manson law, Equation (14-7)], or the total strain correlation, Equation (14-8), that is used to predict the fatigue life in thermal fatigue. This is because the elastic strain is generally small and is therefore a poor parameter to use. Also, stress relaxation with a fixed total strain (such as that occurring during a hold time at the maximum temperature or during slow temperature changes) causes the load to drop and the plastic strain to replace the elastic strain. In the limit, all the elastic strain will disappear and the total strain will be all plastic. In this case, to be discussed later, the Coffin-Manson law should be used. Experimentally, it is also better to use the plastic strain limits and the Coffin-Manson law because the flow stress, and hence the elastic strain, is dependent on the strain rate employed and this makes it difficult to correlate laboratory tests run at one strain rate and actual design conditions using another.

The strain developed by thermal cycling [Equation (14-4)] is used with Equation (14-8) to predict the fatigue life. If all of the strain is plastic (as is the case when stress relaxation causes all of the elastic strain to be replaced by plastic strain), Equation (14-7) is used to predict the fatigue life. The use of Equation (14-4), or a more elaborate version of this equation, coupled with Equation (14-7) or (14-8), is the customary method of handling thermal fatigue problems.[9,10] The differences in approach center on the handling of the influence of the definition of the fatigue life, the cycling frequency, hold times, peak temperature, temperature and strain rate dependence of the mechanical properties, and the

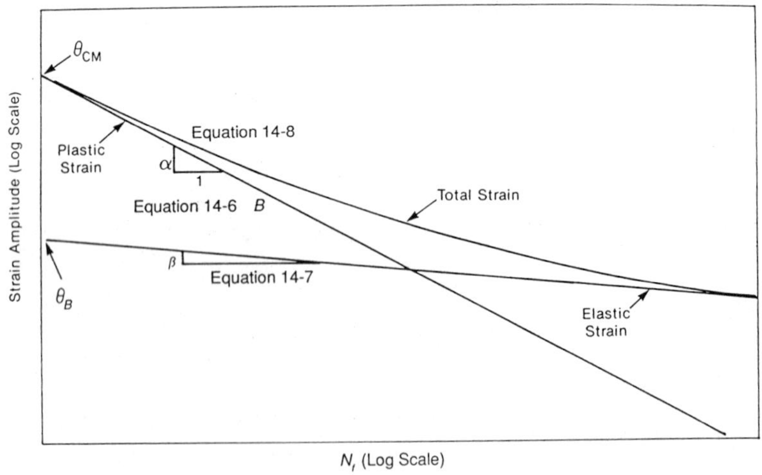

Figure 14-3 Fatigue life as correlated by the elastic, plastic, and total strains.

environmental and metallurgical changes that occur during thermal cycling but not necessarily in isothermal exposures. The following considers these variables in so far as the fatigue of solder is considered. Both eutectic and high Pb solders are considered.

14.2 LOW CYCLE FATIGUE OF SOLDERS—INFLUENCE OF THE DEFINITION FOR FAILURE

The problem of the fatigue failure of solder joints has prompted numerous studies,[11–107] particularly in the past few years. This increased interest is due to the increased problems encountered with surface mounted components. Strain-Fatigue Life data of the sort described by Equations (14-7) and (14-8) have been generated. An example of trend lines from two studies is shown in Figure 14-4. Both data sets were generated with shear loading of solder joints rather than bulk specimens. The Solomon data (the solid lines from the reference[29]) were generated with plastic strain limits, whereas total strain limits were used for the Wild data (the dashed lines from the references[13,14]). For plastic strains that are much larger than the elastic strains, the plastic and total strains will be essentially the same (see Fig. 14-3), enabling both to be compared in Figure 14-4. Solomon used the drop in the hysteresis load to define failure, and Wild used the first appearance of a crack or a joint resistance increase of 0.03 mΩ to define failure. Figure 14-4 highlights two important points: The slopes of the Solomon and Wild curves are different, and they are displaced according to the definition used

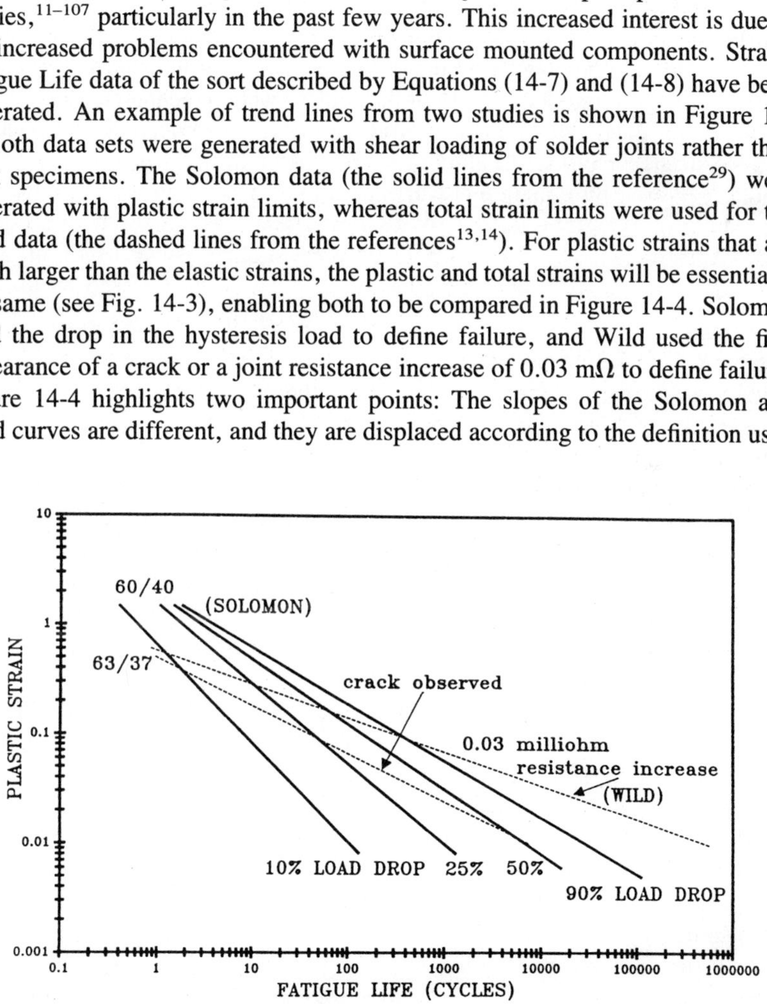

Figure 14-4 Strain life fatigue data for two studies. (From Solomon[29] and Wild[13,14])

for failure. These differences make the use of the data of Figure 14-4 to predict device lives, difficult at best. In this section we shall first consider the influence of the definition of failure; then the differences in the Coffin-Manson slope will be considered.

The difference in the position of the Wild curves is rather straightforward when one considers that, contrary to what one might expect, dc resistance measurements are not very sensitive to the presence of fatigue cracks.[104] Fatigue cycling results in the nucleation and growth of fatigue cracks, which increase the joint resistance. The resistance increase is not, however, a linear function of the crack length.[102] The resistance increase depends on the amount of the measured resistance that is due to the solder layer compared to that which is due to the rest of path being measured and to the fraction of the solder joint that is cracked. The ratio of the thickness of the solder crack to the total resistance path (corrected for conductivity changes between the various materials encompassed by the resistance path) is a scaling factor that determines the sensitivity. This is simply expressed by Equation (14-9), i.e., by

$$\frac{\Delta R}{R_o} = \frac{L_c A_o}{L_o A_c} \tag{14-9}$$

where $\Delta R/R_o$ is the change in resistance for a crack, or cracks, reducing the area conducting the current from A_o to A_c, L_c is the crack thickness (separation of the crack faces), and L_o is the length of the resistive path. As A_c increases so does $\Delta R/R_o$, but the change is scaled by L_c/L_o. When A_c goes to zero, the resistance change becomes infinite, providing the faces do not touch (i.e., L_c is not zero). Even a cracked joint can pass current if nonoxidized crack faces touch. Thus, it is possible to visually observe a fatigue crack while the resistance change might be barely resolvable or even unresolved above the background noise. A larger resistance change might not be produced until the crack is quite large (i.e., A_c is large relative to A_o).[104] Hence, a crack can be observed before a dc resistance increase of 0.03 mΩ is observed.

The growth of fatigue cracks also explains the difference between the Solomon curves. Here, the load required to produce the plastic strain that is being used in the test is the measure of the fatigue crack nucleation and growth. Figure 14-5 (from the reference[25]) shows the hysteresis loops generated in *tests run with a constant plastic strain* (fixed with an analog plastic strain computer). (For simplicity we shall talk of the strain, but it is actually a displacement that is being measured, with the strain being calculated by dividing by the thickness of the solder layer.) The plastic strain includes the anelastic as well as the true plastic strain. It is determined by subtracting out the linear stress-strain portion of the hysteresis loop.

The tilted loops of Figure 14-5 used the signal directly from the extensometer; the vertical loops were drawn with the output from the plastic strain computer,

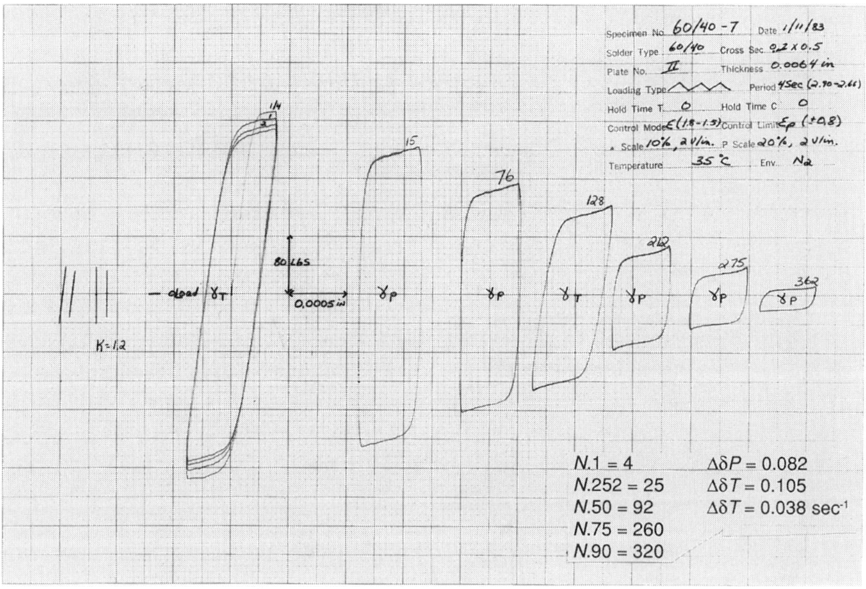

Figure 14-5 Typical hysteresis loops. (From Solomon[25])

where the elastic strain was subtracted out from the total strain signal. The plastic strain computer keeps the loop width (plastic strain) constant throughout the test. As cycling continues, the load required to produce the plastic strain decreases (the cycle number is written above the hysteresis loop). Figure 14-6 plots the load drop parameter $\Phi = 1 - (\Delta P/\Delta P_M)$ versus the cycle number; ΔP is the hysteresis load range at any cycle and ΔP_M is the maximum hysteresis load range (developed on the first complete cycle or after the first few cycles). In this experiment, the test was stopped periodically and the specimen was removed for ultrasonic microscopy.[25,108] This technique allows the cracked areas (the shaded areas in the inserts) to be determined. As can be seen, the load drop is associated with the increase in the cracked areas. Using a larger load drop definition for failure requires more crack growth and more cycles; hence the strain-life curves are displaced to longer lives.

The degree to which the fatigue life is dependent on the load drop definition used for failure is a function of the compliance of the testing machine, the plastic strain employed in the test, and the control limit used in the test.[107] With total displacement testing limits, a relatively soft testing machine, and small plastic strains, the load decrease is abrupt, with little difference between the number of cycles for the initial load drop and larger load drops. This is caused by increases in the plastic strain range with cycling. The data of Figure 14-6 through 14-10 were, however, generated with plastic strain limits. In effect, this is equivalent to using an infinitely stiff testing machine (because the plastic strain computer

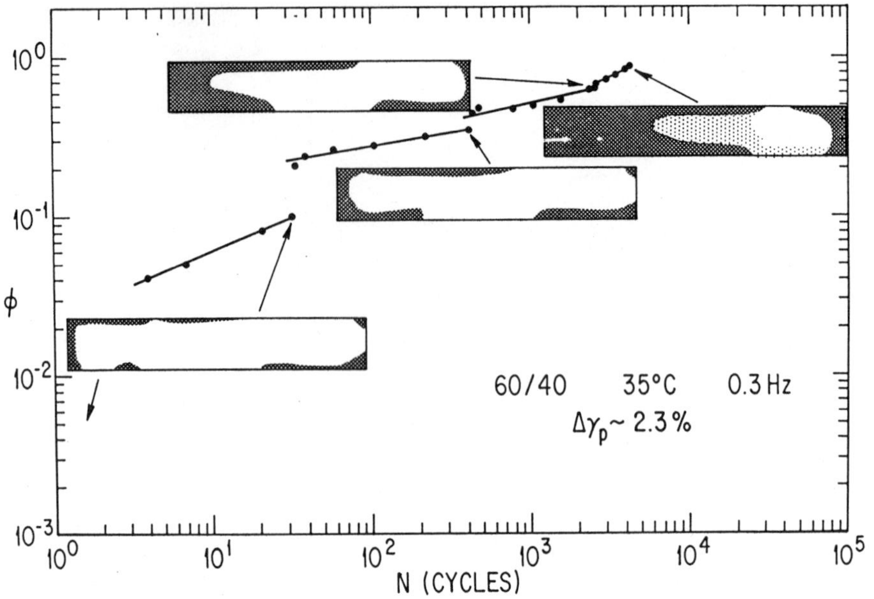

Figure 14-6 Load drop parameter Φ versus the cycle number of an interrupted test. The inserts show the cracked area as measured by ultrasonic microscopy. (From Solomon[25])

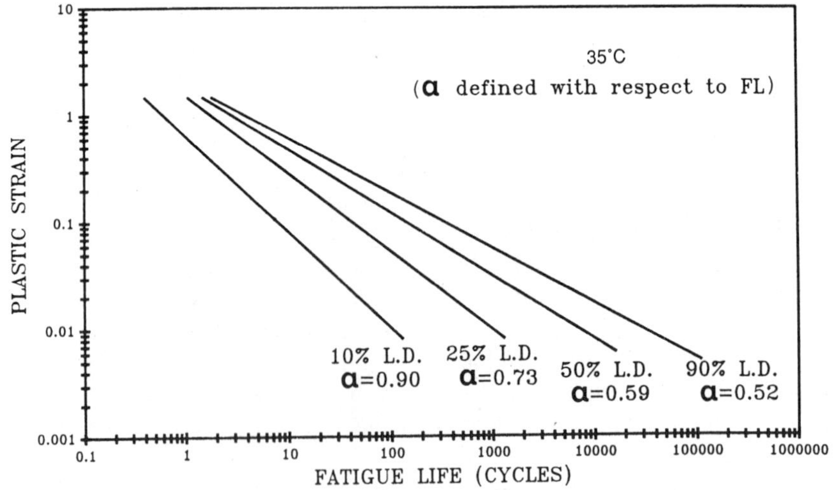

Figure 14-7 Low cycle fatigue life curves for 60% Sn-40% Pb with the fatigue life defined by different load drops. (From Solomon[29])

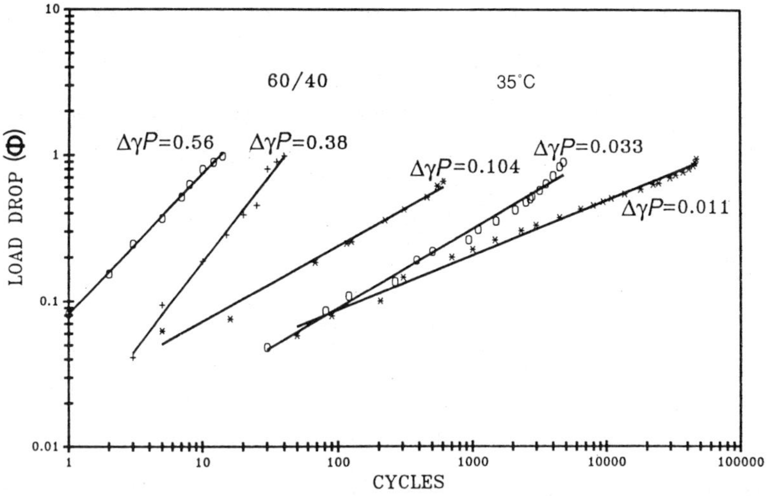

Figure 14-8 Load drop parameter Φ versus the cycle number for different applied plastic strains for 60% Sn-40% Pb solder.

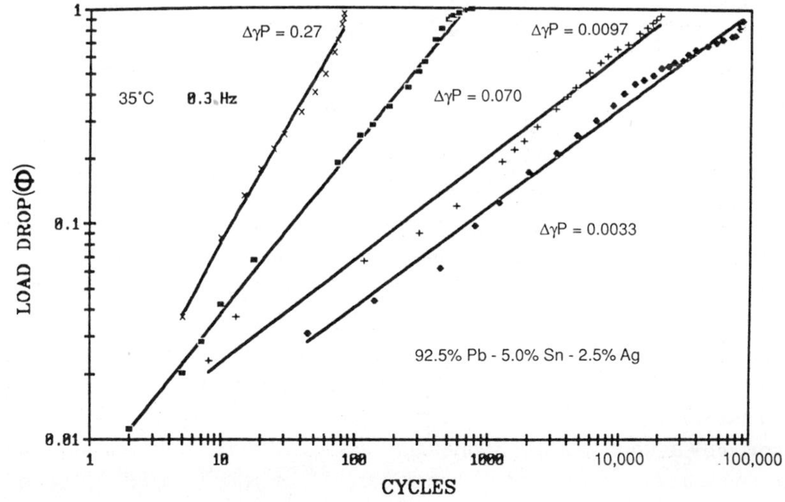

Figure 14-9 Load drop parameter Φ versus the cycle number for different applied plastic strains for 92.5% Pb-5.0% Sn-2.5 Ag solder. (From Solomon[106])

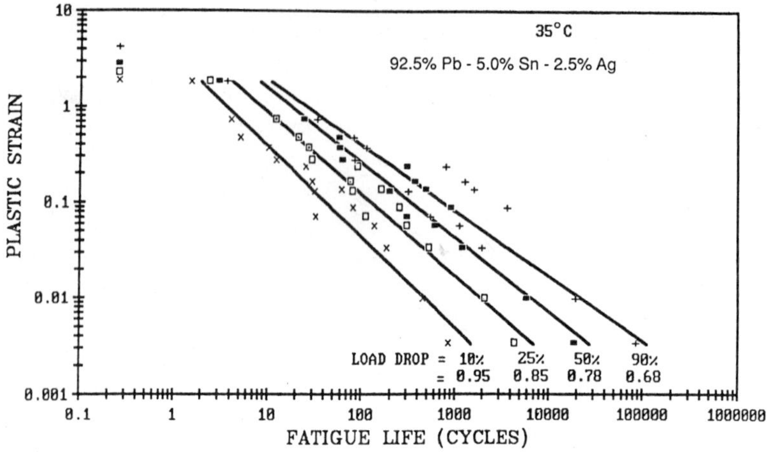

Figure 14-10 Low cycle fatigue life curves for 92.5% Pb-5.0% Sn-2.5% Ag solder with the fatigue life defined by different load drops. (From Solomon[106])

subtracts out the elastic strains and controls only on the plastic strains). The plastic strains are kept constant and the load drop is gradual rather than abrupt. The fatigue life is therefore a strong function of the definition of failure and this must be accounted for before the data can be used in any meaningful manner.

The problem of the dependence of the magnitude of the fatigue life on the definition of failure is inherent in all fatigue tests, but it becomes most critical when one tries to predict the life of solder joints. This is because the actual joints are generally *smaller,* often by an order of magnitude, than the laboratory joints with which the strain life data was generated. With larger equipment, such as turbine discs or aircraft frames, the laboratory specimens are generally *smaller* than the actual device that they are meant to simulate. In these cases the lives determined from laboratory specimens are generally conservative since a crack growing through the laboratory specimen would be contained within the larger device (provided brittle fracture does not intervene, in which case there may be little or no difference between the laboratory test and the part life). This is not the case when the device (in this case a solder joint) is smaller than the laboratory specimen and brittle fracture does not occur.

Another factor that makes the dependence of the fatigue life on the definition of failure critical for solder joint life prediction is the fact that the loading is in shear on a solder layer. In this case, the crack growth rate is, to a first approximation, only a function of the applied plastic shear strain and not also a function of the crack length.[25] This conclusion is based[109] on Rice's analysis[110] of the J integral of a crack in simple shear, with semi-infinite boundary conditions. It is supported[25,29,89,106] by the observation that, for tests run with a constant plastic strain, the load drop (which is proportional to the crack growth rate) does not increase as the cracking gets more severe (for instance, see Figure 14-6). If the applied plastic strain is kept constant, to a first approximation, the crack growth rate will be constant. This means that increasing the specimen size increases the fatigue life because the crack must grow longer to produce failure. It also means that *the life of a test specimen will be longer than that of a smaller solder joint subjected to the same plastic strain.*

With tensile loading (which is more typically used in other design situations), the crack growth rate increases as the crack length gets larger (because the stress intensity and J integral are functions of the crack length as well as the applied plastic strain or stress level). As the crack grows longer, its growth rate increases. Increasing the specimen size produces a less than proportional increase in the fatigue life. For instance, doubling the length to which a crack must grow to produce failure does *not* double the number of cycles required to achieve this crack growth (because the crack growth rate is increasing as the crack length is getting larger). This tends to minimize the influence of the laboratory specimen size relative to that of the device.

With a crack growth rate independent of the crack length, the crack growth rate can be easily determined from the LCF data. For a constant applied plastic strain, which to a first approximation makes the crack growth rate constant, the cyclic crack growth rate $\Delta a/\Delta N$ is just the distance D over which the crack must grow to produce failure, divided by N_f, the number of cycles to produce failure, that is,

$$\frac{\Delta a}{\Delta N} = \frac{D}{N_f} \tag{14-10}$$

As will be seen, D can be thought of as a normalization parameter for the different load drop data. The larger the load drop, the larger N_f and hence the larger D to give a constant crack growth rate for a given applied plastic strain range.

For a real joint subjected to the same plastic shear strain, the crack growth rate should be the same. Failure would then be determined by the length L over which the crack must grow. Equating the crack growth rate in the laboratory test (for the crack growing a distance D in $N_{f(\text{lab})}$ cycles) with that which should occur in a real joint one gets

$$N_f = N_{f(\text{lab})}\frac{L}{D} \tag{14-11}$$

The fatigue life should scale according to the ratio of L/D. The fatigue life, developed in a joint that is smaller than that used in laboratory tests, will be smaller by L/D. The reverse is true if the joint of interest is larger than that of the laboratory joint.

Distance D is determined not only by the geometry of the laboratory specimen but also by the definition used for failure. For instance, for a 50 percent load drop definition for failure, the cracked area will be one-half the joint area (assuming that all of the load drop is due to cracking of the joint and that fatigue softening or hardening can be neglected). For a 25 percent load drop definition for failure, the cracked area must be one-quarter of the joint area. The exact determination of D is difficult because the cracks do not have simple linear fronts that extend across the width of the specimen and grow in the direction of shearing. The crack front is irregular and while it grows in the general direction of shearing, it does not do so in a completely linear manner. However, some simple approximation can be made by assuming that the crack front is linear and goes across the width of the joint.

Figure 14-7 shows the fatigue life curves for 60% Sn-40% Pb solder with failure defined by 10, 25, 50, and 90 percent decreases in the hysteresis load range required to produce a given plastic strain range. Two difficulties are apparent in applying the idea of using a D value to normalize this data. First, the curves are not displaced according to the load drop, that is to say, the 50 percent load drop curve is not at twice the fatigue life of the 25 percent load drop curve or at 5 times the life of the 10 percent load drop data. Secondly, the curves are not parallel; that is to say the Coffin-Manson exponent α is dependent on the load drop definition for failure. These problems can be handled by assuming that a different number of cracks are growing at each load drop level (an assumption that is supported by the ultrasonic microscopy data).

The joint used for the data of Figure 14-6 is 0.1 × 0.5 in (2.54 × 12.7 mm). Assuming that the crack growth is in the 0.5-in (12.7-mm) direction (the direction of shearing), a single crack propagating to reduce the load by one-half would require crack growth of 0.25 in (6.35 mm). Figure 14-6 shows that the actual situation is more complex because there is more than one crack growing. If two cracks were present, the average crack growth of each, D, is 0.25 / 2 = 0.125 in. (3.18 mm). Figure 14-6 also shows that the number of growing cracks decreases, due to crack coalitions, as the test progresses. At the start of the test there may be more than four distinct cracks present, but by the time the load has dropped by 90 percent there are only one or two cracks present. It was found[105] that the fatigue curves for different load drops could be normalized by assuming that for a 25 percent load drop, there were six growing cracks, for a 50 percent load drop there were three growing cracks, and for a 90 percent load drop, there were one and one-half growing cracks.

If the strain range in a joint is $\Delta\gamma_p$ over a length L, the life is given by Equation (14-10) for N_f and D defined by the same load drop and N_f determined by $\Delta\gamma_p$. For instance, in Figure 14-7, for a plastic strain of 10 percent, the fatigue life in this laboratory test will be 9, 50, 170, and 400 cycles for the 10, 25, 50, and 90 percent load drop curves, respectively. The value for D will be different for each load drop, reflecting the different crack lengths required to produce this load drop and differences in the number of cracks that may be present. Assuming that 10, 6, 3, and 1.5 cracks are present at 10, 25, 50, and 90 percent load drop, respectively, D will be $0.05 / 10 = 0.005$, $0.125 / 6 = 0.021$, $0.25 / 3 = 0.083$, and $0.45 / 1.5 = 0.3$, respectively. These values can be thought of as correction factors to normalize the different fatigue curves shown in Figure 14-7, that is, the values of N_f / D are roughly constant (at worst, within a factor of 2 or 3 for different load drops and plastic strain ranges). Furthermore, the exact choices, for the number of cracks growing at any load drop, were made to correlate laboratory data and tests run[105] on actual chip carrier-printed wiring board (CC-PWB) joints. The ratio of the D values normalizes the different load drop curves, but the choice of the exact D value was based on the CC-PWB joint data of the reference.[105] D is not, however, a completely disposable curve fitting parameter since the values used are based on actual joint geometries and the number of cracks assumed to be growing is consistent with the ultrasonic microscopy measurements.

If all the curves were parallel and displaced according to the load drop (i.e., the 50 percent load drop curve was at just twice the fatigue life of the 25 percent load drop curve), the curves would be normalized by assuming that the same number of cracks was growing at different stages of the test (instead of using a different number of cracks) and that a 50 percent load drop required just twice as much crack growth as a 25 percent load drop. This was the sort of behavior observed at 150°C.[29]

The fatigue curves can also be normalized by calculating a crack growth rate directly from the load drop curves. Figure 14-8 shows some of the load drop curves used to develop Figure 14-7. Here the test was not interrupted and a single straight line is developed for each experiment run with a different applied plastic strain range. These curves fit to a simple power law expression, that is,

$$\phi = AN^B \tag{14-12}$$

where the constants A and B are functions of the applied plastic strain range according to

$$A = A_0 \Delta\gamma_P{}^{A_1} \tag{14-13}$$

and

$$B = B_0 \Delta\gamma_P{}^{B_1}. \tag{14-14}$$

Assuming that the crack growth rate da / dN is proportional to the load drop rate $d\phi/dN$, and using Equations 14-12 through 14-14, it can be shown[29] that

$$\frac{da}{dN} \approx A_0 B_0 \Delta\gamma_p (A_1 + B_1)_N (B_0 \Delta\gamma_P{}^{B_1} - 1) \qquad (14\text{-}15)$$

If one further assumes that the fatigue crack extends across the width of the joint, propagating in the length direction, the crack growth rate will be equal to the right-hand side of Equation (14-15) instead of being just proportional to it. Equation (14-15) assumes that a single crack is present and corrects for the presence of multiple cracks or other factors that might influence the load drop curves by altering the constants A_0, A_1, B_0, and B_1. However, Equation (14-15) can be difficult to use because unless $B = 1$ da/dN will be a function of N. Since B is generally less than 1, this reflects a slowdown in the load drop, which was handled in the previous approach by decreasing the number of cracks that were assumed to be growing.

When $B = 1$, Equation (14-15) can be integrated to yield Equation (14-7) with $\alpha = 1/(A_1 + B_1)$ and for $B_1 = 0$ and $B_0 = 1$ (which gives $B = 1$) $\alpha = 1/A_1$. With B not equal to 1, the correlation of α with A and B is more complex but describes the variation in α with the definition of failure.[29] This is the situation that prevails for 60% Sn-40% Pb solder at 35°C. At 125 and 150°C B_1 is small enough to be considered zero and B_0 is close to 1, which keeps $B = 1$ and α independent of plastic strain and the definition of failure.[29]

The problem of the influence of the definition of failure on the fatigue life is not just a problem for 60% Sn-40% Pb solder; it is a general one. This is shown by the load drop data of Figure 14-9, which is for a Pb-rich solder (92.5% Pb-2.5% Ag-5.0% Sn), and by the fatigue life curves that can be determined from this data (i.e., Figure 14-10). As with the 60% Sn-40% Pb data, the fatigue life curves are not parallel (i.e., α decreases with increasing the load drop definition for failure). The same D values will normalize the curves, that is, give roughly (at worst within a factor of 2 or 3) the same values for N_f/D for different load drop definitions of failure, but tests of the sort described in the reference[105] have not been run, so it is not demonstrated that these D values will correlate the laboratory data with tests on actual joints. Also in agreement with the 60% Sn-40% Pb data, at 150°C this high Pb solder exhibited parallel straight strain life Coffin-Manson curves,[106] but again correlations of laboratory and tests on actual joints are lacking.

The important points to remember are

1. In tests using plastic shear strain limits, the rate of drop in the hysteresis load required to produce a given plastic strain does not accelerate as the load drops. This means that the fatigue life to produce a given load drop is highly dependent on the choice of the load drop.

2. This effect should be less in tests using tensile loading because the crack growth rate will accelerate as the crack gets longer, thereby causing the rate of load drop to accelerate. This lessens the differences between the fatigue life defined by different load drops.
3. When the definition of failure has a large influence on the fatigue life, it is necessary to normalize the data to reduce this effect. This was done by treating the LCF data as if it were crack growth data. This necessitated several simplifying assumptions, but at worst the normalization could be done to within a factor of 2 or 3 (generally the normalization was better than a factor of 2, especially when the 10 percent load drop data are disregarded).
4. The question of the influence of the definition of failure is particularly important when considering small solder joints. Here the laboratory test joints are often larger than the joint of interest. The combination of a larger laboratory joint and a dependence of the fatigue life on the definition of failure can make uncorrected laboratory fatigue data highly nonconservative.

Subrahmanyan et al.[95,97] have also treated LCF data in terms of crack growth, only they use a time-based approach rather than a cycle-based approach; therefore a time dependent constitutive equation can be used. They employ a crack length integral that defines the total crack length to failure and the crack growth per cycle that is defined as the damage integral. While their approach uses a different crack growth expression (they use a tensile expression in which the crack growth is related to the crack length and a rate constant), the final equation for the fatigue life is similar to Equation (14-11). They relate failure to the total crack growth (defined by the crack growth integral), divided by the crack growth per cycle (defined by the damage integral). This is in effect what is done by Equations (14-10) and (14-11). Subrahmanyan et al.[95] have successfully used their approach to apply Solomon's data[25,29] to predict the thermal fatigue results of Hall and Sherry[55] for various temperature changes.

14.3 INFLUENCE OF THE TEMPERATURE

The use of isothermal LCF data to predict the behavior of thermally cycled joints hinges on the assumption that the strains developed by thermal cycling can be modeled by mechanical strains developed at a constant temperature. The question arises as to what temperature to use to model cycling over a range of temperature. Taira[9] considers two possibilities: the use of the mean temperature for the thermal cycle or, when creep and high temperature environmental interactions are possible, the use of the highest temperature. Fortunately, for 60% Sn-40% Pb solder and for the high Pb solder discussed here, there is not a great deal of difference in the fatigue behavior at these different temperatures. This is shown in Figures 14-11 and 14-12 for a 50 percent load drop.[24,106] There is a temperature dependence of the fatigue life, but it is not very large one nor, considering the scatter in the data, is it a very significant one. For the 60% Sn-40% Pb solder the fatigue life for a plastic strain range of 5 percent (a middle range strain), the fatigue life decreases from 500 cycles at 35°C to 200 cycles at 125°C and to about 120 at 150°C (not shown in Figure 14-11 but provided in the reference[24]). The high Pb solder shows a somewhat lower temperature dependence of the fatigue life. The 5 percent plastic strain life at 35°C is 700 cycles and decreases to only 300 cycles at 150°C.

Norris and Landzberg[70] also provide data concerning the temperature dependence of the fatigue life. The data they present is for a 95% Pb-5% Sn solder (similar to the high Pb solder of Figure 14-12) but for a very small *total* strain (0.03 percent), which for the high cycling frequency of 1800 cycles/min, will develop even a smaller plastic strain. The fatigue life was reduced from 3.5×10^5 cycles to 10^5 cycles when the temperature was increased from 35 to 150°C. This factor of 3.5 is similar to that shown in Figure 14-12 where the ratio of the fatigue lives is about 2.5. At −50°C, however, Norris and Landzberg show a factor of 200 increase in the fatigue life over that at 150°C. This may be a manifestation of the difference in the slopes of the fatigue life curves shown in Figure 14-12. The lower slope of the −50°C data curve means that when extrapolated to very small plastic strains, the fatigue life will be greatly increased. At large plastic strains, however, the fatigue life at −50°C is actually lower

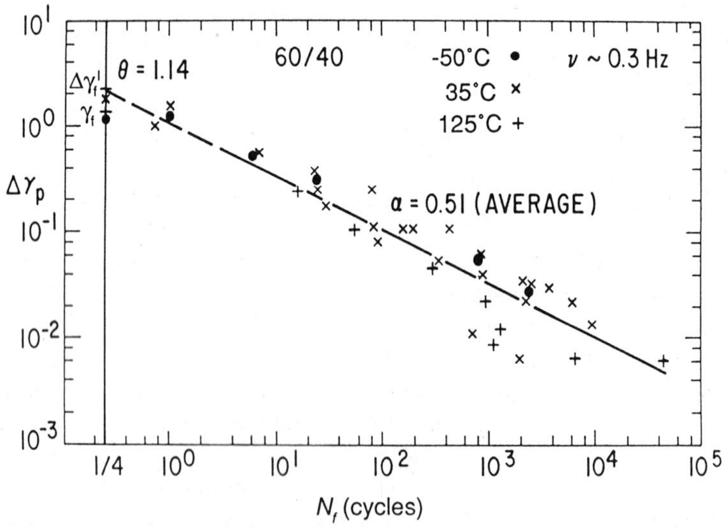

Figure 14-11 Low cycle fatigue curves for a 50 percent load drop definition of failure for tests on 60% Sn-60% Pb solder run at −50, +35, and +125°C. (From Solomon[24])

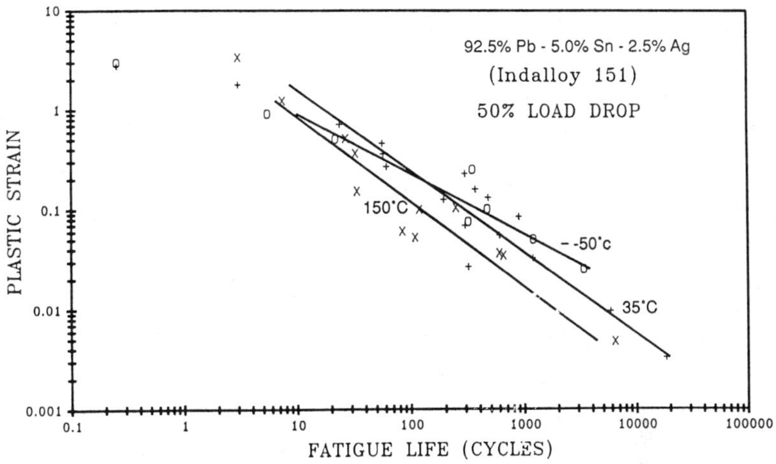

Figure 14-12 Low cycle fatigue curves for a 50 percent load drop definition of failure for tests on 92.5% Pb-2.5% Ag-5.0% Sn solder run at −50, +35, and +150°C. (From Solomon[106])

than that at 150°C. This points up a significant error that can develop by just multiplying the Coffin-Manson fatigue law by a temperature correction factor (as is done by Norris and Landzberg) instead of using the correct Coffin-Manson curve for the temperature of interest.

The data of Figures 14-11 and 14-12 show that the fatigue life is a function of the cycling temperature but that even a factor of 2 or 3 in the relative fatigue life is not very significant with respect to the scatter in the data. The most conservative approach would be to use the peak temperature since this predicts the lowest fatigue life. This is also the temperature to consider when there is a hold time at this peak temperature because as we shall see in the next section, this will cause much of the strain to be developed at this temperature.

14.4 INFLUENCE OF HOLD TIMES AND CYCLING FREQUENCY

The previous discussion has been limited to the results of tests that used a relatively high cycling frequency (0.25 Hz or higher). Cycling at a relatively high frequency speeds up laboratory tests but does not duplicate the actual service behavior of most solder joints.

Service temperature changes seldom ramp up (experience a linear temperature increase) and then ramp down. Rather the temperature is held constant for a period of time before changing. In the case of real service joints this hold time may be a matter of hours, a day, or even longer. To speed up laboratory testing, laboratory tests that incorporate a hold time typically use hold times of an hour or less. Another laboratory approach is to use a slower ramp rate (i.e., a slow linear temperature change of strain rate) with no hold time instead of using a rapid ramp rate and a hold time. The use of a hold time or slow ramp rate is important because hold times or slow ramp rates reduce the fatigue life relative to that achieved with a high ramp rate and no hold time. It is thus necessary to correct the fatigue life measured at a rapid cycling rate when applying this data to a design situation that incorporates a hold time or slow cycling frequency.

The question of the influence of hold times or slow cycling was first studied in Pb[111,112] in the early 1950s. This work was the basis for studies of the influence of hold times and slow cycling on the high temperature fatigue of superalloys[113–116] and more recently of solders.[70] A eutectic solder at 25°C is at 65 percent of its melting point (in degrees K), which is a higher relative use temperature than that of most superalloys. It is therefore not surprising that the fatigue of solders must be handled by using many of the same techniques developed for superalloys and that these techniques were first developed to describe the fatigue of Pb, for which room temperature is also a relatively high homologous temperature.

Two different approaches have been developed to describe frequency and hold time effects. Coffin[113,115] developed a frequency modified version of the Coffin-Manson equation and Manson, et al.[117,118] developed strain range partitioning (SRP). We shall first consider Coffin's frequency-modified Coffin-Manson equation, which is given by

$$(N_f \nu^{(K-1)}) \gamma_p{}^{\alpha} = \theta_{CM} \tag{14-16}$$

This is the same as Equation (14-5) but with the addition of a frequency ν term, $\nu^{(K-1)}$, which modifies the fatigue life. Both ramp and hold time effects are considered because the frequency is just the inverse of the period, which is the sum of the ramp times (i.e., the time to load and unload) and the hold times (i.e., hold times at the maximum and or minimum loads or strains). The effects of the frequency are determined by the magnitude of K. For $K = 1$, there is no dependence of the fatigue life with frequency variations and the frequency term is just equal to 1. For $K = 0$ the fatigue life is modified by $1/\nu$. Since N/ν is just the time to failure, this results in the development of a constant time to failure (for a given applied plastic strain range). When $K = 0$, if the frequency is halved, the number of cycles to failure is halved, to give a constant time to failure.

Figures 14-13 through 14-15 show the fatigue life as a function of the cycling frequency for tests on 60% Sn-40% Pb solder, run at 35 and 150°C.[27] All of the data can reasonably be fit to two curves, one with K equal to or close to 1 and the other with K equal to or close to 0. There is thus a frequency above which there is little or no frequency effect and below which there is a large frequency effect. This behavior is typical of what is often observed in superalloys.[114–116] The existence of two regimes with different K values presents a complication to the use of the frequency-modified fatigue life but does not prevent its use. Over the range of about 10^{-4} to 1 Hz, a single K value can be used

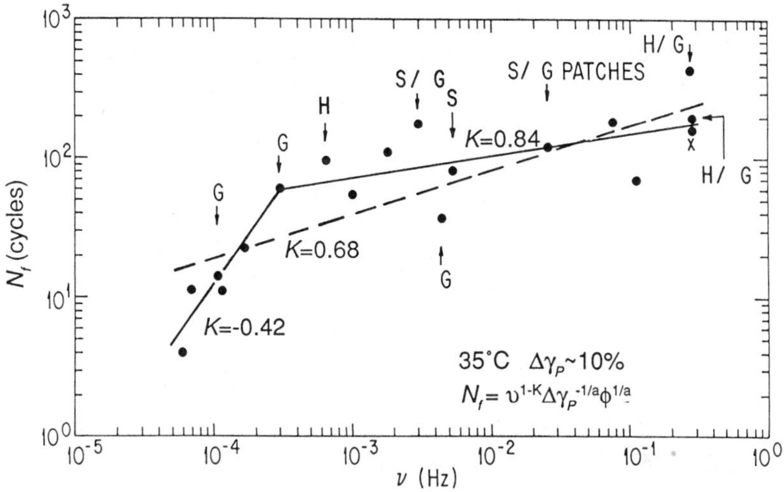

Figure 14-13 Fatigue life versus cycling frequency for 60% Sn-40% Pb solder tested at 35°C and with an applied plastic strain of 10 percent. The notations G, H, S, etc., refer to the nature of the fracture surface. (From Solomon[27])

without too much error. This was the approach taken by Norris and Landzberg.[70] Based on the lead data mentioned earlier, they used a parameter $-k$ instead of $K - 1$ as is used in Equation 14-16. For a 99% Pb-1% Sb alloy, at 30°C, they found $k = 0.33$, which corresponds to $K = 0.67$. This is in good agreement with the average value for 60% Sn-40% Pb solder, shown in Figure 14-13.

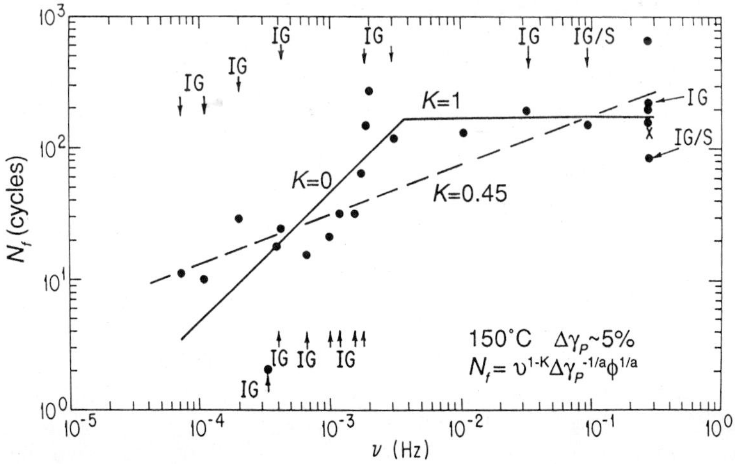

Figure 14-14 Fatigue life versus cycling frequency for 60% Sn-40% Pb solder tested at 150°C and with an applied plastic strain of 5 percent. The notations IG, H, etc., refer to the nature of the fracture surface. (From Solomon[27])

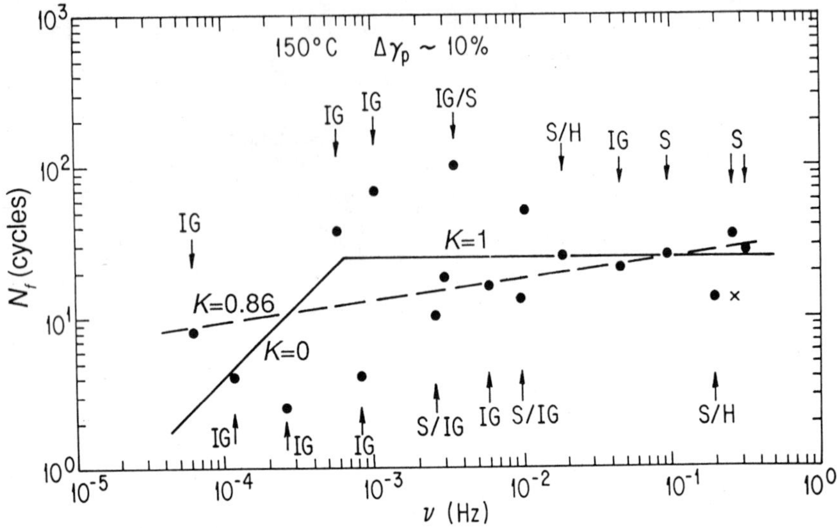

Figure 14-15 Fatigue life versus cycling frequency for 60% Sn-40% Pb solder tested at 150°C and with an applied plastic strain of 10 percent. The notations IG, H, etc., refer to the nature of the fracture surface. (From Solomon[27])

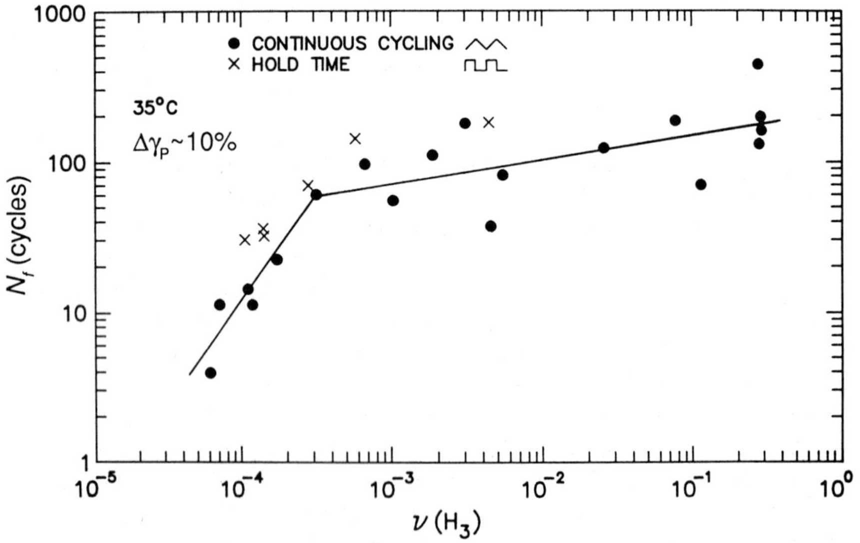

Figure 14-16 Comparison of the 35°C fatigue life N_f versus cycling frequency for continuous cycling and cycling with a hold time for 60% Sn-40% Pb solder. (From Solomon[28])

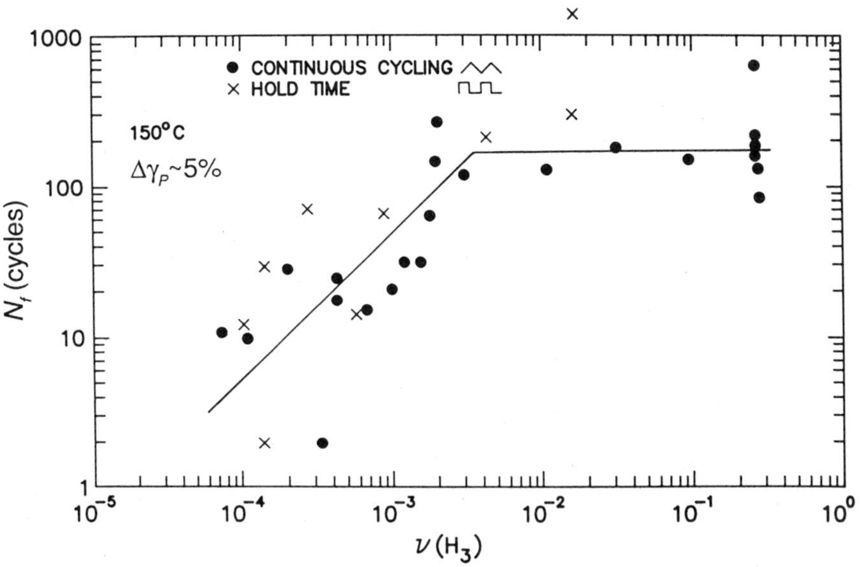

Figure 14-17 Comparison of the 150°C fatigue life N_f versus cycling frequency for continuous cycling and cycling with a hold time for 60% Sn-40% Pb solder. (From Solomon[28])

Figures 14-16 and 14-17 show results for hold time tests[28] superimposed on the continuous cycling data of Figures 14-13 and 14-14. As can be seen, the results are similar, with perhaps a slightly larger fatigue life for the hold time tests (compared at the same frequencies). Vaynman[101] has shown that a high Pb solder (96.5% Pb-3.5% Sn) exhibits similar frequency-life behavior as is shown in Figures 14-13 through 14-17 but with the break in the curve occurring at about 0.01 Hz rather than at the lower frequency exhibited by the 60% Sn-40% Pb solder. This solder also exhibited a more pronounced transition from transgranular to intergranular fracture with increasing hold time.

Figures 14-13 through 14-17 show that the transition from $K = 1$ to $K = 0$ is temperature dependent. With a hold time, one should use the data for the temperature of the hold to determine the appropriate K value to use. There is also a question of how the applied strain (or strain developed during the hold time) and the frequency effect interact. Is K or the transition frequency dependent on the applied plastic strain? Figure 14-18 shows the 35°C strain-life data, with two assumptions regarding the relative interaction between the plastic strain and frequency effect.[24] One set of curves assumes that K and the transition frequency are independent of the applied plastic strain. The dotted line shows another extreme. Here it is assumed that there is a single $K = 0$ curve (as seems to be the case at 150°C,[24]) with the transition frequency given by the intersection of the $K = 0$ and $K = 1$ curves. At frequencies below this transition frequency,

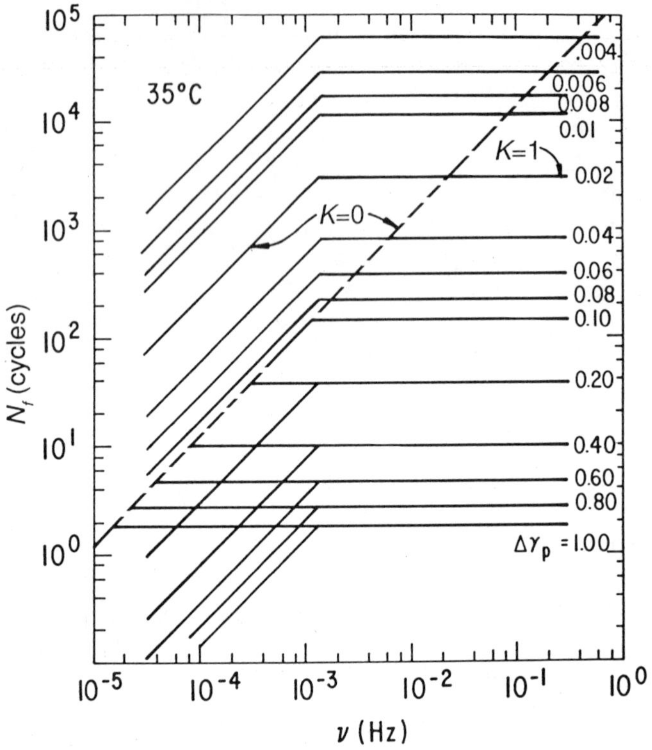

Figure 14-18 Fatigue life as a function of the cycling frequency and applied plastic strain (curves established from 35°C 60% Sn-40% Pb solder data. (From Solomon[24])

the predicted fatigue life ranges between the dotted and solid curves. At higher frequencies it is just given by the solid curve. At low frequencies there can be orders of magnitude between the two predicted extremes. Using the dotted curve is an extremely conservative approach, perhaps overly so.

With a single K = 0 curve (the dotted curve of Figure 14-18) to describe the data, the fatigue life is no longer a function of the applied strain and the Coffin-Manson expression is no longer valid. This represents an extreme limit. In general, however, below the transition frequency, α of the Coffin-Manson expression will be a function of frequency. It will increase,[24] but not necessarily to infinity, as is implied by the dotted curve of Figure 14-18.

The strain range partitioning approach assumes that there are different Coffin-Manson curves for creep and time independent plastic (noncreep) strains. The problem then becomes one of determining the fraction of creep making up the hysteresis. An interactive damage law is used to add up the life determined by the fraction creep, that due to the fraction time independent plastic (noncreep) deformation, and the fraction due to mixed cycling with creep on loading and plastic (noncreep) on unloading or vice versa. The creep Coffin-Manson fatigue life is usually less than that of the noncreep curve. Thus the overall fatigue life, as determined by the interactive damage law, is reduced when the fraction of creep is increased.

The strain range partitioning approach can also be applied to thermal fatigue[117] by developing the stress-strain hysteresis and determining the fraction of creep for each strain increment (each occurring at a different temperature, in the temperature cycling range).

Reducing the cycling frequency, as in done in the data of Figures 14-13 through 14-15, results in more creep deformation. A simple way to determine the amount of creep in the hysteresis is to compare the creep strain with that developed by dynamically straining the test specimen. This comparison is made for 60% Sn-40% Pb solder at 35 and 150°C in Figure 14-19.[26] A single stress-strain rate curve describes the strain rate data, independent of the strain range and cycling frequency employed. This means that the deformation of the 60% Sn-40% Pb solder is Non-Newtonian viscous, in that the flow stress is related to the strain rate raised to a power other than 1. The strain rate and creep curves are roughly parallel, meaning that the fraction creep does not change very much as the cycling frequency is reduced. Figure 14-20 shows computer calculations[26] of the percentage of creep developed from creep curves and strain rate data (a more accurate approach than by just comparing the curves of Figure 14-19). At 35°C, at high strains, there is an increase in the fraction of creep when the frequency is decreased from 0.25 Hz to about 10^{-2} Hz but much less of an increase when the frequency is reduced still further. This increase in the fraction of creep should decrease the fatigue life, but as Figure 14-13 shows, there is no decrease in the fatigue life when the frequency is reduced from 0.25 to 10^{-2} Hz. There is no

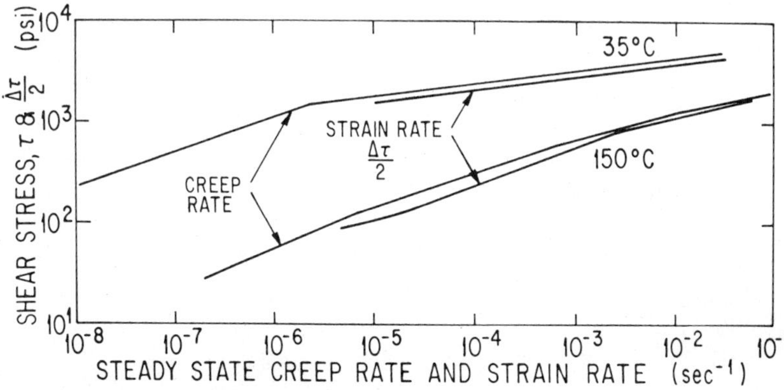

Figure 14-19 Comparison of 60% Sn-40% Pb solder creep and strain rate sensitivity data. (From Solomon[26])

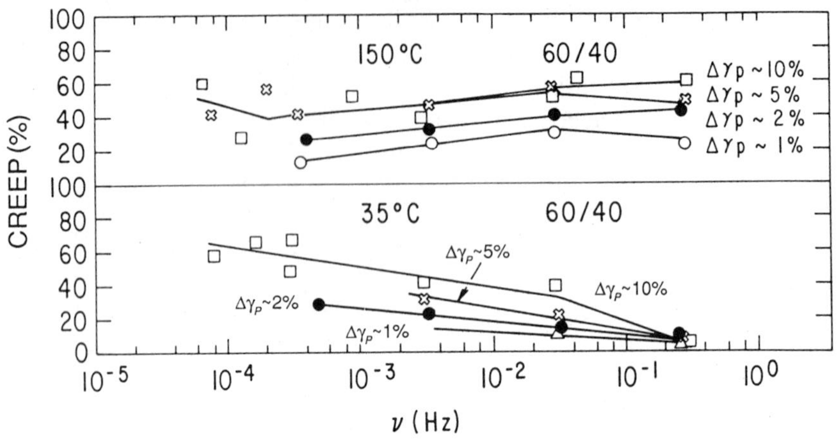

Figure 14-20 Calculated percentage of creep in the hysteresis behavior of 60% Sn-40% Pb solder cycled at 35 and 150°C. (From Solomon[26])

increase in the fraction creep below about 2.5 × 10^{-4} Hz, but there is a large decrease in the fatigue life. For a given applied plastic strain at 150°C there is little or no change in the percentage of creep, even though the fatigue life shows a marked decrease when the frequency is reduced below about 10^{-3} Hz.

It is clear that strain range partitioning in its standard form does not describe the 60% Sn-40% Pb solder data presented here (see the reference[27] for a more detailed comparison of SRP predictions and the test data). This failure stems from the fundamental assumption of SRP that when the deformation is *all* by creep, a characteristic fatigue life is generated for this creep strain. Lowering the cycling frequency (or increasing the hold time) should *not* therefore alter the fatigue life, providing that the deformation remains all by creep and that the applied strain is not altered when this change is made. Figures 14-13 through 14-17 show that this assumption is not valid for 60% Sn-40% Pb solder in the tests described here. SRP can, however, be applied in a slightly modified form if it is assumed that when the deformation is 100 percent creep the time to failure and not the number of cycles to failure reaches a characteristic value t_{cc} for the applied plastic strain. Now, when the frequency is reduced, or the hold time increased, the number of cycles to failure will decrease, so the time to failure t_f does not change (i.e., $N_f\nu = t_f$, which becomes a constant with $t_f = t_{cc}$).

The data of Figures 14-21 through 14-23 show the results of Figures 14-15 through 14-17 in terms of this time-modified SRP approach. The terms f_{cc} and $f_{pp} = 1 - f_{cc}$ follow the SRP nomenclature and refer to the fraction of the applied strain that is creep or noncreep. The fraction creep is determined from Figure 14-20. The N_{pp} term is the fatigue life when the deformation is all noncreep and is approximated by the fatigue life observed when the highest frequency

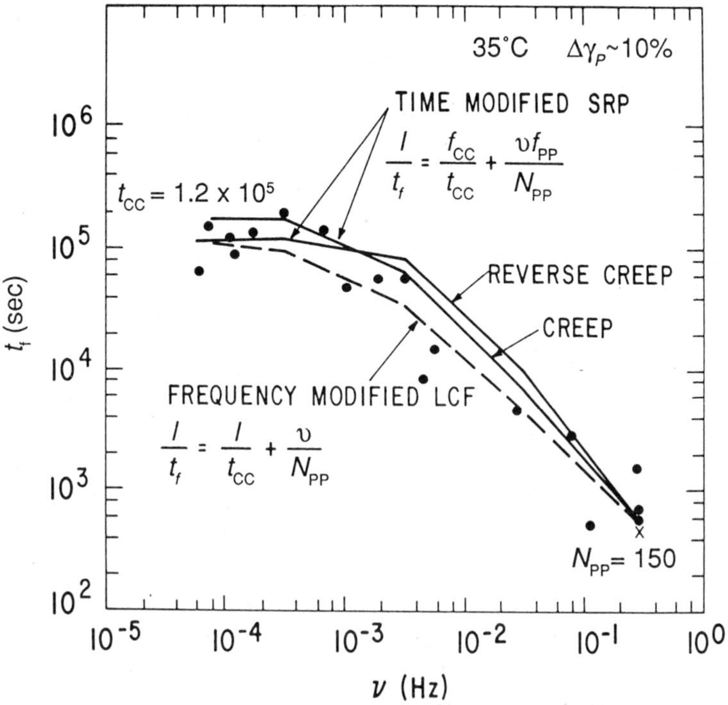

Figure 14-21 Time to failure t_f versus the cycling frequency ν, as correlated by time-modified SRP and frequency-modified LCF, at 35°C for a plastic strain of 10 percent. (From Solomon[27])

(0.25 Hz) was employed. As can be seen, this approach describes the approximately constant time ($t_f = t_{cc}$) to failure that is observed at low cycling frequencies or long hold times. Time-modified SRP is completely analogous to the frequency-modified approach with $K = 1$ at high frequencies and $K = 0$ at low frequencies ($K = 0$ defines a constant time to failure and $K = 1$ defines a constant number of cycles to failure).[27]

Enke et al.[100] have also studied the influence of frequency on the fatigue behavior of 60% Sn-40% Pb solder, using SRP. In agreement with the results presented here they found no influence of cycling frequency when 0.05-Hz tests were compared to those run at 0.017 Hz. The data of Figures 14-13 and 14-16 show that the frequency would have to have been reduced to below 0.0002 Hz before a significant influence of frequency would be noted.

While conventional SRP does not appear to explain the 60% Sn-40% Pb solder data presented here, it may do so for high Pb solders. Figure 14-24[119] shows the creep rate-strain rate comparison for a high Pb solder (the 92.5% Pb-2.5% Ag-5.0% Sn solder described previously). This solder does not display viscous

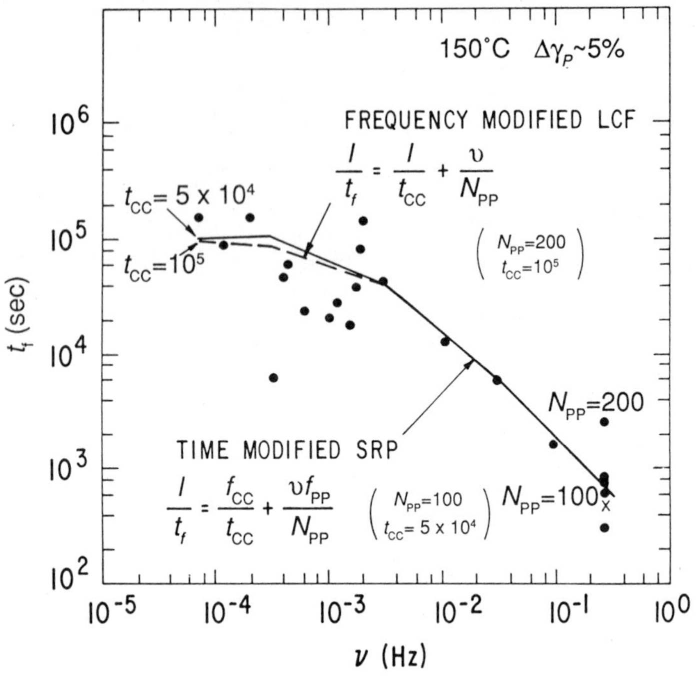

Figure 14-22 Time to failure t_f versus the cycling frequency ν, as correlated by time-modified SRP and frequency-modified LCF, at 150°C for a plastic strain of 5 percent. (From Solomon[27])

behavior. Hence, there are separate curves for each plastic strain that was used in the testing. Here, the creep curves diverge from all of the strain rate curves, meaning that the fraction creep will increase as the cycling frequency decreases. In agreement with SRP, the fatigue life decreases as the cycling frequency decreases and the fraction creep increases. Work is currently going on to see if conventional SRP indeed describes these results.

Fox, Shine, and co-workers[32,120,121] have developed yet another approach to describe the influence of cycling frequency and hold times. They correlate the matrix creep that is developed during cycling or during a hold time in a Coffin-Manson expression that has an α of 1.0. This means that the product of the matrix creep and the fatigue life is a constant. This is also a version of the Monkman-Grant model,[122] in which the product of the time to failure and the creep rate is a constant. (The number of cycles to failure times the time per cycle is the time to failure, and the matrix creep strain per cycle divided by the time per cycle is the matrix creep rate. Thus, if the product of the matrix creep per cycle and the fatigue life is a constant, the product of the time to failure and the matrix creep rate per cycle will also be a constant.)

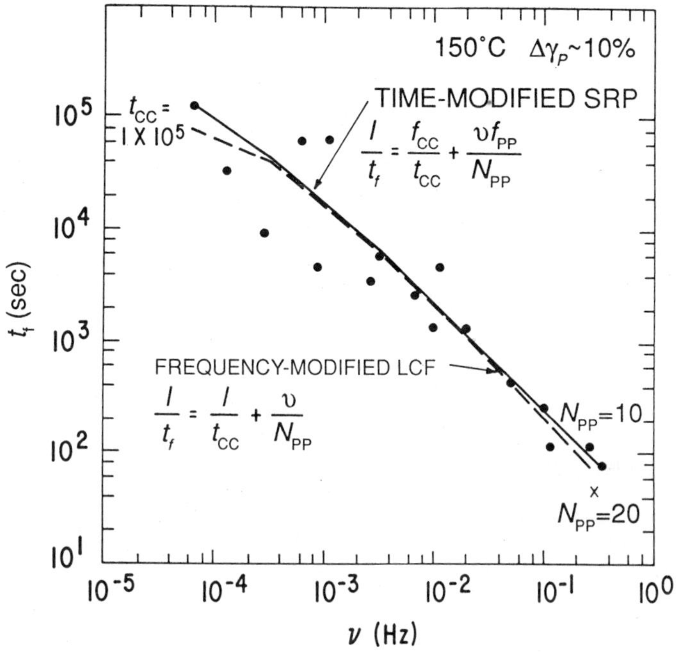

Figure 14-23 Time to failure t_f versus the cycling frequency ν, as correlated by time-modified SRP and frequency-modified LCF, at 150°C for a plastic strain of 10 percent. (From Solomon[27])

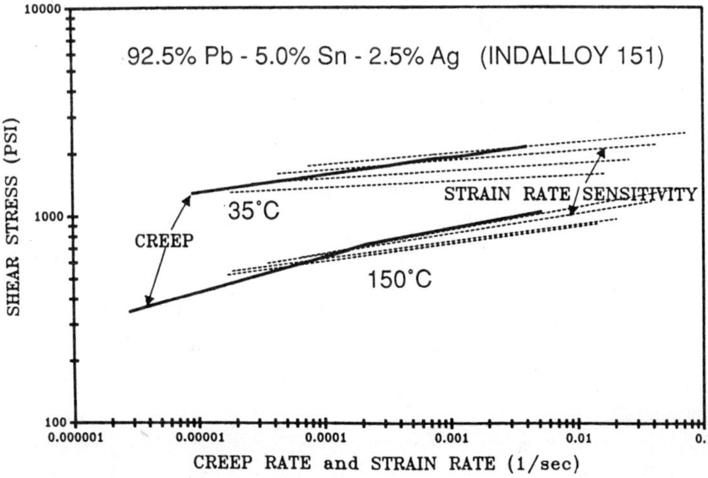

Figure 14-24 Calculated percentage of creep in the hysteresis behavior of 92.5% Pb-5.0% Sn-2.5% Ag solder cycled at 35 and 150°C. (From Solomon[119])

The matrix creep was determined from the upper branch (higher stress region) of the creep curve (see Figure 14-19). This model has been shown to describe some 60% Sn-40% Pb and 63% Sn-37% Pb fatigue data taken at about room temperature at relatively high strain rates. It is not clear if this approach will work at higher temperatures or at very low strain rates. Under these conditions, the deformation is largely by creep, and the fraction of creep (and presumably the amount of matrix creep) does not change with changes in the frequency or hold time, but the fatigue life does. It is further not clear why only the matrix creep correlates with the fatigue life. The value of α of unity is not, however, surprising. Coffin[113,115] has shown this to be the case for superalloys tested at elevated temperatures (although the applied plastic strain was used rather than the matrix creep strain as was done by Fox and Shine).

14.5 INFLUENCE OF THE ENVIRONMENT

Time dependent plastic flow (creep) is not the only time dependent phenomenon that can be the cause of the observed time dependence of the fatigue life. The environment can also produce this effect. Snowden[123,124] has shown that oxygen contents as little as 3×10^{-2} mm Hg will reduce the fatigue life on polycrystalline Pb by about two orders of magnitude, compared to that measured in tests run in a vacuum. This effect, which produced intergranular fractures, saturated at 3×10^{-2} mm Hg, with tests run with higher oxygen contents having much the same fatigue life. The tests described in Figures 14-13 through 14-17 were run in a nitrogen filled box, but penetrations for the loading pull rods and grips allowed oxygen to enter, making it probable that the oxygen content was well in excess of 3×10^{-2} mm Hg.

Neugebauer et al.[76,78] have shown that destroying the hermeticity of some electronic devices greatly reduces their thermal fatigue life compared to that measured in identical tests but on devices whose hermeticity was not destroyed.

Superalloys have been subjected to numerous tests of the influence of the environment on the high temperature fatigue life. The influence of the environment in lowering the fatigue life at low cycling frequencies is well documented.[115,116,125] Oxygen and water vapor[126,127] produce a strong frequency and hold time dependence of the fatigue life.

It is quite possible that oxygen and/or water vapor are responsible for the frequency and hold time dependence of the fatigue life of solder. This would explain why, even though there is a strong frequency and hold time dependence of the fatigue life of 60% Sn-40% Pb solder (see data of Figures 14-13 through 14-17), it does not correlate well with the creep behavior. Vacuum fatigue data will be required, however, to demonstrate that the environment is indeed the cause of the observed frequency and hold time effects. This is an important point since if the environment is the cause, using hermetic joints will greatly improve the fatigue life.

14.6 MICROSTRUCTURAL CHANGES

In assuming that isothermal results can be used to predict thermal fatigue, one is disregarding possible differences in the microstructure that are developed during thermal fatigue to a much greater extent than during isothermal fatigue, even isothermal fatigue at the peak temperature of the thermal range. Microstructural instability is greater for thermally cycled solder because of the local thermal stresses developed between the Pb- and Sn-rich material. These stresses are developed because of the difference in the thermal expansion between these two phases. Regions of coarsening of the Sn- and Pb-rich phases have been observed.[34] in thermally fatigued specimens of 60% Sn-40% Pb solder. Fatigue cracks developed in the region where this coarsening developed, and it is believed[34] that this coarsening plays an important role in the fatigue process. Coarsening is less prevalent in the thermal fatigue of high Pb content solders.[128] Coarsening also seems to be less prevalent when isothermal cycling is used and was not readily apparent in the initially coarse grained solder 60% Sn-40% Pb solder employed to generate much of the data used to illustrate this article. Thus, coarsening, while an important factor in influencing the fatigue process, is not the cause per se.

The development of a coarsened zone is important because this zone is softer than the noncoarsened zone and will thus become more highly strained. This leads to the development of cracks in this zone. Recrystallization and grain growth can eliminate some fatigue damage. Thus, interactions between metallurgical changes and the fatigue process may be important. In general, the number of thermal cycles required to develop a coarsened zone may be dependent on any or all of the following factors: the magnitude of the temperature change, the mean temperature, the rate of temperature change or magnitude of any hold times that are employed, the gross thermal strain being applied, the local thermal strain associated with the expansion difference between the Sn-rich and Pb-rich phases, the magnitude of any prior strain that has not been removed by recrystallization, and, lastly but very importantly, the size and distribution of the Sn- and Pb-rich phases at the start of thermal cycling.

The structure of eutectic solder is unstable even at room temperature,[91] with grain coarsening and precipitation occurring even as a specimen sits at that temperature. The question therefore arises as to which structure to correlate with fatigue tests, a structure observed when the joint is made, after the test, after waiting for a given period of time after producing the joint, etc.? There is even a question as to what is the structure of an actual joint of interest, the initial structure, one developed after a burn in, after standing prior to operation, after operation, etc.? Correlations of the change in structure with ageing and the influence of any changes on the fatigue behavior have not been published. Additionally, the influence of any change will be a function of the type of test that is performed. Softening due to grain coarsening will reduce the life of a load controlled test or a test using the total strain as the limiting variable (because in both cases the plastic strain will increase). Softening will, however, have much less of an impact on plastic strain limited tests because the plastic strain is being kept constant. This is the type of test whose results are shown in the figures of this chapter.

One approach (as was done here) to applying isothermal data to the problem of thermal fatigue is to use a coarse microstructure, which will be more stable than a fine one. There is also some evidence[129,130] that a coarser structure will lead to a shorter fatigue life. Thus, if the isothermal data is obtained on solder with a structure as coarse as that which may be developed during thermal fatigue, or after some time at temperature, some of the microstructural change during thermal fatigue or after time at temperature, may be accounted for. If, however, the coarsening during thermal fatigue is localized, the resulting strain localization will not be accounted for. Clearly, more work in this area is required.

14.7 DETERMINATION OF THE DISPLACEMENT AND STRAIN DISTRIBUTION IN A SOLDER JOINT

In order to apply the isothermal fatigue data to the problem of thermal fatigue, it is necessary to first calculate the thermal displacement and from this displacement to then calculate the strain distribution in the solder joint. We have considered the idealized problem of simple solder posts. Now we shall consider more complex cases.

There are two parts to the problem of calculating the displacement that a solder joint experiences. First one must perform a global analysis of the complete circuit board, complete with components, stiffeners, heat sinks, etc. In this analysis, the components need not be considered in detail; only their influence on the overall compliance is considered. When the whole board is considered, factors such as board bending and resonance can be calculated. This global analysis will show that there are components on the circuit board whose joints experience more displacements than do the joints of other components. These are the joints that are most likely to fail and therefore must be examined in more detail. The displacements determined in this analysis serve as the boundary conditions for the calculation of the displacement that a particular joint experiences. To do this the exact details of a specific single component must be considered.

The global problem does not consider the solder joint or component in detail and is beyond the scope of what is to be discussed here. These problems are also extremely specific to the exact board and component population and are therefore not amenable to a general discussion. They are more properly the subject of a discussion of finite element techniques.

We have already considered the simple case of solder posts attaching a chip carrier to a PWB. Now we shall consider the more complicated case of a fillet joint subjected to a thermal excursion. The most simplified approach would be to use Equation (14-2) to determine the displacement that the joint experiences. This approach does not, however, consider the global strains discussed above or the possibility that the PWB will bend, relieving some of the strain that the joint must absorb.

Hall and co-workers[40,57,58] and others[51] have determined by both experimental and analytical means the magnitude of this bending. The critical variables determining the magnitude of this bending are the elastic properties and geometry of the chip carrier, board, and solder joint. In addition, the temperature and the rate at which the solder is strained are of critical importance in determining the amount of bending. The mechanical properties of solder are determined by the temperature and by the rate at which the solder is strained. The yield strength is increased at low temperatures and high strain rates, and the increased stiffness, resulting from a lack of yielding, will produce more bending. Likewise, at high temperatures or if very low strain rates are imposed, the yield strength will be so low that the solder will impose no forces on the board or chip carrier, and no bending will be developed. Bending has the effect of reducing the displacement that must be accommodated by the solder joint because some of the displacement is taken up by board and chip carrier bending. The nature of the displacement is also a function of the rate of solder straining. At high temperatures or at low enough rates to suppress bending, only shear displacements are developed. At low temperatures or at higher rates, there are increasingly larger amounts of rotational and tensile displacements, with increasing amounts of bending. There is thus a critical interplay between solder properties and the nature and magnitude of the displacements that a solder joint will experience.

The temperature dependence of the flow stress also gives rise to a highly asymmetric thermal fatigue hysteresis loop.[57] For thermal cycling between two temperatures, lower stresses are required at the higher temperature than at the lower temperature. This asymmetry was successfully modeled by the constitutive law developed by Stone et al.,[46] enabling one to calculate the flow stress and strains[94] that are developed in simple joints that are thermally cycled. The calculated thermal strains are then used in conjunction with the techniques described above to determine the fatigue life. More complex calculations are required when the joints are more complex and have a nonuniform strain distribution.

The strain distribution can be calculated from the displacement. For the case of a simple solder pillar, the strain is simply calculated by dividing the displacement by the height of the solder joint, as was done in Equation (14-4). Thermally cycling a simple joint requires a constitutive relationship of the sort mentioned

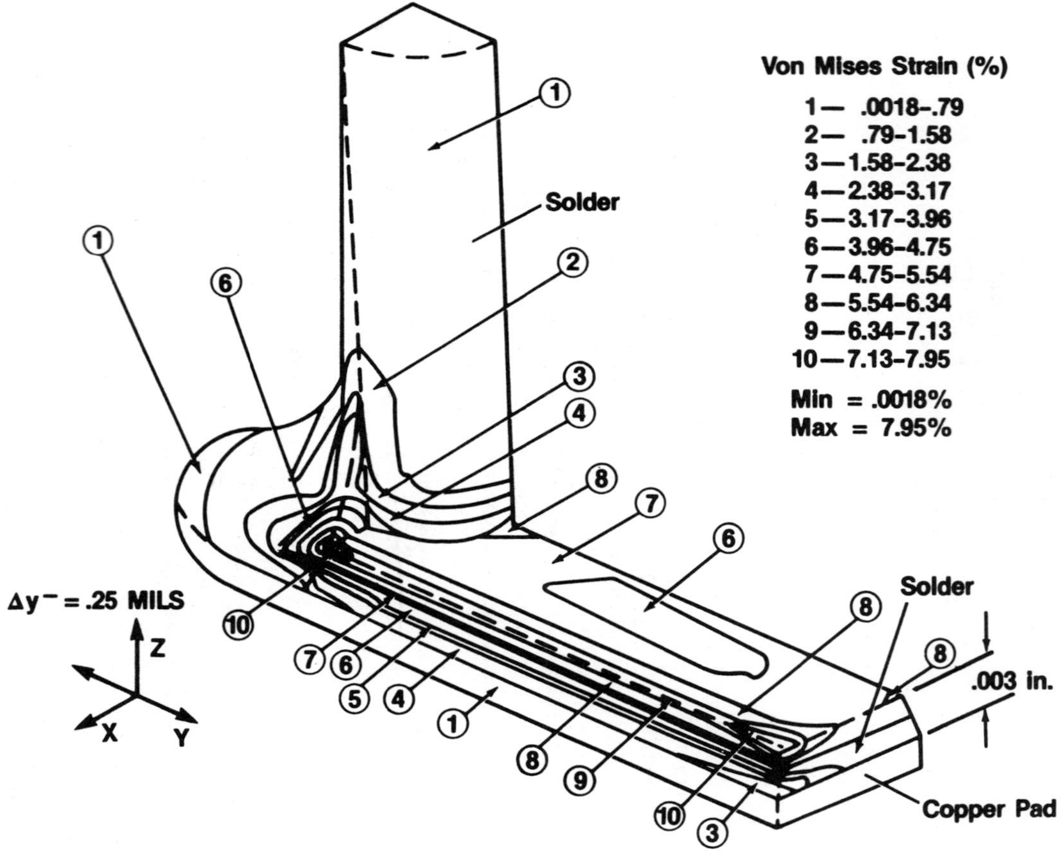

Figure 14-25 Calculated von Mises strain for a leadless CC-PWB joint for a castillated chip carrier subjected to a displacement of 0.25 mils. (From Solomon[105])

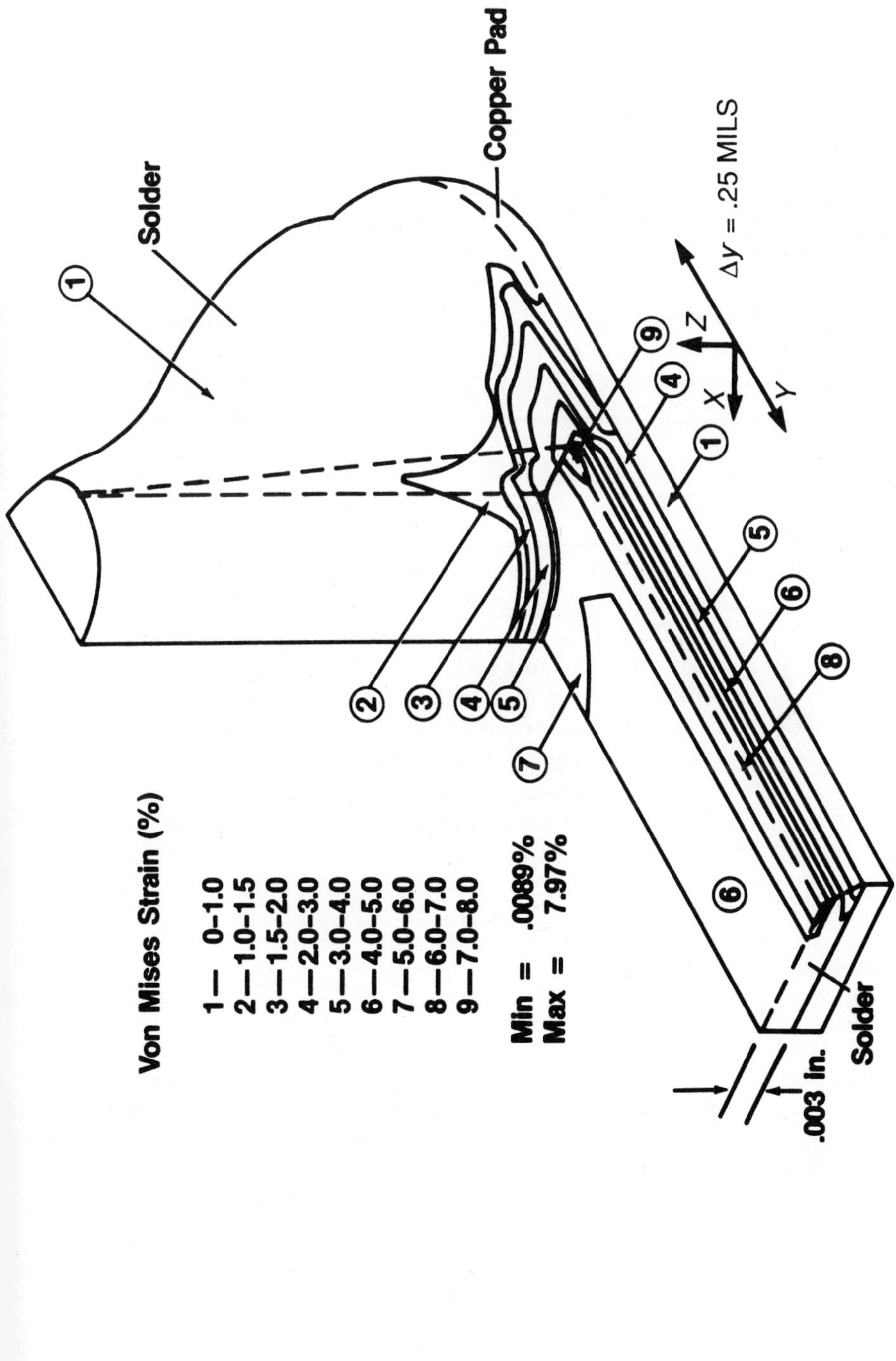

Figure 14-25 *(continued)*

above. More complex joints require a finite element analysis (FEA) to translate the displacement that a joint experiences into a strain field distribution. This has been done for both leadless[60,131] and leaded[41] chip carriers. For the joints containing fillets that attach a leadless chip carrier to a PWB, the highest strains are developed under the chip carrier and at the edge of the chip carrier. This is illustrated in Figure 14-25. This figure shows the Von Mises strain developed in joints that experienced a shear displacement of 2.5×10^{-4} in. The calculations were made for two different joint configurations—a bulbous joint and a thin joint—that represent extremes in the shape of the fillet. As can be seen, the shape of the fillet has an influence on the magnitude of the strain that is developed from a given displacement. It was also found that increasing the thickness of the solder between the chip carrier and the PWB reduced the strains, in accordance with Equation (14-4). These calculations were made to enable the correlation (discussed in the next section) of isothermal, mechanical joint cycling, with predictions made from generic isothermal data. For thermal fatigue cycling the calculations would have to account for the temperature dependence of yielding.

The Von Mises stress distribution (and thus the strain distribution because a linear analysis was used) found by Clatterbaugh and Charles[60] is similar to that shown in Figure 14-25 in that the highest stresses (and strains) were located under the chip carrier, at the edge of the chip carrier. Lau et al.[41] also used a linear analysis to calculate the Von Mises stress but performed the analysis for a J-leaded chip carrier to PWB joint. They found the maximum stress (and thus strain) to be located at the tip of the outer solder fillet radius. This was the point where failures were observed to start in laboratory tests of this type of joint.

The use of the Von Mises strains in these calculations emphasizes an important point. The preceding LCF law was formulated in terms of the shear strain. The more general approach is to describe the local stress state in terms of a correlating parameter and to use this parameter to correlate fatigue test data. For instance, the strains can be described in terms of an equivalent or Von Mises strain, that is, in terms of the principal strains ε_1, ε_2, and ε_3 according to

$$\varepsilon_{VM} = \frac{\sqrt{2}}{3}[(\varepsilon_1 - \varepsilon_2)^2 + (\varepsilon_1 - \varepsilon_3)^2 + (\varepsilon_2 - \varepsilon_3)^2]^{1/2} \qquad (14\text{-}17)$$

This is what was done in Figure 14-25. It is then necessary to convert the strain used in the strain-life data to Von Mises strains. The equivalent Von Mises strain for a test run with a shear strain range $\Delta\gamma$ is $1/\sqrt{3}(\Delta\gamma)$ (see the reference[105]). For a test run with a tensile strain range $\Delta\varepsilon$ the equivalent Von Mises strain is equal to this tensile strain range. Correlations[134] have also been made for the octahedral shearing strain, which is equal to $\sqrt{2}$ times the Von Mises strain. Other similar approaches are possible; these examples are merely to illustrate a technique, not necessarily to describe the best one.

14.8 PREDICTION OF THE FATIGUE LIFE OF SOLDER JOINTS

Isothermal displacement controlled LCF tests were performed on leadless CC-PWB joints,[102] and the results were compared to predictions based on isothermal fatigue data gathered on laboratory solder joints.[105] Figure 14-25 shows the calculated strain distribution that was employed. The calculation was done for extreme variations in the joint geometry, and therefore an average of these extremes was used to describe an average joint. The FEA calculation was done in terms of the total Von Mises strain [Equation (14-5)]. It was therefore necessary to cast the LCF data into this form. The elastic strain was added to the plastic strain of the Coffin-Manson data to develop the total strain-life curves that are shown in Figure 14-26. The plastic strain curves are those of Figure 14-7. Alternatively the plastic strain in the joint could be calculated and Figure 14-7 used directly. Since the calculation is given in terms of a Von Mises strain, it was necessary to convert it into an equivalent shear strain. This was done via Equation (14-7), which shows that a pure shear strain is equivalent to $\sqrt{3}$ times the Von Mises strain.[105]

Figure 14-25 shows that there is a distribution of shear strains in the joint. To simplify the problem, it was assumed that only three strain regions existed: a high strain region under the chip carrier (averaging 0.182 total shear strain), a transition region where the crack entered the fillet (with a 0.095 total shear strain), and the fillet region, which had the lowest strain (only 0.028 total shear

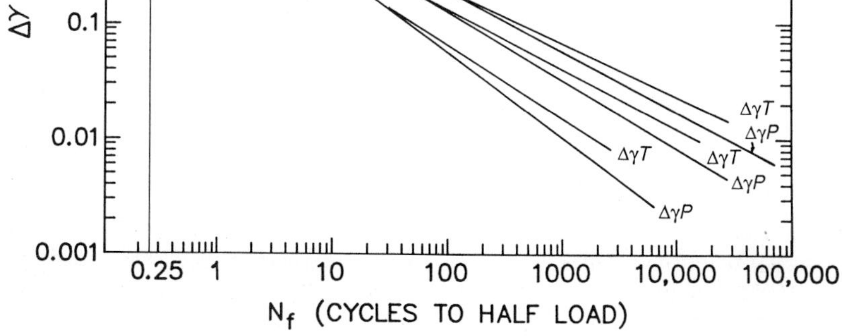

Figure 14-26 LCF curves for the plastic and total strain for failure defined by a 25, 50, and 90 percent drop in the hysteresis load. (From Solomon[105])

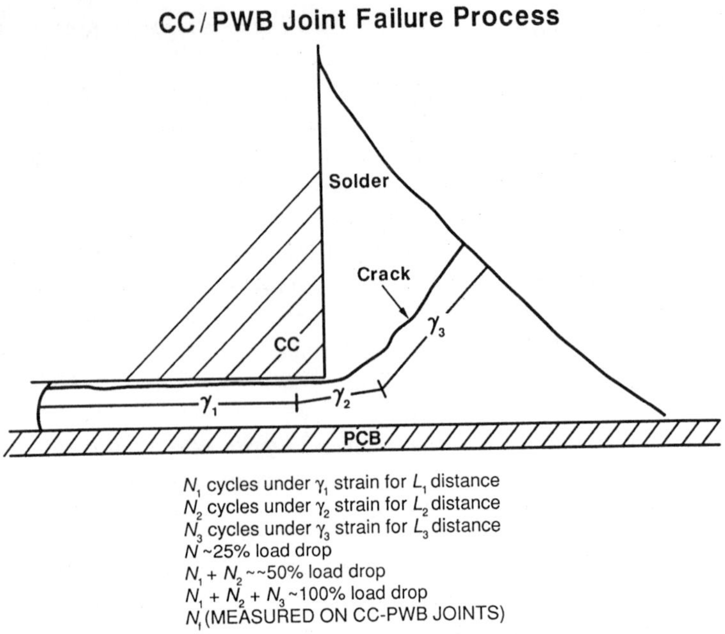

Figure 14-27 Schematic representation of cracking in CC-PWB joints. (From Solomon[105])

strain). This is illustrated schematically in Figure 14-27. Now Figure 14-26 can be used to calculate a fatigue life for each of these strains, providing that the size effect is corrected by Equation (14-11). The length of region 1 was measured at 0.040 in (1.01 mm), that of region 2 was estimated as 0.01 in (0.25 mm), and that of region 3, at 0.015 in (0.38 mm). Table 14-1 lists the calculated cycles to failure for each region using the total strain corrected 25, 50, and 90% load drop curves of Figure 14-7. The *D* values that were used are also given in

Table 14-1 Calculation of N_i for N_f Defined by Different Drops in Load and Values for *D*

	$N_1[L_1 = 0.04$ in V (1.01 mm)]	$N_2[L_2 = 0.01$ in V (0.25 mm)]	$N_3[L_3 = 0.015$ in V (0.38 mm)]
For N_f (25%) $D = 0.021$ in (0.53 mm)	38	26	250
For N_f (50%) $D = 0.083$ in (2.10 mm)	25	22	343
For N_f (90%) $D = 0.30$ in (7.62 mm)	15	15	300

the table. The values N_1, N_2, and N_3 given in the table are the number of cycles to failure for each region. As can be seen, Equation (14-7) and the appropriate value for D allows us to use any of the load drop curves to calculate the fatigue life. In applying the results of Table 14-1, we are guided by the failure process observed in the joints being considered. It was found that cracks were first observed under the chip carrier and then grew into the fillet to failure. This first stage should be determined by N_1. The number of cycles to grow the crack into the fillet is $N_1 + N_2$. We are assuming that the cracking must grow in from a free surface, where there can be an environmental interaction so a crack is not considered to have nucleated in the second region; rather it grows in from under the chip carrier. Hence, N_1 and N_2 are summed to get the life to grow the crack into the fillet (i.e., N_1 and N_2 are assumed to be sequential, not concurrent). For final failure (defined by a 90 percent drop in the hysteresis load) we use $N_1 + N_2 + N_3$. We are neglecting concurrent cracking of the fillet but since it is a free surface, it should be occurring. Since N_3 is so much larger than $N_1 + N_2$, there is little difference between defining failure by just N_3 or by the sum of $N_1 + N_2 + N_3$. Figure 14-28 compares these calculated values and that measured on actual joints subjected to the displacement used in the FEA calculation. The three boxes represent the calculations for the three regions. The bands represent the different results obtained with different load drop laboratory curves and

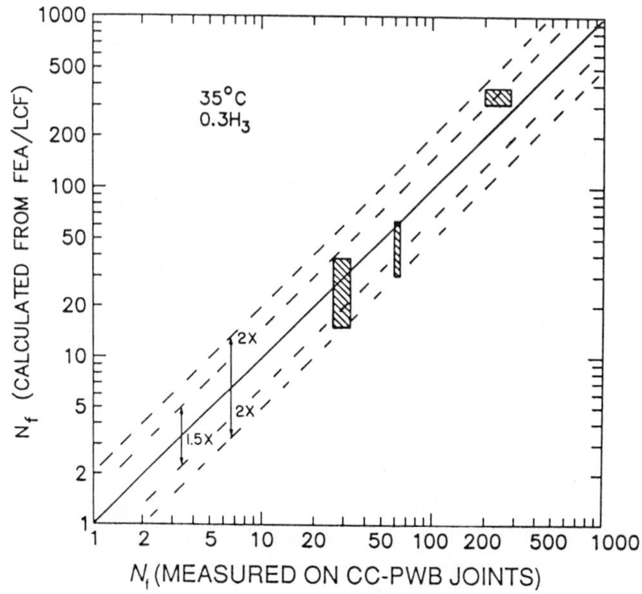

Figure 14-28 Comparison between the predictions made from FEA and LCF data and measurements on CC-PWB joints. (From Solomon[105])

different joint data curves. As can be seen, the prediction and results generally fall within a factor of 2. Some refinements could improve the agreement. For instance, if the final failure were defined by N_3 instead of by $N_1 + N_2 + N_3$, there would be some improvement between the calculation and the prediction for this last box.

The preceding analysis takes account of the geometry of the solder joint but does *not* account for such factors as the influence of the cycling frequency or hold time. The tests on both the GC-PWB and laboratory joints were run without hold times and at a relatively high cycling frequency. As such, the results *cannot* be applied to situations where long hold times or low cycling frequencies are employed. Corrections of the sort described in the previous sections of this chapter must be employed before a more general life prediction can be made.

Clech, Engelmaier, and co-workers[133–136] have developed a "figures of merit" approach, which systematizes the influence of many of the factors, discussed in this chapter, that alter the fatigue life. They provide expressions for the influence of joint height, distance between joints, differences in the coefficient of expansion, temperature cycling range, dwell time, and lead compliance (for leaded joints). These expressions give relative fatigue lives based on tests run on FR-4-type joints, which do not have fillets of the sort shown in Figure 14-25. For this type of joint, the techniques they outline allow one to estimate the effect of changing the solder joint height, varying the hold time, varying the cycling temperature range, etc. However, since the data used to determine the correlations of their model were generated on joints that do not contain large fillets, corrections must be made when applying it to joints where much of the life is spent growing a fatigue crack through a fillet.

14.9 INHERENT LIMITATIONS TO FATIGUE LIFE PREDICTIONS

It is clear from the previous discussion that the calculation of the fatigue life is a complex process, requiring a knowledge of the behavior of the total CC-PWB population, the geometry and mechanical properties of individual joints and the components that they join, the fatigue properties of solder (including the influences of stress state, hold time, cycling frequency, temperature, and even the definition used to define failure), fatigue crack propagation, solder creep, and the influence of microstructural changes. Even if one knew the influence of all of these variables, it would still not be possible to *exactly* determine the number of cycles to a fatigue failure. This is because the fatigue life must be expressed in terms of the statistical variance associated with each of the aforementioned processes. This is especially true for solder joints that are in effect small unmachined castings, for which the variability in the fatigue properties is greater than that for wrought and carefully machined specimens. The general approach is to determine the standard deviation of the fatigue life curve, taking into account all the factors that influence the fatigue life, and then design to only a fraction of this life, say with fatigue curves drawn at the mean life *minus* one or more standard deviations. The Clech, Engelmaier figures of merit approach incorporates failure rates as an integral part of their methodology.

The time or number of cycles to the first joint failure can be estimated by identifying the joint subjected to the highest displacement. Then it can be assumed that the joint has the least favorable geometry (i.e., one for which the strain distribution is a maximum). The appropriate fatigue life curve or crack growth curve, corrected for the stress state, must then be considered statistically. A fatigue life at minus one or more standard deviations, depending upon the degree of conservatism that is desired, should be used (or if the analysis is based upon crack growth data, an upper limit curve should be used). More complex statistical approaches to estimate the number of cycles to the first failure can also be employed.[139] Finally, accelerating factors[140,141] such as the cycling frequency, hold time, mean temperature, grain size, and grain growth should be considered and used to further reduce the fatigue life.

What is meant by failure and the consequence of a joint failure must also be considered. A fatigue crack in a solder joint does not produce a significant overall increase in resistance until the crack is almost completely through the joint.[142,102,104] Even then, if the broken halves contact each other, the increase in resistance may be not be detrimental to the overall circuit. The problem is that vibrations or other forces may separate the broken halves and then produce an open. The point at which the open is actually observed may thus be a function of more than just the fatigue behavior. This further adds to the variability in observed device failure times. It is also possible that the failed joint may not always be called upon to carry current, and thus the observation of a device failure would be further postponed until a failure to carry current was noted.

14.10 NECESSARY FURTHER WORK

Comparisons have been made between the fatigue life measured on actual joints subjected to isothermal cycles and the life calculated from generic isothermal fatigue data. This showed how to account for the joint shape and the size effect that was created because the generic data was developed on joints that were larger than the actual solder joints of interest. The influence of other variables was also discussed but not specifically incorporated in the life prediction analysis. Similar predictions, based upon generic data, should therefore be made for thermally cycled joints, with cycling between different temperatures, with different cycling rates, hold times, etc., and comparisons made with the behavior of actual thermally cycled joints.

A considerable amount of thermal cycle data already exists, but generally there is insufficient data on all of the thermal displacements that were developed (i.e., all the translations and rotations about all three axis) to enable an *accurate* determination of all the strains that were developed in the joint. The calculation of these strains requires not only better input displacement data than currently exists but also requires a temperature and strain rate dependent finite element analysis. With a knowledge of the joint strain state, developed by a given thermal cycle, it should be possible to determine the fatigue life based upon general fatigue data of the sort discussed here and compare these predicted results to those measured on the thermally cycled joints.

The aforementioned approaches have been discussed in the context of leadless chip carrier joints. They must also be extended to leaded joints where the joint compliance is much larger and the stresses developed much lower.

The figures of merit technique developed by Clech, Engelmaier, and co-workers (which is discussed at length in this volume), provides a systematized approach to many of the factors discussed here, for both leaded and leadless joints. It is a relative approach, however, that is best applied when doing a sensitivity analysis of the influence of various variables on the behavior of FR-4 joints. More work is needed to extend it to other types of joints, particularly where large fillets play a dominate role in determining the fatigue life.

The phenomena discussed at length in this chapter are largely due to crack propagation, but instead of measuring the crack growth rate directly, it was inferred from LCF results. While this is a reasonable approach because LCF is dominated by fatigue crack growth, approximations were required to account for multiple cracking and the irregular nature of the crack front that is developed. With specific crack nucleation and crack propagation measurements one could use the more refined crack growth design approaches that have been developed for the design of aircraft engines and air frames. These allow for the calculation of the remaining life in a joint containing a fatigue crack as well as for the prediction of the overall life.

The design and life prediction of solder joints will also have to account for cracking of metallized layers, intermetallic layers, and the strain concentration and subsequent cracking that may be developed in grain coarsened zones. More work is needed to determine the extent to which these factors are life limiting. Tests run on leadless CC-PWB joints[105] have shown that metallization cracking was *not* a major factor in determining the overall life because of the much larger number of cycles spent propagating a crack through the fillet. With joints without fillets it is expected that metallization cracking would play a larger role in determining the joint life.

Of all the necessary future work nothing is more important than the systemization of a design approach that will incorporate all the factors discussed above in a manner that can be used by the average design engineer. Some progress has been made in this area, but all too often successful approaches have been used beyond their "zones of applicability," say by neglecting the influence of a long hold time or by neglecting the fact that the joint of interest has a radically different geometry from the one used to develop the design data. Much of the data exists to develop this systemization; what is needed is a more concerted effort to organize it, do blind predictions of actual thermal cycling tests to compare the results of various approaches, and a standardization body to codify the approach that does the best job. Any resulting code must also define, and clearly state, its zone of applicability. Such future work will not only result in more accurate life predictions, and therefore more reliable solder joints, but will also clearly point to the directions that should be taken to *improve* the joint life, either though improved solder properties or improved joint designs.

14.11 ACKNOWLEDGMENTS

The author would like to acknowledge the USAF materials laboratory for their support (through contract no. F33615-85-C-5065) of the work associated with the references.[102,104,105] The author would also like to acknowledge the use of the section on "Inherent Limitations to Fatigue Life Prediction" and parts of the sections on "Microstructural Changes" and the "Determination of the Displacement and Strain Distribution in a Solder Joint" which appeared in the author's review paper[12] that was published by The Materials Research Society as part of their 1989 spring meeting.

REFERENCES

1. Manson, S. S., "Thermal Stress and Low Cycle Fatigue," New York, McGraw Hill, 1965.
2. ASME Code case N-47, loc. cit. G. R. Halford et al. ASTM-STP 942, 1988, pp. 625–636.
3. Basquin, O. H., *Proc. ASTM* **10,** 1910, pp. 625–630.
4. Coffin, L. F., Jr., *Trans. ASME,* **76,** 1954, pp. 931–950.5.
5. Manson, S. S., "Behavior of Materials Under Conditions of Thermal Stress," *Heat Transfer Symposium,* Univ. of Michigan, June 27–28, 1952, Univ. of Mich. Press; also NACA TN2933, July 1953.
6. Manson, S. S., *Exp. Mech.,* **5**(7), 1965, pp. 193–226.
7. Coffin, L. F., Jr., in "Fracture," London: Chapman and Hall, 1969.
8. Morrow, J., "Cyclic Plastic Strain Energy and Fatigue of Metals," in "Internal Friction, Damping and Cyclic Plasticity," ASTM-STP 378, 1965, pp. 45–87.
9. Taira, S., "Relationship Between Thermal Fatigue and Low-Cycle Fatigue at Elevated Temperature," in ASTM-STP 520, 1973, pp. 80–101.
10. Spera, D. A., and D. F. Mowbray, eds. "Thermal Fatigue of Materials and Components," ASTM-STP 612, 1976.
11. Lau, J. H., and D. W. Rice, *Solid State Technology,* **28**(10), 1985, pp. 91–104.
12. Solomon, H. D., "Review of Critical Variables Determining Solder Fatigue Lives," *Mat. Res. Soc. Symp. Proc.,* **154,** 1989, pp. 441–451.

13. Wild, R. N., *Welding J.*, **51,** 1972, pp. 521s–526s.
14. ______, *Internepcon,* October 1975 (IBM/74Z000481).
15. Jarboe, D. M., "Thermal Fatigue of Solder Alloys," NTIS-BDX-613-2341, February 1980.
16. Shamash, M. B., "Development of Highly Reliable Soldered Joints for Printed Wiring Boards," NTIS N69-25697, Baltimore, MD, Westinghouse Defense and Space Center, 1968.
17. Rathore, H. S., R. C. Yih, and A. R. Edenfeld, *J. of Testing and Evaluation,* **1**(2), 1973, pp. 170–178.
18. Denlinger, M. C., R. W. Korb, and V. F. Lardenoit, "Development of Improved Solders for Electronic Reliability," AFML-TR 77-93, June 1977.
19. Inoue, H., Y. Kurihara, and H. Hachino, *IEEE Trans. CHMT-9,* no. 2, 1986, pp. 190–194.
20. Vaynman, S., M. E. Fine, and D. A. Jeannotte, "Isothermal Fatigue Failure Mechanisms in Low Tin Lead Based Solder," in *TMS-AIME Symposium on Effect of Load and Thermal Histories on Mechanical Behavior of Materials,* Denver, February 1987.
21. ______, *37th Electronic Components Conference,* May 1987, pp. 598–603.
22. ______, "Effect of Hold Time on Fatigue of 96.5Pb-3.5Sn Solder," in *Fatigue 87.*
23. Lawson, L. R., M. E. Fine, and D. A. Jeannotte, "Thermomechanical Fatigue of Lead-Base Low Tin Solder," in *Fatigue 87.*
24. Solomon, H. D., *IEEE Trans. CHMT-9,* no. 4, 1986, pp. 423–432.
25. ______, in "Electronic Packaging—Materials and Processes," J. A. Sartell, Ed., ASM, 1986, pp. 29–47.
26. ______, *Brazing and Soldering,* no. 11, Autumn 1986, pp. 68–75.
27. ______, in "Low Cycle Fatigue," ASTM-STP 942, H. D. Solomon, G. R. Halford, L. R. Kaisand, and B. N. Leis, Eds., ASTM, January 1988, pp. 342–371.
28. ______, *38th Electronic Components Conference,* May 1988, pp. 7–13.
29. ______, *ASME J. of Electronic Packaging,* **111,** 1989, pp. 75–82.
30. Engberg, G., L. E. Larsson, M. Nylen, and H. Steen, *Brazing and Soldering,* no. 11, Autumn 1986, pp. 62–65.
31. Shine, M. C., L. R. Fox, and L. W. Sofia, *4th International Electronics Packaging Conference,* October 1984; also in *Brazing and Soldering,* no. 9, Autumn 1985, pp. 11–14.
32. Shine, M. C., and L. R. Fox, in "Low Cycle Fatigue," ASTM-STP 942, H. D. Solomon, G. R. Halford, L. R. Kaisand, and B. N. Leis, Eds., ASTM, January 1988, pp. 588–611.
33. Frost, H. J., P. R. Lavery, S. D. Lutender, and R. T. Howard, *38th Electronic Components Conference,* May 1988, pp. 13–22.
34. Frear, D., D. Grivas, M. McCormack, D. Tribula, and J. W. Morris, Jr., "Fatigue and Thermal Fatigue of Pb-Sn Joints," in *AIME Spring Conference Proc.,* 1987.
35. Frear, D., D. Grivas, and J. W. Morris, Jr., "The Effect of Cu_6Sn_5 Whisker Precipitates in Bulk 60Sn-40Pb Solder," to be published.
36. Frear, D., D. Grivas, L. Quan, and J. W. Morris, Jr., *Mat. Res. Soc. Proc.* **72,** 1986, pp. 181–186.
37. Weinbel, R. C., J. K. Tien, R. A. Pollak, and S. K. Kang, *J. of Materials Science* **22,** 1987, pp. 3901–3906.

38. Weinbel, R. C., E. A. Schwartzkopf, and J. K. Tien, *Scripta Metallurgicia* **21,** 1987, pp. 1165–1168.
39. Hendrix, B. C., J. K. Tien, S. K. Kang, and T. Reiley, "Stress and Strain Controlled Low Cycle Fatigue of Pb-Sn Solder for Electronic Packaging Applications," to be published.
40. Hall, P. M., *37th Electronic Components Conference,* May 1987, pp. 579–588.
41. Lau, J., G. Harkins, D. Rice, J. Kral, and B. Wells, *37th Electronic Components Conference,* May 1987, pp. 589–597.
42. ______, *IEEE/IRPS,* 1987, pp. 250–259.
43. Lau, J., D. Rice, and P. A. Avery, IEEE CH2295, 1986, pp. 173–184.
44. Hagge, J. K., *Proc IEPS,* 1986, pp. 199–208.
45. Stone, D., H. Wilson, R. Subrahmanyan, and C.-Y. Li, *36th Electronic Components Conference,* May 1986, pp. 630–635.
46. Stone, D., S.-P. Hannula, and C.-Y. Li, *35th Electronic Components Conference,* May 1985, pp. 46–51.
47. Engelmaier, W., IEPS, October 1984, pp. 360–369.
48. ______, *IEEE Trans. CHMT-6,* no. 3, 1983, pp. 232–237.
49. ______, *Circuit World,* **11**(3), 1985, pp. 61–72.
50. Hieber, H. *IEEE/IRPS,* 1988, pp. 234–238.
51. Desaulnier, W. E., Jr., T. E. LaFlamme, W. B. Ammerman, and M. C. Binder, *Proc. Institute of Environmental Sciences,* 1986, pp. 6–19.
52. Marshall, J. L., and S. R. Walter, *Int. J. for Hybrid Microelectronics,* **10,** 1987, pp. 11–17.
53. Tanaka, T., S. Nishijima, and M. Ichikawa, *Current Japanese Materials Research,* **2,** 1987, pp. 235–250.
54. Schmitt-Thomas, Kh. G., and S. Wege, *Brazing and Soldering,* no. 11, Autumn 1986, pp. 27–33.
55. Hall, P. M., and W. M. Sherry, *Interconnection Technology in Electronics, Proc. 3d International Conference,* February 1986, pp. 47–61.
56. Sherry, W. M., J. S. Erich, K. Bartschat, and F. B. Orinz, *IEEE Trans. CHMT-8,* 1985, pp. 417–426.
57. Hall, P. M., *IEEE Trans CHMT-7,* 1984, pp. 314–327.
58. Hall, P. M., T. D. Dudderar, and J. F. Argyle, *IEEE Trans CHMT-6,* 1983, pp. 544–552.
59. O'Clock, G. D., M. S. Peters, G. A. Kleese, and R. V. Martini, *IEEE Trans CHMT-10,* 1987, pp. 82–88.
60. Clatterbaugh, G. V., and H. K. Charles, IEEE 0569-5503, 1985, pp. 60–72.
61. Bangs, E. R., and R. E. Beal, *Welding J.,* **54,** 1975, pp. 377s–383s.
62. Dunn, B. D., "The Resistance of Space-Quality Solder Joints to Thermal Fatigue," ESA STM-207, September 1978.
63. Coombs, V. D., in ASTM-STP 515, 1972, pp. 3–21.
64. Olsen, D. R., and H. M. Berg, *IEEE Trans CHMT-2,* no. 2, 1979, pp. 257–263.
65. Korb, R. W., and D. O. Ross, *IEEE Trans CHMT-6,* no. 3, 1983, pp. 227–797.
66. Pankratz, J.M ., and H. R. Plumlee, *IEEE Trans ED-17,* no. 9, 1970, pp. 793–797.
67. Kong, S. K., N. D. Zommer, D. L. Feutcht, and R. W. Heckel, *IEEE Trans. PHP-13,* no. 3, 1977, pp. 318–321.

68. Zommer, N. D., D. L. Feutch, and R. W. Heckel, *IEEE Trans. ED-23,* no. 8, 1976, pp. 843–850.
69. Levine, E., and J. Ordonez, *IEEE Trans. PHP-13,* no. 3, 1977, pp. 318–321.
70. Norris, K. C., and A. H. Landzberg, *IBM J. Res. Dev.,* May 1969, pp. 266–271.
71. Goldmann, L. S., *IBM J. Res. Dev.,* **13,** 1969, p. 251.
72. Goldmann, L. S., R. D. Herdzik, N. G. Koopman, and V. C. Marcotte, *IEEE Trans PHP-13,* no. 3, 1977, pp. 194–198.
73. Agarwala, B. N., *IEEE/IRPS,* 1985, pp. 198–205.
74. Taylor, J. R., and D. J. Pedder, *Int. J. Hybrid Microelectronics,* **5**(2), 1982, pp. 209–214.
75. Lang, G. A., B. J. Fehder, and W. D. Williams, *IEEE Trans ED-17,* no. 9, 1970, pp. 787–793.
76. Burgess, J. F., R. O. Carson, H. H. Glascock, C. A. Neugebauer, and H. F. Webster, *IEEE Trans. CHMT-7,* 1984, pp. 405–410.
77. Burgess, J. F., R. O. Carson, H. H. Glascock, C. A. Neugebauer, and H. F. Webster, *IEEE 34th Electronic Components Conference,* May 1984, pp. 144–148.
78. Neugebauer, C. A., H. F. Webster, H. D. Solomon, and R. O. Carlson, "The Measurement of Fatigue Suppression in Electronic Solder Joints," in the *8th Soldering Technology Seminar,* February 1984, Naval Weapons Center, China Lake, CA.
79. Miyazaki, M., S. Yoshioka, and A. Aijikata, *Zairyo (Materials)* **30**(331), 1981, pp. 8–13.
80. Berriche, R., S. Vaynman, M. E. Fine, and D. A. Jeannotte, in *3d Conference on Electronic Packaging: Materials and Processes & Corrosion in Microelectronics,* Minneapolis, MN, April 1987, pp. 169–174.
81. Homa, T. R., *ibid.,* pp. 209–216.
82. O'Clock, G. D., M. S. Peters, G. A. Kleese, and R. V. Martini, *ibid.,* pp. 251–257.
83. Frost, H. J., P. R. Lavery, and S. D. Lutender, *ibid.,* pp. 259–267.
84. Frear, D., D. Grivas, M. McCormack, D. Tirbula, J. D. Morris, Jr., *ibid.,* pp. 269–274.
85. Frost, H. J., R. T. Howard, P. R. Lavery, and S. D. Lutender, in *38th IEEE-ECC Conf. Proc.,* May 1988, pp. 13–23.
86. Bae, K., A. F. Sprecher, Z. Guo, H. Conrad, and Y. Jung, in *Micro. Electronic Packaging Technology—Materials and Processes,* ASM, 1989, pp. 109–121.
87. Frost, H. G., G. J. Stone, and R. T. Howard, *ibid.,* pp. 121–128.
88. Hendrix, B. C., P. L. Bretz, J. K. Tien, and S. K. Kang, *ibid.,* pp. 129–134.
89. Solomon, H. D., *ibid.,* pp. 135–146.
90. Fine, M. E., and D. A. Jeannotte, *ibid.,* pp. 147–158.
91. Chilton, A. C., M. A. Whitmore, and W. Hampshire, *ibid.,* pp. 159–179.
92. Lau, J. H., and A. H. Jeans, *ibid.,* pp. 177–188.
93. Oberin, F. W., *ibid.,* pp. 189–202.
94. Wilcox, J. R., R. Subrahmanyan, and C.-Y. Li, *ibid.,* pp. 203–212.
95. Subrahmanyan, R., J. R. WIlcox, and C.-Y. Li, *ibid.,* pp. 213–222.
96. Vaynman, S., and M. E. Fine, *ibid.,* pp. 255–260.

97. Subrahmanyan, R., J. R. Wilcox, and C.-Y. Li, in *39th IEEE-ECC Conf. Proc.,* May 1989, pp. 240–252; also *IEEE Trans. CHMT-12,* no. 4, 1989, pp. 480–492.
98. Schmidt, C. G., *ibid.,* pp. 253–258.
99. Tien, J. K., B. C. Hendrix, P. L. Bretz, and A. I. Attarwala, *ibid.,* pp. 259–263; also *IEEE Trans. CHMT-12,* no. 4, 1989, pp. 502–506.
100. Enke, N. F., T. J. Kilinski, S. A. Schroeder, and J. R. Lesniak, *ibid.,* pp. 264–272; also *IEEE Trans. CHMT-12,* no. 4, 1989, pp. 459–469.
101. Vaynman, S., *ibid.,* pp. 273–276; also *IEEE Trans. CHMT-12,* no. 4, 1989, pp. 469–473.
102. Solomon, H. D., *ibid.,* pp. 277–292; also *IEEE Trans. CHMT-12,* no. 4, 1989, pp. 473–480; also GE report #87CRD185.
103. Frear, D. R., *ibid.,* pp. 293–300, also *IEEE Trans. CHMT-12,* no. 4, 1989, pp. 492–502.
104. Solomon, H. D., "Influence of Temperature on the Low Cycle Fatigue of Surface Mounted Chip Carrier/Printed Wiring Board Joints," presented at the 35th Annual Technical Meeting, Institute of Environmental Sciences, Anaheim, CA, May 1989; also GE report #87CRD186.
105. Solomon, H. D., V. Brzozowski, and D. G. Thompson, "Prediction of Solder Joint Fatigue Life," GE report #88CRD101; also *40th ECTC,* May 1990, pp. 351–360.
106. Solomon, H. D., "High and Low Temperature Strain-Life Behavior of a Pb Rich Solder," 89-WA/EEP-33, presented at the 1989 Winter ASME Meeting, December 1989, San Francisco; *ASME J. Electronic Packaging,* **112,** June 1990, pp. 123–128.
107. Wilcox, J. R., R. Subrahmanyan, and C.-Y. Li, "Assembly Stiffness and Failure Criterion Considerations in Solder Joint Fatigue," presented at the 1989 Winter ASME Meeting, December 1989, San Francisco; to be published in *ASME J. Electronic Packaging,* June 1990.
108. Gilmore, R. S., K. C. Tam, J. D. Young, and D. R. Howard, *Phil. Trans. R. Soc.,* London, **A320,** 1986, pp. 215–235.
109. Wilkening, W. W., private communication.
110. Rice, J. R., in *Treatise on Fracture,* H. Liebowitz, Ed., New York, Academic Press, 1971, V2, Chap. 3, pp. 230–232.
111. Gohn, G. R., and W. C. Ellis, *Proc ASTM,* **51,** 1951, pp. 721–744.
112. Eckel, J. F., *Proc. ASTM,* **51,** 1951, pp. 745–760.
113. Coffin, L. F., Jr., *Met. Trans.,* **2,** 1971, pp. 3105–3113.
114. Solomon, H. D., *Met. Trans.,* **4,** 1973, pp. 341–347.
115. Coffin, L. F., Jr., ASTM-STP 520, 1973, pp. 5–34.
116. Solomon, H. D., and L. F. Coffin, Jr., ASTM-STP 520, 1973, pp. 112–122.
117. Manson, S. S., G. R. Halford, and H. M. Hirshberg, in *Design for Elevated Temperature Environment,* ASME, 1971, pp. 25–28.
118. Manson, S. S., ASTM-STP 520, 1973, pp. 744–782.
119. Solomon, H. D., "The Creep and Strain Rate Sensitivity of a High Pb Content Solder," presented at the AIME symposium on *Deformation in Electronic Materials and Devices,* October 1989, Indianapolis, IN: proceedings to be published by AIME.
120. Knect, S., and L. R. Fox., to be published.

121. Fox, L., this volume.
122. Monkman, F. C., and N. J. Grant, *Proc. ASTM,* **56,** 1956, pp. 593–620.
123. Snowden, K. R., *Acta Met.,* **12,** 1964, pp. 295–303.
124. Snowden, K. U., *Phil. Mag.,* **10,** 1964, pp. 435–440.
125. Gabreilli, F., and R. M. Pelloux, *Met. Trans.,* **13A,** 1982, pp. 1083–1090.
126. Bradshaw, F. J., and C. Wheeler, *Applied Materials Research,* **5,** 1966, pp. 112–120.
127. Bradshaw, F. J., and C. Wheeler, *Intl. J. Fracture Mechanics,* **5,** 1969, pp. 255–268.
128. Vaynman, S., private communication.
129. Wild, R. N., *NEPCON,* 1974, pp. 105–117.
130. Kinser, D. L., J. G. Vaugn, and S. M. Graff, *Electronic Packaging and Production,* May 1976, pp. 61–68.
131. Brzozowski, V., et al., "FEA Analysis of Solder Joints," *Man. Tech. for ADSP,* monthly report March 20, 1987, Martin Marrietta Aerospace.
132. Miller, K. J., and M. W. Brown, Eds., "Multiaxial Fatigue," ASTM-STP 853, 1985.
133. Clech, J.-P., this volume.
134. Clech, J.-P., W. Engelmaier, R. W. Kotlowitz, and J. A. Augis, "Surface Mount Solder Attachment Reliability Figures of Merit—'Design for Reliability' Tools," presented at the 1989 SMART V Conference, New Orleans, LA.
135. ———, *IEEE Trans. CHMT,* **12,** no. 4, 1989, pp. 449–459.
136. Clech, J.-P., M. J. Tervalon, and J. A. Augis, *Proc. IEPS,* San Diego, CA, 1989, **2,** pp. 754–768.
137. Clech, J.-P., and J. A. Augis, *Proc. IEPS,* 1987, pp. 385–410.
138. ———, *Proc. IEPS,* 1988.
139. Lambert, R. G., "Mechanical Durability Prediction Methods," in *Proc. 1989 Annual Reliability and Maintainability Symposium,* Atlanta, GA, 1989.
140. Howard, R. T., IEEE CH1781-4/82/0000/0376, 1982, pp. 376–384.
141. Kotlowitz, R. W., and W. Engelmaier, *6th Intl. Electronic Packaging Conf. Proc.,* 1986, pp. 841–865.
142. Weller, D., L. Fox, and R. Hannemann, *33d ECC Proc.* IEEE, 0569-5503/83/000-0534, 1983, pp. 534–545.

15

Static and Dynamic Analyses of Surface Mount Component Leads and Solder Joints

John H. Lau

Surface mount technology (SMT) is currently one of the strongest trends in electronic packaging and interconnect assembly. This technology offers opportunities for significant cost reduction and for efficient production line automation.

A surface mount assembly (SMA) is basically a composite structure consisting of three major parts: the printed circuit board (PCB), the solder joints, and the surface mount component (SMC). In most consumer products, computers, peripherals, and automotive electronics, the PCB is made of FR-4 epoxy-glass and the solder joint is made of 60% Sn-40% Pb or 63% Sn-37% Pb solder alloys. Unlike traditional through-hole technology, SMT uses smaller SMCs that are soldered directly to the pad surface of a PCB. Consequently, the solder joint is the only mechanical means of attaching the SMC to the PCB. Thus, the integrity of solder joints is one of the most critical issues in SMT developments.[1–202] For a list of solder-joint papers published before 1985, see reference [185].

Basically, there are two groups of SMCs, leadless SMCs (e.g., leadless ceramic chip carriers, epoxy-glass chip carriers, chip resistors, chip capacitors, etc.) and leaded SMCs (e.g., small outline integrated circuits (SOICs), plastic leaded chip carrier (PLCCs), plastic quad flat pack (PQFPs), etc.). It is generally assumed that: (1) the limited compliance of the solder joints of leadless ceramic chip carriers on FR-4 PCB leads to serious reliability problems, and (2) the characteristically free-standing leads of SOICs, PLCCs, and PQFPs have sufficient compliance to virtually preclude any attachment reliability hazard. Unfortunately, the second assumption is not always correct. It is true that the stresses and strains acting at the solder joints are reduced substantially due to the compliance of the leads. However, because of solder's low resistance to fatigue, it is not uncommon to observe solder-joint cracking in SMT assemblies with leaded SMCs.

A variety of environmental stress factors such as temperature and humidity, shock and vibration, thermal and functional cycles, corrosion, etc., may lead to solder-joint failure. The most commonly observed failure modes in practice are *overload* and *fatigue*. Overload failure occurs whenever the stress or strain in the solder joint, brought about by the imposed stress factors, is greater than the strength or fracture toughness of the solder alloy. An example of overload failure condition is extensive bending and twisting of a PCB with SMCs soldered to its surface. On the other hand, fatigue failure of a solder joint takes place via the initiation and slow propagation of a crack until it becomes unstable.[185] The stress factors that typically cause failure by fatigue are far below the overload failure levels. Examples of fatigue failure include the vibration of a PCB with SMCs soldered to its surface and the functional power cycling of a SMC that is soldered to the surface of a PCB. In this chapter, both the static (including thermal effects) and dynamic (including shock and vibration) stress factors will be considered.

There are two parts in this chapter. In Part I, **"Stiffness of Gull-Wing and J Leads and Solder Joints for Surface Mounted Chip Carriers,"** two 12×12 stiffness matrices of the two most common leads and solder joints in SMA are presented. One is for the gull-wing lead and solder joint in a PQFP SMA, and the other is for the J lead and solder joint in a PLCC SMA. Deformation of these composite structures for each imposed unit displacement (or rotation) is also presented to provide a better understanding of their mechanical characteristics. Finally, the whole-field stress distributions in the leads and solder joints for the case of a unit displacement in the 1-direction are given. This case is similar to the global thermal expansion mismatch between the chip carriers and the PCB.

In Part II, **"Solder Joint Reliability Under Vibration Conditions,"** both local and global vibrations of an SMA are considered. For local vibration of the SMA, the frequencies of soldered and unsoldered leads of surface mount wide and narrow SOICs, PLCCs, and PQFPs are presented. It is found that the fundamental frequency of all the soldered leads is at least five times larger than that of the unsoldered leads. For global vibration of the SMA, the frequency equation of a constrained PCB with a SMC is given. With this equation, vibration frequencies of a PCB with a SMC can be ready determined. The results presented should be useful for theoretically determining the reliability of solder joints under shock and vibration conditions and for developing a shock and vibration experiment.

Stiffness of Gull-Wing and J Leads and Solder Joints for Surface Mounted Chip Carriers

Based on the matrix structural analysis of solid mechanics,[203–219] the stiffness matrices of the gull-wing lead and solder joint in a PQFP SMA and the J lead and solder joint in a PLCC SMA are calculated.

The PQFP and PLCC, Figures 15-1*a* and *b*, are package families that meet the Joint Electronic Device Engineering Council (JEDEC) package registration specification. The PQFP has gull-wing shaped leads that are spaced on a 20-mil (0.02-in.) pitch. The PLCC has specially designed J-shaped leads that are spaced on a 50-mil (0.05-in.) pitch. These leads are the electrical and mechanical connections of the packages to the outside world that are soldered to the surface of a PCB. Consequently, the stiffness of the leads plays a major role in transmitting the stress to the solder joints. In general, the larger the lead stiffness (less compliance), the more the joint stress. Therefore, in order to predict the solder-joint integrity, it is necessary to characterize the gull-wing (or J) lead and solder joint as a system.

In this chapter, the finite element method (see for examples, references [203–219]) was used to calculate the stiffness of the leads and solder joints. In order to take all the stress factors into account, all the possible deformations of the leads and solder joints are considered, (i.e., a total of 12 degrees of freedom have been used). It should be noted that these stiffnesses are independent of the geometry and material of the chip carriers and the PCB.

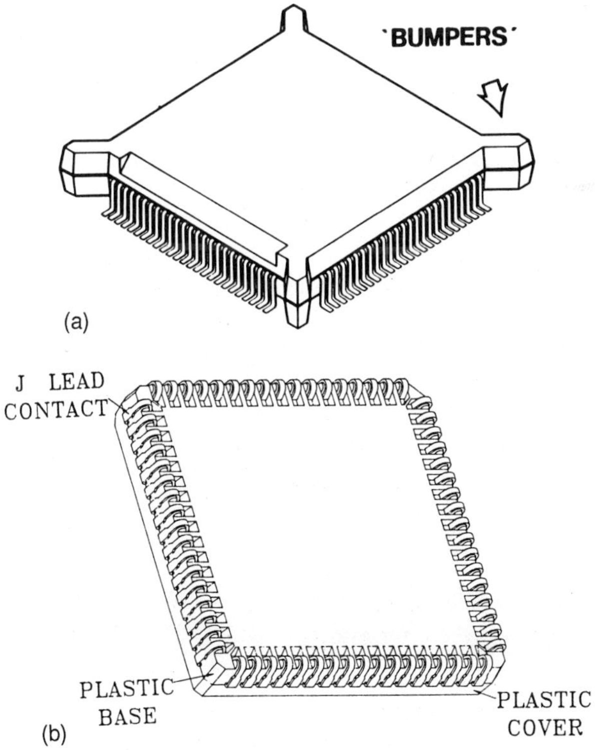

Figure 15-1 (*a*) Plastic quad flat pack (PQFP). (*b*) Plastic leaded chip carrier (PLCC).

15.1 BOUNDARY-VALUE PROBLEM

The composite structures under consideration are shown in Figure 15-2*a* and *b*. For the PQFP, the thickness of the gull-wing lead is 0.005 in., and the radius of the inner elbow is 0.005 in. The overall standoff height of the lead is 0.06 in., and the length of the horizontal part of the lead is 0.012 in. The minimum height of the solder joint is 0.003 in.

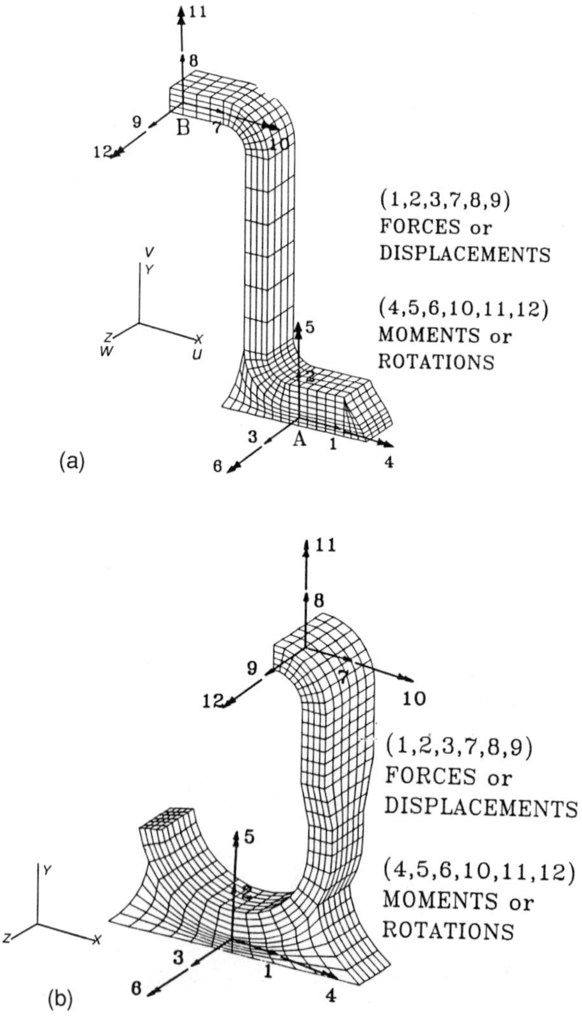

Figure 15-2 (*a*) FE model and coordinate system for PQFP Gull-Wing lead and solder joint. (*b*) Finite-element model and coordinate system for PLCC J-lead and solder joint.

For the PLCC, the thickness of the J lead is 0.008 in., the minimum height of the solder joint is 0.005 in., and the overall standoff height of the lead is 0.097 in. The width of the shoulder of the lead is 0.028 in., and the width of the lower part is 0.018 in. The radius of the centroidal axis of the shoulder is 0.014 in., and the radius of the centroidal axis of the lower curved part is 0.029 in.

The partial differential equations governing the 3-D structures are[220–221]

$$\frac{\partial^2 u}{\partial x^2} + \frac{\partial^2 v}{\partial x \partial y} + \frac{\partial^2 w}{\partial x \partial z} + (1 - 2\mu)\nabla^2 u = 0 \tag{15-1}$$

$$\frac{\partial^2 v}{\partial y^2} + \frac{\partial^2 u}{\partial x \partial y} + \frac{\partial^2 w}{\partial y \partial z} + (1 - 2\mu)\nabla^2 v = 0 \tag{15-2}$$

$$\frac{\partial^2 w}{\partial z^2} + \frac{\partial^2 u}{\partial x \partial z} + \frac{\partial^2 v}{\partial y \partial z} + (1 - 2\mu)\nabla^2 w = 0 \tag{15-3}$$

where

$$\nabla^2 = \frac{\partial^2}{\partial x^2} + \frac{\partial^2}{\partial y^2} + \frac{\partial^2}{\partial z^2} \tag{15-4}$$

In Equations (15-1) through (15-3), u, v, and w are the displacement components in the x, y, and z directions, respectively, and μ is Poisson's ratio. The relationships between the stress components σ_x (normal stress acting in the x direction), σ_y (normal stress acting in the y direction), σ_z (normal stress acting in the z direction), τ_{xy} (shear stress acting in xy plane), τ_{yz} (shear stress acting in the yz plane), and τ_{zx} (shear stress acting in zx plane), and the displacement components u, v, and w are

$$\sigma_x = Z\left[(1 - \mu)\frac{\partial u}{\partial x} + \mu\left(\frac{\partial v}{\partial y} + \frac{\partial w}{\partial z}\right)\right] \tag{15-5}$$

$$\sigma_y = Z\left[(1 - \mu)\frac{\partial v}{\partial y} + \mu\left(\frac{\partial w}{\partial z} + \frac{\partial u}{\partial x}\right)\right] \tag{15-6}$$

$$\sigma_z = Z\left[(1 - \mu)\frac{\partial w}{\partial z} + \mu\left(\frac{\partial u}{\partial x} + \frac{\partial v}{\partial y}\right)\right] \tag{15-7}$$

$$\tau_{xy} = G\left(\frac{\partial u}{\partial y} + \frac{\partial v}{\partial x}\right) \tag{15-8}$$

$$\tau_{yz} = G\left(\frac{\partial v}{\partial z} + \frac{\partial w}{\partial y}\right) \tag{15-9}$$

$$\tau_{zx} = G\left(\frac{\partial w}{\partial x} + \frac{\partial u}{\partial z}\right) \tag{15-10}$$

where

$$Z = \frac{E}{(1 + \mu)(1 - 2\mu)} \tag{15-11}$$

$$G = \frac{E}{2(1 + \mu)} \tag{15-12}$$

and E is Young's modulus for the material.

The statement of the present boundary-value problem is to find a general solution for u, v, and w from Equations (15-1) through (15-3), which satisfies a given set of boundary conditions. There are 12 different sets of displacement boundary conditions in the present study and they will be stated in the results section. The stresses σ_x, σ_y, σ_z, τ_{xy}, τ_{yz}, and τ_{zx} can then be calculated from Equations (15-5) through (15-10) and the Mises stress intensity is determined by the following equation

$$\bar{\sigma} = \frac{\sqrt{2}}{2}\sqrt{(\sigma_x - \sigma_y)^2 + (\sigma_y - \sigma_z)^2 + (\sigma_z - \sigma_x)^2 + 6(\tau_{xy}{}^2 + \tau_{yz}{}^2 + \tau_{zx}{}^2)} \tag{15-13}$$

Obviously, closed-form solutions for the present boundary-value problem would be very difficult to obtain. Therefore, numerical schemes such as the finite element method[203–219] were adopted.

15.2 FINITE ELEMENT METHODS

The finite element method is the principal means of solving boundary-value problems. Part of its advantage stems from its ability to take care of irregular boundary shapes and mixed material structures. Due to the generality and richness of the concepts underlying this method, it has been employed with remarkable success in solving problems in virtually all areas of engineering and mathematical physics.

The basic concept of the finite element method is that a structure can be decomposed into a finite number of elements. For each element, trial function approximations of displacement components are used in conjunction with the variational principle or Galerkin's approximation and matrix methods to transform the boundary-value problem, Equations (15-1) through (15-10), and the boundary conditions into a system of simultaneous algebraic equations.

In formulating the finite element method by the principle of minimum potential energy, the displacement field is represented by interpolation functions together with generalized displacements at a finite number of nodal points in each element. In matrix form the assumed displacement may be written as (Fig. 15-3)

$$[u] = [A][s] \tag{15-14}$$

where $[u]$ is a displacement column matrix, $[s]$ is a nodal displacement column matrix, and $[A]$ is a shape function matrix. In Figure 15-3, ① is the element for one kind of material and ② is the element for the other.

The corresponding strain column matrix is

$$[\varepsilon] = [L][u] \tag{15-15}$$

where $[L]$ is a differential operator matrix. The corresponding stress column matrix is

$$[\sigma] = [D][\varepsilon] - [D][\varepsilon_o] + [\sigma_o] \tag{15-16}$$

where $[D]$ is a material matrix, and $[\varepsilon_o]$ and $[\sigma_o]$ are column matrices of initial strains and initial stresses, respectively. Equation (15-16) degenerates to Equations (15-5) through (15-10) if $[\varepsilon_o] = [\sigma_o] = 0$.

The total potential energy functional (V_p) for any elastic solid, which is divided into a finite number (n) of discrete elements (V_n), is given by

$$\begin{aligned} V_p = \sum_n \left(- \int_{S_n} [\mathrm{u}]^T[\mathrm{T}]dS - \int_{V_n} [\mathrm{u}]^T[\mathrm{F}]dV \right) \\ + \frac{1}{2} \sum_n \int_{V_n} [\sigma]^T[\varepsilon]dV \end{aligned} \tag{15-17}$$

where $[T]$ is a traction column matrix acting on a surface S_n, and $[F]$ is a body force column matrix. Substituting Equations (15-14) through (15-16) into Equation (15-17) leads to

$$V_p = \sum_n \left(-[s]^T[\mathrm{S}] + \frac{1}{2}[\mathrm{s}]^T[k][s] \right) \tag{15-18}$$

where

$$[k] = \int_{V_n} ([L][A])^T[D]\,([L][A])dV \tag{15-19}$$

and

$$\begin{aligned}[S] = &\int_{S_n} [A]^T[T]dS + \int_{V_n} [A]^T[F]dV \\ &+ \int_{V_n} ([L][A])^T[D][\varepsilon_o]dV \\ &- \int_{V_n} ([L][A])^T[\sigma_o]dV\end{aligned} \tag{15-20}$$

are the element stiffness matrix and the nodal force column matrix, respectively. The nodal displacements $[s]$ for different elements are not completely independent; a transformation is needed to relate the element nodal displacements to the independent generalized global displacements. Then the compatibility equations of the assembled structure are written in matrix form as (Fig. 15-3)

$$[s] = [a][r] \tag{15-21}$$

where $[a]$ is a compatibility transformation matrix and $[r]$ is a global displacement column matrix.

Substituting Equation (15-21) into Equation (15-18), yields

$$V_p = -[r]^T[R] + \frac{1}{2}[r]^T[K][r] \tag{15-22}$$

where

$$[K] = \sum_n [a]^T[k][a] \tag{15-23}$$

and

$$[R] = \sum_n [a]^T[S] \tag{15-24}$$

are the global stiffness matrix and the global force column matrix, respectively.

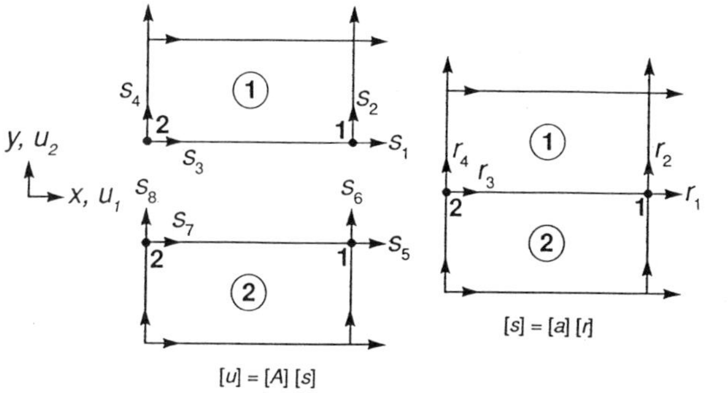

Where [u] = Displacement Column Matrix
[s] = Nodal Displacement Column Matrix
[A] = Shape Function Matrix
[a] = Compatibility Transformation Matrix
[r] = Global Displacement Column Matrix

Figure 15-3 Nodal points, nodal displacements, and global nodal displacements.

The principle of minimum potential energy requires that $\delta V_p = 0$, that is,

$$\frac{\partial V_p}{\partial [r]} \delta[r] = (-[R] + [K][r])\delta[r] = 0 \tag{15-25}$$

For arbitrary $\delta[r]$,

$$[K][r] = [R] \tag{15-26}$$

Thus, for a given force vector $[R]$ the corresponding displacement vector $[r]$ can be obtained by the inversion of $[K]$ and the nodal displacements and the corresponding strains within each element may be found by Equations (15-21) and (15-15), respectively. Most of the commercially available finite element codes calculate the stresses at the nodal points of each element by Equation (15-16) and average the stress values at the interface between elements. In that case, for a composite structure special care should be noted.[203]

Figure 15-2*a* depicts the three-dimensional finite element model used for the determination of the stiffness of the gull-wing lead and solder joint. It consists of 832 3D solid elements. Each element has 20 nodal points. Each nodal point has 3 degrees of freedom. The Young's modulus used for the copper gull-wing lead is 17,500 ksi, and that for the solder alloy is 1500 ksi. The Poisson's ratio for the lead is 0.35, and for the solder it is 0.40.

Figure 15-2*b* depicts the three-dimensional finite element model used for the determination of the stiffness of the J lead and solder joint for the PLCC SMA.

It consists of 1296 3D solid elements. The element type and material properties are exactly the same as those for the gull-wing lead and solder joint for the PQFP SMA.

The coordinate system for measuring forces, moments, displacements, and rotations is the same for both the PQFP and PLCC SMAs and is also shown in Figures 15-2*a* and *b*. This system consists of 12 coordinates that are identified by 12 numbered arrows shown at specific directions and locations on the composite structure. Single arrows (1, 2, 3, 7, 8, 9) represent forces and the corresponding displacements, while double arrows (4, 5, 6, 10, 11, 12) represent moments and the corresponding rotations. All the arrows point in the positive direction of the forces and moments (right-hand rule). Coordinates 1 through 6 are specified at the center of the bottom surface of the solder joint (point A in Fig. 15-2*a* and *b*). Coordinates 7 through 12 are specified at the center of the end surface of the leads (point B in Fig. 15-2*a* and *b*). These two surfaces are assumed to be rigid. There are 12 different sets of displacement boundary conditions imposed at these two surfaces. Therefore, there are 12 analyses of each model and their results are summarized in the next two sections.

15.3 STIFFNESS OF GULL-WING LEAD AND SOLDER JOINT

The 12 loading components (forces and moments) and the 12 corresponding deformation components (displacements and rotations) of the gull-wing lead and solder joint are related to each other through the symmetrical 12 × 12 stiffness matrix [*K*] as shown in Table 15-1. In this matrix, the unit for the forces is lb/in. and the unit for the moments is in. – lb/rad. It should be noted that this stiffness matrix [*K*] is not the same as the one in Equation (15-25).

The definition of a typical stiffness coefficient is denoted by K_{ij}, which means the value of the action (force or moment) at restraint (coordinate) *j* induced by a unit displacement (or rotation) at restraint *i*.[209–219]

For example, the elements in the first row are: 1 = 165, 2 = –325, 3 = 0, 4 = 0, 5 = 0, 6 = –3, 7 = –165, 8 = 325, 9 = 0, 10 = 0, 11 = 0, and 12 = 2. This means that for a unit displacement in the 1 direction (Fig. 15-2*a*) with all the other directions (2 through 12) fixed, the force applied at C-1 (from now on, C- will be the abbreviation for coordinate-) is 165 lb/in., the force applied at C-2 is 325 lb/in. (in the opposite direction of Fig. 15-2*a*), the force applied at C-3 and the moments applied at C-4 and C-5 are zero, the moment applied at C-6 is 3 in. – lb/rad (in the opposite direction of Fig. 15-2*a*), the force applied at C-7 is 165 lb/in. (in the opposite direction of Fig. 15-2*a*), the force applied at C-8 is 325 lb/in., the force applied at C-9 and the moments applied at C-10 and C-11 are zero, and the moment applied at C-12 is 2 in. – lb/rad.

Table 15-1 Stiffness Matrix (Force in lb–in.; Moment in in./lb/rad) for Gull-Wing Lead and Solder Joint

K =

165	−325	0	0	0	−3	−165	325	0	0	0	2
−325	1648	0	0	0	−6	325	−1648	0	0	0	21
0	0	138	6.1	1.3	0	0	0	−138	2.3	2.6	0
0	0	6.1	.3	.1	0	0	0	−6.1	0	.1	0
0	0	1.3	.1	.1	0	0	0	−1.3	0	0	0
−3	−6	0	0	0	.2	3	6	0	0	0	.1
−165	325	0	0	0	3	165	−325	0	0	0	−2.1
325	−1648	0	0	0	6	−325	1648	0	0	0	21
0	0	−138	−6.1	−1.3	0	0	0	138	−2.3	−2.6	0
0	0	2.3	0	0	0	0	0	−2.3	.1	.1	0
0	0	2.6	.1	0	0	0	0	−2.6	.1	.1	0
2	21	0	0	0	.1	−2.1	21	0	0	0	.3

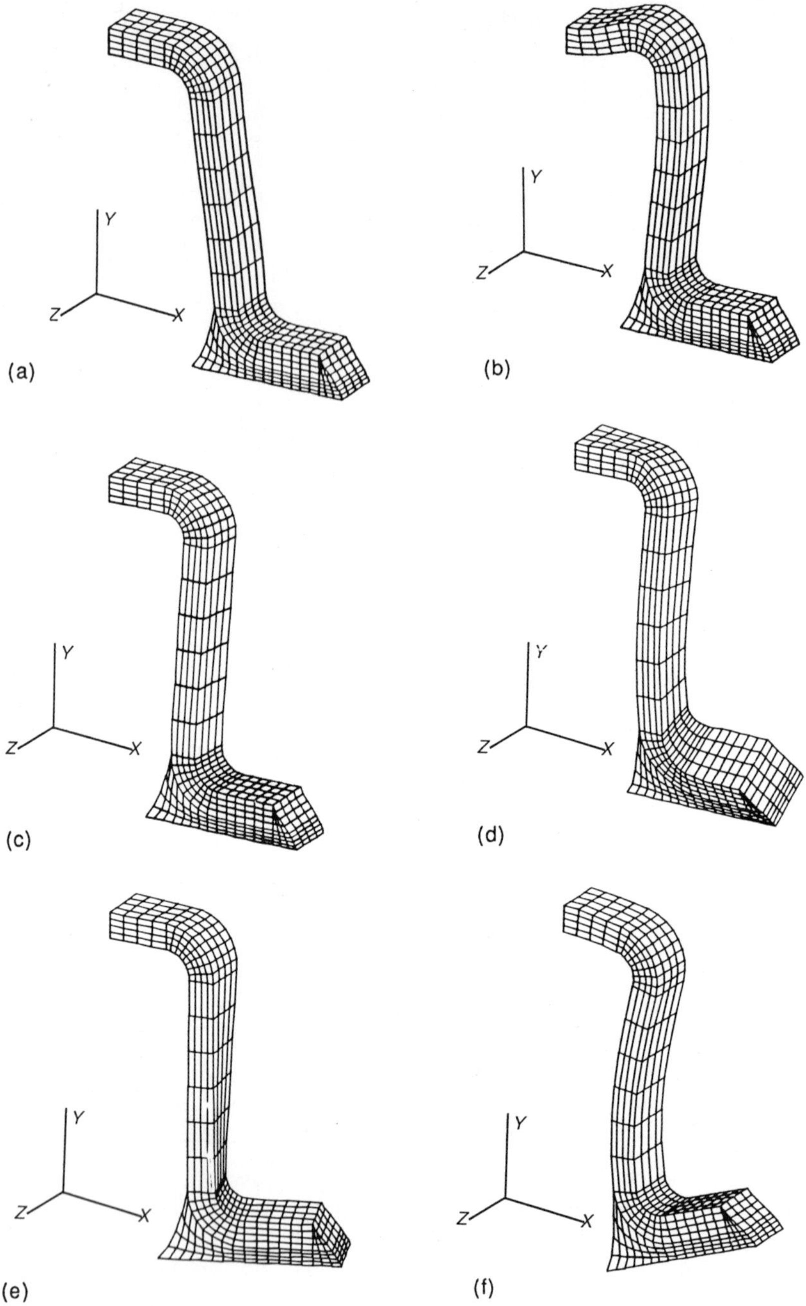

Figure 15-4 Deformation. (*a*) Unit displacement in 1-direction. (*b*) Unit displacement in 2-direction. (*c*) Unit displacement in 3-direction. (*d*) Unit rotation in 4-direction. (*e*) Unit

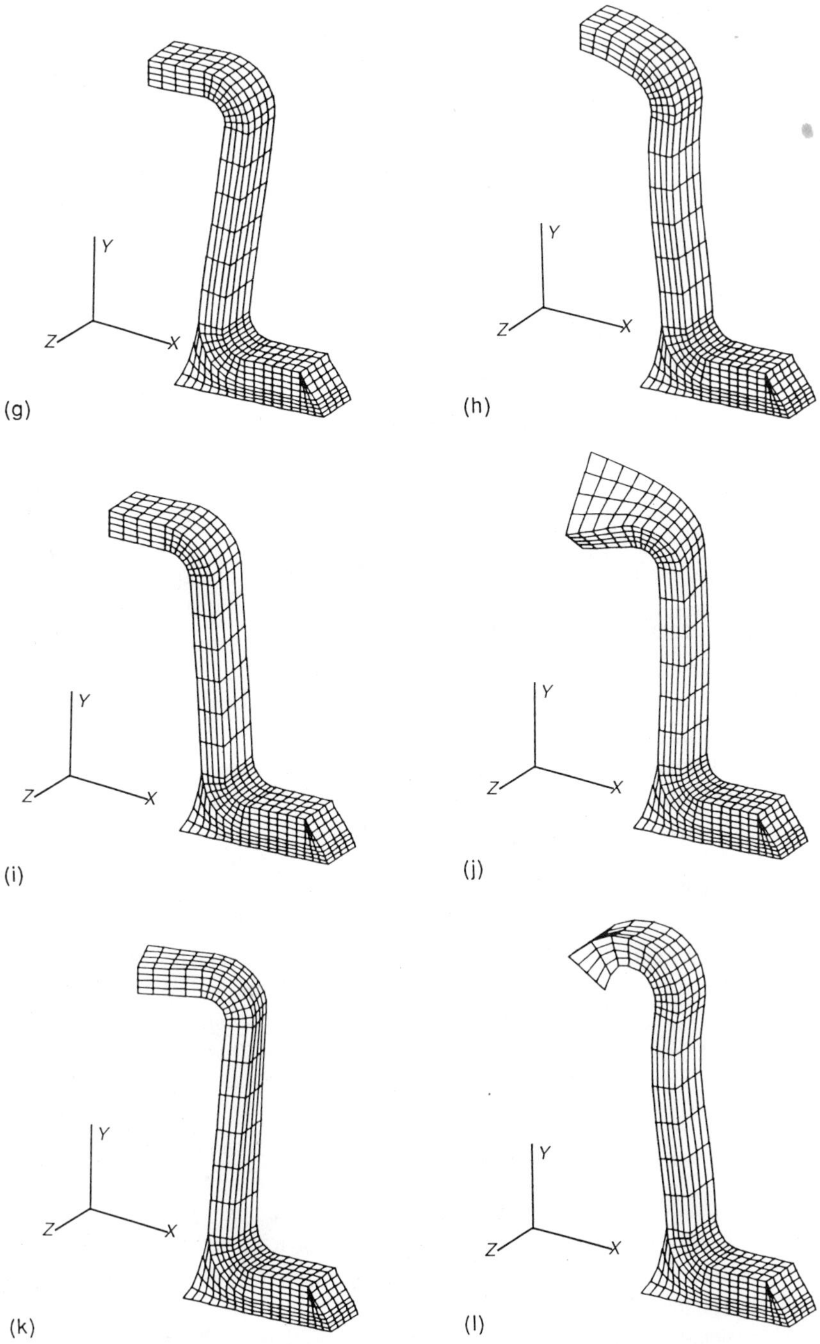

rotation in 5-direction. (*f*) Unit rotation in 6-direction. (*g*) Unit displacement in 7-direction. (*h*) Unit displacement in 8-direction. (*i*) Unit displacement in 9-direction. (*j*) Unit rotation in 10-direction. (*k*) Unit rotation in 11-direction. (*l*) Unit rotation in 12-direction.

The calculated deformations of the composition structure (gull-wing lead and solder joint) for each imposed unit displacement and rotation are shown in Figures 15-4*a* through 4*l*. The stress contours acting in the gull-wing lead and solder joint for a unit displacement (0.0001 in.) in the C-1 direction are shown in Figure 15-5*a* and *b*, respectively. Some significant results are summarized as follows (the unit for forces is lb/in. and the unit for moments is in. – lb/rad):

1. Unit displacement (0.0001 in.) in the 1-direction. The boundary conditions for this case are: at point A, $u = 0.0001$ in., $v = 0$, $w = 0$, $(\partial w / \partial y - \partial v / \partial z)/2 = 0$, $(\partial u / \partial z - \partial w/\partial x) / 2 = 0$, $(\partial v/\partial x - \partial u / \partial y) / 2 = 0$ and at point B, $u = 0$, $v = 0$, $w = 0$, $(\partial w / \partial y - \partial v / \partial z) / 2 = 0$, $(\partial u / \partial z - \partial w / \partial x) / 2 = 0$, $(\partial v / \partial x - \partial u / \partial y) / 2 = 0$.

   The finite element results for the forces and moments are shown in the first row of Table 15-1. The deformed shape of the composite structure is shown in Figure 15-4*a*. The Mises stress contours in the gull-wing lead and solder joint are shown in Figure 15-5*a* and *b*, respectively. It can be seen that the maximum stress (9218 psi) acting in the gull-wing lead occurs at the inner elbows of the gull-wing lead due to stress concentration and bending. Higher stress also occurs at the outer shoulder of the gull-wing lead because of bending. Because of the residual manufacturing stresses, it is not uncommon to observe cracks at the shoulder of the gull-wing leads. In that case, further crack propagation in the gull-wing lead under operation conditions is likely. The maximum Mises stress (1245 psi) acting in the solder joint occurs near the inside upper-tip of the solder fillet, Figure 15-4*b*. The stresses acting on the outer toe of the solder joint are very small. The results of this case can be used as an analog for estimation of the thermal stresses (due to the thermal expansion mismatch between the SMC and PCB) in the lead and solder joint.
2. Unit displacement (0.0001 in.) in the 2-direction. The boundary conditions for this case are: at point A, $u = 0$, $v = 0.0001$ in., $w = 0$, $(\partial w / \partial y - \partial v / \partial z) / 2 = 0$, $(\partial u / \partial z - \partial w / \partial x)/2 = 0$, $(\partial v / \partial x - \partial u / \partial y) / 2 = 0$ and at point B, $u = 0$, $v = 0$, $w = 0$, $(\partial w / \partial y - \partial v / \partial z) / 2 = 0$, $(\partial u / \partial z - \partial w / \partial x) / 2 = 0$, $(\partial v / \partial x - \partial u / \partial y) / 2 = 0$.

   The finite element results are shown in the second row of the stiffness matrix (Table 15-1). The deformed shape is shown in Figure 15-4*b*. The Mises stress contours in the gull-wing lead and solder joint for cases 2 through 12 are not shown in this chapter but can be found in references.[189,190] The maximum stress (32,479 psi) occurs at the upper end-surface of the gull-wing lead at the package. The maximum stress (1062 psi) acting at the solder joint also occurs near the inner solder fillet as in case 1.
3. Unit displacement (0.0001 in.) in the 3-direction. The boundary conditions for this case are: at point A, $u = 0$, $v = 0$, $w = 0.0001$ in., $(\partial w / \partial y$

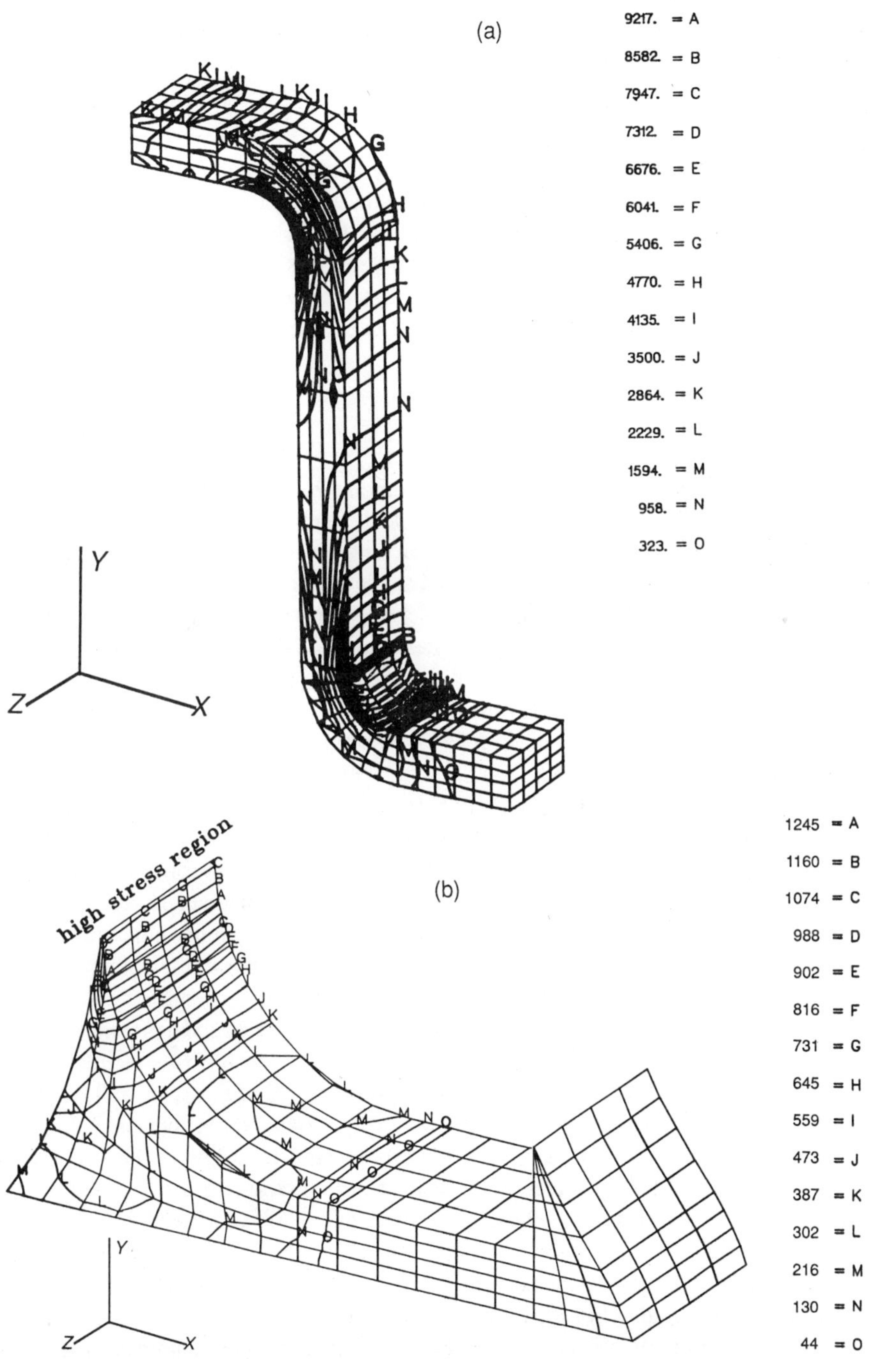

Figure 15-5 (*a*) Mises stress contours in the gull-wing. (*b*) Mises stress contours in the PQFP solder joint. (Unit displacement, 0.0001 in., in 1 direction).

$- \partial v / \partial z) / 2 = 0$, $(\partial u / \partial z - \partial w / \partial x) / 2 = 0$, $(\partial v / \partial x - \partial u / \partial y) / 2 = 0$, and at point B, $u = 0$, $v = 0$, $w = 0$, $(\partial w / \partial y - \partial v / \partial z) / 2 = 0$, $(\partial u / \partial z - \partial w / \partial x) / 2 = 0$, $(\partial v / \partial x - \partial u / \partial y) / 2 = 0$.

The finite element results for the forces and moments are shown in the third row of the stiffness matrix (Table 15-1). Figure 15-4*c* shows the deformed shape. The maximum stress (8255 psi) occurs at both sides of the lower part of the lead near the inner elbow. The maximum stress (1388 psi) in the solder joint occurs at both sides near the tip of the inner solder fillet.

4. Unit rotation (0.0001 rad) in the 4-direction. The boundary conditions for this case are: at point A, $u = 0$, $v = 0$, $w = 0$, $(\partial w / \partial y - \partial v / \partial z) / 2 = 0.0001$ rad, $(\partial u / \partial z - \partial w / \partial x) / 2 = 0$, $(\partial v / \partial x - \partial u / \partial y) / 2 = 0$, and at point B, $u = 0$, $v = 0$, $w = 0$, $(\partial w / \partial y - \partial v / \partial z) / 2 = 0$, $(\partial u / \partial z - \partial w / \partial x) / 2 = 0$, $(\partial v / \partial x - \partial u / \partial y) / 2 = 0$.

   The finite element results are shown in the fourth row of the stiffness matrix (Table 15-1). The deformed shape of this case is shown in Figure 15-4*d*. The maximum stress (4718 psi) occurs near the inner elbow of the lower part of the lead. The maximum stress (778 psi) in the solder joint occurs at both sides near the tip of the inner solder fillet.
5. Unit rotation (0.0001 rad) in the 5-direction. The boundary conditions for this case are: at point A, $u = 0$, $v = 0$, $w = 0$, $(\partial w / \partial y - \partial v / \partial z) / 2 = 0$, $(\partial u / \partial z - \partial w / \partial x) / 2 = 0.0001$ rad, $(\partial v / \partial x - \partial u / \partial y) / 2 = 0$, and at point B, $u = 0$, $v = 0$, $w = 0$, $(\partial w / \partial y - \partial v / \partial z) / 2 = 0$, $(\partial u / \partial z - \partial w / \partial x) / 2 = 0$, $(\partial v / \partial x - \partial u / \partial y) / 2 = 0$.

   The finite element results are shown in the fifth row of the stiffness matrix (Table 15-1). Figure 15-4*e* shows the deformed shape. The maximum stress (1279 psi) occurs near both sides of the inner elbow of the lower part of the lead. The maximum stress (131 psi) in the solder joint occurs near the inner solder fillet.
6. Unit rotation (0.0001 rad) in the 6-direction. The boundary conditions for this case are: at point A, $u = 0$, $v = 0$, $w = 0$, $(\partial w / \partial y - \partial v / \partial z) / 2 = 0$, $(\partial u / \partial z - \partial w / \partial x) / 2 = 0$, $(\partial v / \partial x - \partial u / \partial y) / 2 = 0.0001$ rad, and at point B, $u = 0$, $v = 0$, $w = 0$, $(\partial w / \partial y - \partial v / \partial z) / 2 = 0$, $(\partial u / \partial z - \partial w / \partial x) / 2 = 0$, $(\partial v / \partial x - \partial u / \partial y) / 2 = 0$. This set of boundary conditions means there is a unit rotation in the 6-direction (Figure 15-2*a*) with all the other directions (1 through 5 and 7 through 12) fixed.

   The finite element results are: the forced applied at C-1 is 3 lb/in. (in the opposite direction of Figure 15-2*a*), the force applied at C-2 is 6 lb/in. (in the opposite direction of Figure 15-2*a*), the force applied at C-3 is 0, the moments applied at C-4 and C-5 are 0, the moment applied at C-6 is 0.2 in. – lb/rad, the force applied at C-7 is 3 lb/in., the force applied at C-8 is 6 lb/in., the force applied at C-9 is 0, the moments applied at C-10 and C-11 are 0, the moment applied at C-12 is 0.1 in. – lb/rad. All of these values are shown in the sixth row of the stiffness matrix (Table 15-1). The deformed shape is shown in Figure 15-4*f*. The maximum stress (2999 psi) occurs at the inner elbow of the lower part of the lead. The maximum stress (595 psi) in the solder joint occurs at the tip of the inner solder fillet.

7. Unit displacement (0.0001 in.) in the 7-direction. The boundary conditions for this case are: at point A, $u = 0$, $v = 0$, $w = 0$, $(\partial w / \partial y - \partial v / \partial z) / 2 = 0$, $(\partial u / \partial z - \partial w / \partial x) / 2 = 0$, $(\partial v / \partial x - \partial u / \partial y) / 2 = 0$, and at point B, $u = 0.0001$ in., $v = 0$, $w = 0$, $(\partial w / \partial y - \partial v / \partial z) / 2 = 0$, $(\partial u / \partial z - \partial w / \partial x) / 2 = 0$, $(\partial v / \partial x - \partial u / \partial y) / 2 = 0$.

   The finite element results are shown in the seventh row of the stiffness matrix (Table 15-1). The deformed shape of this case is shown in Figure 15-4*g*. The Mises stress contours acting at the gull-wing lead and solder joint are the same as those for a unit displacement in 1-direction.

8. Unit displacement (0.0001 in.) in the 8-direction. The boundary conditions for this case are: at point A, $u = 0$, $v = 0$, $w = 0$, $(\partial w / \partial y - \partial v / \partial z) / 2 = 0$, $(\partial u / \partial z - \partial w / \partial x) / 2 = 0$, $(\partial v / \partial x - \partial u / \partial y) / 2 = 0$, and at point B, $u = 0$, $v = 0.0001$ in., $w = 0$, $(\partial w / \partial y - \partial v / \partial z) / 2 = 0$, $(\partial u / \partial z - \partial w / \partial x) / 2 = 0$, $(\partial v / \partial x - \partial u / \partial y) / 2 = 0$.

   The finite element results are shown in the eighth row of the stiffness matrix (Table 15-1). Figure 15-4*h* shows the deformed shape of the composite structure. The Mises stress contours in the gull-wing lead and solder joint are the same as those for a unit displacement in 2-direction.

9. Unit displacement (0.0001 in.) in the 9-direction. The boundary conditions for this case are: at point A, $u = 0$, $v = 0$, $w = 0$, $(\partial w / \partial y - \partial v / \partial z) / 2 = 0$, $(\partial u / \partial z - \partial w / \partial x) / 2 = 0$, $(\partial v / \partial x - \partial u / \partial y) / 2 = 0$, and at point B, $u = 0$, $v = 0$, $w = 0.0001$ in., $(\partial w / \partial y - \partial v / \partial z) / 2 = 0$, $(\partial u / \partial z - \partial w / \partial x) / 2 = 0$, $(\partial v / \partial x - \partial u / \partial y) / 2 = 0$.

   The finite element results are shown in the ninth row of the stiffness matrix (Table 15-1). Figure 15-4*i* shows the deformed shape. The Mises stress contours in the composite structure are the same as those for a unit displacement in the 3-direction.

10. Unit rotation (0.0001 rad) in the 10-direction. The boundary conditions for this case are: at point A, $u = 0$, $v = 0$, $w = 0$, $(\partial w / \partial y - \partial v / \partial z) / 2 = 0$, $(\partial u / \partial z - \partial w / \partial x) / 2 = 0$, $(\partial v / \partial x - \partial u / \partial y) / 2 = 0$, and at point B, $u = 0$, $v = 0$, $w = 0$, $(\partial w / \partial y - \partial v / \partial z) / 2 = 0.0001$ rad., $(\partial u / \partial z - \partial w / \partial x) / 2 = 0$, $(\partial v / \partial x - \partial u / \partial y) / 2 = 0$.

    The finite element results are shown in the tenth row of the stiffness matrix (Table 15-1). The deformation shape is shown in Figure 15-4*j*. The maximum stress (2562 psi) occurs near the upper end-surface. The maximum stress (65 psi) in the solder joint occurs near the inner solder fillet.

11. Unit rotation (0.0001 rad) in the 11-direction. The boundary conditions for this case are: at point A, $u = 0$, $v = 0$, $w = 0$, $(\partial w / \partial y - \partial v / \partial z) / 2 = 0$, $(\partial u / \partial z - \partial w / \partial x) / 2 = 0$, $(\partial v / \partial x - \partial u / \partial y) / 2 = 0$, and at point B, $u = 0$, $v = 0$, $w = 0$, $(\partial w / \partial y - \partial v / \partial z) / 2 = 0$, $(\partial u / \partial z - \partial w / \partial x) / 2 = 0.0001$ rad, $(\partial v / \partial x - \partial u / \partial y) / 2 = 0$.

    The finite element results are shown in the eleventh row of the stiffness matrix (Table 15-1). The deformed shape is shown in Figure 15-4*k*. The maximum stress (1525 psi) occurs near the elbow of the lower part of the lead and at the upper end-surface of the lead. The maximum stress (347 psi) in the solder joint occurs at both sides of the tip of the inner solder fillet.
12. Unit rotation (0.0001 rad) in the 12-direction. The boundary conditions for this case are: at point A, $u = 0$, $v = 0$, $w = 0$, $(\partial w / \partial y - \partial v / \partial z) / 2 = 0$, $(\partial u / \partial z - \partial w / \partial x) / 2 = 0$, $(\partial v / \partial x - \partial u / \partial y) / 2 = 0$, and at point B, $u = 0$, $v = 0$, $w = 0$, $(\partial w / \partial y - \partial v / \partial z) / 2 = 0$, $(\partial u / \partial z - \partial w / \partial x) / 2 = 0$, $(\partial v / \partial x - \partial u / \partial y) / 2 = 0.0001$ rad.

    The finite element results are shown in the twelfth row of the stiffness matrix (Table 15-1). Figure 15-4*l* shows the deformed shape. The maximum stress (5264 psi) occurs at the upper end-surface of the lead. The maximum stress (152 psi) in the solder joint occurs at both sides of the tip of the inner solder fillet.

In all the twelve cases considered, the location of the maximum Mises stress acting at the solder joint is near the tip of the inner solder fillet. The stresses in the outer solder fillet are very small. Thus, from a failure mechanism standpoint, solder joint fatigue cracking should initiate at the tip of the inner solder fillet.

15.4 STIFFNESS OF J LEAD AND SOLDER JOINT

Based on the same definition of stiffness, analysis procedure, material property, boundary conditions, etc., for the gull-wing lead and solder joint, the stiffness matrix relating the 12 loading components (forces and moments) and the 12 corresponding deformation components (displacements and rotations) of the J lead and solder joint is determined and is shown in Table 15-2. The deformed shapes for all 12 cases are shown in Figure 15-6*a* through *l*.

15.4.1 Unit Displacement (0.0001 in.) in the 1-Direction

The finite element results for the forces and moments are shown in the first row of Table 15-2. It can be seen that for a unit displacement in the 1 direction (Figure 15-2*b*) with all the other direction fixed, the force applied at C-1 is 298

Table 15-2 Stiffness Matrix (Force in lb/in.; Moment in in.–lb/rad) for J-Lead and Solder Joint

K =

298	105	0	0	0	−13	−298	−105	0	0	0	−14
105	8973	0	0	0	233	−105	8973	0	0	0	−109
0	0	665	26	−21	0	0	0	−665	39	11	0
0	0	26	1.7	−.8	0	0	0	−26	.9	.5	0
0	0	−21	−.8	.9	0	0	0	21	−1.2	−.6	0
−13	233	0	0	0	7	13	−233	0	0	0	−2.3
−298	−105	0	0	0	13	298	105	0	0	0	14
−105	−8973	0	0	0	−233	105	8973	0	0	0	109
0	0	−665	−26	21	0	0	0	665	−39	−11	0
0	0	39	.9	−1.2	0	0	0	−39	3	.6	0
0	0	11	.5	−.6	0	0	0	−11	.6	.4	0
−14	−109	0	0	0	−2.3	14	109	0	0	0	2.1

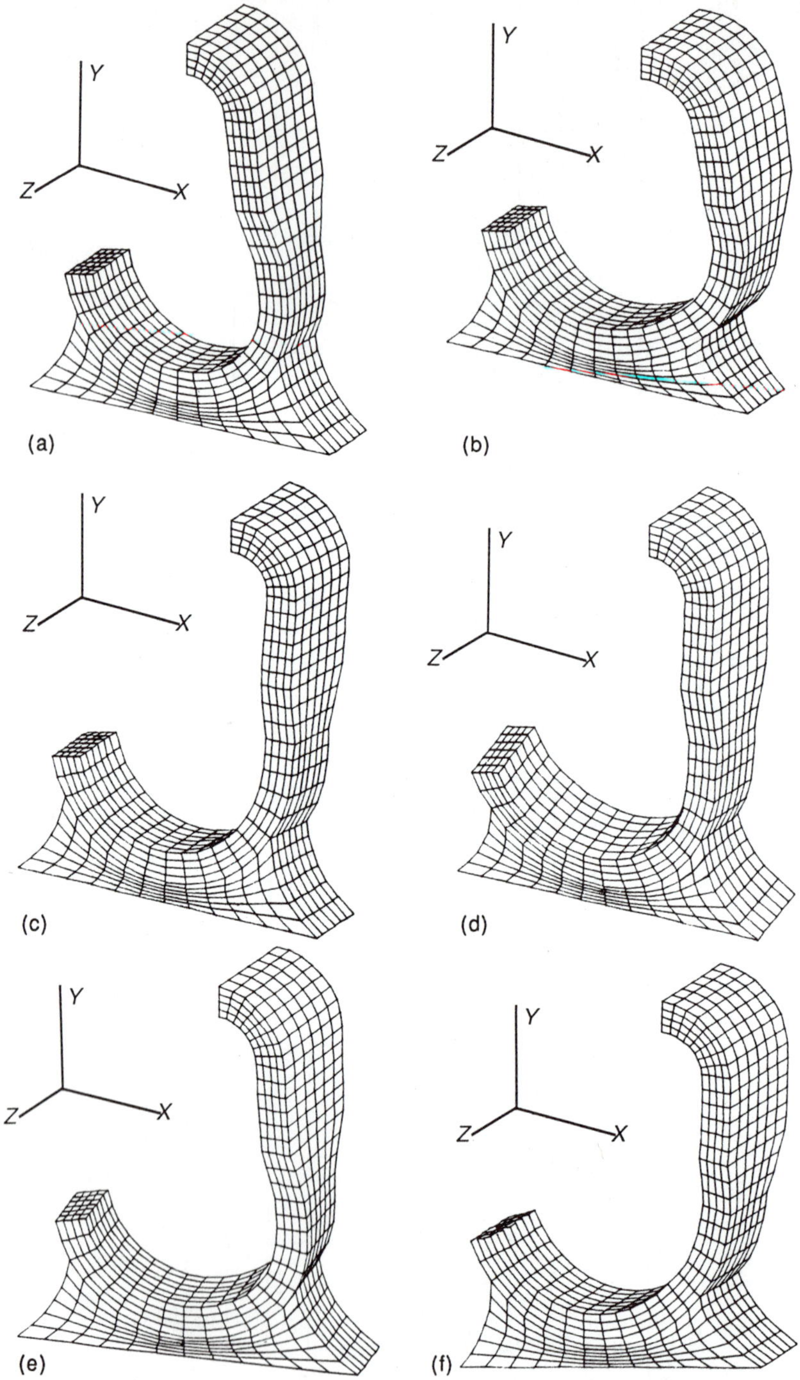

Figure 15-6 Deformation. (*a*) Unit displacement in 1 direction. (*b*) Unit displacement in 2 direction. (*c*) Unit displacement in 3 direction. (*d*) Unit rotation in 4 direction. (*e*) Unit

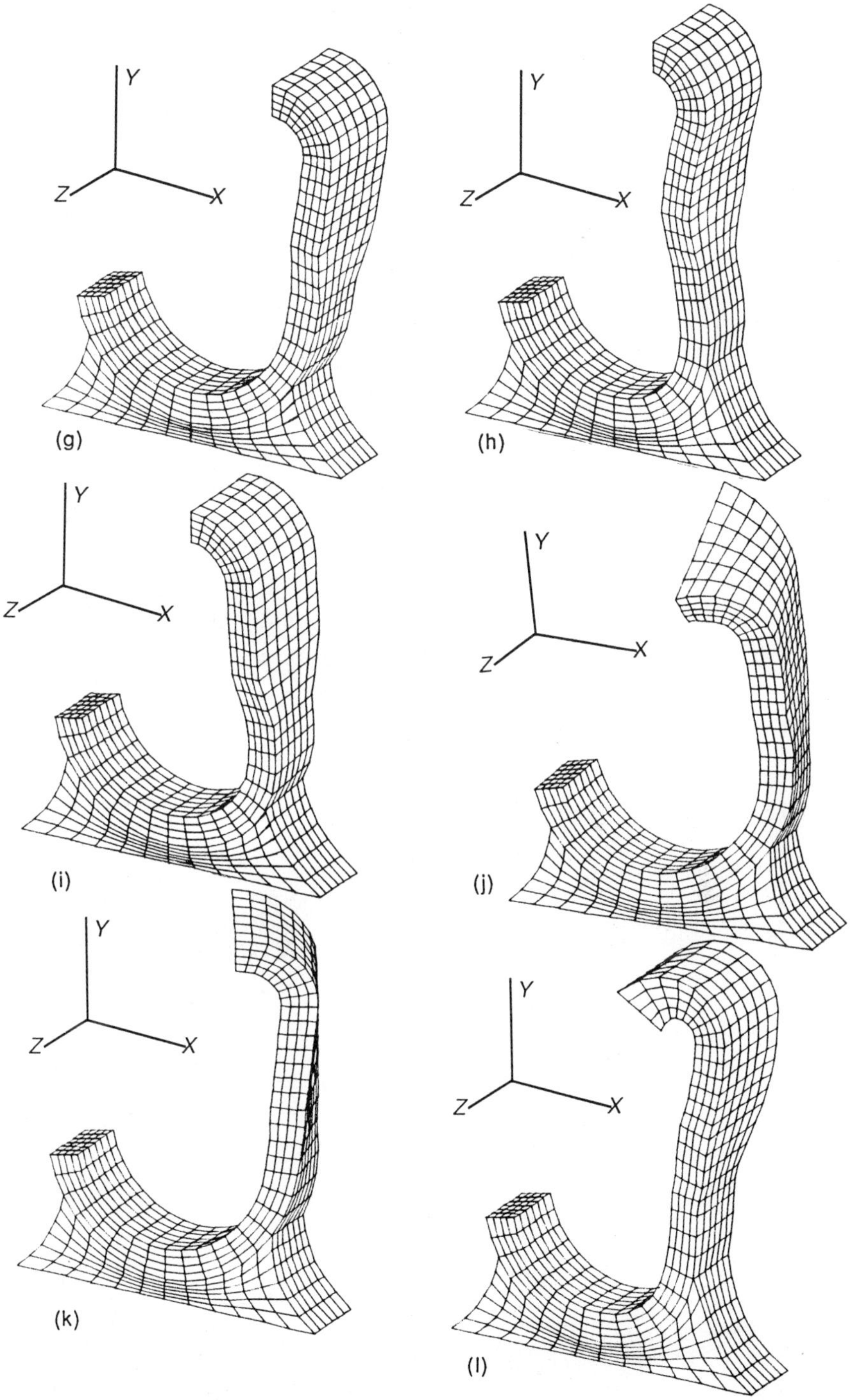

rotation in 5 direction. (*f*) Unit rotation in 6 direction. (*g*) Unit displacement in 7 direction. (*h*) Unit displacement in 8 direction. (*i*) Unit displacement in 9 direction. (*j*) Unit rotation in 10 direction. (*k*) Unit rotation in 11 direction. (*l*) Unit rotation in 12 direction.

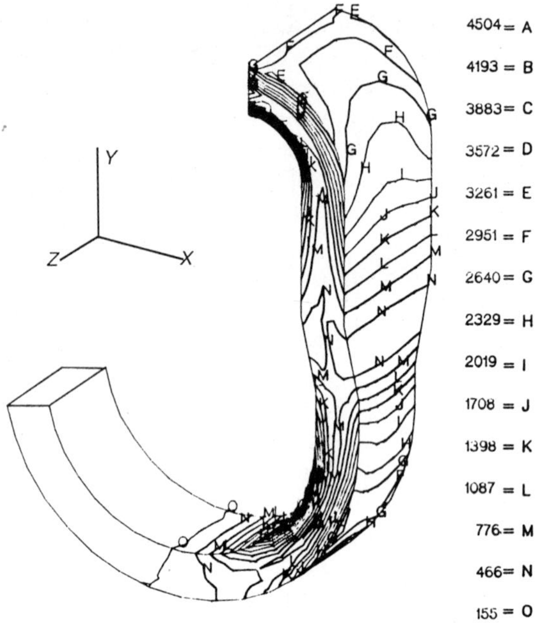

(a)

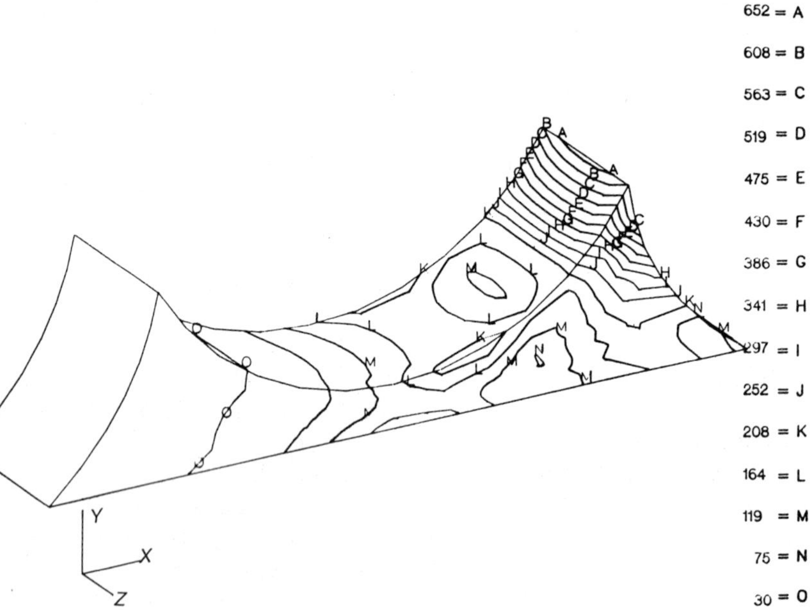

(b)

Figure 15-7 (*a*) Mises stress contours in the J lead. (*b*) Mises stress contours in the PLCC solder joint. (Unit displacement, 0.0001 in., in 1 direction).

lb/in., the force applied at C-2 is 105 lb/in., the force applied at C-3 and the moments applied at C-4 and C-5 are zero, the moment applied at C-6 is 13 in. – lb/rad (in the opposite direction of Figure 15-2*b*), the force applied at C-7 is 298 lb/in. (in the opposite direction of Figure 15-2*b*), the force applied at C-8 is 105 lb/in. (in the opposite direction of Figure 15-2*b*), the force applied at C-9 and the moments applied at C-10 and C-11 are zero, and the moment applied at C-12 is 14 in. – lb/rad (in the opposite direction of Figure 15-2*b*).

The deformed shape of the composite structure is shown in Figure 15-6*a*. The Mises stress contours in the J lead and solder joint are shown in Figure 15-7*a* and *b*, respectively. It can be seen that the maximum stress (4504 psi) acting in the J lead occurs at the inner shoulder of the J lead due to stress concentration and bending. Just like the PQFP's lead, higher stress also occurs at the outer shoulder of the J lead because of bending. Because of residual manufacturing stresses, it is not uncommon to observe cracks at the shoulder of the J lead. In that case, further crack propagation in the J lead under operation conditions is likely. The maximum Mises stress (2083 psi) acting in the solder joint occurs near the tip of the outer solder fillet (Fig. 15-7*b*). The stresses acting on the other tip of the solder joint are very small. Thus, cracking of the solder joint should start near the tip of the outer solder fillet.

15.4.2 Unit Displacement and Rotation in Other Directions

For the other 11 cases the stress contours in the J lead and solder joint are not shown in this chapter but can be found in the reference.[191] However, the calculated forces and moments are summarized in Table 15-2 and the deformed shapes are shown in Figure 15-6*b* through 6*l*.

15.4.3 Comparison of the Stiffness Matrices between the PQFPs and PLCCs

It is important to compare the stress contours in the PLCC solder joint (Figure 15-7*b*) with those in the PQFP solder joint (Figure 15-4*b*). It can be seen that the values of the PLCC solder joint are larger than those of the PQFP solder joint. Furthermore, by comparing the elements of the stiffness matrix of the J lead and solder joint (Table 15-2) with those of the PQFP gull-wing lead and solder joint (Table 15-1), most of the elements of the PQFP are smaller than those of the PLCC. Consequently, the gull-wing leads are more compliant than those of the J leads, and the stresses in the PQFP solder joints are less than those in the PLCC. Restated, the fatigue life of the PQFP solder joints should be longer than those of PLCC solder joints.

Solder Joint Reliability Under Shock and Vibration Conditions

Vibration analysis of structures consists of four major steps. First, the structure of interest is identified, its boundary conditions are estimated, and its interfaces with other structures are understood. Second, the natural frequencies and mode shapes of the structure are determined by analysis or direct experimental measurement. Third, the time history of the loads on the structure are estimated. Fourth, these loads are applied to an analytical model of the structure to determine its response. The first two steps are called "free vibration analysis." The last two steps are called "forced vibration analysis" and are usually executed by modal superposition, direct integration, or frequency domain methods.[222–234]

Free vibration analysis is the most crucial step in structural dynamics. Many vibration problems can be minimized by avoiding coincident resonances that can magnify the dynamic loads and stresses in the structure.

All problems of free vibration of undamped linear systems are eigenvalue problems, see references [222–226,231–234] for examples. In most of continuum mechanics the problem is to determine the natural mode of vibration for which all points in the continuum perform simple harmonic oscillations that are in phase with each other and that have the same frequency. Accordingly, all points in the continuum obtain their maximum velocities simultaneously as they pass through their neutral positions, and they all reach their extreme displacement simultaneously. A frequency for which natural vibration is possible is an eigenvalue of the boundary-value problem and is called a natural frequency. The number of the natural frequencies is equal to the number of degrees of freedom of the continuum. Theoretically speaking, it is infinite.

In this part, we will first present the free vibration of soldered and unsoldered leads of SOICs, PLCCs, and PQFPs. Then we will present the free vibration of a constrained PCB with a SMC.

15.5 FREE VIBRATION OF SOLDERED AND UNSOLDERED LEADS

The structures under consideration are shown in Figure 15-8 for SOICs and Figure 15-2*a* and *b* for PQFPs and PLCCs, respectively. All the leads are assumed to be clamped (fixed) at the upper end due to the larger stiffness of the package bodies compared with the leads. The lower end of the lead is assumed to be clamped for a *good solder joint* (or soldered lead) and free to move for a *bad solder joint* (or unsoldered lead). The partial differential equation governing the 3D vibration of these leads is:[220–226]

$$\nabla(\nabla \cdot \mathbf{Q}) + (1 - 2\mu)\nabla^2\mathbf{Q} = 2(1 + \mu)(1 - 2\mu)\frac{\rho}{E}\frac{\partial^2\mathbf{Q}}{\partial t^2} \qquad (15\text{-}27)$$

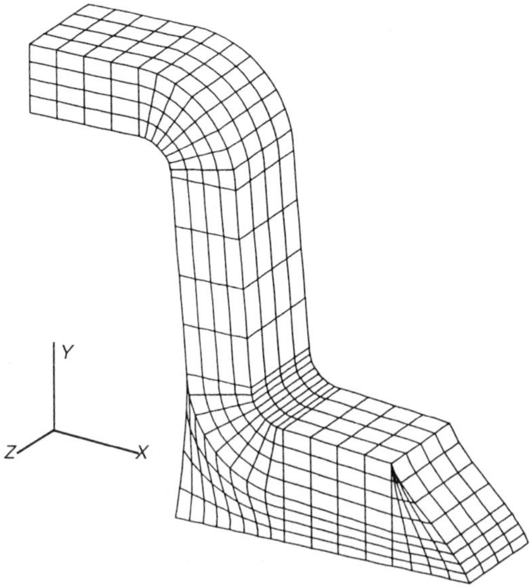

Figure 15-8 Finite-element model for SOIC lead and solder joint.

where ∇ = gradient operator, $\nabla^2 = \nabla \cdot \nabla$, $\mathbf{Q}$ = displacement vector, μ = Poisson's ratio, ρ = mass density, and E = Young's modulus.

The natural modes are characterized by the equation

$$\mathbf{Q} = \sum_{i=1}^{\infty} c_i \mathbf{A}_i \cos(\omega_i t - \gamma_i) \tag{15-28}$$

in which c_i and γ_i are arbitrary constants, ω_i is the angular frequency, and $\mathbf{A}_i(x, y, z)$ is a vector field that is independent of time. Substitution of Equation (15-28) into Equation (15-27) yields

$$\nabla(\nabla \cdot \mathbf{A}_i) + (1 - 2\mu)\nabla^2 \mathbf{A}_i + 2(1 + \mu)(1 - 2\mu)\frac{\rho\omega_i^2}{E}\mathbf{A}_i = 0 \qquad i = 1, 2, \ldots, \infty. \tag{15-29}$$

For free vibration, the equations of the boundary conditions are homogeneous.[220–226] Consequently, Equation (15-29), in conjunction with the boundary conditions, possesses a solution other than $\mathbf{A}_i = 0$ if, and only if, ω_i belongs to one of an infinite sequence of discrete values $\omega_1, \omega_2, \omega_3 \ldots$. These are the natural frequencies of the structure. To any natural frequency ω_i, there corresponds a nonzero vector field $\mathbf{A}_i$ that satisfies Equation (15-29) and the boundary conditions. This vector field contains an arbitrary constant factor; that is, $c_i\mathbf{A}_i$ is also a solution. Thus, the amplitudes of the oscillations of the particles are not determinate, but the ratio of the amplitudes for any two particles is fixed. Each particle executes a simple harmonic oscillation on the straight line in which the vector $\mathbf{A}_i$ lies.

Exact solutions of Equation (15-29) for the structures shown in Figures 15-2*a* and *b* and 15-8 are very difficult to obtain. Therefore, the finite element method was adopted to transform the boundary-value problem, Equation (15-29) and the boundary conditions, into a system of simultaneous algebraic equations, that is,

$$[\boldsymbol{K} - \omega^2 \boldsymbol{M}]\boldsymbol{A} = 0 \tag{15-30}$$

where $\boldsymbol{K}$ = stiffness matrix, $\boldsymbol{M}$ = mass matrix, and $\boldsymbol{A}$ = displacement components matrix. The size of these matrices ($\boldsymbol{K}$ and $\boldsymbol{M}$ are square matrices, and $\boldsymbol{A}$ is a column matrix) equals the number of degrees of freedom of the finite element model. Equation (15-30) can be easily solved for the frequencies (ω_i) and the corresponding mode shapes ($\boldsymbol{A}_i$) by a computer ($i = 1, 2, \ldots, N$, where N equals the number of degrees of freedom).

15.5.1 Vibration Results for Wide SOICs

Figure 15-8 shows the finite element model of the wide SOIC gull-wing lead and solder joint. The length of the horizontal parts of the lead was 0.017 in. The radius of the centroidal axis of the elbows was 0.005 in. and the width of the lead was 0.016 in. The thickness of the gull-wing lead was treated as a variable. Specifically, three different thicknesses were considered, 0.008, 0.01, and 0.012 in. Consequently, the overall standoff heights of the lead modeled were 0.057, 0.059, and 0.061 in.

Table 15-3 Vibration Frequencies for Soldered and Unsoldered Wide SOIC Leads

	Vibration Frequency (kHz)					
	Wide-SOIC Lead Thickness (in./mm)					
Mode	.008/.203		.010/.254		.012/.305	
	Soldered lead	Unsoldered lead	Soldered lead	Unsoldered lead	Soldered lead	Unsoldered lead
1	143.6	24.6	175.8	30.7	205.8	36.7
2	379.4	122.1	440.5	147.1	488.4	169.0
3	747.6	152.4	840.9	185.6	896.8	216.5

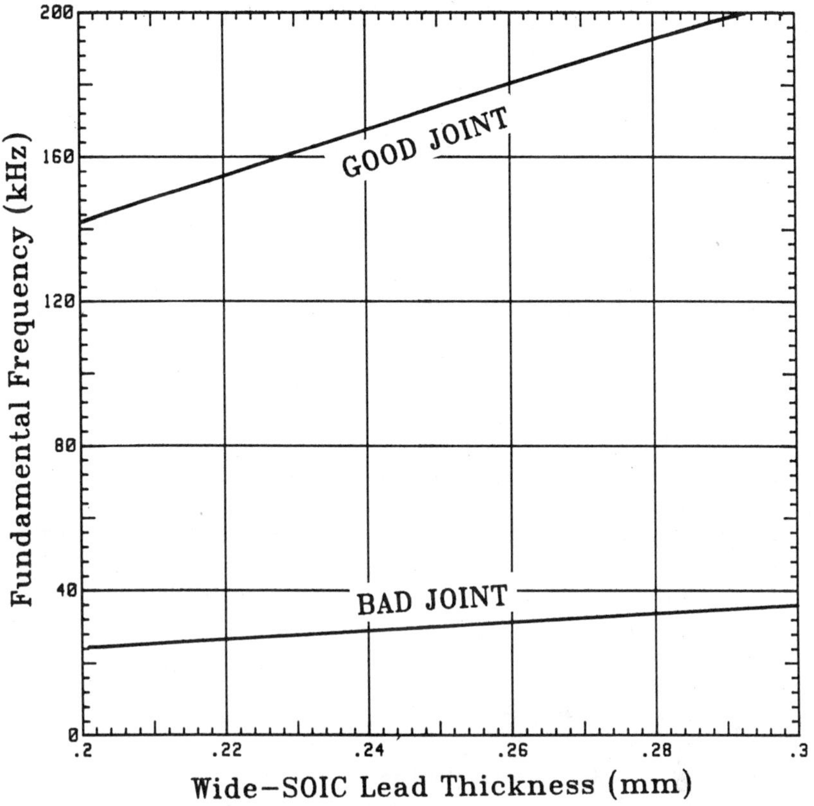

Figure 15-9 Frequencies for various wide-SOIC lead thickness.

Table 15-3 summarizes the first three natural frequencies for all the wide SOIC cases studied. The first (fundamental) frequency of the soldered leads (good joints) is at least 5 times larger than that of the unsoldered leads (bad joints). This is true for all the thicknesses under consideration, as is shown in Figure 15-9. Consequently, it is possible to identify the soldered and unsoldered leads by examining the frequency of vibration of the leads. Figure 15-9 also shows that the fundamental frequency does not vary significantly with the lead thickness.

15.5.2 Vibration Results for Narrow SOICs

The finite element model for the narrow SOIC is basically the same as for the wide SOIC. The lead thicknesses considered in these case were 0.006, 0.008, and 0.01 in. The overall standoff heights of the leads were 0.036, 0.038, and 0.04 in. The length of the horizontal part of the lead was 0.016 in.

Table 15-4 Vibration Frequencies for Soldered and Unsoldered Narrow SOIC Leads

Mode	Vibration Frequency (kHz)					
	Narrow SOIC Lead Thickness (in./mm)					
	.006/.152		.008/.203		.010/.254	
	Soldered lead	Unsoldered lead	Soldered lead	Unsoldered lead	Soldered lead	Unsoldered lead
1	191.9	35.1	248.6	46.6	300.2	57.9
2	564.2	156.8	679.8	199.3	759.8	235.2
3	1032.0	202.5	1218.7	261.3	1250.3	314.9

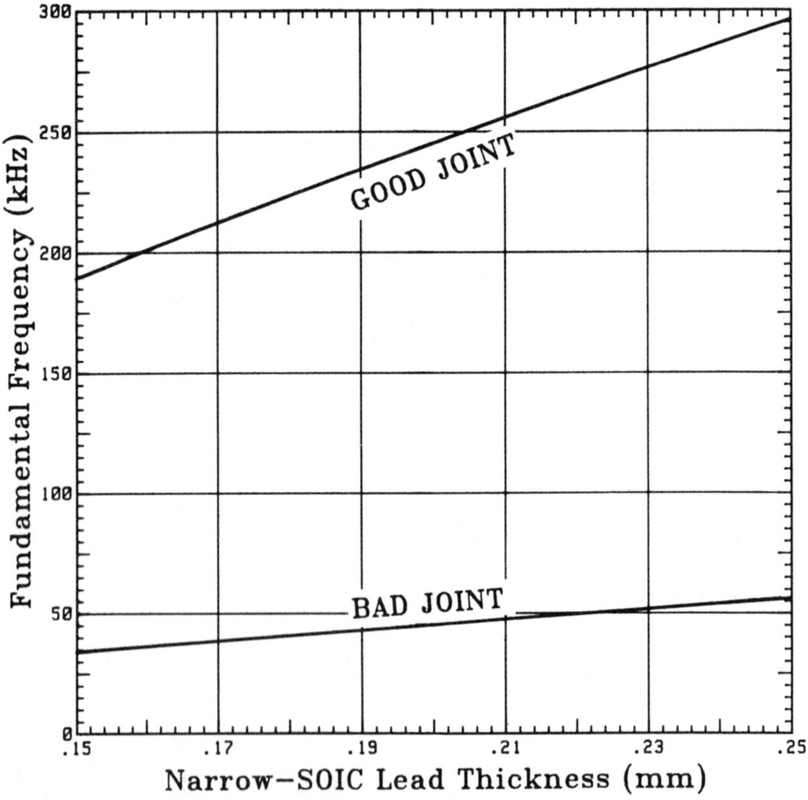

Figure 15-10 Frequencies for various narrow-SOIC lead thickness.

The first three frequencies for all the cases calculated are summarized in Table 15-4. It can be seen in the table that the fundamental frequency of the narrow SOIC leads is higher than that of the wide SOIC leads. This is because the narrow SOIC lead is shorter than the wide SOIC lead (i.e., stiffer in resisting vibration). The fundamental frequency of the narrow SOIC leads with various lead thickness is shown in Figure 15-10. It can be seen that, just like the wide SOIC cases, the fundamental frequency does not vary much with the lead thickness.

15.5.3 Vibration Results for PLCCs

Figure 15-2*b* shows the finite element model for the PLCC J lead and solder joint. The thicknesses of the J lead considered were 0.006, 0.008, and 0.01 in. The overall standoff heights of the lead were 0.095, 0.097, and 0.099 in. The radius of the centroidal axis of the shoulder was 0.014 in., and the radius of the centroidal axis of the lower curved part was 0.029 in. The width of the shoulder of the lead was 0.028 in., and the width of the lower part was 0.018 in.

Table 15-5 shows the first three frequencies of the soldered and unsoldered leads for three different thicknesses. It can be seen that the fundamental frequency of the soldered lead is more than 7 times larger than that of the unsoldered lead. Figure 15-11 shows the plot of fundamental frequency versus J lead thickness. It is again to be noted that the effect of thickness of the lead on the fundamental frequency is negligible and that the frequencies for the soldered lead and the unsoldered lead are very different.

15.5.4 Vibration Results for PQFPs

The finite element model of a PQFP lead and solder joint is shown in Figure 15-2*a*. The thickness of the gull-wing lead were 0.003, 0.005, and 0.007 in. The overall standoff heights of the lead were 0.056, 0.06, and 0.064 in. The

Table 15-5 Vibration Frequencies for Soldered and Unsoldered PLCC Leads

Mode	Vibration Frequency (kHz)					
	PLCC Lead Thickness (in./mm)					
	.006/.152		.008/.203		.010/.254	
	Soldered lead	Unsoldered lead	Soldered lead	Unsoldered lead	Soldered lead	Unsoldered lead
1	67.5	9.0	88.7	12.0	108.6	14.9
2	146.9	24.4	190.6	32.4	229.9	40.2
3	207.5	80.6	255.9	106.2	291.9	130.7

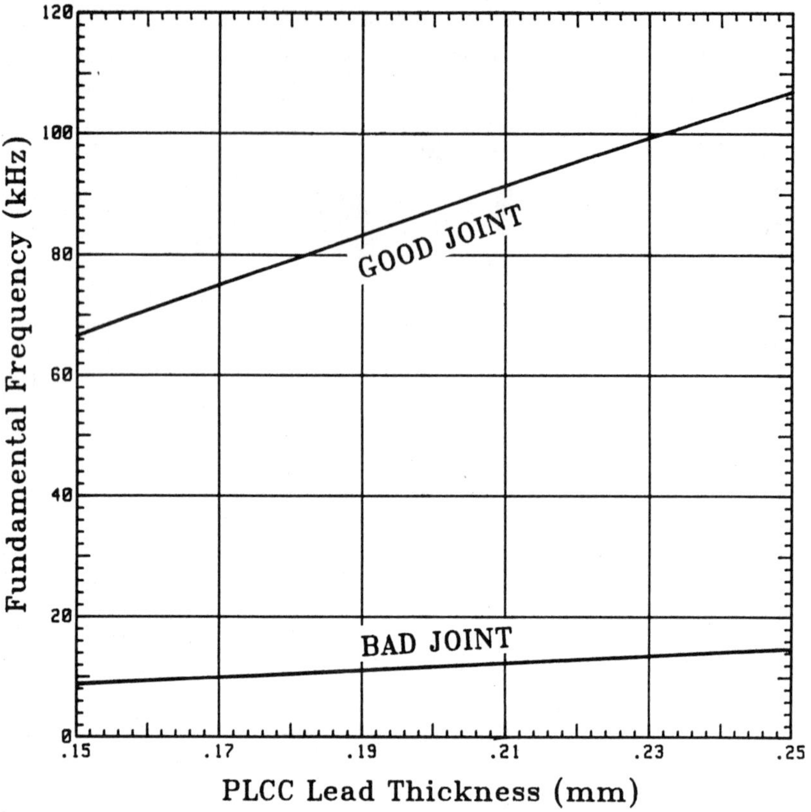

Figure 15-11 Frequencies for various PLCC lead thickness.

length of the horizontal part of the lead was 0.012 in. The radius of the centroidal axis of the elbow was 0.008 in.

The fundamental and the next two higher frequencies of vibration of the soldered and unsoldered leads are shown in Figure 15-12 and Table 15-6, respectively. Again, it can be seen that the frequency difference between the soldered and unsoldered leads is very large. Consequently, a good or bad solder joint can be detected by measuring its natural frequencies.

15.5.5 Experimental Verification

A laser Doppler vibrometer (LDV) system has been developed for measuring the vibration frequencies of all these SMC leads and for solder joint inspection.[200] The experimental results and the finite element solutions presented herein are in close agreement.[200]

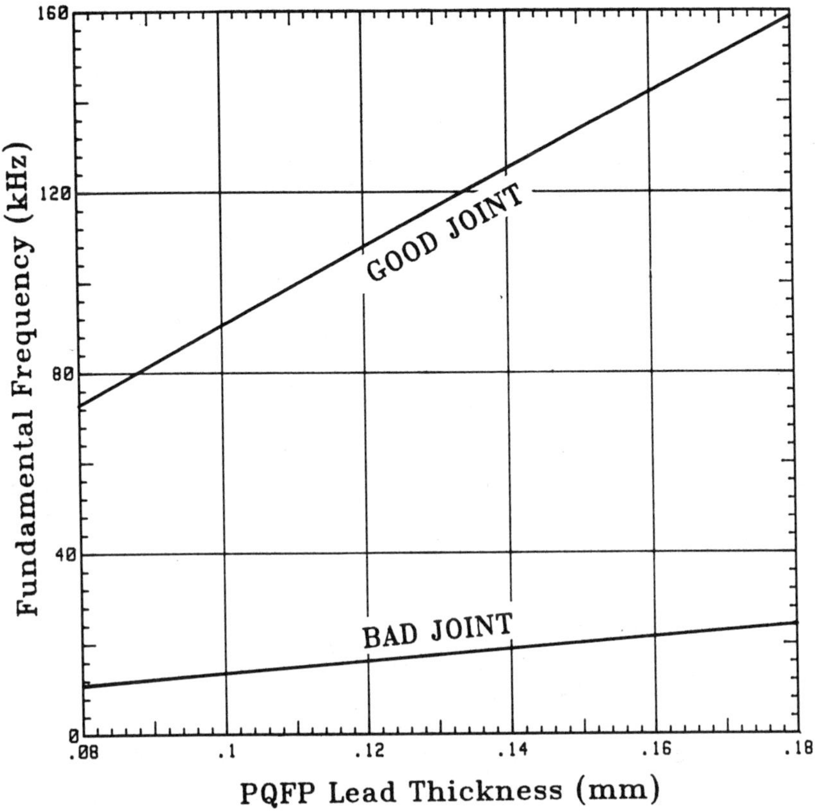

Figure 15-12 Frequencies for various PQFP lead thickness.

Table 15-6 Vibration Frequencies for Soldered and Unsoldered PQFP Leads

Mode	Vibration Frequency (kHz)					
	PQFP Lead Thickness (in./mm)					
	.003/.076		.005/.127		.007/.178	
	Soldered lead	Unsoldered lead	Soldered lead	Unsoldered lead	Soldered lead	Unsoldered lead
1	69.4	10.4	114.3	17.4	157.1	24.3
2	165.0	62.6	261.7	102.3	342.7	138.8
3	335.6	76.3	521.5	123.8	666.3	167.6

15.6 FREE VIBRATION OF A CONSTRAINED PCB WITH A SMC

Figure 15-13 shows a slender PCB with L as its largest dimension. It is supported by linear and rotational springs at both ends. A SMC is attached at the center of the PCB.

The Bernoulli-Euler equation of motion for small amplitude, free transverse vibrations of the structure shown in Figure 15-13 is

$$\mathrm{EI}\,\frac{\partial^4 y}{\partial x^4} + [M\delta(x - l) + m]\,\frac{\partial^2 y}{\partial t^2} = 0 \qquad (15\text{-}31a)$$

where $\delta(x)$ is the Dirac delta function and has the properties

$$\delta(x) = 0,\ x \neq 0 \qquad (15\text{-}31b)$$

$$\int_{-\infty}^{+\infty} \delta(x)dx = 1 \qquad (15\text{-}31c)$$

$$Ł[\delta(x)] = 1 \qquad (15\text{-}31d)$$

and Ł is the Laplace transformation. In Equation (15-31a), E is the Young's modulus, I is the area moment of inertia of beam (PCB) cross section about the neutral axis, EI is the flexural rigidity of the PCB, y is the deflection of the PCB, M is the concentrated mass of a SMC, and m is the mass per unit length of the PCB.

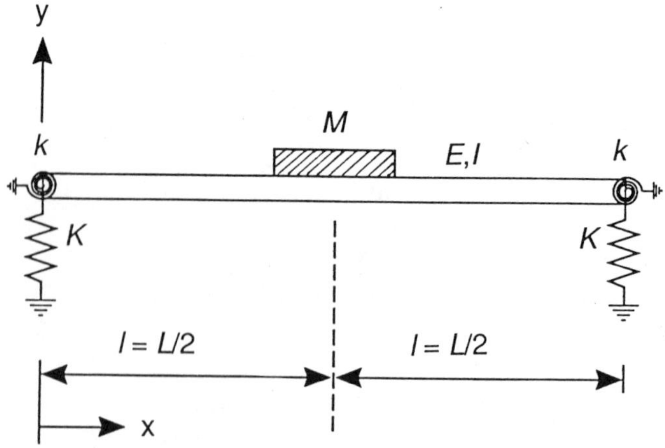

Figure 15-13 Transverse vibration of a PCB with a SMC.

The boundary conditions for the structure shown in Figure 15-13 are
At $x = 0$:

$$\mathrm{EI}\,\frac{\partial^3 y}{\partial x^3} = -Ky \tag{15-32}$$

$$\mathrm{EI}\,\frac{\partial^2 y}{\partial x^2} = k\frac{\partial y}{\partial x} \tag{15-33}$$

At $x = L = 2l$:

$$\mathrm{EI}\,\frac{\partial^3 y}{\partial x^3} = Ky \tag{15-34}$$

$$\mathrm{EI}\,\frac{\partial^2 y}{\partial x^2} = -k\frac{\partial y}{\partial x} \tag{15-35}$$

where K is the stiffness of the translational spring and k is the stiffness of the rotational spring.

For simple harmonic motion with frequency ω, we have

$$y(x, t) = Y(x)e^{iwt} \tag{15-36}$$

Then the equation of motion, Equation (15-31a) becomes

$$\frac{d^4 Y}{dx^2} - \varphi\,[\overline{M}L\delta(x - l) + 1]Y = 0 \tag{15-37}$$

where

$$\varphi^4 = \frac{m\omega^2}{\mathrm{EI}} \tag{15-38}$$

$$\overline{M} = \frac{M}{mL} \tag{15-39}$$

And the boundary conditions of the present problem, Equations (15-32) through (15-35) become

At $x = 0$:

$$L^3\,\frac{d^3 Y}{dx^3} = -\overline{K}Y \tag{15-40}$$

$$L \frac{d^2Y}{dx^2} = \bar{k} \frac{dY}{dx} \tag{15-41}$$

At $x = L = 2l$:

$$L^3 \frac{d^3Y}{dx^3} = \bar{K}Y \tag{15-42}$$

$$L \frac{d^2Y}{dx^2} = -\bar{k} \frac{dY}{dx} \tag{15-43}$$

where

$$\bar{K} = \frac{K L^3}{\text{EI}} \tag{15-44}$$

$$\bar{k} = \frac{kL}{\text{EI}} \tag{15-45}$$

Equations (15-37) and (15-40) through (15-43) define an eigenvalue problem that is not self-adjoint, the solution of which can be most conveniently obtained by Laplace transforms. After some algebraic manipulation, one obtains the following frequency equation:

$$\begin{aligned} &\bar{M}\,(a_{12}B - a_{22}A)\left[\frac{\bar{K}}{\beta^3}(\sinh \omega l - \sin \omega l) - (\cos \omega l + \cosh \omega l)\right] \\ &\quad - \bar{M}\,(a_{21}A - a_{11}B)\left[\frac{\bar{k}}{\beta^2}(\cosh \omega l - \cos \omega l) + \frac{1}{\beta}(\sin \omega l + \sinh \omega l)\right] \\ &\quad - \bar{M}\beta(\sinh \omega l - \sin \omega l)\,\Delta + 2\,\Delta = 0 \end{aligned} \tag{15-46}$$

where

$$\Delta = a_{12}a_{21} - a_{11}a_{22}$$

$$a_{11} = \beta^3 (\sin \omega L + \sinh \omega L) + \frac{\bar{k}^2}{\beta^3} (\sinh \omega L - \sin \omega L)$$

$$- 2\overline{K}\,(\cos \omega L + \cosh \omega L)$$

$$a_{12} = -\frac{\overline{K}}{\beta}(\sin \omega L + \sinh \omega L) + \beta\overline{k}(\sinh \omega L - \sin \omega L)$$

$$-\left(\beta^2 - \frac{\overline{k}\overline{K}}{\beta^2}\right)(\cos \omega L - \cosh \omega L)$$

$$a_{21} = a_{12}$$

$$a_{22} = \beta(\sinh \omega L - \sin \omega L) + \frac{\overline{K}^2}{\beta}(\sin \omega L + \sinh \omega L)$$

$$+ 2\overline{k}(\cos \omega L + \cosh \omega L)$$

$$A = \beta\overline{K}\,(\sinh \omega l - \sin \omega l) - \beta^4\,(\cosh \omega l + \cos \omega l)$$

$$B = \beta^2\overline{k}\,(\cos \omega l - \cosh \omega l) - \beta^3\,(\sin \omega l + \sinh \omega l)$$

$$\beta = \varphi L \tag{15-47}$$

It can be seen from Equation (15-46) that the frequency equation is a combination of the trigonometric and hyperbolic functions. It oscillates very rapidly and attains very large values between successive roots. It can be shown that the function of the frequency equation is monotonic and has no points of inflection between roots. Once the root (frequency parameter, $\beta = \varphi L$) has been obtained, the frequency (f) of the structural system can be determined by Equation (15-38) as

$$f = \frac{\omega}{(2\pi)} = \frac{(\beta)^2}{2\pi L^2}\sqrt{\frac{\mathrm{EI}}{m}} \tag{15-48}$$

The first two roots for a wide range of values of PCB support conditions (spring constants) are shown in Tables 15-6 through 15-10 for $\overline{M} = 0, 0.01, 0.05, 0.1, 1$, respectively. In [199], these tables have been used to calculate the natural frequencies of a PCB with various SMCs. These natural frequencies are very important for designing SMA vibration experiments and for design for reliability of solder joints.

Corresponding to each frequency, the mode shape can be easily obtained from the following expression

$$\frac{y(x)}{y(l)} = \frac{\overline{M}}{2\Delta}\Bigg((-a_{12}B + a_{22}A)\left[\frac{\overline{K}}{\beta^3}(\sinh \varphi x - \sin \varphi x) - (\cos \varphi x + \cosh \varphi x)\right]$$
$$+ (a_{21}A - a_{11}B)\left[\frac{\overline{k}}{\beta^2}(\cosh \varphi x - \cos \varphi x) + \frac{1}{\beta}(\sin \varphi x + \sinh \varphi x)\right]$$
$$+ \beta\Delta\,[\sinh \varphi(x - l) - \sin \varphi(x - l)]U(x - l)\Bigg) \qquad (15\text{-}49)$$

where $y(l)$ is the deflection at $x = l$, and $U(x - l) = 0$ for $x < l$, $U(x - l) = 1/2$ for $x = l$, and $U(x - l) = 1$ for $x > l$.

Table 15-6a Values of β for $\overline{M} = 0$ (Mode 1)

$\overline{K}$ \ $\overline{k}$	0.0	0.01	0.05	0.1	0.5	1.0	100.0
0.00	4.73004	0.69950	1.04348	1.23722	1.80856	2.09598	3.11092
0.01	0.37605	0.37605	0.37605	0.37605	0.37605	0.37605	0.37606
0.05	0.56222	0.56222	0.56223	0.56223	0.56224	0.56226	0.56232
0.10	0.66846	0.66846	0.66847	0.66847	0.66851	0.66854	0.66869
0.50	0.99792	0.99793	0.99797	0.99801	0.99827	0.99850	0.99962
1.00	1.18429	1.18430	1.18438	1.18448	1.18510	1.18564	1.18830
100.00	2.87675	2.87877	2.88671	2.89636	2.96383	3.02968	3.49775

Table 15-6b Values of β for $\overline{M} = 0$ (Mode 2)

$\overline{K}$ \ $\overline{k}$	0.0	0.01	0.05	0.1	0.5	1.0	100.0
0.00	7.85320	4.73411	4.75016	4.76972	4.90881	5.04878	6.22286
0.01	0.49492	0.73969	1.05639	1.24500	1.81100	2.09750	3.11126
0.05	0.74006	0.85688	1.10377	1.27474	1.82064	2.10353	3.11261
0.10	0.88006	0.95699	1.15553	1.30921	1.83247	2.11099	3.11430
0.50	1.31568	1.34095	1.42895	1.51833	1.91971	2.16798	3.12769
1.00	1.56415	1.57924	1.63500	1.69626	2.01415	2.23331	3.14418
100.00	4.66378	4.66379	4.66380	4.66381	4.66388	4.66396	4.66473

Table 15-7a Values of β for $\overline{M} = 0.01$ (Mode 1)

$\overline{K}$ \ $\overline{k}$	0.0	0.01	0.05	0.1	0.5	1.0	100.0
0.00	4.71088	0.69950	1.04348	1.23722	1.80856	2.09598	3.11092
0.01	0.37511	0.37511	0.37511	0.37511	0.37511	0.37511	0.37512
0.05	0.56083	0.56083	0.56083	0.56083	0.56085	0.56086	0.56092
0.10	0.66680	0.66680	0.66680	0.66681	0.66684	0.66688	0.66703
0.50	0.99542	0.99543	0.99546	0.99550	0.99577	0.99600	0.99713
1.00	1.18129	1.18131	1.18139	1.18148	1.18211	1.18266	1.18533
100.00	2.86437	2.86638	2.87431	2.88393	2.95123	3.01695	3.48488

Table 15-7b Values of β for $\overline{M}$ = 0.01 (Mode 2)

$\overline{K}$ \ $\overline{k}$	0.0	0.01	0.05	0.1	0.5	1.0	100.0
0.00	11.00527	4.71491	4.73082	4.75022	4.88801	5.02654	6.17803
0.01	0.49492	0.73969	1.05639	1.24500	1.81100	2.09750	3.11126
0.05	0.74006	0.85688	1.10377	1.27474	1.82064	2.10353	3.11261
0.10	0.88006	0.95699	1.15553	1.30921	1.83247	2.11099	3.11430
0.50	1.31568	1.34095	1.42895	1.51833	1.91971	2.16798	3.12769
1.00	1.56415	1.57924	1.63500	1.69626	2.01415	2.23331	3.14418
100.00	4.66378	4.66379	4.66380	4.66381	4.66388	4.66396	4.66473

Table 15-8a Values of β for $\overline{M}$ = 0.05 (Mode 1)

$\overline{K}$ \ $\overline{k}$	0.0	0.01	0.05	0.1	0.5	1.0	100.0
0.00	4.58791	0.69950	1.04348	1.23722	1.80856	2.09598	3.11092
0.01	0.37148	0.37148	0.37148	0.37149	0.37149	0.37149	0.37150
0.05	0.55540	0.55540	0.55540	0.55541	0.55542	0.55544	0.55550
0.10	0.66034	0.66034	0.66035	0.66035	0.66039	0.66042	0.66058
0.50	0.98570	0.98571	0.98575	0.98579	0.98606	0.98630	0.98746
1.00	1.16964	1.16966	1.16974	1.16984	1.17048	1.17105	1.17379
100.00	2.81414	2.81612	2.82392	2.83339	2.89965	2.96437	3.42732

Table 15-8b Values of β for $\overline{M}$ = 0.05 (Mode 2)

$\overline{K}$ \ $\overline{k}$	0.0	0.01	0.05	0.1	0.5	1.0	100.0
0.00	5.69410	4.59143	4.60528	4.62209	4.73879	4.85026	5.37269
0.01	0.49492	0.73969	1.05639	1.24500	1.81100	2.09750	3.11126
0.05	0.74006	0.85688	1.10377	1.27474	1.82064	2.10353	3.11261
0.10	0.88006	0.95699	1.15553	1.30921	1.83247	2.11099	3.11430
0.50	1.31568	1.34095	1.42895	1.51833	1.91971	2.16798	3.12769
1.00	1.56415	1.57924	1.63500	1.69626	2.01415	2.23331	3.14418
100.00	4.66378	4.66379	4.66380	4.66381	4.66388	4.66396	4.66473

Table 15-9a Values of β for $\overline{M}$ = 0.1 (Mode 1)

$\overline{K}$ \ $\overline{k}$	0.0	0.01	0.05	0.1	0.5	1.0	100.0
0.00	4.32649	0.69950	1.04348	1.23722	1.80856	2.09598	3.11092
0.01	0.36719	0.36719	0.36719	0.36719	0.36719	0.36719	0.36720
0.05	0.54897	0.54897	0.54897	0.54898	0.54899	0.54901	0.54907
0.10	0.65269	0.65269	0.65269	0.65270	0.65274	0.65277	0.65293
0.50	0.97417	0.97418	0.97422	0.97426	0.97454	0.97479	0.97598
1.00	1.15580	1.15582	1.15590	1.15600	1.15667	1.15725	1.16008
100.00	2.75095	2.75287	2.76043	2.76961	2.83381	2.89647	3.34313

Table 15-9b Values of β for $\overline{M}$ = 0.1 (Mode 2)

$\overline{K}$ \ $\overline{k}$	0.0	0.01	0.05	0.1	0.5	1.0	100.0
0.00	5.17591	4.32860	4.33685	4.34668	4.41004	4.46262	4.64755
0.01	0.49492	0.73969	1.05639	1.24500	1.81100	2.09750	3.11126
0.05	0.74006	0.85688	1.10377	1.27474	1.82064	2.10353	3.11261
0.10	0.88006	0.95699	1.15553	1.30921	1.83247	2.11099	3.11430
0.50	1.31568	1.34095	1.42895	1.51833	1.91971	2.16798	3.12769
1.00	1.56415	1.57924	1.63500	1.69626	2.01415	2.23331	3.14418
100.00	4.66378	4.66379	4.66380	4.66381	4.66388	4.66396	4.66473

Table 15-10a Values of β for $\overline{M}$ = 1 (Mode 1)

$\overline{K}$ \ $\overline{k}$	0.0	0.01	0.05	0.1	0.5	1.0	100.0
0.00	3.00108	0.69950	1.04348	1.23722	1.80856	2.09598	3.07129
0.01	0.31620	0.31620	0.31620	0.31620	0.31620	0.31620	0.31621
0.05	0.47265	0.47265	0.47265	0.47265	0.47267	0.47269	0.47277
0.10	0.56181	0.56181	0.56182	0.56182	0.56187	0.56191	0.56211
0.50	0.83693	0.83694	0.83698	0.83704	0.83738	0.83768	0.83914
1.00	0.99064	0.99066	0.99076	0.99089	0.99168	0.99239	0.99582
100.00	2.05921	2.06027	2.06443	2.06947	2.10395	2.13640	2.33266

Table 15-10b Values of β for $\overline{M} = 1$ (Mode 2)

$\overline{K}$ \ $\overline{k}$	0.0	0.01	0.05	0.1	0.5	1.0	100.0
0.00	4.90867	3.00145	3.00290	3.00463	3.01592	3.02569	3.11092
0.01	0.49492	0.73969	1.05639	1.24500	1.81100	2.09750	3.07139
0.05	0.74006	0.85688	1.10377	1.27474	1.82064	2.10353	3.07182
0.10	0.88006	0.95699	1.15553	1.30921	1.83247	2.11099	3.07236
0.50	1.31568	1.34095	1.42895	1.51833	1.91971	2.16798	3.07667
1.00	1.56415	1.57924	1.63500	1.69626	2.01415	2.23331	3.08207
100.00	3.63194	3.63261	3.63523	3.63845	3.66173	3.68580	3.89504

15.7 ACKNOWLEDGMENTS

The author wants to thank Drs. Girvin Harkins, Albert Jeans, Cathy Keely, Larry Moresco, and Donald Rice for their contributions. He also wants to thank Dr. Donald Rice for his support, guidance, and encouragement.

REFERENCES

1. Wilcox, J. R., R. Subrahmanyan, and C.-Y. Li, "Thermal Stress Cycles and Inelastic Deformation in Solder Joints," *Proceedings of ASM 2nd Electronic Materials and Processing Congress,* 1989, pp. 203–211.
2. Lee, S-M, and D. S. Stone, "Grain Boundary Sliding in Surface Mount Solders During Thermal Cycling," *Proceedings of the IEEE 40th Electronic Components and Technology Conference,* 1990, pp. 491–495.
3. Stone, D., H. Wilson, R. Subrahmanyan, and C.-Y. Li, "Mechanisms of Damage Accumulation in Solders During Thermal Fatigue," *Proceedings of the IEEE 36th Electronic Components Conference,* 1986, pp. 630–635.
4. Stone, D., S. P. Hanula, and C.-Y. Li, "The Effects of Device and Materials Variables on the Fatigue Behavior of Solder Joints During the Thermal Cycle," *Proceedings of the 35th IEEE Electronic Components Conference,* Washington, DC, May 1985, pp. 46–51.
5. Stone, D. S., "Creep Fatigue Interaction in Solders and Solder Joints, *ASME Winter. Annual Meeting,* Paper no. 89-WA/EEP-36, San Francisco, CA, December 1989.
6. Chen, W. T., L. C. Lee, C. K. Lim, and D. P. Seraphim, "Mechanical Modelling for Printed Circuit Boards," *Circuit World,* **11** (3), 1985, pp. 68–72.
7. Seraphim, D. P., R. Lasky, and C.-Y. Li, *Principles of Electronic Packaging,* New York, McGraw-Hill, 1989.
8. Ingraham, A. P., J. M. McCreary, and J. A. Varcoe, "Flip-Chip Soldering to Bare Copper Circuits," *Proceedings of the IEEE 40th Electronic Components and Technology Conference,* 1990, pp. 333–337.

9. Homa, T. R., "Cracking of Electronic Packaging Joints Due to Creep-Fatigue," *Proceedings of ASM's 3d Conference on Electronic Packaging and Corrosion in Microelectronics,* 1987, pp. 209–216.
10. Khutoretsky, M., "An Evaluation of Solder Balls on Printed Wiring Boards with Dry Film Solder Resist," *Proceedings of ASM's 3d Conference on Electronic Packaging and Corrosion in Microelectronics,* 1987, pp. 225–233.
11. Luke, D. A., "Electroplating Copper for Printed Circuit Manufacture," *Circuit World,* **13**(1), 1986, pp. 18–23.
12. ———, "Tin-Lead Composition—Its Control and Effect on Solderability," *Circuit World,* **15**(4), 1989, pp. 15–18.
13. Deane, P. A., J. S. Erich, and T. F. Marinis, "A Procedure for Measuring Lead Planarity of Surface Mount Components," *Proceedings of the IEEE 37th Electronic Components Conference,* 1987, pp. 56–63.
14. Warrior, M., "Reliability Improvements in Solder Bump Processing for Flip Chips," *Proceedings of the IEEE 40th Electronic Components and Technology Conference,* 1990, pp. 460–469.
15. Weinhold, M., "Designing for Zero-Defect Soldering," *Circuit World,* **12**(4), 1986, pp. 5–8.
16. Gunter, I. A., "The Solderability Testing of Surface Mount Devices Using the GEC Meniscograph Solderability Tester," *Circuit World,* **13**(1), 1986, pp. 8–12.
17. Steen, H. A. H., "Metallographic Examination of Soldered Joints," *Circuit World,* **14**(3), 1988, pp. 37–40.
18. Landis, R. C., "Plastic Chip Carriers—A Question of Reliability," *Proceedings of NEPCON West,* 1985, pp. 756–767.
19. Hutchins, C. L., "SMT Solder Yield and Reliability Test," *Texas Instruments Technical Papers on Surface Mount Technology,* 1986, pp. 39–43.
20. Tien, J. K., B. C. Hendirx, and A. I. Attarwala, "Damage Storage Measurement in Pb-Sn Solder," *Proceedings of the IEEE 40th Electronic Components and Technology Conference,* 1990, pp. 477–481.
21. Tien, J. K., B. C. Hendrix, P. L. Bretz, and A. I. Attarwala, "Creep-Fatigue Interactions in Solders," *IEEE Transactions on Components, Hybrids, and Manufacturing Technology,* **12**(4), December 1989, pp. 502–505.
22. Hendrix, B. C., P. L. Bretz, J. K. Tien, and S. K. Kang, "Methods for the Assessment of Damage in Cast Lead-Tin Solders Subjected to Accelerated Low Cycling Fatigue," *Proceedings of the 2nd ASM International Electronic Materials and Processing Congress,* 1989, pp. 129–133.
23. Enke, N. F., and B. I. Sandor, "A Generic Approach to Predicting the Fatigue Life of Surface Mount Interconnections," *Proceedings of the NEPCON West,* 1988, pp. 683–687.
24. Kilinski, T. J., D. P. Goetsch, and B. I. Sandor, "Straddle-Board Modeling of SMT Joint Behaviors," *Proceedings of the IEEE 40th Electronic Components and Technology Conference,* 1990, pp. 981–988.
25. Enke, N. F., T. J. Kilinski, S. A. Schroeder, and J. R. Lesniak, "Mechanical Behaviors of 60/40 Tin-Lead Solder Lap Joints," *IEEE Transactions on Components, Hybrids, and Manufacturing Technology,* **12**(4), December 1989, pp. 459–468.

26. Williams, R. L., and A. W. Noblett, "Tailored Coefficient of Thermal Expansion Printed Wiring Boards to Improve the Solder Joint Life of Leadless Ceramic Chip Carriers," *Circuit World,* **15**(2), 1989, pp. 30–32.
27. Riemer, D. E., "Prediction of Temperature Cycling Life for SMT Solder Joints on TCE-Mismatched Substrates," *Proceedings of the IEEE 40th Electronic Components and Technology Conference,* 1990, pp. 418–425.
28. Fine, M. E., and D. A. Jeannotte, "Soft Solder-General Principles and Fatigue Lifetime Estimation," *Proceedings of the 2d ASM International Electronic Materials and Processing Congress,* 1989, pp. 147–157.
29. Zubelewicz, A., R. Berriche, L. M. Keer, and M. E. Fine, "Lifetime Prediction of Solder Materials," *Journal of Electronic Packaging, Transactions of ASME,* **111**(3), September 1989, pp. 179–182.
30. Zubelewicz, A., Q. Guo, E. C. Cutiongco, M. E. Fine, and L. M. Keer, "Micromechanical Method to Predict Fatigue Life of Solder," *ASME Winter Annual Meeting,* Paper no. 89-WA/EEP-44, San Francisco, CA, December 1989.
31. Vaynman, S., "Fatigue Life Prediction of Solder Material: Effect of Ramp Time, Hold Time and Temperature," *Proceedings of the IEEE 40th Electronic Components and Technology Conference,* 1990, pp. 505–509.
32. Vaynman, S., M. E. Fine, and D. A. Jeannotte, "Prediction of Fatigue Life of Lead-Base Low Tin Solder," *Proceedings of the 37th IEEE Electronic Components Conference,* 1987, pp. 598–603.
33. Cutiongco, E. C., S. Vaynman, and M. E. Fine, "Isothermal Fatigue of 63Sn-37Pb Solder," *ASME Winter Annual Meeting,* Paper no. 89-WA/EEP-37, San Francisco, CA, December 1989.
34. Vaynman, S., "Effect of Strain Rate on Fatigue of Low-Tin Lead-Base Solder," *IEEE Transactions on Components, Hybrids, and Manufacturing Technology,* **12**(4), December 1989, pp. 469–472.
35. Berriche, R., S. Vaynman, M. E. Fine, and D. A. Jeannotte, "The Effect of Environment on Fatigue of Low Tin Lead Base Alloys," *Proceedings of ASM's 3d Conference on Electronic Packaging and Corrosion in Microelectronics,* 1987, pp. 169–174.
36. Vaynman, S., and M. E. Fine, "Fatigue of Low-Tin Lead-Based and Tin-Lead Eutectic Solders," *Proceedings of the 2nd ASM International Electronic Materials and Processing Congress,* 1989, pp. 255–259.
37. Feldin, M., C. Zevnik, and I. Kovac, "A New Approach to Outgassing Study of Plated Through-Hole Printed Boards," *Circuit World,* **15**(1), 1988, pp. 22–27.
38. Ellis, J. R., and G. Y. Masada, "Dynamic Behavior of SMT Chip Capacitors During Solder Reflow," *Proceedings of the IEEE International Electronic Manufacturing Technology Symposium,* 1989, pp. 23–29.
39. Lynch, J. T., "Surface Mount—Where Are The Standards?," *Proceedings of the IEEE 37th Electronic Components Conference,* 1987, pp. 64–69.
40. Geis, H., and M. Kottke, "The Role of Surface Corrosion in Degrading Solderability of Tin Coated Copper Leads," *Proceedings of the IEEE 36th Electronic Components Conference,* 1986, pp. 636–640.
41. Liljestrand, L.-G., and L.-O. Andersson, "Accelerated Thermal Fatigue Cycling

of Surface Mounted PWB Assemblies in Telecom Equipment," *Circuit World,* **14**(3), 1988, pp. 69–73.
42. Turnbull, R. B., "Blowholes in PCB Soldering—Their Causes and Cures," *Circuit World,* **12**(2), 1986, pp. 59–61.
43. Scharr, T. A., and M. D. Nagarkar, "Outer Lead and Die Bond Reliability in High Density TAB," *Proceedings of the IEEE 39th Electronic Components Conference,* 1989, pp. 177–183.
44. Iannuzelli, R., "Validation of Module Assembly Physical Models," *Proceedings of the IEEE 40th Electronic Components and Technology Conference,* 1990, pp. 1071–1076.
45. Davy, J. G., and R. Skold, "Computer-Aided Solderability Testing for Receiving Inspection," *Circuit World,* **12**(1), 1985, pp. 34–41.
46. Caswell, G., "Vapour Phase Soldering of SMDs-Reliability Characteristics," *Circuit World,* **11**(4), 1985, pp. 17–23.
47. Engel, P. A., "Torque Stress Analysis for Printed Circuit Boards Carrying Peripherally Leaded Modules," *ASME* Paper no. 86-WAM/EEP-2, December 1986.
48. ———, "Structural Analysis for Circuit Card Systems Subjected to Bending," *Journal of Electronic Packaging, Transactions of ASME,* **112**(1), March 1990, pp. 2–10.
49. Chen, W., "Analyzing the Mechanical Strength of SMT Attached Solder Joints," *Proceedings of the IEEE International Electronic Manufacturing Technology Symposium,* 1989, pp. 61–69.
50. Rafanelli, A. J., "Failure Analysis of a 132-Pin Leaded. Chip Carrier Gate Array," *ASME* Paper no. 86-WAM/EEP-6, December 1986.
51. ———, "Solder Fillet Height Criteria for Surface Mounted Chip Components," *Journal of Electronic Packaging, Transactions of ASME,* **111**(4), December 1989, pp. 299–302.
52. Brierley, C. J., and J. P. McCarthy, "The Reliability of Surface Mounted Solder Joints Under PWB Cylic Mechanical Stresses," *Circuit World,* **12**(3), 1986, pp. 16–19.
53. Hong, J., and M. Chason, "Gold Removal From Component Terminations Via Solder Dipping," *Proceedings of the IEEE 39th Electronic Components Conference,* 1989, pp. 351–354.
54. Chason, M., and J. Hong, "Kinetic Interactions of Copper Lead and Tin on Solder Coated PC Boards Studied Using X-Ray Diffraction," *Proceedings of the IEEE International Electronic Manufacturing Technology Symposium,* 1989, pp. 294–299.
55. Suhir, E., "Can Power Cycling Life of Solder Joint Interconnections Be Assessed on the Basis of Temperature Cycling Test?," *Journal of Electronic Packaging, Transactions of ASME,* **111**(4), December 1989, pp. 310–312.
56. ———, "Could Compliant External Leads Reduce the Strength of a Surface-Mounted Device?" *Proceedings of the 37th IEEE Electronic Components Conference,* May 1987, pp. 589–597; also, *Journal of Applied Mechanics, Transactions of ASME,* **55,** December 1988, pp. 818–821.
57. ———, "Twist-Off Testing of Solder Joint Interconnections," *Journal of Electronic Packaging, Transactions of ASME,* **111**(3), September 1989, pp. 165–171.
58. ———, "Die Attachment Design and Its Influence on Thermal Stresses in the Die

and the Attachment," *Proceedings of the IEEE 37th Electronic Components Conference,* 1987, pp. 508–517.

59. Frost, H. J., and R. T. Howard, "Creep-Fatigue Modelling for Solder-Joint Reliability Predictions Including the Microstructural Evolution of the Solder," *Proceedings of the IEEE 40th Electronic Components and Technology Conference,* 1990, pp. 482–490.
60. Frost, H. J., R. T. Howard, P. R. Lavery, and S. D. Lutender, "Creep and Tensile Behavior of Lead-Rich, Lead-Tin Solder Alloys," *IEEE Transactions on Components, Hybrids, and Manufacturing Technology,* **11**(4), December 1988, pp. 371–379.
61. Frost, H. J., P. R. Lavery, and S. D. Lutender, "Microstructure of Mechanical Properties of Lead-Tin Solder Alloys," *Proceedings of ASM's 3d Conference on Electronic Packaging and Corrosion in Microelectronics,* 1987, pp. 259–267.
62. Frost, H. J., G. J. Stone, and R. T. Howard, "The Effects of Thermal History on the Microstructure and Mechanical Properties of Solder Alloys," *Proceedings of the 2d ASM International Electronic Materials and Processing Congress,* 1989, pp. 121–127.
63. Suryanarayana, D., R. Hsiao, T. P. Gall, and J. M. McCreary, "Flip-Chip Solder Bump Fatigue Life Enhanced by Polymer Encapsulation," *Proceedings of the IEEE 40th Electronic Components and Technology Conference,* 1990, pp. 338–344.
64. Panousis, N. T., J. F. Prosser, A. T. Wang, and Y. N. Yong, "Strength and Reliability of Ceramic Modules Soldered to Flexible Cables," *Proceedings of the IEEE 40th Electronic Components and Technology Conference,* 1990, pp. 377–382.
65. Sherry, W. M., and P. M. Hall, "Materials, structures, and mechanics of solder-joints for surface-mount microelectronics technology," in *Proc. Conf. Interconnection Technology in Electronics,* Fellbach, W. Germany, Feb. 1986, Deutscher Verband für Schweisstechnik, pp. 47–61.
66. Hall, P. M., "Creep and Stress Relaxation in Solder Joints in Surface-Mount Chip Carriers," *IEEE Transactions on Components, Hybrids, and Manufacturing Technology,* **CHMT-10**(4), December 1987, pp. 556–565.
67. Tong, H. M., K. R. Mok, K. R. Grebe, H. L. Yeh, K. K. Srivastava, and J. T. Coffin, "Parylene Encapsulation of Ceramic Packages for Liquid Nitrogen Application," *Proceedings of the IEEE 40th Electronic Components and Technology Conference,* 1990, pp. 345–350.
68. Engelmaier, W., and A. I. Attarwala, "Surface-Mount Attachment Reliability of Clip-Leaded Ceramic Chip Carriers on FR-4 Circuit Boards," *IEEE Transactions on Components, Hybrids, and Manufacturing Technology,* **12**(2), June 1989, pp. 284–296.
69. Oberin, F. W., "Solder Joint Cracking in Potted High Voltage Modules," *Proceedings of the 2d ASM International Electronic Materials and Processing Congress,* 1989, pp. 189–202.
70. Bae, K., A. F. Sprecher, Z. Guo, and H. Conrad, "Effect of Compliance on the Fatigue of Solder Joints in Surface-Mounted Electronic Packages," *Proceedings of the 2d ASM International Electronic Materials and Processing Congress,* 1989, pp. 109–119.

71. ———, "Effect of Composition on the Low-Cycle Fatigue of Pb Alloy Solder Joints," *Proceedings of the IEEE 40th Electronic Components and Technology Conference,* 1990, 496–504.
72. Lauer, L. D., "Dynamic Mechanical Testing of Solder and Solder Joints," *Circuit World,* **13**(1), 1986, pp. 13–17.
73. Andrews, G. J., "Evaluation of the Corrosion and Reliability Aspects of Printed Circuit Boards Flow Soldered Using 'No Clean' Fluxes," *Circuit World,* **16**(3), 1990, pp. 11–13.
74. Mirman, B., and Knecht, S., "Creep Strains in an Elongated Bond Layer," *Proceedings of the 2d IEEE I-THERM,* Las Vegas, NV, 1990, pp. 21–32.
75. Darveaux, R., and I. Turlik, "Shear Deformation of Indium Solder Joints," *Proceedings of the 2d IEEE I-THERM,* Las Vegas, NV, 1990, pp. 40–52.
76. Saint-Martin, X., and Joly, J., "Plastic Package Moisture Absorption Hazards in SMT Board Assembly," *Proceedings of the IEEE International Electronic Manufacturing Technology Symposium,* 1989, pp. 38–42.
77. Hagge, J. K., and G. J. Davis, "Ageing, Solder Thickness and Solder Alloy Effects on Circuit Board Solderability," *Circuit World,* **11**(3), 1985, pp. 8–15.
78. DeVore, J. A., "Solderability and Surface Mounting," *Circuit World,* **14**(4), 1990, pp. 37–41.
79. Wild, R. N., "Some Factors Affecting Leadless Chip Carrier Solder Joint Fatigue Life," *Circuit World,* **14**(4), 1988, pp. 29–36.
80. Smeby, J. M., "Solder Joint Behavior in HCC/PWB Interconnections," *IEEE Transactions on Components, Hybrids, and Manufacturing Technology,* **CHMT-8**(3), September 1985, pp. 391–396.
81. Haramake, T., T. Funamoto, and T. Shida, "Characteristics of Press-Soldered Joints by Using Resistance Heating," *IEEE Transactions on Components, Hybrids, and Manufacturing Technology,* **13**(2), June 1990, pp. 452–456.
82. Lodge, K. J., and D. J. Pedder, "The Impact of Packaging on the Reliability of Flip Chip Solder Bonded Devices," *Proceedings of the IEEE 40th Electronic Components and Technology Conference,* 1990, pp. 470–476.
83. Knecht, S., and L. R. Fox, "Constitutive Relation and Creep-Fatigue Life Model for Eutectic Tin-Lead Solder," *IEEE Transactions on Components, Hybrids, and Manufacturing Technology,* **13**(2), June 1990, pp. 424–433.
84. Shine, M. C., L. R. Fox, and J. W. Sofia, "A Strain Range Partitioning Procedure for Solder Fatigue," *Brazing and Soldering,* no. 9, 1985, pp. 11–20.
85. Fox, L. R., J. W. Sofia, and M. C. Shine, "Investigation of Solder Fatigue Acceleration Factors," *IEEE Transactions on Components, Hybrids, and Manufacturing Technology,* **CHMT-8**(2), 1985, pp. 275–281.
86. Spitz, D. M., B. Cobb, D. L. Yenawine, and W. M. Wolverton, "Solderability Optimization for Vapor Phase Reflowed, Surface Mount, Electronic Assemblies," *Proceedings of ASM's 3d Conference on Electronic Packaging and Corrosion in Microelectronics,* 1987, pp. 239–245.
87. Molverton, M., W. R. Russell, B. Waller, and P. Barry, "Improving Control of the Steam Age Solderability Test for Component Leads," *Proceedings of the 2d ASM International Electronic Materials and Processing Congress,* 1989, pp. 37–39.
88. Harada, M., and R. Satoh, "Mechanical Characteristics of 96.5Sn/3.5Ag Solder in

Micro-Bonding," *Proceedings of the IEEE 40th Electronic Components and Technology Conference,* 1990, pp. 510–517.

89. Mitomi, O., T. Nozawa, and K. Kawano, "Effects of Solder Creep on Optical Component Reliability," *IEEE Transactions on Components, Hybrids, and Manufacturing Technology,* **CHMT-9**(3), September 1986, pp. 265–271.
90. Pinamaneni, S., and D. A. Solomon, "Rapidly Solidified Soft Solder Die-Attach Technology," *IEEE Transactions on Components, Hybrids, and Manufacturing Technology,* **CHMT-9**(4), December 1986, pp. 416–422.
91. Keller, H. N., "Solder Connections with a Ni Barrier," *IEEE Transactions on Components, Hybrids, and Manufacturing Technology,* **CHMT-9**(4), December 1986, pp. 433–439.
92. Wassink, R. J. K., *Soldering in Electronics,* Electrochemical Publications, Scotland, 1989.
93. Lea, C., and F. H. Howie, "Blowholing in PTH Solder Fillets, Part 1: Assessment of the Problem," *Circuit World,* **12**(4), 1986, pp. 14–19.
94. Howie, F. H., and C. Lea, "Blowholing in PTH Solder Fillets, Part 2: The Nature, Origin and Evolution of Gas," *Circuit World,* **12**(4), 1986, pp. 20–25.
95. Seah, M. P., F. H. Howie, and C. Lea, "Blowholing in PTH Solder Fillets, Part 3: Moisture and the PCB," *Circuit World,* **12**(4), 1986, p. 14.
96. Lea, C., M. P. Seah, and F. H. Howie, "Blowholing in PTH Solder Fillets, Part 4: The Plated Copper Barrel," *Circuit World,* **13**(1), 1986, pp. 28–34.
97. Lea, C., and F. H. Howie, "Blowholing in PTH Solder Fillets, Part 5: The Role of the Electroless Copper," *Circuit World,* **13**(1), 1986, pp. 35–42.
98. ———, "Blowholing in PTH Solder Fillets, Part 6: The Laminate, The Drilling and the Hole-Wall Preparation," *Circuit World,* **13**(1), 1986, pp. 43–50.
99. Howie, F. H., D. Tilbrook, and C. Lea, "Blowholing in PTH Solder Fillets, Part 7: Optimising the Soldering," *Circuit World,* **13**(2), 1986, pp. 42–45.
100. Lea, C., F. H. Howie, and M. P. Seah, "Blowholing in PTH Solder Fillets, Part 8: The Scientific Framework Leading to Recommendations for Its Elimination," *Circuit World,* **13**(3), 1986, pp. 11–20.
101. Lea, C., "Residues on Soldered Printed Circuit Assemblies," *Circuit World,* **16**(1), 1989, pp. 44–46.
102. ———, "The Effect of Solder Mask on PCB Solderability," *Circuit World,* **15**(1), 1988, pp. 12–21.
103. ———, "Solderability and Joint Strength of Plated Component Leads," *Proceedings of DVS Conference,* 1986, pp. 180–185.
104. ———, *A Scientific Guide to Surface Mount Technology,* Electrochemical Publications, Scotland, 1988.
105. Lea, C., and F. H. Howie, "Controlling the Quality of Soldering of PTH Solder Joints," *Circuit World,* **11**(3), 1985, pp. 5–7.
106. Pawling, J. F., *Surface Mounted Assemblies,* Electrochemical Publications, Scotland, 1987.
107. Sinnadurai, F. N., *Handbook of Microelectronics Packaging and Interconnection Technologies,* Electrochemical Publications, Scotland, 1985.
108. Chilton, A. C., M. A. Whitmore, and W. Hampshire, "Isothermal Mechanical Fatigue of a Model SMD Joint," *Proceedings of the 2d ASM International Electronic Materials and Processing Congress,* 1989, pp. 159–176.

109. Manko, H. H., *Soldering Handbook for Printed Circuits and Surface Mounting,* New York, Van Nostrand Reinhold, 1986.
110. Johnson, C., and J. Kevra, *Solder Paste Technology: Principles and Applications,* Blue Ridge Summit, PA, Tab Books, 1989.
111. Kevra, J., and D. Mohoric, "Solder Paste: Tack and Cure," *Proceedings of the 2d ASM International Electronic Materials and Processing Congress,* 1989, pp. 313–319.
112. Love, G. F., "Packages for High-Density Surface-Mount Applications," *Circuit World,* **13**(3), 1987, pp. 36–39.
113. Ainsworth, P. A., "Measurement of the Tackiness of Solder Paste," *Hybrid Circuits,* no. 11, September 1986, pp. 27–29.
114. Ciulik, J., and M. R. Notis, "Phase Equilibria and Physical Properties in the Au-Sn System," *Proceedings of the 2d ASM International Electronic Materials and Processing Congress,* 1989, pp. 57–61.
115. Tummala, R. R., and E. Rymaszewski, *Microelectronics Packaging Handbook,* Van Nostrand Reinhold, New York, 1989.
116. Hwang, J. S., "Consideration for Surface Mounting Solder Paste," *Brazing and Soldering,* no. 14, 1988, pp. 13–21.
117. ———, *Solder Paste in Electronics Packaging,* New York, Van Nostrand Reinhold, 1989.
118. Hwang, J. S., and R. M. Vargas, "Solder Joint Reliability—Can Solder Creep?," *Proceedings of the International Symposium on Microelectronics,* 1989, pp. 513–524.
119. Verguld, M. M. F., "Wave Soldering. The Efficiency of Debridging with a Hot-Air Knife," *Circuit World,* **13**(2), 1987, pp. 14–18.
120. Kojima, Y., M. Kikuchi, K. Matsunaga, and N. Yamazaki, "The Reliability of Reflow Soldering by Hot Air Reflow," *Proceedings of the IEEE International Electronic Manufacturing Technology Symposium,* 1989, pp. 43–47.
121. Yamada, S. E., "The Inverted T-Joint for Surface Mount Connectors," *IEEE Transactions on Components, Hybrids, and Manufacturing Technology,* **13**(12), June 1990, pp. 397–400.
122. ———, "A Fracture Mechanics Approach to Soldered Joint Cracking," *IEEE Transactions on Components, Hybrids, and Manufacturing Technology,* **12**(no. 1) Mar. 1989, pp. 99–104.
123. ———, "Stress Analysis of Partially Yielded Soldered Joint for Surface Mount Connectors," *IEEE Transactions on Components, Hybrids, and Manufacturing Technology,* **CHMT-10**(2), June 1987, pp. 236–241.
124. ———, "Time Dependent Crack Growth in Soldered Joint," *Engineering Fracture Mechanics,* **28**(1), 1987, pp. 85–91.
125. ———, "The J-Integral for Augmented Double Cantilever Beams and Its Application to Bonded Joint," *Engineering Fracture Mechanics,* **29**(6), 1988, pp. 673–682.
126. ———, "A Mechanism for Board Warpage by Thermal Expansion of Surface Mounted Connector," *IEEE Transactions on Components, Hybrids, and Manufacturing Technology,* **CHMT-9**(4), December 1986, pp. 508–512.
127. Schoenbaum, G. L., and E. Socha, "Selecting the Accept/Reject Criteria for an

Automated Solder Process Inspection Machine," *Circuit World,* **13**(2), 1987, pp. 19–21.

128. Val, C., J. Rakotonarivo, and A. Gerard, "Plasma Inspection of Visible or Hidden Solder Joints," *Proceedings of the International Symposium on Microelectronics,* 1989, pp. 525–536.
129. ———, "Automatic Inspectiron of Visible or Hidden Solder Joints—A New Concept," *Proceedings of the IEEE 39th Electronic Components Conference,* 1989, pp. 623–628.
130. O'Clock, G. D., M. S. Peters, J. R. Pater, G. A. Kleese, and R. V. Martini, "Pb-Sn Alloy Microstructure: Potential Reliability Indicator for Interconnects," *IEEE Transactions on Components, Hybrids, and Manufacturing Technology,* **CHMT-10**(1), March 1987, pp. 82–88.
131. ———, "The Use of Microstructure as a Reliability Indicator for Solder Interconnects," *Proceedings of ASM's 3d Conference on Electronic Packaging and Corrosion in Microelectronics,* 1987, pp. 251–256.
132. Mizuishi, K., and T. Mori, "Thermal Stability of Various Ball-Limited-Metal Systems Under Solder Bumps," *IEEE Transactions on Components, Hybrids, and Manufacturing Technology,* **11**(4), December 1988, pp. 481–484.
133. Chiou, B.-S., K. C. Liu, J.-G. Duh, and P. S. Palanisamy, "Intermetallic Formation on the Fracture of Sn/Pb Solder and Pd/Ag Conductor Interfaces," *IEEE Transactions on Components, Hybrids, and Manufacturing Technology,* **13**(2), June 1990, pp. 267–274.
134. Driels, M. R., and D. J. Nolan, "Automatic Defect Classification of Printed Wiring Board Solder Joints," *IEEE Transactions on Components, Hybrids, and Manufacturing Technology,* **13**(2), June 1990, pp. 331–340.
135. Wong, B., D. E. Helling, and R. W. Clark, "A Creep-Rupture Model for Two-Phase Eutectic Solders," *IEEE Transactions on Components, Hybrids, and Manufacturing Technology,* **11**(3), September 1988, pp. 284–290.
136. Inaba, M., K. Yamakawa, and N. Iwase, "Solder Bump Formation Using Electroless Plating and Ultrasonic Soldering," *IEEE Transactions on Components, Hybrids, and Manufacturing Technology,* **13**(1), March 1990, pp. 119–123.
137. Matsui, N., S. Sasaki, and T. Ohsaki, "VLSI Chip Interconnection Technology Using Attached Solder Bumps," *IEEE Transactions on Components, Hybrids, and Manufacturing Technology,* **CHMT-10**(4), December 1987, pp. 566–570.
138. Kousaka, M., H. Yanai, T. Kobayashi, and R. K. Selwood, "A Unique Material Solution of Fine Line Printing of Solder Paste," *Proceedings of the International Symposium on Microelectronics,* 1989, pp. 537–543.
139. Lin, K. M., and F. H. Friend, "Wetting Balance Methods for Component and Board Solderability Evaluation," *Circuit World,* **16**(1), 1989, pp. 24–28.
140. Puttlitz, K. J., "Preparation, Structure, and Fracture Nodes of Pb-Sn and Pb-In Terminated Flip Chips Attached to Gold-Capped Microsockets," *Proceedings of the IEEE 40th Electronic Components and Technology Conference,* 1990, pp. 360–367.
141. ———, "Corrosion of Pb-50In Flip-Chip Interconnections Exposed to Harsh Environment," *IEEE Transactions on Components, Hybrids, and Manufacturing Technology,* **13**(1), March 1990, pp. 188–193.

142. Forshaw, M., "Application of Non-Destructive Testing to Inspection of Soldered Joints," *Circuit World,* **15**(3), 1989, pp. 14–17.
143. Agarwala, B. N., "Thermal Fatigue Damage in Pb-In Solder Interconnection," *Proceedings of the 23d IEEE International Reliability Physics Symposium,* 1985, pp. 198–205.
144. Marcotte, V. C., C. Vincent, N. G. Koopman, and P. A. Totta, "Review of Flip Chip Bonding," *Proceedings of the 2d ASM International Electronic Materials and Processing Congress,* 1989, pp. 73–81.
145. Fujiwara, K., M. Asahi, S. Tsurumi, and Y. Takeuchi, "Water-Soluble Flux for Pb-Alloy Josephson Device Packaging," *IEEE Transactions on Components, Hybrids, and Manufacturing Technology,* **CHMT-10**(2), June 1987, pp. 258–262.
146. Fujiwara, K., and M. Asahi, "Characterization of Intermetallic Compound Formation on In/Bi/Sn Solder Bumps Used in Pb-Alloy Josephson Chip Packaging," *IEEE Transactions on Components, Hybrids, and Manufacturing Technology,* **CHMT-10**(2), June 1987, pp. 263–266.
147. Evans, J. W., and J. K. Beddow, "Characterization of Particle Morphology and Rheological Behavior in Solder Paste," *IEEE Transactions on Components, Hybrids, and Manufacturing Technology,* **CHMT-10**(2), June 1987, pp. 224–231.
148. Bowen, R. C., and D. M. Peterson, "A Comparison of Rapid Solidification Cast versus Conventional Die Attach Soft Solders," *IEEE Transactions on Components, Hybrids, and Manufacturing Technology,* **CHMT-10**(3), September 1987, pp. 341–345.
149. Balde, J. W., "The IEEE Compliant Lead Task Force," *IEEE Transactions on Components, Hybrids, and Manufacturing Technology,* **CHMT-10**(3), September 1987, pp. 463–469.
150. Matsui, A., K. Kawai, and E. Asada, "Experimental Research on Fatigue and Fracture of Solder," *Proceedings of the International Symposium on Microelectronics,* 1989, pp. 544–551.
151. Tervalon, M. J., "Local Mismatched in SM Solder Joint FE Analysis," *Proceedings of the IEEE 40th Electronic Components and Technology Conference,* 1990, pp. 954–957.
152. Clech, J.-P., F. M. Langerman, and J. A. Augis, "Local CTE Mismatch in SM Leaded Packages: A Potential Reliability Concern," *Proceedings of the IEEE 40th Electronic Components and Technology Conference,* 1990, pp. 368–376.
153. Clech, J.-P., and J. A. Augis, "Engineering Analysis of Thermal Cycling Accelerated Tests for Surface-Mount Attachment Reliability Evaluation," *Proceedings of the 7th Annual International Electronics Packaging Society Conference,* Boston, MA, November 1987, pp. 385–410.
154. Clech, J.-P., W. Engelmaier, R. W. Kotlowitz, and J. A. Augis, "Surface Mount Solder Attachment Reliability Figures of Merit (Design for Reliability) Tools," *Proceedings of NEPCON West,* March 1989, pp. 206–224.
155. ———, "Reliability Figures of Merit for Surface-Soldered Leadless Chip Carriers Compared to Leaded Packages," *IEEE Transactions on Components, Hybrids, and Manufacturing Technology,* **12**(4), December 1989, pp. 449–458.
156. Frear, D. R., "Microstructural Evolution During Thermomechanical Fatigue of 62Sn-36Pb-2Ag and 60Sn-40Pb Solder Joints," *Proceedings of the IEEE 40th Electronic Components and Technology Conference,* 1990, pp. 518–524.

157. ______, "Thermomechanical Fatigue of Solder Joints: A New Comprehensive Test Method," *IEEE Transactions on Components, Hybrids, and Manufacturing Technology,* **12**(4), December 1989, pp. 492–501.
158. Frear, D., D. Grivas, M. McCormack, D. Tribula, and J. W. Morris, "Fatigue and Thermal Fatigue Testing of Pb-Sn Solder Joints," *Proceedings of ASM's 3d Conference on Electronic Packaging and Corrosion in Microelectronics,* 1987, pp. 269–274.
159. Tribula, D., and J. W. Morris, Jr., "Creep in Shear of Experimental Solder Joints," *ASME Winter Annual Meeting,* Paper no. 89-WA/EEP-30, San Francisco, CA, December 1989.
160. Summers, T. S. E., and J. W. Morris, Jr., "Isothermal Fatigue Behavior of Sn-Pb Solder Joints," *ASME Winter Annual Meeting,* Paper no. 89-WA/EEP-39, San Francisco, CA, December 1989.
161. Tribula, D., D. Grivas, D. R. Frear, and J. W. Morris, "Observations on the Mechanisms of Fatigue in Eutectic Pb-Sn Solder Joints," *Journal of Electronic Packaging, Transactions of ASME,* **111**(2), June 1989, pp. 83–89.
162. Royce, B. S. H., "Differential Thermal Expansion in Microelectronic Systems," *IEEE Transactions on Components, Hybrids, and Manufacturing Technology,* **11**(4), December 1988, pp. 454–463.
163. Schmidt, C. G., "A Phenomenological Constitutive Model for Type Sn62 Solder Wire," *Proceedings of the IEEE 39th Electronic Components Conference,* 1989, pp. 253–258.
164. MacKay, C. A., "Some Causes of Problems with Existing SMT Solder Creams and Possible Improvements," *Circuit World,* **13**(3), 1987, pp. 4–7.
165. ______, "Flue Mechanisms From Solderability Measurements," *Proceedings of the 2d ASM International Electronic Materials and Processing Congress,* 1989, pp. 13–16.
166. Sherman, K., and C. A. MacKay, "A Thermoanalytical Study of the Components and Formulation of a Resin Based Flux," *Proceedings of the 2d ASM International Electronic Materials and Processing Congress,* 1989, pp. 17–30.
167. Roche, N. G., and C. A. MacKay, "An Activity Number for Fluxes," *Proceedings of the 2d ASM International Electronic Materials and Processing Congress,* 1989, pp. 31–35.
168. MacKay, C. A., "Solder Creams for SMT," *Proceedings of the 2d ASM International Electronic Materials and Processing Congress,* 1989, pp. 405–417.
169. Charles, H. K., Jr., and G. V. Clatterbaugh, "Solder Joint Reliability—Design Implications from Finite Element Modeling and Experimental Testing," *ASME Winter Annual Meeting,* Paper no. 89-WA/EEP-22, San Francisco, CA, December 1989.
170. Pan, T. Y., and Y. H. Pao, "Stress Analysis of Multilayer Solder Assemblies," *ASME Winter Annual Meeting,* Paper no. 89-WA/EEP-3, San Francisco, CA, December 1989.
171. Shah, M., "Analysis of Parameters Influencing Stresses in the Solder Joints of Leadless Chip Capacitors," *ASME Winter Annual Meeting,* Paper no. 89-WA/EEP-31, San Francisco, CA, December 1989.
172. Solomon, H. D., V. Brzozowski, and D. G. Thompson, "Predictions of Solder Joint Fatigue Life," *Proceedings of the IEEE 40th Electronic Components and Technology Conference,* 1990, pp. 351–359.

173. Solomon, H. D., "Fatigue of 60/40 Solder," *IEEE Transactions on Components, Hybrids, and Manufacturing Technology,* **CHMT-9**(4), December 1986, pp. 423–432.
174. ———, "Strain-Life Behavior in 60/40 Solder," *Journal of Electronic Packaging, Transactions of ASME,* **111,** June 1989, pp. 75–82.
175. ———, "Room Temperature Low Cycle Fatigue of a High Pb Solder (Indalloy 151)," *Proceedings of the 2d ASM International Electronic Materials and Processing Congress,* April 1989, pp. 135–146.
176. ———, "Low Cycle Fatigue of Surface Mounted Chip Carrier/Printed Wiring Board Joints," *IEEE Transactions on Components, Hybrids, and Manufacturing Technology,* **12**(4), December 1989, pp. 473–479.
177. ———, "High and Low Temperature Strain-Life Behavior of a Pb Rich Solder," *ASME Winter Annual Meeting,* Paper no. 89-WA/EEP-33, San Francisco, CA, December 1989.
178. ———, "The Influence of Hold Time and Fatigue Cycle Wave Shape on the Low Cycle Fatigue of 60/40," *Proceedings of the IEEE 38th Electronic Components Conference,* 1988, pp. 7–12.
179. Huffman, B., R. Brownfield, and B. Natarajan, "Mechanical Characterization of J-Bend Leads and Solder Joints for Surface Mounted PLCC's," *Proceedings of IEEE International Electronic Manufacturing Technology Symposium,* 1986, pp. 218–227.
180. Gee, S. A., and C. G. M. Van Kessel, "Stiffness and Yielding in the PLCC 'J' Lead," *IEEE Transactions on Components, Hybrids, and Manufacturing Technology,* **CHMT-10**(3), September 1987, pp. 379–390.
181. Jahsman, W. E., and P. Jain, "Comparison of Predicted and Measured Lead Stiffnesses of Surface Mounted Packages," *Proceedings of IEEE Electronic Components and Technology Conference,* May 1990, pp. 926–932.
182. Pope, D. E., and K. Byrd, "Experimental Measurements of Surface Mount Component Lead Stiffness," *Proceedings of IEEE Electronic Components and Technology Conference,* May 1990, pp. 934–941.
183. Alger, C. L., D. E. Pope, and P. M. Rehm, "Solderability Test Requirements for Plastic Surface Mount Packages," *Circuit World,* 1989, pp. 30–37.
184. Kotlowitz, R. W., "Compliance Metrics for Surface Mount Component Lead Design," *Proceedings of IEEE Electronic Components and Technology Conference,* May 1990, pp. 926–932.
185. Lau, J. H., and D. W. Rice, "Solder Joint Fatigue in Surface Mount Technology: State of the Art," *Solid State Technology,* **28,** October 1985, pp. 91–104.
186. Lau, J. H., D. W. Rice, and P. A. Avery, "Elasto-Plastic Analysis of Surface-Mount Solder Joints," *IEEE Transactions on Components, Hybrids, and Manufacturing Technology,* **CHMT-10**(3), September 1987, pp. 346–357.
187. Lau, J. H., D. W. Rice, and G. Harkins, "Thermal Stress Analysis of TAB Packages and Interconnections," *IEEE Transactions on Components, Hybrids, and Manufacturing Technology,* **13**(1), March 1990, pp. 183–188.
188. Lau, J. H., and D. W. Rice, "Thermal Stress/Strain Analyses of Ceramic Quad Flat Pack Packages and Interconnections," *Proceedings of the IEEE 40th Electronic Components and Technology Conference,* May 1990, pp. 824–834.

189. Lau, J. H., and G. Harkins, "Stiffness of PQFP 'Gull-Wing' Lead and Its Effect on Solder Joint Reliability," *Proceedings of the 5th IEEE International Electronic Manufacturing Technology Symposium,* October 1988, pp. 131–142.
190. ______, "Stiffness of 'Gull-Wing' Lead and Solder Joint for Plastic Quad Flat Pack," *IEEE Transactions on Components, Hybrids, and Manufacturing Technology,* **13**(1), March 1990, pp. 124–130.
191. Lau, J. H., "Stiffness of PLCC 'J' Lead and Its Effect on Solder Joint Reliability," *ASME Winter Annual Meeting,* Paper no. EEP-3, Boston, December 13–15, 1987,
192. ______, "Thermal Stress Analysis of SMT PQFP Packages and Interconnections," *Journal of Electronic Packaging, Transactions of ASME,* **111,** March 1989, pp. 2–8.
193. Lau, J. H., and G. Harkins, "Thermal Stress Analysis of SOIC Packages and Interconnections," *IEEE Transactions on Components, Hybrids, and Manufacturing Technology,* **CHMT-11**(4), December 1988, pp. 380–389.
194. Lau, J. H., G. Harkins, D. Rice, J. Kral, and B. Wells, "Experimental and Statistical Analyses of Surface-Mount Technology PLCC Solder-Joint Reliability," *IEEE Transactions on Reliability,* **37**(5), December 1988, pp. 524–530.
195. Lau, J. H., and G. H. Barrett, "Stress and Deflection Analysis of Partially Routed Panels for Depanelization," *IEEE Transactions on Components, Hybrids, and Manufacturing Technology,* **CHMT-10**(3), September 1987, pp. 411–419.
196. Lau, J. H., S. J. Erasmus, and D. W. Rice, "Overview of Tape Automated Bonding Technology," *Circuit World,* **16**(2), 1990. pp. 5–24.
197. Lau, J. H., and A. H. Jeans, "Effects of Voids on SMT Solder Joint Reliability," *Proceedings of the 2d ASM International Electronic Materials and Processing Congress,* April 1989, pp. 177–187.
198. Lau, J. H., R. Subrahmanyan, D. Rice, S. Erasmus, and C. Li, "Fatigue Analysis of a Ceramic Pin Grid Array Soldered to an Orthotropic Epoxy Substrate," ASME Winter Annual Meeting Paper no. 90-WA/EEP-47, Dallas, TX, November 1990.
199. Lau, J. H., L. Powers, J. Baker, D. Rice, and W. Shaw, "Solder Joint Reliability of Fine Pitch Surface Mount Technology Assemblies," *IEEE Transactions on Components, Hybrids, and Manufacturing Technology,* **13**(3), September 1990, pp. 534–544.
200. Lau, J. H., and C. A. Keely, "Dynamic Characterization of Surface Mount Component Leads for Solder Joint Inspection," *IEEE Transactions on Components, Hybrids, and Manufacturing Technology,* **12**(4), December 1989, pp. 594–602.
201. Lau, J. H., G. Harkins, D. Rice, and J. Kral, "Thermal Fatigue Reliability of SMT Packages and Interconnections," *Proceedings of IEEE 25th International Reliability Physics Symposium,* San Diego, April 1987, pp. 250–259.
202. Lau, J. H., and D. Rice, "Failure Analysis of Surface Mounted Interconnection," *Proceedings of the 12th ASM International Symposium for Testing and Failure Analysis,* Los Angeles, October 1986, pp. 73–82.
203. Lau, J. H., "A Note on the Calculation of Thermal Stresses in Electronic Packaging by Finite Element Methods," *Journal of Electronic Packaging, Transactions of the ASME,* **111,** December 1989, pp. 313–320.
204. ______, *Elasto-Plastic Finite Element Analysis with Special Reference to Reinforced Concrete,* M.A.Sc. Thesis, University of British Columbia, 1973.

205. Lau, J. H., and A. H. Jeans, "Nonlinear Analysis of Elastomeric Keyboard Domes," *Journal of Applied Mechanics, Transactions of the ASME,* **56,** December 1989, pp. 751–755.
206. Lau, J. H., and L. L. Moresco, "Mechanical Behavior of Microstrip Structures Made from YBaCuO Superconducting Ceramics," *IEEE Transactions on Components, Hybrids, and Manufacturing Technology,* **11**(4), December 1988, pp. 419–426.
207. Lau, J. H., and L. B. Lian-Mueller, "Finite Element Modeling for Optimizing Hermetic Package Reliability," *Journal of Electronic Packaging, Transactions of ASME,* **111,** December 1989, pp. 255–260.
208. Lau, J. H., "Thermal Stress Analysis of Plastic Leaded Chip Carriers," *Proceedings of the IEEE/ASME/ISHM/NIST InterSociety Conference on Thermal Phenomena in Electronic Systems,* May 1990, pp. 57–66.
209. Zienkiewicz, O. C., *The Finite Element Method in Engineering Science,* London, McGraw-Hill, 1971.
210. Burnett, D. S., *Finite Element Analysis,* Reading, MA, Addison-Wesley Publishing Company, 1987.
211. Strang, G., *An Analysis of the Finite Element Method,* Englewood Cliffs, NJ, Prentice-Hall, 1973.
212. Huston, R. L., and C. E. Passerello, *Finite Element Methods,* New York, Marcel Dekker, 1984.
213. Bathe, K. J., *Finite Element Procedures in Engineering Analysis,* Englewood Cliffs, NJ, Prentice-Hall, 1982.
214. Eisley, J. G., *Mechanics of Elastic Structures,* Englewood Cliffs, NJ, Prentice-Hall, 1989.
215. Humar, J. L., *Dynamics of Structures,* Englewood Cliffs, NJ, Prentice-Hall, 1990.
216. Heins, C. P., and K. N. Derucher, *Structural Analysis and Design,* New York, Marcel Dekker, 1980.
217. Ross, C. T. F., *Computational Methods in Structural and Continuum Mechanics,* New York, John Wiley, 1982.
218. McGuire, W., and R. H. Gallagher, *Matrix Structural Analysis,* New York, John Wiley, 1979.
219. Meek, J. L., *Matrix Structural Analysis,* New York, McGraw-Hill, 1971.
220. Fung, Y. C., *Foundations of Solid Mechanics,* Englewood Cliffs, NJ, Prentice-Hall, 1965.
221. Timoshenko, S. P., and J. N. Goodier, *Theory of Elasticity,* New York, McGraw-Hill, 1970.
222. Harris, C. M., and C. E. Crede, *Shock and Vibration Handbook,* New York, McGraw-Hill, 1976.
223. Thomson, W. T., *Theory of Vibrations with Applications,* Englewood Cliffs, NJ, Prentice-Hall, 1972.
224. Meirovitch, L., *Analytical Methods in Vibrations,* London, Macmillan, 1967.
225. Clough, R. W., and J. Penzien, *Dynamics of Structures,* New York, McGraw-Hill, 1976.
226. Langhaar, H. L., *Energy Methods in Applied Mechanics,* New York, John Wiley, 1962.

227. Steinberg, D. S., *Vibration Analysis for Electronic Equipment,* New York, John Wiley, 1973.
228. Steinberg, D. S., "Protecting PCBs From Shock," *Machine Design,* March 1983, pp. 189–192.
229. Markstein, H. W., "Designing Electronics for High Vibration and Shock," *Electronic Packaging and Production,* April 1987, pp. 40–43.
230. Markstein, H. W., "Design With Vibration in Electronics," *Electronic Packaging and Production,* June 1989, pp. 30–34.
231. Lau, J. H., "Vibration Frequencies of Tapered Bars With End Mass," *Journal of Applied Mechanics, Transactions of ASME,* **51,** March 1984, pp. 179–181.
232. ———, "Vibration Frequencies and Mode Shape for a Constrained Cantilever," *Journal of Applied Mechanics, Transactions of ASME,* **51**, March 1984, pp. 182–187.
233. ———, "Vibration Frequencies for a Non-Uniform Beam with End Mass," *Journal of Sound and Vibration,* **97,** December 1984, pp. 513–521.
234. Lau, J. H., and R. E. Miller, "Vibration Analysis of Orthotropic Shells of Revolution," *Proceedings of AIAA/ASME/ASCE 22d Structures, Structural Dynamics and Materials Conferences,* 1981, pp. 122–131.

16

Integrated Matrix Creep: Application to Accelerated Testing and Lifetime Prediction

Sheera Knecht and Les Fox

One of the major goals of mechanical analysis in the electronics industry is to be able to predict, before extensive testing, the reliability of printed circuit boards with surface mounted components under thermal or mechanical cyclic loading. Analysis cannot be entirely substituted for testing, but if it can be used as a guide to testing, in determining the minimum tests necessary and in understanding the results of these tests, the cost of ensuring a reliable product will be substantially reduced. It is not surprising, therefore, that a great deal of effort is currently being applied to mechanical modeling[1] of electronic packages with surface mounted devices with the aim of understanding and improving fatigue life of solder joints.

Practical fatigue life predictions are based on the calculation of certain failure indicators that correlate known lifetimes under one set of conditions (e.g., geometry, loading, materials) with projected lifetimes under another set of conditions. Possible failure indicators might be maximum stress, plastic strain, or some kind of "damage integral." To calculate any of these indicators, a constitutive or stress-strain relation is needed for the materials involved in the analysis. In the case of surface mounted components, a constitutive law for solder is a critical part of mechanical modeling.

Solder material behavior is complex. Eutectic 63% Sn-37% Pb solder (the most commonly used solder) melts at 183°C. The ratio T/T_{melt}, where the temperature is expressed in degrees absolute, is known as the homologous temperature; therefore the homologous temperature of solder at room temperature is about 0.65. At homologous temperatures greater than about 0.5 it is well known that metals exhibit significant creep and stress relaxation. Such behavior is easily observed for solder. However, time dependent creep effects do not account for all of the observed inelastic solder behavior; for the rapid cycling used in accelerated fatigue tests, the noncreep plasticity can be significant. In order for mechanical modeling to provide a link between test results and real world applications, the solder constitutive equation needs to consider all important aspects of material behavior. For example, if short time plasticity is ignored in favor of long-term creep, results of very fast accelerated fatigue tests may be misleading when applied to the long-term stress relaxation experienced under service life conditions.

Although a great deal of published work on mechanical properties of solder is available, this broad range of solder behavior is usually not presented in a form that is convenient for mechanical analysis or for fatigue life calculations. Moreover, existing solder constitutive equations have often not been used to predict experimental results other than those upon which they are based. When it can be seen that it is possible to predict results of testing under different conditions from those used to derive an equation, we can have more confidence that the equation is useful for mechanical modeling.

The first purpose of this study, therefore, is to derive a stress-strain relation for eutectic or near-eutectic Sn-Pb solder that is simple enough for practical use and yet is complex enough to represent the range of solder behavior that is important for electronic applications. The development of the equation is based on the data described in reference 2 for solder joints in both steady-state creep and constant strain rate cycling. Some parameters are difficult to extract from this data and these are taken from other published results. The equation is then shown to predict observed data, also from reference 2, for stress relaxation tests.

The equation is developed from a continuum mechanics and not a material science point of view; the purpose is to determine a relation that is useful for predicting the mechanical behavior of solder joints used in different structures (e.g., leaded and leadless chip carriers) and under different loadings. Since the subject of solder metallurgy is complex and not well understood, no attempt has been made to include microstructural effects in the equation. Although microstructure can affect mechanical properties, it is not possible to quantify microstructural effects at the present time. In accordance with our approach, the solder joints used to generate the data used here are typical joints that would be found in electronic devices. Their microstructure is not specially controlled and is therefore difficult to quantify; however, we feel that using "real" joints will generate practical results. Whenever possible, we are guided in our analysis by knowledge of solder metallurgy, but the general approach is a continuum mechanics one.

An adequate constitutive law allows the calculation of stresses and strains for arbitrary geometries and loading histories. The second part of any reliability analysis is to relate these calculated parameters or "failure indicators" to expected lifetime. The matrix creep criterion developed in reference 2 and explained below is proposed as a failure indicator that is simple enough to be calculated relatively easily, yet complex enough to take into account the time dependent nature of solder fatigue. It has already been shown in reference 2 that the matrix creep failure criterion proposed there gives good correlation with the fatigue life of the solder joints of reference 2. This correlation was achieved using a steady-state creep law with different, experimentally determined, coefficients for each sample considered; the matrix creep was then integrated over the actual measured stress history (see reference 2). It is often necessary, however, to make fatigue life predictions before any testing can be done; therefore the second purpose of this study is to show that the constitutive equation developed here can be used along with the matrix creep failure indicator to predict fatigue life adequately for the solder joints of reference 2. The equation is also applied to fatigue tests described by other authors, and reported lifetimes are shown to correlate with matrix creep.

Finally, the fatigue failure indicator is itself extended to the case of leaded solder joints where stresses are multiaxial and vary with position in the joint. Application of the extended matrix creep failure indicator to a thermal cycle fatigue example from the literature shows the current status of the method.

16.1 GENERAL FORM OF THE CONSTITUTIVE RELATION

In order to be able to predict the loading and deformation histories encountered in electronic applications, the constitutive relation for eutectic or near-eutectic Sn-Pb solder is developed considering three modes of solder deformation. The total shear strain rate $\dot{\gamma}$ is assumed to be the sum of time independent elastic, time independent plastic, and time dependent secondary creep strain rates:

$$\dot{\gamma} = \dot{\gamma}_{el} + \dot{\gamma}_{pl} + \dot{\gamma}_{cr} \tag{16-1}$$

It is difficult to separate fast primary creep from time independent plasticity; therefore $\dot{\gamma}_{pl}$ may include such effects. It is also possible to include a time dependent (transient), slower primary creep strain rate term, but for simplicity this is not done. Similarly, any anelastic (time dependent elastic) effects are also ignored.

The data used to develop the constitutive law consists of stable cycles of fully reversed loading; the initial loading cycles of the test (typically fewer than five) are not considered. As cycling is continued beyond these first few hysteresis loops, the stress-strain curve stabilizes; the number of stable cycles depends on the details of the loading. As cycling continues further, the load needed to maintain a fixed opening of the hysteresis loop decreases (probably due to fatigue cracking) and therefore the relationship between stress and strain changes; these changes are not considered either. It is felt that the most useful information for lifetime prediction can be obtained in the simplest manner from an understanding of the initial stable part of the deformation history.

16.2 DEVELOPMENT OF THE CONSTITUTIVE RELATION

16.2.1 Description of Data

The data used in this study are the same as used for the work of reference [2]. A torsional apparatus was employed to obtain isothermal hysteresis loops and creep data on a 16-I/O solder joint array. The 7.6- by 7.6-mm chip carrier had solder pads approximately 0.64 by 1.0 mm. The 16 pads were arranged in a square array of four per side at an average distance from the center of 3.43 mm, which was used to calculate the average stress on the pads due to the applied torque. These chip carriers were surface mounted to a stainless steel plate 45 by 45 mm. The stainless steel plate had electroplated copper footprints patterned to those of the chip carrier. The assembly was soldered together by 63% Pb-37% Sn solder paste: The joint thicknesses ranged from 0.102 to 0.305 mm (0.004 to 0.12 in.). Additional data was obtained in the same manner, but using G10 substrate material instead of stainless steel. Further details of the experimental procedure can be obtained in reference.[2]

16.2.2 Steady-State Creep Strain Component

When the rate of shear stress $\dot{\sigma}$ is zero, $\dot{\gamma}_{el}$ and $\dot{\gamma}_{pl}$ are also zero and therefore, according to Equation (16-1), the time dependent part of the strain rate is completely described by the steady-state creep strain rate determined from constant stress tests. Details of the steady state creep tests are given in reference 2.

Data used here were taken from samples A, A1, A2, A3, and C (stainless steel substrates) of reference 2. The creep rates were obtained by reanalyzing the raw data, so some variation from reference 2 can be noted. Shear stress versus shear strain rate at 25, 60, and 100°C are shown in Figure 16-1. Two points can be noted:

- Several authors[3,4,8] have noted a grain size dependence in steady-state creep data for eutectic or near-eutectic solder. The joints used here are typical solder joints that would be found in actual devices; therefore, the grain size has not been carefully controlled and varies within each joint. Still, the thinner joints tend to have smaller grain sizes than the thicker ones, although the amount of this difference cannot be quantified (see reference 2). Ob-

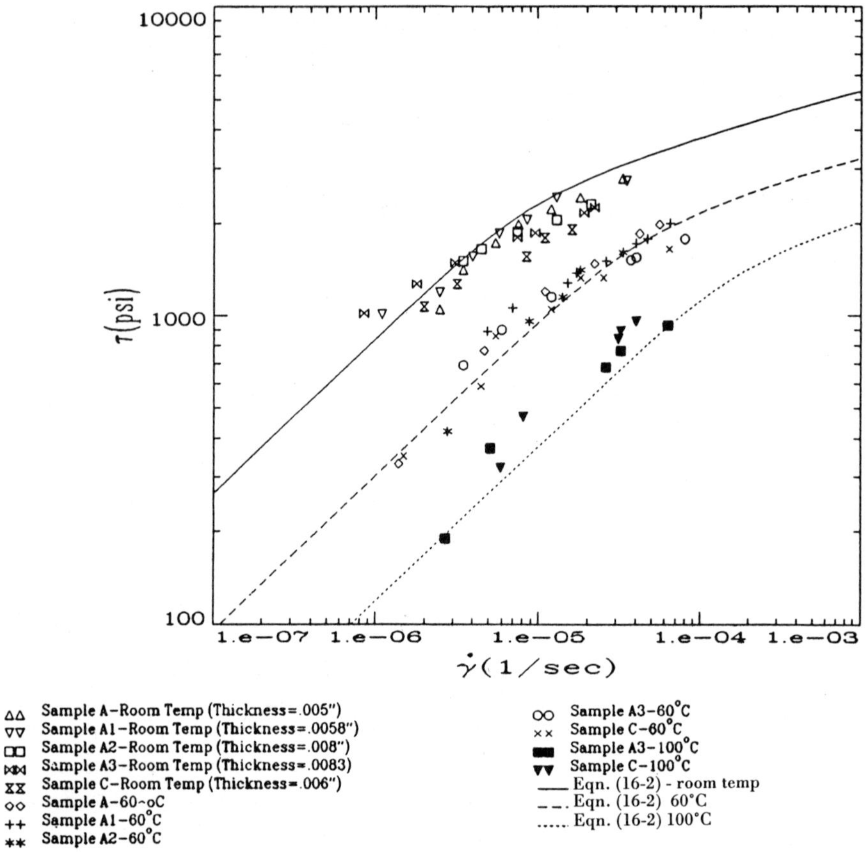

Figure 16-1 Steady-state creep data from (2).

servation of Figure 16-1 shows that thickness effects are not evident here.

- Most recent experimental work on solder[3,4,8–10] shows two (or possibly three) regimes of secondary (steady-state) creep with distinctly different slopes in the stress-strain rate curve. The range of the data from reference 2 is not really large enough to confirm this hypothesis absolutely, but the data in Figure 16-1 do suggest it.

In light of these points, the following approach is taken: Two regimes of steady-state creep are assumed. For convenience, these are called "grain boundary" and "matrix" creep as in reference 2. However, in keeping with the continuum mechanics (rather than metallurgical) approach taken here we do not need to determine any particular creep mechanisms for these regions. We simply assume,

Table 16-1 Comparison of Creep Parameters from Literature

Author	ΔH_{GB} (eV)	n_{GB}	ΔH_{MC} (eV)	n_{MC}	Ref. No.
Grivas et al.	0.5	2.0	0.84	7.1	3
Kashyap et al.	0.46	1.7	1.0	11.1	4
Solomon	—	3.1	—	7.3	5
Baker	0.69	3.4	—	—	7
Lam et al.	0.45	2.4	0.9	5.7	8
Wong et al.	0.47	3.0	0.47	7.0	9

on the basis of empirical data for creep in metals, that each regime has a power law dependence of strain rate on stress and each follows an Arrhenius model for temperature dependence with a different activation energy.

The form of the equation for steady-state creep rate as a function of stress and temperature is

$$\dot{\gamma} = \dot{\gamma}_{er} = C_1 \exp\left(\frac{-\Delta H_{GB}}{kT}\right)\tau^{n_{GB}} + C_2 \exp\left(\frac{-\Delta H_{MC}}{kT}\right)\tau^{n_{MC}} \qquad (16\text{-}2)$$

As can be seen from Figure 16-1, the data are concentrated in the region where both steady-state creep terms contribute significantly to the stress; it is difficult, therefore, to isolate the two regimes and to find values of the two exponents and two activation energies. Instead these parameters are taken from published data.

Representative values of the activation energies (ΔH_{GB} and ΔH_{MC}) and exponents (n_{GB} and n_{MC}) found in the literature are shown in Table 16-1. In this study, values of ΔH_{GB}, ΔH_{MC}, n_{GB}, and n_{MC} are taken from Grivas, Murty, and Morris.[3] Each term is multiplied by a constant that is determined by a least squares fit to the data in Figure 16-1. As can be seen from Table 16-1, the values assumed for ΔH_{GB}, ΔH_{MC}, n_{GB}, and n_{MC} lie within the range of results reported by other authors; these values also fit our data quite well (see Figure 16-1). Other values were considered, but no improvement in fit resulted.

The values of the parameters used in Equation (16-2) are as follows:

$C_1 = 8.31$

$C_2 = 1.12$

ΔH_{GB} = activation energy for grain boundary creep = 0.5 eV

ΔH_{MC} = activation energy for matrix creep = 0.84 eV

k = Boltzmann's constant = $8.63 \times 10^{-5} \dfrac{\text{eV}}{\text{K}}$

Table 16-2 Values of Parameters for Creep Equation (16-3)

	25°C	60°C	100°C
C_0	$1.17 \times 10^{-5} \frac{1}{\text{sec}}$	$5.24 \times 10^{-5} \frac{1}{\text{sec}}$	$2.06 \times 10^{-4} \frac{1}{\text{sec}}$
τ_0	19.8 MPa (2870 psi)	15.0 MPa (2180 psi)	11.7 MPa (1700 psi)

This equation and the data points are shown in Figure 16-1.

T = temperature, K

τ = shear stress, MPa

$\dot{\gamma}$ = shear strain rate, $\frac{1}{\text{sec}}$

n_{GB} = grain boundary creep exponent = 2

n_{MC} = matrix creep exponent = 7.1

The following equivalent form of this equation is more convenient for numerical analysis; values of C_0 and τ_0 for the temperatures used in the tests of reference 2 are listed in Table 16-2:

$$\dot{\gamma} = \dot{\gamma}_{cr} = C_0\left[\left(\frac{\tau}{\tau_0}\right)^2 + \left(\frac{\tau}{\tau_0}\right)^{7.1}\right] \tag{16-3}$$

This equation and the data points are shown in Figure 16-1.

16.2.3 Elastic Strain Component

Considerable variation exists in published values of the elastic modulus of solder. For example, reference 11 quotes a value of 14.8 GPa (2.15 Mpsi) for room temperature tests of bulk samples of eutectic solder. In reference 12 a value of 30.0 GPa (4.35 Mpsi) is reported for bulk samples of 60% Sn-40% Pb alloy at room temperature. In reference 13, 43.4 GPa (6.3 Mpsi) is determined for elastic modulus of 60% Sn-40% Pb alloy while reference 14 shows a range from about 12.4 GPa to 17.3 GPa (1.8 Mpsi to 2.5 Mpsi) increasing with frequency. Solomon[15] discusses in detail how structural compliance in test fixtures influences the measurement of elastic properties of thin joints.

No attempt has been made here to separate the effect of apparatus and substrate compliance of reference 2 on the overall elastic properties. Instead we note that the experiment measures an effective shear modulus and assume the following:

$$\frac{1}{G_{\text{eff}}} = \frac{1}{G_{\text{solder}}} + \frac{1}{G_{\text{structure}}} \tag{16-4}$$

Because of uncertainty involved in calculating the compliance of the fixtures used to obtain the data analyzed here, determination of the correct value of solder modulus would best be done by independent means. References[9,10] present temperature dependent elastic modulus of eutectic solder as $E = -0.088T(°C) + 32$ GPa; these values are used in Figure 16-3*b*.

The value of Poisson's ratio for solder at room temperature is generally taken to be about 0.4.[11]

16.2.4 Time Independent Plastic Strain Component

Hysteresis test data that give stable loops of stress versus strain under steady cycling at different temperatures and frequencies are available for some of the samples described in reference 2. The tests that give data with least noise are those in which the total strain is a ramp function with time (total strain rate is known). The magnitude of the total strain range is automatically adjusted in successive cycles to be that necessary to maintain a particular value of inelastic strain (opening of the hysteresis loop, see reference 2), but the samples are observed to reach a stable loop in less than 10 cycles and not to show significant softening for a reasonably large number of cycles. It is not possible to be more precise since the complete history of deformation of the samples is not available. The time independent (noncreep) part of the constitutive law is derived from these stable ramp cycles as follows:

Using Equations (16-1) and (16-3), and knowing the stress-time history for the hysteresis cycle, the creep strain can be subtracted from the total strain at all stress values of the loop. In this way the value of time independent strain (elastic plus plastic) can be calculated. Typical such curves are shown in Figure 16-2.

During unloading the time independent stress strain curve is assumed to be purely elastic with slope G_{eff}. In the loading regime, the form of the constitutive law is taken to be a Ramberg-Osgood-type relation:

$$\gamma^{\text{time-independent}} = \frac{\tau}{G_{\text{eff}}} + \left(\frac{\tau}{\tau_p}\right)^n \tag{16-5}$$

We determine G_{eff} from the slope of the linear unloading regime; for the data analyzed here, G_{eff} varies from about 350,000 to 550,000 psi at room temperature for the stainless steel substrates and depends on details of sample preparation. At higher temperatures and for G10 substrates, the values of G_{eff} are lower.

Using the proper value for G_{eff} and a linear least squares technique for Equation (16-5), the exponent n was found to be 2 to the nearest whole number for the range of data available. Using $n = 2$, τ_p is determined as a function of temperature

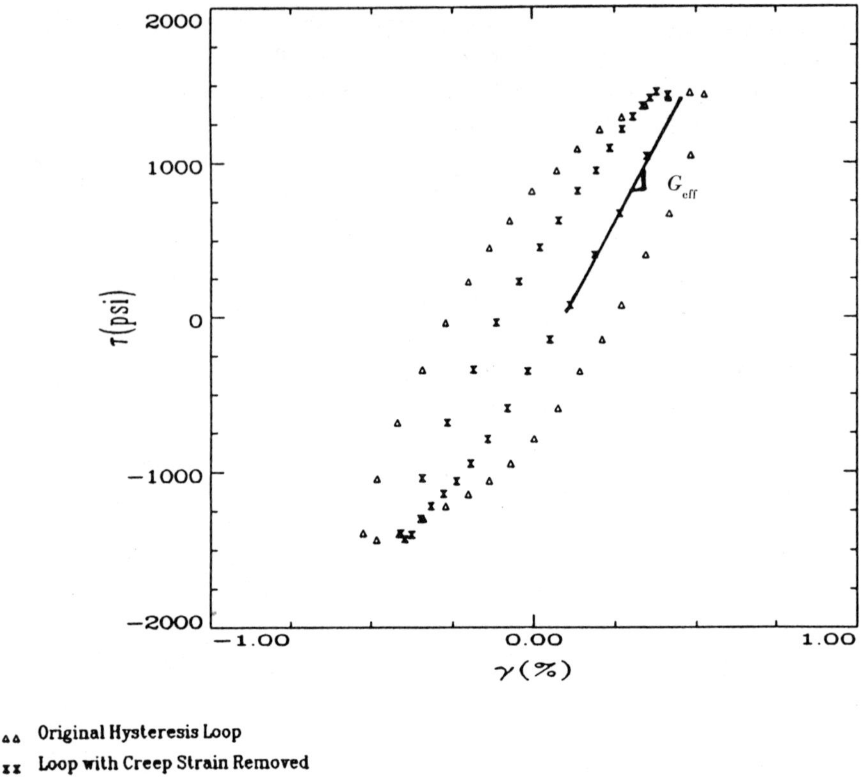

Figure 16-2 Hysteresis loop for sample A3 10-min cycle at 60°C.

by an additional least squares fit. Values of τ_p versus temperature are given in Table 16-3.

The parameter τ_p does not relate to any definite deformation mode. It simply includes all effects not covered by the steady-state creep and elastic terms. As such, it may be considered to be something similar to a "plastic modulus," which is consistent with its high value relative to actual solder stresses. The strong temperature dependence of τ_p may be due to the primary creep effect that it encompasses or perhaps to anelastic effects.

Table 16-3 Values of Plasticity Parameter for Equation (16-5)

Temperature	τ_p
25°C	289 MPa (41,900 psi)
60°C	240 MPa (34,700 psi)
100°C	135 MPa (19,500 psi)

16.3 SUMMARY OF CONSTITUTIVE EQUATION

The above analysis leads to the following form of the constitutive equation (in shear only) expressed in terms of strain rate:

$$\dot{\gamma} = \dot{\gamma}_{\text{tot}} = \left[\frac{1}{G_{\text{solder}}} + \frac{2|\tau|}{\tau_p^2}\left(\frac{1 + \text{sgn}(\dot{\tau}\tau)}{2}\right)\right]\dot{\tau} + C_0\left[\left(\frac{|\tau|}{\tau_0}\right)^2 + \left(\frac{|\tau|}{\tau_0}\right)^{7.1}\right]\text{sgn}(\tau) \qquad (16\text{-}6)$$

where $\text{sgn}(x) = 1$ for $x > 0$, $= -1$ for $x < 0$, and $= 0$ for $x = 0$. Values for the parameters τ_0 and C_0 are given in Table 16-2 and values of τ_p are given in Table 16-3.

Note that for mechanical analysis where the compliance of nonsolder components is not separately accounted for, G_{solder} will be replaced by G_{eff}. This approach is used here for analysis of the available test data.

Since the equation was derived only from hysteresis loops of stable cycling, no yield criterion can be obtained.

16.4 COMPARISON OF THE STEADY-STATE CREEP EQUATION TO PUBLISHED DATA

A great deal of published data exists on steady-state creep of solder joints and much of it is presented in the form of Equation (16-2). Comparison of creep exponents and activation energies has been made in Table 16-1. It is also useful to compare some overall equations to each other and to Equation (16-2).

- Solomon[5]: Solomon's results for steady-state creep rate in shear at 35°C are shown in Figure 16-3*a*. A transition from a regime of slope approximately 7 to one of slope approximately 3 does occur but at a lower value of strain rate than for the data of reference 2.
- Grivas et al.[3]: This study shows a grain size dependence for a "superplastic" region but not for the high stress region. Figure 16-3*a* shows that agreement is obtained with Equation (16-2) using an average grain size of 60-μm. This value is much larger than the grain sizes estimated for the samples used in reference 3. The question of the conditions under which true superplasticity can exist is controversial. Here we avoid the term "superplasticity" and approach the problem from a continuum rather than a metallurgical perspective.

  The 60-μm grain size, however, is in reasonable agreement with the values estimated in reference 2, although it is probably high. Calculation of grain size for ordinary solder joints is difficult because of the wide variation observed in size in a single joint. For comparison, the equation from reference 3 is also plotted for a grain size of 40 μm.

- Wong et al.[9]: These authors have taken data from references[3–5,11] and several others and fit a relation of the same form as Equation (16-2). As is done here, they neglect any grain size dependence since they feel that experimental data are inconclusive on this point. Upper and lower bounds for this data are given by Clech and Augis;[10] these bounds are shown in Figure 16-3*b* where $E = -0.088T(°C) + 32$ GPa (4.3 Mpsi at 25°C); $D = \exp(-5413/T)$ with T in K. For comparison, the data from Figure 16-1 and equation (16-2) are also shown. Note that the data from reference 2 have been converted to tensile from shear using the equivalent transformations given in reference 10 or in Equation (16-12) of this paper.

It can be seen from Figure 16-3*b* that the data on which this study is based lie within or close to the bounds given in reference 10. Since the

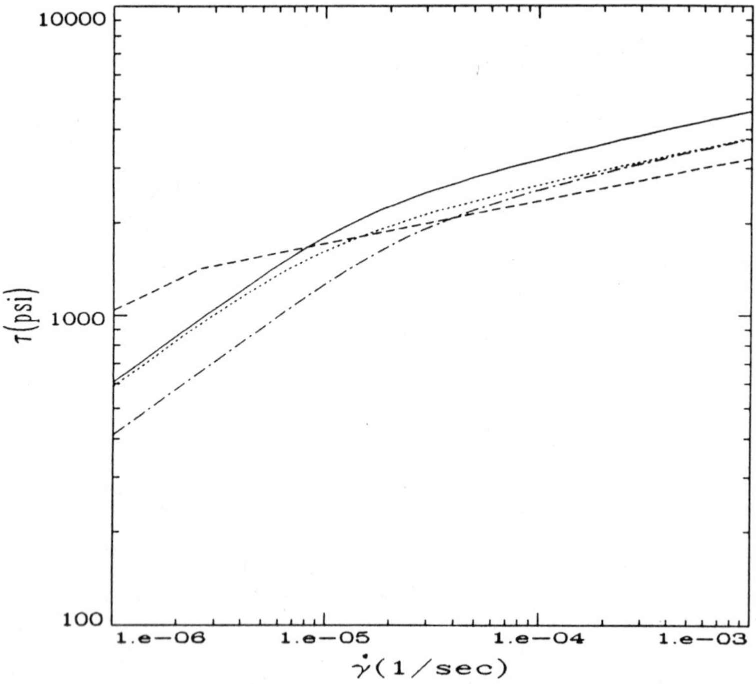

— Equation (16-2)
--· Solomon Data[5]
..... Grivas et al.[3] 60 -μm grains
—.. Grivas et al.[3] 40 -μm grains

Figure 16-3 (*a*) Comparison of creep data in shear (35°C).

steady-state creep rates of reference 2 have been recalculated from the raw data for the purpose of this study, the data in Figure 16-3*b* is slightly different than the data attributed to Shine and Fox in,[10] which was taken from reference 2.

Table 16-1 and Figure 16-3*a* and *b* indicate the difficulties inherent in developing a constitutive law for an arbitrary solder joint: it is evident that substantially different results are reported for what seems to be the same test. It is possible that some part of these differences may be attributed to testing problems such as non-steady-state conditions, loading eccentricities or variations in test temperature;[10] another reason for the different results may be microstructural differences in the solders due to differing solder technologies. At the present time, these differences are not understood well enough to be quantified. It is the

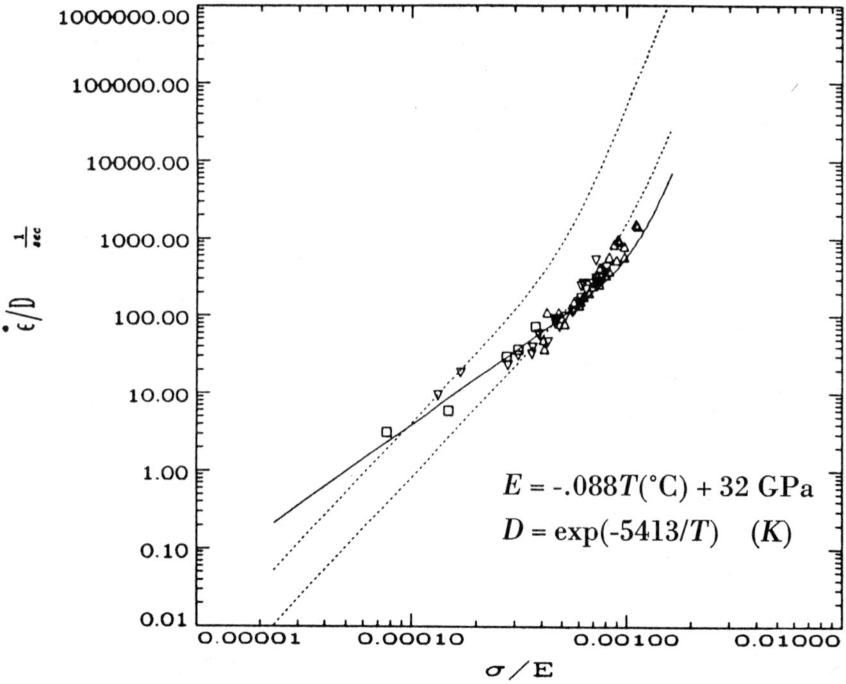

.... ,Wong reference 2 - least squares fit to various data, upper & lower bounds

—— Equation (16-2) converted to tension

△ Room Temp data from reference 2 -(converted to tension)

▽ 60 °C data from reference 2 - (converted to tension)

□ 100 °C data from reference 2 - (converted to tension)

Figure 16-3 (*b*) Comparison of creep data in tension.

contention of this work, however, that useful information can be obtained using a constitutive law for a typical solder joint as long as all relevant modes of solder behavior are taken into account.

16.5 APPLICATION OF CONSTITUTIVE EQUATION TO DATA OF REFERENCE 2

16.5.1 Description of Numerical Procedures

To compare the prediction of the constitutive equation to experimental results, Equation (16-6) will be solved for two cases. First is for a triangle wave strain history, where $\dot{\gamma}$ is constant (equal to $\dot{\gamma}_0$) except that it changes sign. Second is for a square wave where $\dot{\gamma}$ is zero except during transition from positive to negative strain values.

For these cases, Equation (16-6) can be expressed by the following integral:

$$\Delta t = \int_{\tau_i}^{\tau_f} \frac{\left(\dfrac{1}{G_{\text{eff}}} + \dfrac{2|\tau|}{\tau_p^2}\right)\left(\dfrac{1 + \text{sgn}(\dot{\tau}\tau)}{2}\right)}{\dot{\gamma}_0 - C_0\left[\left(\dfrac{|\tau|}{\tau_0}\right)^2 + \left(\dfrac{|\tau|}{\tau_0}\right)^{7.1}\right]\text{sgn}(\tau)} d\tau \qquad (16\text{-}7)$$

where τ_i is the initial value of stress for the time increment considered and τ_f is the final value.

The square wave is an idealization of an actual cycle since it is assumed that loading and unloading are accomplished infinitely fast and, therefore, that the value of strain is always $+\gamma_{\max}$ or $-\gamma_{\max}$ and strain rate is always zero (or infinite for zero time). Any actual test will have finite rise time, but for a cycle consisting of a fast ramp and hold when the ratio of the transition time to the hold time is much less than 1, the cycle may be reasonably approximated by a square wave. Equation (16-6) may also be solved for any other more complex strain history if necessary.

Triangle Wave Cycle

Considering one half-cycle and assuming $\dot{\gamma}_0$ is negative: $\tau_i = +\tau_{\max}$ and $\tau_f = -\tau_{\max}$. Let γ^{tot} = total strain range and let t_{cycle} = time for entire cycle (loading and unloading). Then, $\dot{\gamma}_0 = \dfrac{2\gamma^{\text{tot}}}{t_{\text{cycle}}}$ and the following equation can be written to solve for $\tau_{\max}$:

$$\frac{t_{cycle}}{2} = \int_0^{\tau_{max}} \frac{\dfrac{1}{G_{eff}}}{\dfrac{2\gamma_{tot}}{t_{cycle}} + C_0\left[\left(\dfrac{\tau}{\tau_0}\right)^2 + \left(\dfrac{\tau}{\tau_0}\right)^{7.1}\right]} d\tau + \int_0^{\tau_{max}} \frac{\dfrac{1}{G_{eff}} + \dfrac{2\tau}{\tau_p^2}}{\dfrac{2\gamma_{tot}}{t_{cycle}} - C_0\left[\left(\dfrac{\tau}{\tau_0}\right)^2 + \left(\dfrac{\tau}{\tau_0}\right)^{7.1}\right]} d\tau \quad (16\text{-}8)$$

The equation was solved using a trapezoidal rule procedure, increasing the limit of the integrals until the correct value of time was obtained. The first integral is stable, but the second integral becomes infinitely large as

$$C_0\left[\left(\frac{\tau}{\tau_0}\right)^2 + \left(\frac{\tau}{\tau_0}\right)^{7.1}\right] \rightarrow \frac{2\gamma_{tot}}{t_{cycle}}.$$

This asymptotic value of τ is calculated before the numerical integration so that it can be approached slowly and not exceeded. The value of τ as a function of time is also obtained.

Square Wave Cycle

In the case of the (idealized) square wave, the general form of the hysteresis loop is shown in Figure 16-4 where γ_0 is known, and τ_i and τ_f must be calculated.

In regions B-C and D-A of the idealized cycle of Figure 16-4, deformation takes place infinitely fast, so no creep strain occurs; therefore the following equation holds:

$$\gamma_0 = \frac{\tau_i + \tau_f}{G_{eff}} + \left(\frac{\tau_i}{\tau_p}\right)^2 \quad (16\text{-}9)$$

In regions A-B (and C-D) unloading occurs due to relaxation:

$$t_f = \frac{t_{cycle}}{2} = \frac{1}{G_{eff}\, C_0} \int_{\tau_f}^{\tau_i} \frac{d\tau}{\left[\left(\dfrac{\tau}{\tau_0}\right)^2 + \left(\dfrac{\tau}{\tau_0}\right)^{7.1}\right]} \quad (16\text{-}10)$$

This gives two equations for the two unknowns τ_i and τ_f (initial stress and final stress). There are two extreme cases: first, where there is no relaxation (t_f =

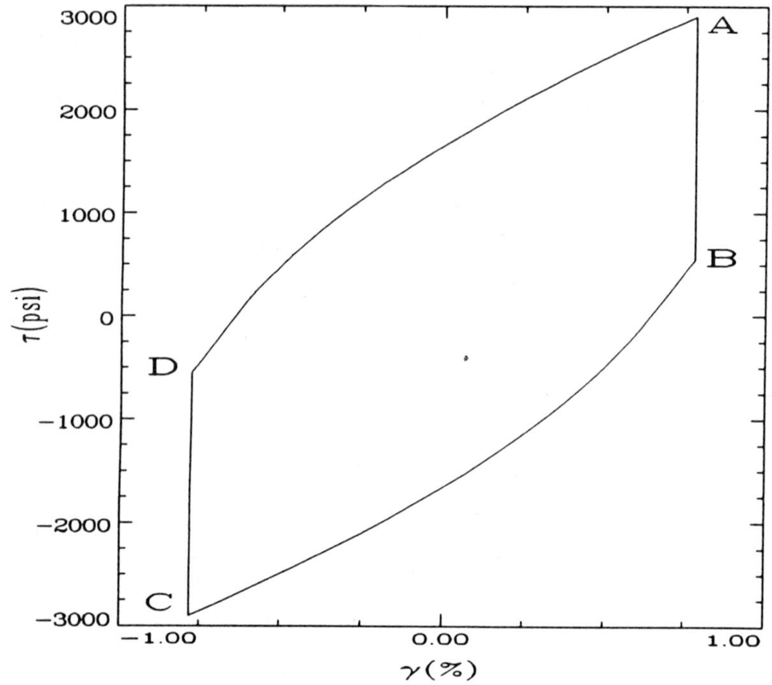

Figure 16-4 Example of hysteresis loop for square wave.

0) and $\tau_f = \tau_i$ can be calculated from Equation (16-9); second, where there is complete relaxation (t_f very large), $\tau_f = 0$, and τ_i is again calculated from Equation (16-9). Knowing the range of τ_f, the two equations may be easily solved by an iterative procedure.

16.5.2 Results

Comparison of the results of solutions to Equation (16-8) with the experimental data for triangle wave tests shows excellent agreement over the range of temperatures (25, 60 and 100°C) and cycle times (from 1- to 20-min cycles) available. Typical results are shown in Figure 16-5*a* and *b* where $G_{\text{eff}} = 450{,}000$ psi and $\gamma_{\text{tot}} = 1.04$ percent. Comparison of Equation (16-3) to the steady-state creep results has already been shown in Figure 16-1.

In addition to the ramp and steady-state creep tests, another set of experimental data is available for these solder joints, namely stress relaxation tests where the displacement-time history approximates a square wave. This data is important because it was not used in the development of the constitutive equation. The test results for the square wave hysteresis loops are quite noisy, so any data

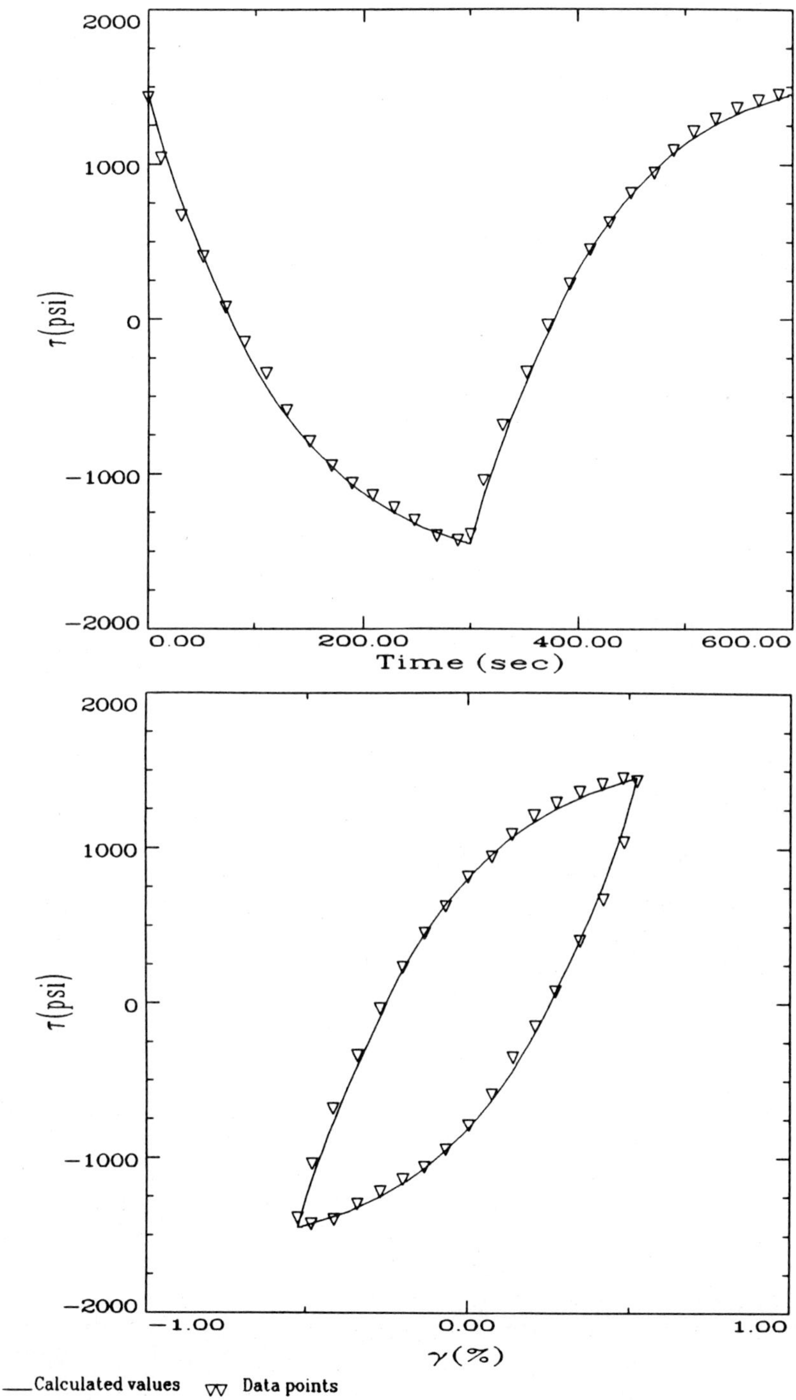

Figure 16-5 (*a* and *b*) Sample A3: 60°C-10-min cycle.

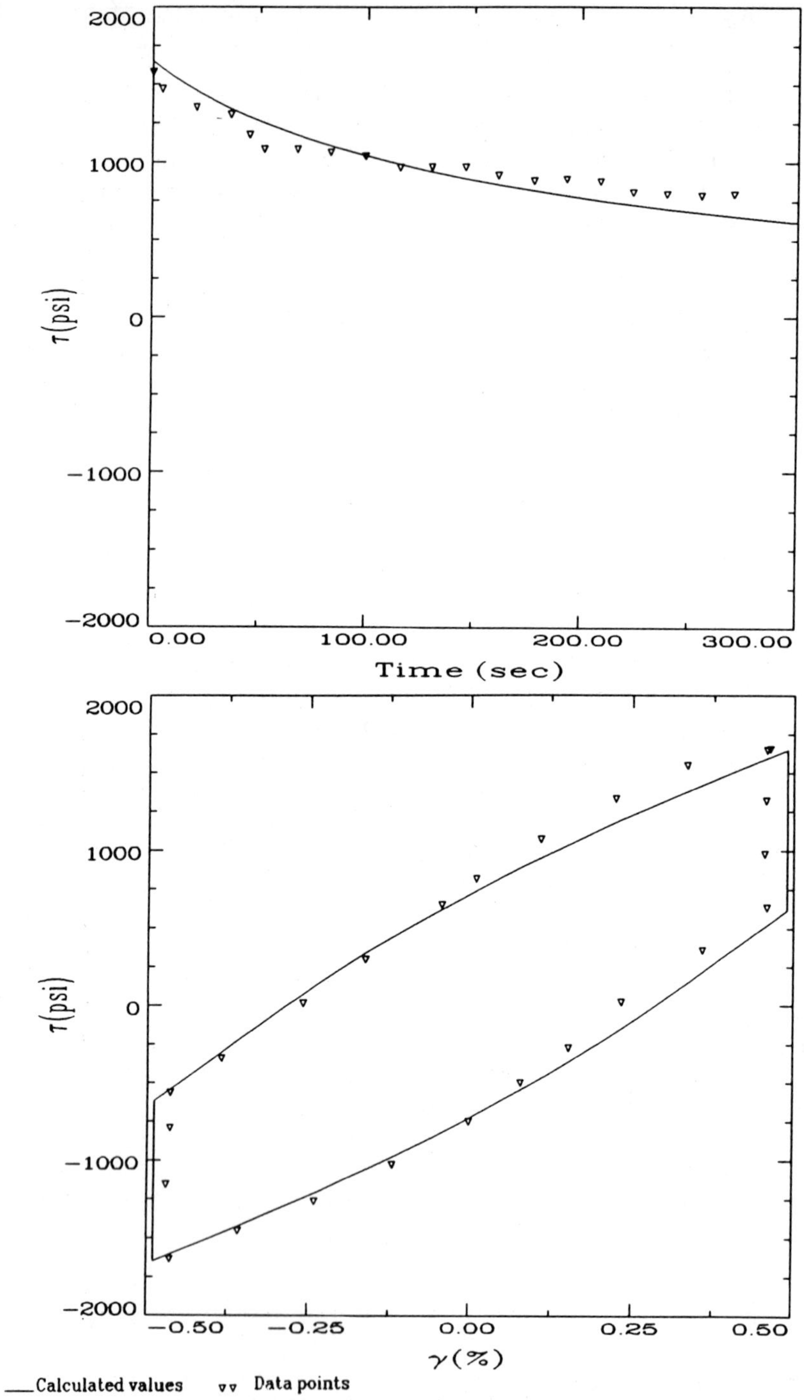

Figure 16-6 (*a* and *b*) Sample A: 60°C-10-min cycle.

points shown are somewhat approximate. The strain history is of course not a perfect square wave, although the ratio of the transition time to the hold time is much less than 1; given the uncertainty in the data, it was felt that further refinements in the numerical procedure to account for the transition period were not warranted. Still, the agreement of the calculations with the approximate data is good. Typical results are shown in Figures 16-6*a* and *b*; for this data, $G_{eff} =$ 300,000 psi and $\gamma_{tot} = 0.98$ percent.

Solutions to the constitutive equation indicate that for fast cycling (less than about 1 min) at room temperature or lower, the equation describes the solder hysteresis loop even if the creep term is neglected. For cycling of more than a few hours at constant strain rate at room temperature or higher, the instantaneous plastic strain is negligible compared to the creep strain. For intermediate cases (for example, 10-min cycles at 60°C), both creep and time independent plasticity contribute significantly to deformation.

16.6 MULTIAXIAL STRESS STATES

16.6.1 Derivation of Constitutive Equation in Three Dimensions

Uniaxial tests to determine the relation between stress and strain in one dimension are fairly easy to conduct. For most cases of interest, however, the actual state of stress is much more complex. It is common in plasticity theory to determine experimentally the behavior of a material in one dimension and then to generalize this behavior to three dimensions using some simple assumptions.

The following discussion assumes that the material is isotropic. For thin solder joints with relatively large grain sizes there often are only a few grains (possibly as few as one or two) through the thickness of the joint. In this case, the assumption of isotropy is not adequate; however, in accordance with an empirical approach (and due to lack of data) we proceed in the traditional way.

Idealizing from experimental observation, it is assumed that the hydrostatic pressure has no effect on the inelastic strains, and thus, that these strains depend only on the history of the stress deviator, $S_{ij} \equiv \sigma_{ij} - 1/3\sigma_{kk}\delta_{ij}$ (sum over repeated indices). Furthermore, zero plastic volume change is assumed. The generalization of a uniaxial relation to multiaxial states is then conventionally given in terms of the second (J_2) invariant of the stress tensor. Equivalently, the Mises or "effective stress" (σ_e) can be used where:

$$\sigma_e \equiv \sqrt{\frac{3}{2}S_{ij}S_{ij}} \equiv \sqrt{3J_2} \tag{16-11}$$

Using this approach and Hooke's law for generalization of the linear term, the complete equation for strain rate becomes

$$\dot{\varepsilon}_{ij} = \frac{1+\nu}{E}\dot{\sigma}_{ij} - \frac{\nu}{E}\dot{\sigma}_{kk}\delta_{ij} + \alpha\left(\frac{\dot{\sigma}_e}{\sqrt{3}\tau_p}\right)\frac{S_{ij}}{\tau_p}$$

$$+ \frac{C_0}{2\tau_0}\left[\left(\frac{\sigma_e}{\sqrt{3}\tau_0}\right) + \left(\frac{\sigma_e}{\sqrt{3}\tau_0}\right)^{6.1}\right]S_{ij} \quad (16\text{-}12)$$

where $\alpha = 1$ for loading ($\dot{J}_2 > 0$), $\alpha = 0$ for unloading ($\dot{J}_2 < 0$), and δ_{ij} is the Kronecker delta function. Note that $\dot{\sigma}_e = (3/2)(\dot{J}_2/\sigma_e)$. It is easily shown that this equation reduces to Equation (16-6) for pure shear since in that case, $\sigma_e = \sqrt{3}|\tau|, S_{12} = S_{21} = \tau$, and $\dot{\gamma} = 2\dot{\varepsilon}_{12}$.

In the case of pure tension the only nonzero stress component is $\sigma_{11} \equiv \sigma$. From the definition of the stress deviator, $S_{11} = 2/3\sigma, S_{22} = S_{33} = \frac{1}{3}\sigma$. Therefore,

$$\sigma_e = \sqrt{\frac{3}{2}\left(\frac{4}{9}\sigma^2 + \frac{1}{9}\sigma^2 + \frac{1}{9}\sigma^2\right)} = \sigma$$

As an example of the use of Equation (16-12), for loading in pure tension the equation becomes

$$\sqrt{3}\dot{\varepsilon} = \sqrt{3}\frac{\dot{\sigma}}{E} + \frac{d}{dt}\left(\frac{\sigma}{\sqrt{3}\tau_p}\right)^2 + C_0\left[\left(\frac{\sigma}{\sqrt{3}\tau_0}\right)^2 + \left(\frac{\sigma}{\sqrt{3}\tau_0}\right)^{7.1}\right]$$

As yet, no experimental evidence exists to confirm the validity of this generalization. For a joint with only a few grains through its thickness, the assumption of isotropy is almost certainly not warranted. However, a generalization of the constitutive equation to three dimensions is necessary for the analysis of any solder joint under thermal loading conditions where stresses are always multiaxial. The generalization presented here allows the constitutive relation to be used with commercial general purpose finite element codes as will be seen later.

16.7 FATIGUE CALCULATIONS AND MECHANICAL SHEAR TESTS

Calculation of stresses, strains, or other parameters for solder joints is not meaningful unless the values obtained can be used to predict structural reliability. The purpose of this section is to show that when the solder constitutive equation is solved for the different test conditions and solder joints used by various authors, calculated parameters can be correlated with fatigue life.

The failure indicator used here is the one proposed by Shine and Fox,[2] which correlates fatigue life with the component of creep strain due to matrix creep,

$$N_f = \frac{C}{\gamma_{MC}} \tag{16-13}$$

$$\gamma_{MC} = \int C_0 \left(\frac{\tau}{\tau_0}\right)^{n_{MC}} dt \qquad \text{(per cycle)} \tag{16-14}$$

In reference 2, C is given as 125 percent while different values of n_{MC}, C_0, and τ_0 are determined experimentally for each sample. In this work, a general constitutive law for eutectic or near-eutectic solder is determined; to be most useful for mechanical modeling, fixed values of n_{MC}, C_0, and τ_0 must often be known before testing can be done and this is the approach taken here. The value of C will then vary depending on factors such as (perhaps) solder microstructure and the way in which failure is determined to have occurred (e.g., electrical resistance increase or change in structural compliance). The effects of such parameters on fatigue life are not well understood. In this context, C can be regarded as an experimentally determined constant. Some idea of the range of values to be expected will be given below.

Parallels can be seen between Equation (16-13) and the Manson-Coffin fatigue relation for materials exhibiting no time dependent effects. The expressions are somewhat similar except that in the Manson-Coffin relation, the total plastic strain (opening of the hysteresis loop) is raised to an empirically determined power, which is usually about -2. With the Shine-Fox matrix creep approach, an inverse linear relation between fatigue life and the matrix creep part of the strain has also been determined experimentally.

Accommodation of time dependent effects within the Manson-Coffin framework remains problematic. It is generally accepted that for materials that exhibit significant time dependence, the traditional Manson-Coffin law must be modified to account for frequency effects; the exact nature of this frequency dependence has not been definitely established. The premise here is that Equation (16-13) is a good failure criterion for solder as demonstrated in reference 2. Because no special modifications need be made to account for effects such as loading history (frequency) or temperature, the equation is very general and may be considered a more fundamental approach to the problem of solder fatigue.

Similarities also exist between Equation (16-13) and the Monkman-Grant relation, which states that time to failure multiplied by steady-state creep rate for one time loading is equal to a constant. In this sense, the constant C can be thought of as the creep ductility exhaustion factor. Again, Equation (16-13) is more general than the Monkman-Grant expression.

16.7.1 Correlation of the Data of Reference 2

In the fatigue tests of reference 2, the solder joints were cycled at a particular temperature with essentially a square wave strain-time history. Opening of the hysteresis loop was kept constant by automatic adjustment of the total strain amplitude. Failure was determined when the load to maintain this opening was reduced by 50 percent.

In reference 2, a different steady-state creep relation in the form of Equation (16-3) was determined for each sample. Then the matrix creep term was integrated over the actual measured stress history to calculate the amount of matrix creep per cycle as defined in Equation (16-14). This method was the basis for the fatigue relation of Equation (16-13) with C determined as 125 percent.

To show that Equations (16-13) and (16-14) combined with the constitutive law are still a useful failure indicator for cases where test results are not yet available, the fatigue tests of reference 2 were analyzed using Equations (16-9) and (16-10) for the square wave. A value of total strain range was chosen from plots of the raw data, cycle time and temperature were known, and a value for G_{eff} was determined from a ramp test on that sample if available or from the square wave plot if not. Input data for equation solutions are shown in Table 16-4.

Matrix creep strain is easily calculated using Equations (16-9) and (16-10). Results are shown in Figure 16-7. Note that these results are somewhat (but not very) different from the values of matrix creep per cycle obtained in reference 2. A least squares fit to this data of an equation of the form of Equation (16-3) gives C equal to 170 percent, in good agreement with the value obtained by Shine and Fox. The data show the scatter that is typical in fatigue testing.

Table 16-4 Summary of Shine-Fox Data: 60°C, Square Wave τ Versus Time Relation-Raw Data for Figure 16-7

Sample Name	G_{eff} (psi)	γ_{tot} (%)	Period (min)	Substrate Material	N_f
A11	70,000	5.8	10	G10	486
A10	200,000	2.1	10	G10	526
A12	110,000	3.0	10	G10	1200
3	125,000	1.6	30	G10	3436
2	100,000	2.5	30	G10	4262
1	95,000	3.3	2	G10	5000
A1	400,000	1.6	10	Stainless Steel (SS)	320
A2	415,000	1.1	2	SS	880
BO	350,000	1.1	10	SS	1250
A	300,000	.98	10	SS	1870

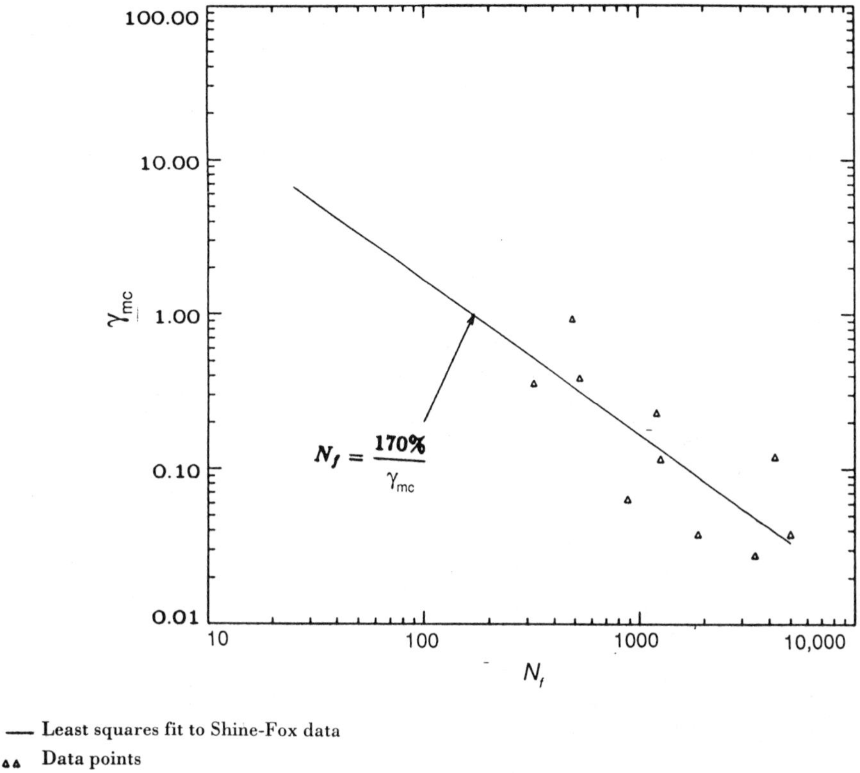

Figure 16-7 Cycles to failure versus matrix creep strain for data of Equation (16-2).

16.7.2 Correlation of the Data of Wild and Solomon

Wild[16] reports fatigue life of solder joints as a function of total strain varied as a sinusoidal function of time; each data point is an average of 30 to 50 tests, so the scatter always present in fatigue tests is minimized. Failure was determined when electrical resistance increased by 10 percent. The joints tested were lap-shear configurations using steel specimens to reduce bending. The joint sizes were 2.9 by 1.1 by 0.1 mm thick (0.115 by 0.045 by 0.004 in.).

To correlate the Wild data with the matrix creep failure indicator, it is necessary to solve the constitutive equation for each reported cycle and calculate the resulting matrix creep strain. All necessary parameters are available except for G_{eff}; somewhat arbitrarily, but based on values from other fatigue testing of solder joints, the value is taken to be 2.41 GPa (350 ksi) at room temperature and 1.72 GPa (250 ksi) at 100°C. Due to the approximate nature of these

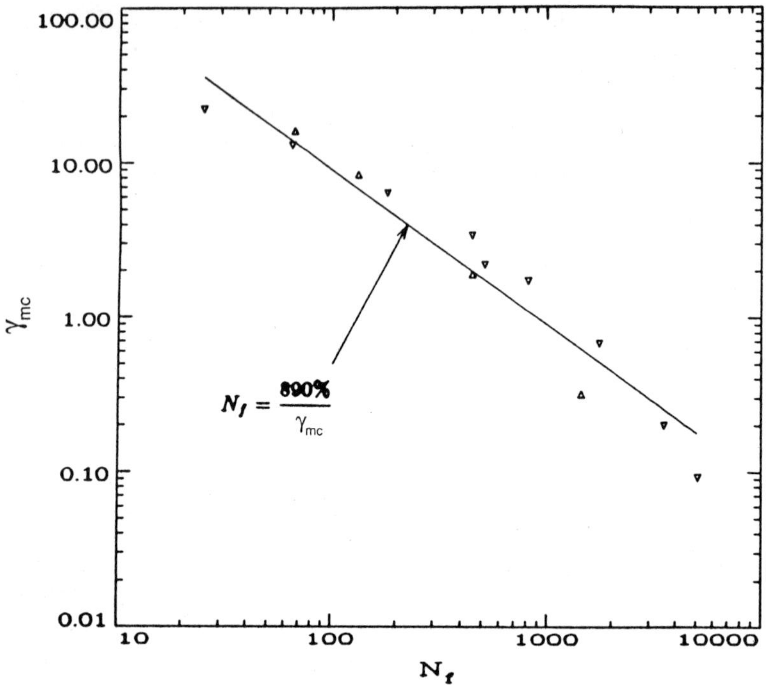

— Least squares fit to Solomon and Wild-Engelmaier Data

▲▲ Solomon Data

▼▼ Wild-Engelmaier Data

Figure 16-8 Cycles to failure versus matrix creep strain for Equations (16-5) and (16-15).

calculations, representation of the sine wave history by a triangle wave is felt to be adequate. For the longest cycles at high strain rates, numerical problems were encountered and no results are available. Engelmaier[17] has derived an expression to relate total strain to cycles to failure by fitting data from reference 16. This expression was used to determine lifetime versus total strain. Input data for data points of Figure 16-8 are given in Table 16-5.

Solomon[6] presents data giving fatigue life versus plastic strain (opening of hysteresis loop) at 35°C for rapid cycling (triangle wave at 1200 cycles/h). The cross sections of the joints are much larger than those used by the other authors considered here (2.4 by 13 mm and 0.13 to 0.23 mm thick). It is not clear what effect this different geometry may have on the fatigue life. Failure is defined in the same way as in reference 2 (50 percent load reduction). A value of G_{eff} is reported as approximately 1.38 GPa (200,000 psi) in reference 15. Equation (16-

Table 16-5 Wild Data: Triangle Wave γ Versus *Time* Relation-Raw Data for Figure 16-8

G_{eff} (psi)	γ_{tot} (%)	Period	Temperature (°C)	N_f
350,000	20	12 sec	25	25
350,000	15	12 sec	25	64
350,000	8	12 sec	25	509
250,000	8	12 sec	100	208
250,000	5	12 sec	100	804
250,000	3	12 sec	100	3500
350,000	5	15 min	25	447
350,000	3	15 min	25	1730
350,000	2	15 min	25	5070

8) can be solved for the test conditions of reference 6 using an estimated value of total strain to obtain a value of plastic strain for a cycle. Iteration gives the correct value of total strain for a desired plastic strain. Matrix creep strain for the cycle can then be calculated. The input data used is given in Table 16-6.

Figure 16-8 shows the results of correlating calculated matrix creep with experimentally determined number of cycles to failure for the Solomon and Wild data. A least squares fit to this data of an equation of the form of Equation (16-13) gives a value of C of 890 percent, higher than the value obtained for the Shine-Fox data.

It can be seen from Figure 16-3*a* that the values of the parameters C_0 and τ_0 are different for the Solomon data than they are for the Shine-Fox data and we would expect (if it were to be calculated) that this would also hold true for τ_p due possibly to different solder microstructure or geometry. In fact, Solomon, using the relation between steady-state creep rate and stress determined for his samples, has calculated the amount of creep strain in his measured hysteresis loops.[6] Using Equation (16-8) the same calculation gives the percentage of creep per cycle to be a factor of 2 higher. Using Solomon's creep calculations, Shine[18]

Table 16-6 Solomon Data: Triangle Wave γ Versus *Time* Relation-Raw Data for Figure 16-8

G_{eff} (psi)	Period	Temperature (°C)	$\Delta\gamma_{pl}$ (%)	N_f
200,000	3 sec	35	15	66
200,000	3 sec	35	10	131
200,000	3 sec	35	5.5	446
200,000	3 sec	35	3.0	1430

has shown that the Solomon data fall on the same line as the data from reference[2]. No steady-state creep data are available in reference[16].

Solomon[6] also has experimental data for lifetime versus frequency at high fixed values of plastic strain which show a sharp drop-off in lifetime as frequency becomes very low. As Solomon points out,[19] the matrix creep failure indicator predicts the opposite of this effect. As cycle period becomes very long, to maintain a fixed opening of the hysteresis loop, maximum stress goes down; since matrix creep varies as the seventh power of stress, the amount of matrix creep per cycle actually begins to decrease with decreasing frequency despite the longer time available for creep strain to accumulate, thereby predicting an increase in lifetime rather than Solomon's observed decrease. It may not be valid to apply the equations of this chapter to directly compare Solomon's data: strain values for available data that fall into the temperature range considered here (25°C $< T <$ 100°C) are much larger than those covered by this work. The lifetimes where the drop-off occurs are generally less than 100 cycles. We infer that effects other than those covered by this work are responsible for Solomon's observations. The matrix creep failure indicator was based on data at the relatively moderate temperature and strain values often found in many applications. Care must be taken in use of the matrix creep approach with due consideration given to its range of applicability.

16.8 ANALYSIS OF LEADED SOLDER JOINTS

This section applies the constitutive equation and fatigue criterion to the analysis of solder joints of a leaded chip carrier. For leadless carriers, it may be possible to assume a pure shear stress state and to use the results of simple tests to get an estimate of fatigue life. For leaded joints, however, it is known from experiments[20] that the compliance of the lead is one of the most important factors influencing lifetime. The interaction of the lead and solder is difficult to simplify and thus an in-depth stress analysis may be warranted. The leaded case differs from the simple shear results cited previously for the following reasons:

- Stresses and strains in a leaded solder joint are multiaxial: In general there are six components of stress and strain, three tension-compression and three shear. Each component may be expected to contribute in some way to damage accumulation and ultimate failure.
- The stresses and strains vary with position throughout the structure.
- The definition of lifetime in fatigue tests of actual components is, for practical reasons, almost always the onset of electrical opens and not a percentage load drop or resistance change as was the case in the pure shear mechanical tests.

At first glance, failure of the leaded solder joint may seem to be a candidate for analysis by fracture mechanics. However, the problem is extremely complex from a fracture point of view: Creep plays an important role, stresses are cyclic, and effects of different nearby materials on the thin solder layers always introduce a mixed-mode fracture problem. Each of these situations separately is difficult to treat using classical fracture mechanics, and together they are more a subject for academic research than practical design engineering.

It is, therefore, our contention that a more empirical approach will be most effective in extending previous results to the leaded case: We will define and then calculate certain "failure indicators." These indicators will be correlated with experimental fatigue data in order to determine a relation to predict expected lifetime. We will be guided in our efforts to choose failure indicators by the following:

- Fatigue failure is known to correlate with calculated strain. For low temperature fatigue calculations of more conventional structures (e.g., automobile components) some form of the empirical Manson-Coffin relation[23] is often used successfully. For the frequency dependent creep case, the relation must be modified.
- Consideration of the use of total strain as a failure indicator leads to the following paradox: For practical cases of leaded chip carriers, total displacement is imposed on the lead-joint by the deformation of the board and chip carrier and relaxation times are very long. Therefore, for stiff or compliant leads, final strains per cycle should be approximately the same. However, it is known that more compliant leads result in longer fatigue life.
- The matrix creep failure indicator described earlier provides *one* resolution to this paradox. Even though total strain per cycle may be the same for both types of leads, the strain will be accumulated at lower stress in the case of the compliant lead resulting in less matrix creep strain than for the stiff lead case. Therefore, we hope to find correlation between matrix creep strain and lifetime for the leaded case as we did for the leadless case.

Engelmaier[20] has published solder fatigue results for thermal cycling of 100-I/O ceramic chip carriers with two types of clip leads having significantly different compliance. In this section, we will attempt to model this structure and calculate failure indicators using the constitutive law and finite element analysis. In order to confidently relate these indicators to fatigue life, much more data should be analyzed than is possible within the scope of this work. The example presented here is offered as a beginning of this process.

16.8.1 Extension of Matrix Creep Failure Indicator to General Case

Multiaxial Stress States

The matrix-creep failure criterion, Equations (16-13) and (16-14), is given for the case where the only stresses or strains are presumed to be pure shear in a single direction. For practical cases, multiaxial stress-strain states will be present. The law may be generalized to three dimensions using the effective strain:

$$N_f = \frac{C}{\sqrt{3}\varepsilon^{MC}} \tag{16-15}$$

where

$$\varepsilon^{MC} \equiv \sqrt{\frac{2}{3}\varepsilon_{ij}^{MC}\,\varepsilon_{ij}^{MC}} \tag{16-16}$$

This criterion reduces to Equation (16-13) for the case of pure shear but other than that is arbitrary; however, if the material is presumed to be isotropic, the generalization should be in terms of a quantity that is invariant with respect to change in choice of coordinate axes. Instead of effective strain, maximum principal strain or another similar quantity might be substituted. It should be kept in mind that the solder joint may not be isotropic if only a few grains are present through the thickness of the joint. For simplicity, isotropy is assumed throughout this chapter.

Variation of Strains with Position

For leaded joints, not only are the strains multiaxial, but they vary with position in the structure; at some points "effective" strain may be high while at others it may be almost zero.

1. The traditional approach to this problem is to use maximum strain values as failure indicators. Maximum strains may be expected to correlate with crack initiation; however, failure would be expected to correlate also with crack growth that might be a function of the geometry and stress state of the entire structure. Using maximum values would be expected to be a conservative approach.
2. Another method to address this question has been suggested.[24] In this approach the total matrix creep to be used in Equation (16-15) is defined

as a volume weighted-average of the matrix creep strain throughout the structure:

$$\varepsilon_{avg}^{MC} = \frac{1}{V}\int \varepsilon^{MC}(x,y,z)dV \tag{16-17}$$

As before, another strain invariant could be used instead of effective strain.

The difficulty with this approach is that it may overpredict lifetimes because it takes into account the entire structure whether or not it is needed to maintain structural integrity. It is easy to imagine a structure with a part of the volume under low stress and a smaller part under high stress where a crack through the high stress region would result in failure. It may not make sense to average the strains over the entire volume in this case; however, in some cases volume average matrix creep may be found to correlate with lifetime and thereby find use for practical design engineering.

3. To eliminate volume averaging over low stress regions, one might confine the integration to regions where the stress exceeds some "trigger" level. The form of Equation (16-6) suggests that τ_0 be considered as such a threshold since it sets the stress above which matrix creep dominates total creep. With this consideration, we also carry out lifetime calculations using Mises stress $> \tau_0$. This approach might be expected to be invalid for lifetime estimates in very low stress cases, where the error in neglecting finite contributions of low stress regions becomes significant. Indeed, we see this in the following example.

Definition of Lifetime

In the tests discussed previously, Shine and Fox[2] and Solomon[5,6] have used 50 percent load reduction to define failure while Wild[16] has used 10 percent resistance change. It is necessary to relate these to the electrical open criterion (the only important one from a practical standpoint) used for the Engelmaier tests. Solomon[21] has studied this problem in depth: he concludes that a resistance increase of 0.1 percent is observed with load drops of 60 to 90 percent. In general it takes a load drop of about 90 percent before the first open circuit is reached. The difference in number of cycles between 50 percent and 90 percent load drop varies with plastic strain range and can be more than an order of magnitude in reference 21. The difference seems to be larger at smaller values of plastic strain range.

Here, we assume that the lifetime definition is taken into account by the constant C in Equation (16-13). This constant may be calculated for the electrical open case by curve fitting to experimental data as was done for the pure shear mechanical tests. The constant will also depend on which definition is used for the matrix creep parameter.

Geometrical factors related to specific joint design, affecting crack initiation and growth, may also be accommodated by the empirical determination of C. It is possible microstructural effects might also be accounted for.

We suggest that, once the effective matrix creep fatigue ductility coefficient C is determined experimentally, further lifetime calculations and sensitivity studies may be carried out on a family of similar designs without further arbitrary adjustments. Additional experimental investigation is required to fully support this assertion.

16.8.2 Description of Model

The finite element analysis of the Engelmaier[20] tests was done using the ABAQUS[25] general purpose finite element code. The finite element mesh is shown in Figure 16-9; elements are isoparametric eight-node bricks.

Most stress analysis is subject to limited computer resources and therefore problems must be simplified to get meaningful results in a timely fashion. The following simplifications were made in the finite element models used here:

- The geometry of the leads and joints is simplified. The configurations are shown in Figure 16-9. Lead geometries are taken from Kotlowitz's analysis of lead compliance for these tests[22] but modified somewhat. The lead is assumed to be placed at the middle of the pad with 3 mil of solder under it. The geometry of the joints is very approximate. Fillets in the foot-toe direction are included but those in the lateral direction are ignored (Figure 16-9).
- Effects of the leads and solder joints on the deformation of the board and chip carrier are ignored. Therefore, the displacements imposed on the solder joints are due only to the lateral deflection of the board and chip carrier caused by thermal mismatch between them. Analysis of the conservative problem of board, chip carrier, and leads with infinitely stiff solder joints shows this to be a reasonable assumption. Imposed displacements are those that would be experienced by a lead on the package corner since this is the most highly stressed lead. These displacements are shown in Figure 16-10*a* and *b* where, $\delta_y = -0.39$ mil and $\delta_x = 0.35$ mil.
- The temperature-time profile has been simplified: The structure is assumed to go instantaneously from 25 to 85°C and to hold at this higher temperature for 450 sec. The second part of the cycle where the structure cools and then holds at low temperature is not included in the analysis since it is assumed that the amount of creep strain accumulated there is relatively low. Two dimensional studies of a structure similar to the type A clip lead

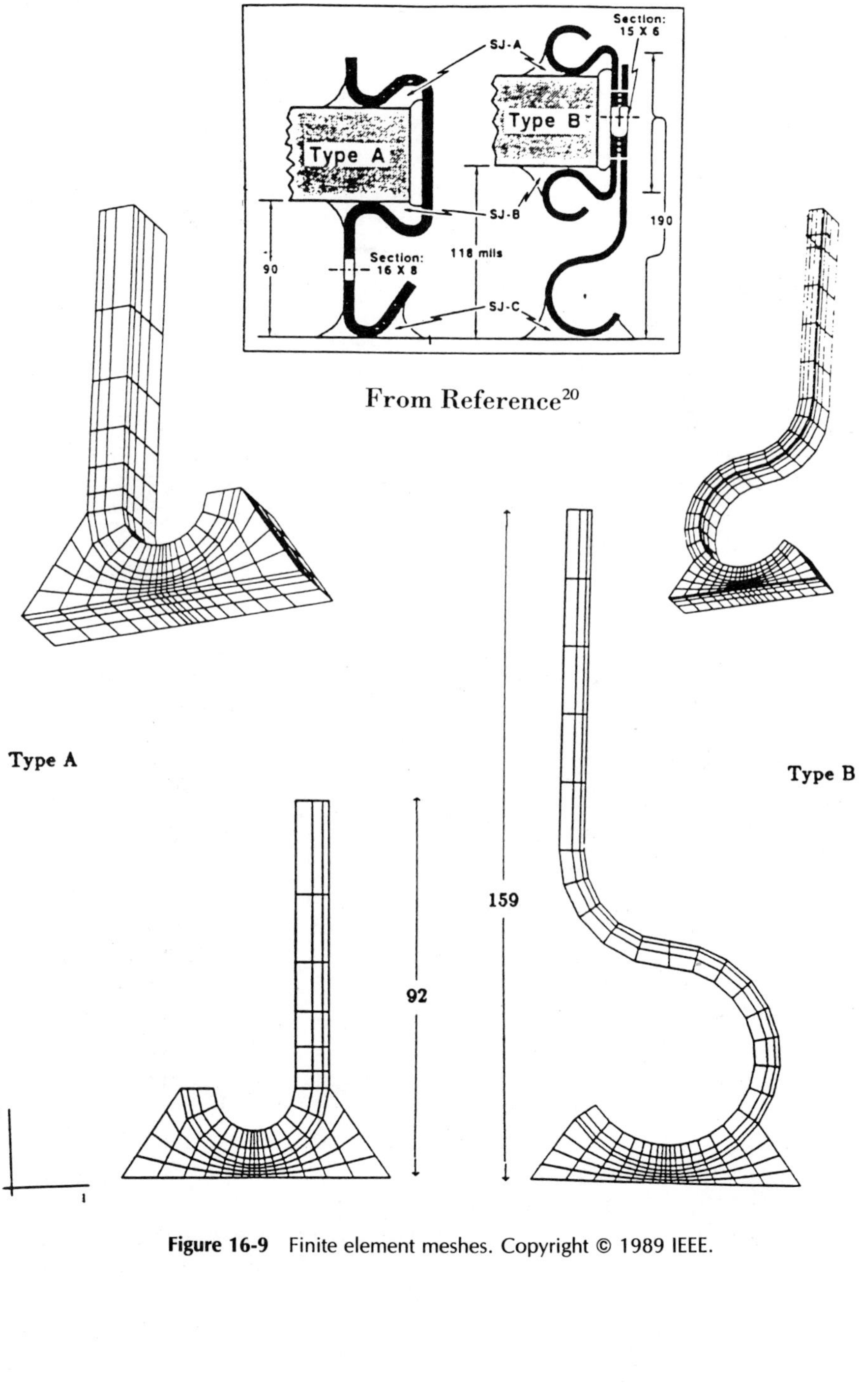

Figure 16-9 Finite element meshes. Copyright © 1989 IEEE.

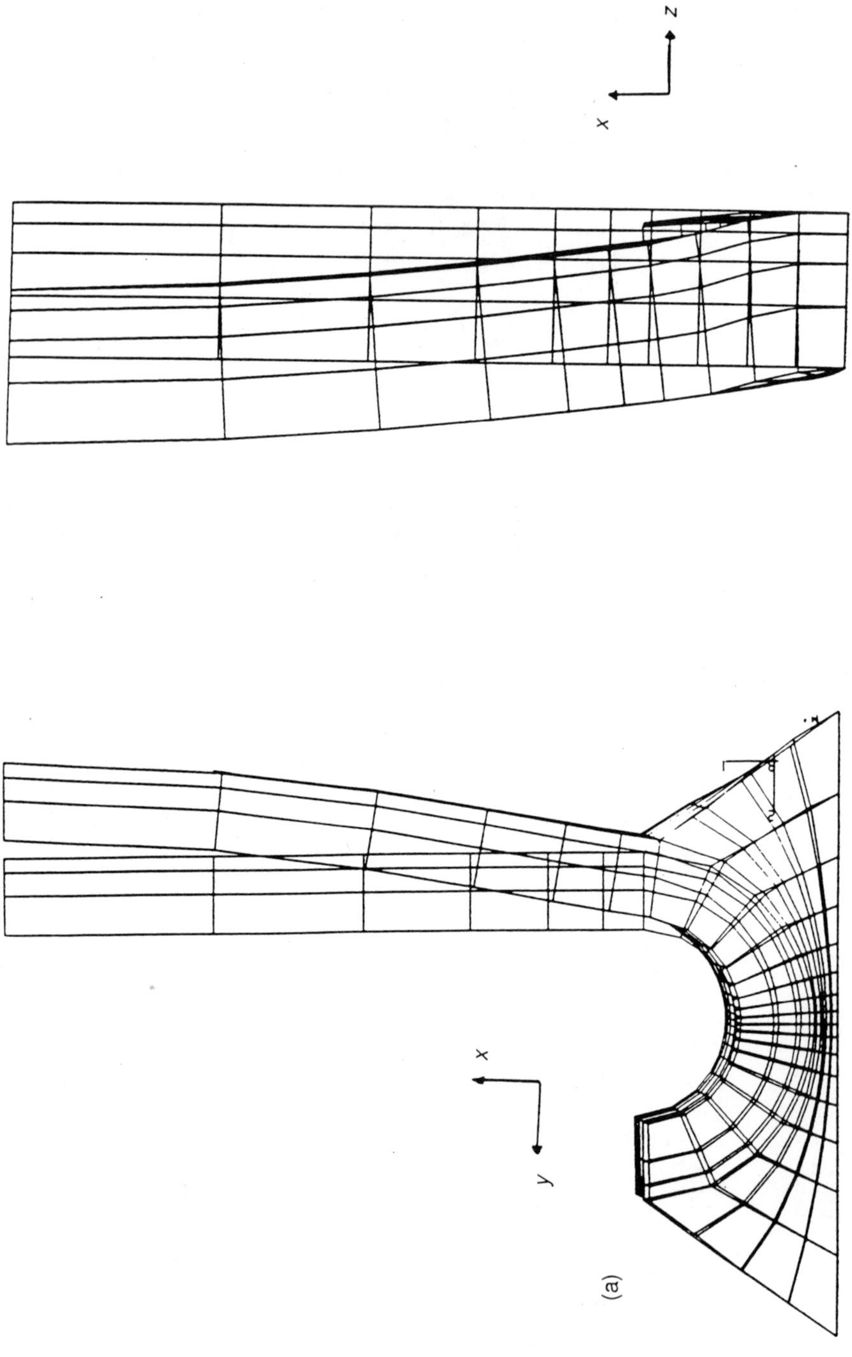
z
x
x
y
(a)

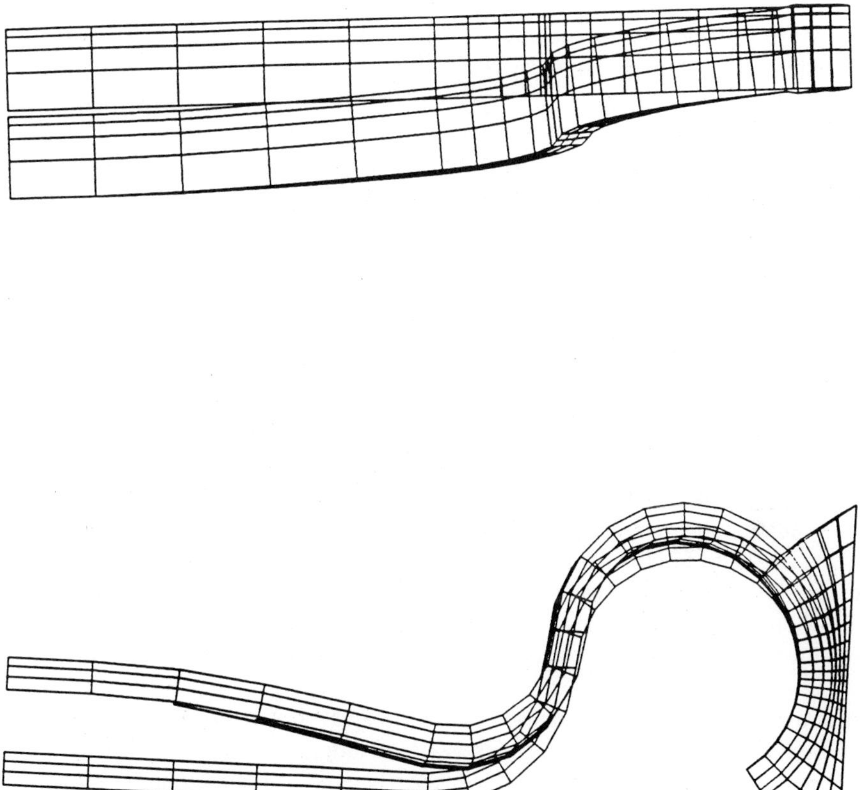

(b)

Figure 16-10 (*a*) Imposed displacements type A. (*b*) Imposed displacements type B.

and solder joint were performed to try to give a simple rough estimate of the effect of this assumption; these studies show that lifetime is decreased by about 10 to 20 percent when this part of the cycle is included. Therefore, to reduce computational run-time, only the warm half of the cycle was analyzed. This approach is consistent with the determination of an effective ductility C.

- The finite element meshes for the two cases studied by Engelmaier[20] are shown in Figure 16-9. It is well known that mesh density can have a large effect on calculated strains and that the effect of mesh refinement should always be checked. Unfortunately, limited computer resources make this difficult for 3D highly nonlinear problems. Mesh studies were carried out with a 2D mesh analogous to the type A lead-joints to guide the choice used here. Results of these 2D models show the level of mesh refinement used for the 1-2 plane in Figure 16-9 to give a maximum strain approximately within 20 percent of the asymptotic value that is obtained from further mesh refinement. This is only a rough estimate for the 3D case.

16.8.3 Results

As explained above, we regard C as an experimentally determined parameter that will vary depending on the definition of failure and on the method of extension of the matrix creep strain in pure shear to a structure with a complex stress-strain state. For the Engelmaier[20] tests, the failure event or criterion is electrical opens. The mean fatigue life is the number of cycles when 50 percent of the population has failed. The joints of the type A (stiff) leads show a mean of 20,000 cycles to failure with a Weibull slope of 2 as shown in Figure 16-11. All failures are assumed to occur at the "SJ-C" location between the lead and the FR4 board. The test was discontinued at 100,000 cycles; at that time the board with type B (compliant) leads had not experienced any failures.

- It is assumed that the calculated lifetime is equal to the mean population lifetime (20,000 cycles). Using this value of fatigue life and the "volume average matrix creep" defined above, the value of C for the type A clip lead is determined to be 1.7 (or 170 percent), which is the same as for the data of reference 2. Using this value of C and calculating volume averaged matrix creep for the type B leads, lifetime is predicted to be 860,000 cycles. This gives a value of the ratio of lifetimes between the type B and type A lead cases of 42. Since no failures occurred in the type B lead test before it was stopped, actual lifetime is not available. Assuming a failure in the next cycle after the end of the test and a Weibull slope of 2 equal to the type A clip lead case, Engelmaier estimates the minimum lifetime to be 600,000 cycles.

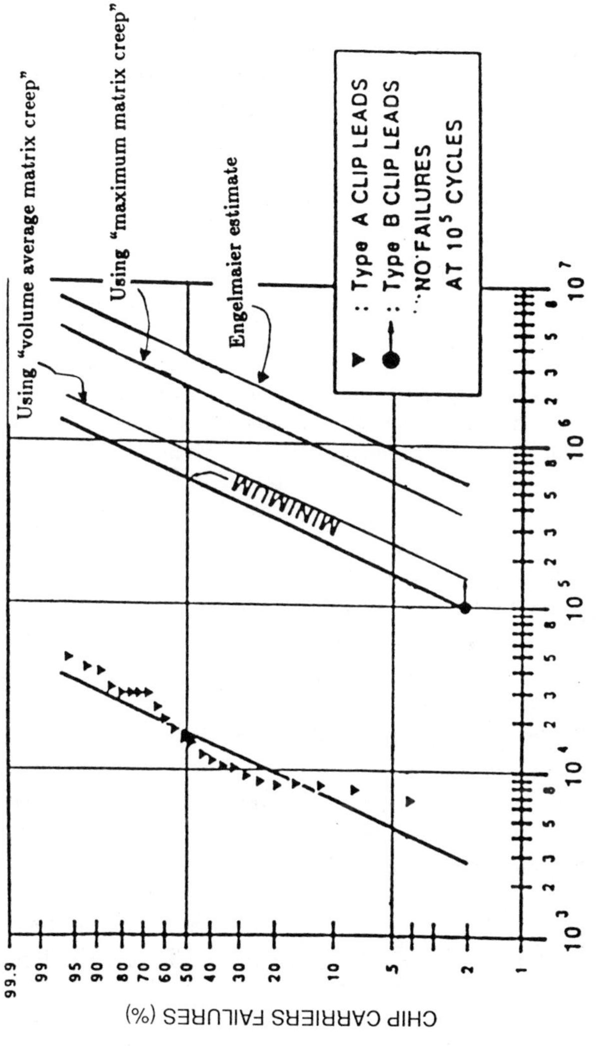

Figure 16-11 Predicted Weibull plots for type B leads. Copyright © 1989 IEEE.

Table 16-7 Summary of Results for Leaded Chip Carriers

Method	N_f Type A (data)	N_j Type B (calculated)	C	Ratio Type B / Type A
Vol. Avg. matrix creep	20,000	860,000	1.7	42
Max. matrix creep	20,000	2,300,000	137	115
Vol. Avg. MC, $\sigma_e \geqslant \tau_0$	20,000	∞	36	∞
Engelmaier	20,000	3,500,000	—	175
Actual	20,000	>600,000	—	>30

- Using matrix creep strain at that location in the structure that experiences the maximum value of effective matrix creep strain, C for the type A lead case is found to be 137. It is to be expected that this number is much higher than for the pure shear case. Stresses for the "pure shear" tests are calculated using measured total force; for an elongated joint actual stress will not be constant and therefore the value used to determine C for the pure shear mechanical tests is really an average itself. Using $C = 137$ and maximum effective creep strain calculated for the joints of the type B lead, the resulting lifetime is 2,300,000 cycles, or a ratio of 115. For comparison, Engelmaier estimates this ratio at 175 using a ratio of the relative compliances of the leads and a relationship between total strain and fatigue life based on the simple shear mechanical tests of Wild.[16] The Weibull plots of fatigue failure for the type A solder joints and projections for the type B joints are shown in Figure 16-11.
- Using volume averaging over that volume whose stress is about τ_0 results in a value of $C = 36$ for the type A clip lead. For the type B lead, calculated lifetime is infinite since there is no region where Mises stresses exceed τ_0.

Each of these methods give results that are consistent with available test data. More information is needed to choose the most appropriate failure indicator. Results are summarized in Table 16-7.

16.9 CONCLUSIONS

A stress-strain relation for a typical solder joint has been developed from shear tests and generalized to three dimensions. The equation accounts for elastic, time independent plastic and steady-state creep components of solder strain. Solutions to the equation have been developed for triangular and square wave strain versus time profiles; these solutions have shown that the equation gives a very good description of the Shine-Fox[2] data over a range of conditions.

The constitutive equation has been solved for pure shear mechanical test conditions where fatigue life has been measured. The matrix creep failure indicator proposed in reference 2 (Equations 16-13 and 16-14) has been calculated for three independent sets of such solder fatigue data; it has been shown that the slope of the logarithm of cycles to failure versus the logarithm of matrix creep curve is about -1 for all data considered. The effective matrix creep ductility varies depending on the data being considered. While it is necessary to determine this constant experimentally, relative design comparisons can be made on this basis before testing. Since it has been shown that the slope of the N_f versus γ_{MC} curve does not vary, fewer tests are necessary to predict fatigue life for a given construction.

Finally, some possible extensions of the matrix creep failure indicator to the case of leaded chip carriers are proposed. As an example, these criteria are applied to the analysis of thermal tests of clip leaded ceramic chip carriers conducted by Engelmaier. The example shows that unresolved issues exist in the application of matrix creep to the more complex stress states of leaded solder joints. With more data, some resolution of the questions posed can be expected.

16.10 ACKNOWLEDGMENTS

The authors wish to thank B. Mirman and M. C. Shine for their helpful discussions of this work.

REFERENCES

1. Lau, J. H., D. W. Rice, and P. A. Avery, "Elastoplastic Analysis of Surface-Mount Solder Joints," *IEEE Trans. Comp., Hybrids, Manuf. Technol.*, **CHMT-10,** 1987, pp. 346–357.
2. Shine, M. C., and L. R. Fox, "Fatigue of Solder Joints in Surface Mount Devices," *Low Cycle Fatigue—ASTM Special Technical Publication 942,* 1987, pp. 588–610.
3. Grivas, D., K. L. Murty, and J. W. Morris, "Deformation of Pb/Sn Eutectic Alloys at Relatively High Strain Rates," *Acta Met.,* **27,** 1979, pp. 731–737.
4. Kashyap, B. P., and G. S. Murty, "Experimental Constitutive Relations for High Temperature Deformation of a Pb/Sn Eutectic Alloy," *Materials Science and Engineering,* **50,** 1981, pp. 205–213.
5. Solomon, H. D., "Creep, Strain Rate Sensitivity and Low Cycle Fatigue of 60/40 Solder," *Brazing and Soldering,* no. 11, 1986, pp. 68–75.
6. ———, "Low-Frequency, High Temperature, Low-Cycle Fatigue of 60Sn-40Pb Solder," *Low Cycle Fatigue—ASTM Special Technical Publication 942,* 1987, pp. 342–370.
7. Baker, E., "Stress Relaxation in Tin-Lead Solders," *Materials Science and Engineering,* **38,** 1979, pp. 241–247.

8. Lam, S. T., A. Arieli, and A. K. Mukherjee, "Superplastic Behavior of Pb-Sn Eutectic Alloy," *Materials Science and Engineering,* **40,** 1979, pp. 241–247.
9. Wong, B., D. Helling, and R. W. Clark, "A Creep-Rupture Model for Two Phase Eutectic Solders," *IEEE Trans. Comp., Hybrids, Manuf. Technol.,* **CHMT-11,** 1988, pp. 305–325.
10. Clech, J. P., and J. A. Augis, "Temperature-Cycling, Structural Response and Attachment Reliability of Surface-Mounted Leaded Packages," *1988 IEPS,* Dallas, November 1988, pp. 305–325.
11. Westinghouse Defense and Space Center, "Development of Highly Reliable Soldered Joints for Printed Circuit Boards," Aug. 1968.
12. International Tin Research Institute, "Solder Alloy Data," I.T.R.I. Publication No. 656.
13. Adams, P. J., "Thermal Fatigue of Solder Joints in Micro-Electronic Devices," M.S. Thesis, MIT, 1986.
14. Engelmaier, W., "Test Method Considerations of SMT Solder Joint Reliability," *IEPS* October 1984, pp. 360–369.
15. Solomon, H. D., "Low Cycle Fatigue of 60/40 Solder-Plastic Strain Limited vs Displacement Limited Testing," *Electronic Packaging: Materials and Processes,* J. A. Sartell, Eds., ASM, 1989, pp. 29–47.
16. Wild, R. N., "Some Fatigue Properties of Solders and Solder Joints," IBM Report No. 74z000481; also INTERNEPCON, Brighton, England, October 1975.
17. Engelmaier, W., "Fatigue of Leadless Chip Carrier Solder Joints During Power Cycling," *IEEE Trans. Comp., Hybrids, Manuf. Technol.,* **CHMT-6,** 1983, pp. 232–237.
18. Shine, M. C., private communication.
19. Solomon, H. D., private communication.
20. Engelmaier, W., and A. I. Attarwala, "Surface-Mount Attachment Reliability of Clip-Leaded Ceramic Chip Carriers on FR-4 Circuit Boards," *IEEE Trans. Comp., Hybrids, Manuf. Technol.,* **CHMT-12,** 1989, pp. 284–296.
21. Solomon, H. D., "Low Cycle Fatigue of Surface Mounted Chip Carrier/Printed Wiring Board Joints," *1989 ECC,* pp. 277–292.
22. Kotlowitz, R. W., "Comparative Compliance of Representative Lead Designs for Surface Mounted Components," *1989 ECC,* pp. 791–831.
23. Knott, J. F., *Fundamentals of Fracture Mechanics,* New York, John Wiley, 1973.
24. Iannuzzelli, R., "Validation of Module Assembly Physical Models," *40th Electronic Components and Technology Conference,* May 21–23, 1990.
25. Hibbitt, Carlsson and Sorensen Inc., ABAQUS V 4.8, 1989.

17

Solder Attachment Reliability, Accelerated Testing, and Result Evaluation

Werner Engelmaier

17.1 THE RELIABILITY OF ELECTRONIC ASSEMBLIES AND SOLDER JOINT RELIABILITY

The reliability of electronic assemblies depends on the reliability of their individual elements—e.g., integrated circuit (ICs); active components: chip carriers (CCs), small-outline integrated circuits (SOICs), small-outline transistors (SOTs), dual in-line packages (DIPs), small-outline J-leaded devices (SOJs), etc.); passive components (chip resistors, chip capacitors); printed wiring boards (PWBs); connectors; etc.—and the reliability of the mechanical, thermal, and electrical interfaces (or attachments) between these elements. These interfaces can be connector contacts, wire-wraps, thermal shunts, bonding layers between a printed circuit board and a heat sink, leads soldered into plated through holes (PTHs), chip solder attachments (flip-chip or C4), or surface mount (SM) solder joints.

In this context reliability is defined as: *the ability to function as expected for an expected period of time without exceeding an expected acceptable failure probability*.

It should be noted that the chapter title refers to the reliability of solder attachments, not the reliability of solder joints. A solder joint in isolation is neither reliable nor unreliable; it becomes so only in the context of the individual elements that are connected via the solder joints and the characteristics of these elements together with the use environment, the design life, and the acceptable failure probability for the electronic assembly.

Among the different solder connections, surface mount solder attachment is unique since the solder joints not only provide the electrical interconnection but are also the sole mechanical attachment of the electronic components to the printed wiring board. Thus, surface mount solder attachment requires that special reliability concerns be addressed and that explicit design attention is provided to assure long-term reliability.

An additional complicating factor is the fact that at typical operating temperatures solder behaves in a fashion more akin to a very viscous fluid than a typical structural material. It creeps and stress-relaxes readily and can be thought of as having a load-bearing capability on a strictly temporary basis only. Solder joint strength, beyond a rudimentary minimum level easily obtained with all but grossly defective solder joints, is not of concern for the long-term reliable functioning of most surface mount solder attachments.

Surface mount solder joints subjected to cyclic thermomechanical strains and stresses in operation, accumulate fatigue damage in the solder which, in time, results in fatigue cracks and eventually in mechanical and electrical failure of the solder attachment. Thus, surface mount solder joints raise the concern of premature wear-out failure for inadequate designs.

17.1.1 "Bathtub" Reliability Curve—Hazard Rate Model

Reliability experiences with electronic components have given rise to the "bathtub" reliability curve as shown in Figure 17-1, where the instantaneous failure rate (hazard rate) in FITs (failures in 10^9 device hours) is plotted against the operating time in hours.[1] The bathtub curve has three failure regions with distinct characteristics, which are associated with failures due to "infant mortality," random failures in "steady state" operation, and failures due to "wearout." Normal operation of mature product defines the low steady-state (random) failure region. The failure rate in this region is in part dependent on the complexity of the product. In the wearout phase the accumulated damage from normal operation becomes large enough to cause rising failure rates, signaling the approaching end of useful life. The wearout phase is primarily dependent on the choice of the design parameters and the use conditions.

Infant mortality comprises the operational phase from start-up to typically no longer than one year of operation and is characterized by an initially high but rapidly decreasing failure rate. An early high infant-mortality failure rate is typically associated with a small fraction of the population which consists of "weak" or "defective" "freaks" and is dependent on the process controls during manufacture and the quality of the resulting product. Appropriate burn-in or environmental stress screening (ESS) procedures can reduce these "infant mortality" failures. The infant mortality failure rates are characterized by[1,2]

$$\lambda(t) = \lambda_1 t^{-\alpha} \qquad (17\text{-}1)$$

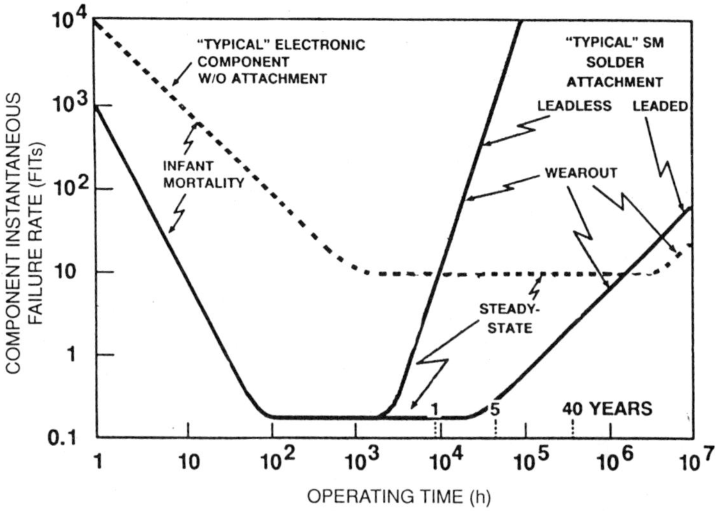

Figure 17-1 Reliability "bathtub" curves for "typical" SM solder attachments and a "typical" electronic component. The curve for the "typical" component is based on test and field failure data. For the solder attachments the infant-mortality and steady-state phases are "best, educated" guesses, while the wearout phases are based on the material in the section "Analytical Model of Solder Shear Fatigue." The location of the wearout curves is highly dependent on the choice of design parameters and the use environment to the extent that the wearout region bands shown are not wide enough to include even all likely let alone possible wearout scenarios.

where $\lambda(t)$ is the instantaneous failure rate in FITs, λ_1 the failure rate at $t = 1$ hour, t the operating time in hours, and $-\alpha$ the Weibull slope. MIL Handbook 217D[3] gives for reflowed solder joints $\lambda_1 = 18$ FITs per joint and $\alpha = 0.6$. It should be noted that these values, used to plot the infant mortality region for a "typical" surface mount solder attachment (56 solder joints) in Figure 17-1, while widely used,[3] are rather "soft." The low failure rates preclude experimental determination and hard field data in adequate quantities do not exist for either the component or solder attachment infant mortality curves.

The random steady-state region is characterized by a constant, or near constant, failure rate of mature assemblies in normal operation. The failures are of course, neither random nor steady state and have meaning only in a purely statistical sense; they cannot be characterized mechanistically, but failure rates tend to be higher for more complex components. Chip components characteristically fall well below the indicated bathtub curve in Figure 17-1, whereas very complex, high I/O chip carriers could fail at higher rates. For solder joints no evidence is available that shows that the "random steady-state" region even exists; it is entirely possible that "infant mortality" and "wearout" form the entire active failure rate history. Nevertheless, Reference 3 gives a steady-state failure rate for reflowed solder joints of $\lambda_L = 0.07$ FITs per joint; given the experience with chip carriers having surface mount solder attachments consisting of 44 to 100 individual solder joints, this value for λ_L appears too high by perhaps as much as one or two orders of magnitude. Large electronic systems can contain hundreds of thousands of solder joints; a steady-state failure rate of that magnitude would make such systems impossible. For Figure 17-1, $\lambda_L = 0.0035$ FITs per joint was assumed.

The most important aspect illustrated in Figure 17-1, however, is the difference in the "wearout" regions of the "typical" electronic component without solder attachment and the "typical" surface mount solder attachment. Wearout is characterized by a failure rate that increases monotonically with time. It results from the accumulation of damage from some mechanism and represents the point in the operational life where the amount of accumulated damage has reached levels that cause failure. For electronic components, wearout during the intended service life is encountered only rarely; of course, given the steady-state component failure rate, the number of components in a system, and the years of service, system failures caused by component failures even at low failure rates could become rather significant. Whereas, for electronic components from reputable suppliers and which are operated in recommended fashion "wearout" failures are not expected during the intended operational life, this is not necessarily true for surface mount solder attachments. Without appropriate design attention, surface mount solder attachments, particularly of the type without compliant leads, can easily be the cause of premature wearout failures. The wearout regions in Figure 17-1 are based on the experimental results from accelerated reliability tests that have given rise to a solder joint reliability model described later. The appropriate design steps which can assure that significant wearout of surface mount solder attachments does not occur during the intended operational life of the electronic assembly, are outlined here and in Chapter 18, "Surface Mount — Attachment Reliability."

Consistent good quality solder joints should result in significantly lower infant mortality and (quasi-) steady-state failure rates. However, depending on the choice of attachment technology (leadless versus leaded), the design parameters, and the severity of the use environment, wearout failures can start to assert themselves about the same time as the components reach the steady-state phase. Depending on the rigidity of the SM attachment (leadless or leaded), the wearout failure rates will rise to exceed the steady-state failure rate of the component. While, to perhaps "mid-life," components might dominate systems' failures, solder joint failures could be the determining failure mode toward the end of the design life.

From Figure 17-2, which shows the information from Figure 17-1 in terms of the cumulative component failure probability versus operating time, it is clear that for the first couple of years of operation the reliability, or more accurately any lack of reliability, is dominated by component failures, albeit at low failure

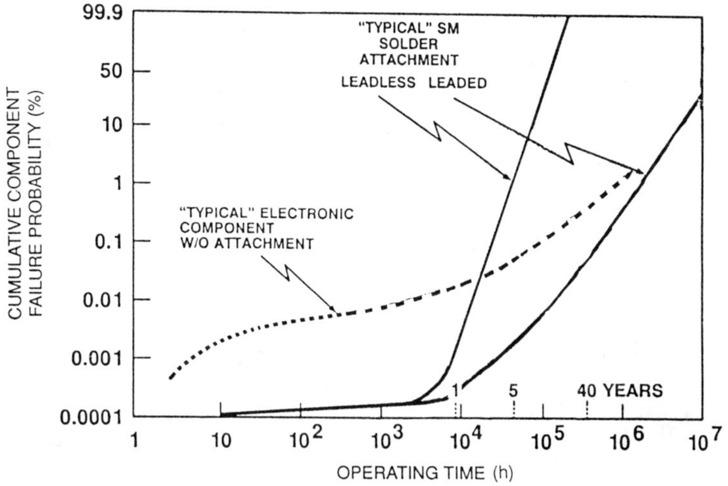

Figure 17-2 Cumulative failure probabilities for a "typical" electronic component and a "typical" surface mount solder attachment. The cumulative failure probability for solder-attached components is the sum of the components and attachment curves.

probabilities of less than about 0.5 percent. If premature wearout is not prevented by an adequate "design for reliability" that takes into account the use environment and the intended life, then failures due to wearout of the surface mount solder attachments can cause rapidly increasing failure probabilities signifying the end of useful life before reaching the desired product life. The cumulative failure probability of a surface-mount-solder-attached electronic component is the sum of the instantaneous failure probabilities of the component without attachment and of the surface mount solder attachment. The level of failure probability that is acceptable depends on the criticality of function of the system and the number of components in the system.

17.2 SOLDER JOINT LOADING CONDITIONS AND RELIABILITY

The reliability of the solder joint attachments of electronic components surface mounted to circuit board substrates requires explicit attention in the design phase as well as during manufacturing. Surface mount (SM) solder joints in use can be subjected to a variety of loading conditions that can lead to premature failure in inadequate designs. The following loading conditions could exist either singly, sequentially, or simultaneously: (1) cyclic differential thermal expansion, (2) vibration (transport and mobile use), (3) thermal shock (rapid temperature change causing transient differential warpages) during cooling from soldering (reflow) operations or from severe use environments, and (4) mechanical shock (high acceleration) from accidental misuse or severe use conditions.

While vibration, thermal shock, and mechanical shock do occur in practical applications, they are the exceptions rather than the norm in terms of the main cause of failure; the failure mechanism of primary concern for SM solder joint reliability is cyclic differential thermal expansion that results in fatigue damage. It should be noted that the resistance against failures induced by vibration, thermal shock, and mechanical shock, as well as low operating temperature cycles depends primarily on solder joint strength because these failures are stress-driven. This is not the case, however, for fatigue failure resistance at typical operating temperatures. Fatigue failures, which are primarily strain-driven, are prevented by minimizing the cyclic fatigue damage by reducing or/and accommodating the cyclic expansion differences.

Consequently, this chapter deals primarily with the long-term fatigue reliability of SM solder joints. While these different loading conditions—and the associated damage mechanisms—need to be considered separately, their cumulative damage needs to be combined for complete reliability estimates (see the section "Multiple Cyclic Load Histories").

SM solder joints can be subjected to considerable cyclic strains during operation caused by different thermal expansions of:

1. The surface mounted components and the substrates to which they are solder-attached
2. The solder and the materials (component, lead, substrate) to which the solder is bonded

The thermal expansion differences for the first cause result from differences in temperature due to power dissipation internal to the component in combination with system internal or external temperature variations. These temperature variations are caused by component or system load fluctuations or on/off cycles, by diurnal cycles, and by seasonal changes. Differences in the coefficients of thermal expansion (CTE) of the materials involved in the solder attachment system together with these temperature variations result in the thermal expansion differences.

Appropriate design for reliability measures for the first cause can take one of two forms, which are best employed in combination for improved reliability margins. These measures are:

1. Tailoring the CTEs of substrate or/and component
2. Increasing the attachment compliancy

The thermal expansion differences for the second cause result from the difference in the CTEs of the solder itself and the material(s) to which it is bonded.

While this cause can have a significant contributory effect to cause 1, by itself it becomes a reliability threat only when the ΔCTE is large and the use environment is very severe. The appropriate design for reliability measure, besides reducing ΔCTE, is keeping the solder joint attachment (interface) dimensions as small as possible. It should be noted that this failure cause is stress-driven (i.e. it results from overstress) and is compounded by the fact that the highest stresses from this cause occur at the solder joint-metallization pad interface that typically consists of brittle intermetallic compounds of tin and the metallization metal.

Assessing the reliability potential of a design and assuring the reliability of an electronic assembly in use requires knowledge of the use environment(s), the resulting loading conditions and the accumulating fatigue damage. It further requires an analytical relationship that relates the accumulating fatigue damage to cyclic fatigue life in general, generic terms with the option of including specific characteristics of the design in question that might also affect reliability. This empirical model and, especially, the non-generic design-specific characteristics need to be verified or established, which requires a well-designed and coordinated accelerated reliability test program. Lastly, an acceleration transform is necessary that allows valid comparisons of results from different accelerated test conditions and that, more importantly, allows the extrapolation of these accelerated test results to predict the reliability of the product in use. Given the time, temperature, and stress dependent behavior of solder, these relationships are anything but straightforward. Furthermore, some of the reliability and accelerated testing consequences are not intuitively obvious and can be rather subtle. In particular, the connection between accelerated test results and field reliability prediction is frequently misunderstood. These aspects will be discussed in this chapter.

17.3 RELIABILITY AND ACCELERATED TESTING—OVERVIEW

In Figure 17-3 the steps necessary for SM solder attachment reliability assurance are given in the form of a flow chart.

A generic understanding of reliability and the underlying failure mechanisms is the first step toward designing and assuring product reliability. For this, a generic database is required, one that is based on a combination of low-acceleration (acceleration factors of ~ 10 to 20 in terms of mean-time-to-failure) fatigue tests that closely mimic expected thermal cyclic field conditions and high-acceleration (acceleration factors ~ 100 to 500) "quick and dirty" fatigue tests. Thus, in the low-acceleration tests all effects that have an impact on long-term product reliability should be present and only the effects of the reduced dwell times at the cycle extremes need to be scaled to estimate product reliability from the accelerated test results. The high-acceleration tests are a compromise between the realities of insufficient time and limited resources that do not permit the more

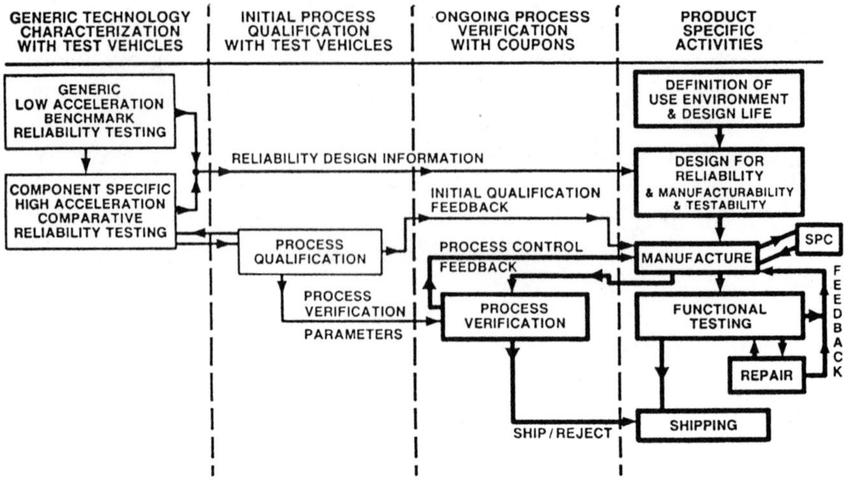

Figure 17-3 Flow chart for surface mount solder attachment reliability assurance, indicating the need for a generic (thin outline) technical understanding to develop the tools for a design for reliability, and the need for product specific (heavy outline) design for reliability as well as for ongoing process verification and product control.

accurate low-acceleration tests to be performed on all product and the need, sometimes based more on emotional rather than technically justifiable arguments, to have reliability data on all possible product variations. However, the test limitations and the much higher potential for inaccurate, even misleading, results must be clearly kept in mind.

For the accelerated tests, it is imperative that the failures occur as the result of accelerating the proper failure mechanisms and not as the result of a failure mechanism unique to the test conditions. Thermal shock testing, in particular, does *not* accelerate the long-term fatigue mechanisms of concern for most SM applications and can produce misleading results. In this context thermal shock is defined as a temperature change rapid enough to cause transient unequal or even opposite warpages of components and substrate due to the transient temperature gradients caused by the rapid transfer of the test vehicles from hot to cold test environments and vice versa. The underlying driving mechanism causing most of the long-term solder joint fatigue damage in product in operation results from the creep and stress relaxation behavior of solder. Therefore, accelerated fatigue testing must include well-defined dwell times at the cycle extremes to allow for some, albeit incomplete, creep and stress relaxation. The incomplete cyclic creep and stress relaxation inherent in accelerated test cycles is in contrast to actual field operational environments, where sufficient time is usually available for essentially complete creep and stress relaxation. Thus, the significantly smaller fatigue damage resulting from accelerated test cycles must be accounted for with a fatigue acceleration transform.

It also needs to be understood that the strongly temperature, time, and stress-level dependent material behavior of many of the materials involved in electronic packaging makes accelerated tests with temperature extremes substantially different from actual cyclic use conditions questionable at best. Temperatures near the dielectric glass transition temperatures ($> +100°C$ for FR-4) or low temperatures where stress relaxation for solder becomes negligible ($< -20°C$) render results from tests employing temperatures outside this range useless and frequently misleading for most actual applications.

The generic understanding generated from the accelerated fatigue tests must be translated into reliability design information that can be used for design for reliability. The use environment, the expected design life, and the acceptable probability of failure at the end of this design life must be established to set the reliability goals for the design.

17.4 THE THERMAL EXPANSION MISMATCH PROBLEM

The reliability problem caused by the mismatch in thermal expansion of component and substrate, given the cyclic variations in the thermal conditions during functional use, results not so much from the cyclically recurring strains in the solder joints but from the fact that solder responds to applied strains and stresses by time dependent, albeit relatively rapid, plastic deformation. Metals undergoing cyclic strains sustain cyclically accumulating fatigue damage. Fatigue damage due to plastic strains is significantly larger than the damage from elastic strains. Solder at typical operating temperatures is being used significantly in excess of 60 percent of its absolute melting point. Thus, solder is subject to significant stress relaxation and creep, processes that convert elastic strains to plastic strains but which are strongly time, temperature, and stress dependent. Thus, in all but very rapid cycling without dwell or hold times and/or operation at low temperatures ($< -20°C$), plastic deformation dominates and determines the fatigue life of solder joints. The cyclic fatigue damage is proportional to the area circumscribed by the cyclic hysteresis loop in a stress-strain diagram.[4–9] The reliability hazard can be reduced by decreasing the hysteresis loop size; reducing the total stress range, the total strain range, and the magnitude of the cyclic plastic strains accomplishes this reduction. The importance of the various design parameters affecting reliability to first-order and the necessary trade-offs between sometimes interdependent and competing design parameters are shown in the section "Analytical Model of Solder Shear Fatigue."

Methods to reduce fatigue damage are CTE-tailoring and increased attachment compliancy. CTE-tailoring reduces the cyclic hysteresis loop primarily by reducing the total cyclic strain range, while attachment compliancy primarily works by reducing the cyclic stress level at the solder joint. These methods are described below.

17.4.1 Solution 1: CTE-Tailoring to Reduce Expansion Mismatch

In CTE-tailoring, the coefficients of thermal expansion of substrate and/or components are tailored, by deliberate design or material choice, to result in the minimum steady-state component-substrate expansion differential resulting from the expected temperature variations and temperature gradients.[9] This is significantly different from CTE-matching—attempting to make the CTEs of substrate and components equal—which was forced upon vendors in order to meet unrealistically extreme MIL-spec temperature cycling acceptance criteria (e.g., −55 to +125°C cycling)[10] and which results in less than optimum reliability for actual use conditions. It also needs to be realized, however, that the tailored component-substrate CTE difference can only be optimum for one component—preferably the component representing the largest reliability hazard due to large size, leadless attachment, large power dissipation, etc.—and has to be suboptimum for all others. For these other components, the reliability hazard should be lower or equal due to other design features, such as smaller size and compliant leads. Solder joints of even perfectly CTE-tailored component-substrate combinations will ultimately fail in fatigue because:

1. Material variations (patch-to-patch, x-y directionality) in product produce ranges of CTE that can result in significant CTE-mismatches.
2. CTE-tailoring does not eliminate transient cyclic strains.[6]
3. The expansion mismatch between the solder (~25 ppm/°C) and the material to which it is attached[11]
4. The expansion mismatch of the α and β phases within the solder itself

While these causes of fatigue damage can result in sizeable damage contributions, they are not likely to cause failures by themselves within the design life of a typical product. This is true with the exception of very large CTE mismatches between the solder and the attachment material in combination with severe use environments.

For the more severe use environments (e.g., Use Category 9, Transportation—Under Hood, see section "Testing Considerations"), it is possible that the reliability threat comes from the expansion mismatch between the solder itself and the materials to which it is bonded. The reliability problem needs to be assessed using an interfacial stress analysis[12] and appropriate accelerated testing.

For ceramic components mounted to ceramic substrates, the steady-state CTE-tailoring occurs virtually automatically since the CTEs for ceramic chip carriers and ceramic substrates are typically in the ranges of 5.4 to 6.7 and 6.0 to 8.3

ppm/°C, respectively.[6] Active components, whether forced air or conduction cooled, always operate at a temperature higher than the substrate; therefore, the substrate CTE should be somewhat higher than the component CTE. Fine CTE-tailoring and CTE-matching, while theoretically possible, are in practice very difficult due to the inherent CTE variations in manufactured parts. However, the limitations of ceramic substrates—small size due to brittleness, electrical losses due to dielectric properties—provide strong incentives for organic-based substrates.[13] Using constraining layers (e.g., copper-clad Invar, copper-clad molybdenum, copper-infused molybdenum,[14] graphite fibers, Kevlar fibers), the CTE for FR-4 epoxy-glass substrates (which can range from 16 to 24 ppm/°C, depending primarily on resin content) has been reduced to as low as 8.6 ppm/°C by substituting copper-clad Invar layers for the copper power and ground layers[6,15] and 6 ppm/°C by substituting Kevlar for the glass fiber reinforcements.[16]

Conversely, plastic leaded components have a CTE range (20 to 23 ppm/°C) falling within the CTE range of FR-4 substrates. Therefore, like ceramic-on-ceramic assemblies, plastic components are inherently CTE-tailored. Plastic leaded components in addition have leads that provide varying degrees (dependent on design and material) of compliancy[16–21] to accomodate expansion mismatches. Compliant leaded attachments are described in depth below.

17.4.2 Solution 2: Attachment Compliancy to Accommodate Expansion Mismatch

Attachment compliancy[11,16,19,21] comes primarily from suitably designed component leads, and to a much smaller degree can be assisted by compliant surface pad layers on the substrate.[22] For designs requiring hermetic (ceramic) chip carriers mounted directly to conventional (not CTE-tailored) FR-4 substrates, the large differential expansions must be accommodated by the compliancy of the attachment design. The solution of choice is the attachment of the ceramic chip carrier to the substrate using edge clip-leads, but even highly compliant lead designs are inadequate for large expansion mismatches and long cyclic lives.[7,11] In addition, highly compliant leads can present manufacturability problems due to easy accidental bending of the leads and functional problems due to high electrical losses at high signal speeds from the long lead length.

Leads reduce the cyclic fatigue damage in the solder joints by reducing the cyclic stresses in the solder. A differential expansion between component and substrate causes strains and stresses in the substrate, the component body, the leads, and the solder joints. The strains in the substrate, component, and leads are primarily elastic with the elastic energies stored by warpage and bending.[23,24] In solder, the elastic strains are converted to plastic strains with time via stress relaxation and creep, processes that are accelerated at higher temperatures and stress levels. At typical use temperatures, the solder rapidly responds to stresses and displacements by visco-plastic flow and plastically deforms not only by converting the elastic strains within the solder joint but also by yielding to the elastic energies stored in substrate, component body, and leads.

The primary purpose of a compliant lead is to reduce the stress level at the solder joint significantly below the solder yield strength to reduce the cyclic hysteresis loop area and thus the cyclic fatigue damage. It needs to be emphasized that compliant leads do not reduce the expansion mismatch. The compliant leads elastically store strain energy, which over time unloads into the solder joints due to creep and stress relaxation. Thus, the total strain at the solder joint is not affected; the maximum stresses, however, are significantly reduced. At lower stress levels, the visco-plastic flow slows significantly to a degree that the contention has been advanced that a threshold stress level exists below which creep and stress relaxation stop. Unfortunately, the often quoted threshold stress level for solder creep of ~300 psi can readily be demonstrated by simple experiments to be nonexistent. An added reliability benefit can result, however, if stress levels are reduced to a point where the conversion of the elastically stored strain energies to plastic deformations in the solder, due to stress relaxation or creep, is retarded to such an extent that stress relaxation is still significantly incomplete even for operational cycles.

17.5 ANALYTICAL MODEL OF SOLDER SHEAR FATIGUE

Analytical formulations for solder joint fatigue are developed in this section. The parameter nomenclature and units, where appropriate, are provided in conjunction with the equations where they first occur.

17.5.1 Solder Joint Fatigue

The long-term reliability of a SM solder attachment is governed by the difference between the required design life and the cyclic fatigue life of the solder joints as determined by the combined effects of component design, assembly design, and use environment. The cyclic fatigue life of a solder joint is determined by the amount of cyclically accumulating fatigue damage.[6] Solder joint rupture occurs when the total accumulated damage exceeds the capability of solder to sustain such damage.

The solder joint response to cyclic displacements is characterized by a hysteresis loop in the shear stress-strain plane as shown in Figure 17-4 in stylized form. The area of this hysteresis loop is representative of the visco-plastic strain energy density per cycle ΔW and thus the fatigue damage accumulating per cycle. For leadless SM solder attachments, the hysteresis loop is limited by a constant stress envelope—independent of the thermal expansion mismatch—determined by the solder yield strength τ_y during the initial elastic loading and

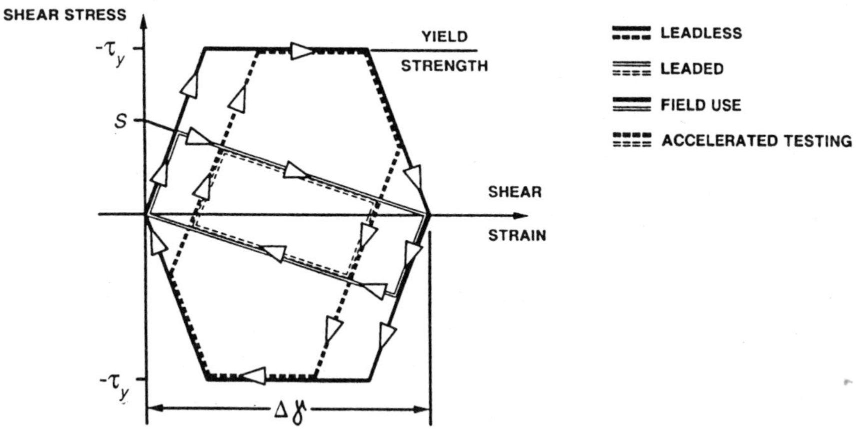

Figure 17-4 Stress-strain response (stylized) of solder joints in both leadless and leaded surface mount attachments to identical total cyclic displacements. The areas of the cyclic hysteresis loops are indicative of the cyclically sustained fatigue damage. Thus, the difference between leadless and compliant leaded SM attachments as well as between field use cycles and accelerated test cycles with incomplete creep-stress relaxation is illustrated (from references[8,9]).

plastic yielding and by the stress reduction lines during creep and stress relaxation. It should be noted that unlike the yield strengths of common engineering metals, the yield strength of solder is highly dependent on temperature and strain rate. For typical leaded SM solder attachments, the maximum solder joint stresses, S are significantly below the solder yield strength and depend on the thermal expansion mismatch. Thus, the stress reduction lines, except for some initial elastic loading, determine the hysteresis loop in the stress direction. For both leadless and leaded SM solder attachments, the maximum shear strains $\Delta\gamma$ are determined by the displacements resulting from the thermal expansion mismatch between component and substrate. Due to stress relaxation and creep, which allow the elastically stored strain energy in the compliant leads to unload into the solder joints with time, there is no difference in the total displacements at the solder joints. These hysteresis loops are illustrated in Figure 17-4. It should be noted that symmetric hysteresis loops as shown in the figure result only if the cyclic strains are mechanically applied at isothermal conditions or if the dwell time at T_o is longer than at $T_o + \Delta T_e$ to compensate for the slower stress relaxation at the lower temperature.

A generalized fatigue damage law for metals has been proposed on the basis of cumulative stored visco-plastic strain energy density.[4] The cyclic shear fatigue life $\overline{N}_f$ is related to ΔW in a stabilized fatigue cycle by the equation

$$\overline{N}_f = \frac{1}{2}\left[\frac{\Delta W}{W_f'}\right]^{\frac{1}{c}} \qquad (17\text{-}2)$$

where c = fatigue ductility exponent

$\overline{N}_f$ = mean fatigue life [$N_f(50\%)$], cycles-to-failure

W_f' = intercept energy term, a material constant

ΔW = visco-plastic strain energy density per cycle

The exponent c is in the range of -0.5 to -0.7 for most metals. The well-known Manson-Coffin plastic strain-fatigue life relationship[5] can be directly derived from—and is a special stress limited case of—this generalized fatigue damage function and is

$$\overline{N}_f = \frac{1}{2}\left[\frac{\Delta\gamma_p}{2\varepsilon_f'}\right]^{\frac{1}{c}} \tag{17-3}$$

where $\Delta\gamma_p$ = cyclic plastic shear strain range

ε_f' = fatigue ductility coefficient

For metals used in temperature ranges where time and temperature dependent creep and stress relaxation become significant relative to the initial plastic strains caused by yielding (typically in excess of 60 percent of the absolute melting temperature), the determination of either ΔW or Δy_p is not straightforward. The total plastic strains will increase with time as strain energy elastically stored in the component-lead-solder joint-substrate structure system will be converted to cumulative irrecoverable visco-plastic strain energy in the structure member (solder) that creeps/stress relaxes. With sufficient time, which for solder at operating temperatures can be quite short, virtually all the elastically stored strain energy accumulates as visco-plastic strain energy in the solder joints. This maximizes the cyclically circumnavigated hysteresis loop in the shear stress-strain plane and thus maximizes the cyclic fatigue damage to the solder joint (see Fig. 17-4).

The fatigue life of solder joints is not dependent on cyclic frequency per se; frequency can, however, appear as a correlative variable if data from cyclic tests with identical cycle shapes (sinusodial, saw-tooth, trapezoidal, square, etc.) are compared, since in these cases the cycle half-period and the time available for stress relaxation/creep are directly related.

In operational use the cycles typically consist of a ramp-up, an operational dwell period, a ramp-down, and a dwell period, which depending on the application can be operational or non-operational. Superimposed on the system operational cycles might be cyclic temperature variations due to daily and seasonal outside environmental conditions. The ramp-up will depend on the heat transfer and power dissipation conditions of the system and is of reliability concern only if it leads to temperature changes rapid enough to cause thermal shock (see the section "Reliability and Accelerated Testing—Overview") or if it is slow enough for significant stress relaxation to take place during this transient period. The operational and non-operational dwell periods depend on the operational characteristics of the system for both its duration and constancy; it can have superimposed on it more rapid cyclic thermal variations due to system load fluctuations. The down-ramp is a near-exponential decay depending on the system heat transfer parameters.

Thus, frequency is not necessarily a measure of the dwell time and using frequency instead of dwell time in correlating accelerated test data can result in significant errors in the analysis. Frequency also does not provide any indication of the rate of change of the thermal conditions; solder can be quite strain-rate sensitive at higher rates of change.

For accelerated fatigue testing, the dwell times during the cycle half-periods are insufficient for full stress relaxation to take place (see Figure 17-4); it is the shortened dwell times that provide the test acceleration but that also require that the reduced cyclic fatigue damage during accelerated testing be accounted for.

Test acceleration by increasing the thermal cyclic loading is only possible to a limited extent since the material characteristics of the solder limit both the maximum and minimum test temperatures that can be applied without introducing damage mechanisms that do not correspond to the field operational damage mechanisms. For leadless SM solder attachments, analysis is relatively simple since, unlike the response of leaded attachments, it is not complicated by interacting effects between the solder and the compliant lead structures.

17.5.2 Leadless Solder Attachments

A model for the mean cyclic fatigue life for the solder joints of a leadless surface mounted component (or a component with very stiff leads) has been derived from isothermal experimental data[25] (see Fig. 11-5). This model is expressed by the relationship[6]

$$\overline{N}_f = \frac{1}{2}\left[\frac{\Delta\gamma}{2\varepsilon_f'}\right]^{\frac{1}{c}} \tag{17-4}$$

where for eutectic and 60% Sn-40% Pb solder[6]

$$2\varepsilon_f' \approx 0.65 \tag{17-5}$$

and

$$c = -0.442 - 6 \times 10^{-4}\overline{T}_{SJ} + 1.74 \times 10^{-2}\ln\left(1 + \frac{360}{t_D}\right) \tag{17-6}$$

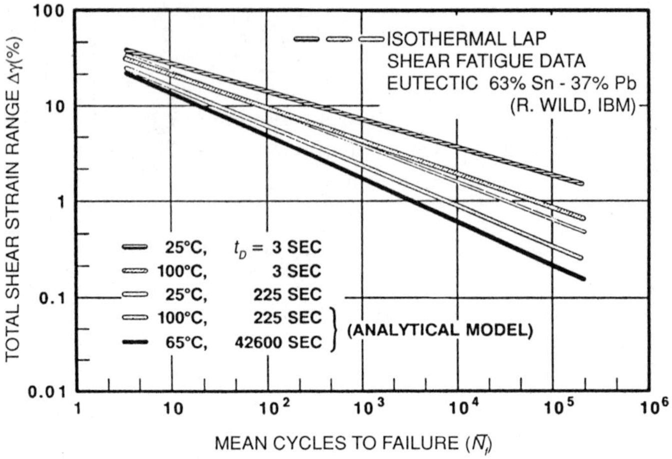

Figure 17-5 Manson-Coffin plot showing the results from reference[25] for isothermal fatigue cycling of solder joints at different temperatures and cyclic dwell times and some extrapolated results based on Equation (17-4).

and where T_S, T_C = steady-state operating temperature for substrate, component ($T_C > T_S$ for power dissipation in component)

$\overline{T}_{SJ} = \frac{1}{4}(T_C + T_S + 2T_o)$, mean cyclic solder joint temperature

T_o = temperature during off half-cycle

t_D = half-cycle dwell time (min), average time available for stress relaxation at T_S/T_C and T_o

$\Delta\gamma$ = cyclic total plastic shear strain range in solder after complete stress relaxation

ε_f' = fatigue ductility coefficient, defined for solder in Equation (17-5)

While Equation (17-4) has the same form as Equation (17-3), it overcomes the impractical difficulty of determining the time and temperature dependent plastic shear strain range Δy_p by using the maximum available displacement shear and accounting for the incomplete degree of stress relaxation/creep with a time and temperature dependent exponent.

The effect of time and temperature can be clearly seen in Figure 17-6 where the experimental results from isothermal mechanically applied strain cycling[17–19] are shown for temperatures of 23 and 81°C and cyclic dwell times ranging from 3 to 180 sec. The cycle shape for these data was trapezoidal with constant strain rates.

The total cyclic strain range is given by

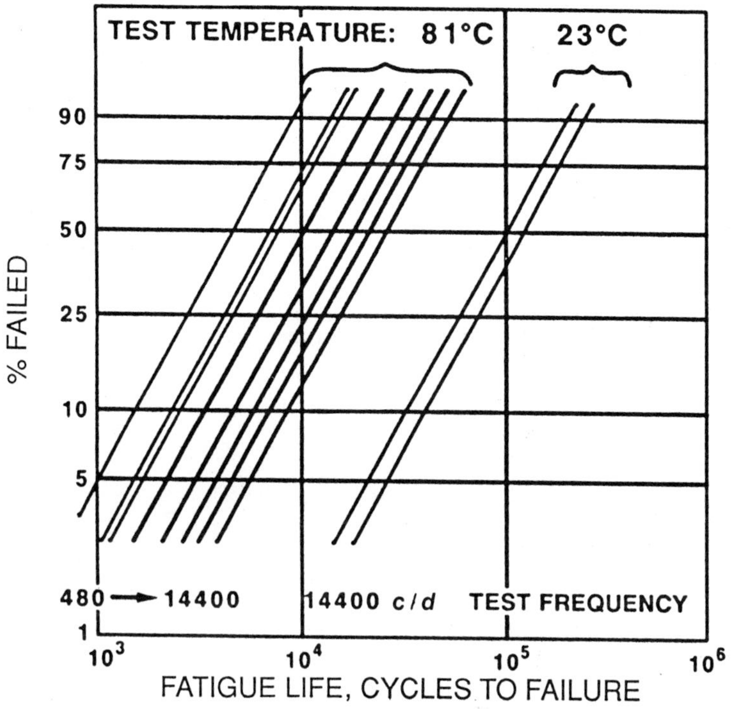

Figure 17-6 Weibull statistics plot of test results from accelerated isothermal mechanical cycling tests of plastic leaded chip carriers employing two temperatures and near square wave cycle shapes. The tests were carried out with three similar components indistinguishable in their fatigue lives over cyclic dwell times ranging from 90 to 3 sec (f = 480 to 14,400 cycles/day). The test data have similar Weibull slopes indicating the same failure mechanism and show the effect of different degrees of the time and temperature dependent cyclic stress relaxation in the solder joints on the cyclic fatigue lives (from references[17–19]).

$$\Delta\gamma = F\frac{L_D}{h}\Delta(\alpha\Delta T) \tag{17-7}$$

where the thermal expansion mismatch is

$$\Delta(\alpha\Delta T) = |\alpha_S\Delta T_S - \alpha_C\Delta T_C| \tag{17-8}$$

and where F = empirical "nonideal" factor indicative of deviations of real solder joints from idealizing assumptions and accounting for secondary and frequently untractable effects such as cyclic warpage, cyclic transients, nonideal solder joint geometries, solder microstructure, brittle intermetallic compounds, Pb-rich boundary layers, and solder/bonded material expansion differences, as well as inaccuracies and uncertainties in the parameters in Equations (17-7) and (17-8); $1.5 > F > 1.0$ for column-like leadless solder attachments, $1.2 > F > 0.7$ for castellated and chip component leadless solder attachments with fillets, $F \approx 1$ for solder attachments utilizing compliant leads

h = solder joint height, for leaded attachments $h \equiv 1/2$ of solder paste stencil depth

$2L_D$ = maximum distance between component solder joints, see Equations (17-13) and (17-14)

α_S, α_C = coefficient of thermal expansion (CTE) for substrate, component

ΔT_C = cyclic temperature swing for component

ΔT_S = cyclic temperature swing for substrate (at component)

$\Delta(\alpha\Delta T)$ = absolute cyclic differential thermal expansion between component and substrate, see Equation (17-8)

Some values for coefficients of thermal expansion of materials common is electronic packaging are given in Table 17-1.

The right side in Equation (17-8) can be simplified to

$$\Delta(\alpha\Delta T) = \Delta\alpha\Delta T_e \tag{17-9}$$

where

$$\Delta T_e = \left| \frac{\alpha_S \Delta T_S - \alpha_C \Delta T_C}{\Delta\alpha} \right| \tag{17-10}$$

and where ΔT_e = equivalent cyclic temperature swing, accounting for component power dissipation effects as well as component-external temperature variations

$\Delta\alpha$ = absolute difference in coefficients of thermal expansion of component and substrate, CTE-mismatch, ΔCTE

Equation (17-10), of course, simplifies in the case of no power dissipation in the component to

$$\Delta T_e = \Delta T_S = \Delta T_C \tag{17-11}$$

Thus, from combining Equations (17-4), (17-7), and (17-9), the mean cyclic life of a leadless surface mount solder attachment is obtained from

$$\overline{N}_f = \frac{1}{2}\left[\frac{F}{2\varepsilon_f'} \frac{L_D \Delta\alpha\Delta T_e}{h}\right]^{\frac{1}{c}} \tag{17-12}$$

Table 17-1 Some Measured Planar Linear Thermal Expansion Coefficients

Component, Substrate,* Material	Planar Linear Thermal Expansion Coefficients,** α (ppm/°C)
Plastic chip carrier	20–23
Ceramic chip carrier	5.4–6.7
FR-4 laminate	12–24
FR-4 multilayer board	16–24
CTE-modified FR-4 MLB	7.2–14
Ceramic multilayer board	6.0–8.3
Solder	23–25

*The specific construction (number of layers, resin/glass/metal content, etc.) has a major impact on α.
**Most values were measured between room temperature and 110°C.

which contains all the essential first-order design parameters. Figure 17-7 shows in a Manson-Coffin plot the results from three different test programs that correlate well with the prediction from Equation (17-12) using $F = 1.0$. A value of $F = 1.27$ would virtually match the model prediction and the experimental data.

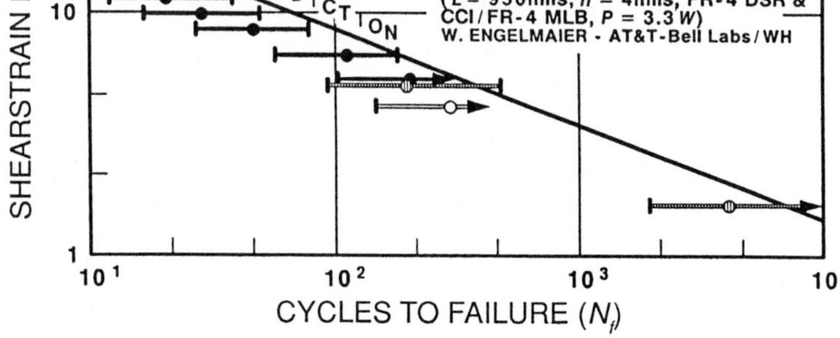

Figure 17-7 Manson-Coffin plot showing the results (mean and range) from three accelerated functional and thermal cycling test programs for leadless ceramic chip carriers of different sizes on different substrates over widely varying test conditions. To facilitate comparison, the test results are normalized to a common condition of $\overline{T}_{SJ} = 45°C$ and $t_D = 6.2$ min ($f = 58$ cycles/day) using the acceleration transform [Eq. (17-18)]. The model prediction [Eq. (17-4)] for this common condition indicates excellent agreement between model and experimental data with a nonideal factor [$1.12 \leq F \leq 1.27$, Eq. 17-7] offset (from reference[6]).

The maximum distance between solder joint centers for rectangular or square components with a multiplicity of solder joints, (e.g., chip carriers, SOICs, etc.) is approximately given by

$$L_D = \frac{1}{2}\sqrt{L^2 + W^2} \qquad (17\text{-}13)$$

where L = longest component dimension between component edges with solder joints

W = L for square components

= for rectangular components, the shorter component dimension between component edges with solder joints or the component dimension between component edges without solder joints

For components with two solder joints, such as discrete chip components and metal electrode face bonded components (MELFs) the formulation

$$L_D = \frac{L + L_M}{4} \qquad (17\text{-}14)$$

can be used

where L_M = distance between metallizations for solder joints

For temperature cycling between low temperatures (<0°C), where stress relaxation and creep are insignificant, the cyclic life can be determined from Equation (17-3). Of course, the plastic strain range $\Delta\gamma_p$ is in this case significantly less than the total strain range [Eq. (17-7)]; the elastic strain range, determinable from the low temperature stress-strain behavior of solder, needs to be subtracted.

17.5.3 Leaded Solder Attachments

The effects of lead compliancy on the fatigue life make the analytical determination of the cyclic life of leaded solder attachments significantly more complicated and these effects had until very recently[7] not been fully quantified. Techniques to determine the cyclic strain energy density in the solder joints of leaded SM attachments have been developed.[26] From the results of accelerated tests that mimic functional use conditions, the cyclic life of the product in the field has been estimated[11] by extrapolation using the acceleration transform in Equation (7-18). For leadless SM attachments, the high component and substrate stiffnesses ensure rapid and near complete stress relaxation at most use temperatures, even at the cyclic low temperature excursions (provided the temperatures are not too low and the cycles too short). Stresses are large but are limited by the solder yield strength. In contrast, as shown in Figure 17-4, for leaded SM attachments with adequate lead compliancy, the stresses at the solder joints are reduced to levels at which solder creep is significantly retarded even at the cyclic high temperature excursions and for long operational dwell times. Thus, the stress level effects on the solder joint creep become a determining factor for SM solder joint reliability.

The solder joint response is characterized by a hysteresis loop in the shear stress-strain plane between the stress reduction (combination of creep and stress relaxation) lines at the temperature extremes. The product of the largest applied stress and the total cyclic shear strain range is the maximum available cyclic strain energy density ΔW, which is proportional to the product $S\Delta\gamma$ (see Fig. 17-4). The largest applied stress is proportional to

$$S \propto \frac{K_D L_D \Delta\alpha \Delta T_e}{A} \tag{17-15}$$

where A = effective solder joint area ($\equiv$ 2/3 solder-wetted lead area projecting to solder pad)

K_D = "diagonal" flexural stiffness of unconstrained component lead determined by strain energy methods or finite element analysis

S = largest stress applied to a leaded attachment solder joint

The magnitude of S controls the stress reduction rates during the dwells at constant temperature. From Equations (17-7), (17-9), and (17-15), the cyclic strain energy density (a measure of the cyclic fatigue damage) is, for surface mount solder attachments with compliant leads, proportional to

$$\Delta W \propto \frac{K_D \left(L_D \Delta\alpha \Delta T_e \right)^2}{Ah} \tag{17-16}$$

A plot of the cyclic strain energy density per cycle ΔW versus mean cyclic life $\overline{N}_f$ is shown in Figure 17-8.[8,9] The experimental data are for five different leaded assemblies subjected to accelerated testing in functional[11] or temperature cycling.[27] The slope of the regression line ΔW versus $\overline{N}_f$ is $|c| < 0.64$ [see Eq. 17-2)]. For the three assemblies with the longest lives, lowest bound estimates for $\overline{N}_f$ had to be used since no failures had been observed at the indicated numbers of cycles. Fatigue testing to failure undoubtedly will result in larger values of $\overline{N}_f$ resulting in a reduced slope in Figure 17-8 and a value for the exponent c

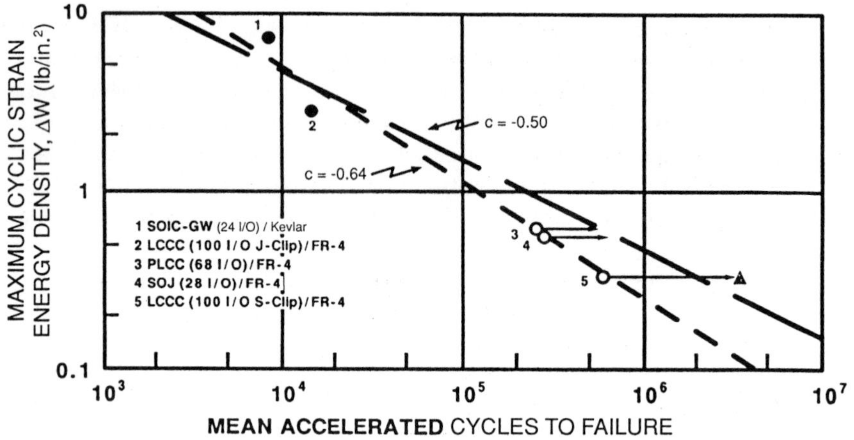

Figure 17-8 Plot of test results from accelerated functional and thermal cycling tests for five different leaded surface mount component assemblies. Since for three tests no failures occurred during testing, only the N_f (50%) lives are plotted. For tests 3 to 5 the projected lives are calculated from Equation (17-19) under the assumption that the first failure occurs on the very next cycle after test termination. For test 5 a life projection based on the test results from test 2 and Equation (17-18) is also shown (from references[8,9,11,27]).

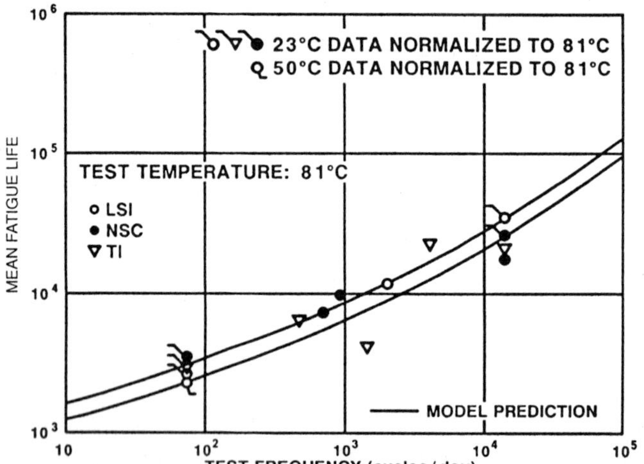

Figure 17-9 Plot of test results from accelerated isothermal mechanical cycling tests of plastic leaded chip carriers employing three test temperatures and near square wave cycle shapes. The tests were carried out with three similar components indistinguishable in their fatigue lives over cyclic dwell times ranging from 10 min to 3 sec (f = 72 to 14,400 cycles/day). The test data are normalized to a comparison temperature of 81°C using the acceleration transform [Equation (17-18)]; the test results and the model prediction [Eq. (17-17)] show excellent correlation (from references[17–19]).

closer to the more commonly observed values ($c \sim -0.5$ to -0.4) for solder. This is clearly shown to be the case in Figure 17-9, where the results from the IEEE Compliant Lead Task Force[17–19,28] are plotted against the model prediction resulting in excellent correlation between the data and the model. Thus, from combining Equations (17-4) and (17-16) and benchmarking the proportionality in Equation (17-16) with the results in Figure 17-8, the mean cyclic life of surface mount solder attachments with compliant leads is given by the equation

$$\overline{N}_f = \frac{1}{2}\left[\frac{F}{2\varepsilon_f'}\,\frac{K_D(L_D\Delta\alpha\Delta T_e)^2}{(200\text{ psi})\,Ah}\right]^{\frac{1}{c}} \tag{17-17}$$

which contains all the essential first-order design parameters.

17.5.4 Acceleration Transform

The differences in the cyclic hysteresis loops—and thus the cyclic fatigue damage—resulting from accelerated test cycles with insufficient dwell times for complete stress relaxation and product functional cycles resulting in essentially complete stress relaxation are shown in Figure 17-4. This difference in fatigue damage per cycle needs to be accounted for with an acceleration transform in order to make product reliability estimates based on accelerated test results.

From Equations (17-2) through (17-6) an acceleration transform[6] results with which accelerated test results, denoted by (t), can be extrapolated to use conditions, denoted by (u):

$$\overline{N}_f(u) = \frac{1}{2}\left[[2\,\overline{N}_f(t)]^{c_t}\,\frac{\Delta D_u}{\Delta D_t}\right]^{\frac{1}{c_u}} \tag{17-18}$$

where

$$\Delta D = \Delta\gamma \tag{17-19}$$

from Equation (17-7) leadless attachments and

$$\Delta D = \Delta W \tag{17-20}$$

from Equation (17-16) for leaded attachments. It should be noted that (u) and (t) can also denote different accelerated test conditions; transforming the results obtained at conditions (t) to equivalent (u) condition results permits meaningful comparisons of test data. In Equation (17-18), the cyclic fatigue damage is in ratio form; thus, it is not necessary that the values of ΔD, the measure of cyclic fatigue damage, be explicitly known. Only the ratio relationship between test and use fatigue damage is needed. This makes it much easier to transform data for surface mount solder attachments since many and in the case of functional accelerated testing (see section "Accelerated Fatigue Reliability Testing") all parameters going into the measure of cyclic fatigue damage ΔD are identical and cancel each other in the $\Delta D_u/\Delta D_t$ ratio.

The validity of this acceleration transform has been demonstrated by comparing and correlating the results of accelerated fatigue tests carried out at different temperatures and different cycle dwell times (see Figs. 17-5 through 17-9). In fact, Figures 17-7 and 17-9 required the use of the acceleration transform in Equation (17-18); without it the test data resulting from different test conditions could not be meaningfully compared. For leadless attachments such correlations can be found in References [6–10,15,22,29] and for leaded attachments in References [11,16–21,28]. Correlations have been established over dwell times ranging from 3 sec to 16 h and temperatures from −20 to +150°C.

It needs to be emphasized that the acceleration transform cannot be used for results from tests that either have temperature cycles that traverse over or even come close to the glass transition temperature of any of the organic-based materials in the components or substrates,[6] or that have low temperature (< −20°C) extremes that do not permit significant stress relaxation of the solder.[9]

Further, it has been shown that for cycles with large temperature swings ΔT, where the lower temperature extreme permits no significant stress relaxation of

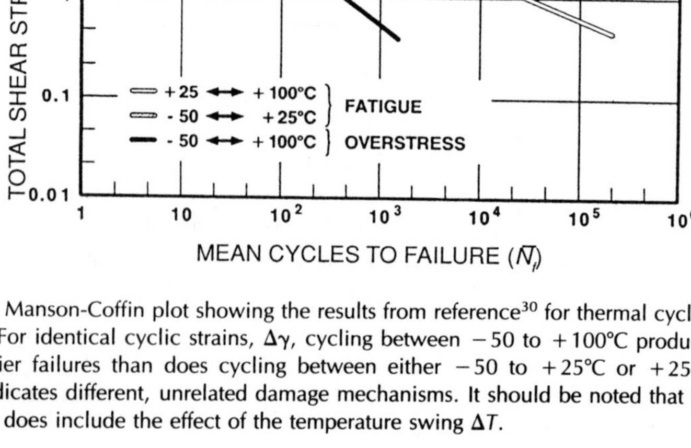

Figure 17-10 Manson-Coffin plot showing the results from reference[30] for thermal cycling of solder joints. For identical cyclic strains, $\Delta\gamma$, cycling between −50 to +100°C produces significantly earlier failures than does cycling between either −50 to +25°C or +25 to +100°C; this indicates different, unrelated damage mechanisms. It should be noted that the total strain range does include the effect of the temperature swing ΔT.

the solder and the upper temperature extreme is well in the temperature range of readily stress-relaxing solder, different damage mechanisms are responsible for failure. Figure 17-10 shows experimental results[30] of thermal cycling over typical temperature swings (ΔT = 75°C) and a temperature swing (ΔT = 150°C) that is not typically encountered by electronic assemblies (see section "Testing Conditions"). The results for the ΔT = 75°C thermal cycling—even from −50 to +25°C—follow the damage mechanism described by Equation (17-12), for ΔT = 150°C they clearly do not and the damage mechanism is likely dependent on overstress and stress-related recrystallisation[31] considerations.

17.5.5 Failure Statistical Considerations

Solder joint fatigue failures are the result of a wearout mechanism (Fig. 17-1) and as such have a statistical distribution around the mean cyclic life. This statistical distribution is illustrated in Figure 17-11 using the data from a number

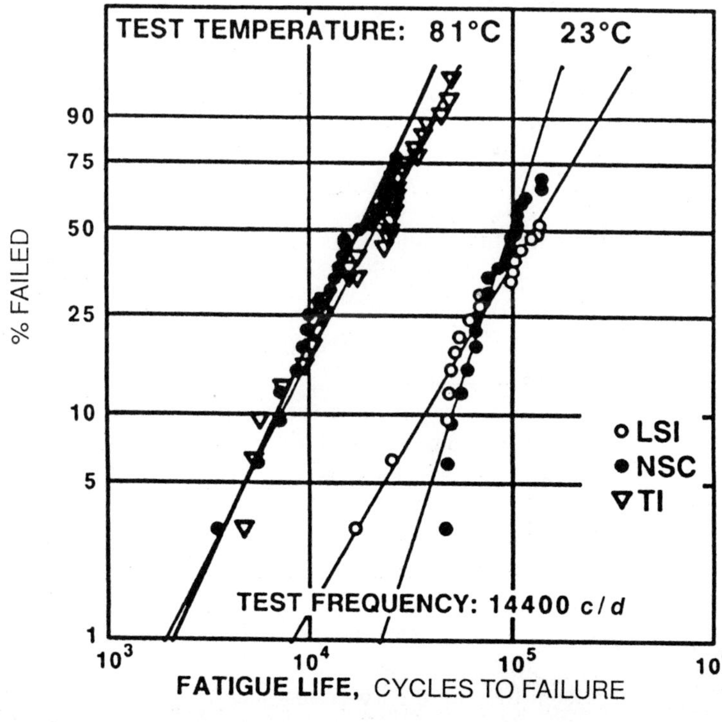

Figure 17-11 Raw data Weibull statistics plot from accelerated isothermal mechanical cycling tests of plastic leaded chip carriers employing near square wave cycle shapes with dwell times of 3 sec (f = 14,400 cycles/day) at two temperatures (from references[17–19]).

of fatigue experiments,[17–19]; these results are also included in Fig. 17-6 as linear regression plots. These statistical distributions are Weibull distributions with typical Weibull slopes $\beta \simeq 2$[10,11,16–21] for attachments via compliant leads and $\beta \simeq 4$[9,10,15,29] for leadless or stiff leaded attachments.

The Weibull cumulative failure distribution can be described in terms of the mean cyclic life $\overline{N}_f$ by the relationship

$$F(N) = 1 - 0.5^{\left[\frac{N}{\overline{N}_f}\right]^{\beta}} \tag{17-21}$$

where $F(N)$ = acceptable cumulative failure probability for the component under consideration after N cycles

N = expected product service life, cycles of operation

β = Weibull shape parameter, slope of Weibull probability plot

Equation (17-21) gives the probability of failure for a single component; for a system consisting of m different component types each having a population of $n(m)$ components, the system probability of failure, $F_{\Sigma}(N)$, is

$$F_{\Sigma}(N) = 1 - \prod_{m=1}^{m} [1 - F_m(N)]^n \tag{17-22}$$

which reduces for n identical components to

$$F_{\Sigma}(N) = 1 - 0.5^{n\left[\frac{N}{\overline{N}_f}\right]^{\beta}} \tag{17-23}$$

where n = number of the component under consideration in system, number of individually monitored components on test

In statistical reliability evaluations it should be recognized, however, that the confidence in the reliability projections resulting from relationships such as Equations (17-21) to (17-23) depends on the confidence established for the constituent parameters and reliability evaluations for whole systems as shown in Equation (17-22) will have significantly lower confidence levels than are established for the individual constituent components.

The product design life and the expected cyclic frequency in operation give the expected number of operating cycles N. Of course, given the statistical distribution of failures in a wearout mechanism like fatigue, the design mean fatigue life $\overline{N}_f(D)$ has to be larger than N,[8,9,16] such that

$$\bar{N}_f(D) = N\left[\frac{n\ \ln(0.5)}{\ln[1 - F(N)]}\right]^{\frac{1}{\beta}} \tag{17-24}$$

where $\bar{N}_f(D)$ = design mean fatigue life [N_f(50%)], cycles-to-failure

for a design with *n* identical components. A mix of diverse components requires a more sophisticated statistical analysis. While the effects of the number of components and the component mix can be accounted for in the choice of the acceptable failure probability *F(N)*, the primary considerations affecting the choice of *F(N)* are dictated by the criticality of function of the system and any available redundancies.

It should be noted that Equation (17-24) deals with components, *not* individual solder joints. On components with a multitude of I/Os, only the "corner" solder joints (eight on chip carriers, four on small outline packages, two on chip resistors) have the highest failure probability; all other solder joints have lower failure probabilities since they experience significantly less cyclic fatigue damage. From a strict point of view, the statistical liabilities of these less strained solder joints also need to be accounted for; however, this is a rather cumbersome task and requires accuracy to a degree which neither the analytical modeling efforts nor accelerated testing can provide.

17.6 ACCELERATED FATIGUE RELIABILITY TESTING

17.6.1 Testing Considerations

The general considerations necessary for valid accelerated fatigue testing of SM solder joints have been discussed at length in the literature.[6,10,21,32] There are two primary purposes for accelerated testing: (1) to obtain a measure of long-term reliability of the SM solder attachments and (2) to provide a process verification test. The latter test purpose, however, is not served by an accelerated test for long-term solder fatigue. Rather, it is a combination overstress-tensile fatigue-shear fatigue test leading to rapid failure of substandard solder joints with reduced strength; it assures consistent quality and verifies the manufacturing process (see Fig. 17-3); it is a processing quality screen [e.g., environmental stress screening (ESS)], not a reliability test. The former test purpose requires that the tests, while carried out on an accelerated basis, have to simulate use conditions without introducing extraneous failure mechanisms or significant non-representative physical and/or material behavior. The closer the test conditions and the test vehicles are to operational reality, the more accurately the test results will represent the reliability of the product. This also implies that the higher the test acceleration—ratio of mean-time-to-failure (MTTF) for product in the field and MTTF in accelerated testing—the less accurate and detailed can be the results from the accelerated tests; that is, one gets what one pays for.

Table 17-2 lists realistic, but worst-case, use environments that can be experienced by electronic assemblies. These use environments are put into nine use categories representing some of the most common applications and are

Table 17-2 Realistic Worst-Case Use Environments and Appropriate Accelerated Testing for Surface Mounted Electronics by Use Categories

	Worst Use Environment					Years	Accelerated Testing			
Use Category	T_{min} (°C)	T_{max} (°C)	ΔT* (°C)	t_D (h)	Cycles/ year	of service	T_{min} (°C)	T_{max} (°C)	ΔT** (°C)	t_D (min)
1–Consumer	0	+60	35	12	365	1–3	+25	+100	75	15
2–Computers	+15	+60	20	2	1460	~5	+25	+100	75	15
3–Telecomm	−40	+85	35	12	365	7–20	+25	+100	75	15
4–Commercial aircraft	−55	+95	20	2	3000	~10	0	+100	100	15
5–Industrial &			20	12	185					
automotive—	−55	+65	&40	12	100	~10	0	+100	100	15
passenger			&60	12	60					
compartment			&80	12	20			& "Cold"†		
6–Military	−55	+95	40	12	100		0	+100	100	15
ground & ship			&60	12	265	~5		& "Cold"†		
7–Space: leo				1	8760		0	+100	100	15
	−40	+85	35			5–20				
geo				12	365			& "Cold"†		
8–Military: a			40	2	500					
avionics: b	−55	+95	60	2	500		0	+100	100	15
c			80	2	500	~5				
			&20	1	1000			& "Cold"†		
9–Automotive—			60	1	1000		0	+100	100	15
under hood	−55	+125	&100	1	300	~5				
			&140	2	40		& "Cold"† & "Large ΔT"††			

& = in addition

*ΔT represents the maximum temperature swing but does not include power dissipation effects; for power dissipation see Equation (17-10); power dissipation can make pure temperature cycling accelerated testing significantly inaccurate.

**All accelerated test cycles shall have temperature ramps <20°C/min and dwell times at temperature extremes shall be 15 min measured on the test boards. This will give ~ 24 test cycles/day.

†The failure/damage mechanism for solder changes at lower temperatures; for assemblies seeing significant cold environment operations, additional "Cold" cycling from perhaps −40 to 0°C with dwell times long enough for temperature equilibration and for a number of cycles equal to the "Cold" operational cycles in actual use is recommended.

††The failure/damage mechanism for solder is different for large cyclic temperature swings traversing the stress-to-strain −20 to +20°C transition region; for assemblies seeing such cycles in operation, additional appropriate testing with cycles similar in nature and number to actual use is recommended.

ordered roughly by increasing severity. It should be noted that the cyclic temperature swings ΔT are not the difference between the maximum and minimum temperatures that can be experienced; they are significantly smaller and move between the extremes primarily due to seasonal, and perhaps geographic, temperature variations. For some use categories the severity of the cyclic environment can change seasonally leading to multiple cyclic temperature ranges; for others, different ranges result from multiple use conditions (see the section "Multiple Cyclic Load Histories"). It should be noted that the listed maximum temperature swings are for component-external causes only; they do not include any effects due to component-internal power dissipation. These effects need to be evaluated using Equation (17-10); depending on the values of the parameters, the effective temperature swing ΔT_e can be either smaller or larger than ΔT. Also included in Table 17-2 are estimates of the effective half-cycle dwell times t_D as well as of the design service lives. It should be noted that the Automobile-Under Hood (Use Category 9) service life is shorter than the life for Automobile-Passenger Compartment (Use Category 5) electronics. It is anticipated that because of the severity of the under-the-hood environment reliable electronic packaging for the entire life of the automobile is not possible. Under-the-hood electronics need to be placed either into environmentally controlled enclosures or need to be viewed (and accordingly designed) as auto parts requiring periodic replacement. Use Category 7, space, requires different environments because of the large differences between low earth orbit ("leo") and geo-synchronous orbit ("geo") satellites; the three use environments for Use Category 8, Military Avionics, account for differences in the severity of environmental mission profiles due to the type of aircraft or the location on the aircraft.

Table 17-2 also lists recommendations for the test conditions for accelerated thermal cycling. For all test cycles, dwells at the temperature extremes of 15-min duration, measured on the test vehicles, are recommended. These test recommendations are designed to assure that failures occur in accelerated testing as the result of the same failure mechanisms as they would for product in use. It should be noted that for electronic assemblies seeing significant cold environment operations (Use Categories 5 through 9) or large temperature swings traversing the damage mechanism region (-20 to $+20$°C) in which the change from stress-to-strain control takes place, additional representative testing is recommended (see section "Solder Joint Fatigue" and Fig. 17-10). The total damage from the multiple cyclic tests can be assessed using the procedure outlined in section "Multiple Cyclic Load Histories."

Table 17-3 lists the number of failure-free test cycles (as described in Table 17-2) required to assure that the failure risk per component is below the stated cumulative failure probabilities at the end of the stated service lives, given that 32 components are on test. The minimum number of failure-free test cycles $N(t)$

Table 17-3 Accelerated Testing Required for Surface Mounted Electronics by Use Categories to Empirically Assure Failure Risk Levels at End of Service Life

		Service Life								
		1 year			3 years			5 years		
		Failure-Free Test Cycles, $N(t)$, Required* with 32 Components on Test to Assure Cumulative Failure Probability, $F(N)$ in ppm at End of Service Life								
Use Category	Attachment Type	1 ppm	100 ppm	10,000 ppm	1 ppm	100 ppm	10,000 ppm	1 ppm	100 ppm	10,000 ppm
1	Leadless	1600	470	140	5200	1500	440	9000	2600	760
	Leaded**	4500	390	33	14,700	1300	107	25,400	2200	184
2	Leadless	1400	420	127	4400	1300	400	7500	2300	680
	Leaded**	950	86	25	3000	270	25	5100	460	42
3	Leadless	1800	530	150	6000	1700	500	10,500	3000	870
	Leaded**	5200	430	36	17,200	1400	118	29,900	2600	210
4	Leadless	1800	520	154	5700	1700	500	9800	2900	850
	Leaded**	540	47	25	1700	150	25	3000	290	25

5	Leadless	1200	340	100	3900	1100	320	6800	2000	560
	Leaded**	6500	540	44	21,200	1800	146	36,900	2800	250
6	Leadless	3200	890	204	10,600	3000	690	18,600	5200	1200
	Leaded**	16,000	1300	100	53,900	4300	330	94,700	7500	590
7	Leadless	15,600	4700	1420	48,800	14,800	4500	82,900	25,100	7600
leo	Leaded**	19,000	1740	159	59,500	5400	500	101,000	9,200	840
7	Leadless	1090	310	86	3700	1030	290	6400	1800	510
geo	Leaded**	1650	130	25	5500	440	34	9700	770	60
8	Leadless	2600	740	210	8700	2500	700	15,200	4300	1210
$\Delta T = 20\&40$	Leaded**	4700	370	28	15,900	1250	99	27,900	2200	173
8	Leadless	5000	1430	410	16,200	4700	1350	28,100	8100	2300
$\Delta T = 20\&60$	Leaded**	23,500	1950	160	77,100	6400	530	134,000	11,100	920
8	Leadless	7900	2300	680	25,300	7400	2200	43,500	12,700	3800
$\Delta T = 20\&80$	Leaded**	71,200	6200	530	229,000	19,800	1700	394,000	34,200	3000
9	Leadless	13,600	4100	1200	42,600	12,900	3900	72,400	22,200	6600
	Leaded**	177,000	16,200	1500	553,000	50,600	4600	940,000	90,500	7800

8700 cycles ≈ 1 year test time

*The required number of test cycles is obtained for an acceptable cumulative failure probability $F(N)$ at the end of the service life and a number of components n on test from Equation (17-25).

**These numbers apply to compliant leads (diagonal lead stiffness < 100 lb/in.); stiffer leads will move the required test cycles towards the leadless cycle numbers.

assuring a cumulative failure probability *F(N)* with *n* individually monitored components on test, is obtained from

$$N(t) = \bar{N}(t)\left[\left[\frac{\ln(0.5)}{\ln[1 - F(N)]}\right]^{\frac{c(u)}{c(t)}}\left[\frac{\ln[1 - 1/n]}{\ln(0.5)}\right]\right]^{\frac{1}{\beta}} \tag{17-25}$$

where $\bar{N}(t)$ is obtained by rearranging Equation (17-18) to

$$\bar{N}(t) = \frac{1}{2}\left[[2N]^{c_u}\frac{\Delta D_t}{\Delta D_u}\right]^{\frac{1}{c_t}} \tag{17-26}$$

and *N* is the product of service life and the expected number of annual cycles.

Increasing the number of components on test reduces the required number of failure-free test cycles. However, this reduction of the test duration is limited by both the number of components that can realistically be put on test and the slow response of the required test cycles to even substantial increases in the number of components on test. It should be noted that for the more severe use environments, assuring very low failure risks (1 and 100 ppm) can take longer than the service life, which is of course absurdly impractical. However, it illustrates that because of the unique properties of the materials involved in electronics packaging (see the section "Thermal Expansion Mismatch Problem") there are definite limitations to what can be obtained by accelerated testing.

Another notable feature brought out in Table 17-3 is the difference between stiff leadless and compliant leaded SM solder attachments in terms of evaluating the results of accelerated testing and designing for low acceptable failure probabilities. The two attachment types have different Weibull statistics slopes β (see section "Failure Statistical Considerations") and different dependencies on the cyclic temperature swings ΔT [see Eqs. (17-12 and (17-17)]; this results in the values of *N(t)* at the different failure probabilities, service lives, and use categories, which at first glance certainly can appear peculiar.

These points are illustrated in Figure 17-12 where the cumulative failure probability curves resulting from the different use environments described in Table 17-2 are shown for one specific chip carrier. It is quite obvious that this chip carrier is not appropriate for some of the more severe use environments in its leadless version, but to illustrate the effect of the different use environments it is necessary to have a common basis for comparison. The nine use environments result in reliable operating lives that are as much as a factor of 250 apart for stiff leadless solder attachments; for compliant leaded attachments this difference shrinks to about a factor of 15. In Figure 17-12 the effect of the different Weibull failure statistical slopes is also evident; this together with the effect of the lead

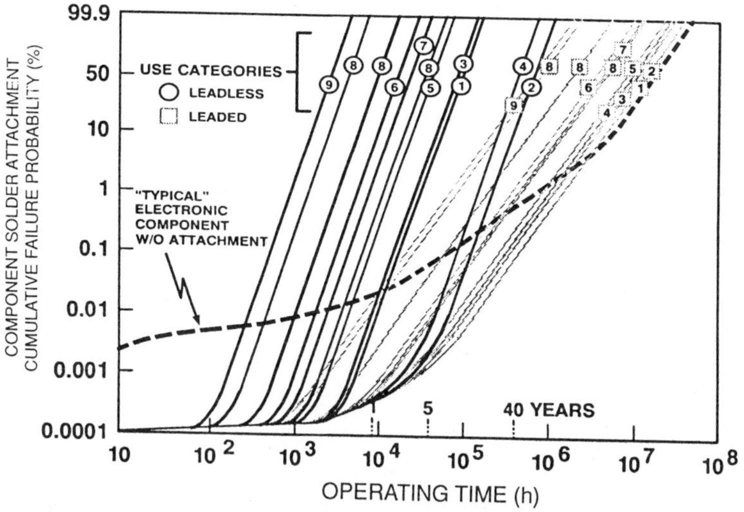

Figure 17-12 Cumulative failure probabilities for a typical electronic component and for the surface mount solder attachment of a 68-I/O chip carrier (L_D = 620 mils, $T_C - T_S$ = 5°C) subjected to the use environments described in Table 17-2. For stiff leadless attachment h = 3 mils, α_S = 8 ppm/°C, α_C = 6 ppm/°C; for the compliant leaded attachment h = 5 mils, A = 800 mils^2, α_S = 17.5 ppm/°C, α_C = 20.5 ppm/°C, K_D = 55 lb/in. The Use Category 7 geo curves are not shown.

compliancy produces significantly lower failure probabilities for leaded attachments.

The primary parameter to accelerate a SM reliability test is the cyclic dwell time. In order to translate the results from an accelerated test to product reliability information, it is essential that the effects of reduced dwells be understood and that the resulting reduced cyclic fatigue damage due to the incomplete stress relaxation and creep of the solder joints can be accounted for (see the section "Analytical Model of Solder Shear Fatigue" and Fig. 17-4). For product in operation, the dwells are measured in hours or days; for functional and thermal accelerated cycling the dwell durations can be as low as 3 to 5 min[11,29] and for mechanical cycling as low as 3 sec.[18,19] The presence of an elastic strains[34] sets limits to the minimum dwell times that can be correlated with use environments.

Another way to accelerate tests is to increase the applied strains. Frequently, this has been attempted by extending the applied test temperatures significantly beyond realistic use temperatures. From section "Analytical Model of Solder Shear Fatigue" it is clear that even large increases in ΔT produce only relatively modest test accelerations; this has led to extreme test conditions such as cycling from -65 to $+150$°C, perhaps even in thermal shock, to force early but unrepresentative failures, rather than accelerated failures. Given the time and temperature dependence of many of the materials involved in surface mounting,[10,32] extending the test temperature ranges significantly is not a viable approach (see the section "Acceleration Transform"). Small extensions of the test temperature ranges are possible (see Table 17-2). However, for assemblies designed for high reliability ($\Delta\alpha \sim 2$ppm/°C) very little cyclic fatigue damage occurs in the solder joints during thermal cycling, where the whole assembly is always at uniform temperature. For these assemblies valid thermal cycling tests become impractically long.

There is, however, another way to increase the applied strains. For accelerated test purposes the CTEs of component and test vehicle substrate can be deliberately mismatched[6,16,32] thereby significantly increasing the applied cyclic strain range. Examples of such deliberate mismatches would be plastic leaded components on epoxy-Kevlar substrates or CTE-constrained FR-4 substrates using copper-clad Invar or molybdenum; ceramic components could be tested mounted on FR-4 substrates. The acceleration transform given in Equation (17-18) can then be used to obtain the equivalent cyclic lives for actual product.

With mechanical cycling, increasing the applied strains is a simple matter since the strains are applied by mechanical deformation of the test vehicle substrate or external loading of the components. However, caution needs to be exercised to prevent a change in failure mechanism due to overstressing, high strain rates, or changing from predominantly shear to tension fatigue.

17.6.2 Accelerated Test Conditions

In Table 17-4 typical accelerated fatigue tests are listed together with some common test parameters, acceleration factors, and comments. It should be noted that temperature cycling could be performed at the higher frequencies possible with functional cycling, if specially designed facilities allowing rapid uniform temperature changes without even localized thermal shock are used instead of standard commercial environmental chambers. In either test well-defined dwells at the temperature extremes are necessary and should be at least 25 percent of the cycle period in duration. Reference should also be made to the discussion in the section "Reliability and Accelerated Testing—Overview."

Functional accelerated testing, which combines both external temperature variations and component-internal power dissipation cycling, provides the most representative test environment without introducing extraneous and uncontrolled differences between accelerated testing and product functional use. In functional accelerated testing, the conditions and test vehicles mimic the product use conditions and the product as closely as possible,[11,15] with the exception of the test dwell times and thus the test frequency. While a shortening of the dwell times leads to less complete conversion of the elastically stored strain energies to plastic deformations in the solder via stress relaxation and creep—and thus to

Table 17-4 Recommended Accelerated Fatigue Test Parameters

Test	Temperature	Frequency	Warpage	Acceleration Factors*	Comments
Functional cycling	Realistic	<100 cycles/ day; $\dot{T} < 30°C/min$; defined dwells	Realistic	~ 10 to 20	Realistic power dissipation and heat transfer
Temperature cycling	$0 \leftrightarrow 100°C$ $25 \leftrightarrow 100°C$ $-40 \leftrightarrow 0°C$	$\leqslant 24$ cycles/day; $\dot{T} < 30°C/min$; defined dwells	Somewhat higher than realistic	<1 to ~ 10; $\leqslant \sim 30$ with deliberate test vehicle CTE-mismatch	Very long test durations for $\Delta CTE < 5$ ppm/°C
Mechanical cycling	Isothermal**	$<10{,}000$ cycles/day; near square wave	Bend radius $\pm$ 40 to 80 in. (± 1 to 2 m)	$\leqslant 500$	Excludes all CTE-mismatch effects; beware of overstressing; somewhat higher tensile strains

*Mean-time-to-failure acceleration factors for ΔT (test) $\approx \Delta T$ (use).

**At 0°C, room temperature, 60°C, or 80°C.

less fatigue damage per cycle and a longer fatigue life in terms of cycles-to-failures—the total time necessary to produce failures is still significantly reduced.

Thermal cycling provides reliability data with significantly simpler test arrangements, albeit at the price of reduced accuracy and increased uncertainty. While in functional cycling the shape of the cyclic waveform is defined by the thermal environment of the product, special care is necessary for thermal cycling (temperature variations in a thermal test chamber) to achieve proper test conditions. The cyclic waveform needs to have heating and cooling ramps for which the rates of temperature change are below ~30°C/min to avoid thermal shock conditions. In addition, the waveform can have a significant effect on the (effective) dwell times at temperature. The preferred waveform is trapezoidal because it is closest to actual use conditions and has well-defined dwells at the temperature extremes, but sinusoidal and even sawtooth waveform have been used. For sinusoidal waveforms there arises the question of what the effective dwell time is. In Reference[6] where the solder fatigue model underlying the section "Analytical Model of Solder Fatigue" was first developed from isothermal sinusoidal mechanical cycling results,[25] the assumption was made that the effective dwell time is one-quarter of the cycle period. While for sawtooth waveforms some stress relaxation and creep occurs during the up and down ramps, it is so small that for all practical purposes $t_D = 0$ for accelerated tests of this nature. Thus, sawtooth thermal cycling is more akin to low frequency vibration than to thermal expansion mismatch fatigue. The limits on the rate of temperature change virtually prevent the use of dual-temperature chambers where the test vehicles are moved between two chambers that remain at the two temperature extremes. Without complex special arrangements it is virtually impossible to avoid at least local thermal shock conditions and putting the entire test results in question.

For the "Cold" temperature cycling, shown in Tables 17-2 and 17-4, well-defined dwells are not important (since stress relaxation and creep are not significant factors at those temperatures) as long as the whole assembly reaches temperature equilibrium. The temperature ramp rates are, however, important both to avoid thermal shock and to stay within representative strain rate ranges.

The "quick and dirty" isothermal mechanical cycling tests provide coarse comparative data that can, however, be "benchmarked" using the more accurate low acceleration tests. The most extensive isothermal mechanical cycling program, reported in References 17–19 and 28, involved the controlled deflection of the test vehicle circuit board. The equipment developed for the IEEE Compliant Lead Task Force is commercially available. Other mechanical cycling tests involve test vehicle board twisting or some tensile arrangements.[34,35] It should be noted that isothermal cycling excludes any effects of thermal cycling, such as the thermal expansion differences between the solder and the materials to which it is bonded and between the lead frame and component body materials. While these are secondary effects in most applications, they clearly make important additional cyclic strain contributions in product use.[11] It should be noted, however, that isothermal mechanical cycling with well-controlled applied strains can give highly accurate results. There is no fundamental difference between the primary damage mechanisms operative in thermal cycling and isothermal mechanical cycling. Without the isothermal data from Wild[25] the solder fatigue model described earlier could not have been developed.

17.6.3 Test Vehicle Design

The test vehicles must be designed to allow electrical monitoring for continuity of the solder joints. This requires specially designed test boards as well as specially prepared components. In order to capture a significant number of solder joints, the solder joints should be connected into a continuity daisy-chain test net containing six or more solder joints.[10] Components with solder joints on four sides can be divided into four daisy-chains, one for each side. For components with an odd number of solder joints on a side it is best to skip the center-most joint in the daisy-chains; the corner joints, because they are the most distant from the component center (neutral point) and experience the highest deformations and thus the most fatigue damage, should always be included in the daisy-chains. These center-most joints not connected into the continuity daisy-chain test net can serve to bring power to a resistor chip simulating functional power dissipation and to monitor test "junction" temperatures with thermistors.[11]

The test vehicles should include provisions to monitor and record the cyclic temperature profiles (see, e.g., Reference 11) at various locations on the test vehicles. Accurate knowledge of the actual cyclic temperature histories is essential for the evaluation of the test results and the reliability extrapolation to use conditions.

17.6.4 Sample Statistics and Solder Joint Defects

As has been discussed in the section "Failure Statistical Considerations," the failures in a population of SM solder attached components will follow a Weibull statistical distribution. It is therefore important for the test plan to include provisions to obtain the results from a failure population large enough to provide this statistical information. The less there is known about the assembly being tested and the less well controlled the test setup, the larger is the number of data points required. In order to obtain a statistically adequate number of data points (one per daisy-chain test net) for a well-controlled and understood test setup, a minimum of 24, but preferably 32, daisy-chains should be on test.

While it is possible, but of significant difficulty, to produce test vehicles with intentional variations and "defects" (e.g., misalignments, solder volume, lead nonplanarities, voids, etc), it is perhaps better to subject test vehicles to the accelerated testing in quantities that capture typical, but acceptable, product variations and defects. These variations and defects need to be identified and recorded prior to testing to allow correlation between any early failures that might occur and these defects. It needs to be said, however, that all evidence so far indicates no detrimental effects of so-called defects, (i.e., pad/lead-misalignments, small solder volume, lead nonplanarities, gas-bubble or shrinkage voids, nonshiny solder, etc.) on long-term reliability. Gross defects are likely to cause yield or infant-mortality problems due to inadequate solder joint strength to withstand the rigors of manufacturing, assembly, and transport.

17.6.5 Test Vehicle Assembly and Conditioning

The test vehicles should be assembled and processed as similarly as possible to actual product. Product-similar test vehicle assemblies should reflect product solder joint characteristics such as grain structure, composition, and intermetallic compound layers and their thickness. Solder grain structure is inherently unstable; it will coarsen with time and temperature,[36,37] particularly in the presence of cyclic strains.[37–40] Equivalent ageing in terms of grain growth can be estimated from

$$\ln\left(\frac{t_2}{t_1}\right) = \frac{E_A}{k}\left[\frac{1}{T_2} - \frac{1}{T_1}\right] \tag{17-27}$$

where the activation energy for grain coarsening $E_A = 0.42$ eV and Boltzman's constant $k = 8.62 \times 10^{-5}$eV/K.[41]

Fine grain structure in the solder joints will result in longer fatigue lives in accelerated testing, which does not allow the time for these time dependent processes to occur to the same degree as they would in product use over many years. The changes in the structure of solder, which undergoes and accumulates fatigue damage, appear to follow the sequence of grain structure coarsening,[37–40] micro-voiding/cavitation formation at grain boundaries,[40] micro-cracking with ~50 to 70 percent of life remaining, coalescence of micro-cracks, and fracture. Therefore, it stands to reason that accelerated test vehicles should start with a solder grain structure that is more representative of the solder grain structure in product after some period of operation.

After assembly, the test vehicles should be subjected to an accelerated thermal aging (e.g., 300 h at 100°C[11]) in air to simulate a reasonable use period and to accelerate such possible processes as solder grain growth, intermetallic compound growth, and oxidation. Storing the test vehicles after this artificial ageing for some additional time at room temperature before commencing with the fatigue testing serves to further stabilize the solder structure.

17.6.6 Failure Definition and Detection

Solder joint fractures manifest themselves in product predominantly by intermittant electrical failures during either thermal transients or dynamic loading conditions. These intermittents can, at least initially, be of exceedingly short duration (<1 μs). Only after many cycles of the solder joint fracture surfaces working against each other will electrical contact be broken during steady-state conditions at one of the cyclic tempertures extremes.[10,29]

For multiple solder joints daisy-chained into continuity nets, the optimum failure definition is the first confirmed interruption of electrical continuity even of very short duration.[10] This failure definition not only closely corresponds to the failure indication during product field operation, but it also does not require test interruptions with time consuming and uncertain failure searches. Since full solder joint fractures manifest themselves—at least initially—as high electrical resistance spikes of exceedingly short duration and almost exclusively during the cyclic transients, the continuity nets require continuous, *not* multiplexed (i.e., switched) monitoring.

For practical and economic reasons also, continuous monitoring for electrical discontinuity appears to be the most viable option given the large number of test nets required to produce statistically adequate databases. Equipment to monitor definite changes in resistance is significantly more expensive than the electronics necessary to detect essentially open nets, without providing more consistent and more relevant data.

Some previous studies have been carried out by monitoring with the sensing current limited to 1 mA and 5 V and resistance spikes of 1 μs or longer detectable.[11,15] For the IEEE Compliant Lead Task Force study[17–19,28] failure detection was carried out on 256 continuously and simultaneously monitored channels. The sensing electronics could detect 1000-Ω resistance spikes of 0.2-μs duration with a maximum sensing current at 1.25 mA at zero net resistance decreasing asympotically to zero current at 10 V potential across a full net open. The argument has been advanced that even at these currents and voltages, arcing across the short-duration gap is possible. Commercial equipment has recently become available that can detect resistance spikes of 0.2-μs duration with a 300-Ω threshold limit with a sensing current maximum of 1.25 mA at zero net resistance declining to 0 mA at 0.6-V maximum voltage at infinite net resistance.[42]

17.7 PREDICTION OF SM SOLDER JOINT RELIABILITY

17.7.1 Simple Cyclic Load Histories

Surface mount solder joint reliability prediction, within the stated caveats and limitations in the section "Analytical Model of Solder Shear Fatigue," involves primarily combining and rearranging some of the equations in these sections. From Equations (17-12) and (17-24) the expected cyclic life with a failure probability of $x = 100\ F(N)$ is for leadless SM solder attachments

$$N_f(x\%) = \frac{1}{2}\left[\frac{F}{2\varepsilon_f'}\frac{L_D\Delta\alpha\Delta T_e}{h}\right]^{\frac{1}{c}}\left[\frac{\ln(1 - 0.01x)}{\ln(0.5)}\right]^{\frac{1}{\beta}} \tag{17-28}$$

where $N_f(x\%)$ = number of operating cycles at x percent failure probability

For leaded solder attachments, the cyclic life prediction

$$N_f(x\%) = \frac{1}{2}\left[\frac{F}{2\varepsilon_f'}\frac{K_D(L_D\Delta\alpha\Delta T_e)^2}{(200\text{ psi})\,Ah}\right]^{\frac{1}{c}}\left[\frac{\ln(1 - 0.01x)}{\ln(0.5)}\right]^{\frac{1}{\beta}} \tag{17-29}$$

results from combining Equations (17-17) and (17-24). Equation (17-29) is in English units; the scaling coefficient for metric dimension is 1.38 MPa instead of 200 psi in the denominator in the first bracket.

It should be noted that these equations apply only to solder joints consisting of 60% Sn-40% Pb or eutectic solder. Further, for leaded components with lead materials that have CTEs significantly lower than copper alloy materials, the results from Equation (17-29) will be optimistic since the fatigue damage contributions from the solder-lead material CTE-mismatch are not included.

For high-frequency applications, $t_D < 3$ sec, and low temperature applications, $T_0 < -20°C$, for which the stress relaxation and creep in the solder joint is substantially incomplete, if not non-existent, the direct application of the Manson-Coffin fatigue relationship (see Eq. (17-3) in the section "Analytical Model of Solder Shear Fatigue" and Reference 5) is advised.

17.6.2 Multiple Cyclic Load Histories

Multiple cyclic load histories [e.g., "Cold" temperature fatigue cycles combined with higher temperature creep-fatigue cycles (see Table 17-2) combined with vibration] all make their contributions to the cumulative fatigue damage in solder joints. Under the assumption that these damage contributions are linearly cumulative—this assumption underlies Equation (17-1) as well—and that the simultaneous occurrence or the sequencing order of these load histories makes no significant difference, the Palmgren-Miner's rule[43] can be applied

$$\sum\frac{N_i}{N_{f,i}} \leq 1 \tag{17-30}$$

where N_i = actually applied number of cycles at a specific cyclic load level

$N_{f,i}$ = fatigue life at a specific cyclic load level alone

Equation (17-30) can be used with the allowable sum of the fatigue damage fractions significantly less than unity to provide margins of safety.

17.8 ACKNOWLEDGMENT

Most of this work was done while the author was with AT&T Bell Laboratories.

REFERENCES

1. Klinger, D. J., Y. Nakada, and M. A. Menendez, *AT&T Reliability Manual* Eds., New York, Van Nostrand Reinhold, 1990.
2. Holcomb, D. P., and J. C. North, "An Infant Mortality and Long-Term Reliability Failure Rate Model for Electronic Equipment," *AT&T Technical Journal V.* **64** *No.* (1), 1985.
3. *MIL-HDBK-217D, Military Handbook - Reliability Prediction of Electronic Equipment,* January 15, 1982, Department of Defense, Washington, D.C. 20301.
4. Morrow, J. D., "Cyclic Plastic Strain Energy and Fatigue of Metals," *ASTM STP 378*, ASTM, Philadelphia, PA, 1964, pp. 45–87.
5. Manson, S. S., *Thermal Stress and Low Cycle Fatigue,* New York, McGraw-Hill, 1966.
6. Engelmaier, W., "Functional Cycling and Surface Mounting Attachment Reliability," *ISHM Technical Monographic Series 6984-002,* Int. Soc. for Hybrid Microelectronics, Silver Spring, MD, October 1984, pp. 87–114; also in *Circuit World,* **11**(3), spring 1985, pp. 61–72.
7. ———, "Fatigue Life of Leadless Chip Carrier Solder Joints During Power Cycling," *IEEE Trans. CHMT,* **6**(3) pp. 232–237.
8. Clech, J.-P., W. Engelmaier, R. W. Kotlowitz, and J. A. Augis, "Surface Mount Solder Attachment Reliability Figures of Merit—'Design for Reliability' Tools," *Proc. 1989 SMART V Conf.,* New Orleans, LA, January 9–12, 1989.
9. Engelmaier, W., "Surface Mount Solder Joint Long-Term Reliability: Design, Testing, Prediction," *Soldering & Surface Mount Technology,* no. 1, February 1989, pp. 14–22.
10. ———, "Is Present-Day Accelerated Cycling Adequate for Surface Mount Attachment Reliability Evaluation?," *IPC-TP-653,* Institute for Interconnecting and Packaging Electronic Circuits, Lincolnwood, IL. September 1986.
11. Engelmaier, W., and A. I. Attarwala, "Surface Mount Attachment Reliability of Clip-Leaded Ceramic Chip Carriers on FR-4 Circuit Boards," *IEEE Trans. CHMT,* **12**(2), June 1989, pp. 284–296; also partially in *IEPS Journal,* **9**(4), January 1988, pp. 3–11.

12. Suhir, E., "Axisymmetric Elastic Deformations of a Finite Circular Cylinder with Application to Low Temperature Strains and Stresses in Solder Joints," *J. Appl., Mech.,* **56**(2), June 1989, pp. 328–333.
13. "Guidelines for Surface Mounting and Interconnecting Chip Carriers," *IPC-CM-78,* Institute for Interconnecting and Packaging Electronic Circuits, Lincolnwood, IL, November 1983.
14. Klemencic, R., E. Kny, and W. Schmidt, "Multilayer Circuit Boards with Molybdenum-Copper Metal Cores," *Circuit World,* **15**(4), July 1989, pp. 5–8.
15. Engelmaier, W., "Surface Mount Solder Joint Reliability of Leadless Ceramic Chip Carriers on Different Substrates," to be published.
16. Kotlowitz, R. W., and W. Engelmaier, "Impact of Lead Compliance on the Solder Attachment Reliability of Leaded Surface Mounted Devices," *Proc. Int. Electronics Packaging Soc. Conf.,* San Diego, CA, November 17–19, 1986, pp. 841–865.
17. Engelmaier, W., "Solder Joint Reliability and Testing Considerations for Leaded Chip Carriers," *Nat. Electronic Packaging and Production Conf.,* NEPCON West, Anaheim, CA, February 26–28, 1985, pp. 884–885 (extended abstract only); Presentation available on IEPS Videotape VR 001, "Lead Compliance and Solder Integrity of Surface Mount Leaded Packages," Int. Electronics Packaging Society, Wheaton, IL.
18. ———, "Interim Report on the Test Results of IEEE Test Program," IEPS Videotape VR 004, "IEEE Compliant Lead Task Force Results of Phase A," Int. Electronics Packaging Society, Wheaton, IL.
19. ———, "IEEE Compliant Lead Task Force—Phase I: Correlation and Analysis of Solder Joint Fatigue Results Obtained at Different Cyclic Frequencies and Temperatures," *Proc. Int. Electronics Packaging Soc., Conf.,* Orlando, FL, October 21–23, 1985, p. 279 (extended abstract only); also, presentation available on IEPS Videotape VR 002, "Second Interim Report on the IEEE Complaint Lead Task Force," Int. Electronics Packaging Society, Wheaton, IL.
20. Kotlowitz, R. W., "Comparative Compliance of Generic Lead Designs for Surface Mounted Components," *Proc. Int. Electronics Packaging Soc. Conf.,* Boston, MA, November 9–11, 1987, pp. 965–984.
21. ———, "Comparative Compliance of Representative Lead Designs for Surface Mounted Components," *Proc. 1988 IEPS Conf.,* Dallas, TX, November 7–9, 1988, pp. 908–948.
22. Liljestrand, L.-G., and L.-O. Anderson, "Accelerated Thermal Fatigue Cycling of Surface Mounted PWB Assemblies in Telecom Equipment," *Proc. Int. Packaging Soc. Conf.,* Boston, MA, November 9–11, 1987, pp. 965–984; also in *Circuit World* **14**(3), March 1988, pp. 69–73.
23. Hall, P. M., "Strain Measurements During Thermal Chamber Cycling of Leadless Ceramic Chip Carriers Soldered to Printed Boards," *Proc. 34th Electronic Components Conf.,* New Orleans, LA, May 1984, p. 107.
24. ———, "Creep and Stress Relaxation in Solder Joints in Surface-Mounted Chip Carriers," *Proc. 37th Electronic Component Conf.,* Boston, MA, May 1987, p. 579.
25. Wild, R. N., "Some Fatigue Properties of Solders and Solder Joints," IBM Tech. Rep. 73Z000421, January 1973.

26. Clech, J.-P. and J. A. Augis, "Temperature Cycling, Structural Response and Attachment Reliability of Surface-Mounted Leaded Packages," *Proc. 1988 IEPS Conf.*, Dallas, TX, November 7–9, 1988, pp. 305–325.
27. Hines, L. L., AT&T Bell Laboratories, Columbus, OH, private communication, July 1988.
28. Balde, J. W. "The IEEE Compliant Lead Task Force," *IEEE Trans CHMT-10,* Sept. 1987, pp. 463–469.
29. Ng, K. G., "Assessing Surface Mount Attachment Reliability of Ceramic Cylindrical Resistors on FR-4 Circuit Boards," *Proc. Expo SMT '88,* Las Vegas, NV, November 1988, pp. 1–6.
30. Wild, R. N., "1974 IRAD Study—Fatigue Properties of Solder Joints," IBM Report No. M45-74-002, Contract No. IBM 4A69, Jan. 5, 1975.
31. ———, IBM, Owego, NY, private communication, May 1989.
32. Engelmaier, W., "Test Method Considerations for SMT Solder Joint Reliability," *Proc. Int. Electronics Packaging Soc. Conf.,* Baltimore MD, Oct. 29–31, 1984, pp. 360–369; also in *Brazing & Soldering,* no. 9, autumn 1985, pp. 40–43.
33. Tien, J. K., B. C. Hendrix, B. L. Bretz, and A. I. Attarwala, "Creep-Fatigue Interactions in Solders" *Proc. 39th Electronics Components Conf.,* Houston, TX, May 1989, pp. 259–263.
34. Solomon, H. D., "Low Cycle Fatigue of Surface Mounted Chip Carrier/Printed Wiring Board Joints," *Proc 39th Electronics Components Conf.,* Houston, TX, May 1989, pp. 277–292.
35. Enke, N. F., T. J. Kilinski, S. A. Schroeder, and J. R. Lesniak, "Mechanical Behaviors of 60/40 Tin-Lead Solder Lap Joints," *Proc. 39th Electronics Components Conf.,* Houston, TX, May 1989, pp. 264–272.
36. Lampe, B. T., *Welding Research Supplement* No. 10, 1976, p. 330.
37. Karjalainen, L. P., *et al.,* "Relation Between Microstructure and Fracture Behavior in Surface Mounted Joints," *Brazing & Soldering,* no. 15, autumn 1988, pp. 37–42.
38. Wolverton, W. M., "The Mechanisms and Kinetics of Solder Joint Degradation," *Brazing & Soldering,* no. 13, autumn 1987, pp. 33–38.
39. Frear, D., D. Grivas, and J. W. Morris, "A Microstructural Study of the Thermal Fatigue Failures of 60Sn-40Pb Solder Joints," *J. Electron. Mater. (USA),* **17**(2), March 1988, pp. 171–180.
40. Greenwood, P. J., T. C. Reiley, J. K. Tien, and V. Raman, "Cavitation in a Pb-Low Sn Solder During Low Cycle Fatigue," *Scripta Met.,* **22**(9), September 1988, pp. 1465–1468.
41. Evans, J. W., "Grain Growth in Eutectic Solder: Impact on Accelerated Testing," *Proc. 33rd Ann. IPC Meeting,* Boston, MA, April 1–6, 1990.
42. Sofia, J. W., Anatech, Inc., Wakefield, MA, Private communications, November 1988.
43. Miner, M. A., "Cumulative Damage in Fatigue," *J. Applied Mechanics,* vol. **12,** 1945.

18

Surface Mount Attachment Reliability and Figures of Merit for Design for Reliability

J-P M. Clech and J. A. Augis

Surface Mount (SM) assemblies in the field are subject to complex thermal fluctuations. The cyclic thermal expansion mismatch between SM components and the substrates to which they are attached may lead to fatigue or wearout of solder joints during the useful life of a product. Attachment reliability is the ability of solder joints to provide electrical and mechanical connections under stated conditions and for a specified duration. Packaging engineers and physical designers thus bear the responsibility to design SM attachments that will survive the desired product life. This is achieved by proper design of the attachment and soldering according to recommended Workmanship Requirements (e.g., reference[1]). While several companies have gathered attachment reliability data for generic SM components, the data often remain proprietary or are not readily available for less common or emerging packaging technologies. Reliability evaluation can then be an involved process[2,3] including the design and assembly of test vehicles, accelerated testing, data, and failure mode analysis, all of which require the appropriate expertise and adequate commitment of time and resources. This process amounts to a 6-month to 1-year effort while design cycles for SM packages or electronic assemblies are as short as 1 month, sometimes less. Clearly, design tools are needed to build in attachment reliability throughout the design process and provide reliability information quickly, based on critical design and assembly parameters.

Figures of Merit (FMs) have been developed[4,5] that serve as design for reliability tools in the design of solder joint interconnections and account for the primary factors that determine attachment reliability in the field. The FM formulation includes the substrate-component thermal expansion mismatch, the attachment design parameters, the thermal environment, the product service life, and the acceptable cumulative failure probability per device or attachment reliability target. The FMs are derived from simple solder joint theories and empirical fatigue data for the attachment of SM devices. By using the FMs, packaging engineers and physical designers can assess the impact of design parameters on end-use attachment reliability and, if need be, change them to meet a stated reliability target. The FM tools provide guidelines to ensure that no significant wearout of solder joints occurs during the useful life of an electronic assembly and give quantitative estimates of attachment reliability for quality solder joints. Because of simplifications inherent to design tools and tolerances on design parameters, reliability estimates derived from the FM tools should be considered as order of magnitude estimates. End-use reliability for a given product and application can only result from design for reliability, followed by quality assembly.

This chapter focuses on engineering approaches to address SM attachment reliability at the design stage. Fundamentals of solder joint fatigue for leadless and leaded attachments are reviewed in the next section, which provides the theoretical foundation for developing the FM tools. The FMs are then presented and applied to several examples.

18.1 MECHANICS AND FATIGUE OF SM SOLDER JOINTS

This section reviews simplified concepts of the mechanics and fatigue of solder joints in SM leadless and leaded components, providing insight into the theoretical aspects underlying the FM formulation.

18.1.1 SM Leadless Attachments

Figure 18-1 is a schematic of an SM leadless assembly showing the design parameters that govern the solder joint response in thermal cycling:

- L (in.) is the distance from the neutral axis of the component to the center line of the farthest solder joint; for a perimeter chip carrier, L is half the diagonal dimension of the package.
- h (in.) is the solder joint height or component standoff above the substrate.
- $\Delta\alpha$ (in./in.°C) is the absolute mismatch between the coefficients of thermal expansion (CTE) of the component α_c and substrate α_s: $\Delta\alpha = |\alpha_c - \alpha_s|$.

When the assembly is subject to a temperature swing ΔT(°C),* solder joints of leadless components undergo a gross shear strain.[3,6,7] Assuming complete

*In power cycling, an equivalent temperature range of ΔT_e is used to account for component power dissipation.[3]

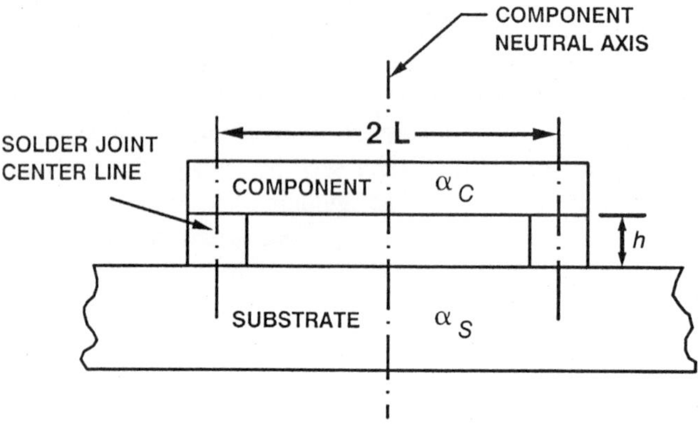

Figure 18-1 Schematic of surface mount leadless component showing the design parameters that affect solder attachment reliability in thermal cycling.

stress relaxation for solder at the temeprature extremes, the amplitude of the applied shear strain is

$$\Delta\gamma = \frac{L}{h}\Delta\alpha\Delta T \tag{18-1}$$

The solder joint response during thermal cycling is characterized by hysteresis loops in the shear stress-strain plane (see Fig. 18-2). Such loops have been obtained experimentally for leadless chip carriers[8,9] and numerical simulations have been proposed by several authors.[10,11,12,13] Their general shape gives insight into the solder joint deformation and stress history. The initial response as the temperature ramps up is *elasto-plastic* (elastic plus time independent plastic deformations) and is limited by the solder yield strength. During the dwell at higher temperature, T_{high}, creep and stress relaxation occur along an isothermal stress reduction line with a steep slope determined by the stiffness of the assembly. In accelerated testing with short dwells, stress reduction may be incomplete. If the dwell at T_{high} is long enough, as in slow temperature cycling or low frequency use conditions (e.g., 1 cycle/day) stress reduction proceeds to zero stress and maximum plastic strain. The stress-strain response is reversed in the next half-cycle with stress reduction proceeding in the opposite direction at low temperature (T_{low}). For leadless or "stiff" SM attachments, the largest cyclic strain energy density associated with the hysteresis loop area is obtained as

$$\Delta W \sim C\Delta\gamma \tag{18-2}$$

where C is a material constant characteristic of solder and independent of the assembly design parameters.

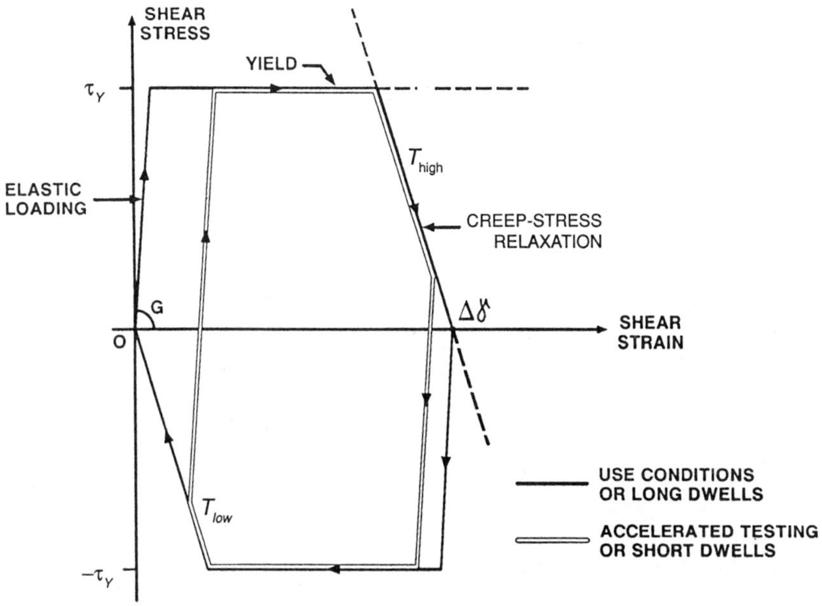

Figure 18-2 Hysteresis loops characterizing the solder joint response in the shear stress-strain plane during thermal cycling; case of SM leadless components.

The plastic part of the cyclic shear strain $\Delta\gamma_p$ leads to solder joint failures in low cycle fatigue. This is described by Coffin-Manson relationships[3,6,14] that give the median cyclic life $\overline{N}_f$ as

$$\overline{N}_f = C_1 \, (\Delta\gamma_p)^{\frac{1}{m}} \tag{18-3}$$

where C_1 is a material constant and the exponent parameter m is about -0.5. The solder joint cyclic life is expected to decrease with increasing CTE mismatch. For large temperature fluctuations and CTE-mismatched substrates and components, the component size may have to be limited or the standoff height be increased for the solder joints to survive the desired cycles of operation.

18.1.2 SM Leaded Attachments

One solution to the problem of solder fatigue in leadless assemblies is to provide a more compliant attachment, using leaded components. Compliant leads take up the component-substrate thermal expansion mismatch,[15,16] thus reducing the cumulative fatigue damage imparted to solder joints. Figure 18-3 is a schematic

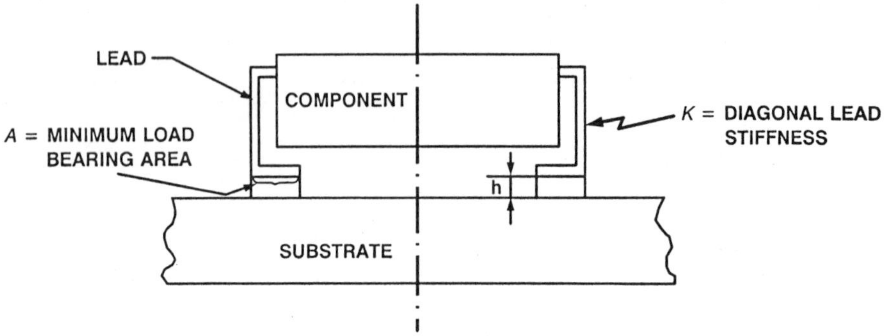

Figure 18-3 Schematic of surface mount leaded component showing the design parameters, other than those defined in Figure 18-1 for leadless assemblies, that affect attachment reliability.

of an SM leaded assembly. In addition to the design parameters defined in the leadless case, the attachment reliability of leaded assemblies is determined by:

- K (lb/in.) the diagonal lead stiffness
- A (in.^{2}) the minimum load bearing area of the solder joint, subjected to a shear force transmitted by the lead to the solder joint
- h (in.) a characteristic solder joint thickness

The diagonal lead stiffness K is the flexural spring constant of the lead in the diagonal direction of the package where the thermal expansion mismatch is the largest. The parameter K is obtained by classical strain energy methods[17–19] or finite element analysis. In the FM formulation for leaded components, the minimum cross-sectional area A is approximated as two-thirds the solder wetted lead area projected to the surface mounting pad. We take the characteristic solder joint thickness h as half the solder paste stencil thickness, a representative dimension for the average solder thickness under the lead foot.

The solder joint response in thermal cycling follows a hysteresis loop as in the leadless case. The initial solder response is *elastic* and, if the leads are compliant enough, stresses remain below the yield strength of solder (see Fig. 18-4). At the temperature extremes, stress reduction proceeds along isothermal lines of much lesser slope than in the leadless case.[20] The product of S, the largest applied stress, and $\Delta\gamma$, the total cyclic shear strain range, gives the maximum available cyclic strain energy density, a measure of cyclic fatigue damage.

The thermal expansion mismatch $L\Delta\alpha\Delta T$ is transmitted to the joint as a force $K(L\Delta\alpha\Delta T)$, and the resulting shear stress

$$S = \frac{K\,(L\Delta\alpha\Delta T)}{A} \qquad (18\text{-}4)$$

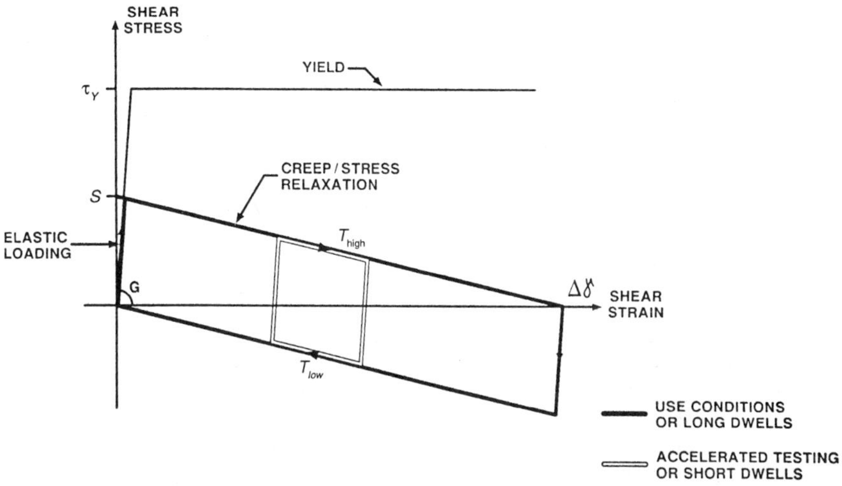

Figure 18-4 Hysteresis loops of solder joint response during thermal cycling; case of a leaded assembly with compliant leads.

determines the initial stress reduction rate during dwells at constant temperature. The maximum cyclic strain energy density is then obtained as

$$\Delta W = S\Delta\gamma = \frac{K(L\Delta\alpha\Delta T)^2}{Ah} \tag{18-5}$$

The median cyclic life for solder joints of leaded components has been found to follow a generalized fatigue law*[21]

$$\overline{N}_f = C_2[\Delta W]^{\frac{1}{m}} \tag{18-6}$$

where C_2 is a material constant and the exponent parameter m is in the range of -0.5 to -0.7. Figure 18-5 shows a plot of ΔW versus $\overline{N}_f$[4,5] for five SM leaded assemblies subjected to accelerated testing in functional[16] or slow temperature cycling: 20 to 100°C, 10 cycles/day.[23] The slope of the regression line, ΔW versus $\overline{N}_f$ is $m = -0.64$. For the three assemblies with the longest lives, lower bounds estimates of $\overline{N}_f$ were used since no failures were observed at the end of tests. Testing to failure would give larger values of $\overline{N}_f$ and a reduced slope for the regression line, closer to the more commonly observed values $m \sim -0.5$ for solder.

*The Coffin-Manson relationship [Eq. (18-3)] is a particular case of and can be derived from Morrow's generalized fatigue law.[22]

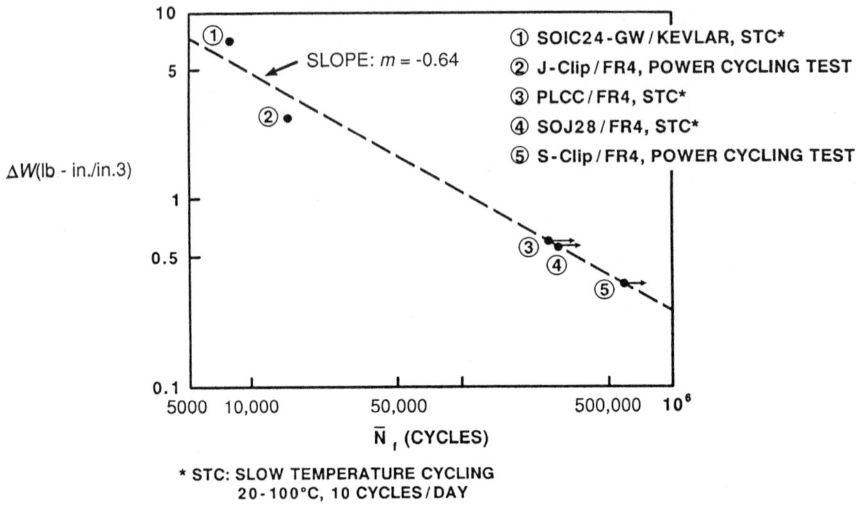

Figure 18-5 Plot of cyclic strain energy density versus median cycles to failure for SM leaded assemblies subjected to accelerated thermal cycling.

Equations (18-5) and (18-6) suggest that the cyclic life can be increased by providing more lead compliance (lower K) or increasing the lead-solder wetting area (larger A). The lead stiffness can be reduced most effectively by increasing the overall lead height[17–19] or by using a thinner lead frame.[24] Often, an increase in the width of the lead foot at the tip of the lead is beneficial to reliability since it increases the lead-solder wetted area with no significant reduction in lead compliance (see Example 18-3).

18.1.3 Failure Distribution

Another important feature of SM reliability studies is the statistical distribution of solder joint failures. The wearout of solder joints follows a Weibull distribution around the median cycles to failure,** $\overline{N}_f$ (see Fig. 18-6). The cumulative distribution function, or cumulative failure probability, at the current cycle n is

$$F(n) = 1 - 0.5^{\left(\frac{n}{\overline{N}_f}\right)^{\beta}} \qquad (18\text{-}7)$$

**Median cycles to failure are used instead of the characteristic scaling factors of Weibull distributions, since fatigue data for solder and SM attachments are most often reported in terms of median life.

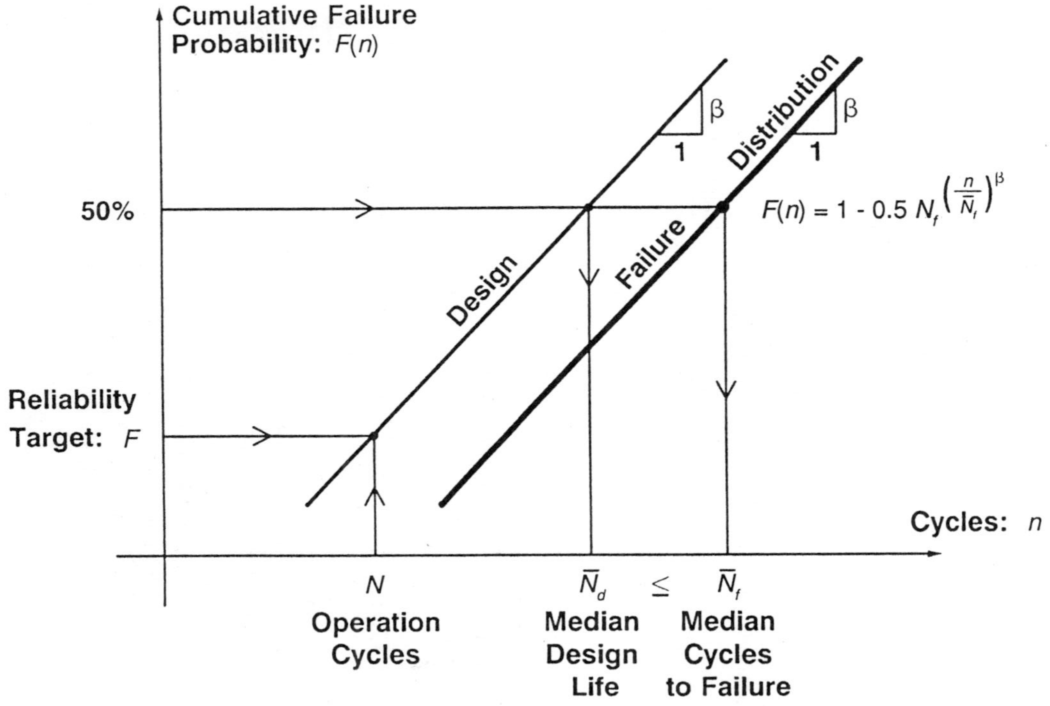

Figure 18-6 Weibull plot: cumulative failure probability versus cycles. The design line must be to the left of the wearout failure distribution for the reliability expectations to be met.

where the Weibull slope, or shape factor β, is about 4 for leadless attachment[3] and 2 to 3 for leaded attachment.[15,16,23] To our knowledge, the difference in β's for leadless and leaded attachments is not well understood.

For practical purposes, reliability requirements are stated in terms of the largest acceptable cumulative failure probability per device,† $F(n) \leq$ F, at the end of the product life (that is, at most a fraction F of the overall population is tolerated to fail). Through the Weibull distribution, we can relate $\overline{N}_f$ to the number of operation cycles N during the product life. Reliability requirements will be met if the median design life $\overline{N}_d$ is less than median cycles to failure (see Fig. 18-6). This condition is written as

$$\overline{N}_d = \left[\frac{\ln(0.5)}{\ln(1 - F)}\right]^{\frac{1}{\beta}} \times N \leq \overline{N}_f \qquad (18\text{-}8)$$

which will account explicitly for the number of operation cycles and the reliability requirements in the formulation of design tools. The fatigue laws and the statistical failure distribution then imply

$$Q = \overline{N}_d^{-m} \times \Delta d \leq \overline{N}_f^{-m} \times \Delta d \qquad (18\text{-}9)$$

where $\Delta d = \Delta\gamma$ for leadless attachment and $\Delta d = \Delta W$ for leaded attachment. The quantity Q on the left-hand side of Equation (18-9) is a relative measure of damage accumulated during the median design life; it should be less than the maximum allowable damage given on the right-hand side of Equation (18-9).

18.2 FIGURES OF MERIT FOR ATTACHMENT RELIABILITY

The above considerations of solder joint design, fatigue, and wearout have been packaged as design for reliability tools. These FM tools address SM attachment reliability at several stages of the design process:

- FM (comp) for *component* design and/or selection
- FM (assy) for component-substrate *assembly* design
- FM (env) for the reliability impact of the thermal *environment*
- FM (rel) for end-use attachment *reliability*.

The FM formulas are summarized hereafter for easy reference. This formulation has been developed for and applies only to SM attachment using 60% Sn – 40% Pb or eutectic solder and leaded assemblies with a copper alloy lead frame.

†One solder joint failure is one device failure.

For leadless (ll) SM solder attachment:

$$\mathrm{FM(comp,ll)} \equiv 20\,\frac{h}{L} \tag{18-10}$$

$$\mathrm{FM(assy,ll)} \equiv 8 \times 10^{-5}\,(^\circ\mathrm{C}^{-1})\,\frac{h}{L\Delta\alpha} \tag{18-11}$$

$$\mathrm{FM(env,ll)} \equiv 3.2 \times 10^{-3}\,\frac{h}{L\Delta\alpha\Delta T} \tag{18-12}$$

$$\mathrm{FM(rel,ll)} \equiv 0.32\,(\mathrm{cycles}^{0.44})\,\frac{h}{L\Delta\alpha\Delta T N^{0.44}}\left[\frac{\ln(1-F)}{\ln(0.5)}\right]^{\frac{0.44}{\beta}} \tag{18-13}$$

For leaded (ld) SM solder attachment:

$$\mathrm{FM(comp,ld)} \equiv 10^7\,(\mathrm{psi})\,\frac{Ah}{KL^2}\;(10^7\ \mathrm{psi\ or\ } 6.9 \times 10^4\ \mathrm{MPa}) \tag{18-14}$$

$$\mathrm{FM(assy,ld)} \equiv 2.5 \times 10^{-4}\,(\mathrm{psi} \times {}^\circ\mathrm{C}^{-2})\,\frac{Ah}{K(L\Delta\alpha)^2}$$
$$(2.5 \times 10^{-4}\ \mathrm{psi} \times {}^\circ\mathrm{C}^{-2}\ \mathrm{or}\ 1.72\ \mathrm{Pa} \times {}^\circ\mathrm{C}^{-2}) \tag{18-15}$$

$$\mathrm{FM(env,ld)} \equiv 0.625\,(\mathrm{psi})\,\frac{Ah}{K(L\Delta\alpha\Delta T)^2}\;(0.625\ \mathrm{psi\ or\ } 4.3 \times 10^3\ \mathrm{MPa}) \tag{18-16}$$

$$\mathrm{FM(rel,ld)} \equiv 87.5\,(\mathrm{psi} \times \mathrm{cycles}^{0.44})\,\frac{Ah}{K(L\Delta\alpha\Delta T)^2 N^{0.44}}\left[\frac{\ln(1-F)}{\ln(0.5)}\right]^{\frac{0.44}{\beta}}$$
$$(87.5\ \mathrm{psi} \times \mathrm{cycles}^{0.44}\ \mathrm{or}\ 0.604\ \mathrm{MPa} \times \mathrm{cycles}^{0.44}) \tag{18-17}$$

The FMs are dimensionless, with increasing values implying higher attachment reliability. There are no set lower bounds for FM (comp), FM (assy) and FM (env); however, for FM (rel), a target of 1 or more is needed to meet reliability requirements. The scaling constants in the FM formulas were selected such as:

1. FM(rel) $\geqslant$ 1 warranties reliability requirements with an acceptable failure probability per device F.

2. $0.7 < \text{FM(rel)} \leqslant 1$ is a region of uncertainty whether these requirements are met.
3. $\text{FM(rel)} \leqslant 0.7$ means that the SM design does not meet the reliability requirements.

If condition 1 is not met, it is recommended to check for condition 2; the SM design may still be acceptable when it falls in the gray area. A more detailed study, including testing and extrapolation of test data to use conditions, is then necessary to establish attachment reliability and capture the effect of design parameters not included in the FM tools. It is also emphasized that the FM formulation assumes complete stress relaxation for solder. In the case of leaded attachment with very compliant leads, stresses in the solder joints are likely to be low and this will be a conservative assumption. For field temperature excursions with incomplete stress relaxation, cycles to failure predicted from the FMs will be conservative; a more accurate prediction is then necessary that includes the effect of dwell times and mean solder joint temperatures.[25] Note also that the above formulation does not address the effect of parameters such as joint shape, mechanical support (pad size, conformal coating), or joint quality. For a lack of data, their impact on attachment reliability is better assessed through a carefully planned test program.

18.2.1 Derivation of FM Formulas

The reliability figures of merit, FM(rel), are the inverse of the cumulative solder joint damage Q [see Equation (18-9)], with appropriate weighting factors D_1 and D_2 that scale the FM (rel) metric to a minimum of 1 for reliable attachment. The FM(rel)'s are defined from Equation (18-9) as

$$\text{FM(rel,ll)} = \frac{D_1}{\overline{N}_{\bar{d}}^{-m} \times \Delta\gamma} \quad \text{for leadless attachment} \qquad (18\text{-}18)$$

and

$$\text{FM(rel,ld)} = \frac{D_2}{\overline{N}_{\bar{d}}^{-m} \times \Delta W} \quad \text{for leaded attachment} \qquad (18\text{-}19)$$

The other formulas are obtained by working backward from Equations (18-18) and (18-19) and preserving in each FM the design parameters relevant to that step of the process. Appendix 18-A gives the details of the FM derivation and of the selection of scaling constants to fit the FM model to empirical reliability data. Again, the end-use FM(rel) should be larger than one for reliability requirements to be met. While the other FMs do not predict reliability in an absolute way, they are indicative of reliability potential and are useful for comparison purposes. For example, since FM(comp) measures the ability of a component to provide for reliable attachment in a given application, a new component design with a larger FM(comp) than the current design is an acceptable alternative when the latter meets expected reliability requirements.

18.2.2 FMs for Multiple Thermal Fluctuations

A design evaluation based exclusively on worst-case thermal fluctuations can be too conservative or lead to rejection of SM attachments that would survive typical operating cycles and occasional worst-case fluctuations. In Appendix 18-B, we show that Miner's rule[26] provides a simple way of extending the FM formulation to thermal cycles of various amplitudes. A generalized formulation of FM(rel) for complex thermal fluctuations with various amplitudes ΔT_i during N_i cycles is obtained as

$$\mathrm{FM(rel,ll)} = \frac{0.32\,h}{L\Delta\alpha \times S_{ll}} \left[\frac{\ln(1 - F)}{\ln(0.5)}\right]^{\frac{1}{a\beta}} \quad \text{for leadless attachment} \tag{18-20}$$

and

$$\mathrm{FM(rel,ld)} = \frac{87.5\,Ah}{K(L\Delta\alpha)^2 \times S_{\mathrm{ld}}} \left[\frac{\ln(1 - \mathrm{F})}{\ln(0.5)}\right]^{\frac{1}{a\beta}} \quad \text{for leaded attachment} \tag{18-21}$$

where

$$S_{\mathrm{ll}} = \left(\sum_i N_i(\Delta T_i)^a\right)^{\frac{1}{a}}$$

and

$$S_{\mathrm{ld}} = \left(\sum_i N(\Delta T_i)^{2a}\right)^{\frac{1}{a}}$$

are measures of the cumulative environmental stress imparted to the SM assembly by multiple thermal fluctuations in the field, and $a = 1 / 0.44$.

18.2.3 Graphical Interpretation and Reliability Charts

For reliability at an acceptable cumulative failure probability F per device, FM(rel) must be larger than or equal to 1. In terms of FM(assy), and for multiple thermal fluctuations, this condition is equivalent to

$$\text{in the leadless case: FM(assy,ll)} \geqslant 2.5 \times 10^{-4} \left[\frac{\ln(0.5)}{\ln(1 - F)}\right]^{\frac{1}{a\beta}} \times S_{\mu} \tag{18-22}$$

$$\text{in the leaded case: FM(assy,ld)} \geqslant 2.86 \times 10^{-6} \left[\frac{\ln(0.5)}{\ln(1 - F)}\right]^{\frac{1}{a\beta}} \times S_{\text{ld}} \tag{18-23}$$

These equations signify that the SM attachment is robust enough to withstand the damage imparted to solder joints throughout the product life. On the left-hand sides of Equations (18-22) and (18-23), the FM(assy)'s group the design parameters that govern attachment reliability and, as such, serve as measures of assembly robustness. The quantities on the right-hand sides are relative measures of the cumulative environmental stress.

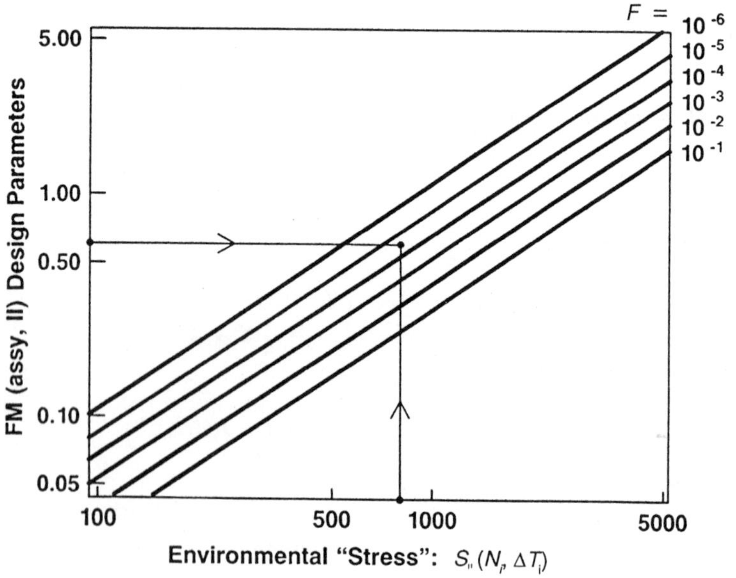

Figure 18-7a Chart for attachment reliability estimates for SM leadless components.

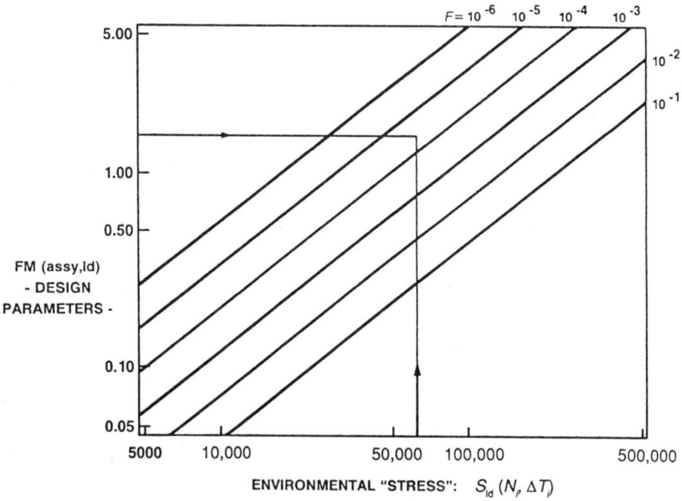

Figure 18-7b Chart for attachment reliability estimates for SM leaded components.

The above interpretation leads to a graphical estimate of attachment reliability using the charts in Figure 18-7*a* and *b*. On the horizontal axis, S_{ll} or S_{ld} combine the cycles of operation and temperature fluctuations for complex thermal histories; on the vertical axis, FM(assy) includes the important design parameters. The lines of constant cumulative failure probability $F = x$ are parallel in the log-log coordinate systems: (S_{ll}, FM(assy,ll)) for leadless devices, and (S_{ld}, FM(assy,ld)) for leadless devices. Given the design parameters and use conditions, a representative point above the line $F = x$ ensures a cumulative failure level less than x, per device, at the end of the design life.

18.3 EXAMPLES

18.3.1 Example 18-1: Chip Components on FR-4 Printed Wiring Boards

The use of FMs is illustrated here for size code 1206 and 1825[1] chip components surface mounted on an FR-4 printed wiring board. The intended service life is 10 years with a reliability target $F < 10^{-4}$ (0.01 percent) at the end of life. The use thermal environment is described by a daily temperature fluctuation of 20°C at the circuit board level. The relevant design data are

$$L = \begin{cases} 0.047 \text{ in for the 1206s} \\ 0.075 \text{ in for the 1825s} \end{cases}$$

$h = 0.005$ in.
$\Delta\alpha = 9 \times 10^{-6}/°\text{C}$

From equation (18-11), we get

$$\text{FM(assy,ll)} = \begin{cases} 0.946 \text{ for the 1206s} \\ 0.593 \text{ for the 1825s} \end{cases}$$

The cumulative environmental stress is described by

$$N_1 = 3650 \text{ cycles (1 cycle/day, 10 years)}, \ \Delta T_1 = 20°\text{C}$$

$$S_{11} = (3650 \times 20^{\frac{1}{0.44}})^{0.44} = 738$$

Figure 18-8 shows that, at 10 years, the representative points A and B are above the line $F = 10^{-4}$ and reliability requirements are met for both the 1206 and the 1825 chips.

For a similar application with an extended service life of 20 years:

$$N_1 = 7300 \text{ cycles (1 cycle/day, 20 years)}, \ \Delta T_1 = 20°\text{C}$$

$$S_{11} = (7300 \times 20^{\frac{1}{0.44}})^{0.44} = 1002$$

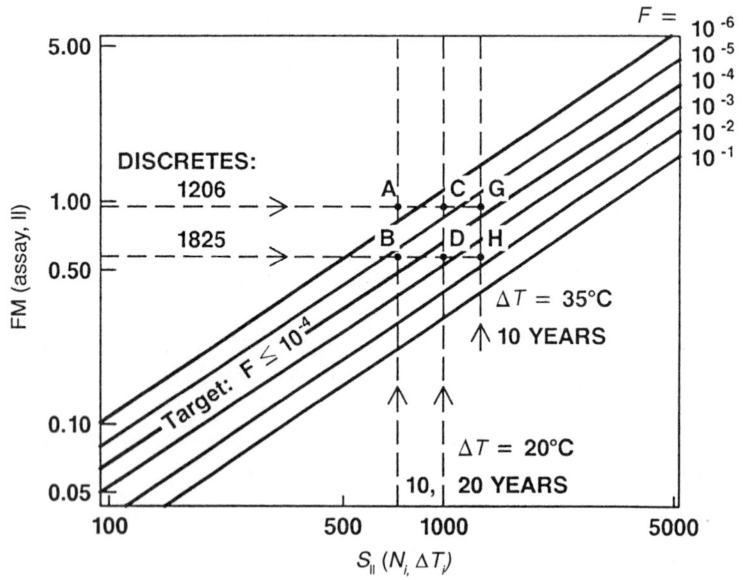

Figure 18-8 Example 18-1: Attachment reliability estimates for size 1206 and 1825 discrete chips on FR-4, using the design-reliability chart for leadless assemblies (Fig. 18-7a).

The data in Figure 18-8 is shifted to the right, from A and B to C and D. Point C for the 1206 chips shows the reliability requirements are still met at 20 years, with a predicted cumulative failure probability of less than 10^{-5} (10 ppm). Point D for the 1825s falls below the line: $F = 10^{-4}$ and reliability is not proven in by the FM tools. Checking for the gray area, we find FM(rel,ll) ~ 0.9, suggesting that the SM design for the larger chips may still be acceptable. Indeed, accelerated testing[27] shows that 180 mil long is an acceptable size for cylindrical resistors on FR-4 in an outside installation with a slightly more severe thermal environment than the one described here.

Finally, for a more severe environment with a daily temperature fluctuation of 35°C for 10 years, the data in Figure 18-8 move from A and B to G and H. Point G is above the target and the 1206s meet reliability requirements. Point H for the 1825s is below target; in this harsher environment, the expected cumulative failure rate at 10 years is between 0.1 and 1 percent.

18.3.2 Example 18-2: 50-Mil Pitch Ceramic Leaded Devices on FR-4

Next, we apply the FMs to 50-mil pitch ceramic leaded devices. The design data for a 100 I/O device with J-clip leads on FR-4 are given as

$$L = 0.919 \text{ in.} \qquad A = 800 \times 10^{-6} \text{ in.}^2 \qquad h = 0.005 \text{ in.}$$

$$\Delta\alpha = 9.5 \times 10^{-6}/°\text{C}, \qquad K = 58 \text{ lb/in.}$$

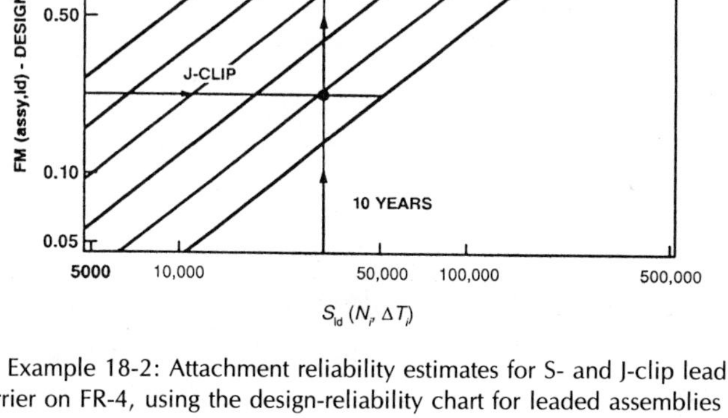

Figure 18-9 Example 18-2: Attachment reliability estimates for S- and J-clip lead ceramic leaded chip carrier on FR-4, using the design-reliability chart for leaded assemblies (Fig. 18-7b).

From Equation (18-15), we get FM(assy,ld) = 0.23. A useful life of 10 years is targeted in a thermal environment described by 3650 thermal fluctuations of 15°C and 120 fluctuations of 60°C:

$$S_{ld} = (3650 \times 15^{\frac{2}{0.44}} + 120 \times 60^{\frac{2}{0.44}})^{0.44} = 3.03 \times 10^4$$

The representative point on Figure 18-9 shows an unacceptable cumulative failure level of about 1 percent at the end of the desired service life. Using a more compliant S-clip lead (e.g., reference 16) for which $K \sim 10$ lb/in., FM(assy,ld) = 1.31 results in a more reliable attachment with $F < 10^{-5}$ at 10 years (see Fig. 18-9).

18.3.3 Example 18-3: 25- and 50-Mil Pitch Plastic Leaded Components on FR-4

For similar substrates and environments, the FMs are a function of package size, lead compliance, and lead-solder wetting area, thus allowing a direct comparison of attachment reliability within a family of devices. In this example, we use the FM approach to compare attachment reliability for 25-mil pitch plastic quad flat packs (PQFPs) and 50-mil pitch plastic leaded chip carriers (PLCCs) on FR-4. The analysis illustrates further the type of information that can be obtained by using the FM tools.

The design data are given in Table 18-1 for 68-I/O PLCCs and PQFPs with 84, 132, and 244 I/Os. The structural models that were used to determine the diagonal lead stiffnesses are shown in Figure 18-10. The PQFP gull-wing lead is compliant with a low $K \sim 12$ lb/in., to be compared to $K \sim 55$ lb/in. for the 68 I/O PLCC J lead. The thinner lead frame of the PQFP gull wings is a significant contributor to their compliance.[24] The FM(assy,ld) data for the PQFPs cover the design range for the lead-solder wetted area. The range of FM(assy,ld) for the 132-I/O PQFP is comparable to that of the 68-I/O PLCC; by design, attachment reliability is expected to be similar for these two packages.

Two thermal environments are considered:

- A controlled ambient, central office with a typical daily fluctuation of 10°C and an occasional surge of 30°C associated with air conditioning failure, say at most once a month. For a 20-year system life, the environmental stress is described by:

$$N_1 = 7300 \text{ cycles (1 cycle/day, 20 years)}, \Delta T_1 = 10°\text{C}.$$

$$N_2 = 240 \text{ cycles (1 cycle/month, 20 years)}, \Delta T_2 = 30°\text{C}$$

$$S_{ld} = (7300 \times 10^{\frac{2}{0.44}} + 240 \times 30^{\frac{2}{0.44}})^{0.44} = 1.09 \times 10^4$$

Table 18-1 Design Data for Plastic Packages on FR-4 Circuit Boards

Package	L (in.)	K (lb/in.)	A (10^{-6} in.^{2})	h (mil)	$\Delta\alpha$ (10^{-6}/°C)	FM (assy,ld)
68-I/O PLCC	0.674	55	810	4	4	2.03
84-I/O PQFP	0.531	12	240–290	3	4	3.32–4.02
132-I/O PQFP	0.743	12	240–290	3	4	1.70–2.05
244-I/O PQFP	1.146	12	240–290	3	4	0.71–0.86

- An outside installation with an averaged daily fluctuation of 30°C and a system life of 20 years for which

$$N_1 = 7300 \text{ cycles (1 cycle/day, 20 years)}, \ \Delta T_1 = 30°\text{C}$$

$$S_{ld} = (7300 \times 30^{\frac{2}{0.44}})^{0.44} = 4.51 \times 10^4$$

The results for both environments are shown in Figure 18-11 where the FM(assy,ld) data for the PQFPs are plotted as bands that cover the design range for the lead-solder wetted area. Also shown are the FM(assy,ld) lines for the 132-I/O PQFP when the lead-solder wetted area is reduced by 50 percent.

In the first scenario of a central office environment, the cumulative failure probability at end of service life is predicted to be smaller than 10^{-5} (10 ppm) for all packages. Even the large 244-I/O PQFPs, over 1.6 in. on the side, have a large reliability margin for this application. When the lead-solder wetted area is reduced by 50 percent, due for example to lead offset or a reduced solder joint length, the 132-I/O PQFPs still have adequate attachment reliability. This suggests that, by design, these devices are robust enough to tolerate minimum assembly defects.

For the more severe environment of an outside plant installation, the cumulative failure probability at 20 years is less than 10^{-5} (10 ppm) for the PLCCs and 84- or 132-I/O PQFPs. For the 244-I/O PQFPs, we predict: $10^{-4} \leqslant F \leqslant 10^{-3}$, which is unacceptable for an application with stringent reliability requirements. Here, the size of fine-pitch PQFPs may have to be limited to no more than 132 I/Os.

Last, some lead design considerations are discussed for the PQFPs as they relate to design choices having reliability implications. The PQFP packages are made to JEDEC standards[28] where the lead width is in the range 8 to 12 mil. In the PQFP model shown in Figure 18-10, the width of the lead below the cross section taper is 8 mil. For a larger width of 12 mil, the diagonal lead stiffness increases by a trivial amount, less than 1 percent,[24] while the lead-solder wetted area increases by 25 percent. FM(assy,ld) increases since it goes as A / K, and the larger width at the tip of the lead is beneficial to attachment reliability. A

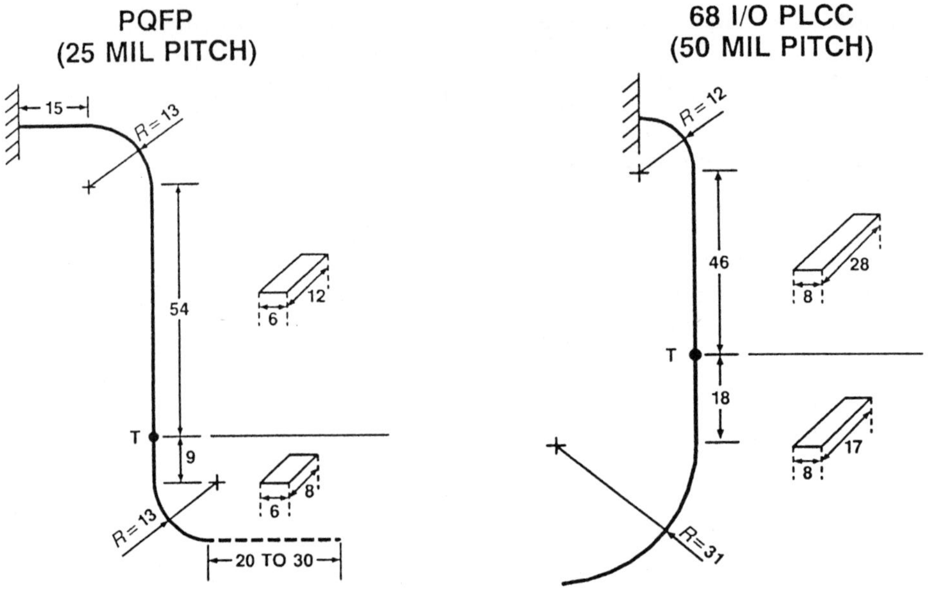

Figure 18-10 Structural models used to determine the diagonal stiffness of PQFP and PLCC corner leads (Example 18-3).

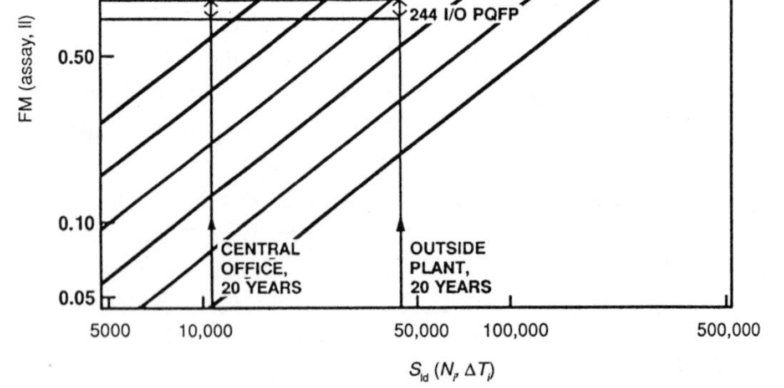

Figure 18-11 Example 18-3: Attachment reliability estimates for PLCCs and PQFPs on FR-4, using the design-reliability chart for leaded assemblies (Fig. 18-7*b*).

uniform, 12-mil-wide cross section would also simplify the lead frame manufacturing; however, the potential for solder bridging during reflow needs to be considered before recommending a lead foot 12 mil wide.

18.4 CONCLUDING REMARKS

The high reliability levels expected of today's electronic assemblies make attachment reliability a critical issue. SMT is a reliable technology, with good design and assembly practices the key to end-use reliability. Upfront design for reliability, using tools such as described in this chapter, can help packaging engineers and physical designers build in attachment reliability.

18.5 ACKNOWLEDGMENTS

We would like to acknowledge the contribution of Werner Engelmaier and Robert Kotlowitz of AT&T Bell Laboratories, Whippany, NJ, to the initial formulation of figures of merit. We are also grateful to Peter M. Hall, Chi K. Ng, and Ephraim Suhir for their time and effort in reviewing this paper.

Appendix 18-A FM Formulas and Scaling Constants

At the component level, the parameters that affect attachment reliability and can be influenced by the package designer are the component dimensions and to some degree the attachment design: solder paste thickness, leadless or leaded attachment, and for leaded packages, the lead compliance. The FMs at the component level are formulated as:

$$\mathrm{FM(comp,ll)} = R\mathrm{(comp,ll)}\,\frac{h}{L} \qquad \text{for leadless} \qquad (18\text{-A1})$$

and

$$\mathrm{FM(comp,ld)} = R\mathrm{(comp,ld)}\,\frac{Ah}{KL^2} \qquad \text{for leaded} \qquad (18\text{-A2})$$

In these and the next FM equations, the $R(.,.)$ factors are scaling constants that lead to dimensionless FMs. The constants R(comp,ll) = 20 in./in. (mm/mm) and R(comp,ld) = 10^7 lb/in.^{2} (6.9×10^4 MPa) are arbitrary; their values reflect the combinations of design parameters for highly reliable SM attachments. For leadless components, h is the component standoff height; for leaded components, h is taken as half the solder paste thickness (e.g., $h = 5 \times 10^{-3}$ in. for 10-mil paste, $h = 3 \times 10^{-3}$ in. for fine-pitch 6-

mil paste. L is the distance from the neutral axis of the component to the center line of the further solder joint. For a large chip carrier, L is the distance from the neutral axis of the package to a corner joint, about half the diagonal dimension of the package. For small, discrete or chip components, L is from the neutral axis to the center line of the end solder joints (e.g., $L \approx 47$ mil for a 120-mil-long 1206-chip resistor).

At the next next stage of component-substrate assembly design, designers select a substrate material. Each component has an FM(assy) that includes the component-substrate CTE mismatch $\Delta\alpha$:

$$\mathrm{FM(assy,ll)} = \mathrm{FM(comp,ll)}\,\frac{R(\mathrm{assy,ll})}{\Delta\alpha} \quad \text{for leadless} \tag{18-A3}$$

and

$$\mathrm{FM(assy,ld)} = \mathrm{FM(comp,ld)}\left(\frac{\mathrm{R(assy,ld)}}{\Delta\alpha}\right)^2 \quad \text{for leaded} \tag{18-A4}$$

with the R constants specified as $\mathrm{R(assy,ll)} = 4 \times 10^{-6}/°\mathrm{C}$ and $\mathrm{R(assy,ld)} = 5 \times 10^{-6}/°\mathrm{C}$.

The impact of power dissipation or nominal temperature fluctuations (ΔT) at the board level is captured in the environment FMs:

$$\mathrm{FM(env,ll)} = \mathrm{FM(assy,ll)}\,\frac{R(\mathrm{env,ll})}{\Delta T} \quad \text{for leadless} \tag{18-A5}$$

and

$$\mathrm{FM(env,ld)} = \mathrm{FM(assy,ld)}\left[\frac{R(\mathrm{env,ld})}{\Delta \mathrm{T}}\right]^2 \quad \text{for leaded} \tag{18-A6}$$

with $R(\mathrm{env,ll}) = 40°\mathrm{C}$, and $R(\mathrm{env,ld}) = 50°\mathrm{C}$.

Last, the reliability FMs include the number of expected cycles of operation N and the acceptable cumulative failure probability F per device at the end of the design life:

$$\mathrm{FM(rel)} = \mathrm{FM(env)}\,\frac{R(\mathrm{rel})}{N^{0.44}}\left[\frac{\ln(1 - F)}{\ln(0.5)}\right]^{\frac{0.44}{\beta}} \tag{18-A7}$$

with $\mathrm{R(rel)} = 100\ \mathrm{cycles}^{1/2}$ in the leadless case and $\mathrm{R(rel)} = 140\ \mathrm{cycles}^{1/2}$ in the leaded case.

The R(rel) constants were selected such that $\mathrm{FM(rel)} \geqslant 1$ when reliability requirements are met, at least to the specified acceptable failure probability. A plot of FM(env) versus predicted median cycles of operations ($N = N_{50\%}$ for $F = 50$ percent) was generated (Fig. 18-A1) using available reliability data for 13 types of leadless and leaded assemblies for which design parameters and predicted field lives have been documented.[3,4,5,16,25,27] Since FM(env) is inversely proportional to $\Delta\gamma$ in the leadless case and ΔW in the leaded case, Figure 18-A1 bears similarity to an inverted Coffin-Manson plot.[14] The slope of the regression lines for either the leadless or the leaded data is 0.44, a more accurate value than the 0.5 that was previously used.[4,5] For applications with short dwell times,

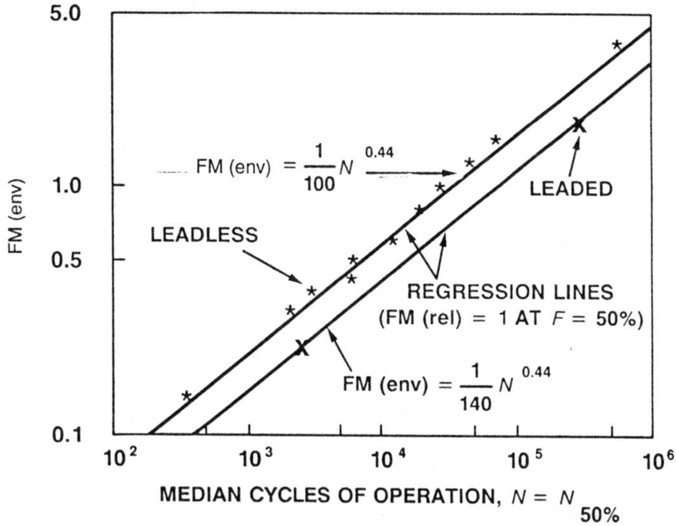

Figure 18-A1 Fit of the leadless and leaded FM models to empirical data, using median cycles to failure for 13 tested assemblies.

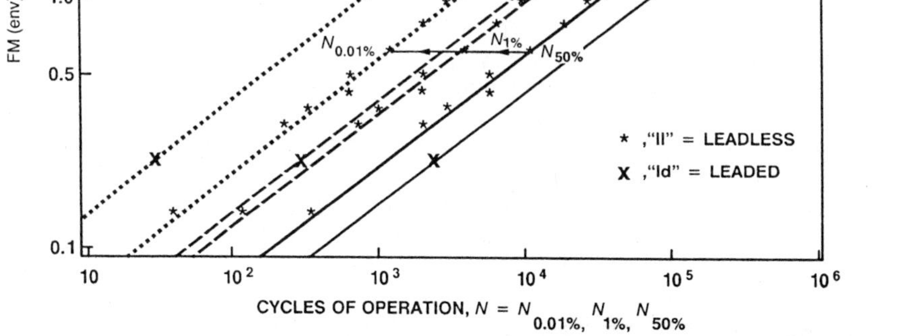

Figure 18-A2 Fit of the FM models to the empirical data at several levels of cumulative failure probabilities. FM (env) versus cycles of operation showing the fit of the FM model: FM (rel) = 1, to the data at acceptable cumulative failure probabilities $F = 10^{-4}$ (0.01%), 10^{-2} (1%), and 50%.

a more sophisticated model and a variable exponent should be used to account for incomplete stress relaxation.[25] The R(rel) constants were obtained by fitting the FM model FM(rel) $= 1$ at $F = 50$ percent, to the regression lines in Figure 18-A1. They are actually the inverse of these line intercepts with the FM(env) axis. Figure 18-A2, where we have plotted the data and the lines FM(rel) $= 1$ at $F = 10^{-4}$, 10^{-2}, and 50 percent, shows that the fit of the FM model to the data is preserved at all failure levels. Because of the simplifications of the FM model and uncertainties in some of the design parameters, a design for which FM(rel) is slightly less than 1 may still be acceptable. A decrease in FM(rel) from 1 to 0.7 corresponds to a change in predicted life by a factor of 2 or an order of magnitude in predicted failure level. In this gray area $0.7 <$ FM(rel) $\leqslant 1$, reliability requirements may still be met; however, a more detailed analysis than provided by the FM tools, and perhaps an accelerated test program, are necessary to prove in reliability. For FM(rel) $\leqslant 0.7$, the SM attachment design does not meet reliability requirements.

Appendix 18-B
FMs for Multiple Thermal Fluctuations

SM assemblies in the field are subject to complex thermal histories. As a first cut, the thermal history is best described by series of cycles of variable amplitudes. For example, the thermal history in a controlled environment for telecommunication equipment consists of typical daily thermal fluctuations and occasional temperature surges that occur when cooling equipment (fans or ambient air conditioning) fails.

Thermal cycles of amplitudes ΔT_i use up a fraction $N_i/N_{f,i}$ of the useful fatigue life, where N_i is the number of operation cycles at amplitude ΔT_i, and $N_{f,i}$ is the number of cycles to failure at the same amplitude ΔT_i. Using Miner's rule,[26] the number of cycles to failure for multiple thermal cycles is then estimated from the relationship

$$\frac{N_1}{N_{f,1}} + \frac{N_2}{N_{f,2}} + \frac{N_3}{N_{f,2}} \cdots = D \leqslant 1 \qquad (18\text{-B1})$$

where D is the cumulative fatigue damage. The formulation of Miner's rule is based on the net energy absorbed at failure, as is the cyclic fatigue life for solder joints. While Miner's rule has not been proven in for solder joint fatigue, it is a reasonable first approach at handling cumulative fatigue damage in SM attachments.

From the reliability condition FM(rel) $= 1$ for failure in the field, we get the cycles to failure at an acceptable failure level F as

$$N_f(F) = \left(\frac{0.32\,h}{L\Delta\alpha\Delta T}\right)^a \left[\frac{\ln(1 - F)}{\ln(0.5)}\right]^{\frac{1}{\beta}} \quad \text{for leadless attachment} \qquad (18\text{-B2})$$

and

$$N_f(F) = \left(\frac{87.5\,Ah}{K(L\Delta\alpha\Delta T)^2}\right)^a \left[\frac{\ln(1 - F)}{\ln(0.5)}\right]^{\frac{1}{\beta}} \quad \text{for leaded attachment} \qquad (18\text{-B3})$$

with $a = 1/0.44$.

Substituting Equations (17-B2) and (17-B3) into Equation (17-B1) for cumulative fatigue damage gives

$$N_1(\Delta T_1)^a + N_2(\Delta T_2)^a + \cdots = D\left(\frac{0.32\,h}{L\Delta\alpha}\right)^a \left[\frac{\ln(1 - F)}{\ln(0.5)}\right]^{\frac{1}{\beta}} \tag{18-B4}$$

for leadless attachment

$$N_1(\Delta T_1)^{2a} + N_2(\Delta T_2)^{2a} + \cdots = D\left(\frac{87.5\,Ah}{K(L\Delta\alpha)^2}\right)^a \left[\frac{\ln(1 - F)}{\ln(0.5)}\right]^{\frac{1}{\beta}} \tag{18-B5}$$

for leaded attachment

The generalized formulation of FM(rel) for multiple thermal loads derives from the above combinations of N_i's and ΔT_i's:

$$\text{FM(rel,ll)} = \frac{0.32\,h}{L\Delta\alpha \times S_{ll}} \left[\frac{\ln(1 - F)}{\ln(0.5)}\right]^{\frac{1}{a\beta}} \quad \text{for leadless attachment} \tag{18-B6}$$

and

$$\text{FM(rel,ld)} = \frac{56.25\,Ah}{K(L\Delta\alpha)^2 \times S_{\text{ld}}} \left[\frac{\ln(1 - F)}{\ln(0.5)}\right]^{\frac{1}{a\beta}} \quad \text{for leaded attachment} \tag{18-B7}$$

where

$$S_{\text{ll}} = \left(\sum_i N_i(\Delta T_i)^a\right)^{\frac{1}{a}}$$

and

$$S_{\text{ld}} = \left(\sum_i N_i(\Delta T_i)^{2a}\right)^{\frac{1}{a}}$$

are measures of the cumulative environmental stress imparted to the SM assembly by multiple thermal fluctuations.

REFERENCES

1. IPC-SM-782, "Surface Mount Land Patterns (Configurations and Design Rules)," Institute for Interconnecting and Packaging Electronic Circuits, Lincolnwood, IL, March 1987.
2. Lau, J. H., and D. W. Rice, "Solder Joint Fatigue in Surface-Mount Technology: State of the Art," *Solid State Technology,* Oct. 1985, pp. 99–100.
3. Engelmaier, W., "Fatigue Life of Leadless Chip Carriers Solder Joints During Power

Cycling," *IEEE Transactions on Components, Hybrids and Manufacturing Technology,* **CHMT-6,** September 1983, pp. 232–237.

4. Clech, J.-P., W. Engelmaier, R. W. Kotlowitz, and J. A. Augis, "Surface Mount Solder Attachment Reliability Figures of Merit—'Design for Reliability' Tools," *Proceedings of the SMART V Conference,* New Orleans, LA, Technical Paper SMT V-48, January 9–12, 1989.
5. ______, "Reliability Figures of Merit for Surface-Soldered Leadless Chip Carriers Compared to Leaded Packages," *IEEE Transactions on Components, Hybrids and Manufacturing Technology,* **12**(4), December 1989, pp. 449–458.
6. Norris, K. C., and A. H. Landzberg, 1969, "Reliability of Controlled Collapse Interconnections," *IBM Journal of Research and Development,* May 1969, pp. 266–271.
7. Howard, R. T., S. W. Sobeck, and C. Sanetra, "A New Package Related Failure Mechanism for Leadless Ceramic Chip Carriers (LC-3s) Solder Attached to Alumina Substrates," *Solid State Technology,* February 1983, pp. 115–122.
8. Hall, P. M., "Forces, Moments, and Displacements During Thermal Chamber Cycling of Leadless Ceramic Chip Carriers Soldered to Printed Boards," *IEEE Transactions on Components, Hybrids and Manufacturing Technology,* **CHMT-7**(4), December 1984, pp. 314–327.
9. ______, "Creep and Stress Relaxation in Solder Joints in Surface Mounted Chip Carriers," *IEEE Transactions on Components, Hybrids and Manufacturing Technology,* **CHMT-10**(4), December 1987, pp. 556–565.
10. Stone, D., S.-P. Hannula, and C.-Y. Li, "The Effects of Service and Materials Variables on the Fatigue Behavior of Solder Joints During the Thermal Cycle," *Proceedings of the 35th Electronic Components Conference,* New York, 1985, pp. 46–51.
11. Hall, P. M., and W. M. Sherry, "Materials, Structures and Mechanics of Solder Joints for Surface Mounted Microelectronics Technology," *Lectures of the 3rd International Conference on Interconnection Technology in Electronics,* Fellbach, W. Germany, Feb. 18–20, 1986, pp. 47–61.
12. Clech, J.-P., and J. A. Augis, "Engineering Analysis of Thermal Cycling Accelerated Tests for Surface—Mount Attachment Reliability Evaluation," *Proceedings of the 1987 International Electronic Packaging Conference,* Boston, MA, November 9–11, 1987, pp. 385–410.
13. Lovasco, F., "Solder Joint Modeling Using Finite Element Analysis," *Proceedings of the 1988 International Electronic Packaging Conference,* Dallas, TX, November 7–10, 1988, pp. 410–415.
14. Manson, S.S., *Thermal Stress and Low-Cycle Fatigue,* McGraw-Hill, N.Y., 1966.
15. Balde, J., "The IEEE Compliant Lead Task Force," *IEEE Transactions on Components, Hybrids and Manufacturing Technology,* **CHMT-10**(3) September 1987, pp. 463–469.
16. Engelmaier, W., "Surface Mount Attachment Reliability of Clip-Leaded Ceramic Chip Carriers on FR-4 Circuit Boards," *International Electronic Packaging Society Journal,* **9**(4), January 1988, pp. 3–11.
17. Kotlowitz, R. W., "Comparative Compliance of Generic Lead Designs For Surface Mounted Components," *Proceedings of the 1987 International Electronic Packaging Conference,* Boston, MA, November 9–11, 1987, pp. 965–984.

18. ______, "Comparative Compliance of Representative Lead Designs for Surface Mounted Components," *Proceedings of the 1988 International Electronic Packaging Society Conference,* Dallas, TX, November 7–10, 1988 pp. 908–948.
19. ______, "Compliance of Surface Mount Component Lead Designs With Rectangular and Circular Cross-Sections," *Proceedings of the 1989 International Electronic Packaging Conference,* vol. II, San Diego, CA, September 11–13, 1989, pp. 1071–1085.
20. Clech, J.-P., and J.A. Augis, "Temperature Cycling, Structural Response and Attachment Reliability of Surface-Mounted Leaded Packages," *Proceedings of the 1988 International Electronic Packaging Conference,* Dallas, TX, November 7–10, 1988, pp. 305–325.
21. Morrow, J.D., "Cyclic Plastic Strain Energy and Fatigue of Metals," *ASTM STP 378,* ASTM, Philadelphia, 1964, pp. 45–87.
22. Lemaitre, J., "Evaluation of Dissipation and Damage in Metals Submitted to Dynamic Loading," *Proceedings of the International Conference on Mechanical Behavior of Materials,* vol. II (Fatigue of Metals), August 15–20, 1971, pp. 540–549.
23. Hines, L. L., 1988, AT&T Bell Laboratories, Columbus, OH, private communication.
24. Clech, J.-P., M. J. Tervalon, and J. A. Augis, "Reliability Evaluation of 25 and 50 Mil Pitch Surface Mounted Plastic Leaded Packages," *Proceedings of the 1989 International Electronic Packaging Conference,* vol. II, San Diego, CA, September 11–13, 1989, pp. 754–758.
25. Engelmaier, W., 1989, AT&T Bell Laboratories, Whippany, NJ, private communications.
26. Miner, M. A., "Cumulative Damage in Fatigue," *Journal of Applied Mechanics,* September 1945, pp. A-159–A-184.
27. Ng, K. K., "Assessing Surface Mount Attachment Reliability of Ceramic Cylindrical Resistors in FR-4 Circuit Boards," *Technical Proceedings, Expo SMT '88,* Las Vegas, NV, November 1988, pp. 1–6.
28. JEDEC, *JEDEC Registered and Standard Outlines for Solid State Products,* JEDEC Publication no. 95, Electronic Industries Association, November 1988.

Authors' Biographies

Abbas I. Attarwala—Research Assistant at the Center Strategic Materials at University of Texas—working on his PhD. He received his BS and MS degrees in metallurgy/material science from Columbia University in New York. During his junior and senior years he worked at AT&T Bell Laboratories as a student employee studying the failure models of fatigue-cycled leadless surface-mount components on circuit boards. He was awarded the Francis Rhode Prize from the Henry Krumb School of Mines at Columbia University. Mr. Attarwala is a member of Tau Beta Pi and the AIME.

Jacques A. Augis—Group Supervisor, Reliability of Surface Mount Technology for AT&T Applications—AT&T. Dr. Augis joined AT&T Bell Laboratories, Columbus, Ohio, in 1968, working on the "Electromechanical Apparatus" to improve the active life and reliability of Wire Spring and Reed Relays. After several other assignments related to device quality and reliability, he was transferred to the Interconnection Laboratory, Whippany, New Jersey, in 1982, working on Printed Wiring Board reliability. Dr. Augis received the Ingénieur degree from the Ecole Superieure de Physique et Chimie Industrielles, Paris, in 1959, and the PhD degree in physics from Case Western Reserve University, Cleveland, Ohio, in 1969. Dr. Augis has published about 25 technical papers, and holds four U.S. patents.

Jean-Paul M. Clech—Member of the Technical Staff—Advanced Computer Research Institute, Lyon, France. He was previously with AT&T Bell Laboratories in Whippany, New Jersey. Within the Interconnection Systems and Assembly Reliability Department, he has responsibility for surface mount assembly and reliability issues.

Dr. Clech graduated from Ecole Centrale de Paris where he received his Diplôme d'Ingénieur in 1981. He also holds an associate degree in economics from the Sorbonne. He received the MS and PhD degrees in mechanical engineering from Northwestern University in 1982 and 1985, respectively.

Dr. Clech has published and presented papers in the areas of engineering mechanics, finite element modeling, thermomechanical fatigue and fracture, surface mount attachment and assembly reliability. An active member of ASME, IEEE and IEPS, he has chaired workshops and technical sessions at numerous conferences.

Werner Engelmaier—President—Engelmaier Associates, Inc. Engelmaier Associates, Inc. is a firm providing consulting services on reliability, manufacturing, and processing aspects of electronic packaging. This consulting company provides services to OEMs, printed circuit board manufacturers, PWB subcontract assembly companies and end users, such as telecommunications and computer companies and the military, in the areas of design for manufacturability and quality, processing effects, yield improvement, environmental stress screening, design for reliability, accelerated reliability testing, reliability prediction, and failure mode analysis for solder joints, plated-through holes, and dynamically used flexible circuitry.

Prior to forming his own company, Mr. Engelmaier was a Distinguished Member of Technical Staff at AT&T Bell Laboratories, Whippany, New Jersey. During his 24 years at Bell Labs, until his retirement in 1990, he dealt with many of the aspects involved in electronic packaging and interconnection technology. His efforts within the Bell System/AT&T ranged from R&D at Bell Labs to 'fire-fighting' at Western Electric, and included the development of specifications and guidelines, as well as technology transfer to production facilities.

Born in Vienna, Austria, Mr. Engelmaier holds the Mechanical Engineering degrees from Technologisches Gewerbe-Museum (TGM), Vienna, Austria, the University of South Carolina, and Massachusetts Institute of Technology. He is a member of the American Society of Mechanical Engineers, the American Society for Testing and Materials, the IEEE Components, Hybrids, and Manufacturing Technology Society, the Institute for Interconnecting and packaging Electronic Circuits, the International Electronics Packaging Society, the Surface Mount Technology Association, Phi Beta Kappa, Pi Mu Epsilon, Sigma Xi, and Tau Beta Pi.

He has conducted numerous tutorial workshops for AT&T, NEPCON, IEPS, IPC, and ISHM. He has contributed to numerous task forces, specifications, standards, and design guides. He is on the Editorial Advisory Board of *Electronic Packaging & Production* magazine and chairs committees at IPC and ASTM. He has been the recipient of many honors and awards, including the *IPC President's Award* in 1988, the *Electronic Packaging Achievement Award* from IEPS/*Electronic Packaging & Production* magazine in 1987, and the *Distinguished Technical Staff Award* from AT&T Bell Laboratories in 1986. For two of his developments he received *R&D 100 Awards* in 1978 and 1981, from *Research & Development* magazine. Mr. Englemaier is the author of over 90 technical publications in a variety of fields related to electronic packaging and interconnection technology, and has been awarded a number of patents on these areas.

Morris E. Fine—Walter P. Murphy Professor Emeritus—Materials Science & Engineering, McCormick School of Engineering and Applied Science of Northwestern University, Evanston, Illinois. Dr. Fine is currently Director of the AISI-sponsored Steel Resource Center at Northwestern. He helped to organize the first Materials Science Department in the United States. He was on the Manhattan Project at both the Chicago and Los Alamos sites, and a member of the Technical Staff of the Bell Telephone Laboratories, Murray Hill, N.J., 1946–54. His formal education was at the University of Minnesota, where he was awarded the BMetEng with distinction in 1940, and a PhD in 1943. His professional society activities include Chairman of the Institute of Metals Division of The Metallurgical Society of AIME, and member of the Board of Directors of AIME and of The Metallurgical Society of AIME. Dr. Fine is a Fellow of five societies: the American Society for Metals, The Metallurgical Society of AIME, the American Physical Society, the American Ceramic Society, and the American Association for the Advancement for Science. He was elected to the National Academy of Engineering in 1973, where he was Chairman of the Membership Committee in 1979. He has won a number of awards, including the ASM Gold Medal, TMS Mathewson Medal, and the AIME James Douglas Gold Medal. Dr. Fine is author or coauthor of over 200 papers. His research interests include alloy development, phase transformations in solids, the theory of strength of metals, alloys and composites, magnetic properties of metals and ceramics, fatigue of metals and tribology of metal and ceramic matrix composites.

Les Fox—Digital Equipment Corporation. Since 1981, Dr. Fox has been working on advanced development of high-density packaging technologies for VLSI. He is especially interested in thermal phenomena, and has developed several proprietary chip cooling technologies while at Digital. Dr. Fox is especially interested in the materials science and reliability physics of advanced electronic packaging and interconnect systems. Dr. Fox holds the ScB in physics from Brown University, and the PhD in low-temperature solid state physics from MIT.

Dr. Fox is a member of IEEE-CHMT, Chairman of IEEE-CHMT Technical

Committee TC-6 on Electronic Packaging, and a member of IEPS. He has published several papers on original investigations on the reliability of solder joint interconnects.

Harold J. Frost—Faculty Member—Thayer School of Engineering. Dr. Frost's primary research interests are in high-temperature deformation of metals and alloys, the structure of defects and grain boundaries, and the evolution of microstructure through phase transformation and grain growth. He received his AB in engineering and applied physics, as well as his PhD in materials science, from Harvard. He has done post-doctoral work at Cambridge University, MIT, and Harvard. Dr. Frost is a member of the Materials Research Society, the American Society of Metals, and the Metallurgical Society of the AIME.

Dennis Grivas—Research Engineer—Center for Advanced Materials of the Lawrence Berkeley Laboratory. Dr. Grivas received the PhD degree in material science and engineering at the University of California at Berkeley in 1978.

Leslie A. Guth—Distinguished Member of Technical Staff—AT&T Bell Laboratories' Engineering Research Center in the Environmental Materials and Processes Group in Princeton, N.J. Dr. Guth's research interests include soldering flux and cleaning technology. She has a BA in physics from the University of North Carolina-Chapel Hill, and a PhD in materials science from the University of Pennsylvania. Dr. Guth is an active member of the Institute for Interconnecting and Packaging Electronic Circuits (IPC), participating on several committees. She chairs the IPC Electromigration Task Group and the IPC Test Monitoring and Validation Committee for the Industry/DoD/U.S.EPA cooperative effort in testing CFC alternatives. For her work in IPC, Dr. Guth received the 1988–1990 IPC President's Award. She was also a member of the 1989 United Nations Environmental Program's Solvents Technical Options Assessment Committee, which evaluated CFC-113 alternative technologies. This report was used in generating additional restrictions and modifications to the 1987 Montreal Protocol phase-down. Dr. Guth holds two patents and has numerous publications.

Peter M. Hall—Distinguished Professor of Physics—Johnson C. Smith University, Charlotte, NC. Dr. Hall teaches physics, is in charge of a pre-engineering program, and serves on the faculty of the Honors College. He has 31 years experience at AT&T Bell Labs in the fields of thin films, superconductivity, electronics packaging, and solder mechanics. Dr. Hall received his BA in physics from Hobart College, and his PhD in physics from Iowa State University. He is a Fellow of IEEE and a Bell Labs Fellow. He serves on the program committee of the Electronic Components and Technology Conference and is program chairman of the 1990 VLSI and GaAs Packaging Workshop. He also serves as Associate Editor of the ASME Transactions of the *Journal of Electronic Pack-*

aging. Dr. Hall has published over 45 papers, and is coauthor of *Thin Film Technology*.

Bryan C. Hendrix—Research Scientist—University of Texas at Austin. In 1981, Dr. Hendrix joined CSTA at the U.S. Army Aberdeen Proving Grounds as a junior engineering fellow. He joined the Center for Strategic Materials at Columbia University in 1985 as a research assistant. Dr. Hendrix received his BS, MS, and DES degrees in metallurgy and materials science from Columbia University in New York. He was awarded the Francis Rhode Prize from the Henry Krumb School of Mines, and the "Pop" Herrington medal from the School of Engineering and Applied Science at Columbia University. Dr. Hendrix is a member of MRS, ASM, TMS/AIME, and Tau Beta Pi.

Charles L. Hutchins—Independent Consultant—Dr. Hutchins is recognized worldwide for his expertise in Surface Mount Technology. His experience includes 22 years at Texas Instruments, where he served in the capacities of Engineering Manager, Quality Department Manager, and Product Manager. Dr. Hutchins was most recently responsible for engineering management at TI's SMT Center for 3 years, specifically process engineering, SPC implementation, yield improvement, product design and development, and the Customer SMT Training Program. Since leaving TI, he has worked as a technical and managerial consultant to several major companies and conducted numerous SMT training sessions. He is also an Adjunct Professor at the National Training Center for Microelectronics, Northampton Community College, Bethlehem, PA. Dr. Hutchins holds PhD and MS degrees in electrical engineering from North Carolina State University.

He is a past Vice President of the Surface Mount Technology Association, with responsibility for the Technical Session of SM '89, the technical conference for the SMTA. Dr. Hutchins is the author of over 25 technical papers, in addition to his textbook *Surface Mount Technology—How to Get Started*. He also writes a monthly column on SMT for *Electronic Engineering Times* and serves on the Editorial Advisory Board of *Surface Mount Technology Magazine*.

Jennie S. Hwang—President—International Electronic Materials Corp. Prior to this appointment, Dr. Hwang held senior managerial and research positions with SCM Corp., Martin Marietta Corp., and Sherwin-Williams Company.

In conjunction with multidisciplinary trainings, Dr. Hwang's 14 years of experience as technology/business manager covering various industries has brought her a unique breadth and depth of knowledge of materials and processes from both the theoretical and practical aspects.

Dr. Hwang received her PhD in materials science and metallurgical engineering from the Engineering School of Case Western Reserve University, an MS in liquid crystal chemistry from the Liquid Crystal Institute of Kent State

University, and a second Master's in inorganic chemistry from Columbia University.

Dr. Hwang's major professional interests are in global technological trends, market development of the eletronics industry, and electronic materials in general. In specific, her interests include interconnection and packaging materials/process, surface mount technology, paste technology, and solder paste.

Dr. Hwang is the top authority in the electronic paste field. She has been an invited lecturer and speaker at numerous major international and national conferences on the subject. Dr. Hwang is a member of the International Society of Hybrid Microelectronics, International Electronics Packaging Society, American Ceramic Society, American Chemical Society, American Welding Society, American Society of Metals, American Society of Testing and Materials, Surface Mount Technology Association, and the Adhesion Society. She has served as the president of ASM Electronic materials-Cleveland and is serving on the membership committee of the American Society of Metals and the IPC committee.

In addition to many publications, she is the author of the book *Solder Paste in Electronics Packaging Technology & Applications for Surface Mount, Hybrid Circuits, and Component Manufacturing*. She holds several U.S. overseas patents in the area of paste technology.

Thomas J. Kilinski—Research Assistant—University of Wisconsin-Madison. Mr. Kilinski has worked with the Electronics Packaging and Interconnection Program since 1987. He is involved with the development of mechanical test procedures for determining the behaviors of SMT interconnections. He presented his work at the IEEE Electronic Components and Technology Conferences in 1989 and 1990. He received the BS and MS degrees in engineering mechanics from the University of Wisconsin-Madison in 1988 and 1990, respectively.

Sheera Knecht—Research Scientist—Digital Equipment Corporation. Ms. Knecht is currently working on the mechanical analysis of high-density electronic packaging, and on the development of analytical methods to assess the reliability of solder joint interconnect. She joined Digital Equipment Corporation in 1986. Previously, she worked at MIT, as a research engineer on problems of hydraulic fracture, and at the Bird-Johnson Company. Ms. Knecht received her BSc degree in engineering from Brown University, and her MS in solid mechanics from Harvard University.

John H. Lau—Senior Engineer—Hewlett-Packard Company, Palo Alto, California. His research activities cover a broad range of electronic packaging and manufacturing technology including MCM, SMT, TAB, FPT, CAE, and DFM.

Prior to joining Hewlett-Packard Laboratories in 1984, Dr. Lau worked for Exxon Production and Research Company and Sandia National Laboratory. He received the BE degree in civil engineering from National Taiwan University, the MASc degree in structural engineering from the University of British Columbia, the MS degree in engineering mechanics from the University of Wisconsin, the MS degree in management science from Fairleigh Dickinson University, and the PhD degree in theoretical and applied mechanics from the University of Illinois. He has more than 20 years of research and development experience in heat transfer, thermal stress, vibration, and reliability analyses of mechanical and electronic components and systems. He has published extensively in these areas, and is an associate editor of the ASME *Journal of Electronic Packaging* as well as an associate editor of the IEEE *Transactions on Components, Hybrids, and Manufacturing Technology*. He has also served as a member of technical committees, session chairman, and workshop speaker of several IEEE, ASME, ASM, and ISHM conferences and symposiums. He is the chairman of the IEEE/CHMT TC-9 (Thermal Management) committee, and the technical program chairman of the 1990 IEEE/CHMT International Electronics Manufacturing Technology Symposium. Dr. Lau is a registered professional engineer in New York and California. He is a senior member of IEEE and is listed in the American Men and Women of Science.

Jon R. Lesniak—Research Assistant—University of Wisconsin-Madison. Mr. Lesniak is involved with the Electronics Packaging and Interconnection Program, where he develops high-precision measurement systems. He presented his work at the IEEE Electronic Components and Technology Conference in 1989. He worked at John Deere Co. developing the SPATE 9000 system, a system that measures stress patterns by thermal emissions. He received the BS in engineering mechanics from the University of Wisconsin-Madison in 1987. He is continuing his work towards the PhD.

Che-Yu Li—Faculty Member—Department of Materials Science and Engineering, Cornell University. The emphasis of Dr. Li research has been on mechanical properties of materials and their application in the energy and electronics industries. For the past decade, he has been actively engaged in electronics packaging research, specializing in solder interconnects, metallization, and TAB. These research activities include mechanical reliability studies and micromechanical testing of electronic components down to submicron dimensions. Dr. Li received his PhD degree from Cornell University in 1960. He is the coeditor of the book *Principles of Electronic Packaging*.

Colin Alan MacKay—Staff Metallurgist—MCC Packaging. As a permanent member of technical staff, Dr. MacKay is involved with the ongoing bonding program, and leads the amalgams development project. In addition he supplies advice and direction on a wide range of bonding and metallurgical subjects to both projects and shareholders. Dr. MacKay has held posts with the British Insulated Callenders Cables as research officer, The International Tin Research Institute (for 16 years), latterly as Director of Research and Head of Metallurgy Department, and on moving to the U.S.A., he was Director of Research at Alpha

Metals Newark, N.J., and Director of Research at the Semi Alloys Subsidiary of Allied-Signal Corporation. He joined MCC from Allied-Signal in 1985. Dr. MacKay earned his BSc (Honours Physics), MSc, and PhD at London University.

Dr. MacKay has served on committees for the British Standards Institute and the European Commission Electrotechnique Internationale (CEI) and is currently on several committees of IPC. He is a Chartered Engineer, a member of the British Institute of Physics, the Institution of Metallurgists, the Institute of Metal Finishing, the British Association of Brazing and Soldering, the International Society of Hybrid Microelectronics (ISHM), and IEEE. He is currently Chairman of the Solders and Brazing Committee of the ASM.

Dr. MacKay has published more than 70 technical papers. He also holds patents in the fields of solder alloys, special solder creams, preforms, packaging, and interconnect and amalgams.

James L. Marshall—Director of the Center for Materials Characterization and Director of the Industry-University Cooperative Research Center—University of North Texas, Denton, Texas. Dr. Marshall has been a member of the faculty of the Department of Chemistry at the University of North Texas (UNT), developing a research program in analytical chemistry. In 1981, he moved on to industry, helping to develop the Advanced Manufacturing Technology program at Motorola, Inc., in Ft. Worth, Texas. In 1987, Dr. Marshall returned to UNT to establish a materials program in the Center for Materials Characterization. Dr. Marshall earned his BS in chemistry at Indiana University, and his PhD in Chemistry at Ohio State University. He cofounded the North Texas Section of the Materials Research Society, and has served as an officer, including President, several times. Dr. Marshall is the author of a reference book on instrumental organic chemistry.

J. W. Morris—Program Leader for Structural Materials—Center for Advanced Materials at the Lawrence Berkeley Laboratory. The research of this group concerns the design of new materials and includes topics that range from fundamental theory to the application of advanced engineering materials.

Professor Morris holds an ScD in materials science from MIT. He is a Fellow in ASM. He was awarded the Hardy Gold Metal of the AIME, the Bradley Stoughton Teaching Award of the ASM, and has authored numerous technical papers.

Bela I. Sandor—Professor—Department of Engineering Mechanics, University of Wisconsin-Madison. Professor Sandor is also Director of the Electronics Packaging and Interconnection Program at this institution. He teaches courses in the fatigue of materials and advanced laboratory techniques in the area of mechanics of materials. He joined the Wisconsin faculty in 1968. His research interests include creep-fatigue interactions, damage evolution, life prediction, and rational test acceleration methods of monolithic and composite materials.

He worked at Bell Telephone Laboratories in the early 1960s, and was a Research Fellow of the Japan Society for the Promotion of Science in 1985. His PhD degree is in theoretical and applied mechanics from the University of Illinois-Urbana. He has published several books and about 70 papers on fatigue and related subjects.

Harvey D. Solomon—Staff Metallurgist—GE Research and Development Center, Schenectady NY, and adjunct faculty member at Union College, Schenectady NY. Dr. Solomon joined the General Electric Corporate Research and Development Center in 1968, with an initial assignment at G.E.'s Nuclear Energy Division in San Jose. He returned to the R&D center at Schenectady in 1969 to continue his research into materials problems encountered by the Nuclear Energy Division. This work centered on the Low Cycle Fatigue of carbon steels and Zr alloys, general physical metallurgy, and stress corrosion cracking of stainless steels. Specific studies were performed on high-strain low-cycle fatigue crack growth, phase transformations in duplex stainless steels, weld sensitization of stainless steels, stress-corrosion cracking in high-temperature water, and the development of corrosion and stress corrosion sensors. His interests have extended beyond the concerns of the Nuclear Energy Division into superalloys for aircraft engines and gas turbines, the influence of creep and environment on high-temperature fatigue, liquid metal surface tension measurements, welding metallurgy, and, in recent years, to the problem of solder fatigue and electrical component problems. This most recent work has included the testing and study of leadless, as well as leaded, chip carrier/printed wiring board joints, as well as the study of special laboratory joints. He currently teaches a course on Welding Metallurgy at Union College. Dr. Solomon received a BS in metallurgical engineering from New York University, and a PhD in metallurgy and materials science from The University of Pennsylvania. Dr. Solomon is the author or coauthor of over 55 articles in technical journals and over 50 additional General Electric Company documents. He was also the coeditor of three conference proceedings and the chairman of the 1985 ASTM symposium on LCF. He has recently completed a chapter on weld embrittlement for Vol. 25 of the Academic Press series *Treatise on Materials Science and Technology,* and an invited review article on solder fatigue for the Materials Research Society. Dr. Solomon has presented conference keynote talks on Stainless Steel Metallurgy, the Physical Metallurgy of Duplex Stainless Steels, and the fatigue of solders and solder joints, as well as over 100 other talks. His work on Duplex Stainless Steels was recognized with the 1979 ASTM Vilella Award (with T.M. Devine) and his work on solder fatigue with the 1986 IEEE-CHMT Transactions Best Paper Award.

Janet R. Sterritt—Manager, Research and Development—Hollis Automation. Hollis Automation is a capital equipment manufacturer for the printed circuit assembly industry. Her expertise is in wave soldering, reflow, and cleaning processes. Ms. Sterritt joined Hollis Automation in 1987, after designing and fabricating boards for 10 years at Xerox Corporation. She holds a BSEE from the University of Rochester. Ms. Sterritt holds the patent for the wave soldering Accuknife. She developed the XL7 and Future1SMT wave soldering processes. She directed the R&D team in the development of various nozzle designs for aqueous cleaning, leading to the development of the Hurricane Jet and the Directed Impact Force nozzles. She has written numerous technical articles which have been published in various trade journals.

Donald S. Stone—Assistant Professor—Materials Science and Engineering, the University of Wisconsin at Madison, WI. At the University of Wisconsin for 3 years, Dr. Stone currently researches gain boundary sliding in solders and solder joints, and the mechanical properties of thin films with applications in tribology and electronic packaging. His expertise is the mechanical properties of solids with emphasis on theory and measurement. He received a BS in physics from the University of Texas at Austin, and a PhD in materials science from Cornell University. Dr. Stone is a member of TMS-AIME and MRS.

Ravi Subrahmanyan—Research Scientist—Motorola, Inc., Dr. Subrahmanyan is currently involved in materials-related research and development in areas such as electronics packaging, device technology, structural design, and reliability. Since 1985, he has been actively involved in solder fatigue research at Cornell University. To date, he has several publications to his credit that describe fundamental aspects of this phenomenon. He has also been responsible for the development of techniques for the fabrication and mechanical testing of solder joints as small as 50 μm. Dr. Subrahmanyan received the BTech degree in metallurgical engineering at the Indian Institute of Technology, Madras, India. He earned his MS and PhD degrees in materials science at Cornell University.

Tammy Summers—University of California, Berkeley. Ms. Summers is currently working toward her PhD Thesis in materials science. Her thesis work involves a microstructural study of fatigue in Sn-Pb solders.

John K. Tien—Director, Strategic Materials R&D Laboratory—University of Texas. Dr. Tien joined Pratt & Whitney Aircraft in 1969 to conduct research on jet engine materials. He joined the Faculty of Metallurgy and Materials Science at Columbia University in 1972. In 1983, he was honored by Columbia University by becoming the Henry Marion Howe Chair Professor of Metallurgy and Materials Science. Dr. Tien received his BS degree in mechanical engineering from Worchester Polytechnic Institute, in Worchester, Massachusetts, MA and MEng degrees from Yale University, and MS and PhD degrees from Carnegie-Mellon University in Pittsburgh, Pennsylvania. He has been elected an Outstanding Young Man of America by the U.S. Chamber of Commerce for "outstanding ability, contributions and service to America." He was awarded the Bradley Stoughton Best Young Professor Award for excellence in teaching and research contributions, by the American Society of Metals. In 1984, Dr. Tien was honored with the K. C. Li Gold Medal for meritorious achievement. He is a member of committees of the National Advisory Board and the Argonne National Laboratory's MST Review Committee. He is a member of Sigma Chi and Tau Beta Pi. He is active on the national level in TMS/AIME and the ASM, and has chaired various committees and many international conferences. Dr. Tien, together with his students and colleagues, has written over 200 technical papers on a broad range of metallurgical and other topics.

Dana Tribula—Lam Research Institute, Fremont, California. Dr. Tribula studied creep and thermal fatigue of near-eutectic Sn-Pb solders as her PhD Thesis work in materials science at the University of California, Berkeley.

James R. Wilcox—Research Scientist—IBM, Endicott, New York. Dr. Wilcox has most recently been studying the deformation and fatigue properties of solder alloys. He joined IBM in 1982, where he worked on various process-development projects for advanced circuit packaging. Previously, he worked for several years at the GE Corporate Research and Development Center in Schenectady, New York, where his research efforts included amorphous metal processing and fatigue performance evaluation of Ni-based superalloys. Dr. Wilcox received his BS and MS degrees in metallurgical engineering from Michigan Technological University, Houghton, Michigan. In 1986, he was awarded an IBM resident study fellowship, and received his PhD degree in materials science from Cornell University.

Index